Lecture Notes in Mathematics

1620

Editors:
A. Dold, Heidelberg
F. Takens, Groningen

Subseries: Fondazione C.I.M.E., Firenze
Advisor: Roberto Conti

T0203137

Springer
Berlin
Heidelberg
New York
Barcelona
Budapest
Hong Kong
London
Milan
Paris
Santa Clara
Singapore
Tokyo

R. Donagi   B. Dubrovin
E. Frenkel   E. Previato

Integrable Systems and Quantum Groups

Lectures given at the 1st Session of the
Centro Internazionale Matematico Estivo
(C.I.M.E.) held in Montecatini Terme, Italy,
June 14–22, 1993

Editors: M. Francaviglia, S. Greco

 Springer

Authors

Ron Donagi
Department of Mathematics
University of Pennsylvania
Philadelphia, PA 19104, USA

Boris Dubrovin
SISSA
Via Beirut, 2– 4
34014 Miramare, Italy

Edward Frenkel
Department of Mathemetics
Harvard University
Cambridge, MA 01610, USA

Emma Previato
Department of Mathemaitcs
Boston University
Boston, MA 022 15, USA

Editors

Mauro Francaviglia
Istituto di Fisica Matematica
"J.-L. Lagrange"
Università di Torino
Via Carlo Alberto, 10
10123 Torino, Italy

Silvio Greco
Dipartimento di Matematica
Politecnico di Torino
Corso Duca degli Abruzzi, 24
10129 Torino, Italy

Cataloging-in-Publication Data applied for
  Die Deutsche Bibliothek - CIP-Einheitsaufnahme

Centro Internazionale Matematico Estivo:
Lectures given at the ... session of the Centro Internazionale
Matematico Estivo (CIME) ... - Berlin ; Heidelberg ; New
York ; London ; Paris ; Tokyo ; Hong Kong : Springer
  Früher Schriftenreihe. - Früher angezeigt u.d.T.: Centro
  Internazionale Matematico Estivo: Proceedings of the ... session of the
  Centro Internazionale Matematico Estivo (CIME)
NE: HST
  1995,1. Integrable systems and quantum groups. - 1995
Integrable systems and quantum groups : held in Montecatini
Terme, Italy, June 14 - 22, 1995 / R. Donagi ... Ed.: M.
Francaviglia ; S. Greco. - Berlin ; Heidelberg ; New York ;
Barcelona ; Budapest ; Hong Kong ; London ; Milan ; Paris ;
Tokyo : Springer, 1995
  (Lectures given at the ... session of the Centro Internazionale
  Matematico Estivo (CIME) ... ; 1995,1)
  (Lecture notes in mathematics ; Vol. 1620 : Subseries: Fondazione
  CIME)
  ISBN 3-540-60542-8
NE: Donagi, Ron; Francaviglia, Mauro [Hrsg.]; 2. GT

Mathematics Subject Classification (1991): 81R50, 58F07

ISBN 3-540-60542-8 Springer-Verlag Berlin Heidelberg New York

© Springer-Verlag Berlin Heidelberg 1996
Printed in Germany
Typesetting: Camera-ready TeX output by the author
SPIN: 10479683   46/3142-543210 - Printed on acid-free paper

FOREWORD

"Integrable Systems" form a classical subject having its origins in Physics and is deeply related with almost all the most important domains of Mathematics. Intriguing and often surprising are in fact the interplays of Integrable Systems with Algebraic Geometry, which started perhaps with Jacobi, to be then forgotten and discovered again in more recent years, with a fantastic impulse to both Theoretical Physics and Pure Mathematics. A very recent related topic is the theory of so-called "Quantum Groups", which is nowadays generating a further incredible amount of fruitful interplay between Physics and Mathematics, especially in the domains proper to Geometry and Algebra.

The scientific organisers of this CIME Session have for several years been enjoing these fascinating interrelations, also promoting and working in a research project supported by CNR and involving geometers and mathematical physicists from Genova, Milano, Torino and Trieste. Thus, the idea of a CIME session on this subject was quite ripe and necessary to the mathematical community when the proposal was worked out and promptly accepted by CIME.

Of course it was immediately clear to us that covering all the subjects was an impossible task. However, we were lucky enough to obtain the collaboration of four outstanding main lecturers, coming from different research experiences and animated by different points of view, but sharing the interest in the crosspoint of these disciplines and extremely able to develop their lectures with the most fascinating interdisciplinary attitude. The Course resulted then in a stimulating and exciting experience, not only for the participants but also for us. We really hope that the reader will find the same excitement as we did in following these lectures, even if a book cannot reproduce the unique atmosphere that existed in the school and surrounded it, in a special place where quiet and friendship were at the basis of a fruitful interaction.

The Session was held in fact in Montecatini Terme, at Villa "La Querceta", from June 14 to June 22, 1993. It consisted of the main four courses of six lectures each, accompanied by a number of interesting seminars concerning special topics and/or recent research announcements.

These proceedings contain the expanded versions of the four main courses. It took some time to collect them and recast them in their final form, but we believe that the importance and completeness of these texts made this longer delay worthwile. Unfortunately the lack of space has made impossible to include also the seminars, which shall be just listed on page VIII. For an outline of the contents of this volume we refer the reader to the Tables of Contents and the Introductions of the four sets of lectures themselves.

We are thankful to CIME for having given us the possibility of living such an exciting experience; we especially acknowledge the patience and the help of our colleagues Roberto Conti and Pietro Zecca. Special thanks are also due to our friend and collaborator Franco Magri, who was a source of inspiration for this session and helped us to construct its scientific structure.

Mauro Francaviglia
Silvio Greco

TABLE OF CONTENTS

R. DONAGI, E. MARKMAN   Spectral covers, algebraically
completely integrable, Hamiltonian
systems, and moduli of bundles 1

B. DUBROVIN   Geometry of 2D topological
field theories 120

B. FEIGIN, E. FRENKEL   Integrals of motion and
quantum groups 349

E. PREVIATO   Seventy years of spectral
curves: 1923-1993 419

SEMINARS

B. DUBROVIN, Integrable Functional Equations

M. BERGVELT, Grassmannians, Heisenberg Algebras and Toda Lattices

M. NIEDERMAIER, Diagonalization Problem of Conserved Charges

C. REINA, A Borel-Weil-Bott Approach to Representations of Quantum Groups

A. STOLIN, Rational Solutions of Classical Yang-Baxter Equations and Frobenius Lie Algebras

M. ADAMS, Bohr-Sommerfeld Quantization of Spectral Curves

G. MAGNANO, A Lie-Poisson Pencil for the KP Equation

K. MARATHE, Geometrical Methods in QFT

L. GATTO, Isospectral Curves for Elliptic Systems

R. SCOGNAMILLO, Prym-Tyurin Varieties and the Hitchin Map

Spectral covers,
algebraically completely integrable,
Hamiltonian systems,
and moduli of bundles

Ron Donagi [1]

University of Pennsylvania

and

Eyal Markman [2]

University of Michigan

Contents

1 Introduction   3

2 Basic Notions   6
  2.1 Symplectic Geometry   6
  2.2 Integrable Systems   9
  2.3 Algebraically Completely Integrable Hamiltonian Systems. 11
  2.4 Moment Maps and Symplectic Reduction 12
  2.5 Finite dimensional Poisson loop group actions 17
      2.5.1 Finite dimensional approximations 17
      2.5.2 Type loci 18

3 Geodesic flow on an ellipsoid   21
  3.1 Integrability 21
  3.2 Algebraic integrability 23
  3.3 The flows 25
  3.4 Explicit parametrization 27

4 Spectral curves and vector bundles   30
  4.1 Vector Bundles on a Curve............................... 30
  4.2 Spectral Curves and the Hitchin System 34
  4.3 Polynomial Matrices 37
      4.3.1 Explicit Equations for Jacobians of Spectral Curves with a Cyclic
            Ramification Point 39
      4.3.2 Geodesic Flow on Ellipsoids via 2 × 2 Polynomial Matrices 40

[1] Partially supported by NSA Grant MDA904-92-H3047 and NSF Grant DMS 95-03249
[2] Partially supported by a Rackham Fellowship, University of Michigan, 1993

5 Poisson structure via levels   43
   5.1 Level structures   43
   5.2 The cotangent bundle   43
   5.3 The Poisson structure   44
   5.4 Linearization   45
   5.5 Hamiltonians and flows in $T^*\mathcal{U}_D$ 47

6 Spectral flows and KP 48
   6.1 The hierarchies   48
   6.2 Krichever maps   50
   6.3 Compatibility of hierarchies   53
        6.3.1 Galois covers and relative Krichever maps   54
        6.3.2 Compatibility of stratifications   56
        6.3.3 The compatibility theorem, the smooth case   57
        6.3.4 The compatibility theorem, singular cases   58
        6.3.5 Elliptic Solitons   61
        6.3.6 Outline of the proof of the compatibility theorem   64

7 The Cubic Condition and Calabi-Yau threefolds   66
   7.1 Families of Tori   66
   7.2 The Cubic Condition   68
   7.3 An Integrable System for Calabi-Yau Threefolds   71

8 The Lagrangian Hilbert scheme and its relative Picard   82
   8.1 Introduction   82
   8.2 Lagrangian Hilbert Schemes   84
   8.3 The construction of the symplectic structure   87
   8.4 Partial compactifications: a symplectic structure on the moduli space of
        Lagrangian sheaves   94
   8.5 Examples   96
        8.5.1 Higgs Pairs   96
        8.5.2 Fano Varieties of Lines on Cubic Fourfolds   97

9 Spectral covers   99
   9.1 Algebraic extensions   99
   9.2 Decomposition of spectral Picards 101
        9.2.1 The question 101
        9.2.2 Decomposition of spectral covers 103
        9.2.3 Decomposition of spectral Picards 104
        9.2.4 The distinguished Prym 105
   9.3 Abelianization 106
        9.3.1 Abstract vs. K-valued objects 106
        9.3.2 The regular semisimple case: the shift 107
        9.3.3 The regular case: the twist along the ramification 109
        9.3.4 Adding values and representations 111
        9.3.5 Irregulars? 112

1 Introduction

The purpose of these notes is to present an algebro-geometric point of view on several interrelated topics, all involving integrable systems in symplectic-algebro-geometric settings. These systems range from some very old examples, such as the geodesic flow on an ellipsoid, through the classical hierarchies of $KP-$ and KdV-types, to some new systems which are often based on moduli problems in algebraic geometry.

The interplay between algebraic geometry and integrable systems goes back quite a way. It has been known at least since Jacobi that many integrable systems can be solved explicitly in terms of *theta functions*. (There are numerous examples, starting with various *spinning tops* and the *geodesic flow on an ellipsoid*.) Geometrically, this often means that the system can be mapped to the total space of a family of Jacobians of some curves, in such a way that the flows of the system are mapped to linear flows along the Jacobians. In practice, these curves tend to arise as the spectrum (hence the name *'spectral'* curves) of some parameter-dependent operator; they can therefore be represented as branched covers of the parameter space, which in early examples tended to be the Riemann sphere $\mathbf{CP}^1$.

In *Hitchin's system*, the base $\mathbf{CP}^1$ is replaced by an arbitrary (compact, non-singular) Riemann surface Σ. The cotangent bundle $T^*\mathcal{U}_\Sigma$ to the moduli space $\mathcal{U}_\Sigma$ of stable vector bundles on Σ admits two very different interpretations: on the one hand, it parametrizes certain *Higgs bundles*, or vector bundles with a (canonically) twisted endomorphism; on the other, it parametrizes certain *spectral data*, consisting of torsion-free sheaves (generically, line bundles) on spectral curves which are branched covers of Σ. In our three central chapters (4,5,6) we study this important system, its extensions and variants. All these systems are linearized on Jacobians of spectral curves.

We also study some systems in which the spectral curve is replaced by a higher-dimensional geometric object: a *spectral variety* in Chapter 9, an algebraic *Lagrangian subvariety* in Chapter 8, and a *Calabi-Yau manifold* in Chapter 7. Our understanding of some of these wild systems is much less complete than is the case for the curve-based ones. We try to explain what we know and to point out some of what we do not. The Calabi-Yau systems seem particularly intriguing. Not only are the tori (on which these systems are linearized) not Jacobians of curves, they are in general not even abelian varieties. There are some suggestive relations between these systems and the conjectural mirror-symmetry for Calabi-Yaus.

The first three chapters are introductory. In Chapter 2 we collect the basic notions of *symplectic geometry* and *integrable systems* which will be needed, including some information about *symplectic reduction*. (An excellent further reference is [AG].) In Chapter 3 we work out in some detail the classical theory of geodesic flow on an ellipsoid, which is integrable via hyperelliptic theta functions. We think of this both as a beautiful elementary and explicit example and as an important special case of the much more powerful results which follow. (Our presentation follows [Kn, Re, D5]). Some of our main algebro-geometric objects of study are introduced in Chapter 4: vector bundles and their moduli spaces, spectral curves, and the *'spectral systems'* constructed from them. In particular, we consider the *polynomial matrix system* [AHH, B1] (which contains the geodesic flow on an ellipsoid as special case) and *Hitchin's system* [H1, H2].

Each of the remaining five chapters presents in some detail a recent or current research topic. Chapter 5 outlines constructions (from [Ma1, Bn, Ty1]) of the Poisson structure on the spectral system of curves. This is possible whenever the twisting line bundle K is a non-negative twist $\omega_\Sigma(D)$ of the canonical bundle ω_Σ, and produces an algebraically completely integrable Hamiltonian system. Following [Ma1] we emphasize the deformation-theoretic construction, in which the Poisson structure on an open subset of the system is obtained via symplectic reduction from the cotangent bundle $T^*\mathcal{U}_{\Sigma,D}$ of the moduli space $\mathcal{U}_{\Sigma,D}$ of stable bundles with a *level-D structure*.

In Chapter 6 we explore the relation between these spectral systems and the KP-hierarchy and its variants (multi-component KP, Heisenberg flows, and their KdV-type subhierarchies). These hierarchies are, of course, a rich source of geometry: The Krichever construction (e.g. [SW]) shows that any Jacobian can be embedded in KP-space, and these are the only finite-dimensional orbits [Mul, AdC, Sh]. Following [AB, LM1] we describe some "multi-Krichever" constructions which take spectral data to the spaces of the KP, $mcKP$ and Heisenberg systems. Our main new result is that the flows on the spectral system which are obtained by pulling back the $mcKP$ or Heisenberg flows via the corresponding Krichever maps are *Hamiltonian* with respect to the Poisson structure constructed in Chapter 5. In fact, we write down explicitly the Hamiltonians for these KP flows on the spectral system, as residues of traces of meromorphic matrices. (Some related results have also been obtained recently in [LM2].)

The starting point for Chapter 7 is an attempt to understand the condition for a given family of complex tori to admit a symplectic structure and thus become an ACIHS. We find that the condition is a symmetry on the derivatives of the period map, which essentially says that the periods are obtained as partials of some field of symmetric cubic tensors on the base. In the rest of this Chapter we apply this idea to an analytically (not algebraically) integrable system constructed from any family of Calabi-Yau 3-folds. Some properties of this system suggest that it may be relevant to a purely hodge-theoretic reformulation of the mirror-symmetry conjectures. (This chapter is based on [DM].)

Chapter 8 is devoted to the construction of symplectic and Poisson structures in some inherently non-linear situations, vastly extending the results of Chapter 5. The basic space considered here is the moduli space parametrizing line-bundle-like sheaves supported on (variable) subvarieties of a given symplectic space X. It is shown that when the subvarieties are Lagrangian, the moduli space itself becomes symplectic. The spectral systems considered in Chapter 5 can be recovered as the case where X is the total space of $T^*\Sigma$ and the Lagrangian subvarieties are the spectral curves. (A fuller version of these results will appear in [Ma2].)

In the final chapter we consider extensions of the spectral system to allow a higher-dimensional base variety S, an arbitrary reductive group G, an arbitrary representation $\rho : G \to AutV$, and values in an arbitrary vector bundle K. (Arbitrary reductive groups G were considered, over a curve $S = \Sigma$ with $K = \omega_\Sigma$, by Hitchin [H2], while the case $K = \Omega_S$ over arbitrary base S is Simpson's [Sim1]). We replace spectral curves by various kinds of spectral covers, and introduce the cameral cover, a version of the Galois-closure of a spectral cover which is independent of K and ρ. It comes with an action of W, the Weyl group of G. We analyze the decomposition, under the action of W, of the cameral and spectral Picard varieties, and identify the distinguished Prym in there. This is shown to correspond, up to certain shifts and twists, to the fiber of the Hitchin map in this general setting, i.e. to moduli of Higgs bundles with a given $\tilde{S}$. Combining this

5

with some obvious remarks about existence of Poisson structures, we find that the moduli spaces of K-valued Higgs bundles support algebraically completely integrable systems. Our presentation closely follows that of [D4]

It is a pleasure to express our gratitude to the organizers, Mauro Francaviglia and Silvio Greco, for the opportunity to participate in the CIME meeting and to publish these notes here. During the preparation of this long work we benefited from many enjoyable conversations with M. Adams, M. Adler, A. Beauville, R. Bryant, C. L. Chai, I. Dolgachev, L. Ein, B. van Geemen, A. Givental, M. Green, P. Griffiths, N. Hitchin, Y. Hu, S. Katz, V. Kanev, L. Katzarkov, R. Lazarsfeld, P. van Moerbeke, D. Morrison, T. Pantev, E. Previato and E. Witten.

6

2 Basic Notions

We gather here those basic concepts and elementary results from symplectic and Poisson geometry, completely integrable systems, and symplectic reduction which will be helpful throughout these notes. Included are a few useful examples and only occasional proofs or sketches. To the reader unfamiliar with this material we were hoping to impart just as much of a feeling for it as might be needed in the following chapters. For more details, we recommend the excellent survey [AG].

2.1 Symplectic Geometry

Symplectic structure

A symplectic structure on a differentiable manifold M of even dimension $2n$ is given by a non-degenerate closed 2-form σ. The non degeneracy means that either of the following equivalent conditions holds.

- σ^n is a nowhere vanishing volume form.

- Contraction with σ induces an isomorphism $\rfloor\sigma : TM \to T^*M$

- For any non-zero tangent vector $v \in T_mM$ at $m \in M$, there is some $v' \in T_mM$ such that $\sigma(v,v') \neq 0$.

Examples 2.1

1. Euclidean space

   The standard example of a symplectic manifold is Euclidean space $\mathbb{R}^{2n}$ with $\sigma = \Sigma dp_i \wedge dq_i$, where $p_1,\cdots,p_n$, $q_1,\cdots,q_n$ are linear coordinates. Darboux's theorem says that any symplectic manifold is locally equivalent to this example (or to any other).

2. Cotangent bundles

   For any manifold X, the cotangent bundle $M := T^*X$ has a natural symplectic structure. First, M has the tautological 1-form α, whose value at $(x,\theta) \in T^*X$ is θ pulled back to T^*M. If $q_1,\cdots,q_n$ are local coordinates on X, then locally $\alpha = \Sigma p_i dq_i$ where the p_i are the fiber coordinates given by $\partial/\partial q_i$. The differential

   $$\sigma := d\alpha$$

   is then a globally defined closed (even exact) 2-form on M. It is given in local coordinates by $\Sigma dp_i \wedge dq_i$, hence is non-degenerate.

3. Coadjoint orbits

   Any Lie group G acts on its Lie algebra $\mathbf{g}$ (adjoint representation) and hence on the dual vector space $\mathbf{g}^*$ (coadjoint representation). Kostant and Kirillov noted that for any $\xi \in \mathbf{g}^*$, the coadjoint orbit $\mathcal{O} = G\xi \subset \mathbf{g}^*$ has a natural symplectic structure. The tangent space to $\mathcal{O}$ at ξ is given by $\mathbf{g}/\mathbf{g}_\xi$, where $\mathbf{g}_\xi$ is the stabilizer of ξ:

   $$\mathbf{g}_\xi := \{x \in \mathbf{g} \mid ad_x^*\xi = 0\} = \{x \in \mathbf{g} \mid (\xi,[x,y]) = 0 \quad \forall\, y \in \mathbf{g}\}.$$

Now ξ determines an alternating bilinear form on $\mathfrak{g}$

$$x, y \longmapsto (\xi, [x, y]),$$

which clearly descends to $\mathfrak{g}/\mathfrak{g}_\xi$ and is non-degenerate there. Varying ξ we get a non-degenerate 2-form σ on $\mathcal{O}$. The Jacobi identity on g translates immediately into closedness of σ.

Hamiltonians

To a function f on a symplectic manifold (M, σ) we associate its *Hamiltonian vector field* v_f, uniquely determined by

$$v_f \rfloor \sigma = df.$$

A vector field v on M is Hamiltonian if and only if the 1-form $v \rfloor \sigma$ is exact. We say v is *locally Hamiltonian* if $v \rfloor \sigma$ is closed. This is equivalent to saying that the flow generated by v preserves σ. Thus on a symplectic surface $(n = 1)$, the locally Hamiltonian vector fields are the area-preserving ones.

Example: (Geodesic flow)

A Riemannian metric on a manifold X determines an isomorphism of $M := TX$ with T^*X; hence we get on M a natural symplectic structure together with a C^∞ function $f =$ (squared length). The geodesic flow on X is the differential equation, on M, given by the Hamiltonian vector field v_f. Its integral curves are the geodesics on M.

Poisson structures

The association $f \mapsto v_f$ gives a map of sheaves

$$v : C^\infty(M) \longrightarrow V(M) \tag{1}$$

from C^∞ functions on the symplectic manifold M to vector fields. Now $V(M)$ always has the structure of a Lie algebra, under commutation of vector fields. The symplectic structure on M determines a Lie algebra structure on $C^\infty(M)$ such that v becomes a morphism of (sheaves of) Lie algebras. The operation on $C^\infty(M)$, called *Poisson bracket*, is

$$\{f, g\} := (df, v_g) = -(dg, v_f) = \frac{n\, df \wedge dg \wedge \sigma^{n-1}}{\sigma^n}.$$

More generally, a *Poisson structure* on a manifold M is a Lie algebra bracket $\{ , \}$ on $C^\infty(M)$ which acts as a derivation in each variable:

$$\{f, gh\} = \{f, g\}h + \{f, h\}g, \quad f, g, h \in C^\infty(M).$$

Since the value at a point m of a given derivation acting on a function g is a linear function of $d_m g$, we see that a Poisson structure on M determines a global 2-vector

$$\psi \in H^0(M, \overset{2}{\wedge} TM).$$

or equivalently a skew-symmetric homomorphism

$$\Psi : T^*M \longrightarrow TM.$$

Conversely, any 2-vector ψ on M determines an alternating bilinear bracket on $C^\infty(M)$, by

$$\{f,g\} := (df \wedge dg, \psi),$$

and this acts as a derivation in each variable. An equivalent way of specifying a Poisson structure is thus to give a global 2-vector ψ satisfying an integrability condition (saying that the above bracket satisfies the Jacobi identity, hence gives a Lie algebra).

We saw that a symplectic structure σ determines a Poisson bracket $\{\ ,\ \}$. The corresponding homomorphism Ψ is just $(\rfloor\sigma)^{-1}$; the closedness of σ is equivalent to integrability of ψ. Thus, a Poisson structure which is (i.e. whose 2-vector is) everywhere non-degenerate, comes from a symplectic structure.

A general Poisson structure can be degenerate in two ways: first, there may exist non-constant functions $f \in C^\infty(M)$, called *Casimirs*, satisfying

$$0 = df\rfloor\psi = \Psi(df),$$

i.e.

$$\{f,g\} = 0 \text{ for all } g \in C^\infty(M).$$

This implies that the rank of Ψ is less than maximal everywhere. In addition, or instead, rank Ψ could drop along some strata in M. For even r, let

$$M_r := \{m \in M | rank(\Psi) = r\}.$$

Then a basic result [We] asserts that the M_r are submanifolds, and they are canonically foliated into *symplectic leaves*, i.e. r-dimensional submanifolds $Z \subset M_r$ which inherit a symplectic structure. (This means that the restriction $\psi_{|z}$ is the image, under the inclusion $Z \hookrightarrow M_r$, of a two-vector ψ_Z on Z which is everywhere nondegenerate, hence comes from a symplectic structure on Z.) These leaves can be described in several ways:

- The image $\Psi(T^*M_r)$ is an involutive subbundle of rank r in TM_r; the Z are its integral leaves.

- The leaf Z through $m \in M_r$ is $Z = \{z \in M_r | f(m) = f(z)$ for all Casimirs f on $M_r\}$.

- Say that two points of M are ψ-connected if there is an integral curve of some Hamiltonian vector field passing through both. The leaves are the equivalence classes for the equivalence relation generated by ψ-connectedness.

Example 2.2 The Kostant-Kirillov symplectic structures on coadjoint orbits of a Lie algebra $\mathbf{g}$ extend to a Poisson structure on the dual vector space $\mathbf{g}^*$. For a function $F \in C^\infty(\mathbf{g}^*)$ we identify its differential $d_\xi F$ at $\xi \in \mathbf{g}*$ with an element of $\mathbf{g} = \mathbf{g}^{**}$. We then set:

$$\{F,G\}(\xi) := (\xi, [d_\xi F, d_\xi G]).$$

This is a Poisson structure, whose symplectic leaves are precisely the coadjoint orbits. The rank of $\mathbf{g}$ is, by definition, the smallest codimension ℓ of a coadjoint orbit. The Casimirs are the ad-invariant functions on $\mathbf{g}^*$. Their restrictions to the largest stratum $\mathbf{g}^*_{\dim \mathbf{g}-\ell}$ foliate this stratum, the leaves being the *regular* (i.e. largest dimensional) coadjoint orbits.

2.2 Integrable Systems

We say that two functions h_1, h_2 on a Poisson manifold (M, ψ) *Poisson commute* if their Poisson bracket $\{h_1, h_2\}$ is zero. In this case the integral flow of the Hamiltonian vector field of each function h_i, $i = 1, 2$ is tangent to the level sets of the other. In other words, h_2 is a conservation law for the Hamiltonian h_1 and the Hamiltonian flow of h_2 is a symmetry of the Hamiltonian system associated with (M, ψ, h_1) (the flow of the Hamiltonian vector field v_{h_1} on M).

A map $f : M \to B$ between two Poisson manifolds is a *Poisson map* if pullback of functions is a Lie algebra homomorphism with respect to the Poisson bracket

$$f^*\{F, G\}_B = \{f^*F, f^*G\}_M.$$

Equivalently, if $df(\psi_M)$ equals $f^*(\psi_B)$ as sections of $f^*(\overset{2}{\wedge} T_B)$. If $H : M \to B$ is a Poisson map with respect to the trivial (zero) Poisson structure on B we will call H a *Hamiltonian map*. Equivalently, H is Hamiltonian if the Poisson structure ψ vanishes on the pullback $H^*(T^*B)$ of the cotangent bundle of B (regarding the latter as a subbundle of (T^*M, ψ)). In particular, the rank of the differential dH is less than or equal to $\dim M - \frac{1}{2} \operatorname{rank}(\psi)$ at every point. A Hamiltonian map pulls back the algebra of functions on B to a commutative Poisson subalgebra of the algebra of functions on M.

The study of a Hamiltonian system (M, ψ, h) simplifies tremendously if one can extend the Hamiltonian function h to a Hamiltonian map $H : M \to B$ of maximal rank $\dim M - \frac{1}{2} \operatorname{rank}(\psi)$. Such a system is called a completely integrable Hamiltonian system. The Hamiltonian flow of a completely integrable system can often be realized as a linear flow on tori embedded in M. The fundamental theorem in this case is Liouville's theorem (stated below).

Definition 2.3 1. Let V be a vector space, $\sigma \in \overset{2}{\wedge} V^*$ a (possibly degenerate) two form. A subspace $Z \subset V$ is called *isotropic (coisotropic)* if it is contained in (contains) its symplectic complement. Equivalently, Z is isotropic if σ restricts to zero on Z. If σ is nondegenerate, a subspace $Z \subset V$ is called *Lagrangian* if it is both isotropic and coisotropic. In this case V is even (say $2n$) dimensional and the Lagrangian subspaces are the n dimensional isotropic subspaces.

2. Let (M, σ) be a symplectic manifold. A submanifold Z is *isotropic* (respectively *coisotropic, Lagrangian*) if the tangent subspaces $T_z Z$ are, for all $z \in Z$.

Example 2.4 For every manifold X, the fibers of the cotangent bundle T^*X over points of X are Lagrangian submanifolds with respect to the standard symplectic structure. A section of T^*X over X is Lagrangian if and only if the corresponding 1-form on X is closed.

We will extend the above definition to Poisson geometry:

Definition 2.5 1. Let U be a vector space, ψ an element of $\overset{2}{\wedge} U$. Let $V \subset U$ be the image of the contraction $\rfloor \psi : U^* \to U$. Let $W \subset U^*$ be its kernel. W is called the null space of ψ. ψ is in fact a nondegenerate element of $\overset{2}{\wedge} V$ giving rise to a symplectic form $\sigma \in \overset{2}{\wedge} V^*$ (its inverse). A subspace $Z \subset U$ is *Lagrangian* with respect to ψ if Z is a Lagrangian subspace of $V \subset U$ with respect to σ. Equivalently, Z is Lagrangian if $(U/Z)^*$ is both an isotropic and a coisotropic subspace of U^* with respect to $\psi \in \overset{2}{\wedge} U \cong \overset{2}{\wedge} (U^*)^*$.

2. Let (M, ψ) be a Poisson manifold, assume that ψ has constant rank (this condition will be relaxed in the complex analytic or algebraic case). A submanifold $Z \subset M$ is *Lagrangian* if the tangent subspaces $T_z Z$ are, for all $z \in Z$. Notice that the constant rank assumption implies that each connected component of Z is contained in a single symplectic leaf.

Theorem 2.6 *(Liouville). Let M be an m-dimensional Poisson manifold with Poisson structure ψ of constant rank $2g$. Suppose that $H : M \to B$ is a proper submersive Hamiltonian map of maximal rank, i.e, $\dim B = m - g$. Then*

i) *The null foliation of M is induced locally by a foliation of B (globally if H has connected fibers).*

ii) *The connected components of fibers of H are Lagrangian compact tori with a natural affine structure.*

iii) *The Hamiltonian vector fields of the pullback of functions on B by H are tangent to the level tori and are translation invariant (linear).*

Remark 2.7 : If H is not proper, but the Hamiltonian flows are complete, then the fibers of H are generalized tori (quotients of a vector space by a discrete subgroup, not necessarily of maximal rank).

<u>Sketch of proof of Liouville's theorem:</u>

i) Since H is a proper submersion the connected components of the fibers of H are smooth compact submanifolds. Since H is a Hamiltonian map of maximal rank $m - g$, the pullback $H^*(T_B^*)$ is isotropic and coisotropic and hence H is a Lagrangian fibration. In particular, each connected component of a fiber of H is contained in a single symplectic leaf.

ii),iii) Let A_b be a connected component of the fiber $H^{-1}(b)$. Let $0 \to T_{A_b} \to T_{M|_{A_b}} \overset{dH}{\longrightarrow} (T_b B) \otimes \mathcal{O}_{A_b} \to 0$ be the exact sequence of the differential of H. Part i) implies that the null subbundle $W_{|A_b} := Ker[\Psi : T^* M \to TM]_{|A_b}$ is the pullback of a subspace W_b of $T_b^* B$. Since H is a Lagrangian fibration, the Poisson structure induces a surjective homomorphism $\phi_b : H^*(T_b^* B) \to T_{A_b}$ inducing a trivialization $\bar{\phi}_b : (T_b^* B / W_b) \otimes \mathcal{O}_{A_b} \overset{\sim}{\longrightarrow} T_{A_b}$.

A basis of the vector space $T_b^* B / W_b$ corresponds to a frame of global independent vector fields on the fiber A_b which commute since the map H is Hamiltonian. Hence A_b is a compact torus.

$\square$

2.3   Algebraically Completely Integrable Hamiltonian Systems

All the definitions and most of the results stated in this chapter for C^∞-manifolds translate verbatim and hold in the complex analytic and complex algebro-geometric categories replacing the real symplectic form by a holomorphic or algebraic $(2,0)$-form (similarly for Poisson structures). The (main) exception listed below is due to the differences between the Zariski topology and the complex or C^∞ topologies. A Zariski open subset is the complement of the zero locus of a system of polynomial equations. It is hence always a dense open subset.

The (local) foliation by symplectic leaves exists only local analytically. For example, a rank 2 translation invariant section $\psi \in H^0(A, \overset{2}{\wedge} TA)$ on a 3 dimensional abelian variety A which is simple (does not contain any abelian subvariety) is an algebraic Poisson structure with a non algebraic null foliation.

We will relax the definitions of a Lagrangian subvariety and integrable system in the algebro-geometric category:

Definition 2.8 Let (M, ψ) be a Poisson smooth algebraic variety. An irreducible and reduced subvariety $Z \subset M$ is *Lagrangian* if the tangent subspace $T_z Z \subset T_z M$ is Lagrangian for a generic point $z \in Z$.

Definition 2.9 An *algebraically completely integrable Hamiltonian system* consists of a proper flat morphism $H : M \to B$ where (M, ψ) is a smooth Poisson variety and B is a smooth variety such that, over the complement $B \smallsetminus \Delta$ of some proper closed subvariety $\Delta \subset B$, H is a Lagrangian fibration whose fibers are isomorphic to abelian varieties.

Multiples of a theta line bundle embed an abelian variety in projective spaces with the coordinates being theta functions. Thus, a priori, the solutions of an algebraically completely integrable Hamiltonian system can be expressed in terms of theta functions. Finding explicit formulas is usually hard. In the next chapter we will study one example, the geodesic flow on ellipsoids, in some detail. Later we will encounter certain equations of Kdv type, the Hitchin system, and a few other examples. Other classical integrable systems include various Euler-Arnold systems, spinning tops, the Neumann system of evolution of a point on the sphere subject to a quadratic potential.

Most of these systems are the complexification of real algebraic systems. Given a real algebraic symplectic variety (M, σ) and an algebraic Hamiltonian h on M we say that the system is *algebraically completely integrable* if its complexification $(M_\mathbf{C}, \sigma_\mathbf{C}, h_\mathbf{C})$ is. A real completely integrable system (M, σ, h) need not be algebraically completely integrable even if (M, σ, h) are algebraic:

A Counter Example: Let (M, σ) be $(\mathbb{R}^2, dx \wedge dy)$ and $h : \mathbb{R}^2 \to \mathbb{R}$ a polynomial of degree d whose level sets are nonsingular. The system is trivially completely integrable, but it is algebraically completely integrable if and only if $d = 3$ because in all other cases the generic fiber of the complexification is a complex affine plane curve of genus $\frac{(d-1)(d-2)}{2} \neq 1$.

Action Angle Coordinates:

Let (M, σ) be a $2n$-dimensional symplectic manifold, $H : M \to B$ a Lagrangian fibration by compact connected tori.

Theorem 2.10 *(real action angle coordinates)*.
In a neighborhood of a fiber of $H : M \to B$ one can introduce the structure of a direct product $(\mathbb{R}^n/\mathbb{Z}^n) \times \mathbb{R}^n$ with action coordinates $(I_1 \cdots I_n)$ on the factor $\mathbb{R}^n$ and angular coordinates $(\phi_1, \cdots \phi_n)$ on the torus $(\mathbb{R}^n/\mathbb{Z}^n)$ in which the symplectic structure has the form $\sum_{k=1}^n dI_k \wedge d\phi_k$.

The Local action coordinates on B are canonical up to affine transformation on $\mathbb{R}^n$ with differential in $SL(n, \mathbb{Z})$. The angle coordinates depend canonically on the action coordinates and a choice of a Lagrangian section of $H : M \to B$.
Remarks:

1. In action angle coordinates the equations of the Hamiltonian flow of a function h on B becomes: $\dot{I}_k = 0$, $\dot{\phi}_k = c_k(I_1, \cdots, I_n)$ where the slopes c_k are $c_k = \frac{\partial h}{\partial I_k}$.
2. In the polarized complex analytic case, we still have local holomorphic action coordinates. They depend further on a choice of a Lagrangian subspace of the integral homology $H_1(A_b, \mathbb{Z})$ with respect to the polarization (a section of $\overset{2}{\wedge} H^1(A_b, \mathbb{Z})$).

2.4   Moment Maps and Symplectic Reduction

Poisson Actions

An action ρ of a connected Lie group G on a manifold M determines an *infinitesimal action*

$$d\rho : \mathbf{g} \longrightarrow V(M),$$

which is a homomorphism from the Lie algebra of G to the Lie algebra of C^∞ vector fields on M. When (M, σ) is symplectic, we say that the action ρ is *symplectic* if

$$(\rho(g))^*\sigma = \sigma, \qquad \text{all} \quad g \in G,$$

or equivalently if the image of $d\rho$ consists of locally Hamiltonian vector fields.

We say that the action ρ is *Poisson* if it factors through the Lie algebra homomorphism
(1) $v : C^\infty(M) \to V(M)$ and a Lie algebra homomorphism

$$H : \mathbf{g} \longrightarrow C^\infty(M).$$

This imposes two requirements on ρ, each of a cohomological nature: the locally Hamiltonian fields $d\rho(X)$ should be globally Hamiltonian, $d\rho(X) = v(H(X))$; and it must be possible to choose the $H(X)$ consistently so that

$$H([X, Y]) = \{H(X), H(Y)\}.$$

(a priori the difference between the two terms is a constant function, since its v is zero, so the condition is that it should be possible to make all these constants vanish simultaneously.)

Moment Maps

Instead of specifying the Hamiltonian lift

$$H : \mathbf{g} \longrightarrow C^\infty(M)$$

for a Poisson action of G on (M, σ), it is convenient to consider the equivalent data of the *moment map*

$$\mu : M \longrightarrow \mathbf{g}^*$$

defined by

$$(\mu(m), X) := H(X)(m).$$

It is a Poisson map with respect to the Kostant-Kirillov Poisson structure on $\mathbf{g}$ (example 2.2), and is G-equivariant.

Examples 2.11   1. Any action of G on a manifold X lifts to an action on $M := T^*X$. This action is Poisson. The corresponding moment map $T^*X \to \mathbf{g}^*$ is the dual of the infinitesimal action $\mathbf{g} \to \Gamma(TX)$. It can be identified with the pullback of differential forms from X to G via the action.

2. The coadjoint action of G on $\mathbf{g}^*$ is Poisson, with the identity as moment map.

Symplectic Reduction

Consider a Poisson action of G on (M, σ) for which a reasonable quotient G/M exists. (We will remain vague about this for now, and discuss the properties of the quotient on a case-by-case basis. A general sufficient condition for the quotient to be a manifold is that the action is proper and free.) The Poisson bracket on M then descends to give a Poisson structure on M/G. The moment map,

$$\mu : M \longrightarrow \mathbf{g}^*,$$

determines the symplectic leaves of this Poisson structure: let $\xi = \mu(m)$, let $\mathcal{O}$ be the coadjoint orbit through ξ and let G_ξ be the stabilizer of ξ. Assume for simplicity that $\mu^{-1}(\xi)$ is connected and μ is submersive at $\mu^{-1}(\xi)$. Then, the leaf through m is

$$\mu^{-1}(\mathcal{O}_\xi)/G \approx \mu^{-1}(\xi)/G_\xi.$$

These symplectic leaves are often called the Marsden-Weinstein reductions M_{red} of M.

As an example, consider a situation where G acts on X with nice quotient X/G. The lifted action of G on $M = T^*X$ is Poisson, and has a quotient M/G which is a vector bundle over X/G. The cotangent $T^*(X/G)$ sits inside $(T^*X)/G$ as the symplectic leaf over the trivial orbit $\mathcal{O}_0 = \{0\} \subset \mathbf{g}^*$.

In contrast, the action of G on $\mathbf{g}^*$ does not in general admit a reasonable quotient. Its action on the dense open subset $\mathbf{g}^*_{reg}$ of regular elements (cf. example 2.2) does have a quotient, which is a manifold. The Poisson structure on the quotient is trivial, so the symplectic leaves are points, in one-to-one correspondence with the regular orbits. We refer to this quotient simply as $\mathbf{g}^*/G$. The map $\pi_{reg} : \mathbf{g}^*_{reg} \to \mathbf{g}^*/G$ extends to $\pi : \mathbf{g}^* \to \mathbf{g}^*/G$, and there is a sense in which $\mathbf{g}^*/G$ really is the quotient of all $\mathbf{g}^*$. Each coadjoint orbit $\mathcal{O}$ is contained in the closure of a unique regular orbit $\mathcal{O}'$ and $\pi(\mathcal{O}) = \pi_{reg}(\mathcal{O}')$.

A Diagram of Quotients

In the general situation of Poisson action (with a nice quotient π) of G on a symplectic manifold (M, σ), there is another, larger, Poisson manifold $\bar{M}$, which can also be considered as a reduction of M by G. Everything fits together in the commutative diagram of Poisson maps:

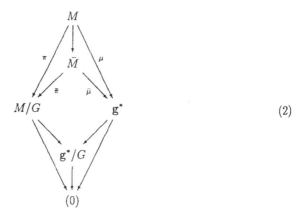

$$(2)$$

$$(0)$$

$\bar{M}$ may be described in several ways:

- $\bar{M}$ is the quotient of M by the equivalence relation $m \sim gm$ if $g \in G_{\mu(m)}$, i.e., if $g(\mu(m)) = \mu(m)$.

- $\bar{M}$ is the fiber product $\bar{M} = (M/G) \times_{(\mathbf{g}^*/G)} \mathbf{g}^*$.

- $\bar{M}$ is the dual realization to the realization $M \to \mathbf{g}^*/G$.

A *realization* of a Poisson manifold P is defined to be a Poisson map from a symplectic manifold M to P (see [We]). The realization will be called *full* if it is submersive. A pair of realizations $P_2 \xleftarrow{f_2} M \xrightarrow{f_1} P_1$ is called a dual pair if functions on one induce vector fields along the fibers of the other (i.e., the two opposite foliations are symplectic complements of each other).

We note that in the diagram of quotients, any two opposite spaces are a dual pair of realizations.

Given a full dual pair with connected fibers, the symplectic leaf foliations on P_1 and P_2 induce the same foliation on M (P_1 and P_2 have the "same" Casimir functions). The bijection between symplectic leaves on P_1 and P_2 is given by

$$P_1 \supset S_1 \mapsto f_2(f_1^{-1}(S_1)) = f_2(f_1^{-1}(x)) \quad \forall\, x \in S_1.$$

Returning to moment maps, we have over a coadjoint orbit $\mathcal{O} \subset \mathbf{g}^*$:

- $\mu^{-1}(\mathcal{O})$ is coisotropic in M

- $\pi(\mu^{-1}(\mathcal{O}))$ is a symplectic leaf M_{red} in M/G

- $\bar{\mu}^{-1}(\mathcal{O})$ is also a symplectic leaf in $\bar{M}$. It is isomorphic to $\mu^{-1}(\mathcal{O})/(\text{null})$, or to $\mu^{-1}(\mathcal{O})/\sim$, or to $M_{red} \times \mathcal{O}$.

Example 2.12 Take M to be the cotangent bundle T^*G of a Lie group G. Denote by $\mu_L : T^*G \to \mathbf{g}^*$ the moment map for the lifted left action of G. The quotient $\pi : M \to M/G$ is just the moment map $\mu_R : T^*G \to \mathbf{g}^*$ for the lifted right action, and $\bar{M}$ is the fiber product $\mathbf{g}^* \times_{(\mathbf{g}^*/G)} \mathbf{g}^*$.

Example 2.13 If G is a connected commutative group T, the pair of nodes $\mathbf{t}^*$ and $\mathbf{t}^*/T$ coincide. Consequently, so do M/T and $\bar{M}$. The diagram of quotients degenerates to

$$
\begin{array}{c}
M \\
\downarrow{\scriptstyle \pi} \\
M/T \\
\downarrow{\scriptstyle \bar{\mu}} \\
\mathbf{t}^* \\
\downarrow \\
(0)
\end{array}
\qquad (3)
$$

Example 2.14 Consider two Poisson actions on (M, σ) of two groups G, T with moment maps μ_G, μ_T with connected fibers. Assume that

i) The actions of G and T commute.

It follows that $\mu_T : M \to \mathbf{t}^*$ factors through M/G and $\mu_G : M \to \mathbf{g}^*$ factors through M/T. Assume moreover

ii) T is commutative,

iii) $M \to \mathbf{g}^*/G$ factors through $\mathbf{t}^*$

Then $\bar{\mu}_G : \bar{M}_G \to \mathbf{g}^*$ factors through M/T and the two quotient diagrams fit nicely together:

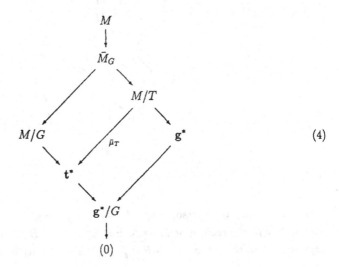

$$(4)$$

Remark 2.15 Note that condition iii in example 2.14 holds whenever $M/G \to \mathbf{t}^*$ is a completely integrable system (with connected fibers). In that case the map $M/G \to \mathbf{t}^*$ pulls back $C^\infty(\mathbf{t}^*)$ to a *maximal* commutative subalgebra $\mathcal{I}_T$ of $(C^\infty(M/G), \{,\})$. The map $M/G \to \mathbf{g}^*/G$ pulls back $C^\infty(\mathbf{g}^*/G)$ to a commutative Lie subalgebra $\mathcal{I}_G$ of $(C^\infty(M/G), \{,\})$. As the two group actions commute so do the subalgebras $\mathcal{I}_G$ and $\mathcal{I}_T$. By maximality, $\mathcal{I}_T$ contains $\mathcal{I}_G$ and consequently $M/G \to \mathbf{g}^*/G$ factors through $\mathbf{t}^*$.

The diagram of quotients for a Poisson action (diagram 2) generalizes to an analogous diagram for any full dual pair of realizations $P_2 \xleftarrow{f_2} M \xrightarrow{f_1} P_1$. Denote by $\bar{M}$ the image of M in the Poisson manifold $P_1 \times P_2$ under the diagonal Poisson map $f_1 \times f_2 : M \to P_1 \times P_2$. The realization dual to $f_1 \times f_2 : M \to \bar{M}$ is the pullback of the symplectic leaf foliations on P_1 or P_2 (they pull back to the same foliation of M).

The following is the analogue of Example 2.14 replacing the commutative T-action by a realization:

Example 2.16 Let $M/G \xleftarrow{\pi} M \xrightarrow{\mu} \mathbf{g}^*$ be the full dual pair associated to a Poisson action of G on M and $N \xleftarrow{\ell} M \xrightarrow{h} B$ a full dual pair of realizations with connected fibers where:

(i) h is G-invariant

(ii) $h : M \to B$ is a Hamiltonian map (B is endowed with the trivial Poisson structure) and

(iii) The composition $M \xrightarrow{\mu} \mathbf{g}^* \longrightarrow \mathbf{g}^*/G$ factors through $h : M \to B$.

Then we get a diagram analogous to the one in example 2.14:

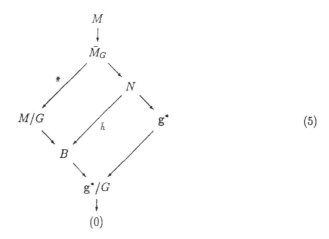

$$(5)$$

It follows that the Poisson map $M \xrightarrow{h \times \mu} B \times_{(\mathbf{g}^*/G)} \mathbf{g}^*$ into the fiber product space factors through the realization $M \xrightarrow{\ell} N$ dual to $h : M \to B$. If, moreover, $M/G \to B$ is a Lagrangian fibration, then $M \xrightarrow{h \times \mu} B \times_{(\mathbf{g}^*/G)} \mathbf{g}^*$ is itself a realization dual to $h : M \to B$.

2.5 Finite dimensional Poisson loop group actions

We present in this section two elementary constructions related to finite dimensional symplectic leaves in the Poisson quotient Q_∞ of an infinite dimensional symplectic space M by subgroups of loop groups. The material in this section will only be used in section 6.3 so the reader may prefer to read it in conjunction with that section.

We will not construct the quotient Q_∞. The spaces involved are constructed independently. Rather, we will analyze the relationship between the Poisson action of the loop group on the infinite dimensional spaces and its descent to the finite dimensional symplectic leaves of Q_∞. In fact, our main purpose in this section is to provide the terminology needed in order to study the Poisson loop group action in the finite dimensional setting (convention 2.20 and corollary 2.22).

In section 2.5.1 we note that the infinitesimal Hamiltonian actions of elements of the loop group descend to Hamiltonian vector fields on finite dimensional symplectic approximations $M_{(l,l)}$. The $M_{(l,l)}$'s dominate finite dimensional Poisson subvarieties Q_l of Q_∞ with positive dimensional fibers. In section 2.5.2 the action of certain maximal tori in the loop group further descends to finite Galois covers of certain (type) loci in Q and we examine the sense in which it is Hamiltonian.

2.5.1 Finite dimensional approximations

The loop group G_∞ is the group $GL(n, \mathbb{C}((z)))$. The level infinity group G_∞^+ is its positive part $GL(n, \mathbb{C}[[z]])$. Let (M, σ) be a symplectic variety with a Poisson loop group action whose moment map is

$$\mu : M \to \mathfrak{g}_\infty^*.$$

In section 6.3 M will be the cotangent bundle of a projective (inverse) limit of finite dimensional smooth algebraic varieties (the cotangent bundles of the moduli spaces of vector bundles with level structure). It is thus the inductive (direct) limit of projective limits of finite dimensional varieties. All constructions (morphisms, group actions, symplectic structures etc ...) can be made precise as limits of the standard constructions on finite dimensional approximations. We will omit the technical details as our point is to transfer the discussion back to the finite dimensional symplectic leaves of the Poisson quotient $Q_\infty := M/G_\infty^+$.

Let $G_\infty^{\geq l}$, $l \geq -1$, be the subgroup of G_∞^+ of elements equal to 1 up to order l. Denote by $\mu_{G_\infty^{\geq l}}$ its moment map. We assume that the subquotients

$$M_{(l,k)} := \mu_{G_\infty^{\geq l}}^{-1}(0)/G_\infty^{>k}, \quad k \geq l,$$

are smooth, finite dimensional and that they approximate M:

$$M = \lim_{l \to \infty} \lim_{\infty \leftarrow k} M_{(l,k)}.$$

Notice that $M_{(l,l)}$ is a symplectic reduction, hence symplectic.

Let a be an element of the loop algebra $\mathfrak{g}_\infty$ with poles of order at most l_0. The Hamiltonian vector field ξ_a on M is an infinite double sequence of Hamiltonian vector fields on $M_{(l,k)}$, $l \geq 0$, $k \geq \max\{l, l_0\}$ compatible with respect to projections and inclusions (by a Hamiltonian vector field on $M_{(l,k)} \subset M_{(k,k)}$ we mean, the restriction of a Hamiltonian vector field on $M_{(k,k)}$ which is tangent to $M_{(l,k)}$).

The quotient $Q_\infty := M/G_\infty^+$ is the direct limit $\lim_{l\to\infty} Q_l$ of the finite dimensional Poisson varieties

$$Q_l := \mu_{G_\infty^{\geq l}}^{-1}(0)/G_\infty^+ = M_{(l,k)}/G_k = M_{(l,l)}/G_l$$

where $G_k := G_\infty^+/G_\infty^{\geq k}$ is the finite dimensional level-k group (we assume that the quotients Q_l are smooth).

Example 2.17 The homogeneous G_∞^+-space $\mathcal{U}_\infty := G_\infty^+/GL(n,\mathbb{C})$ is endowed with a canonical infinitesimal G_∞-action via its embedding as the degree-0 component of the homogeneous G_∞-space $G_\infty/GL(n,\mathbb{C}[[z^{-1}]])$

$$G_\infty^+/GL(n,\mathbb{C}) \hookrightarrow G_\infty/GL(n,\mathbb{C}[[z^{-1}]])$$

(the degree of $a \in G_\infty$ is the signed order of the pole/zero of $\det(a)$). Let M be an open subset of the cotangent bundle $T^*\mathcal{U}_\infty$ for which the regularity assumptions on the approximating quotients $M_{(l,k)}$ hold. This will be made precise in section 4.3 and the quotients Q_l will be the spaces of conjugacy classes of polynomial matrices studied in that section.

Unfortunately, the action of $a \in \mathfrak{g}_\infty$ above is not defined on Q_l. It is well defined only when we retain at least the l_0-level structure, i.e., on $M_{(l,k)}$, $k \geq l_0$. In section 2.5.2 we will see that the action of certain maximal tori in G_∞ descends to *finite* Galois covers of certain loci in Q.

2.5.2 Type loci

Let (M,σ) be a smooth symplectic variety endowed with an infinitesimal Poisson action $\mu_G^* : \mathfrak{g} \to [\Gamma(M,\mathcal{O}_M),\{,\}]$ of a group G. Consider a subgroup $G^+ \subset G$, a commutative subgroup $T \subset G$, and their intersection $T^+ := T \cap G^+$. Assume further that the following conditions hold:

i) The infinitesimal G^+-action integrates to a free action on M,
ii) T^+ is a maximal commutative subgroup whose Weyl group $W_{T^+} := N_{G^+}(T^+)/T^+$ is finite.

Definition 2.18 The *type* τ of T is the class of all commutative subgroups T' of G which are conjugate to T via an element of G^+.

Let $W := [N_{G^+}(T^+) \cap N(T)]/T^+$ be the corresponding subgroup of both W_{T^+} and W_T. Denote by

$$\mathfrak{g}_\tau^* \subset \mathfrak{g}^*$$

(respectively, $\mathfrak{g}_T^* \subset \mathfrak{g}^*$) the subset of elements whose stabilizer (with respect to the coadjoint action) is a torus of type τ (respectively, precisely T).

Example 2.19 Let G be the loop group, G^+ the level infinity group and $T \subset G$ a maximal torus of type $\underline{n}$ determined by a partition of the integer n (see section 6.1). In this case G^+ and T generate G. It follows that $W = W_T = W_{T^+}$ and the type τ is invariant throughout a coadjoint orbit in $\mathbf{g}^*$.

Assume that a "nice" (Poisson) quotient $Q := M/G^+$ exists. Let

$$M^\tau := \mu_G^{-1}(\mathbf{g}_\tau^*), \quad \text{and} \quad Q^\tau := M^\tau/G^+ \subset Q$$

be the loci of type τ. Note that for each T of type τ there is a canonical isomorphism

$$\mu_G^{-1}(\mathbf{g}_T^*)/[N_{G^+}(T^+) \cap N(T)] \overset{\cong}{\to} Q^\tau \subset Q.$$

In particular, a choice of T of type τ determines a canonical W-Galois cover of Q^τ

$$\tilde{Q}^T := \mu_G^{-1}(\mathbf{g}_T^*)/T^+. \tag{6}$$

All the $\tilde{Q}^T$ of type τ are isomorphic (not canonically) to a fixed abstract W-cover $\tilde{Q}^\tau$. Note that $\tilde{Q}^T$ is a subset of M/T^+. We get a canonical "section" (the inclusion)

$$s_T : \tilde{Q}^T \hookrightarrow M/T^+ \tag{7}$$

into a T-invariant subset. Consequently, we get an induced T-action on the Galois cover $\tilde{Q}^T$. The moment map μ_T is T-invariant, hence, descends to M/T^+. Restriction to $s^T(\tilde{Q}^T)$ gives rise to a canonical map

$$\bar{\mu}_T : \tilde{Q}^T \to \mathbf{t}^*. \tag{8}$$

The purpose of this section is to examine *the extent to which $\bar{\mu}_T$ is the moment map of the T-action with respect to the Poisson structure on Q*. In general, the G^+-equivariant projection

$$j : \mathbf{g}^* \twoheadrightarrow (\mathbf{g}^+)^*$$

might *forget the type*. Coadjoint orbits $S \subset (\mathbf{g}^+)^*$ may intersect nontrivially the images $j(\mathbf{g}_\tau^*)$ of several types (e.g., take $S = 0$ in example 2.19 and observe that the kernel of j intersects coadjoint orbits of all types). Consequently, symplectic leaves Q_S of Q would intersect nontrivially several type loci Q^τ. If $Q_S^{\tau^{open}}$ is an open subvariety of Q_S of type τ (e.g., if Q is the disjoint union of finitely many type loci and τ is a *generic type*) then the corresponding open subvariety $\tilde{Q}_S^{\tau^{open}}$ of $\tilde{Q}_S^T$ will be a symplectic variety. In this case the T-action on $\tilde{Q}_S^{\tau^{open}}$ is Poisson whose moment map $\bar{\mu}_T$ is given by (8).

The Galois W-covers $\tilde{Q}_S^T$ of the nongeneric type loci in Q_S are not symplectic. Nevertheless, motivated by the fact that $\bar{\mu}_T$ can be extended canonically to M/T^+

$$
\begin{array}{ccc}
M & \overset{\mu_T}{\longrightarrow} & \mathbf{t}^* \\
\downarrow & \nearrow & \\
& \overset{\bar{\mu}_T}{\nearrow} & \\
M/T^+ & \overset{\supset}{\leftarrow} & \tilde{Q}_S^T \\
\downarrow & & \downarrow \\
Q & \overset{\supset}{\leftarrow} & Q_S^\tau
\end{array}
\tag{9}
$$

we will adopt the:

Convention 2.20 i) Given an element h of t we will say that the corresponding vector field $\tilde{\xi}_h$ on $\tilde{Q}_S^T$ is the *Hamiltonian vector field of h* (even if the type τ of T is not generic in Q_S). ii) We will refer to the pair $(\bar{\mu}_T, \mu_T)$ as the moment map of the T-action on $\tilde{Q}_S^T$.

Remarks 2.21 Let G be the loop group and M, G^+, T as in section 2.5.1,

1. Diagram (9) has an obvious finite dimensional approximation in which Q_S^τ, $\tilde{Q}_S^T$ and T stay the same but with M replaced by $M_{(l,l)}$ and Q by Q_l. By μ_T we mean in this context a linear homomorphism $\mu_T^* : \mathrm{t} \to \mathrm{t}/\mathrm{t}_\infty^{\geq l} \to \Gamma(M_{(l,l)}, \mathcal{O}_{M_{(l,l)}})$.

2. (Relation with the diagram of quotients (2)) Let S be a coadjoint orbit of level l, i.e., $S \subset \mathrm{g}_l^* := (\mathrm{g}_\infty^+/\mathrm{g}_\infty^{\geq l})^* \subset (\mathrm{g}_\infty^+)^*$. There is a rather subtle relationship between the Galois cover $\tilde{Q}_S^T \to Q_S^\tau$ and the space $\bar{M}_{(l,l)}$ dual to g_l^*/G_l from the diagram of quotients (2) of level l. The Galois cover $\tilde{Q}_S^T \to Q_S^\tau$ factors canonically through an intermediate subspace $\tilde{Q}_S^T / \sim$ of $\bar{M}_{(l,l)}$. Note that the loop group moment map μ_{G_∞} descends to a map

$$\bar{\mu}_{G_\infty} : \tilde{Q}_S^T \to (\mathrm{g}_\infty^*)_T \subset \mathrm{g}_\infty^*.$$

Two points $\tilde{x}_1, \tilde{x}_2 \in \tilde{Q}_S^T$ in a fiber over $x \in Q_S^\tau$ are identified in $\tilde{Q}_S^T / \sim$ if and only if $\bar{\mu}_{G_\infty}(\tilde{x}_1)$ and $\bar{\mu}_{G_\infty}(\tilde{x}_1)$ project to the same point in $S \subset (\mathrm{g}_\infty^+)^*$. The relation $\sim$ is a geometric realization of the partial type-forgetfullness of the projection $j : \mathrm{g}_\infty^* \to (\mathrm{g}_\infty^+)^*$. The loci in $\bar{M}_{(l,l)}$ at which the type is not forgotten are precisely the loci to which the moment map of the infinitesimal *loop group* action descends (from $M^T := \mu_{G_\infty}^{-1}((\mathrm{g}_\infty^*)_T)$. Note that the moment map of the level infinity subgroup descends by definition of the quotient $\bar{M}_{(l,l)}$). In particular, the infinitesimal action of the maximal torus t integrates to a Poisson action in these loci. (See section 6.3.2 for examples of such loci.)

Assume further that we have a "nice" quotient $B := M/G$ and that G is generated by T and G^+. We get the type loci $B^\tau := M^\tau/G$. Fixing T of type τ we get the W-cover $\tilde{B}^T := M^T/T$. The restriction of the moment map μ_T to M^T descends further to $\phi_T : \tilde{B}^T \to \mathrm{t}^*$ and we get the commutative diagram

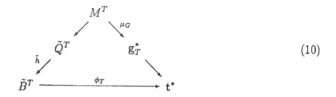

$$(10)$$

Corollary 2.22 i) The T-action on M descends to a canonical action on the W-cover $\tilde{Q}^T$ of the type locus $Q^\tau \subset Q$. ii) Its moment map, in the sense of convention 2.20, is $(\phi_T \circ \tilde{h}, \mu_T)$. iii) If the type τ of T is the generic type in a symplectic leaf $Q_S \subset Q$ and $Q_S^{\tau^{open}} \subset Q_S$ is an open subvariety, then the corresponding W-cover $\tilde{Q}_S^{\tau^{open}}$ is symplectic and $\phi_T \circ \tilde{h}$ is the moment map of the T-action in the usual sense.

3 Geodesic flow on an ellipsoid

Consider the geodesic flow on an ellipsoid

$$E = \{(x_1, \cdots, x_{n+1}) | \sum_{i=1}^{n+1} \frac{1}{a_i} x_i^2 = 1\} \subset \mathbb{R}^{n+1},$$

where the metric is induced from the standard one on $\mathbb{R}^{n+1}$, and where the a_i are distinct positive numbers, say

$$0 < a_1 < \cdots < a_{n+1}.$$

For $n = 1$, the problem is to compute arc length on an ellipse. It amounts to computing the integral

$$s = \int \sqrt{\frac{a_1^2 + (a_2 - a_1)x^2}{a_1(a_1 - x^2)}} \, dx.$$

(Hence the name *elliptic* for this and similar integrals.)

For $n = 2$, the problem was solved by Jacobi. Each geodesic γ on E determines a hyperboloid E', intersecting E in a pair of ovals. The geodesic γ oscillates in the band between these ovals, meeting them tangentially. In fact, each tangent line of γ is also tangent to the hyperboloid E'. The solutions can be parametrized explicitly in terms of hyperelliptic theta functions.

The geodesic flow on an n-dimensional ellipsoid is integrable, in fact algebraically integrable. We will see this, first using some elementary geometric techniques to describe the geodesics concretely, and then again using the algebraic description of hyperelliptic jacobians which will be extended later to all spectral curves.

3.1 Integrability

The geodesic flow on the $2n$-dimensional symplectic manifold $TE \approx T^*E$ is given by the Hamiltonian function h=length square. We need $n - 1$ farther, commuting, independent, Hamiltonians.

Consider the family of quadrics confocal to E:

$$E_\lambda : \sum_{i=1}^{n+1} \frac{x_i^2}{a_i - \lambda} = 1,$$

depending on a parameter λ. (The name makes sense only when $n = 1$: we get the family of ellipses ($\lambda < a_1$), hyperbolas ($a_1 < \lambda < a_2$), and empty (real) conics ($\lambda > a_2$), with fixed foci.)

Here is an intrinsic way to think of this family. Start with a linear pencil

$$Q_\lambda = Q_0 + \lambda Q_\infty, \qquad (\lambda \in \mathbb{P}^1)$$

of quadrics in general position in projective space $\mathbb{P}^{n+1}$. By "general position" we mean that there are exactly $n + 2$ values of $\lambda \in \mathbb{P}^1$ such that Q_λ is singular, and for those λ, Q_λ is a cone (i.e. its singular locus, or vertex, is a single point).

Lemma. *A generic linear subspace $L \approx \mathbb{P}^{k-1}$ in $\mathbb{P}^{n+1}$ is tangent to Q_λ for k values of λ. The points of tangency p_λ are pairwise harmonic with respect to each of the quadrics Q_μ.*

Proof: Four points of $\mathbb{P}^1$ are harmonic if their cross ratio is -1; e.g. $0, \infty, a, -a$. Two points $p_1, p_2 \in \mathbb{P}^1$ are harmonic with respect to a quadric Q if the set $\{p_1, p_2\} \cup (Q \cap \mathbb{P}^1)$ is harmonic. For example, two points on the line at infinity in $\mathbb{P}^2$ (i.e. two directions in the affine plane) are harmonic with respect to some (hence every) circle, iff the directions are perpendicular.

Since Q is tangent to L if and only if $Q \cap L$ is singular, the first part of the lemma follows by restriction of the pencil to L. The second part follows by restricting to the line $p_{\lambda_1}, p_{\lambda_2}$, where in appropriate coordinates $Q_{\lambda_1} = x^2$ and $Q_{\lambda_2} = y^2$, so the points of tangency are $0, \infty$ and the quadric Q_λ vanishes at $\pm a$, where $a^2 = \lambda$.

$\square$

We choose the parameter λ so that Q_∞ is one of the singular quadrics. The dual Q_λ^* of a non-singular Q_λ is a non-singular quadric in $(\mathbb{P}^{n+1})^*$. The dual of Q_∞ is a hyperplane $H_\infty \subset (\mathbb{P}^{n+1})^*$ (corresponding to the vertex of Q_∞), with a non-singular quadric $Q_\infty^* \subset H_\infty$. We get a family of confocal quadrics by restriction to the affine space $\mathbb{R}^{n+1} := (\mathbb{P}^{n+1})^* \setminus H_\infty$:

$$E_\lambda := Q_\lambda^*|_{\mathbb{R}^{n+1}}.$$

If we choose coordinates so that

$$\begin{aligned} Q_\infty &= \sum_{i=1}^{n+1} x_i^2 \\ Q_0 &= \sum_{i=0}^{n} a_i x_i^2 \end{aligned}$$

(where $a_0 = -1$ and the other a_i are as above), we retrieve the original E_λ. (Euclidean geometry in $\mathbb{R}^{n+1}$ is equivalent, in the sense of Klein's program, to the geometry of $\mathbb{P}^{n+1}$ with a distinguished "light-cone" Q_∞. In this equivalence, $E = E_0$ corresponds to Q_0, which determines the pencil $\{Q_\lambda\}$, which corresponds to the confocal family $\{E_\lambda\}$.)

Dualizing the lemma, for $k = n + 1, n$, gives the following properties of the confocal family. (The reader is invited to amuse herself by drawing the case $n = 1$ in the plane.)

(1) Through a generic point x of $\mathbb{R}^{n+1}$ pass $n + 1$ of the E_λ.

(2) These $n + 1$ quadrics intersect perpendicularly at x.

(3) A generic line ℓ in $\mathbb{R}^{n+1}$ is tangent to n of the E_λ.

(4) The tangent hyperplanes to these n quadrics (at their respective points of tangency to ℓ) are perpendicular.

By property (3) we can associate to a generic line ℓ in $\mathbb{R}^{n+1}$ an unordered set of n values λ_i, $1 \le i \le n$, such that Q_{λ_i} is tangent to ℓ. When ℓ comes from a point of TE, one of these, say λ_n, equals 0. The remaining $n - 1$ values λ_i (or rather, their symmetric functions) give $n - 1$ independent functions on TE; in fact, they can take an arbitrary $(n - 1)$-tuple of values. These functions descend to the projectivized tangent bundle $\mathbb{P}(TE)$; so together with the original Hamiltonian h ($=$ length squared) they give n independent functions on TE. The key to integrability is:

Chasles' Theorem. *The λ_i are flow invariants, i.e. they are constant along a geodesic $\gamma = \gamma(t)$.*

Proof. For any curve $\gamma(t)$ in $\mathbb{R}^{n+1}$, the family of tangent lines $\ell(t)$ gives a curve Λ in the Grassmannian $Gr(1, \mathbb{R}^{n+1})$ of affine lines in $\mathbb{R}^{n+1}$. This curve is developable, i.e. its tangent line $T_{\ell(t)}\Lambda$ is given by the pencil of lines through $\gamma(t)$ in the osculating plane of γ at t. When γ is a geodesic, this plane is the span of $\ell(t)$ and the normal vector $n(t)$ to E at $\gamma(t)$. Write $\lambda_i(t)$ for the value of λ_i at $\ell(t)$.

Let Z_i be the hypersurface in $Gr(1, \mathbb{R}^{n+1})$ parametrizing lines tangent to $E_{\lambda_i(t)}$, for some fixed t. The tangent space $T_{\ell(t)}Z_i$ contains all lines through $\gamma(t)$ in the tangent hyperplane $T_{p_i(t)}E_{\lambda_i(t)}$, and this hyperplane contains the normal $n(t)$, by property (4). Hence:

$$T_{\ell(t)}\Lambda \subset \{\text{lines through } \gamma(t), \text{ in } T_{p_i(t)}E_{\lambda_i(t)}\} \subset T_{\ell(t)}Z_i \qquad i = 1, \cdots, n-1.$$

If the family Λ of tangent lines to a geodesic meets Z_i, it must therefore stay in it.

$\square$

3.2 Algebraic integrability

Since a line ℓ determines two (opposite) tangent vectors of given non-zero length, we have identified the fiber of the geodesic flow as a double cover $\tilde{K}$ of

$$K := \{\ell \in Gr(1, \mathbb{R}^{n+1}) | \ell \text{ is tangent to } E_{\lambda_1}, \cdots, E_{\lambda_n}\}.$$

Next we want to interpret this in terms of the real points of a complex abelian variety. We follow Knörrer's approach [Kn], which in turn is based on [Mo],[Re] and [D5].

Start with the pencil of quadrics in $\mathbb{P}^{2n+1}$ (over $\mathbb{C}$):

$$Y_\lambda := Y_0 + \lambda Y_\infty$$

with

$$\begin{aligned} Y_0 &= \textstyle\sum_{i=1}^{2n+1} a_i x_i^2 - x_0^2 \\ Y_\infty &= \textstyle\sum_{i=1}^{2n+1} x_i^2. \end{aligned}$$

The base locus $X = Y_0 \cap Y_\infty$ is non-singular if the a_i are distinct. We set $a_0 = \infty$. The family of linear subspaces $\mathbb{P}^n$ contained in a fixed quadric Y_λ consists of two connected components, or rulings, for the non-singular Y_λ ($\lambda \notin \{a_i\}$), and of a single ruling for $\lambda = a_i$. We thus get a double cover

$$\pi : C \to \mathbb{P}^1$$

of the λ-line, parametrizing the rulings. (More precisely, one considers the variety of pairs

$$\mathcal{P} = \{(A, \lambda) | A \text{ is a } \mathbb{P}^n \text{ contained in } Q_\lambda\},$$

and takes the Stein factorization of the second projection.) Explicitly, C is the hyperelliptic curve of genus n:

$$C: \quad s^2 = \Pi_{i=1}^{2n+1}(t - a_i).$$

Miles Reid [Re] showed that the Jacobian $J(C)$ is isomorphic to the variety

$$F := \{A \in Gr(n-1, \mathbb{P}^{2n+1}) | A \subset X\}$$

of linear subspaces in the base locus. An explicit group law on F is given in [D5], and corresponding results for rank 2 vector bundles on C are in [DR]. Since we are interested in a family of varieties F with varying parameters, we need some information about the isomorphism. Let $Pic^d(C)$ denote the variety parametrizing isomorphism classes of degree-d line bundles on C. Then $Pic^0(C) = J(C)$ is a group, $Pic^1(C)$ is a torser (= principal homogeneous space) over it, but, these two have no natural identification; while $Pic^2(C) \approx J(C)$ canonically, using the hyperelliptic bundle on C. It turns out that F is isomorphic to $Pic^0(C)$ and to $Pic^1(C)$, but neither isomorphism is canonical. Rather, we may think of F as "$Pic^{\frac{1}{2}}(C)$": it is a torser over $J(C)$, and has a natural torser map

$$F \times F \to Pic^1(C).$$

All of this is based on the existence of a natural morphism

$$j : F \times C \to F.$$

The ruling p (on the quadric $Y_{\pi(p)}$) contains a unique subspace $\mathbb{P}^n$ which contains a given $\mathbb{P}^{n-1}$-subspace $A \in F$. This $\mathbb{P}^n$ intersects X in the union of A and another element of F, which we call $j(A, p)$. We can also think of j as a family of involutions of F, indexed by $p \in C$. This extends to a map

$$F \times J(C) \to F$$

which gives the torser structure on F. Once F is thus identified with $J(C)$, the map j becomes

$$j(A, p) = p - A,$$

up to an additive constant. Since this is well defined globally, points $A \in F$ must behave as line bundles on C of "degree $\frac{1}{2}$". In particular, we have for $0 \le i \le 2n+1$ the involution

$$\begin{aligned} j_i : \quad & F \to F \\ & A \mapsto j(A, a_i), \end{aligned}$$

where we set $a_i = \infty$ and identify the $a_i \in \mathbb{P}^1$ with the $2n+2$ Weierstrass points $\pi^{-1}(a_i) \in C$. Explicitly, each j_i is induced by the linear involution

$$\overline{j}_i : \mathbb{P}^{2n+1} \to \mathbb{P}^{2n+1}$$

flipping the sign of the i-th coordinate.

Consider the linear projection

$$\begin{aligned} \rho : \mathbb{P}^{2n+1} \quad & \to \quad \mathbb{P}^{n+1} \\ (x_0, \cdots, x_{2n+1}) \quad & \mapsto \quad (x_0, \cdots, x_{n+1}), \end{aligned}$$

which commutes with the $\overline{j}_i$, $n+2 \le i$. Recall that in $\mathbb{P}^{n+1}$ we have the pencil of quadrics Q_λ, with dual quadrics E_λ in $(\mathbb{P}^{n+1})^*$.

Proposition.

(i) *The projection ρ maps F to*

$$F' := \{B \in Gr(n-1, \mathbb{P}^{n+1}) | B \text{ is tangent to } Q_{n+2}, \cdots, Q_{2n+1}\}.$$

(ii) *The induced $\rho : F \to F'$ is a finite morphism of degree 2^n, and can be identified with the quotient of F by the group $G \approx (\mathbb{Z}/2\mathbb{Z})^n$ generated by the involutions j_i, $n+2 \leq i \in 2n+1$.*

(iii) *Duality takes F' isomorphically to the variety K of lines in $(\mathbb{P}^{n+1})^*$ tangent to E_{λ_i}, $\lambda_i = a_{n+1+i}, 1 \leq i \leq n$.*

We omit the straightforward proof. Let $\tilde{G} \subset G$ be the index-2 subgroup generated by the products $j_{i_1} \circ j_{i_2}$, and set

$$\widetilde{K} := F/\tilde{G}.$$

We obtain natural commuting maps, whose degrees are indicated next to the arrows:

$$
\begin{array}{ccccc}
F & \xrightarrow{2^{n-1}} & \widetilde{K} & \xrightarrow{2^{n+1}} & Pic^1(C) \\
& & \downarrow^{2} & & {}^{2}\downarrow \\
& & K & \xrightarrow{2^{n+1}} & \text{Kummer}^1(C).
\end{array}
$$

Here $\text{Kummer}^d(C)$ stands for the quotient of $Pic^d C$ by the involution

$$L \mapsto dH - L,$$

where H is the hyperelliptic bundle $\in Pic^2(C)$. The composition of the maps in the top row is multiplication by 2:

$$F \approx Pic^{\frac{1}{2}}(C) \xrightarrow{\cdot 2} Pic^1(C).$$

In conclusion, the fiber of the geodesic flow on $E = E_0$ with invariants $h = 1$ (say) and $\lambda_i = a_{n+1+i}, 1 \leq i \leq n$ can be identified with the real locus in $\widetilde{K} = \widetilde{K}(\lambda_1, \cdots, \lambda_n)$. The latter is a 2^{n+1}-sheeted cover of $Pic^1(C)$, so up to translation by some points of order 2, it is an abelian variety, isomorphic to a 2^{n+1} sheeted cover of the hyperelliptic Jacobian $J(C)$.

3.3   The flows

Two details of the above story are somewhat unsatisfactory: First, the asymmetry between the $n-1$ Hamiltonians λ_i and the remaining Hamiltonian H (length squared). And second, the fact that the complexified total space TE of the system is not quite symplectic. Indeed, for an arbitrary algebraic hypersurface $M \subset \mathbb{C}^{n+1}$, given by $f = 0$, the complexified metric on $\mathbb{C}^{n+1}$ induces bundle maps

$$TM \hookrightarrow T\mathbb{C}^{n+1}|_M \overset{\sim}{\to} T^*\mathbb{C}^{n+1}|_M \twoheadrightarrow T^*M,$$

but the composition is not an isomorphism; rather, it is degenerate at points where

$$0 = (\nabla f)^2 = \sum_{k=1}^{n+1} \left(\frac{\partial f}{\partial x_k} \right)^2.$$

For an ellipsoid $\sum x_k^2/a_k = 1$ (other than a sphere) there will be an empty real, but non-empty complex degeneracy locus, given by the equation $\sum (x_k/a_k)^2 = 0$.

Both of these annoyances disappear if we replace the total space by the tangent bundle TS of the sphere

$$S = \{(x_1, \cdots, x_{n+1}) \in \mathbb{C}^{n+1} \mid \sum x_k^2 = 1\},$$

i.e.

$$TS = \{(x, y) \in \mathbb{C}^{2n+2} \mid \sum x_k^2 = 1, \sum x_k y_k = 0\}.$$

This is globally symplectic, and the n (unordered) Commuting Hamiltonians can be taken to be the values λ_i, $1 \leq i \leq n$, such that the line

$$\ell_{x,y} := (\text{line through } y \text{ in direction } x)$$

is tangent to E_{λ_i}. The original system TE can be recovered as a $\mathbb{C}^*$-bundle (where $\mathbb{C}^*$ acts by rescaling the tangent direction x) over the hypersurface $\lambda = 0$ in TS.

Here is the explicit equation of the hypersurface:

$$\lambda = 0 \quad \Leftrightarrow \quad \ell_{x,y} \text{ is tangent to } E = \{\sum \frac{x_k^2}{a_k} = 1\}$$
$$\Leftrightarrow \quad -1 + \sum \frac{(y_k + t x_k)^2}{a_k} = 0 \text{ has a unique solution } t$$
$$\Leftrightarrow \quad (-1 + \sum \frac{y_k^2}{a_k} + 2t(\sum \frac{x_k y_k}{a_k}) + t^2(\sum \frac{x_k^2}{a_k}) = 0 \text{ has a unique solution}$$
$$\Leftrightarrow \quad 0 = (\sum \frac{x_k y_k}{a_k})^2 - (\sum \frac{x_k^2}{a_k})(-1 + \sum \frac{y_k^2}{a_k}).$$

More generally, this computation shows that $\ell_{x,y}$ is tangent to E_λ if and only if

$$0 = (\sum \frac{x_k y_k}{a_k - \lambda})^2 - (\sum \frac{x_k^2}{a_k - \lambda})(-1 + \sum \frac{y_k^2}{a_k - \lambda}) = \sum_k \frac{x_k^2}{a_k - \lambda} + \sum_{k \neq l} \frac{x_k y_k x_l y_l - x_k^2 y_l^2}{(a_k - \lambda)(a_l - \lambda)}.$$

As a function of λ, the last expression has first order poles at $\lambda = a_k$, $1 \leq k \leq n+1$, so it can be rewritten as

$$\sum \frac{1}{a_k - \lambda} F_k(x, y)$$

where the F_k are found by taking residue at $\lambda = a_k$:

$$F_k(x, y) := x_k^2 + \sum_{\ell \neq k} \frac{(x_k y_\ell - x_\ell y_k)^2}{a_k - a_\ell}.$$

We see that fixing the $n + 1$ values $F_k(x, y)$, $1 \leq k \leq n + 1$, subject to the condition

$$\sum_{k=1}^{n+1} F_k = 1,$$

is equivalent to fixing the n (unordered) values λ_i, $1 \leq i \leq n$.

This determines the hyperelliptic curve

$$C : s^2 = \prod_{k=1}^{n+1}(t - a_k) \cdot \prod_{i=1}^{n}(t - \lambda_i),$$

and the corresponding abelian variety

$$\widetilde{K} = \widetilde{K}(\lambda_1, \cdots, \lambda_n) = J(C)/\widetilde{G} \approx \{(x, y) | \ell_{x,y} \text{ is tangent to } E_{\lambda_i}, \quad 1 \le i \le n\}.$$

Theorem.

(1) *Geodesic flow on the quadric E_{λ_i} is the Hamiltonian vector field on TS given by the (local) Hamiltonian λ_i. On $\widetilde{K}$ it is a constant vector field in the direction of the Weierstrass point $\lambda_i \in C$.*

(2) *The Hamiltonian vector field on TS with Hamiltonian F_k is constant on $\widetilde{K}$, in the direction of the Weierstrass point $a_k \in C$.*

The direction at $\ell \in \widetilde{K}$ of geodesic flow on E_{λ_i} was described in the proof of Chasles' theorem. The direction given by the Weierstrass point λ_i is given at $A \in F$ as the tangent vector at λ_i to the curve

$$p \mapsto j(A_i, p) \quad (\text{where } A_i = j(A, \lambda_i)).$$

The proof of (1) amounts to unwinding the definitions to see that these two directions agree. (For details, see [Kn] and [D5].) Since the level sets of the λ_i and the F_k are the same, the Hamiltonian vector field of F_k evolves on the same $\widetilde{K}$, and is constant there. A monodromy argument on the family of hyperelliptic curves then shows that its direction must agree with a_k. Mumford gives an explicit computation for this in [Mum2], Theorem 4.7, following Moser [Mo].

We have identified the flows corresponding to $2n + 1$ of the Weierstrass points. The remaining one, at $\lambda = \infty$, corresponds to the Hamiltonian

$$H = \frac{1}{2}\sum_{k=1}^{n+1} a_k F_k = \frac{1}{2}\sum a_k x_k^2 + \frac{1}{2}\sum y_k^2,$$

giving Neumann's system, which is the starting point for the analysis in [Mo] and [Mum2].

3.4 Explicit parametrization

Fix the hyperelliptic curve of genus n

$$C : s^2 = f(t) := \prod_{i=1}^{2n+1}(t - a_i),$$

with projection

$$\begin{aligned} \pi : C &\to \mathbb{P}^1 \\ (t, s) &\mapsto t \end{aligned}$$

and involution

$$i : C \to C$$
$$(t,s) \mapsto (t,-s).$$

We identify the various components $Pic^d(C)$ by means of the base point ∞, which is a Weierstrass point. The affine open subset $J(C) \smallsetminus \Theta$ can be described geometrically, by Riemann's theorem:

$$J(C) \smallsetminus \Theta \approx$$
$$\{L \in Pic^{n-1}(C) | h^0(L) = 0\} \approx$$
$$\{(p_1, \cdots, p_n) \in Sym^n C | p_i \neq \infty, p_i \neq i(p_j)\},$$

where the last identification sends L to the unique effective divisor of $L(\infty)$. Mumford [Mum2] gives an explicit algebraic parametrization of the same open set, which he attributes to Jacobi: to the n-tuple $D = (p_1, \cdots, p_n)$ he associates three polynomials of a single variable t:

(i) $U(t) := \prod_{i=1}^n (t - t(p_i))$.

(ii) $V(t)$ is the unique polynomial of degree $\leq n-1$ such that the meromorphic function

$$V \circ \pi - s \quad : \quad C \to \mathbf{P}^1$$

vanishes on the divisor $D \subset C$. It is obtained by Lagrange interpolation of the expansions of s at the p_i, e.g. when the p_i are all distinct,

$$V(t) = \sum_{i=1}^n s(p_i) \prod_{j \neq i} \frac{t - t(p_j)}{t(p_i) - t(p_j)}.$$

(iii) $W(t) = \frac{f(t) - V(t)^2}{U(t)}$; the definition of V and the equation $s^2 = f(t)$ guarantee that this is a monic polynomial of degree $n+1$.

Conversely, the polynomials U, V, W determine the values $t(p_i)$, $s(p_i)$, hence the divisor D. By reading off the coefficients, we obtain an embedding:

$$(U, V, W) : J(C) \backslash \Theta \hookrightarrow \mathbf{C}^{3n+1}.$$

The image is

$$\{(U, V, W) | V^2 + UW = f\}.$$

This description fits beautifully with the integrable system on TS representing geodesic flow on the E_λ. We can rephrase our previous computation as:

$$\ell_{x,y} \text{ is tangent to } E_{\lambda_1}, \cdots, E_{\lambda_n} \Leftrightarrow f_1(t)f_2(t) = UW + V^2,$$

where, for $(x,y) \in TS$, we set:

$f_1(t) = \prod_{k=1}^{n+1}(t - a_k)$ (this is independent of x,y)
$f_2(t) = f_1(t) \cdot \sum_k \frac{F_k(x,y)}{t-a_k}$ (this varies with x, y; the roots are the λ_i)
$U(t) = f_1(t)(\sum_k \frac{x_k^2}{t-a_k})$
$W(t) = f_1(t)(1 + \sum_k \frac{y_k^2}{t-a_k})$
$V(t) = \sqrt{-1} \cdot f_1(t) \cdot (\sum_k \frac{x_k y_k}{t-a_k})$

The entire system TS is thus mapped to $\mathbb{C}^{3n+1}$. Each abelian variety $\widetilde{K} = \widetilde{K}(\lambda_1, \cdots, \lambda_n)$ is mapped to $\mathcal{J}(C) \backslash \Theta$ embedded in $\mathbb{C}^{3n+1}$ as before, where C is defined by $s^2 = f(t)$, and $f = f_1 \cdot f_2$, with f_1 fixed (of degree $n+1$) and f_2 variable (of degree n). On each $\widetilde{K}$ the map is of degree 2^{n+1}; the group $(\mathbb{Z}/2\mathbb{Z})^{n+1}$ operates by sending

$$(x_k, y_k) \mapsto (\epsilon_k x_k, \epsilon_k y_k), \quad \epsilon_k = \pm 1.$$

4 Spectral curves and vector bundles

We review in this chapter a general construction of an integrable system on the moduli space of Higgs pairs (E, φ) consisting of a vector bundle E on a curve and a meromorphic 1-form valued endomorphism φ (theorem 4.8). These moduli spaces admit a natural foliation by Jacobians of spectral curves. The spectral curves are branched covers of the base curve arising from the eigenvalues of the endomorphisms φ.

We concentrate on two examples:

- The Hitchin system supported on the cotangent bundle of the moduli space of vector bundles on a curve (section 4.2), and
- An integrable system on the moduli space of conjugacy classes of polynomial matrices (section 4.3).

The latter is then used to retrieve the Jacobi-Moser-Mumford system which arose in chapter 3 out of the geodesic flow on an ellipsoid.

Both examples are endowed with a natural symplectic or Poisson structure. The general construction of the Poisson structure on the moduli spaces of Higgs pairs is postponed to chapter 5.

We begin with a short survey of some basic facts about vector bundles on a curve.

4.1 Vector Bundles on a Curve

We fix a (compact, non-singular) curve Σ of genus g. A basic object in these lectures will be the moduli space of stable (or semistable) vector bundles on Σ of given rank r and degree d. To motivate the introduction of this object, let us try to describe a "general" vector bundle on Σ. One simple operation which produces vector bundles from line bundles is the direct image: start with an r-sheeted branched covering $\pi : C \to \Sigma$, ramified at points of some divisor R in the non-singular curve C. Then any line bundle $L \in Pic\ C$ determines a rank-r vector bundle $E := \pi_* L$ on Σ. As a locally free sheaf of $\mathcal{O}_\Sigma$-modules of rank r, this is easy to describe

$$\Gamma(\mathcal{U}, \pi_* L) := \Gamma(\pi^{-1}\mathcal{U}, L),$$

for open subsets $\mathcal{U} \subset \Sigma$. As a vector bundle, the description is clear only at unbranched points of Σ: if $\pi^{-1}(p)$ consists of r distinct points $p_1, \cdots, p_r$ then the fiber of E at p is naturally isomorphic to the direct sum of the fibers of L:

$$E_p \approx \bigoplus_{i=1}^{r} L_{p_i}.$$

At branch points of π, E_p does not admit a natural decomposition, but only a filtration. This is reflected in a drop in the degree. Indeed, the Grothendieck-Riemann-Roch theorem says in our (rather trivial) case that

$$\chi(\pi_* L) = \chi(L),$$

where χ is the holomorphic Euler characteristic,

$$\chi(E) := \deg E - (g-1)\operatorname{rank}\ E = \deg E + \chi(\mathcal{O}) \cdot \operatorname{rank}\ E.$$

Using Hurwitz' formula:

$$\chi(\mathcal{O}_C) = r\,\chi(\mathcal{O}_\Sigma) - \frac{1}{2}\deg\,R,$$

this becomes:

$$\deg\,E = \deg\,L - \frac{1}{2}\deg\,R.$$

Example 4.1 Consider the double cover

$$\begin{aligned}\pi : \mathbb{P}^1 &\longrightarrow \mathbb{P}^1\\ w &\longrightarrow z = w^2\end{aligned}$$

branched over $0, \infty$. The direct image of the structure sheaf is:

$$\pi_*\mathcal{O} \approx \mathcal{O} \oplus \mathcal{O}(-1).$$

We can think of this as sending a regular function $f = f(w)$ (on some invariant open set upstairs) to the pair $(f_+(z), f_-(z))$ downstairs, where

$$f(w) = f_+(w^2) + w f_-(w^2).$$

In the image we get all pairs with f_+ regular (i.e. a section of $\mathcal{O}$) and f_- regular and vanishing at ∞ (i.e. a section of $\mathcal{O}(-1)$). (Similar considerations show that

$$\pi_*\mathcal{O}(-1) \approx \mathcal{O}(-1) \oplus \mathcal{O}(-1)$$

and more generally:

$$\pi_*\mathcal{O}(d) \approx \mathcal{O}([\frac{d}{2}]) \oplus \mathcal{O}([\frac{d-1}{2}]).$$

Note that this has degree $d-1$, as expected). The structure of π_*L near the branch point $z = 0$ can be described, in this case, by the action on the local basis a, b (of even, odd sections) of multiplication by the section w upstairs:

$$a \mapsto b, \qquad b \mapsto za,$$

i.e. w is represented by the matrix

$$\begin{pmatrix} 0 & z \\ 1 & 0 \end{pmatrix}$$

whose square is $z \cdot I$. At a branch point where k sheets come together, the corresponding action (in terms of a basis indexed by the $k-th$ roots of unity) is given by the matrix:

$$P_k := \begin{pmatrix} 0 & & & 0 & z \\ 1 & & & \cdot & \cdot \\ \cdot & \cdot & & \cdot & \cdot \\ & \cdot & \cdot & \cdot & \cdot \\ & & \cdot & 0 & 0 \\ 0 & & & 1 & 0 \end{pmatrix} \tag{11}$$

whose k-th power is $z \cdot I$.

Example 4.2   Now consider a 2-sheeted branched cover $\pi : C \to \mathbb{P}^1$, where $g(C) > 0$. If we take $\chi(L) = 0$, i.e. $\deg L = g - 1$, we get $\deg(\pi_* L) = -2$. The equality

$$\ell := h^0(C, L) = h^0(\mathbb{P}^1, \pi_* L)$$

implies

$$\pi_* L \approx \mathcal{O}(\ell - 1) \oplus \mathcal{O}(-\ell - 1).$$

In particular, we discover a very disturbing phenomenon: as the line bundle L varies continuously, in $Pic^{g-1} C$, so should presumably $\pi_* L$; but if we consider a 1-parameter family of line bundles L_t such that

$$L_0 \in \Theta$$
$$L_t \notin \Theta, \quad t \neq 0,$$

we see that the vector bundle $\pi_* L_t$ jumps from its generic value, $\mathcal{O}(-1) \oplus \mathcal{O}(-1)$ to $\mathcal{O} \oplus \mathcal{O}(-2)$ at $t = 0$. Similar jumps can clearly be forced on a rank-r bundle by considering r-sheeted branched covers.

The moral of these examples is that if we want a moduli space parametrizing the "general" vector bundle on a curve and having a reasonable (say, separated) topology, we cannot consider *all* bundles. In the case of $\mathbb{P}^1$, we will end up with only the balanced bundles such as $\mathcal{O}(-1) \oplus \mathcal{O}(-1)$, thus avoiding the possibility of a discontinuous jump.

The slope $\mu(E)$ of a vector bundle E is defined by:

$$\mu(E) := \frac{\deg E}{\operatorname{rank} E}.$$

A bundle E is called stable (resp., semistable) if for every subbundle $F \subset E$ (other than $0, E$),

$$\mu(F) < \mu(E), \qquad (resp.\ \mu(F) \leq \mu(E)).$$

The basic result due to Mumford and Seshadri [Se], is that reasonable (coarse) moduli spaces $\mathcal{U}_\Sigma^s(r, d) \subset \mathcal{U}_\Sigma(r, d)$ exist, with the following properties:

- $\mathcal{U}_\Sigma^s(r, d)$ is smooth; its points parametrize isomorphism classes of stable bundles of rank r and degree d on Σ; it is an open subset of $\mathcal{U}_\Sigma(r, d)$.

- $\mathcal{U}_\Sigma(r, d)$ is projective; its points parametrize equivalence classes of semistable bundles, where two bundles are equivalent, roughly, if they admit filtrations by semistable subbundles (of constant slope) with isomorphic graded pieces.

- Both are coarse moduli spaces; this means that any "family", i.e. vector bundle on a product $S \times \Sigma$, where S is any scheme, whose restrictions E_s to copies $s \times \Sigma$ of Σ are (semi) stable of rank r and degree d, determines a unique morphism of S to $\mathcal{U}_\Sigma^s(r, d)$ (respectively, $\mathcal{U}_\Sigma(r, d)$) which sends each $s \in S$ to the isomorphism (resp. equivalence) class of E_s, and has the obvious functoriality properties.

33

Examples

$g = 0$. The stable bundles are the line bundles $\mathcal{O}(d)$. The semi-stable bundles are the balanced vector bundles, $\mathcal{O}(d)^{\oplus r}$. Thus $\mathcal{U}_{P^1}(r, d)$ is a point if $r \mid d$, empty otherwise, while the stable subset is empty when $r \neq 1$.

$g = 1$. Let $h := gcd(r, d)$. Atiyah [At] shows that $\mathcal{U}_\Sigma(r, d)$ is isomorphic to the symmetric product $S^h \Sigma$, and that each semistable equivalence class contains a unique decomposable bundle $E = \oplus_{i=1}^{h} E_i$, where each E_i is stable of rank r/h and degree d/h. (Other bundles in this equivalence class are filtered, with the E_i as subquotients.) Thus when $h = 1$, $\mathcal{U}_\Sigma^s = \mathcal{U}_\Sigma$, and when $h > 1$, $\mathcal{U}_\Sigma^s$ is empty.

The possibilities for semistable bundles are illustrated in the case $r = 2$, $d = 0$: given two line bundles $L_1, L_2 \in Pic^0 \Sigma$, the possible extensions are determined, up to non zero scalars, by elements of

$$Ext^1_{\mathcal{O}_\Sigma}(L_1, L_2) \approx H^1(L_2 \otimes L_1^{-1}).$$

The direct sum is thus the only extension when $L_1 \not\approx L_2$, while if $L_1 \approx L_2 \approx L$ there is, up to isomorphism, also a unique non-trivial extension, say E_L. There is, again, a jump phenomenon: by rescaling the extension class we get a family of vector bundles with generic member isomorphic to E_L and special member $L \oplus L$. This explains why there cannot exist a moduli space parametrizing *isomorphism* classes of semistable bundles; neither $L \oplus L$ nor E_L is excluded, and the point representing the former is in the closure of the latter, so they must be identified, i.e. E_L and $L \oplus L$ must be declared to be equivalent.

Higher Genus.

The only other cases where an explicit description of $\mathcal{U}_\Sigma(r, d)$ is known are when $r = 2$ and $g = 3$ [NR] or $r = 2$ and Σ is hyperelliptic of any genus [DR]. In the latter case, the moduli space $\mathcal{U}_\Sigma(2, \xi)$ of rank 2 vector bundles with a fixed determinant line bundle ξ of odd degree is isomorphic to the family of linear spaces $\mathbb{P}^{g-2}$ in the intersection of the two quadrics in $\mathbb{P}^{2g+1}$ used in Chapter 3. In the even degree case, $\mathcal{U}_\Sigma(2, \xi)$ can also be described in terms of the same two quadrics; when $g = 2$, it turns out to be isomorphic to $\mathbb{P}^3$, in which the locus of semistable but non-stable points is the Kummer surface $K := \mathcal{J}(\Sigma)/ \pm 1$, with its classical embedding in $\mathbb{P}^3$ as a quadric with 16 nodes.

Elementary deformation theory lets us make some general statements about $\mathcal{U} := \mathcal{U}_\Sigma(r, d)$ and $\mathcal{U}^s := \mathcal{U}_\Sigma^s(r, d)$:

Lemma 4.3 *For $g \geq 2$:*

1. *$\dim \mathcal{U} = 1 + r^2(g - 1)$, and $\mathcal{U}^s$ is a dense open subset.*

2. *Stable bundles E are simple, i.e. the only (global) endomorphisms of E are scalars.*

3. *Stable bundles are non-singular points of $\mathcal{U}$.*

4. *At points of $\mathcal{U}^s$ there are canonical identifications*

$$\begin{aligned} T_E \mathcal{U}^s &\approx H^1(End\ E) \\ T_E^* \mathcal{U}^s &\approx H^0(w_\Sigma \otimes End\ E). \end{aligned}$$

The proof of (2) is based on the observation that any nonzero $\alpha : E \to E$ must be invertible, otherwise either ker α or im α would violate stability. Therefore $H^0(End\ E)$ is a finite dimensional division algebra containing $\mathbb{C}$, hence equal to it. Since a vector bundle E on Σ is determined by a 1-cocycle with values in $GL(r, \mathcal{O}_\Sigma)$ (= transition matrices), a first order deformation of E is given by a 1-cocycle with values in the associated bundle of Lie algebras, i.e. (up to isomorphism) by a class in $H^1(End\ E)$. The functoriality property of $\mathcal{U}$ ("coarse moduli space") implies that this is the Zariski tangent space, $T_E\mathcal{U}$. By Riemann-Roch

$$h^1(End\ E) = r^2(g-1) + h^0(End\ E).$$

so the minimal value is obtained at the simple points, and equals $1 + r^2(g-1)$ as claimed in (1). The identification of $T_E^*\mathcal{U}^s$ follows from that of $T_E\mathcal{U}^s$ by Serre duality.

$\square$

4.2 Spectral Curves and the Hitchin System

The relation between vector bundles and finite dimensional integrable systems arises from Hitchin's amazing result.

Theorem 4.4 *[H1, H2] The cotangent bundle to the moduli space of semistable vector bundles supports a natural ACIHS.*

At the heart of Hitchin's theorem is a construction of a spectral curve associated to a 1-form valued endomorphism of a vector bundle. The spectral construction allows a uniform treatment of a wide variety of algebraically completely integrable Hamiltonian systems. We will concentrate in this section on the algebro-geometric aspects of these systems leaving their symplectic geometry to Chapter 8. We work with vector bundles over curves, other structure groups will be treated in Chapter 9. The reader is referred to [BNR] and [H2] for more details.

The total space of the cotangent bundle $T^*\mathcal{U}_\Sigma^s(r, d)$ of the moduli space of stable vector bundles parametrizes pairs (E, φ) consisting of a stable vector bundle E and a covector φ in $H^1(\Sigma, End\ E)^* \simeq H^0(\Sigma, End E \otimes \omega_\Sigma)$, i.e., a 1-form valued endomorphism of E.

Consider more generally a pair (E, φ) of a rank r vector bundle E and a section $\varphi \in Hom(E, E \otimes K)$ where K is a line bundle on Σ. The i-th coefficient b_i of the characteristic polynomial of (E, φ) is a homogeneous polynomial of degree i on K^{-1}, hence a section of $H^0(\Sigma, K^{\otimes i})$. In fact $b_i = (-1)^i \cdot \text{trace}(\overset{i}{\wedge} \varphi)$.

The Hamiltonian map of the Hitchin system is the characteristic polynomial map

$$H : T^*\mathcal{U}_\Sigma(r, d) \longrightarrow B_\omega := \bigoplus_{i=1}^r H^0(\Sigma, \omega^{\otimes i}).$$

The fibers of the Hitchin map H turn out to be Jacobians of curves associated canonically to characteristic polynomials.

Going back to the general K-valued pair (E, φ), notice that a characteristic polynomial char $(\varphi) = y^r - \text{tr}(\varphi)y^{r-1} + \cdots + (-1)^r \det \varphi$ in $B_K := \oplus_{i=1}^r H^0(\Sigma, K^{\otimes i})$ defines a morphism from the line bundle K to $K^{\otimes r}$. The inverse image C of the zero section in $K^{\otimes r}$ under a polynomial P in B_K is called a spectral curve. If P is the characteristic polynomial of a pair (E, φ) then indeed the fibers of $\pi : C \to \Sigma$ consist of eigenvalues of φ. If $K^{\otimes r}$ has a section without multiple zeroes (e.g., if it is very ample) then the generic spectral curve is smooth.

Lagrange interpolation extends a function on the inverse image $\pi^{-1}(U) \subset C$ of an open set U in Σ to a unique function on the inverse image of U in the surface K which is a polynomial of degree $\leq r - 1$ on each fiber. It follows that the direct image $\pi_* \mathcal{O}_C$ is isomorphic to $\mathcal{O}_\Sigma \oplus K^{-1} \oplus \cdots \oplus K^{1-r}$. Assuming that $K^{\otimes i}$ has no sections for $i < 0$, the genus $h^1(C, \mathcal{O}_C) = h^1(\Sigma, \pi_* \mathcal{O}_C)$ of C is equal to $\deg(K) \cdot r(r-1)/2 + r(g-1) + 1$. In particular, when $K = \omega_\Sigma$, the genus of C is equal to half the dimension of the cotangent bundle. The data (E, φ) determines moreover a sheaf L on the spectral curve which is a line bundle if the curve is smooth. Away from the ramification divisor R in C, L is the tautological eigenline subbundle of the pullback $\pi^* E$. More precisely, the homomorphism $(\pi^*(\varphi) - y \cdot I) : \pi^* E \to \pi^*(E \otimes K)$, where $y \in H^0(C, \pi^* K)$ is the tautological eigenvalue section, has kernel $L(-R)$.

Conversely, given a spectral curve C and a line bundle L on it we get a pair $(\pi_* L, \pi_*(\otimes y))$ of a rank r vector bundle on Σ and a K valued endomorphism (see example 4.1). The two constructions are the inverse of each other.

Proposition 4.5 *[H2, BNR] If C is an irreducible and reduced spectral curve there is a bijection between isomorphism classes of*

- *Pairs (E, φ) with spectral curve C.*

- *Rank 1 torsion free sheaves L on C.*

Under this correspondence, line bundles on C correspond to endomorphisms φ which are regular in every fiber, i.e., whose centralizer in each fiber is an r-dimensional subspace of the corresponding fiber of End E. (This notion of regularity agrees with the one in Example 2.2.)

We conclude that the fiber of the Hitchin map $H : T^* U_\Sigma^s(r, d) \to B_\omega$ over a characteristic polynomial $b \in B_\omega$ is precisely the open subset of the Jacobian $J_C^{d+r(1-g_\Sigma)+g_C-1}$ consisting of the line bundles L whose direct image is a stable vector bundle. Moreover, the construction of the characteristic polynomial map and a similar description of its fibers applies to moduli spaces of pairs with K-valued endomorphism where K need not be the canonical line bundle (Theorem 4.8).

The missing line bundles in the fibers of the Hitchin map indicate that we need to relax the stability condition for the pair (E, φ).

Definition: A pair (E, φ) is stable (semistable) if the slope of every φ-invariant subbundle of E is less than (or equal to) the slope of E.

As in the case of vector bundles we can define an equivalence relation for semistable pairs, where two bundles are equivalent, roughly, if they admit φ-invariant filtrations by semistable pairs (of constant slope) with isomorphic graded pieces. Two stable pairs are equivalent if and only if they are isomorphic.

Theorem 4.6 *[H1, Sim1, Nit] There exists an algebraic coarse moduli scheme* $\text{Higgs}_K :=$ $\text{Higgs}_\Sigma(r, d, K)$ *parametrizing equivalence classes of semistable K-valued pairs.*

The characteristic polynomial map $H : \text{Higgs}_K \to B_K$ is a proper algebraic morphism. A deeper reason for working with the above definition of stability is provided by the following theorem from nonabelian Hodge theory:

Theorem 4.7 *[H1, Sim2]* *There is a canonical real analytic diffeomorphism between*

- *The moduli space of conjugacy classes of semisimple representations of the fundamental group $\pi_1(\Sigma)$ in $GL(r, \mathbb{C})$ and*

- *The moduli space of semistable ω-valued (Higgs) pairs (E, φ) of rank r and degree 0.*

In the case of Hitchin's system $(K = \omega_\Sigma)$, the symplectic structure of the cotangent bundle extends to the stable locus of the moduli space of Higgs pairs giving rise to an integrable system $H : \text{Higgs}_\Sigma(r, d, \omega_\Sigma) \to B_\omega$ whose generic fiber is a complete Jacobian of a spectral curve.

We will show in Chapter 6 that the Hitchin system is, in fact, the lowest rank symplectic leaf of a natural infinite dimensional Poisson variety $\text{Higgs}_\Sigma(r, d)$ obtained as an inductive limit of the moduli spaces $\text{Higgs}_\Sigma^{sm}(r, d, \omega(D))$ of $\omega(D)$-valued pairs as D varies through all effective divisors on Σ. The basic fact, generalizing the results of [H2, BNR, B1] is:

Theorem 4.8 *[Bn, Ma1]* *Let D be an effective divisor (not necessarily reduced) on a smooth algebraic curve Σ of genus g. Assume that $[\omega(D)]^{\otimes r}$ is very ample and if $g = 0$ assume further that $\deg(D) > \max(2, \rho)$ where $0 \le \rho < r$ is the residue of $d \bmod r$. Then*

1. *The moduli space $\text{Higgs}_\Sigma^s(r, d, \omega(D))$ of stable rank r and degree d $\omega(D)$-valued Higgs pairs has a smooth component $\text{Higgs}_\Sigma^{sm}(r, d, \omega(D))$ of top dimension $r^2(2g - 2 + \deg(D)) + 1 + \epsilon_{D=0}$, where $\epsilon_{D=0}$ is 1 if $D = 0$ and zero if $D > 0$. $\text{Higgs}_\Sigma^{sm}(r, d, \omega(D))$ is the unique component which contains Higgs pairs supported on irreducible and reduced spectral curves.*

2. *$\text{Higgs}_\Sigma^{sm}(r, d, \omega(D))$ has a canonical Poisson structure.*

3. *The characteristic polynomial map $H : \text{Higgs}_\Sigma^{sm}(r, d, \omega(D)) \to B_{\omega(D)}$ is an algebraically completely integrable Hamiltonian system. The generic (Lagrangian) fiber is a complete Jacobian of a smooth spectral curve of genus $r^2(g-1) + 1 + (\deg D)(\frac{r(r-1)}{2})$.*

4. *The foliation of $\text{Higgs}_\Sigma^{sm}(r, d, \omega(D))$ by closures of top dimensional symplectic leaves is induced by the cosets of*

$$H^0\left(\Sigma, \left[\bigoplus_{i=1}^r \omega_\Sigma(D)^{\otimes i}\right](-D)\right) \text{ in } B_{\omega(D)}.$$

Definition 4.9 *As in the theorem, we will denote by $\text{Higgs}_\Sigma^{sm}(r, d, \omega(D))$ the unique component which contains Higgs pairs supported on irreducible and reduced spectral curves.*

In Chapter 6 we will discuss the relationship of these integrable systems with flows of KdV type. In the next section we will discuss the example of geodesic flow on the ellipsoid as a Hamiltonian flow of a symplectic leaf of one of these spaces. See [B1, Ma1] for more examples.

The Hitchin system has been useful in the study of the geometry of the moduli space of vector bundles. The main technique is to reduce questions about vector bundles to questions about spectral Jacobians. Hitchin used these ideas to compute the cohomology groups $H^i(\mathcal{U}, S^k T)$, $i = 0, 1$, of the symmetric products of the tangent bundle of the moduli space $\mathcal{U}$ of rank 2 and odd degree stable vector bundles. In [BNR] these techniques provided the first mathematical proof that the dimensions of the space of sections of the generalized theta line bundle are

$$h^0(\mathcal{U}_\Sigma(n, n(g-1)), \ \Theta) \ = \ 1,$$
$$h^0(\mathcal{SU}_\Sigma(n), \ \Theta) \ = \ n^g,$$

where $\mathcal{SU}_\Sigma(n)$ denotes the moduli space of vector bundles with trivial determinant line bundle. (This of course is now subsumed in the Verlinde Formula for sections of powers of theta bundles.) These ideas were proven useful in the proof of the existence of a projectively flat connection on the bundles of level k theta sections over the moduli space $\mathcal{M}_g$ of curve of genus g [H3], an important fact in conformal field theory. Kouvidakis and Pantev applied these ideas to the study of automorphisms of the moduli space of vector bundles [Ko-P].

4.3   Polynomial Matrices

Theorem 4.8 has a concrete description when the base curve Σ is $\mathbb{P}^1$. Let K be the line bundle $\mathcal{O}_{\mathbb{P}^1}(d)$. Consider the moduli space $\text{Higgs}_K := \text{Higgs}_K^{sm}(-r, r)$ of pairs (E, φ) consisting of a vector bundle E of rank r and degree $-r$ with a K-valued endomorphism $\varphi : E \to E \otimes K$ (we also follow the notation of definition 4.9 singling out a particular component). Choose a coordinate x on $\mathbb{P}^1 - \{\infty\}$. The space B_K of characteristic polynomials becomes

$$\{P(x, y) = y^r + b_1(x)y^{r-1} + \cdots + b_r(x) \mid b_i(x) \text{ is a polynomial in } x \text{ of degree } \leq i \cdot d\}.$$

The total space of the line bundle $\mathcal{O}_{\mathbb{P}^1}(d \cdot \infty)$ restricted to the affine line $\mathbb{P}^1 - \{\infty\}$ is isomorphic to the affine plane, and under this isomorphism $P(x, y)$ becomes the equation of the spectral curve as an affine plane curve.

Denote by $B^0 \subset B_K$ the subset of smooth spectral curves. Let $Q := Q_r(d)$ be the subset of Higgs_K parametrizing pairs (E, φ) with a smooth spectral curve and a vector bundle E isomorphic to $E_0 := \oplus^r \mathcal{O}_{\mathbb{P}^1}(-1)$. Q is a Zariski open (dense) subset of Higgs_K because:

i) by definition 4.9 $\text{Higgs}_K^{sm}(r, -r)$ is irreducible,

ii) E_0 is the unique semistable rank r vector bundle of degree $-r$ on $\mathbb{P}^1$ and semistability is an open condition.

The bundle $\text{End } E_0$ is the trivial Lie algebra bundle $\mathfrak{gl}_r(\mathbb{C}) \otimes \mathcal{O}_{\mathbb{P}^1}$. Hence, every point in Q is represented by an element $\varphi \in M_r(d) := H^0(\mathbb{P}^1, \mathfrak{gl}_r(\mathbb{C}) \otimes \mathcal{O}_{\mathbb{P}^1}(d \cdot \infty))$, i.e., by an $r \times r$

matrix φ with polynomial entries of degree $\leq d$. Denote the inverse image of B^0 in $M_r(d)$ by $M_r^0(d)$. The subset $Q \subset \text{Higgs}_K$ is simply the quotient of $M_r^0(d)$ by the conjugation action of $PGL_r(\mathbb{C})$.

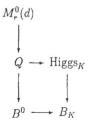

In this setting, Theorem 4.8 specializes to the following theorem of Beauville and Adams-Harnad-Hurtubise-Previato generalizing results of Mumford and Moser [Mum2] in rank 2:

Theorem 4.10 *[B1, AHH]*

1. *The quotient Q of the action of $PGL_r(\mathbb{C})$ by conjugation on $M_r^0(d)$ is a smooth variety.*

2. *The fiber of the characteristic polynomial maps $H : Q \to B^0$ over the polynomial of a spectral curve C is the complement $J_C^{g-1} - \Theta$ of the theta divisor in the Jacobian of line bundles on C of degree $g - 1$ ($g = \text{genus of } C$).*

3. *The choice of $d + 2$ points $a_1, \cdots a_{d+2}$ on $\mathbb{P}^1$ determines a Poisson structure on Q. The characteristic polynomial map $H : Q \to B^0$ is an algebraically completely integrable Hamiltonian system with respect to each of these Poisson structures.*

4. *The symplectic leaves of Q are obtained by fixing the values (of the coefficients) of the characteristic polynomials at the points $\{a_i\}_{i=1}^{d+2}$.*

We note that in [B1] the Poisson structure on Q was obtained as the reduction of a Poisson structure on $M_r(d)$. The latter was the pullback of the Kostant-Kirillov Poisson structure via the embedding

$$M_r(d) \hookrightarrow \mathfrak{gl}_r(\mathbb{C})^{d+2}$$

by Lagrange interpolation at $a_1, \cdots, a_{d+2}$. This embedding will be used in section 4.3.2 where geodesic flow on ellipsoids is revisited.

A choice of a divisor $D = a_1 + \cdots + a_{d+2}$ of degree $d+2$ on $\mathbb{P}^1$ determines an isomorphism of $\mathcal{O}_{\mathbb{P}^1}(d \cdot \infty)$ with $\omega_{\mathbb{P}^1}(D)$. For example, if $a_1, \cdots, a_i$ are finite, $a_{i+1} = a_{a+2} = \cdots = a_{d+2} = \infty$ then we send a polynomial $f(x)$ of degree $\leq d$ to the meromorphic 1-form

$$\frac{f(x)}{\prod_{j=1}^{i}(x - a_j)} dx.$$

When the $d + 2$ points are distinct, Lagrange interpolation translates to the embedding

$$\text{Res} : M_r(d) = H^0(\mathbb{P}^1, \mathfrak{gl}_r(\mathbb{C}) \otimes \omega_{\mathbb{P}^1}(D)) \hookrightarrow \mathfrak{gl}_r(\mathbb{C})^{d+2}$$

via the residues of meromorphic 1-form valued matrices at the points a_i (if a_i has multiplicity 2 or higher, we replace the i-th copy of $\mathfrak{gl}_r(\mathbb{C})$ by its tangent bundle or higher order infinitesimal germs at a_i of sections of the trivial bundle $\mathfrak{gl}_r(\mathbb{C}) \otimes \mathcal{O}_{\mathbb{P}^1}$).

4.3.1 Explicit Equations for Jacobians of Spectral Curves with a Cyclic Ramification Point

A further simplification occurs for matrices with a nilpotent leading coefficient (nilpotent at ∞). The projection $M_r^0(d) \to Q$ has a natural section over the image $N \subset Q$ of this locus. So N can be described concretely as a space of polynomials (rather than as a quotient of such a space).

As a consequence we obtain explicit equations in $M_r(d)$ for the complement $J_C - \Theta$ of the theta divisor in the Jacobian of every irreducible and reduced r-sheeted spectral curve over $\mathbb{P}^1$ which is totally ramified and smooth at ∞ (generalizing the equations for hyperelliptic curve (case $r = 2$) obtained in [Mum2]).

Lemma 4.11 Let $A = A_d x^d + \cdots + A_1 x + A_0$ be an $r \times r$ traceless matrix with polynomial entries of degree $\leq d$

i) whose spectral curve in $\mathcal{O}_{\mathbb{P}^1}(d \cdot \infty)$ is irreducible and reduced and smooth over ∞, and

ii) whose leading coefficient A_d is a nilpotent (necessarily regular) matrix.

Then there exists a unique element $g_0 \in PGL_r(\mathbb{C})$ conjugating A to a matrix $A' = x^d \cdot J + \sum_{i=0}^{d-1} A_i' x^i$ of the form:

$$A' = x^d \begin{pmatrix} 0 & & & 0 & 0 \\ 1 & & & 0 & 0 \\ 0 & \cdot & & 0 & 0 \\ 0 & & \cdot & 0 & 0 \\ 0 & & \cdot & 0 & 0 \\ 0 & & & 1 & 0 \end{pmatrix} + x^{d-1} \begin{pmatrix} \star & \cdots & \star & \beta_r \\ \star & \cdots & \star & 0 \\ \vdots & \vdots & \vdots & \vdots \\ \star & \cdots & \star & 0 \end{pmatrix} + \sum_{i=0}^{d-2} x^i A_i' \qquad (12)$$

where $(-1)^{r+1}\beta_r$ is the (leading) coefficient of x^{dr-1} in the determinant $b_r(x)$ of $A(x)$.

Remark: Notice that the coefficients $b_i(x)$ in the characteristic polynomial $P(x, y) = y^r + b_1(x)y^{r-1} + \cdots + b_r(x)$ of $A(x)$ satisfy degree $b_i(x) \leq d \cdot i - 1$ since A is nilpotent at ∞, and degree $b_r(x) = dr - 1$ since the spectral curve is smooth over ∞. Thus $\beta_r \neq 0$.

Proof (of lemma 4.11): Let J be the nilpotent regular constant matrix appearing as the leading coefficient of $A'(x)$ in the normalized form (12). Let $\mathbb{C}[J]$ be the algebra of polynomials in J with constant coefficients. The proof relies on the elementary fact that $\mathbb{C}^r$, as a left $\mathbb{C}[J]$-module, is free. Any vector with non zero first entry is a generator. A_d is conjugate to J. Thus we may assume that $A_d = J$ and it remains to show that there exists a unique element in the stabilizer of J in $PGL_r(\mathbb{C})$ conjugating $A(x)$ to the normal form (12).

Since $A_d = J$, the first entry in the right column R of A_{d-1} is β_r. Thus R is a generator of $\mathbb{C}^r$ as a $\mathbb{C}[J]$-module. Any element $g \in PGL_r(\mathbb{C})$ in the commutator subgroup of J is an invertible element in $\mathbb{C}[J]$ and can be written (up to scalar multiple) in the form

$g = I + N$, N nilpotent. The right column of $gA_{d-1}g^{-1}$ is $R + NR$ and there exists a unique nilpotent $N \in \mathbb{C}[J]$ such that $NR = \begin{pmatrix} \beta_r \\ 0 \\ \vdots \\ 0 \end{pmatrix} - R$. Thus g is unique up to a scalar factor.

$\square$

Denote the affine subvariety of $M_d^0(r)$ of matrices satisfying the $r^2 + r - 1$ equations (12) by $\tilde{N}$. The subvariety $\tilde{N}$ is a section of the principal $PGL_r(\mathbb{C})$ bundle $M_r^0(d) \to Q$ over the locus N of conjugacy classes of polynomial matrices with a nilpotent leading coefficient. N is a Poisson subvariety of Q with respect to any Poisson structure on Q determined by a divisor D as in theorem 4.10, provided that D contains the point at infinity $\infty \in \mathbb{P}^1$.

Choose a characteristic polynomial $P(x, y) = y^r + b_1(x)y^{r-1} + \cdots + b_r(x)$ in $B_{\mathcal{O}_{\mathbb{P}^1}(d)}$ of a smooth spectral curve C with degree $b_i(x) \leq id - 1$, $b_r(x)$ of degree $rd - 1$ with leading coefficient $(-1)^{r+1}\beta_r$. Theorem 4.10 implies that the equations

a) $A_d = J$,

b) The r-th column of A_{d-1} is $\begin{pmatrix} \beta_r \\ 0 \\ \vdots \\ 0 \end{pmatrix}$,

c) char $(A(x)) = P(x, y)$

define a subvariety of $M_r(d)$ isomorphic to the complement $J_C^{g-1} - \Theta_C$ of the theta divisor in the Jacobian of C.

4.3.2 Geodesic Flow on Ellipsoids via 2×2 Polynomial Matrices

We use polynomial matrices to retrieve the Jacobi-Moser-Mumford system which arose in chapter 3 out of the geodesic flow on an ellipsoid. Our presentation follows [B1].

Consider a spectral polynomial $P(x, y)$ in $B_{\mathcal{O}_{\mathbb{P}^1}(d \cdot \infty)}^0$ of the form

(i) $P(x, y) = y^2 - f(x)$ where f(x) is monic of degree $2d - 1$.

The corresponding spectral curve C is smooth, hyperelliptic of genes $g = d - 1$ and ramified over ∞. Theorem 4.10 implies that the fiber

$$H^{-1}(P(x, y)) = \left\{ \begin{pmatrix} V & U \\ W & -V \end{pmatrix} \mid V^2 + UW = f(x) \right\} / PGL_2(\mathbb{C})$$

of the characteristic polynomial map is isomorphic to the complement $J_C^{g-1} - \Theta$ of the theta divisor.

Lemma 4.11 specializes in our case to the following statement (note that $\beta_r = 1$ since f is taken to be monic):

The $PGL_2(\mathbb{C})$ orbit of a matrix $\begin{pmatrix} V & U \\ W & -V \end{pmatrix}$ over $H^{-1}(P(x,y)) \cong J_C^{g-1} - \Theta$ contains a unique matrix satisfying

(ii) W *is monic of degree d,*
    U *is monic of degree $d-1$ and*
    $\deg V \leq d-2$.

In other words, condition (ii) and

(iii) $V^2 + UW = f(x)$

are the equations of $J_C^{g-1} - \Theta$ as an affine subvariety of the subspace of traceless matrices in $M_2(d)$. In fact, condition (ii) defines a section $\varphi : N \to M_2(d)$ over the locus N in Q of conjugacy classes with characteristic polynomial satisfying condition (i).

Recall that the Jacobi-Moser-Mumford integrable system linearizing the geodesic flow of the ellipsoid $\sum_{i=1}^d a_i^{-1} x_i^2 = 1$ is supported on the tangent bundle TS of the sphere $S \subset \mathbb{R}^d$. Our discussion ended by describing the quotient of TS by the group $G \simeq (\mathbb{Z}/2\mathbb{Z})^d$ of involutions. We will describe in the next three steps an isomorphism of this quotient with a symplectic leaf X of Q.

<u>Step I:</u> (Identification of the symplectic leaf X). Assume that the points $a_1, \cdots, a_d \in \mathbb{P}^1 - \{\infty\}$ are distinct and let $a_{d+1} = a_{d+2} = \infty$. Let $X \subset Q$ be the symplectic leaf over the subspace of characteristic polynomials $P(x,y) = y^2 - f(x)$ satisfying

$$f(a_i) = 0, \ 1 \leq i \leq d, \ \deg f = 2d-1 \text{ and } f \text{ is monic}.$$

The spectral curves of matrices in the leaf X have genus $d-1$, and are branched over the fixed $g+2$ points $a_1, \cdots, a_d, \infty$ and g varying points.

<u>Step II:</u> (Embedding of X in the product $\mathcal{N}^d$ of the regular nilpotent orbit). The isomorphism $\mathcal{O}_{\mathbb{P}^1}(d \cdot \infty) \xrightarrow{\sim} \omega_{\mathbb{P}^1}(\sum_{i=1}^d a_i + 2 \cdot \infty)$ sending $F(x)$ to $\frac{F(x)dx}{\prod_{i=1}^d (x-a_i)}$ translates the matrix $\begin{pmatrix} V(x) & U(x) \\ W(x) & -V(x) \end{pmatrix}$ to a matrix φ of meromorphic 1-forms. The residues of φ satisfy:

$$R_\infty := Res_\infty(\varphi) = \begin{pmatrix} 0 & -1 \\ s & 0 \end{pmatrix} \text{ for some } s \in \mathbb{C} \quad \text{(condition } (ii)),$$

$$R_i := Res_{a_i}(\varphi) = \begin{pmatrix} V(a_i) & U(a_i) \\ W(a_i) & -V(a_i) \end{pmatrix} \frac{1}{\prod_{\substack{j=1 \\ j \neq i}}^d (a_i - a_j)}.$$

The residues at the finite points a_i can be calculated using Lagrange interpolation of polynomials of degree d at the $d+1$ points $a_1, \cdots, a_d, \infty$ given by the formula

$$F(x) = \sum_{i=1}^d F(a_i) \frac{\prod_{\substack{j=1 \\ j \neq i}}^d (x - a_j)}{\prod_{\substack{j=1 \\ j \neq i}}^d (a_i - a_j)} + F(\infty) \prod_{j=1}^d (x - a_j) \tag{13}$$

where $F(\infty)$ is the leading coefficient of $F(x)$.

The residues R_∞, R_i are nilpotent regular 2×2 matrices and the residue theorem implies that $R_\infty = -\sum_{i=1}^d R_i$. The residue map $Res : X \to \mathcal{N}^d$ defines a symplectic embedding $\varphi \mapsto (R_1, \cdots, R_d)$ of the symplectic leaf X of Q in the Cartesian product of d copies of the regular nilpotent orbit $\mathcal{N}$ in $\mathfrak{gl}_2(\mathbb{C})$.

<u>Step III</u> (The 2^d covering $TS \to X$). Endow $\mathbb{C}^2$ with the symplectic structure $2dx \wedge dy$. The map $\mathbb{C}^2 - \{(0,0)\} \to \mathcal{N}$ sending (x,y) to $\begin{pmatrix} xy & -x^2 \\ y^2 & -xy \end{pmatrix}$ is a symplectic $SL_2(\mathbb{C})$-equivariant double cover of the regular nilpotent orbit $\mathcal{N}$ (where $SL_2(\mathbb{C}) \cong Sp_2(\mathbb{C}, 2dx \wedge dy)$ acts on $\mathbb{C}^2$ via the standard representation). We obtain a 2^d-covering $\tau : (\mathbb{C}^2 - \{(0,0)\})^d \to \mathcal{N}^d$. The residue theorem translates to the fact that the image $Res(X) \subset \mathcal{N}^d$ is covered by

$$\left\{ (\bar{x}, \bar{y}) = ((x_1, y_1), \cdots, (x_d, y_d)) \mid \sum x_i y_i = 0 \text{ and } \sum_{i=1}^d x_i^2 = 1 \right\}.$$

This is exactly the tangent bundle $TS \subset (\mathbb{C}^2)^d$ of the sphere $S \subset \mathbb{C}^d$.

5 Poisson structure via levels

We construct a Poisson structure on the moduli space of meromorphic Higgs pairs in two steps (following [Ma1]):

- First we realize a dense open subset of moduli as the orbit space of a Poisson action of a group on the cotangent bundle of the moduli space of vector bundles with level structures (sections 5.1, 5.2 and 5.3).

- Next we exhibit a 2-vector on the smooth locus of moduli, using a cohomological construction (section 5.4). On the above dense open set this agrees with the Poisson structure, so it is a Poisson structure everywhere.

We summarize the construction in section 5.5 in a diagram whose rotational symmetry relates dual pairs of realizations.

5.1 Level structures

Fix a curve Σ, an effective divisor $D = \sum p_i$ in Σ, and a rank r vector bundle E on Σ. A level D structure on E is an $\mathcal{O}_D$-isomorphism $\eta : E \otimes \mathcal{O}_D \xrightarrow{\sim} \mathcal{O}_D^{\oplus r}$. Seshadri [Se] constructs a smooth, quasi-projective moduli space $\mathcal{U}_\Sigma(r, d, D)$ parametrizing stable pairs (E, η). Here stability means that for any subbundle $F \subset E$,

$$\frac{degF - degD}{rankF} < \frac{degE - degD}{rankE}.$$

The level-D group is the projectivized group of $\mathcal{O}_D$-algebra automorphisms,

$$G_D := \mathbb{P}Aut_{\mathcal{O}_D}(\mathcal{O}_D^{\oplus r}).$$

(i.e. the automorphism group modulo complex scalars $\mathbb{C}^*$.) It acts on $\mathcal{U}_\Sigma(r, d, D)$: an element $g \in G_D$ sends

$$[(E, \eta)] \mapsto [(E, \bar{g} \circ \eta)],$$

where $\bar{g} \in Aut_{\mathcal{O}_D}(\mathcal{O}_D^{\oplus r})$ lifts g, and $[\cdot]$ denotes the isomorphism class of a pair. The open set $\mathcal{U}_\Sigma^\circ(r, d, D)$, parametrizing pairs (E, η) where E itself is stable, is a principal G_D-bundle over $\mathcal{U}_\Sigma^s(r, d)$. The Lie algebra $\mathbf{g}_D$ of G_D is given by $\mathbf{gl}_r(\mathcal{O}_D)/$scalars.

5.2 The cotangent bundle

We compute deformations of a pair

$$(E, \eta) \in \mathcal{U}_D := \mathcal{U}_\Sigma(r, d, D)$$

as we did for the single vector bundle

$$E \in \mathcal{U} := \mathcal{U}_\Sigma(r, d).$$

Namely, E is given (in terms of an open cover of Σ) by a 1-cocycle with values in the sheaf of groups $GL_r(\mathcal{O}_\Sigma)$. Differentiating this cocycle with respect to parameters gives a 1-cocycle with values in the corresponding sheaf of Lie algebras, so we obtain the identification

$$T_E \mathcal{U} \approx H^1(EndE).$$

Similarly, the pair (E, η) is given by a 1-cocycle with values in the subsheaf

$$GL_{r,D}(\mathcal{O}_\Sigma) := \{f \in GL_r(\mathcal{O}_\Sigma) | f - 1 \in gl_r(\mathcal{I}_D)\}.$$

Differentiating, we find the natural isomorphism

$$T_{(E,\eta)} \mathcal{U}_D \approx H^1(\mathcal{I}_D \otimes EndE),$$

so by Serre duality,

$$T^*_{(E,\eta)} \mathcal{U}_D \approx H^0(EndE \otimes \omega(D)).$$

(we identify $EndE$ with its dual via the trace.) We will denote a point of this cotangent bundle by a triple (E, φ, η), where $(E, \eta) \in \mathcal{U}_D$ and φ is a D-Higgs field,

$$\varphi : E \to E \otimes \omega(D).$$

5.3 The Poisson structure

The action of the level group G_D on $\mathcal{U}_D$ lifts naturally to an action of G_D on $T^* \mathcal{U}_D$. Explicitly, an element $g \in G_D$ with lift $\bar{g} \in GL_r(\mathcal{O}_D)$ sends

$$(E, \varphi, \eta) \mapsto (E, \varphi, \bar{g} \circ \eta).$$

The lifted action has the following properties:

(1) It is Poisson with respect to the standard symplectic structure on $T^* \mathcal{U}_D$ (holds for any lifted action, see example 2.2).

(2) The moment map

$$\mu : T^* \mathcal{U}_D \to g_D^*$$

is given by

$$\mu(E, \varphi, \eta) : A \mapsto \text{Res Trace } (A \cdot \varphi^\eta), \qquad (14)$$

where

$$A \in g_D = (gl_r(\mathcal{O}_D))/(\text{scalars}) \approx (gl_r(\mathcal{O}_D))_{\text{traceless}},$$
$$\varphi^\eta := \eta \circ \varphi \circ \eta^{-1} \in H^0(gl_r(\omega(D) \otimes \mathcal{O}_D)),$$

and the residue map

$$Res : H^0(\omega(D) \otimes \mathcal{O}_D) \to H^1(\omega) \approx \mathbb{C}$$

is the coboundary for the restriction sequence

$$0 \to \omega \to \omega(D) \to \omega(D) \otimes \mathcal{O}_D \to 0$$

(cf. [Ma1] Proposition 6.12).

(3) G_D acts freely on the open subset $(T^* \mathcal{U}_D)^\circ$ parametrizing triples (E, φ, η) where (E, η) is stable and (E, φ) is a stable Higgs bundle, since such bundles are simple. This makes $(T^* \mathcal{U}_D)^\circ$ into a principal G_D-bundle over an open subset Higgs_D° of Higgs_D'.

We conclude that the symplectic structure on $(T^* \mathcal{U}_D)^\circ$ induces a Poisson structure on Higgs_D°. The symplectic leaves will then be the inverse images under μ of coadjoint orbits in g_D^*.

5.4 Linearization

The main remaining task is to find a two-vector on the non-singular locus Higgs_D^{ns} whose restriction to Higgs_D° is the above Poisson structure. This two vector is then automatically Poisson. The algebraic complete integrability of the component Higgs_D^{sm} (see definition 4.9) would then follow: The spectral curve C_b, for generic $b \in B_D$, is non-singular, so its Jacobian $\mathcal{J}(C)$ is contained in Higgs_D^{sm}. Thus any Hamiltonian vector field must be constant on the generic fiber $\mathcal{J}(C)$, hence on all fibers.

A natural two-vector defined over all of Higgs_D^{ns} can be given in several ways. One [Ma1] is to identify the tangent spaces to Higgs_D (and related spaces) at their smooth points as hypercohomologies, $\mathbb{H}^1$, of appropriate complexes:

space	at	complex
$\mathcal{U}$	E	$EndE$
$\mathcal{U}_D$	(E, η)	$EndE(-D)$
Higgs_D	(E, φ)	$\overline{\mathcal{K}} = [EndE \xrightarrow{ad\varphi} EndE \otimes \omega(D)]$
$T^*\mathcal{U}_D$	(E, φ, η)	$\mathcal{K} := [EndE(-D) \xrightarrow{ad\varphi \otimes i} EndE \otimes \omega(D)]$.

where $i : \mathcal{O}(-D) \hookrightarrow \mathcal{O}$ is the natural inclusion. These identifications are natural, and differentials of maps between these spaces are realized by maps of complexes. For example, the fibration $T^*\mathcal{U}_D \to \mathcal{U}_D$, with fiber $T^*_{(E,\eta)}\mathcal{U}_D$, gives the sequence

$$
\begin{array}{ccccccccc}
0 & \to & T^*_{(E,\eta)}\mathcal{U}_D & \to & T_{(E,\varphi,\eta)}(T^*\mathcal{U}_D) & \to & T_{(E,\eta)}\mathcal{U}_D & \to & 0 \\
 & & \| & & \| & & \| & & \\
0 & \to & H^0(EndE \otimes \omega(D)) & \to & \mathbb{H}^1(\mathcal{K}) & \to & H^1(EndE(-D)) & \to & 0
\end{array}
$$

derived from the short exact sequence of complexes,

$$0 \to EndE \otimes \omega(D)[-1] \to K \to EndE(-D) \to 0,$$

while the (rational) map $T^*\mathcal{U}_D \to \mathrm{Higgs}_D$ gives

$$
\begin{array}{ccccccccc}
0 & \to & \mathfrak{g}_D & \to & T_{(E,\varphi,\eta)}(T^*\mathcal{U}_D) & \to & T_{(E,\varphi)}\mathrm{Higgs}_D & \to & 0 \\
 & & \| & & \| & & \| & & \\
0 & \to & \frac{H^0(EndE \otimes \mathcal{O}_D)}{H^0(EndE)} & \to & \mathbb{H}^1(\mathcal{K}) & \to & \mathbb{H}^1(\overline{\mathcal{K}}) & \to & 0
\end{array}
$$

which derives from:

$$0 \to \mathcal{K} \to \overline{\mathcal{K}} \to EndE \otimes \mathcal{O}_D \to 0.$$

The dual of a complex $\mathcal{L} : A \to B$ of vector bundles is the complex

$$\mathcal{L}^\vee : B^* \otimes \omega \to A^* \otimes \omega.$$

Grothendieck duality in this case gives a natural isomorphism

$$\mathbb{H}^1\,(\mathcal{L}) \approx \mathbb{H}^1\,(\mathcal{L}^\vee)^*.$$

We note that $\mathcal{K}$ is self-dual, in the sense that there is a natural isomorphism of complexes, $J : \mathcal{K}^\vee \overset{\sim}{\to} \mathcal{K}$. For $\overline{\mathcal{K}}$ we obtain a natural isomorphism of complexes, $\overline{\mathcal{K}}^\vee \overset{\sim}{\to} \overline{\mathcal{K}} \otimes \mathcal{O}(-D)$, hence (composing with i) a morphism

$$I : \overline{\mathcal{K}}^\vee \to \overline{\mathcal{K}}.$$

Combining with duality, we get maps

$$\mathbb{H}^1\,(\mathcal{K})^* \approx \mathbb{H}^1\,(\mathcal{K}^\vee) \overset{J}{\approx} \mathbb{H}^1\,(\mathcal{K})$$

and

$$\mathbb{H}^1\,(\overline{\mathcal{K}})^* \approx \mathbb{H}^1\,(\overline{\mathcal{K}}^\vee) \overset{I}{\to} \mathbb{H}^1\,(\overline{\mathcal{K}}).$$

These give elements of $\otimes^2 \mathbb{H}^1\,(\mathcal{K})$ and $\otimes^2 \mathbb{H}^1\,(\overline{\mathcal{K}})$. Both are skew symmetric (since ad_φ, and hence I, J, are), so we obtain global two-vectors on $T^*\mathcal{U}_D^{ns}$ and Higgs_D^{ns}. At stable points these agree with (the dual of) the symplectic form and its reduction modulo G_D, which is what we need.

Another way to find the two-vector on Higgs_D is based on the interpretation of Higgs_D as a moduli space of sheaves on the total space S of $\omega(D)$. At such a simple sheaf $\mathcal{E}$, with support on some spectral curve C, Mukai [Mu1] identifies the tangent space to moduli with

$$Ext^1_{\mathcal{O}_S}(\mathcal{E}, \mathcal{E}),$$

and notes that any two-form $\sigma \in H^0(\omega_S)$ determines an alternating bilinear map:

$$Ext^1_{\mathcal{O}_S}(\mathcal{E}, \mathcal{E}) \times Ext^1_{\mathcal{O}_S}(\mathcal{E}, \mathcal{E}) \to Ext^2_{\mathcal{O}_S}(\mathcal{E}, \mathcal{E}) \overset{tr}{\to} H^2(\mathcal{O}_S) \overset{\sigma}{\to} H^2(\omega_S) \approx \mathbb{C},$$

hence a two-form on moduli. Mukai uses this argument to produce symplectic structures on the moduli spaces of sheaves on $K3$ and abelian surfaces. The same argument works, of course, for sheaves on $T^*\Sigma$; this reconstructs the symplectic form on Hitchin's system.

Our surface S (the total space of $\omega(D)$) is related to $T^*\Sigma$ by a birational morphism $\alpha : T^*\Sigma \to S$. The symplectic form σ does not descend to S, but its inverse σ^{-1} does give a two vector on S which is non-degenerate away from D and is closed there (since it is locally equivalent to the Poisson structure on $T^*\Sigma$).

Tyurin notes [Ty1] that a variant of this argument produces a two-vector on moduli from a two-vector on S. Now the birational morphism $T^*\Sigma \to S$ takes the Poisson structure on $T^*\Sigma$ to one on S, so the Mukai-Tyurin argument gives the desired two-vector on Higgs_D. In chapter 8 this approach is generalized to higher dimensional varieties.

A third argument for the linearization is given by Bottacin [Bn]. He produces an explicit two-vector at stable points using a deformation argument as above, and then makes direct, local computations to check closedness of the Poisson structures and linearity of the flows.

5.5 Hamiltonians and flows in $T^*\mathcal{U}_D$

We saw that the level group G_D acts on $T^*\mathcal{U}_D$, inducing the Poisson structure on the quotient Higgs$_D$, and that the moment map is

$$\mu : T^*\mathcal{U}_D \quad \to \quad \mathfrak{g}_D^*$$
$$\mu(E, \varphi, \eta)(A) \quad := \quad \text{Res Trace } (A \cdot \varphi^\eta).$$

The characteristic polynomial map $\tilde{h} : T^*\mathcal{U}_D \to B_D$ is a composition of the Poisson map $T^*\mathcal{U}_D \to \text{Higgs}_D$ with the Hamiltonian map $h : \text{Higgs}_D \to B_D$. Hence $\tilde{h}$ is also Hamiltonian. Clearly, $\tilde{h}$ is G_D-invariant.

The composition $T^*\mathcal{U}_D \overset{\mu}{\to} \mathfrak{g}_D^* \to \mathfrak{g}_D^*/G_D$ is a G_D-invariant Hamiltonian morphism and hence factors through Higgs$_D$. It follows that it factors also through B_D since $h :$ Higgs$_D \to B_D$ is a Lagrangian fibration whose generic fiber is connected (see remark 2.15). The conditions of example 2.16 in section 2.4 are satisfied and we get a diagram with a 180° rotational symmetry in which opposite spaces are dual pairs of realizations. The realization dual to $T^*\mathcal{U}_D \to \mathfrak{g}_D^*/G_D$ is the rational morphism $T^*\mathcal{U}_D \to G\text{Higgs}_D :=$ Higgs$_D \times_{(\mathfrak{g}_D^*/G_D)} \mathfrak{g}_D^*$ to the fiber product. The one dual to $\tilde{h} : T^*\mathcal{U}_D \to B_D$ is the morphism $T^*\mathcal{U}_D \to GB_D := B_D \times_{(\mathfrak{g}_D^*/G_D)} \mathfrak{g}_D^*$ to the fiber product. We write down the spaces and typical elements in them:

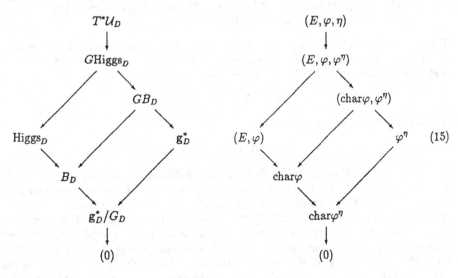

$$\tag{15}$$

6 Spectral flows and KP

Our aim in this section is to relate the general spectral system which we have been considering to the KP and multi-component KP hierarchies. We start by reviewing these hierarchies and their traditional relationship to curves and bundles via the Krichever map. We then reinterpret these flows as coming from Hamiltonians on the limit T^*U_∞ of our previous symplectic spaces. We show that Higgs_∞ can be partitioned into a finite number of loci, each of which maps naturally to one of the $mcKP$-spaces in a way which intertwines isospectral flows with KP flows. As an example we consider the Elliptic solitons studied by Treibich and Verdier.

6.1 The hierarchies

KP

Following the modern custom (initiated by Sato, explained by Segal-Wilson [SW], and presented elegantly in [AdC, Mul, LM1] and elsewhere), we think of the KP hierarchy as given by the action of an infinite-dimensional group on an infinite-dimensional Grassmannian: set

$$
\begin{aligned}
K &:= \mathbb{C}((z)) = \text{field of formal Laurent series in a variable } z \\
Gr &:= \{\text{subspaces } W \subset K | \text{projection } W \to K/\mathbb{C}[[z]]z \text{ is Fredholm}\} \\
&= \{\text{subspaces "comparable to } \mathbb{C}[z^{-1}]"\}.
\end{aligned}
$$

This can be given an algebraic structure which allows us to talk about vector fields on Gr, finite-dimensional algebraic subvarieties, etc. Every $a \in K$ determines a vector field KP_a on Gr, whose value at $W \in Gr$ is the map

$$
W \hookrightarrow K \xrightarrow{a} K \to K/W,
$$

considered as an element of

$$
Hom(W, K/W) \approx T_W Gr.
$$

The (double) KP hierarchy on Gr is just this collection of commuting vector fields. For $a \in \mathbb{C}[[z]]$, this vector field comes from the action on Gr of the one-parameter subgroup $exp(ta)$ in $\mathbb{C}[[z]]^*$, which we consider trivial. The KP hierarchy itself thus consists of the vector fields KP_a, for $a \in \mathbb{C}[z^{-1}]z^{-1}$, on the quotient $Gr/(\mathbb{C}[[z]]^*)$. This quotient is well-behaved: the action of $\mathbb{C}^*$ is trivial, and the unipotent part $1 + z\mathbb{C}[[z]]$ acts freely and with transversal slices. One restricts attention to the open subset of this quotient ("the big cell") parametrizing W of fixed index (the index of W is the index of the Fredholm projection) and satisfying a general position condition with respect to the standard subspace $W_0 := \mathbb{C}[z^{-1}]$. Sato's construction identifies this subset with the space Ψ of pseudo differential operators of the form

$$
\mathcal{L} = D + \sum_{i=1}^{\infty} u_i D^{-i}
$$

where

$$
u_i = u_i(t_1, t_2, \cdots)
$$

and $D = \partial/\partial t_1$. The resulting flows on Ψ have the familiar Lax form:

$$\frac{\partial \mathcal{L}}{\partial t_i} = [(\mathcal{L}^i)_+, \mathcal{L}],$$

where $(\mathcal{L}^i)_+$ is the differential operator part of $\mathcal{L}^i$.

multi component KP

The kth multi-component KP hierarchy $(mcKP)$ is obtained by considering instead the Grassmannian Gr_k of subspaces of $K^{\oplus k}$ comparable to $(\mathbb{C}[z^{-1}])^{\oplus k}$. The entire "loop algebra" $gl(k, K)$ acts here, but to obtain commuting flows we need to restrict to a commutative subalgebra. For the k-th multi-component KP we take the simplest choice, of diagonal matrices, i.e. we consider the action of $(\mathbb{C}[z^{-1}]z^{-1})^{\oplus k}$ on the quotient $Gr_k/(\mathbb{C}[[z]]^*)^k$. There is a big cell $\Psi_k \subset Gr_k/(\mathbb{C}[[z]]^*)^k$, consisting as before of subspaces in general position with respect to a reference subspace W_0, on which the flow is given by a Lax equation (for vector-valued operators).

Heisenberg flows

More generally, for a partition

$$\underline{n} = (n_1, \cdots, n_k)$$

of the positive integer n, we can consider, following [AB], the maximal torus $Heis_{\underline{n}}$ of type $\underline{n}$ in $GL(n, K)$, as well as $heis_{\underline{n}}$, the corresponding Lie subalgebra in $gl(n, K)$. These consist of matrices in block-diagonal form, where the ith block is a formal power series in the $n_i \times n_i$ matrix

$$P_{n_i} := \begin{pmatrix} 0 & & & 0 & z \\ 1 & & & 0 & 0 \\ 0 & \cdot & & 0 & 0 \\ 0 & & \cdot & 0 & 0 \\ 0 & & \cdot & 0 & 0 \\ 0 & & & 1 & 0 \end{pmatrix} \tag{16}$$

We recall that this matrix arises naturally when we consider a vector bundle which is the direct image of a line bundle, near a point where n_i sheets come together: in terms of a natural local basis of the vector bundle, it expresses multiplication by a coordinate upstairs (see (11)). The $\underline{n}^{th}$ $mcKP$ (or "Heisenberg flows" of type $\underline{n}$) lives on the quotient of Gr_n by the non-negative powers of the P_{n_i}, and a basis for the surviving flows is indexed by k-tuples $(d_1, \cdots, d_k)$, $d_i > 0$. Again, this can all be realized by Lax equations on an appropriate space $\Psi_{\underline{n}}$ of pseudo differential operators. When $\underline{n} = (1, \cdots, 1)$ we recover the nth $mcKP$. When $\underline{n} = (n)$, the flows are pulled back from the standard KP flows on Gr, via the mixing map

$$m_n : Gr_n \to Gr$$

sending

$$\widetilde{W} \subset \mathbb{C}((\tilde{z}))^{\oplus n}$$

to

$$W := \{\sum_{i=0}^{n-1} a_i(z^n)z^i | (a_0(\tilde{z}), \cdots, a_{n-1}(\tilde{z})) \in \widetilde{W}\} \subset \mathbb{C}((z)).$$

An arbitrary k-part partition $\underline{n}$ of n determines a map

$$m_{\underline{n}} : Gr_n \to Gr_k,$$

and the $\underline{n}^{th}$ Heisenberg flows are pullbacks of the kth $mcKP$. The natural big cell in this situation is determined by the cartesian diagram:

$$
\begin{array}{ccccc}
Gr_n & \longrightarrow & Gr_n/Heis_{\underline{n}}^+ & \hookleftarrow & \Psi_{\underline{n}} \\
\downarrow m_{\underline{n}} & & \downarrow & & \downarrow \\
Gr_k & \longrightarrow & Gr_k/(C[[z]]^*)^k & \hookleftarrow & \Psi_k
\end{array}
$$

6.2 Krichever maps

The data

A basic Krichever datum (for the KP hierarchy) consists of a quintuple

$$(C, p, z, L, \eta)$$

where:

C is a (compact, non-singular) algebraic curve
$p \in C$
z is a local (analytic or formal) coordinate at p
$L \in PicC$
$\eta : L \otimes \hat{\mathcal{O}}_p \xrightarrow{\sim} \hat{\mathcal{O}}_p \approx \mathbb{C}[[z]]$ is a (formal) trivialization of L near p.

If we fix C, p and z, we think of the Krichever datum as giving a point of

$$\mathcal{U}_C(1, \infty p) := \varprojlim_{\ell} \mathcal{U}_C(1, \ell p).$$

The Krichever map

$$\{\text{Krichever data}\} \to Gr$$

sends the above datum to the subspace

$$W := \eta(H^0(C, L(\infty p))) = \bigcup_k \eta(H^0(C, L(kp))) \subset \mathbb{C}((z)).$$

This subspace is comparable to $\mathbb{C}[z^{-1}]$, since it follows from Riemann-Roch that the dimension of $H^0(L(kp))$ differs from k by a bounded (and eventually constant) quantity.

The flows

Let's work with a coordinate z which is analytic, i.e. it actually converges on some disc. A line bundle L on C can be trivialized (analytically) on the Stein manifold $C \setminus p$. We can think of (L, η) as being obtained from $\mathcal{O}_{C \setminus p}$ by glueing it to $\hat{\mathcal{O}}_p$ via a 1-cocycle, or transition function, which should consist of an invertible function g on a punctured neighborhood of p in C. Conversely, we claim there is a map:

$$\exp : K \longrightarrow \mathcal{U}_C(1, \infty p),$$

$$f \longmapsto (L, \eta).$$

For $f \in \mathbb{C}(z)$, this is defined by the above analytic gluing, using $g := \exp f$, which is indeed analytic on a punctured neighborhood. For $f \in \mathbb{C}[[z]] \approx \hat{\mathcal{O}}_p$, on the other hand, we take $(L, \eta) := (\mathcal{O}, \exp f)$. These two versions agree on the intersection, $f \in \mathbb{C}[z]_{(0)}$, so the map is uniquely defined as claimed. (The bundles we get this way all have degree 0, but we can also obtain maps

$$\exp_{g_0} : K \longrightarrow \mathcal{U}_C(1, d, \infty p)$$

to the moduli space of level-∞p line bundles of degree d, simply by fixing a meromorphic function g_0 on a neighborhood of p which has order d at p, and replacing the previous g by $g_0 \exp f$. We will continue to suppress the degree d in our notation.)

Any $a \in K$ gives an additive flow on K, which at $f \in K$ is

$$t \longmapsto f + ta.$$

Under the composed map

$$K \overset{\exp_{g_0}}{\longrightarrow} \mathcal{U}_C(1, \infty p) \overset{Krichever}{\longrightarrow} Gr,$$

this is mapped to the double KP flow KP_a on Gr. For $a \in \mathbb{C}[[z]]$ this flow does not affect the isomorphism class of L, and simply multiplies η by $exp(ta)$. On the other hand, the ith KP flow, given by $a = z^{-i}$, changes both L and η if $i > 0$. The projection to $Pic\ C$ is a linear flow, whose direction is the ith derivative at p, with respect to the coordinate z, of the Abel-Jacobi map $C \to Pic\ C$. Dividing out the trivial flows corresponds to suppressing η, so we obtain, for each C, p, z and degree $d \in \mathbb{Z}$, a finite-dimensional orbit of the KP flows in $Gr/\mathbb{C}[[z]]^*$, isomorphic to $Pic^d C$.

Multi-Krichever data

Several natural generalizations of the Krichever map to the multi-component KP can be found in [AB, LM1] and elsewhere. Here are some of the possibilities. We can consider "multi-Krichever" data

$$(C, D, z_i, L, \eta)$$

involving a curve C with a divisor D consisting of k distinct points $p_i (1 \le i \le k)$, a coordinate z_i at each p_i, a line bundle L, and a formal trivialization η_i at each p_i. Fixing C, p_i and z_i, we have a multi-Krichever map

$$\{\text{multi-Krichever data}\} \approx \mathcal{U}_C(1, \infty D) \longrightarrow Gr_k$$

sending

$$(L, \eta_i) \mapsto W := (\eta_1, \cdots, \eta_k)(H^0(C, L(\infty D))) \subset \mathbb{C}((z))^{\oplus k}.$$

The k-component KP flow on the right hand side given by $a = (a_1, \ldots, a_k) \in K^k$ restricts to the flow on the multi-Krichever data which multiplies the transition function at p_i (for an analytic trivialization of L on $C \setminus D$) by $\exp a_i$.

We can also consider "vector-Krichever" data (C, p, z, E, η) where the line bundle L is replaced by a rank n vector bundle E, and

$$\eta : E \otimes \hat{\mathcal{O}}_p \overset{\sim}{\to} (\hat{\mathcal{O}}_p)^n \approx (\mathbb{C}[[z]])^n$$

is now a (formal) trivialization of E near p. Not too surprisingly, the vector-Krichever map

$$\{\text{vector-Krichever data}\} \to Gr_n$$

sends the above datum to the subspace

$$W := \eta(H^0(C, E(\infty p))) \subset (\mathbb{C}((z)))^n.$$

In the next subsection we will see that the interesting interaction of these two types of higher Krichever maps occurs not by extending further (to objects such as (C, D, z_i, E, η_i)), but by restricting to those vector data on one curve which match some multi-data on another.

KdV-type subhierarchies

Among the Krichever data one can restrict attention to those quintuples where z^{-n} (for some fixed n) happens to extend to a regular function on $C \setminus p$, i.e. gives a morphism

$$f = z^{-n} : C \to \mathbb{P}^1$$

of degree n, such that the fiber $f^{-1}(\infty)$ is $n \cdot p_0$. The Krichever map sends such data to the nth KdV hierarchy, the distinguished subvariety of Gr (invariant under the (double) KP flows) given by

$$\text{KdV}_n := \{W \in Gr \mid z^{-n}W \subset W\}.$$

The corresponding subspace of Ψ is

$$\{\mathcal{L} \mid \mathcal{L}^n = \mathcal{L}_+^n \text{ is a } \underline{\text{differential}} \text{ operator}\}.$$

Fixing a partition $\underline{n} = (n_1, \cdots, n_k)$ of n, we can similarly consider the covering data of type $\underline{n}$, consisting of the multi-Krichever data (C, p_i, z_i, L, η_i) plus a map $f : C \to \Sigma$ of degree n to a curve Σ with local coordinate z at a point $\infty \in \Sigma$, such that

$$f^{-1}(p) = \Sigma n_k p_i, \quad f^{-1}(z) = z_i^{n_i} \text{ at } p_i.$$

Such a covering datum clearly gives a multi-Krichever datum on C, but it also determines a vector-Krichever datum (E, η) on Σ: The standard m-sheeted branched cover

$$\begin{aligned} f_m : \mathbb{C} &\to \mathbb{C} \\ \tilde{z} &\mapsto z = \tilde{z}^m \end{aligned}$$

of the z-line determines an isomorphism

$$s_m : (f_m)_* \mathcal{O} \overset{\sim}{\to} \mathcal{O}^{\oplus m}$$

given by

$$\sum_{i=0}^{m-1} a_i(\tilde{z}^m)\tilde{z}^i \mapsto (a_0(z), \cdots, a_{m-1}(z)).$$

To the covering datum above we can then associate the rank-n vector bundle $E := f_* L$ on Σ, together with the trivialization at p obtained by composing

$$\oplus_i f_{*, p_i}(\eta_i) : (f_* L)_p \overset{\sim}{\to} \oplus_i f_{*, p_i}(\mathcal{O}_{p_i})$$

with the isomorphisms

$$f_{*,p_i}(\mathcal{O}_{p_i}) \xrightarrow{\sim} \mathcal{O}_p^{\oplus n_i}$$

which are conjugates of the standard isomorphisms s_{n_i} by the chosen local coordinates z, z_i. Finally, we note that there are obvious geometric flows on these covering data: L and η_i flow as before, while everything else stays put. The compatibility of the two types of higher Krichever data is expressed by the commutativity of the diagram:

$$
\begin{array}{ccc}
\{\underline{n} - \text{covering data}\} & \approx & \{f : C \to \Sigma; \ p_i, z_i, L, \eta_i; \ z \mid \ldots\} \\
\downarrow & & \downarrow \\
\{\text{vector Krichever data on } \Sigma\} & & \{\text{multi Krichever data on } C\} \\
\| & & \| \\
\cup_{\Sigma,p,z}\mathcal{U}_\Sigma(n,\infty p) & & \cup_{C,D,z_i}\mathcal{U}_C(1,\infty D) \qquad (17)\\
\downarrow & & \downarrow \\
Gr_n & \xrightarrow{m_{\underline{n}}} & Gr_k \\
\downarrow & & \downarrow \\
Gr_n/Heis_{\underline{n}}^+ & \longrightarrow & Gr_k/(\mathbb{C}[[z]]^\times)^k.
\end{array}
$$

The mcKP flows on the bottom right pull back to the Heisenberg flows on the bottom left, and to the geometric flows on the $\underline{n}$-covering data.

6.3 Compatibility of hierarchies

Fix a smooth algebraic curve Σ of arbitrary genus and a point P in it. The moduli space $\text{Higgs}_D := \text{Higgs}_\Sigma^{sm}(n, d, \omega(D))$ (see definition 4.9) can be partitioned into type loci. We consider the Zariski dense subset consisting of the union of finitely many type loci $\text{Higgs}_D^{\underline{n}}$ indexed by partitions $\underline{n}$ of n. A Higgs pair in $\text{Higgs}_D^{\underline{n}}$ has a spectral curve $C \to \Sigma$ whose ramification type over $P \in \Sigma$ is $\underline{n} = (n_1, \ldots, n_k)$.

Fix a formal local parameter z on the base curve Σ at P. A Higgs pair in $\text{Higgs}_D^{\underline{n}}$ (or rather its spectral pair (C, L), see proposition 4.5) can be completed to an $Heis_{\underline{n}}^+$-orbit of an $\underline{n}$-covering data $(C, P_i, z_i, L, \eta_i) \to (\Sigma, P, z, E, \eta)$ in finitely many ways. These extra choices form a natural finite Galois cover $\widetilde{\text{Higgs}_D^{\underline{n}}}$ of each type locus $\text{Higgs}_D^{\underline{n}}$. We obtain Krichever maps (see diagram 17) from the Galois cover $\widetilde{\text{Higgs}_D^{\underline{n}}}$ to the quotients $Gr_n/Heis_{\underline{n}}^+$ and $Gr_k/(\mathbb{C}[[z]]^\times)^k$. Both the mcKP and Heisenberg flows pull back to the same geometric flow on the Galois cover. It is natural to ask:

Question 6.1 (The compatibility question) *Is the Heisenberg flow Poisson with respect to the natural Poisson structure on Higgs_D?*

The Compatibility Theorem 6.5 and its extension 6.13 provide an affirmative answer. We factor the moment map of the Heisenberg action through natural finite Galois covers of the ramification type loci in the space of characteristic polynomials (equations (20) and (21)).

The compatibility naturally follows from the construction of the Poisson structure via level structures. Recall the birational realization of the moduli space $\text{Higgs}_\Sigma^{sm}(n, d, \omega(lP +$

D)) as a quotient of the cotangent bundle $T^*\mathcal{U}_{lP+D}$ of the moduli space $\mathcal{U}_{lP+D} := \mathcal{U}_\Sigma(n,d,lP+D)$ of vector bundles with level structures (Chapter 5). This realization is a finite dimensional approximation of the limiting realization of the moduli space

$$\mathrm{Higgs}_{\infty P+D} := \varprojlim_{l\to\infty} \mathrm{Higgs}_{lP+D}$$

as a quotient of (a subset of) the cotangent bundle $T^*\mathcal{U}_{\infty P+D}$. The ramification type loci Higgs^n_D, their Galois covers $\widetilde{\mathrm{Higgs}}^n_D$ and the infinitesimal Heis_n-action on $\widetilde{\mathrm{Higgs}}^n_D$ become special cases of those appearing in the construction of section 2.5.2. The compatibility theorem follows from corollary 2.22 accompanied by the concrete identification of the moment maps in our particular example.

The rest of this section is organized as follows. In section 6.3.1 we emphasize the ubiquity of the setup of relative Krichever maps. They can be constructed for any family $\mathcal{J} \to B$ of Jacobians of branched covers of a fixed base curve Σ. The analogue of the compatibility question 6.1 makes sense whenever the family $\mathcal{J} \to B$ is an integrable system (see for example question 6.2).

Starting with section 6.3.2 we concentrate on the moduli spaces of Higgs pairs. Sections 6.3.2 and 6.3.3 consider the case of smooth spectral curves. Especially well behaved is the case where the point $P \in \Sigma$ of the $\underline{n}$-covering data is in the support of the polar divisor D. In this case the symplectic leaves foliation of the moduli space of Higgs pairs is a refinement of the type loci partition (lemma 6.3).

Section 6.3.4 is a generalization to singular cases. As an example, we consider in section 6.3.5 the Elliptic Solitons studied by Treibich and Verdier. We conclude with an outline of the proof of the compatibility theorem in section 6.3.6.

Note: Type-$(1,1,\ldots,1)$ relative Krichever maps were independently considered by Y. Li and M. Mulase in a recent preprint [LM2].

6.3.1 Galois covers and relative Krichever maps

Let $B_D := \oplus_{i=1}^n H^0(\Sigma,(\omega_\Sigma(D))^{\otimes i})$ be the space of characteristic polynomials. For simplicity, we restrict ourselves to the Zariski open subset Bsm_D of reduced and irreducible n-sheeted spectral curves $\pi : C \to \Sigma$ in $T^*_\Sigma(D)$ whose fiber over $P \in \Sigma$ consists of *smooth* points of C. Denote by Higgsm_D the corresponding open subset of Higgs_D. The ramification type stratification of Higgsm_D is induced by that of Bsm_D

$$Bsm_D = \cup_{\underline{n}} Bsm^{\underline{n}}_D.$$

Given a Higgs pair (E,φ) in $\mathrm{Higgsm}^{\underline{n}}_D$ corresponding to a torsion free sheaf L on a spectral cover $C \to \Sigma$ we can complete it to an $\underline{n}$-covering data

$$(C,P_i,z_i,L,\eta_i) \to (\Sigma,P,z,E,\eta)$$

by choosing i) a formal local parameter z at P, ii) an n_i-th root z_i of π^*z at each point P_i of C over P and iii) formal trivializations η_i of the sheaf L at P_i. The Heis^+_n-orbit of an $\underline{n}$-covering data consists precisely of all possible choices of the η_i's. Thus, for fixed P and z only a finite choice is needed in order to obtain the points of the quotients of the Grassmannians $Gr_n/\text{Heis}_{\underline{n}}^+$ and $Gr_k/(\mathbb{C}[[z]]^\times)^k$ (see diagram 17). These choices are independent of the sheaf L. The choices are parametrized by the Galois cover $\widetilde{Bsm}_D^{\underline{n}} \to Bsm_D^{\underline{n}}$ consisting of pairs (C, λ) of a spectral curve C in $Bsm^{\underline{n}}$ and the discrete data λ which amounts to:

i) (Parabolic data) An ordering $(P_1, P_2, \ldots, P_k)$ of the points (eigenvalues) in the fiber over P compatible with the fixed order of the ramification indices $(n_1, \ldots, n_k)$ (say, $n_1 \leq n_2 \leq \ldots \leq n_k$).

ii) A choice of an n_i-th root z_i of $\pi^* z$ at P_i.

Denote by $\widetilde{\text{Higgsm}}_D^{\underline{n}}$ the corresponding Galois cover of $\text{Higgsm}_D^{\underline{n}}$. We get a canonical relative Krichever map

$$\kappa_{\underline{n}} : \widetilde{\text{Higgsm}}_D^{\underline{n}} \to Gr_n/\text{Heis}_{\underline{n}}^+ \tag{18}$$

from the Galois cover to the quotient Grassmannian.

The Galois group of $\widetilde{Bsm}_D^{\underline{n}} \to Bsm_D^{\underline{n}}$ is the Weyl group $W_{\underline{n}} := N(\text{Heis}_{\underline{n}}^+)/\text{Heis}_{\underline{n}}^+$ of the maximal torus of the level infinity group G_∞^+. For example, $W_{(1,\ldots,1)}$ is the symmetric group S_n, while $W_{(n)}$ is the cyclic group of order n. The discrete data $\lambda = [(P_1, P_2, \ldots, P_k), (z_1, \ldots, z_k)]$ of a point (C, λ) in $\widetilde{Bsm}_D^{\underline{n}}$ is equivalent to a commutative $\mathbb{C}[[z]]$-algebras isomorphism

$$\lambda : \text{heis}_{\underline{n}}^+ \overset{\cong}{\to} \oplus_{i=1}^k \widehat{\mathcal{O}_{C,(P_i)}} \tag{19}$$

of the torus algebra with the formal completion of the structure sheaf $\mathcal{O}_C$ at the fiber over P. The inverse λ^{-1} sends z_i to the generator of the i-th block of the torus $\text{heis}_{\underline{n}}^+$ given by (16). The finite Weyl group $W_{\underline{n}}$ acts on $\text{heis}_{\underline{n}}^+$, hence on λ, introducing the $W_{\underline{n}}$-torsor structure on $\widetilde{Bsm}_D^{\underline{n}}$. (See also lemma 6.17 part 2 for a group theoretic interpretation.)

We note that a Galois cover $\tilde{B} \to B$ as above can be defined for any family $\mathcal{J} \to B$ of Jacobians of a family $\mathcal{C} \to B$ of branched covers with a fixed ramification type $\underline{n}$ of a fixed triple (Σ, P, z). We obtain a relative Krichever map

$$\kappa_{\underline{n}} : \tilde{\mathcal{J}} \to Gr_n/\text{Heis}_{\underline{n}}^+$$

as above. The Heisenberg flow pulls back to an infinitesimal action, i.e., a Lie algebra homomorphism

$$d\rho : \text{heis}_{\underline{n}} \to V(\tilde{\mathcal{J}})$$

into a commutative algebra of vertical tangent vector fields. When $\mathcal{J}$ (and hence $\tilde{\mathcal{J}}$) is an integrable system we are led to ask the compatibility question 6.1: *is the action Poisson?* i.e., *can $d\rho$ be lifted to a Lie algebra homomorphism*

$$\mu_{\text{heis}_{\underline{n}}}^* : \text{heis}_{\underline{n}} \to \Gamma(\mathcal{O}_{\tilde{B}}) \hookrightarrow [\Gamma(\mathcal{O}_{\tilde{\mathcal{J}}}), \{,\}]?$$

A priori, the vector fields $d\rho(a)$, $a \in \text{heis}_{\underline{n}}$ may not even be *locally* Hamiltonian.

Inherently nonlinear examples arise from the Mukai-Tyurin integrable system of a family of Jacobians of a linear system $B := \mathbb{P}H^0(S, \mathcal{L})$ of curves on a symplectic or Poisson surface S (see chapter 8). Consider for example the:

Question 6.2 *Let* $\pi : S \to \mathbb{P}^1$ *be an elliptic K3 surface and* $\mathcal{L}$ *a very ample line bundle on* S. *Fix* $P \in \mathbb{P}^1$ *and a local parameter* z *and consider the Galois cover* $\widetilde{B}^{(1,1,\cdots,1)}$ *of the generic ramification type locus. Is the Heisenberg action Poisson on* $\widetilde{\mathcal{J}} \to \widetilde{B}^{(1,1,\cdots,1)}$ *(globally over* $\widetilde{B}^{(1,1,\cdots,1)}$*)?*

The compatibility question has an intrinsic algebro-geometric formulation: The j-th KP flow of the P_i-component is the vector field whose direction along the fiber over $(b, \lambda) \in \widetilde{B}^{(1,1,\cdots,1)}$ is the j-th derivative of the Abel-Jacobi map at P_i. Using the methods of chapter 8 it is easy to see that the Heisenberg action is symplectic. As we move the point $P(0) := P$ in $\mathbb{P}^1$ and its (Lagrangian) fiber in S we obtain an analytic (or formal) family of *Lagrangian* sections $\mathcal{AJ}(P_i(z))$ of $\widetilde{\mathcal{J}} \to \widetilde{B}^{(1,1,\cdots,1)}$ (see corollary 8.13). Translations by the sections $\mathcal{AJ}(P_i(z)) - \mathcal{AJ}(P_i(0))$ is a family of symplectomorphisms of $\widetilde{\mathcal{J}}$. Thus, the vector field corresponding to its j-th derivative with respect to the local parameter z is locally Hamiltonian. It seems unlikely however that the Heisenberg flow integrates to a global Poisson action for a general system as in question 6.2. It is the *exactness* of the symplectic structure in a neighborhood of the fiber over P in $T^*\Sigma$ which lifts the infinitesimal symplectic Heisenberg action to a Poisson action in the Hitchin's system case (see equation (21)).

6.3.2 Compatibility of stratifications

Prior to stating the compatibility theorem 6.5 we need to examine the Poisson nature of the Galois covers $\widetilde{\mathrm{Higgsm}}_D^n$. $\mathrm{Higgsm}_D^{(1,1,\cdots,1)}$ is a Zariski *open Poisson* subvariety of Higgsm_D. Hence, the unramified Galois cover $\widetilde{\mathrm{Higgsm}}_D^{(1,1,\cdots,1)}$ is endowed with the canonical pullback Poisson structure.

Though non generic, the other type strata are as important. The cyclic ramification type (n), for example, corresponds to the single component KP-hierarchy (see [SW]). When the point P of the $\underline{n}$-covering data is in the support of the divisor D, we obtain a *strict compatibility* between the P-ramification type stratification of Higgsm_D and its symplectic leaves foliation. All Galois covers $\widetilde{\mathrm{Higgsm}}_D^n$, $P \in D$ are thus endowed with the canonical pullback Poisson structure:

Lemma 6.3 *(conditional compatibility of stratifications) When the point* $P \in \Sigma$ *is in the support of* D, *the symplectic leaves foliation is a refinement of the ramification type stratification* $\mathrm{Higgsm}_D = \cup_{\underline{n}} \mathrm{Higgsm}_D^{\underline{n}}$.

Proof: We need to show that the ramification type $\underline{n}$ of the spectral cover $\pi : C \to \Sigma$ over $P \in D$ is fixed throughout the symplectic leaf $\mathrm{Higgsm}_S \subset \mathrm{Higgsm}_D$ of a Higgs pair (E_0, φ_0). The symplectic leaves of Higgsm_D are determined by coadjoint orbits of the level D algebra $\mathbf{g}_D$. The coadjoint orbit S is determined by the residue of the Higgs field, namely, the infinitesimal data $(E_{|D}, \varphi_{|D})$ encoded in the value of φ at D (see [Mal] Remark 8.9 and Proposition 7.17)). Thus, the Jordan type of the Higgs field φ at P is fixed throughout Higgsm_S. In general, (allowing singularities over P), the Jordan type depends both on the ramification type of $C \to \Sigma$ and on the sheaf L on C corresponding to the Higgs pair. For Higgs pairs (E, φ) in Higgsm_D, the spectral curve C is smooth over P, hence, its ramification type coincides with the Jordan type of φ at P.

$\square$

6.3.3   The compatibility theorem, the smooth case

We proceed to introduce the moment map of the infinitesimal Poisson action

$$d\rho : \text{heis}_{\underline{n}} \to V(\widetilde{\text{Higgsm}}_D^{\underline{n}}).$$

Throughout the end of this subsection we will assume the

Condition 6.4 *Ramification types $\underline{n}$ other than $(1, 1, \ldots, 1)$ are considered only if P is in the support of D.*

This condition will be relaxed later by conditions 6.11 or 6.12.

Let $b \in Bsm_D^{\underline{n}}$ be the polynomial of the spectral curve $\pi_b : C_b \to \Sigma$. Recall that spectral curves are endowed with a tautological meromorphic 1-form $y_b \in H^0(C_b, \pi_b^* \omega_\Sigma(D))$ with poles over $D \subset \Sigma$ (see section 4.2). Let $\phi_{P_i}^j$ be the function on $\widetilde{Bsm}_D^{\underline{n}}$ given at a pair $(b, \lambda) \in \widetilde{Bsm}_D^{\underline{n}}$ by

$$\phi_{P_i}^j(b, \lambda) := Res_{P_i}((z_i)^{-j} \cdot y_b). \tag{20}$$

The Lie algebra homomorphism $\mu^*_{\text{heis}_{\underline{n}}}$ sends the inverse (in $\text{heis}_{\underline{n}}$) of the generator of the i-th block of the torus $\text{heis}_{\underline{n}}^+$ given in (16) to the function $\phi_{P_i}^1 \circ \widetilde{\text{char}}$ on $\widetilde{\text{Higgsm}}_D^{\underline{n}}$. In other words, $\mu^*_{\text{heis}_{\underline{n}}} : \text{heis}_{\underline{n}} \to [\Gamma(\mathcal{O}_{\widetilde{\text{Higgsm}}_D^{\underline{n}}}), \{,\}]$ factors as a composition $\phi \circ \widetilde{\text{char}}^*$ through $\Gamma(\mathcal{O}_{\widetilde{Bsm}_D^{\underline{n}}})$. If we regard λ also as an isomorphism from $\text{heis}_{\underline{n}}$ to $\oplus_{i=1}^k \mathcal{O}_{C,(P_i)}$ (eq. (19)), then $\phi : \text{heis}_{\underline{n}} \to \Gamma(\mathcal{O}_{\widetilde{Bsm}_D^{\underline{n}}})$ is given by

$$(\phi(a))(b, \lambda) = \sum_{\{P_i\}} Res_{P_i}(\lambda(a) \cdot y_b), \quad a \in \text{heis}_{\underline{n}}. \tag{21}$$

Theorem 6.5 *(The Compatibility Theorem, smooth case) (Assuming condition 6.4) The relative Krichever map*

$$\kappa_{\underline{n}} : \widetilde{\text{Higgsm}}_D^{\underline{n}} \to Gr_n/\text{Heis}_{\underline{n}}^+$$

intertwines the Heisenberg flow on $Gr_n/\text{Heis}_{\underline{n}}^+$ (and the mcKP flow on $Gr_k/(\mathbb{C}[[z]]^\times)^k$) with an infinitesimal Poisson action of the maximal torus $\text{heis}_{\underline{n}}$ on $\widetilde{\text{Higgsm}}_D^{\underline{n}}$. The latter is induced by the Lie algebra homomorphism

$$\mu^*_{\text{heis}_{\underline{n}}} = \phi \circ \widetilde{\text{char}}^* : \text{heis}_{\underline{n}} \xrightarrow{\phi} \Gamma(\mathcal{O}_{\widetilde{Bsm}_D^{\underline{n}}}) \xrightarrow{\widetilde{\text{char}}^*} [\Gamma(\mathcal{O}_{\widetilde{\text{Higgsm}}_D^{\underline{n}}}), \{,\}]$$

which factors through the homomorphism ϕ given by (21).

Remarks 6.6   1. The subalgebra $\text{heis}_{\underline{n}}^+$ acts trivially on $Gr_n/\text{Heis}_{\underline{n}}^+$ hence also on $\widetilde{\text{Higgsm}}_D^{\underline{n}}$. This corresponds to the fact that the functions $\phi_{P_i}^j$, $j \leq 0$ are Casimir. Indeed, if P is not contained in D, the 1-form y_b is holomorphic at the fiber over P and $\phi_{P_i}^j$ is identically zero for $j \leq 0$. If P is in D then the finite set of non-zero $\phi_{P_i}^j$, indexed by finitely many non-positive integers j, are among the Casimirs that induce the highest rank symplectic leaves foliation (see [Ma1] Proposition 8.8).

2. The multi-Krichever map $\kappa_{\underline{n}}$ depends on auxiliary parameters P and z. In other words, it lives naturally on an infinite dimensional space $\cup_{P,z} \widetilde{\mathrm{Higgsm}}_{D,P,z}^{\underline{n}}$ in which P and z are allowed to vary. This is not as bad as it might seem, since the j-th flow on $\widetilde{\mathrm{Higgsm}}_D^{\underline{n}}$ really depends only on our finite dimensional choices of P and the j-th order germ of z there. Similarly, our Hamiltonians $\phi_{P_i}^j$ depend at most on the $(j + n \deg D)$-th germ of z, the shift arising, as in part 1 of this remark, from the possible poles above P of the tautological 1-form on the spectral curve. So we may think of $\cup_{P,z} \widetilde{\mathrm{Higgsm}}_{D,P,z}^{\underline{n}}$ as the inverse limit of a family of finite dimensional moduli spaces, indexed by the level. Each KP flow or Hamiltonian is defined for sufficiently high level.

6.3.4   The compatibility theorem, singular cases

The condition that the fiber over P of the embedded spectral curve be smooth is too restrictive. The embedded spectral data $(\bar{C} \subset T_\Sigma^*(D), \bar{L})$ of a Higgs pair (E, φ) may have singularities over P which are canonically resolvable. The point is that the rank 1 torsion free sheaf $\bar{L}$ on $\bar{C}$ determines a partial normalization $\nu : C \to \bar{C}$ and a unique rank 1 torsion free sheaf L on C such that i) $\bar{L}$ is isomorphic to the direct image $\nu_* L$, and ii) L is locally free at the fiber over P. We are interested in those Higgs pairs for which the fiber of C over P is smooth. Such data may also be completed in finitely many ways to an $\underline{n}$-covering data as in section 6.3.1.

Definition 6.7 The singularities over P of a spectral pair $(\bar{C} \subset T_\Sigma^*(D), \bar{L})$ are said to be *resolved by the spectral sheaf* $\bar{L}$ if i) $\bar{C}$ is irreducible and reduced. ii) The sheaf $\bar{L}$ is the direct image $\nu_* L$ of a rank 1 torsion free sheaf L on the normalization $\nu : C \to \bar{C}$ of the fiber of $\bar{C}$ over P.

Fixing a symplectic leaf Higgs_S we may consider the type sub-loci in the locus of Higgs pairs whose spectral curve has at worst singularities over P which are resolved by the spectral sheaf. The topology of these type loci is quite complicated. As a result, the Galois covers of these type loci may not have a symplectic structure. Nevertheless, the construction of section 2.5.2, as used in section 6.3.6, provides *canonical embeddings* of the Galois covers of these type loci in (finite dimensional) symplectic varieties. These embeddings realize the Heisenberg flow as a Hamiltonian flow.

Control over the topology is regained below by restraining the singularities. In the smooth case (section 6.3.3) it is the smoothness which assures that the generic ramification type locus in a symplectic leaf Higgs_S is open (rather than only Zariski dense). The point is that degenerations from a ramification type $\underline{n}$ through other types back to type $\underline{n}$ must end with a singular fiber over P (and are thus excluded). If $P \in D$, there are symplectic leaves Higgs_S, $S \subset \mathbf{g}_D^*$ of Higgs_D for which the singularities over P are encoded in the infinitesimal data associated to the coadjoint orbit S and shared by the generic Higgs pair in Higgs_S. Often, this is an indication that the Poisson surface $T_\Sigma^*(D)$ is not the best to work with. Moreover, a birational transformation $T_\Sigma^*(D) \to X_S$, centered at points of the fiber over P and encoded in S, can simultaneously resolve the singularities (over P) of the spectral curve of the generic point in Higgs_S (see example 6.9). In such a case, smoothness of the proper transform of the spectral curve in X_S at points of the fiber over P is an *open* condition and the corresponding locus in Higgs$_S$ with generic ramification type is *symplectic*. When the multiplicity of P in D is greater than 1 the correspondence between the coadloint orbits $S \subset g_D^*$ and their surfaces X_S can be quite complicated. Instead of working the correspondence out, we will use the following notion of S-smoothness to assure (see condition 6.11) that the generic ramification type locus in a symplectic leaf Higgs$_S$ is open (rather than only Zariski dense).

Definition 6.8 Let S be a coadjoint orbit of g_D. An irreducible and reduced spectral curve $\pi : \bar{C} \to \Sigma$ is *S-smooth over* D if i) a line bundle L on the resolution of the singularities $\nu : C \to \bar{C}$ of the fibers over D results in a Higgs pair $(E, \varphi) := \pi \circ \nu_*(L, \otimes \nu^*(y))$ in Higgs$_S$. and ii) the arithmetic genus of the normalization C above is equal to half the dimension of the symplectic leaf Higgs$_S$.

If $\bar{C}$ is an irreducible and reduced spectral curve then, by the construction of [Sim1], the fiber of the characteristic polynomial map in Higgs$_\Sigma^{sm}(n, d, \omega(D))$ is its compactified Jacobian, the latter being the moduli space of all rank 1 torsion free sheaves on $\bar{C}$ with a fixed Euler characteristic. The compactified Jacobian is known to be irreducible for irreducible and reduced curve on a surface (i.e., with planar singularities, [AIK]). Moreover, the symplectic leaf Higgs$_S$ intersects the compactified Jacobian of $\bar{C}$ in a union of strata determined by partial normalizations. If (E, φ) in Higgs$_S$ corresponds to $(\bar{C}, \bar{L})$ and $\nu : (C, L) \to (\bar{C}, \bar{L})$ is a partial normalization where L is a locally free sheaf on C, then any twist F of L by a locally free sheaf in $Pic^0(C)$ (the component of $\mathcal{O}_C$) will result in a Higgs pair $(E', \varphi') := (\pi \circ \nu_*(F), \pi_*(\otimes y))$ in Higgs$_S$. The point is that the residue of (E', φ') (with respect to any level-D structure) will be in the same coadjoint orbit as that of (E, φ). S-smoothness of $\bar{C}$ over D is thus equivalent to the geometric condition:

The fiber of the characteristic polynomial map over $\bar{C}$ intersects Higgs$_S$ in a Lagrangian subvariety isomorphic to the compactified Jacobian of the resolution C of the singularities of $\bar{C}$ over D.

The following example will be used in section 6.3.5 to describe a symplectic leaf which parametrizes Elliptic solitons.

Example 6.9 Let $D = P$, $S \subset g_D^* \cong gl(n)^* \cong gl(n)$ be the coadjoint orbit containing the diagonal matrix

$$A = \begin{pmatrix} -1 & 0 & \cdots & & 0 \\ 0 & -1 & & & \\ & & \ddots & & \vdots \\ \vdots & & & -1 & 0 \\ 0 & \cdots & & 0 & n-1 \end{pmatrix}$$

S is the coadjoint orbit of lowest dimension with characteristic polynomial $(x+1)^{n-1}(x-(n-1))$. Its dimension $2n-2$ is $(n-2)(n-1)$ less than the generic rank of the Poisson structure of $gl(n)^*$. If non-empty, each component of Higgs$_S$ is a smooth symplectic variety of dimension dim Higgs$_P - (n-1) - (n-2)(n-1) = [n^2(2g-1)+1] - (n-1)^2$ (see theorem 4.8 and [Ma1] proposition 7.17). The spectral curves $\bar{C} \subset T_\Sigma^*(P)$ which are S-smooth over P will have two points in the fiber over P, one smooth at the eigenvalue with residue $n-1$ and one (singular if $n \geq 3$) at the eigenvalue with residue -1. Assume $n \geq 3$. The resolution C of the singularity of a typical (though not all) such $\bar{C}$ will be unramified over P with $n-1$ points collapsed to one in $\bar{C}$. The sheaf $\bar{L}$ (of a Higgs pair in Higgs$_S$ with spectral curve $\bar{C}$) will be a pushforward of a torsion free sheaf L on C. In contrast, line bundles on that $\bar{C}$ will result in Higgs pairs in another symplectic leaf Higgs$_{S_{reg}}$ corresponding to the *regular* coadjoint orbit S_{reg} in $\mathrm{gl}(n)^* \cong \mathrm{gl}(n)$ with characteristic polynomial $(x+1)^{n-1}(x-(n-1))$. These Higgs pairs will *not* be S_{reg}-smooth. S_{reg}-smoothness coincides with usual smoothness of the embedded spectral curve which is necessarily ramified with ramification index $n-1$ at the point with residue -1 over P.

The S-smooth spectral curves will be smooth on the blowup $\widehat{X_S}$ of $T^*_\Sigma(P)$ at residue -1 over P. If we blow down in $\widehat{X_S}$ the proper transform of the fiber of $T^*_\Sigma(P)$ we get a surface X_S with a marked point x_{n-1} over P. An S-smooth spectral curve $\bar{C}$ will correspond to a curve on X_S through x_{n-1}. It will be smooth at x_{n-1} if in addition it is of ramification type $(1, 1, \ldots, 1)$ over P. Consider the compactification $\mathbb{P}(T^*_\Sigma(P) \oplus \mathcal{O}_\Sigma)$ of $T^*_\Sigma(P)$. Blowing it up and down as above we get a ruled surface $\bar{X}_S$ over Σ which is isomorphic to the projectivization $\mathbb{P}W$ of the unique nontrivial extension

$$0 \to \omega_\Sigma \to W \to \mathcal{O}_\Sigma \to 0.$$

In particular, the surface $\bar{X}_S$ is *independent of the point* P. The point is that blowing up and down the ruled surface $\mathbb{P}V := \mathbb{P}(T^*_\Sigma(P) \oplus \mathcal{O}_\Sigma)$ at residue -1 over P results with the ruled surface of a rank 2 vector bundle W whose sheaf of sections is a subsheaf of $V := T^*_\Sigma(P) \oplus \mathcal{O}_\Sigma$. This subsheaf consists of all sections which restrict at P to the subspace of the fiber $V_{|P}$ spanned by $(-1, 1)$ (i.e., W is a Hecke transform of V, see [Ty2]). Clearly, ω_Σ is a subsheaf of W and the quotient W/ω_Σ is isomorphic to $\mathcal{O}_\Sigma$. The resulting extension is non-trivial because $H^0(\Sigma, W)$ and $H^0(\Sigma, \omega_\Sigma)$ are equal as subspaces of $H^0(\Sigma, V)$.

Remark 6.10 S-smoothness over D is stronger than having singularities over D which are resolved by the spectral sheaf. They differ when the singularity appears at an infinitesimal germ of too high an order to be detected by S. E.g., take $n = 3$ in example 6.9 and consider a pair $(\bar{C}, \bar{L})$ with a tacnode at residue -1 over P (two branches meet with a common tangent). The arithmetic genus of the normalization $\nu : C \to \bar{C}$ of the fiber over P will drop by 2 while the pushforward $\bar{L} := \nu_*(L)$ of a line bundle L on C will belong to a symplectic leaf Higgs$_S$ whose rank is 2 less than the maximal rank (rather than 4). Hence $(\bar{C}, \bar{L})$ is S-singular.

We denote by Higgsm$_{S/D}$ the subset of Higgs$_S \subset$ Higgs$_D$ parametrizing Higgs pairs whose spectral curve is S-smooth over D.

Unfortunately, the compatibility of stratifications (lemma 6.3) does not extend to the S-smooth case. To overcome this inconvenience we may either assume condition 6.11 or condition 6.12.

Condition 6.11 *Consider the ramification locus* Higgsm$^{\underline{n}}_{S/D}$ *in a component of* Higgsm$_{S/D}$ *only if it is the generic ramification type in this component.*

Note that Higgsm$_{S/D}$ parametrizes only Higgs pairs whose spectral curve is S-smooth over D. S-smoothness over D assures that if the type $\underline{n}$ is a generic ramification type in a component of Higgsm$_{S/D}$, then the corresponding component of Higgsm$_{S/D}^{\underline{n}}$ is an *open* subset of Higgsm$_{S/D}$ (i.e., it excludes degenerations of type $\underline{n}$ Higgs pairs through other types back to type $\underline{n}$). In particular, these components of Higgsm$_{S/D}^{\underline{n}}$ are *symplectic*. Alternatively, we may relax condition 6.11 even further at the expense of losing the symplectic nature of the loci and having to resort to convention 2.20:

Condition 6.12 *Consider only Higgs pairs with a spectral curve whose singularities over P are resolved by its spectral sheaf. Adopt convention 2.20.*

Theorem 6.13 *The compatibility theorem 6.5 holds for: i) The $W_{\underline{n}}$-Galois covers of the type loci in Higgsm$_{S/D}$ satisfying the genericity condition 6.11 (instead of condition 6.4). ii) The $W_{\underline{n}}$-Galois covers of the locus in Higgs$_{D}^{\underline{n}}$ consisting of Higgs pairs satisfying condition 6.12. In ii) however we adopt convention 2.20.*

6.3.5 Elliptic Solitons

In this subsection we illustrate the possibilities in the singular case with a specific example. Let Σ be a smooth elliptic curve. A Σ-*periodic Elliptic KP soliton* is a finite dimensional solution to the KP hierarchy, in which the orbit of the first KP equation is isomorphic to Σ. Its Krichever data $(C, \tilde{P}, \frac{\partial}{\partial z}, L)$ consists of a reduced and irreducible curve C, a smooth point $\tilde{P}$, a nonvanishing tangent vector at $\tilde{P}$ and a rank 1 torsion free sheaf L on C of Euler characteristic 0 (we suppress the non essential formal trivialization η and consider only the first order germ of z which is equivalent to choosing a nonzero tangent vector at P). We will denote the global vector field extending $\frac{\partial}{\partial z}$ also by $\frac{\partial}{\partial z}$. The periodicity implies that the image of the Abel Jacobi map $AJ : C \hookrightarrow J_C$, $Q \mapsto Q - \tilde{P}$ is tangent at 0 to a subtorus isomorphic to Σ. Composing the Abel Jacobi map with projection to Σ we get a *tangential morphism* $\pi : (C, \tilde{P}) \to (\Sigma, P)$. Its degree n is called the *order* of the Elliptic soliton. In general, a tangential morphism $\pi : (C, \tilde{P}) \to (\Sigma, P)$ is a morphism with the property that $AJ(C)$ is tangent at $AJ(\tilde{P})$ to $\pi^* J_{\Sigma}^0$. Notice that composing a tangential morphism $\pi : (C, \tilde{P}) \to (\Sigma, P)$ with a normalization $\nu : \tilde{C} \to C$ results in a tangential morphism. A tangential morphism is called *minimal* if it does not factor through another tangential morphism.

The KP elliptic solitons enjoyed a careful and detailed study by A. Treibich and J.-L. Verdier in a series of beautiful papers (e.g., [TV1, TV2]). Their results fit nicely with our picture:

Theorem 6.14 *The variety of Krichever data of Elliptic KP solitons of order n with a fixed pointed elliptic curve and a tangent vector $(\Sigma, P, \frac{\partial}{\partial z})$ is canonically birational to the divisor of traceless Higgs pairs in the symplectic leaf Higgs$_S$ of Higgs$_{\Sigma}^{sm}(n, 0, \omega_{\Sigma}(P))$ corresponding to the coadjoint orbit S of example 6.9. The KP flows are well defined on Higgsm$_{S/P} \subset$ Higgs$_S$ as the Hamiltonian vector fields of the functions $\phi_{\tilde{P}}^j$ given in (20).*

Note that if non-empty (which is the case) Higgs$_S$ is $2n$-dimensional (see example 6.9). The correspondence between tangential covers and spectral covers is a corollary of the following characterization of tangential covers due to I. M. Krichever and A. Treibich. For simplicity we consider only the case in which the tangency point $\tilde{P}$ is not a ramification point of π.

Theorem 6.15 *[TV2] Assume that $\pi : C \to \Sigma$ is unramified at $\tilde{P}$. Then π is tangential if and only if there exists a section $y \in H^0(C, \pi^*[\omega_\Sigma(P)])$ satisfying:*
a) Near a point of $\pi^{-1}(P) - \tilde{P}$ (away from the tangency point $\tilde{P}$), $y - \pi^*(dz/z)$ is a holomorphic multiple of $\pi^*(dz)$, where z is a local parameter at P. (If $\pi : C \to \Sigma$ is unramified over P, this is equivalent to saying that the residues $\operatorname{Res}_{P_i}(y)$ are the same at all P_i other than $\tilde{P}$ in the fiber over $P \in \Sigma$).
b) The residue $\operatorname{Res}_{\tilde{P}}(y)$ at $\tilde{P}$ does not vanish if $n \geq 2$.

It follows by the residue theorem that there is a unique such section which has residue $n - 1$ at $\tilde{P}$ and which is moreover traceless $tr(y) = 0 \in H^0(\Sigma, \omega_\Sigma(P))$. Let dz be a global holomorphic non zero 1-form on Σ. The function $k := y/\pi^*(dz)$ is called a *tangential function* in [TV2]. It was also proven that a tangential morphism of order n has arithmetic genus $\leq n$ and is minimal if and only if its arithmetic genus is n. ([TV1] Corollaire 3.10).

Sketch of proof of theorem 6.15: (for C smooth, $\pi : C \to \Sigma$ unramified over P.)
Step 1: (Cohomological identification of the differential of the Abel-Jacobi map)
    The differential $dAJ : T_Q C \cong H^0(Q, \mathcal{O}_Q(Q)) \to H^1(C, \mathcal{O}_C)$ of the Abel-Jacobi map at $Q \in C$ is identified as the connecting homomorphism of the short exact sequence

$$0 \to \mathcal{O}_C \to \mathcal{O}_C(Q) \to \mathcal{O}_Q(Q) \to 0.$$

Similarly, the differential $d(AJ \circ \pi^{-1})$ of the composition $\Sigma \overset{\pi^{-1}}{\hookrightarrow} \operatorname{Sym}^n C \overset{AJ}{\to} J_C^n$ is given by

$$T_P\Sigma \cong H^0(P, \mathcal{O}_P(P)) \overset{\pi^*}{\hookrightarrow} H^0(\pi^{-1}(P), \mathcal{O}_{\pi^{-1}(P)}(\pi^{-1}(P))) \overset{\cong}{\to} T_{[\pi^{-1}(P)]}\operatorname{Sym}^n C \overset{dAJ}{\to} H^1(C, \mathcal{O}_C),$$

where the composition of the last two arrows is the connecting homomorphism of the short exact sequence

$$0 \to \mathcal{O}_C \to \mathcal{O}_C(\pi^{-1}(P)) \to \mathcal{O}_{\pi^{-1}(P)}(\pi^{-1}(P)) \to 0.$$

Step 2: (residues as coefficients in a linear dependency of tangent lines)
    Clearly, the tangent line to $(AJ \circ \pi^{-1})(\Sigma)$ at the image of P is in the span of the tangent lines to $AJ(C)$ at the points P_i. If, in addition, π is tangential with tangency point $\tilde{P} \in C$, then the tangent lines to $AJ(C)$ at the points in the fiber over P are linearly dependent. If $\pi : C \to \Sigma$ is unramified over P we can write these observations in the form of two linear equations:

$$\sum dAJ_{P_i}(\frac{\partial}{\partial z_i}) = d(AJ \circ \pi^{-1})_P(\frac{\partial}{\partial z}), \tag{22}$$

and

$$\sum a_i dAJ_{P_i}(\frac{\partial}{\partial z_i}) = 0 \quad \text{linear dependency.} \tag{23}$$

Above, $\frac{\partial}{\partial z_i}$ is the lift of $\frac{\partial}{\partial z}$, i.e., $d\pi(\frac{\partial}{\partial z_i}) = \frac{\partial}{\partial z}$. We claim that the coefficients a_i in (23) are residues of a meromorphic 1-form y at the points of the fiber. More precisely we have:

Lemma 6.16 *Assume that $\pi : C \to \Sigma$ is unramified over P. There exists a section $y \in H^0(C, \pi^*\omega_\Sigma(P))$ with residues $(a_1, a_2, \ldots, a_n)$ at the fiber over P if and only if $(a_1, a_2, \ldots, a_n)$ satisfy equation (23).*

Proof: The global tangent vector field $\frac{\partial}{\partial z} \in H^0(\Sigma, T\Sigma)$ gives rise to the commutative diagram:

$$
\begin{array}{ccccc}
H^0(C, \pi^* \omega_\Sigma(P)) & \longrightarrow & H^0(\pi^{-1}(P), \pi^* \omega_\Sigma(P)|_{\pi^{-1}(P)}) & \longrightarrow & H^1(C, \pi^* \omega_\Sigma) \\
\cong \Big\downarrow \scriptstyle{\int \frac{\partial}{\partial z}} & & \cong \Big\downarrow \scriptstyle{\int \frac{\partial}{\partial z}} & & \cong \Big\downarrow \scriptstyle{\int \frac{\partial}{\partial z}} \\
H^0(C, \mathcal{O}_C(\pi^{-1}(P))) & \rightarrow & H^0(\pi^{-1}(P), \mathcal{O}_{\pi^{-1}(P)}(\pi^{-1}(P))) & \xrightarrow{\ dAJ\ } & H^1(C, \mathcal{O}_C).
\end{array}
\tag{24}
$$

The middle contraction $\int \frac{\partial}{\partial z}$ maps residues $(a_1, a_2, \ldots, a_n) \in H^0(\pi^{-1}(P), \pi^* \omega_\Sigma(P)|_{\pi^{-1}(P)})$ to $(a_1 \frac{\partial}{\partial z_1}, a_2 \frac{\partial}{\partial z_2}, \ldots, a_n \frac{\partial}{\partial z_n})$. The lemma follows by the exactness of the horizontal sequences in the diagram.

$\square$

Step 3: We conclude that $\pi : C \to \Sigma$ is tangential if and only if there is a 1-form y as in the theorem. If $\pi : C \to \Sigma$ is tangential then using lemma 6.16 we see that equation (22) gives rise to a 1-form y with residues $(-1, -1, \ldots, n - 1)$ as required. Conversely, given a 1-form y with residues $(-1, -1, \ldots, n - 1)$ lemma 6.16 and equations (23) and (22) imply the tangentiality. This completes the proof of theorem 6.15 in the generic case considered.

$\square$

Theorem 6.14 would follow once the existence of either n-sheeted spectral covers S-smooth over P, or degree n tangential covers smooth and unramified over P is established for every choice of (Σ, P). This was done in ([TV1] Theorem 3.11) by studying the linear system of transferred spectral curves on the surface $\tilde{X}_S$ of example 6.9 and applying Bertini's theorem to show that the generic transferred spectral curve is smooth in X_S. It follows that the blow up of the point with residue -1 over P resolves the generic n-sheeted spectral curve of Higgs pairs in Higgs$_S$ to a smooth curve of genus n unramified over P.

Any S-smooth spectral curve $\bar{C}$ in $T^*\Sigma(P)$ of a Higgs pair (E, φ) in Higgs$_S$ admits a unique partial normalization C of arithmetic genus n by the spectral sheaf $\bar{L}$ corresponding to (E, φ). The tautological 1-form $\bar{y}$ pulls back to a 1-form on C of the type which characterize tangential covers by theorem 6.15.

Conversely, a degree n tangential cover $\pi : (C, \tilde{P}) \to (\Sigma, P)$ of arithmetic genus n which is smooth and unramified over P is sent to the spectral curve $\bar{C}$ in $T^*\Sigma(P)$ of the Higgs pair

$$(E, \varphi) := (\pi_*(L), [\otimes y : \pi_*(L) \to \pi_*(L) \otimes \omega_\Sigma(P)])$$

for some, hence every, choice of a line bundle L on C. Note that $\bar{C}$ is reduced since it is irreducible and the branch through residue $n - 1$ over P is reduced. The canonical morphism $\nu : C \to \bar{C}$ is the resolution by the spectral sheaf $\nu_* L$. Hence $(\bar{C}, \nu_* L)$ is S-smooth. (The arithmetic genus of $\bar{C}$ is $\frac{1}{2}(n^2 - n + 2)$, the common arithmetic genus to all n-sheeted spectral curves in $T^*\Sigma(P)$.)

Finally we note that, as the tangency point $\tilde{P}$ over P is marked by having residue $n - 1$, it does not have monodromy and all the KP flows corresponding to it are well define on Higgsm$_{S/P}$ as the Hamiltonian vector fields of the functions $\phi_{\tilde{P}}^j$ given in (20).

64

6.3.6 Outline of the proof of the compatibility theorem

For simplicity we assume that $D = lP$, $l \geq 0$. The general case is similar replacing $\mathcal{U}_\Sigma(n,d,\infty P)$ by $\mathcal{U}_\Sigma(n,d,\infty P + D)$. Let $G_\infty := GL(n,K)$ be the loop group and G_∞^+ the level infinity subgroup. (More canonically, $K \cong \mathbb{C}((z))$ should be thought of as the completion of the function field of Σ at P, and we may postpone the choice of a coordinate z until we need to choose generators for a maximal torus $\mathrm{Heis}_{\underline{n}}$ of G_∞.)

Denote by $M_{l,k}$, $k \geq l$ the pullback of $T^*\mathcal{U}_\Sigma(n,d,lP)$ to $\mathcal{U}_\Sigma(n,d,kP)$ via the rational forgetful morphism. $M := T^*\mathcal{U}_\Sigma(n,d,\infty P)$ is defined as the limit of finite dimensional approximations (see 2.5.1 for the terminology)

$$T^*\mathcal{U}_\Sigma(n,d,\infty P) := \varprojlim_{l \to l} \varprojlim_{\infty \leftarrow k} M_{l,k}.$$

Denote by M^s (resp. $M^s_{l,k}$) the subset of $T^*\mathcal{U}_\Sigma(n,d,\infty P)$ (resp. $M_{l,k}$) consisting of triples (E,φ,η) with a *stable* Higgs pair (E,φ). We arrive at the setup of section 2.5.1 with the Poisson quotient $Q_\infty := M^s/G^+$ being the direct limit

$$Higgs_\infty := \varinjlim_{l \to \infty} \mathrm{Higgs}_\Sigma^{sm}(n,d,lp).$$

We emphasize that the stability condition is used here for the morphism

$$M^s_{l,k} \to \mathrm{Higgs}_\Sigma^{sm}(n,d,lp)$$

between the two *existing* coarse moduli spaces to be well defined, and *not* to define the quotient.

The *infinitesimal* loop group action on $\mathcal{U}_{\infty P} := \mathcal{U}_\Sigma(n,d,\infty P)$ (the derivative of the action defined in section 6.2 on the level of Čech 1-cocycles) may be lifted to an infinitesimal action on its cotangent bundle. The point is that the infinitesimal action of $a \in \mathfrak{g}_\infty$, with poles of order $\leq l_0$, is well defined on the finite dimensional approximation $\mathcal{U}_\Sigma(n,d,lP)$ for $l \geq l_0$. Thus, it lifts to all cotangent bundles $M_{l,l}$, $l \geq l_0$. $M_{l,k}$ embeds naturally in $M_{k,k}$ as an invariant subvariety. This defines the action on the limit $T^*\mathcal{U}_\Sigma(n,d,\infty P)$. As a lifted action it is automatically Poisson. Its moment map

$$\mu_\infty^* : \mathfrak{g}_\infty \to \Gamma[\mathcal{O}_{T^*\mathcal{U}_{\infty P}}, \{,\}]$$

(the limit of the moment maps for the finite dimensional approximations) is given by the same formula that we have already encountered for the level groups (see 14):

$$(\mu_\infty^*(a))(E,\varphi,\eta) = Res_P trace(a \cdot (\varphi)^n), \quad a \in \mathfrak{g}_\infty. \tag{25}$$

Choosing a maximal torus $\mathrm{heis}_{\underline{n}} \subset \mathfrak{g}_\infty$ of type $\underline{n}$ we arrive at the setup of section 2.5.2. In particular, we obtain the type locus $\mathrm{Higgs}_\Sigma^{\underline{n}}(n,d,lp)$ in $\mathrm{Higgs}_\Sigma^{sm}(n,d,lp)$.

Lemma 6.17 *1. The algebro-geometric definition of the ramification type loci coincides with the group theoretic definition 2.18 when $char(E,\varphi)$ is an integral (irreducible and reduced) spectral curve.*

2. A choice of generators for a maximal torus $\mathrm{heis}_{\underline{n}}$ as in (16) determines a canonical isomorphism between the group theoretic and the algebro-geometric $W_{\underline{n}}$-Galois covers.

Proof: 1) The stabilizer $\mathbf{t} \subset \mathbf{g}_\infty$ of $(\varphi)^\eta \in \mathbf{g}_\infty \otimes_{\mathcal{O}_{(P)}} \omega_{\Sigma,(P)} \cong \mathbf{g}_\infty^*$ with spectral curve $\pi : C = char(E, \varphi) \to \Sigma$ is precisely U^η where U is the stalk of $Ker[ad\varphi : EndE \to EndE \otimes \omega_\Sigma(\text{l}P)]$ at the formal punctured neighborhood of P. In addition, U is canonically isomorphic to the stalk of the structure sheaf at the formal punctured neighborhood of the fiber of C over P via the completion of the canonical embedding:

$$\pi_* \mathcal{O}_C \hookrightarrow \pi_* EndL \hookrightarrow EndE.$$

Hence, the level infinity structure η provides a *canonical* isomorphism

$$\lambda : \mathbf{t} \stackrel{\cong}{\to} U \tag{26}$$

from the stabilizer algebra $\mathbf{t}$ to the structure sheaf at the formal punctured neighborhood of the fiber of C over P.

2) As the types coincide, we may choose the level infinity structure η so that the stabilizer $\mathbf{t}$ coincides with the fixed torus $\mathbf{heis}_{\underline{n}}$. We may further require that the isomorphism λ given by (26) coincides with the one in (19). This determines the $\text{Heis}_{\underline{n}}^+$-orbit of η uniquely, i.e., a point in the group theoretic Galois cover.

□

Theorems 6.5 and 6.13 would now follow from corollary 2.22 provided that we prove that the homomorphism ϕ of the theorems (given by 21) is indeed the factorization of the $\mathbf{heis}_{\underline{n}}$-moment map through the characteristic polynomial map. (Note that the existence of this factorization follows from diagram (10)). In other words, we need to prove the identity

$$\sum_{\{P_i\}} Res_{P_i}(\lambda(a) \cdot y_b) = Res_P trace(a \cdot (\varphi)^\eta) \quad a \in \mathbf{heis}_{\underline{n}}, \ b = char(E, \varphi) \tag{27}$$

as functions on the set of all Higgs pairs (E, φ, λ) in $\widetilde{Higgs}_{l_p}^n$ for which the spectral sheaf resolves the singularities of their spectral curve over P (see definition 6.7). Above, η is any level infinity structure in the $\text{Heis}_{\underline{n}}^+$-orbit as in the proof of lemma 6.17 or, equivalently, $\lambda(a) = \eta^{-1} \circ a \circ \eta$ where we identify the structure sheaf of the formal punctured neighborhood with U of that lemma. If the embedded spectral curve $\bar{C}_b$ is singular, the P_i are the points over P of its resolution $\nu : C_b \to \bar{C}_b$, and the tautological meromorphic 1-form y_b should be replaced by the pullback $\nu^*(y)$ of the tautological 1-form y on the surface $T_\Sigma^*(\text{l}P)$.

Conjugating the right hand side of (27) by η, we get

$$\sum_{\{P_i\}} Res_{P_i}(A \cdot \nu^*(y)) = Res_P trace((\pi \circ \nu)_*[A \cdot \nu^*(y)])$$

for A a (formal) meromorphic function at the fiber over P. Working formally, we can consider only the "parts" with first order pole $r_i dlog z_i$ of $A \cdot \nu^*(y)$ at P_i. The equality follows from the identity $dlog z_i = (\pi \circ \nu)^*[\frac{1}{n_i} dlog z]$ which imply (projection formula) that $(\pi \circ \nu)_*(\otimes dlog z_i)$ acts as $\frac{1}{n_i} e_{P_i} \otimes dlog z$ were e_{P_i} is the projection onto the eigenspace of the point P_i.

□

66

7 The Cubic Condition and Calabi-Yau threefolds

We pose in section 7.1 the general question: when does a family of polarized abelian varieties or complex tori support a completely integrable system? In section 7.2 we describe a general necessary infinitesimal symmetry condition on the periods of the family (the cubic condition of lemmas 7.1 and 7.2) and a sufficient local condition (lemmas 7.4 and 7.5).

In section 7.3 we use the Yukawa cubic to construct a symplectic structure (and an ACIHS) on the relative intermediate Jacobian over the moduli space of gauged Calabi-Yau threefolds (theorem 7.7). The symplectic structure extends to the bundle of Deligne cohomologies and we show that the image of the relative cycle map as well as bundles of sub-Hodge-structures are isotropic (corollary 7.11).

7.1 Families of Tori

Consider a Poisson manifold (X, ψ) together with a Lagrangian fibration

$$\pi : \mathcal{X} \longrightarrow B$$

over a base B, whose fibers

$$X_b := \pi^{-1}(b), \qquad b \in B$$

are tori. (We say π is Lagrangian if each fiber X_b is a Lagrangian submanifold of some symplectic leaf in $\mathcal{X}$.) All these objects may be C^∞, or may be equipped with a complex analytic or algebraic structure.

On B we have the tangent bundle $\mathcal{T}_B$ as well as the vertical bundle $\mathcal{V}$, whose sections are vector fields along the fibers of π which are constant on each torus. The pullback $\pi^*\mathcal{V}$ is the relative tangent bundle $\mathcal{T}_{\mathcal{X}/B}$; in the analytic or algebraic situations, we can define $\mathcal{V}$ as $\pi_*\mathcal{T}_{\mathcal{X}/B}$. The data π and ψ determine an injection

$$i : \mathcal{V}^* \hookrightarrow \mathcal{T}_B$$

or, equivalently, a surjection

$$i' : \mathcal{T}_B^* \twoheadrightarrow \mathcal{V}$$

sending a 1-form α on B to the vertical vector field

$$i'(\alpha) := \pi^*\alpha \,\lrcorner\, \psi.$$

The image $i(\mathcal{V}^*) \subset \mathcal{T}_B$ is an integrable distribution on B. Its integral manifolds are the images in B of symplectic leaves in $\mathcal{X}$.

In this section we start with a family of tori $\pi : \mathcal{X} \to B$ and ask whether there is a *Lagrangian structure* for π, i.e. a Poisson structure on $\mathcal{X}$ making the map π Lagrangian. More precisely, we fix $\pi : \mathcal{X} \to B$ and an injection $i : \mathcal{V}^* \hookrightarrow \mathcal{T}_B$ with integrable image, and ask for existence of a Lagrangian structure ψ on $\mathcal{X}$ inducing the given i.

In the C^∞ category there are no local obstructions to existence of a Lagrangian structure: the fibration π is locally trivial, so one can always find action-angle coordinates near each fiber, and use them to define ψ. In the analytic or algebraic categories, on the other hand, the fibers X_b (complex tori, or abelian varieties) have invariants, given essentially by their *period matrix* $p(X_b)$, so the fibration may not be analytically locally trivial. We will see that there is an obstruction to existence of a Lagrangian structure for $\pi : \mathcal{X} \to B$, which we formulate as a symmetry condition on the derivatives of the period map p. These derivatives can be considered as a linear system of quadrics, and the condition is, roughly, that they be the polars of some cubic (= section of $\text{Sym}^3\mathcal{V}$).

Let X be a g-dimensional complex torus, and $\gamma_1, \cdots, \gamma_{2g}$ a basis of the integral homology $H_1(X, \mathbb{Z})$. There is a unique basis $\alpha_1, \cdots, \alpha_g$ for the holomorphic differentials $H^0(X, \Omega^1_X)$ satisfying

$$\int_{\gamma_{g+i}} \alpha_j = \delta_{ij}, \qquad 1 \leq i, j \leq g,$$

so we define the period matrix $P = p(X, \gamma)$ by

$$p_{ij} := \int_{\gamma_i} \alpha_j, \qquad 1 \leq i, j \leq g.$$

Riemann's first and second bilinear relations say that X is a principally polarized abelian variety (PPAV) if and only if P is in *Siegel's half space*:

$\mathbb{H}_g :=$ {symmetric $g \times g$ complex matrices whose imaginary part is positive definite}.

In terms of a dual basis $\gamma_1^*, \cdots, \gamma_{2g}^*$ of $H^1(X, \mathbb{Z})$, the integral class $\omega := \sum_{i=1}^g \gamma_i^* \wedge \gamma_{g+i}^* \in H^2(X, \mathbb{Z})$ is a Kähler class if and only if P is in Siegel's half space. In this case we call ω a *principal polarization*.

Given a family $\pi : \mathcal{X} \to B$ of PPAVs together with a continuously varying family of symplectic bases $\gamma_1, \cdots, \gamma_{2g}$ for the fiber homologies, we then get a period map

$$p : B \longrightarrow \mathbb{H}_g.$$

If we change the basis γ by a symplectic transformation

$$\begin{pmatrix} A & B \\ C & D \end{pmatrix} \in Sp(2g, \mathbb{Z}),$$

the period matrix P goes to $(AP + B)(CP + D)^{-1}$. So given a family π without the choice of γ, we get a multi-valued map of B to $\mathbb{H}_g$, or a map

$$p : B \longrightarrow \mathcal{A}_g$$

to the moduli space of PPAV. The latter is a quasi projective variety, which can be described analytically as the quotient

$$\mathcal{A}_g = \mathbb{H}_g / \Gamma$$

of $\mathbb{H}_g$ by the action of the modular group

$$\Gamma := Sp(2g, \mathbb{Z})/(\pm 1).$$

A PPAV X determines a point $[X]$ (or carelessly, X) of $\mathcal{A}_g$. This point is non-singular if X has no automorphisms other than ± 1, and then the tangent space $T_{[X]}\mathcal{A}_g$ can be identified with $\text{Sym}^2 V_X$, where V_X is the tangent space (at $0 \in X$) to X. This can be seen by identifying $T_{[X]}\mathcal{A}_g$ with $T_{[X]}\mathbb{H}_g$ and recalling that $\mathbb{H}_g$ is an open subset of $\mathrm{Sym}^2 V_X$. More algebraically, this follows from elementary deformation theory: all first-order deformations of X are given by

$$H^1(X, \mathcal{T}_X) \approx H^1(X, V_X \otimes_c \mathcal{O}_X) \approx V_X \otimes H^1(X, \mathcal{O}_X) \approx \otimes^2 V_X,$$

and in there the deformations as abelian variety, i.e., the deformations preserving the polarization bilinear form on $H_1(X, \mathbb{Z})$, are given by the symmetric tensors $\mathrm{Sym}^2 V_X$.

7.2 The Cubic Condition

Our condition for an analytic or algebraic family $\pi : \mathcal{X} \to B$ of PPAVs, given by a period map $p : B \to \mathcal{A}_g$, to have a Lagrangian structure ψ inducing a given $i : \mathcal{V}^* \hookrightarrow \mathcal{T}_B$, can now be stated as follows. The differential of p is a map of bundles:

$$dp : \mathcal{T}_B \longrightarrow \mathrm{Sym}^2 \mathcal{V},$$

so the composite

$$dp \circ i \; : \; \mathcal{V}^* \longrightarrow \mathrm{Sym}^2 \mathcal{V}$$

can be considered as a section of $\mathcal{V} \otimes \mathrm{Sym}^2 \mathcal{V}$, and the condition is that it should come from the subbundle $\mathrm{Sym}^3 \mathcal{V}$. In other words, there should exist a cubic $c \in H^0(B, \, \mathrm{Sym}^3 \mathcal{V})$ whose polar quadrics give the directional derivatives of the period map: if the tangent vector $\partial/\partial b \in T_b B$ equals $i(\beta)$ for some $\beta \in \mathcal{V}^*$, then:

$$\frac{\partial p}{\partial b} = \beta \rfloor c.$$

We give two versions of this cubic condition. In the first, we check the existence of a two vector ψ, not necessarily satisfying the Jacobi identity, for which π is Lagrangian, and which induces a given injection $i : \mathcal{V}^* \hookrightarrow \mathcal{T}_B$. (Note that neither the definition of the map i induced by the two-vector ψ, nor the notion of π being Lagrangian, require ψ to be Poisson.)

Lemma 7.1 *(Weak cubic condition, Poisson form).* *A family* $\pi : \mathcal{X} \to B$ *of polarized abelian varieties has a two vector* ψ *satisfying*

a) $\pi : \mathcal{X} \to B$ *is Lagrangian*
b) ψ *induces a given* $i : \mathcal{V}^* \hookrightarrow \mathcal{T}_B$

if and only if

$$dp \circ i \in \mathrm{Hom}(\mathcal{V}^*, \mathrm{Sym}^2 \mathcal{V})$$

comes from a cubic

$$c \in H^0(B, \, \mathrm{Sym}^3 \mathcal{V}).$$

Moreover, in this case there is a unique such 2-vector ψ *which satisfies also*

c) The zero section $z : B \to \mathcal{X}$ *is Lagrangian, i.e.,* $(T^*_{\mathcal{X}/B})|_{z(B)}$ *is* ψ*-isotropic (here we identify the conormal bundle of the zero section with* $(T^*_{\mathcal{X}/B})|_{z(B)}$*.)*

Proof: (Note: we refer below to the vertical bundle $\mathcal{T}_{\mathcal{X}/B}$ by its, somewhat indirect, realization as the pullback $\pi^*\mathcal{V}$.) The short exact sequence of sheaves on $\mathcal{X}$:

$$0 \to \pi^*\mathcal{V} \to \mathcal{T}_{\mathcal{X}} \to \pi^*\mathcal{T}_B \to 0$$

determines a subsheaf $\mathcal{F}$ of $\overset{2}{\wedge}\mathcal{T}_{\mathcal{X}}$ which fits in the exact sequences:

$$0 \to \mathcal{F} \to \overset{2}{\wedge}\mathcal{T}_{\mathcal{X}} \to \pi^*\overset{2}{\wedge}\mathcal{T}_B \to 0$$
$$0 \to \pi^*\overset{2}{\wedge}\mathcal{V} \to \mathcal{F} \to \pi^*(\mathcal{V}\otimes\mathcal{T}_B) \to 0.$$

The map π is Hamiltonian with respect to the two-vector $\psi \in H^0(B_{\mathcal{X}}, \overset{2}{\wedge}\mathcal{T}_{\mathcal{X}})$ if and only if ψ goes to 0 in $\overset{2}{\wedge}\mathcal{T}_B$, i.e., if and only if it comes from $H^0(\mathcal{F})$. The question is therefore whether $i \in H^0(B, \mathcal{V}\otimes\mathcal{T}_B) \subset H^0(\mathcal{X}, \pi^*(\mathcal{V}\otimes\mathcal{T}_B))$ is in the image of $H^0(\mathcal{X},\mathcal{F})$. Locally in B, this happens if and only if i goes to 0 under the coboundary map

$$\begin{array}{ccc}
\pi_*\pi^*(\mathcal{V}\otimes\mathcal{T}_B) & \longrightarrow & R^1\pi_*\pi^*\overset{2}{\wedge}\mathcal{V} \\
\| & & \| \\
\mathcal{V}\otimes\mathcal{T}_B & \longrightarrow & \overset{2}{\wedge}\mathcal{V}\otimes\mathcal{V}.
\end{array}$$

This latter map factors through the period map

$$1\otimes dp : \mathcal{V}\otimes\mathcal{T}_B \longrightarrow \mathcal{V}\otimes\operatorname{Sym}^2\mathcal{V}$$

and a Koszul map

$$\mathcal{V}\otimes\operatorname{Sym}^2\mathcal{V} \longrightarrow \overset{2}{\wedge}\mathcal{V}\otimes\mathcal{V}.$$

Now exactness of the Koszul sequence

$$0 \to \operatorname{Sym}^3\mathcal{V} \to \mathcal{V}\otimes\operatorname{Sym}^2\mathcal{V} \to \overset{2}{\wedge}\mathcal{V}\otimes\mathcal{V}$$

shows that the desired ψ exists if and only if

$$dp \circ i = (1\otimes dp)(i) \in \mathcal{V}\otimes\operatorname{Sym}^2\mathcal{V}$$

is in the subspace $\operatorname{Sym}^3\mathcal{V}$. (The Hamiltonian map π will automatically be Lagrangian, since i is injective.)

We conclude that, locally on B, i lifts to a 2-vector ψ satisfying conditions a), b), if and only if $dp \circ i$ is a cubic. If ψ_1, ψ_2 are two such lifts then $\psi_1 - \psi_2 \in H^0(\mathcal{X}, \overset{2}{\wedge}\pi^*\mathcal{V})$. Moreover, $\psi_1 - \psi_2$ is determined by its restriction to the zero section because $\overset{2}{\wedge}\pi^*\mathcal{V}$ restricts to a trivial bundle on each fiber. The zero section induces a splitting $\mathcal{T}_{\mathcal{X}|z(B)} \simeq \pi^*\mathcal{T}_B \oplus (\pi^*\mathcal{V})_{|z(B)}$ and hence a well defined pullback $z^*(\psi) \in H^0(B, \overset{2}{\wedge}\mathcal{V})$ (locally on B). The normalizations $\psi - \pi^*(z^*(\psi))$ patch to a unique global section satisfying a), b) c).

$\square$

The symplectic version of this lemma is:

Lemma 7.2 : *(Weak cubic condition, quasi-symplectic form). A family $\pi : \mathcal{X} \to B$ of principally polarized abelian varieties has a 2-form σ satisfying*
a) $\pi : \mathcal{X} \to B$ has isotropic fibers,
b) σ induces a given (injective) homomorphism $j : \mathcal{T}_B \hookrightarrow \mathcal{V}^*$,
if and only if

$$(1 \otimes j^*) \circ dp \in \mathrm{Hom}(\mathcal{T}_B, \mathcal{T}_B^* \otimes \mathcal{V}) \cong \mathcal{T}_B^* \otimes \mathcal{T}_B^* \otimes \mathcal{V} \text{ is in } \mathrm{Sym}^2 \mathcal{T}_B^* \otimes \mathcal{V}.$$

Moreover, in this case, there exists a unique 2-form σ satisfying a), b), and the additional condition
c) the zero section is isotropic $(z^*\sigma = 0)$.

Remark 7.3 Riemann's first bilinear condition implies further that $(1 \otimes j^*) \circ dp$ maps to $\mathrm{Sym}^3 \mathcal{T}_B^*$, i.e., $(\mathrm{Sym}^2 j^*) \circ dp \in \mathrm{Hom}(\mathcal{T}_B, \mathrm{Sym}^2 \mathcal{T}_B^*)$ comes from a cubic $c \in H^0(B, \mathrm{Sym}^3 \mathcal{T}_B^*)$.

The cubic condition for an embedding $j : \mathcal{T}_B \hookrightarrow \mathcal{V}^*$ does not guarantee that the induced 2-form σ on $\mathcal{X}$ is closed. In that sense, the cubic condition is a necessary condition for j to induce a symplectic structure while the following condition is necessary and sufficient (but, in general, harder to verify).

Closedness Criterion for a Symplectic Form: *Given a family $\pi : \mathcal{X} \to B$ of polarized abelian varieties and a surjective $j' : \mathcal{V} \to \mathcal{T}_B^*$, there exists a closed 2-form σ on $\mathcal{X}$ satisfying conditions a), b), c) of Lemma 7.2 if and only if $j'(\mathcal{H}_1(\mathcal{X}/B, \mathbb{Z})) \subset \mathcal{T}_B^*$ is a Lagrangian lattice in T^*B, i.e., if locally on B it consists of closed 1-forms. Moreover, the 2-form σ is uniquely determined by j'.*

Proof: $j'(\mathcal{H}_1(\mathcal{X}/B, \mathbb{Z}))$ is Lagrangian $\Longleftrightarrow$ the canonical symplectic structure $\tilde{\sigma}$ on T^*B is translation invariant under $j'(\mathcal{H}_1(\mathcal{X}/B, \mathbb{Z})) \Longleftrightarrow (j')^*(\tilde{\sigma})$ descends to the unique 2-form σ on $\mathcal{X} = \mathcal{V}/\mathcal{H}_1(\mathcal{X}/B, \mathbb{Z})$ satisfying conditions a), b), c) of Lemma 7.2.

$\square$

It is instructive to relate the cubic condition to the above criterion. This is done in lemma 7.4 in a down to earth manner and is reformulated in lemma 7.5 as a coordinate free criterion.

Lemma 7.4 *("Strong Cubic Condition") Let V be a g-dimensional vector space, $\{e_1, \cdots, e_g\}$ a basis, $B \subset V^*$ an open subset, $p : B \to \mathbb{H}_g \hookrightarrow \mathrm{Sym}^2 V$ a holomorphic map ($\mathbb{H}_g$ is embedded in $\mathrm{Sym}^2 V$ via the basis $\{e_j\}$), $\pi : \mathcal{X} \to B$ the corresponding family of principally polarized abelian varieties. Then the following are equivalent:*
(i) There exists a symplectic structure σ on $\mathcal{X}$ such that $\pi : (\mathcal{X}, \sigma) \to B$ is a Lagrangian fibration and σ induces the identity isomorphism

$$\mathrm{id} \in \mathrm{Hom}(\mathcal{T}_{\mathcal{X}/B}, \pi^* \mathcal{T}_B^*) \simeq \mathrm{Hom}(\pi^*\mathcal{V}, \pi^*\mathcal{V}).$$

(ii) $p : B \to \mathrm{Sym}^2 V$ is, locally in B, the Hessian of a function on B,
(iii) $dp \in \mathrm{Hom}(\mathcal{T}_B, \mathrm{Sym}^2\mathcal{V}) \simeq (\mathcal{V} \otimes \mathrm{Sym}^2\mathcal{V})$ is a section of $\mathrm{Sym}^3\mathcal{V}$.

Proof: Let $\{e_j^*\}$ be the dual basis of V.

(i) $\Leftrightarrow$ (iii): By the closedness criterion above, there exists σ as in (i) if and only if the subsheaf of lattices $\mathcal{H}^1(\mathcal{X}/B, \mathbf{Z}) \subset T^*B$ is Lagrangian, i.e., if and only if its basis

$$\{e_1, \cdots, e_g, (p \rfloor e_1^*), \cdots, (p \rfloor e_g^*)\}$$

consists of closed 1-forms. The e_i's are automatically closed. If we regard the differential dp as a section of $T_B^* \otimes \mathrm{Sym}^2 V$, then the two-form $d(p \rfloor e_j^*)$ is equal to the anti-symmetric part of the contraction $dp \rfloor e_j^* \in T_B^* \otimes V \cong V \otimes V$. Hence, closedness of $(p \rfloor e_j^*)$, $1 \le j \le g$, is equivalent to the symmetry of $dp \in V \otimes \mathrm{Sym}^2 V$ also with respect to the first two factors, i.e., to dp being a section of $\mathrm{Sym}^3 V$.

(ii) $\Rightarrow$ (iii). Clear.

(iii) $\Rightarrow$ (ii). Follows from the Poincare lemma.

$\square$

The additional information contained in the "Strong Cubic Condition" and lacking in the "Weak Cubic Condition" is that a Lagrangian sublattice (with respect to the polarization) $\mathcal{L} \subset \mathcal{H}_1(\mathcal{X}/B, \mathbf{Z})$ is mapped via $j' : V \xrightarrow{\sim} T_B^*$ to a sublattice $j'(\mathcal{L}) \subset T^*B$ Lagrangian with respect to the holomorphic symplectic structure on T^*B. (In the above lemma, $\mathcal{L} = \mathrm{Sp}\{e_1, \cdots, e_g\}$). The coordinate free reformulation of lemma 7.4 is:

Lemma 7.5 *("Strong cubic condition") Let $j' : V \xrightarrow{\sim} T_B^*$ be an isomorphism of the vertical bundle $V = R_{\pi_*}^0(T_{\mathcal{X}/B})$ of the family $\pi : \mathcal{X} \to B$ of polarized abelian varieties with the cotangent bundle of the base. Assume only that j' maps a sublattice $\mathcal{L} \subset \mathcal{H}_1(\mathcal{X}/B, Z)$ Lagrangian with respect to the polarization to a sublattice $j'(\mathcal{L}) \subset T^*B$ Lagrangian with respect to the holomorphic symplectic structure on T^*B. Then there exists a symplectic structure σ on $\mathcal{X}$ s.t. $\pi : \mathcal{X} \to B$ is a Lagrangian fibration and inducing j' if and only if j' satisfies the weak cubic condition, i.e.*

$$dp \circ i \in H^0(B, \mathrm{Sym}^3 V) \qquad \text{where } i = (j')^{*^{-1}}.$$

Remark 7.6 In most cases however, $j'(\mathcal{L})$ being Lagrangian implies $j'(\mathcal{H}_1(\mathcal{X}/B, \mathbf{Z}))$ being Lagrangian via the global monodromy action and without reference to the weak cubic condition.

Finally we remark that the above discussion applies verbatim to the case of polarized complex tori (not necessarily algebraic) since only the first Riemann bilinear condition was used.

7.3 An Integrable System for Calabi-Yau Threefolds

The *Hodge group* $H^{p,q}$ of an n-dimensional compact Kähler manifold X is defined as the space of harmonic forms on X of type (p, q), i.e. involving p holomorphic and q antiholomorphic differentials. Equivalently, $H^{p,q}$ is isomorphic to the $q - th$ cohomology $H^q(X, \Omega^p)$ of the sheaf of holomorphic p-forms on X. The *Hodge theorem* gives a natural decomposition of the complex cohomology,

$$H^k(X, \mathbf{C}) \approx \oplus_{p+q=k} H^{p,q} \approx \oplus_{p+q=k} H^q(\Omega^p).$$

The *Hodge number* $h^{p,q}$ is the complex dimension of $H^{p,q}$. The *Hodge filtration* of $H^k(X, \mathbb{C})$ is defined by

$$F^i H^k(X, \mathbb{C}) := \oplus_{\substack{p+q=k \\ p \geq i}} H^{p,q}.$$

The *k-th intermediate Jacobian* of X [CG] is:

$$J^k(X) := H^{2k-1}(X, \mathbb{C})/(F^k H^{2k-1}(X, \mathbb{C}) + H^{2k-1}(X, \mathbb{Z}))$$

$$\approx (F^{n-k+1} H^{2n-2k+1}(X, \mathbb{C}))^*/H_{2n-2k+1}(X, \mathbb{Z}).$$

Elementary properties of the Hodge filtration imply that this is a complex torus, but generally not an abelian variety unless $k = 1$ or $k = n$: it satisfies Riemann's first bilinear condition (which expresses the skew symmetry of the cup product on H^{2k-1}), but not the second, since the sign of the product (on primitive pieces) will vary with the parity of p. The extreme cases correspond to the connected component of the Picard ($k = 1$) and Albanese ($k = n$) varieties.

The Hodge decomposition does not depend holomorphically on parameters, since both holomorphic and antiholomorphic differentials are involved. The advantage of the Hodge filtration is that it does vary holomorphically and even algebraically when X is algebraic. The F^p can be defined algebraically, as the hypercohomology of the complex

$$0 \to \Omega^p \to \Omega^{p+1} \to \cdots \to \Omega^n \to 0. \tag{28}$$

In particular, the intermediate Jacobian $J^k(X)$ varies holomorphically with X. This means that a smooth analytic family $\mathcal{X} \to B$ of compact Kähler manifolds gives rise to analytic vector bundles $F^i \mathcal{H}^k(\mathcal{X}/B)$ and to smooth analytic families $\mathcal{J}^k(\mathcal{X}/B) \longrightarrow B$ of intermediate Jacobians of the fibers.

The bundle $\mathcal{H}^k(\mathcal{X}/B)$ is the complexification of a bundle $\mathcal{H}^k(\mathcal{X}/B, \mathbb{Z})$ of discrete groups. In particular, it has a natural local trivialization. In other words, it admits a natural flat connection, called the *Gauss-Manin* connection. The holomorphic subbundles $F^i \mathcal{H}^k(\mathcal{X}/B)$ are in general not invariant with respect to this connection, since the Hodge decomposition and filtration do change from point to point. *Griffiths' transversality* says that when a holomorphic section of $F^i \mathcal{H}^k$ is differentiated, it can move at most one step:

$$\nabla(F^i \mathcal{H}^k) \subset F^{i-1} \mathcal{H}^k \otimes \Omega^1_B.$$

An n-dimensional compact Kähler manifold X is called *Calabi-Yau* if it has trivial canonical bundle,

$$\omega_X = \Omega^n_X \approx \mathcal{O}_X,$$

and satisfies

$$h^{p,0} = 0 \text{ for } 0 < p < n.$$

A *gauged Calabi-Yau* is a pair (X, s) consisting of a Calabi-Yau manifold X together with a non-zero volume form

$$s : \mathcal{O}_X \xrightarrow{\approx} \omega_X.$$

A theorem of Bogomolov, Tian and Todorov [Bo, Ti, To] says that X has a smooth (local analytic) universal deformation space M_X. We say that a family $\chi : \mathcal{X} \to \mathcal{M}$ of Calabi-Yaus $X_t, t \in \mathcal{M}$, is *complete* if the local classifying map $\mathcal{M} \supset U_t \to M_{X_t}$ is an isomorphism for some neighborhood of every point $t \in \mathcal{M}$. It follows that $\mathcal{M}$ is smooth and that the tangent space at t to $\mathcal{M}$ is naturally isomorphic to $H^1(X, \mathcal{T}_X)$. Typically, such families might consist of all Calabi-Yaus in some open subset of moduli, together with some "level" structure.

The choice of gauge s gives an isomorphism

$$\rfloor s : \mathcal{T}_X \longrightarrow \Omega_X^{n-1},$$

hence an isomorphism

$$T_X \mathcal{M} \approx H^{n-1,1}(X).$$

Starting with a complete family $\chi : \mathcal{X} \to \mathcal{M}$, we can construct

- The bundle $\mathcal{J}^k \to \mathcal{M}$ of intermediate Jacobians of the Calabi-Yau fibers.

- The space $\tilde{\mathcal{M}}$ of gauged Calabi-Yaus, a $\mathbb{C}^*$-bundle over $\mathcal{M}$ obtained by removing the 0-section from the line bundle $\chi_*(\omega_{\mathcal{X}/\mathcal{M}})$.

- The fiber product

$$\tilde{\mathcal{J}}^k := \mathcal{J}^k \times_{\mathcal{M}} \tilde{\mathcal{M}},$$

which is an analytic family of complex tori $\pi : \tilde{\mathcal{J}}^k \to \tilde{\mathcal{M}}$.

Theorem 7.7 *Let $\mathcal{X} \to \mathcal{M}$ be a complete family of Calabi-Yau manifolds of odd dimension $n = 2k - 1 \geq 3$. Then there exists a canonical closed holomorphic 2-form σ on the relative k-th intermediate Jacobian $\pi : \tilde{\mathcal{J}} \to \tilde{\mathcal{M}}$ with respect to which π has maximal isotropic fibers. When $n = 3$, the 2-form σ is a symplectic structure and $\pi : \tilde{\mathcal{J}} \to \tilde{\mathcal{M}}$ is an analytically completely integrable Hamiltonian system.*

Proof.

Step I. There is a canonical isomorphism

$$T_{(X,s)}\tilde{\mathcal{M}} \approx F^{n-1}H^n(X, \mathbb{C}).$$

Indeed, the natural map $p : \tilde{\mathcal{M}} \to \mathcal{M}$ gives a short exact sequence

$$0 \to T_{(X,s)}(\tilde{\mathcal{M}}/\mathcal{M}) \to T_{(X,s)}\tilde{\mathcal{M}} \to T_X \mathcal{M} \to 0,$$

in which the subspace can be naturally identified with $H^0(\omega_X) = H^0(\Omega_X^n)$, and the quotient with $H^1(\mathcal{T}_X)$, which goes isomorphically to $H^1(\Omega_X^{n-1})$ by $\rfloor s$. What we are claiming is that this sequence can be naturally identified with the one defining $F^{n-1}H^n$:

$$0 \to H^0(\omega_X) \to F^{n-1} \to H^1(\Omega_X^{n-1}) \to 0,$$

i.e., that the extension data match, globally over $\tilde{\mathcal{M}}$. To see this we need a natural map $T_{(X,x)}\tilde{\mathcal{M}} \to F^{n-1}H^n$ inducing the identity on the sub and quotient spaces.

Over $\tilde{\mathcal{M}}$ there is a tautological section s of $F^n\mathcal{H}^n(\tilde{\mathcal{X}}/\tilde{\mathcal{M}}, \mathbb{C})$. The Gauss-Manin connection defines an embedding

$$\nabla_{(\cdot)}s : T_{(X,s)}\tilde{\mathcal{M}} \longrightarrow H^n(X, \mathbb{C}).$$

Griffiths' transversality implies that the image is in $F^{n-1}H^n(X,\mathbb{C})$. Clearly $\nabla_{(\cdot)}s$ has the required properties.

We will need also a description of the isomorphism in terms of Dolbeault cohomology. We think of a 1-parameter family $(X_t, s_t) \in \tilde{\mathcal{M}}$, depending on the parameter t, as living on a fixed topological model X on which there are families $\bar{\partial}_t$ of complex structures (given by their $\bar{\partial}$-operator) and s_t of C^∞ n-forms, such that s_t is of type $(n,0)$ with respect to $\bar{\partial}_t$, all t. Since the s_t are now on a fixed underlying X, we can differentiate with respect to t:

$$s_t = s_0 + ta \qquad (\text{mod } t^2).$$

Griffiths transversality now says that a is in $F^{n-1}H^n(X_0)$. It clearly depends only on the tangent vector to $\tilde{\mathcal{M}}$ along (X_t, s_t) at $t = 0$, so we get a map $T_{(X,s)}\tilde{\mathcal{M}} \longrightarrow F^{n-1}H^n$ with the desired properties.

Step II. Let $\mathcal{V}$ be the vertical bundle on $\tilde{\mathcal{M}}$ coming from $\pi : \tilde{\mathcal{J}} \to \tilde{\mathcal{M}}$. It is isomorphic to

$$F^k \mathcal{H}^n(\tilde{\mathcal{X}}/\tilde{\mathcal{M}})^*$$

(recall $n = 2k - 1$). Combining with Step I, we get a natural injection

$$j : T_{\tilde{\mathcal{M}}} \hookrightarrow \mathcal{V}^*,$$

which above a given (X,s) is the inclusion of $F^{n-1}H^n(X)$ into $F^k H^n(X)$. Its transpose

$$j' : \mathcal{V} \twoheadrightarrow T^*\tilde{\mathcal{M}}$$

determines a closed 2-form $\sigma := (j')^*\tilde{\sigma}$ on $\mathcal{V}$, where $\tilde{\sigma}$ is the standard symplectic form on $T^*\tilde{\mathcal{M}}$ (see example 2.1). By construction, the fibers of $\mathcal{V}$ over $\tilde{\mathcal{M}}$ are maximal isotropic with respect to this form.

Step III. We need to verify that $\tilde{\sigma}$ descends to

$$\mathcal{J}^k(\tilde{\mathcal{X}}/\tilde{\mathcal{M}}) = \mathcal{V}/\mathcal{H}_n(\tilde{\mathcal{X}}/\tilde{\mathcal{M}}, \mathbb{Z}).$$

Equivalently, a locally constant integral cycle

$$\gamma \in \Gamma(B, \mathcal{H}_n(\tilde{\mathcal{X}}/\tilde{\mathcal{M}}, \mathbb{Z})),$$

defined over some open subset B of $\tilde{\mathcal{M}}$, gives a section of $\mathcal{V}$ on B; hence through j', a 1-form ξ on B, and we need this 1-form to be closed. Explicitly, if a is a section of $T_{\tilde{\mathcal{M}}}$ over B, we have

$$a \rfloor s \in \Gamma(B, F^{n-1}\mathcal{H}^n(\tilde{\mathcal{X}}/\tilde{\mathcal{M}})) \subset \Gamma(B, \mathcal{H}^n(\tilde{\mathcal{X}}/\tilde{\mathcal{M}}))$$

and ξ is defined by:

$$\langle \xi, a \rangle := \int_\gamma (a \rfloor s).$$

Consider the function

$$g : B \longrightarrow \mathbb{C}$$

$$g(X,s) := \int_\gamma s.$$

If we set

$$a = \frac{\partial}{\partial t}\Big|_{t=0} (X_t, s_t)$$

as in Step I, we get:

$$\langle dg, a\rangle = \frac{\partial}{\partial t}\Big|_{t=0} g(X_t, s_t) = \frac{\partial}{\partial t}\Big|_{t=0} \int_\gamma s_t = \int_\gamma (a \lrcorner s) = \langle \xi, a\rangle,$$

so $\xi = dg$ is closed.

$\square$

Remark 7.8 (1) The most interesting case is clearly $n = 3$, when $\tilde{\mathcal{J}}$ has an honest symplectic structure. The cubic field on $\tilde{\mathcal{M}}$ corresponding to this structure by lemma 7.1 made its first appearance in [BG] and is essentially the *Yukawa coupling*, popular among physicists and mirror-symmetry enthusiasts. At $(X, x) \in \tilde{\mathcal{M}}$ there is a natural cubic form on $H^1(\mathcal{T}_x)$:

$$c : \otimes^3 H^1(\mathcal{T}_X) \to H^3(\overset{3}{\wedge} \mathcal{T}_X) = H^3(\omega_X^{-1}) \overset{\cdot s^2}{\to} H^3(\omega_X) \overset{\int}{\to} \mathbb{C},$$

which pulls back to the required cubic on $\mathcal{T}_{(X,s)}\tilde{\mathcal{M}}$. Hodge theoretically, this cubic can be interpreted as the third iterate of the infinitesimal variation of the periods, or the Hodge structure, of X c.f. [IVHS] and [BG]. By Griffiths transversality, each tangent direction on $\mathcal{M}$, $\theta \in H^1(\mathcal{T}_X)$, determines a linear map

$$\theta_i : H^{i,3-i} \longrightarrow H^{i-1,4-i} \qquad i = 3, 2, 1,$$

and clearly the composition

$$\theta_1 \circ \theta_2 \circ \theta_3 : H^{3,0} \longrightarrow H^{0,3}$$

becomes $c(\theta)$ when we use s to identify $H^{3,0}$ and its dual $H^{0,3}$ with $\mathbb{C}$.

(2) For $n = 2k - 1 \geq 5$, we get a closed 2-form on $\tilde{\mathcal{J}}$ which is in general not of maximal rank. The corresponding cubic is identically 0. Hodge theoretically, the "cubic" multiplies the gauge $s \in H^0(\omega_X)$ by two elements of $H^1(\mathcal{T}_X)$ (landing in $H^{n-2,2}$) and then with an element of $F^k H^n$. When $k > 2$ there are too many $dz's$, so the product vanishes.

(3) The symplectic form σ which we constructed on $\tilde{\mathcal{J}}$ is actually exact. Recall that the natural symplectic form $\tilde{\sigma}$ on $T^*\tilde{\mathcal{M}}$ is exact: $\tilde{\sigma} = d\tilde{\alpha}$, where $\tilde{\alpha}$ is the action 1-form. We obtained σ by pulling $\tilde{\sigma}$ back to $(j')^*\tilde{\sigma}$ on $\mathcal{V}$, and observing that the latter is invariant under translation by locally constant integral cycles γ, hence descends to $\tilde{\mathcal{J}}$. Now a first guess for the anti-differential of σ would be the 1-form $(j')^*\tilde{\alpha}$; but this is *not* invariant under translation: if the cycle γ corresponds, as in Step III of the proof, to a 1-form ξ on $\tilde{\mathcal{M}}$, then the translation by γ changes $(j')^*\tilde{\alpha}$ by $\pi^*\xi$, where $\pi : \mathcal{V} \to \tilde{\mathcal{M}}$ is the projection. To fix this discrepancy, we consider the tautological

function $f \in \Gamma(\mathcal{O}_{\mathcal{V}})$ whose value at a point $(X, s, v) \in \mathcal{V}$ (where $(X, s) \in \tilde{\mathcal{M}}$ and $v \in F^k H^n(X)^*$) is given by

$$f(X, s, v) = v(s). \tag{29}$$

This f is linear on the fibers of π, so df is constant on these fibers, and therefore translation by γ changes df by π^* of a 1-form on the base $\tilde{\mathcal{M}}$. This 1-form is clearly ξ, so we conclude that

$$(j')^* \tilde{\alpha} - df \tag{30}$$

is a global 1-form on $\mathcal{V}$ which is invariant under translation by each γ, hence descends to a 1-form α on $\tilde{\mathcal{J}}$. It satisfies $d\alpha = \sigma$, as claimed.

(4) Another way to see the exactness of σ on $\tilde{\mathcal{J}}$ is to note that it comes from a *quasi-contact structure* κ on $\mathcal{J}$. By a quasi-contact structure we mean a line subbundle κ of $T^*\mathcal{J}$. It determines a tautological 1-form on the $\mathbb{C}^*$-bundle $\tilde{\mathcal{J}}$ obtain from κ by omitting its zero section. Hence, it determines also an exact 2-form σ on $\tilde{\mathcal{J}}$. We refer to the pair $(\tilde{\mathcal{J}}, \sigma)$ as the *quasi-symplectification* of $(\mathcal{J}, \kappa)$. Conversely, according to [AG], page 78, a 2-form σ on a manifold $\tilde{\mathcal{J}}$ with a $\mathbb{C}^*$-action ρ is the quasi-symplectification of a line subbundle of the cotangent bundle of the quotient $\mathcal{J}$ if and only if σ is homogeneous of degree 1 with respect to ρ (and the contraction of σ with the vector field generating ρ is nowhere vanishing).

In our case, there are two independent $\mathbb{C}^*$-actions on the total space of $T^*\tilde{\mathcal{M}} \simeq [F^{n-1}\mathcal{H}^n(\tilde{\mathcal{X}}/\tilde{\mathcal{M}}, \mathbb{C})]^*$: the $\mathbb{C}^*$-action on $\tilde{\mathcal{M}}$ lifts to an action $\bar{\rho}'$ on $T^*\tilde{\mathcal{M}}$, and there is also the action $\bar{\rho}''$ which commutes with the projection to $\tilde{\mathcal{M}}$ and is linear on the fibers. The symplectic form $\tilde{\sigma}$ is homogeneous of weight 0 with respect to $\bar{\rho}'$ and of weight 1 with respect to $\bar{\rho}''$, hence of weight 1 with respect to $\bar{\rho} := \bar{\rho}' \cdot \bar{\rho}''$. Hence, $\tilde{\sigma}$ is the symplectification of a contact structure on $T^*\tilde{\mathcal{M}}/\bar{\rho} \simeq [F^{n-1}\mathcal{H}^n(\mathcal{X}/\mathcal{M}, \mathbb{C})]^*$ (suppressing the gauge).

Denote a point in $[F^{n-1}\mathcal{H}^n(\tilde{\mathcal{X}}/\tilde{\mathcal{M}}, \mathbb{C})]^*$ by (X, s, ξ). The actions, for $t \in \mathbb{C}^*$, are given by:

$$\begin{aligned} \bar{\rho}' &: (X, s, \xi) \longmapsto (X, ts, t^{-1}\xi) \\ \bar{\rho}'' &: (X, s, \xi) \longmapsto (X, s, t\xi) \\ \bar{\rho} &: (X, s, \xi) \longmapsto (X, ts, \xi). \end{aligned}$$

The function f on $\mathcal{V}$, given by (29), is the pullback $(j')^*(\bar{f})$ of the function $\bar{f}$ on $T^*\tilde{\mathcal{M}}$ given by

$$\bar{f}(X, s, \xi) = \xi(s).$$

The symplectic structure $\tilde{\sigma}$ on $T^*\tilde{\mathcal{M}}$ takes the vector fields generating the actions $\bar{\rho}', \bar{\rho}''$, and $\bar{\rho}$ to the 1-forms $-d\bar{f}, \tilde{\alpha}$ and $\tilde{\alpha} - d\bar{f}$, respectively. The 1-form $\tilde{\alpha} - d\bar{f}$, which is homogeneous of degree 1 with respect to $\bar{\rho}$, is the 1-form canonically associated to the contact structure on $T^*\tilde{\mathcal{M}}/\bar{\rho} \simeq [F^{n-1}\mathcal{H}^n(\mathcal{X}/\mathcal{M}, \mathbb{C})]^*$ (namely, the contraction of $\tilde{\sigma}$ with the vector field of $\bar{\rho}$.)

Similarly, we have three action ρ', ρ'' and $\rho = \rho' \cdot \rho''$ on the total space of $\mathcal{V} \simeq [F^k\mathcal{H}^n(\tilde{\mathcal{X}}/\tilde{\mathcal{M}}, \mathbb{C})]^*$. The surjective homomorphism $j' : \mathcal{V} \to T^*\tilde{\mathcal{M}}$ is $(\rho', \bar{\rho}'), (\rho'', \bar{\rho}'')$, and $(\rho, \bar{\rho})$-equivariant. The 1-form $\tilde{\alpha} - d\bar{f}$ pulls back to the 1-form $(j')^*(\tilde{\alpha}) - df$ given by (30). Clearly, the action ρ commutes with translations by $\mathcal{H}_n(\tilde{\mathcal{X}}/\tilde{\mathcal{M}}, \mathbb{Z})$. Since the 2-form $(j')^*\tilde{\sigma}$ is also $\mathcal{H}_n(\tilde{\mathcal{X}}/\tilde{\mathcal{M}}, \mathbb{Z})$-equivariant, $(j')^*(\alpha) - df$ descends to a

1-form α on $\tilde{\mathcal{J}}$. Clearly, $d\alpha = \sigma$ and α is homogeneous of degree 1 with respect to the $\mathbb{C}^*$-action on $\tilde{\mathcal{J}}$. Hence α comes from a quasi-contact structure κ on $\mathcal{J}$.

The Abel-Jacobi map of a curve to its Jacobian has an analogue for intermediate Jacobians. Let Z be a codimensional-k cycle in X, i.e. a formal linear combination $Z = \sum m_i Z_i$, with integer coefficients, of codimension k subvarieties $Z_i \subset X$. If Z is homologous to 0, we can associate to it a point $\mu(Z) \in \mathcal{J}^k(X)$, as follows. Choose a real $(2n - 2k + 1)$-chain Γ in X whose boundary is Z, and let $\mu(Z)$ be the image in

$$\mathcal{J}^k(X) \approx (F^{n-k+1}H^{2n-2k+1}(X,\mathbb{C}))^* / H_{2n-2k+1}(X,\mathbb{Z})$$

of the linear functional

$$\int_\Gamma \in (F^{n-k+1}H^{2n-2k+1}(X,\mathbb{C}))^*$$

sending a cohomology class represented by a harmonic form α to $\int_\Gamma \alpha$. Changing the choice of Γ changes $\int_\Gamma$ by an integral class, so $\mu(Z)$ depends only on Z. This construction depends continuously on its parameters: given a family $\pi : \mathcal{X} \to B$ and a family $\mathcal{Z} \to B$ of codimension-k cycles in the fibers which are homologous to 0 in the fibers, we get the *normal function*, or Abel-Jacobi map

$$\mu : B \longrightarrow \mathcal{J}^k(\mathcal{X}/B)$$

to the family of intermediate Jacobians of the fibers.

Abstractly, a normal function $\nu : B \to \mathcal{J}^k(\mathcal{X}/B)$ is a section satisfying the infinitesimal condition:
Any lift

$$\tilde{\nu} : B \longrightarrow \mathcal{H}^n(\mathcal{X}/B,\mathbb{C})$$

of

$$\nu : B \to J^k(\mathcal{X}/B) \simeq \mathcal{H}^n(\mathcal{X}/B,\mathbb{C})/[F^k\mathcal{H}^n(\mathcal{X}/B,\mathbb{C}) + \mathcal{H}^n(\mathcal{X}/B,\mathbb{Z})]$$

satisfies

$$\nabla\tilde{\nu} \in F^{k-1}\mathcal{H}^n(\mathcal{X}/B,\mathbb{C}) \otimes \Omega^1_B \qquad (31)$$

or equivalently

$$(\nabla\tilde{\nu}, s) = 0 \text{ for any section } s \text{ of } F^{k+1}\mathcal{H}^n(\mathcal{X}/B,\mathbb{C})$$

where $\nabla\tilde{\nu}$ *is the Gauss-Manin derivative of* $\tilde{\nu}$.

This condition is independent of the choice of the lift $\tilde{\nu}$ by Griffiths' transversality. It is satisfied by the Abel-Jacobi image of a relative codimension k-cycle (see [Gr]). More generally, we can consider maps

$$B \xrightarrow{\mu} \mathcal{J}^k$$
$$\searrow_q \quad \swarrow$$
$$M$$

The pullback $\mathcal{J}^k(\mathcal{X}/B) \to B$ of the relative intermediate Jacobian to B has a canonical section $\nu : B \to \mathcal{J}^k(\mathcal{X}/B)$. We will refer to the subvariety $\mu(B)$ as a *multivalued normal function* if $\nu : B \to \mathcal{J}^k(\mathcal{X}/B)$ is a normal function.

78

Theorem 7.9 *Let* $\tilde{\mathcal{X}} \to \tilde{\mathcal{M}}$ *be a complete family of gauged Calabi-Yau manifolds of dimension* $n = 2k - 1 \geq 3$, $\tilde{\mathcal{J}} \to \tilde{\mathcal{M}}$ *the relative intermediate Jacobian,* $B \xrightarrow{q} \tilde{\mathcal{M}}$ *a base of a family* $\mathcal{Z} \to B$ *of codimension-k cycles homologous to 0 in the fibers of* $q^*\tilde{\mathcal{X}} \to B$. *Then i) the Abel-Jacobi image in* $\tilde{\mathcal{J}}$ *of* B *is isotropic with respect to the quasi-symplectic form* σ *of theorem 7.7. ii) Moreover, the Abel-Jacobi image is also integral with respect to the 1-form* α *given by (30).*

Proof. i) We follow Step III of the proof of theorem 7.7. We thus think, locally in B, of X as being a fixed C^∞ manifold with variable complex structure $\bar{\partial}_b$, n-form s_b, and cycle Z_b, subject to the obvious compatibility. We choose a family Γ_b, $b \in B$ of n-chains whose boundary is Z_b, and consider the 1-form ξ on B given at $b \in B$ by $\int_{\Gamma_b}$; we need to show that ξ is closed. (The new feature here is that instead of the cycles $\gamma_b \in H_n(X_b, \mathbb{Z})$ we have chains, or relative cycles $\Gamma_b \in H_n(X_b, |Z_b|, \mathbb{Z})$, where $|Z_b|$ is the support of Z_b, which varies with b.)

As before, we consider the function

$$g : B \longrightarrow \mathbb{C}$$
$$g(X, s, Z, \Gamma) := \int_\Gamma s,$$

and we claim $\xi = dg$. This time, in the integral $\int_{\Gamma_b} s_b$, both the integrand and the chain depend on b. So if we take a normal vector v to the supports $|Z_b|$ along Γ_b, we obtain two terms:

$$\frac{\partial}{\partial b} \int_{\Gamma_b} s_b = \int_{\Gamma_b} \frac{\partial s}{\partial b} + \int_{\partial \Gamma_b} (v \rfloor s_b).$$

In the second term, however, s_b is of type $(n, 0)$ with respect to the complex structure $\bar{\partial}_b$, so the contraction $v \rfloor s_b$ is of type $(n-1, 0)$ regardless of the type of v. Since $\partial \Gamma_b = Z_b$ is of the type $(k-1, k-1)$, the second term vanishes identically, so we have $dg = \xi$ as desired.

ii) Integration $\int_\Gamma(\cdot)$ defines a section of $\mathcal{V} \simeq [F^k \mathcal{H}^n]^*$. The function g on B is the pullback via $\int_\Gamma(\cdot)$ of the function f on $\mathcal{V}$ given by the formula (29). Similarly, integration $\int_\Gamma(\cdot)$ defines the section ξ of $T^*\tilde{\mathcal{M}} \simeq [F^{n-1}\mathcal{H}^n]^*$. The pullback of the tautological 1-form $\tilde{\alpha}$ by ξ is ξ itself. The equation $\xi - dg = 0$ translates to the statement that the 1-form $(j')^*\tilde{\alpha} - df$ vanishes on the section $\int_\Gamma(\cdot)$ of $\mathcal{V}$ (see formula (30)). In particular, its descent α vanishes on the Abel-Jacobi image of $\mathcal{Z} \to B$.

□

Again, the most interesting case is $n = 3$. When B dominates the moduli space $\tilde{\mathcal{M}}$, i.e. for a multivalued choice of cycles on the general gauged Calabi-Yau of a given type, the normal function produces a Lagrangian subvariety of the symplectic $\tilde{\mathcal{J}}$, generically transversal to the fibers of the completely integrable system.

Remark 7.10 (1) The result of Theorem 7.9 holds for every multi-valued normal function $\mu : B \to \tilde{\mathcal{J}}^k(\tilde{\mathcal{X}}/\tilde{\mathcal{M}})$, not only for those coming from cycles. Given a vector field $\frac{\partial}{\partial b}$ on B, a lift $\tilde{\nu} : B \to \mathcal{H}^n$, and any section s of $F^{k+1}\mathcal{H}^n(\tilde{\mathcal{X}}/\tilde{\mathcal{M}}, \mathbb{C})$, the infinitesimal condition for normal functions (31) becomes

$$0 = \left(\nabla_{\frac{\partial}{\partial b}}\tilde{\nu}, s\right) = \frac{\partial}{\partial b}(\tilde{\nu}, s) - \left(\tilde{\nu}, \nabla_{\frac{\partial}{\partial b}}s\right). \tag{32}$$

When s is the tautological gauge, $\frac{\partial}{\partial b}(\tilde{\nu}, s)$ is the pullback of df by the projection of $\tilde{\nu}$ to $\mathcal{V} \cong \mathcal{H}^n/F^k\mathcal{H}^n$ (where f is defined by the equation (29)). Similarly, $\left(\tilde{\nu}, \nabla_{\frac{\partial}{\partial b}}s\right)$ is the contraction $\xi \rfloor \frac{\partial}{\partial b}$ of the pullback ξ of the tautological 1-form $\tilde{\alpha}$ on $T^*\tilde{\mathcal{M}}$ by the composition

$$\tilde{\mu} : B \to \mathcal{H}^n(\tilde{\mathcal{X}}/\tilde{\mathcal{M}}) \to \mathcal{H}^n/F^2 \simeq [F^{n-1}\mathcal{H}^n]^* \simeq T^*\tilde{\mathcal{M}}.$$

Thus, the infinitesimal condition for a normal function (32) implies that the image $\mu(B) \subset \tilde{\mathcal{J}}^k(\tilde{\mathcal{X}}/\tilde{\mathcal{M}})$ is integral with respect to the 1-form α (defined in (30)).

In the case of CY 3-folds ($n = 3, k = 2$) we see that the Legendre subvarieties of $\mathcal{J}^2 \to \mathcal{M}$ (i.e., the κ-integral subvarieties of maximal dimension $h^{2,1}$ where κ is the contact structure of Remark 7.8(4)) are precisely the multivalued normal functions.

(2) Both the infinitesimal condition for a normal function (31) and the (quasi) contact structure κ on the relative Jacobian $\mathcal{J}^k \to \mathcal{M}$ (see Remark 7.8(4)) are special cases of a more general filtration of Pfaffian exterior differential systems on the relative intermediate Jacobian $\mathcal{J}^k \to \mathcal{M}$ of any family $\mathcal{X} \to \mathcal{M}$ of $n = 2k - 1$ dimensional projective algebraic varieties.

The tangent bundle $T\mathcal{J}^k$ has a canonical decreasing filtration (defined by (34) below)

$$T\mathcal{J}^k = F^0 T\mathcal{J}^k \supset F^1 T\mathcal{J}^k \supset \cdots \supset F^{k-1}T\mathcal{J}^k \supset 0.$$

The quotient $T\mathcal{J}^k/F^1 T\mathcal{J}^k$ is canonically isomorphic to the pullback of the Hodge bundle $\mathcal{H}^n/F^i\mathcal{H}^n$. The $F^{k-1}T\mathcal{J}^k$ integral subvarieties are precisely the multi-valued normal functions.

When $\mathcal{J}^k$ is the relative intermediate Jacobian of a family of CY n-folds, the subbundle $F^1 T\mathcal{J}^k$ is a hyperplane distribution on $\mathcal{J}^k$ which defines the (quasi) contact structure κ of Remark 7.8(4).

When $n = 3$, $k = 2$, the filtration is a two step filtration

$$T\mathcal{J}^2 \supset F^1 T\mathcal{J}^2 \supset 0$$

and the $F^1 T\mathcal{J}^2$-integral subvarieties are precisely the normal functions.

The filtration $F^i T\mathcal{J}^k$, $0 \leq i \leq k - 1$ is defined at a point $(b, y) \in \mathcal{J}^k$ over $b \in \mathcal{M}$ as follows: Choose a section $\tilde{\nu} : \mathcal{M} \to \mathcal{J}^k$ through (b, y) with the property that any lift $\tilde{\nu} : \mathcal{M} \to \mathcal{X}^n(\mathcal{H}/\mathcal{M}, \mathbb{C})$ of ν satisfies the horizontality condition

$$\nabla\tilde{\nu} \in F^i\mathcal{H}^n(\mathcal{X}/\mathcal{M}, \mathbb{C}) \otimes \Omega^1_{\mathcal{M}} \tag{33}$$

The section ν defines a splitting

$$T_{(b,y)}\mathcal{J}^k = T_b\mathcal{M} \oplus \left[H^n(X_b, \mathbb{C})/F^k H^n(X_b, \mathbb{C})\right]$$

and the i-th piece of the filtration is defined by

$$F^i T_{(b,y)} \mathcal{J}^k := T_b \mathcal{M} \oplus \left[F^i H^n(X_b, \mathbb{C})/F^k H^n(X_b, \mathbb{C}) \right]. \tag{34}$$

The horizontality condition (33) implies that the subspace $F^i T \mathcal{J}^k_{(b,y)}$ is independent of the choice of the section ν through (b, y). Moreover, the subbundle $F^i T \mathcal{J}^k$ is invariant under translations by its integral sections, namely, by sections $\nu : \mathcal{M} \to \mathcal{J}^k$ satisfying the i-th horizontality condition.

We noted above that when $k = 1$ the intermediate Jacobian $\mathcal{J}^k(X)$ becomes the connected component $\mathrm{Pic}^0(X)$ of the Picard variety. The generalization of the Picard variety itself is the *Deligne cohomology* group $D^k(X)$, cf. [EZ]. This fits in an exact sequence

$$0 \to \mathcal{J}^k(X) \to D^k(X) \xrightarrow{p} H^{k,k}(X, \mathbb{Z}) \to 0, \tag{35}$$

where the quotient is the group of Hodge (k, k)-classes,

$$H^{k,k}(X, \mathbb{Z}) := H^{k,k}(X, \mathbb{C}) \cap H^{2k}(X, \mathbb{Z}).$$

Any codimension-k cycle Z in X has an Abel-Jacobi image, or cycle class $\mu(Z)$ in $D^k(X)$. Its image $p(\mu(Z))$ is the cycle class of Z in ordinary cohomology.

Formally, $D^k(X)$ is defined as the hypercohomology $\mathbb{H}^{2k}$ of the following complex of sheaves on X starting in degree 0.

$$0 \to \mathbb{Z} \to \mathcal{O}_X \to \Omega^1_X \to \cdots \to \Omega^{k-1}_X \to 0.$$

The forgetful map to $\mathbb{Z}$ is a map of complexes, with kernel the complex

$$0 \to \mathcal{O}_X \to \Omega^1_X \to \cdots \to \Omega^{k-1}_X \to 0.$$

The resulting long exact sequence of hyper cohomologies gives (35).

Let $H^{k,k}_{alg}$ be the subgroup of $H^{k,k}(X, \mathbb{Z})$ of classes of algebraic cycles. (The Hodge conjecture asserts that $H^{k,k}_{alg}$ is of finite index in $H^{k,k}(X, \mathbb{Z})$.) The inverse image

$$D^k_{alg}(X) := p^{-1}(H^{k,k}_{alg})$$

has an elementary description: it is the quotient of

$$\mathcal{J}^k(X) \times \{\text{codimension}-k \text{ algebraic cycles}\}$$

by the subgroup of codimension-k cycles homologous to 0, embedded naturally in the second component and mapped to the first by Abel-Jacobi.

As X varies in a family, the rank of $H^{k,k}(X, \mathbb{Z})$ can jump up (at those X for which the variable vector subspace $H^{k,k}(X, \mathbb{C})$ happens to be in special position with respect to the "fixed" lattice $H^{2k}(X, \mathbb{Z})$). To obtain a well-behaved family of Deligne cohomology groups, we require that

$$H^{k,k}(X, \mathbb{C}) = H^{2k}(X, \mathbb{C}).$$

For example, this holds for $k = 1$ or $k = n - 1$ if $h^{2,0} = 0$. In this case we also have $H^{k,k}_{alg} = H^{k,k}(X, \mathbb{Z})$ and hence $D^k_{alg}(X) = D^k(X)$, by the Lefschetz theorem on $(1, 1)$-classes [GH].

Corollary 7.11 *Let $\mathcal{X} \to \mathcal{M}$ be a complete family of 3-dimensional Calabi-Yau manifolds, $\tilde{\mathcal{X}} \to \tilde{\mathcal{M}}$ the corresponding gauged family. Let $\mathcal{D} \to \mathcal{M}$, $\tilde{\mathcal{D}} \to \tilde{\mathcal{M}}$, be their families of (second) Deligne cohomology groups, $\mathcal{J}$, $\tilde{\mathcal{J}}$ their relative intermediate Jacobians. Then there is a natural contact structure κ on $\mathcal{D}$ with symplectification $\sigma = d\alpha$ on $\tilde{\mathcal{D}}$ with the following properties:*

(a) *σ, α, and κ restrict to the previously constructed structures on $\tilde{\mathcal{J}}$ and $\mathcal{J}$.*

(b) *The fibration $\tilde{\mathcal{D}} \to \tilde{\mathcal{M}}$ is Lagrangian.*

(c) *The multivalued normal functions of $\mathcal{D}$ (resp. $\tilde{\mathcal{D}}$) are precisely the κ-integral (resp. α-integral) subvarieties. In particular, all multi-valued normal functions in $\mathcal{D}$ are isotropic.*

Proof: The contact structure κ on $\mathcal{J}$ defines one on $\mathcal{J} \times \{cycles\}$, which descends to $\mathcal{D}$ since the equivalence relation is κ-integral by remark 7.10.

$\square$

The mirror conjecture of conformal field theory predicts that to a family $\mathcal{X} \to \mathcal{M}$ of Calabi-Yau three folds, with some extra data, corresponds a "mirror" family $\mathcal{X}' \to \mathcal{M}'$, cf. [Mor] for the details. A first property of the conjectural symmetry is that for $X \in \mathcal{M}$, $X' \in \mathcal{M}'$,
$$h^{2,1}(X) = h^{1,1}(X'), h^{1,1} = h^{2,1}(X').$$
The conjecture goes much deeper, predicting a relation between the Yukawa cubic of $\mathcal{M}$ and the numbers of rational curves of various homology classes in a typical $X' \in \mathcal{M}'$. This has been used spectacularly in [CdOGP] and subsequent works, to predict those numbers on a non-singular quintic hypersurface in $\mathbb{P}^4$ and in a number of other families.

We wonder whether the conjecture could be reformulated and understood as a type of Fourier transform between the integrable systems on the universal Deligne cohomologies $\tilde{\mathcal{D}}$ and $\tilde{\mathcal{D}}'$ of the mirror families $\tilde{\mathcal{M}}$ and $\tilde{\mathcal{M}}'$. Note that the dimensions $h^{2,1}$ and $h^{1,1}$ which are supposed to be interchanged by the mirror, can be read off the continuous and discrete parts of the fibers of $\pi : \tilde{\mathcal{D}} \to \tilde{\mathcal{M}}$, respectively. One may try to imagine the mirror as a transform, taking these Lagrangian fibers over $\tilde{\mathcal{M}}$ (which encode the Yukawa cubic, as in Section 7.2) to Lagrangian sections over $\tilde{\mathcal{M}}'$, which should somehow encode the numbers of curves in X' via their Abel-Jacobi images.

8 The Lagrangian Hilbert scheme and its relative Picard

8.1 Introduction

The Lagrangian Hilbert scheme of a symplectic variety X parametrizes Lagrangian subvarieties of X. Its relative Picard parametrizes pairs (Z, L) consisting of a line bundle L on a Lagrangian subvariety Z. We use the cubic condition of chapter 7 to construct an integrable system structure on components of the relative Picard bundle over the Lagrangian Hilbert scheme.

We interpret the generalized Hitchin integrable system, supported by the moduli space of Higgs pairs over an algebraic curve (see Ch V), as a special case of this construction. Other examples discussed include:

a) Higgs pairs over higher dimensional base varieties (example 8.5.1), and

b) Fano varieties of lines on hyperplane sections of a cubic fourfold (example 8.5.2).

Understanding the *global* geometry of such an integrable system requires a compactification and a study of its boundary. Our compactifications of the relative Picard are moduli spaces of sheaves and we study the symplectic structure at (smooth, stable) points of the boundary.

Let X be a smooth projective symplectic algebraic variety, σ an everywhere non degenerate algebraic 2-form on X. A smooth projective Lagrangian subvariety Z_0 of X determines a component $\bar{B}$ of the Hilbert scheme parametrizing deformations of Z_0 in X. The component $\bar{B}$ consists entirely of Lagrangian subschemes. Its dense open subset B, parametrizing smooth deformations of Z_0, is a smooth quasi-projective variety [Ra, V].

Choose a very ample line bundle $\mathcal{O}_X(1)$ on X and a Hilbert polynomial p. The relative Picard $h : \mathcal{M}^p \to B$, parametrizing line bundles with Hilbert polynomial p which are supported on Lagrangian subvarieties of X, is a quasi-projective variety (see [Sim1]). If the Chern class $c_1(L_0) \in H^2(Z_0, \mathbb{Z})$ of a line bundle on Z_0 deforms as a $(1,1)$-class over the whole of B, then L_0 belongs to a component $\mathcal{M}$ of $\mathcal{M}^p$ which *dominates* the Hilbert scheme B. (By Griffiths' and Deligne's Theorem of the Fixed Part, [Sch] Corollary 7.23, this is the case for example, if $c_1(L_0)$ belongs to the image of $H^2(X, \mathbb{Q})$). Such components $\mathcal{M}$ are integrable systems, in other words:

Theorem 8.1 *There exists a canonical symplectic structure $\sigma_{\mathcal{M}}$ on the relative Picard bundle $\mathcal{M} \xrightarrow{h} B$ over the open subset B of the Hilbert scheme of smooth projective Lagrangian subvarieties of X. The support map $h : \mathcal{M} \to B$ is a Lagrangian fibration.*

The relative Picard over the Hilbert scheme of curves on a $K3$ or abelian surface is an example [Mu1]. In example 8.5.2, X is a symplectic fourfold.

Remark 8.2 Theorem 8.1 holds in a more general setting where X is a smooth projective algebraic variety, σ is a meromorphic, generically non degenerate closed 2-form on X. We let D_0 denote its degeneracy divisor, D_∞ its polar divisor, and set $D = D_0 \cup D_\infty$. Let Z_0 be a smooth projective Lagrangian subvariety of X which does not intersect D. Denote by B

the open subset of a component of the Hilbert scheme parametrizing smooth deformations of Z_0 which stay in $X - D$. Then B is smooth and Theorem 1 holds. A special case is when X has a generically non-degenerate Poisson structure ψ. In this case D_∞, the polar divisor of the inverse symplectic structure, is just the degeneracy locus of ψ, while D_0 is empty. The case where the subvariety Z_0 does intersect the degeneracy locus D_∞ of the Poisson structure is also of interest. It is discussed below under the category of Poisson integrable systems.

The moduli space of 1-form valued Higgs pairs is related to the case where $X = \mathbb{P}(\Omega_Y^1 \oplus \mathcal{O}_Y)$ is the compactification of the cotangent bundle of a smooth projective algebraic variety Y, and $D = \mathbb{P}\Omega_Y^1$ is the divisor at infinity (see example 8.5.1).

The relative Picard bundle $\mathcal{M}$ is in fact also a Zariski open subset of a component of the moduli space of stable coherent sheaves on X (see [Sim1] for the construction of the moduli space). Viewed in this way, Theorem 8.1 extends a result of Mukai [Mu1] for sheaves on a $K3$ or abelian surface.

Theorem [Mu1]: *Any component of the moduli space of simple sheaves on X is smooth and has a canonical symplectic structure.*

Kobayashi [Ko] generalized the above theorem to the case of simple vector bundles on a (higher dimensional) compact complex symplectic manifold (X, σ):
The smooth part of the moduli space has a canonical symplectic structure.

In view of Theorem 8.1 and Kobayashi's result one might be tempted to speculate that every component of the moduli space of (simple) sheaves on a symplectic algebraic variety has a symplectic structure. This is *false*. In fact, some components are odd dimensional (see example 8.19).

Returning to our symplectic relative Picard $\mathcal{M}$, it is natural to ask whether its *compactification* is symplectic. More precisely:

(i) *Does the symplectic structure extend to the smooth locus of the closure of the relative Picard $\mathcal{M}$ in the moduli space of stable (Lagrangian) sheaves?*

(ii) *Which of these components $\bar{\mathcal{M}}$ admits a smooth projective symplectic birational model?*

A partial answer to (i) is provided in Theorem 8.18. We provide a cohomological identification of the symplectic structure which extends as a 2-form $\sigma_{\bar{\mathcal{M}}}$ over the smooth locus of $\bar{\mathcal{M}}$. We do not know at the moment if the 2-form $\sigma_{\bar{\mathcal{M}}}$ is *non-degenerate* at every smooth point of $\bar{\mathcal{M}}$. The cohomological identification of $\sigma_{\bar{\mathcal{M}}}$ involves a surprisingly rich *polarized Hodge-like structure* on the algebra $\text{Ext}^*_X(L, L)$ of extensions of a Lagrangian line bundle L by itself as an $\mathcal{O}_X$-module.

Much of the above generalizes to Poisson integrable systems. Tyurin showed in [Ty1] that Mukai's theorem generalizes to Poisson surfaces:
The smooth part of any component of the moduli space of simple sheaves on a Poisson surface has a canonical Poisson structure.

When the sheaves are supported as line bundles on curves in the surface, we get an integrable system. More precisely:

Theorem 8.3 *Let (X, ψ) be a Poisson surface, D_∞ the degeneracy divisor of ψ. Let B be the Zariski open subset of a component of the Hilbert scheme of X parametrizing smooth irreducible curves on X which are not contained in D_∞. Then*

i) *B is smooth,*

ii) *the relative Picard bundle $h : \mathcal{M} \to B$ has a canonical Poisson structure $\psi_\mathcal{M}$,*

iii) *The bundle map $h : \mathcal{M} \to B$ is a Lagrangian fibration and*

iv) *The symplectic leaf foliation of $\mathcal{M}$ is induced by the canonical morphism $B \to$ Hilb$_{D_\infty}$ sending a curve Z to the subscheme $Z \cap D_\infty$ of D_∞.*

The generalization to higher dimensional Poisson varieties is treated here under rather restrictive conditions on the component of the Lagrangian Hilbert scheme (see condition 8.9). These restrictions will be relaxed in [Ma2].

The rest of this chapter is organized as follows: In section 8.2 we review the deformation theory of Lagrangian subvarieties. The construction of the symplectic structure is carried out in section 8.3 where we prove Theorems 8.1 and 8.3. In section 8.4 we outline the extension of the symplectic structure to the smooth locus of the moduli space of Lagrangian sheaves (Theorem 8.18). We discuss the examples of Higgs pairs and of Fano varieties of lines on cubics in section 8.5.

8.2 Lagrangian Hilbert Schemes

Let X be a smooth n-dimensional projective algebraic variety, $Z \subset X$ a codimension q subvariety and $\mathcal{O}_X(1)$ a very ample line bundle. The Hilbert polynomial p of Z is defined to be

$$p(n) := \chi\big(\mathcal{O}_Z(n)\big) := \sum (-1)^i \dim H^i\big(Z, \mathcal{O}_Z(n)\big).$$

Grothendieck proved in [Gro] that there is a projective scheme Hilb$_X^p$ parametrizing all algebraic subschemes of X with Hilbert polynomial p and having all the expected functoriality and naturality properties.

The Zariski tangent space $T_{[Z]}$Hilb$_X^p$ at the point $[Z]$ parametrizing a subvariety Z is canonically identified with the space of sections $H^0(Z, N_{X/Z})$ of the normal bundle (normal sheaf if Z is singular).

The scheme Hilb$_X^p$ may, in general, involve pathologies. In particular, it may be non-reduced. A general criterion for the smoothness of the Hilbert scheme at a point $[Z]$ parameterizing a locally complete intersection subscheme Z is provided by:

Definition 8.4 *The semi-regularity map $\pi : H^1(Z, N_{X/Z}) \longrightarrow H^{q+1}(X, \Omega_X^{q-1})$ is the dual of the natural homomorphism*

$$\pi^* : H^{n-q-1}\big(X, \Omega_X^{n-q+1}\big) \longrightarrow H^{n-q-1}\big(Z, \omega_Z \otimes N_{Z/X}^*\big).$$

Here $\omega_Z \cong \overset{q}{\wedge} N_{Z/X} \otimes \omega_X$ is the dualizing sheaf of Z and the homomorphism π^* is induced by the sheaf homomorphism

$$\Omega_X^{n-q+1} \cong \omega_X \otimes \overset{q-1}{\wedge} T_X \longrightarrow \omega_X \otimes \overset{q-1}{\wedge} N_{Z/X} \cong \omega_Z \otimes N_{Z|X}^*. \tag{36}$$

Theorem *(Severi-Kodaira-Spencer-Bloch [Ka]) If the semi-regularity map π is injective, then the Hilbert scheme is smooth at $[Z]$.*

Together with a result of Ran it implies:

Corollary 8.5 *Let (X, ψ) be a Poisson surface with a degeneracy divisor D_∞ (possibly empty). Let $Z \subset X$ be a smooth irreducible curve which is not contained in D_∞. Then the Hilbert scheme Hilb_X^p is smooth at $[Z]$.*

Proof: The Poisson structure induces an injective homomorphism $\phi : N^*_{Z/X} \hookrightarrow T_Z$. If Z intersects D_∞ non-trivially then $N_{Z/X} \simeq \omega_Z(Z \cap D_\infty)$ and hence $H^1(Z, N_{X/Z}) = (0)$ and the semi-regularity map is trivially injective.

Note that in our case $n = 2$, $q = 1$ and the dual of the semi-regularity map

$$\pi^* : H^0\big(X, \omega_X\big) \longrightarrow H^0\big(Z, \omega_Z \otimes N^*_{Z/X}\big)$$

is induced by the sheaf homomorphism

$$\omega_X \to \omega_{X_{|Z}} \to \omega_Z \otimes N^*_{Z/X}$$

given by (36). If $D_\infty = \emptyset$ (X is symplectic) then ω_X, $\omega_{X_{|Z}}$ and $\omega_Z \otimes N^*_{Z/X}$ are all trivial line bundles and hence both π^* and the semi-regularity map are isomorphisms. If $D_\infty \cap Z = \emptyset$ but $D_\infty \neq \emptyset$ then π fails to be injective but the result nevertheless holds by a theorem of Ran which we recall below (Theorem 8.7).

□

The condition that the curve Z is not contained in D_∞ is necessary as can be seen by the following counterexample due to Severi and Zappa:

Example 8.6 ([Mum1] Section 22) Let C be an elliptic curve, E a nontrivial extension $0 \to E_1 \to E \to E_2 \to 0$, $E_i \simeq \mathcal{O}_C$ and $\pi : X = \mathbb{P}(E) \to C$ the corresponding ruled surface over C. Denote by Z the section $s : C \to X$ given by the line subbundle $E_1 \subset E$. Let $\mathcal{O}_X(-1)$ be the tautological subbundle of $\pi^* E$. Then $\mathcal{O}_X(1)$ is isomorphic to the line bundle $\mathcal{O}_X(Z)$ and the canonical bundle ω_X is isomorphic to $\pi^*\big(\omega_C\big) \otimes \mathcal{O}_X(-2) \simeq \mathcal{O}_X(-2)$. $H^0(X, \overset{2}{\wedge} T_X)$ is thus isomorphic to $H^0(C, \mathrm{Sym}^2 E^*)$ which is one dimensional. It follows that X has a unique Poisson structure ψ up to a scalar factor. The divisor $D_\infty = 2Z$ is the degeneracy divisor of ψ.

Clearly, $N_{Z/X} \simeq \pi^* T_C \simeq T_Z$ and hence $H^0(Z, N_{Z/X})$ is one dimensional. On the other hand, Z has no deformations in X (its self intersection is 0 and a deformation Z' of Z will contradict the nontriviality of the extension $0 \to E_1 \to E \to E_2 \to 0$).

□

A curve Z on a symplectic surface X is automatically Lagrangian. In the higher dimensional case we replace the curve Z by a Lagrangian subvariety. Lagrangian subvarieties of symplectic varieties have two pleasant properties:

i) *The condition of being Lagrangian is both open and closed,*

ii) *Their deformations are unobstructed.*

More precisely, we have:

Theorem 8.7 *(Voisin [V], Ran [Ra]) Let X be a smooth projective algebraic variety, σ a generically non degenerate meromorphic closed 2-form, D_∞ its polar divisor, D_0 its degeneracy divisor. Assume that $Z_0 \subset X - D_\infty - D_0$ is a smooth projective Lagrangian subvariety. Then*

(i) *The subset of the Hilbert scheme Hilb_X^p parametrizing deformations of Z_0 in $X - D_\infty$ consists entirely of Lagrangian subvarieties.*

(ii) *The Hilbert scheme is smooth at $[Z_0]$.*

Sketch of Proof: (i) The Lagrangian condition is closed. Thus, it suffices to prove that the open subset of smooth deformations of Z_0 is Lagrangian. If $Z \subset X - D_\infty$ then $\sigma_{|_Z}$ is a closed holomorphic 2-form and the cohomology class $[\sigma_{|_Z}]$ in $H^{2,0}(Z)$ vanishes if and only if $\sigma_{|_Z}$ is identically zero. Since σ induces a *flat* section of the Hodge bundle of relative cohomology with $\mathbb{C}$-coefficients, then $[\sigma_{|_Z}] = 0$ is an open and closed condition.

(ii). The symplectic structure σ induces a canonical isomorphism $N_{Z/X} \simeq \Omega_Z^1$ for any Lagrangian projective smooth subvariety $Z \subset X - D_\infty - D_0$. Ran proved a criterion for unobstructedness of deformations: the T^1-lifting property (see [Ra, Ka]). Let $S_n = \mathrm{Spec}(\mathbb{C}[t]/t^{n+1})$. Any flat $(n+1)$-st order infinitesimal embedded deformation $Z_{n+1} \to S_{n+1}$ of $Z_0 = Z$ restricts canonically to an n-th order deformation $Z_n \to S_n$. In our context, the T^1-lifting property amounts to the following criterion:
Given any $(n+1)$-st order flat embedded deformation $Z_{n+1} \to S_{n+1}$, every extension
(a) of $Z_n \to S_n$ to a flat embedded deformation $\tilde{Z}_n \to S_n \times_{\mathbb{C}} S_1$
lifts to an extension
(b) of $Z_{n+1} \to S_{n+1}$ to $\tilde{Z}_{n+1} \to S_{n+1} \times_{\mathbb{C}} S_1$.

Extensions in (a) and (b) are classified by $T^1(Z_i/S_i) \cong H^0(Z_i, \mathcal{N}_{\varphi_i/S_i})$ where $\varphi_i : Z_i \to S_i \times X$ is the canonical morphism and $\mathcal{N}_{\varphi_i/S_i}$ is the relative normal sheaf. Recall that the De Rham cohomology and its Hodge filtration can be computed using the algebraic De Rham complex (28). Consequently, the discussion of part (i) applies in the infinitesimal setting to show that T_{Z_i/S_i} is *Lagrangian* as a subbundle of the pullback $(\varphi^i)^* T_X$ with respect to the non-degenerate 2-form $(\varphi^i)^*(\sigma)$ on $(\varphi^i)^* T_X$. The relative normal sheaf is the quotient

$$0 \to T_{Z_i/S_i} \to (\varphi^i)^* T_X \to \mathcal{N}_{\varphi_i/S_i} \to 0.$$

Hence the symplectic structure induces an isomorphism $\mathcal{N}_{\varphi_i/S_i} \simeq \Omega_{Z_i/S_i}^1$. By a theorem of Deligne, $H^0\big(\Omega_{Z_i/S_i}^1\big)$, and hence also $H^0\big(\mathcal{N}_{\varphi_i/S_i}\big)$, is a free $\mathcal{O}_{S_i}$-module [Del]. Thus, $H^0\big(\mathcal{N}_{\varphi_{n+1}/S_{n+1}}\big) \longrightarrow H^0\big(\mathcal{N}_{\varphi_n/S_n}\big)$ is surjective and the T^1-lifting property holds.

□

Note that the naive analogue of the above theorem fails for Poisson varieties. In general, deformations of Lagrangian subvarieties need not stay Lagrangian. Consider for example $(\mathbb{P}^{2n}, \psi)$ where the Poisson structure ψ is the extension of the standard (non degenerate) symplectic structure on $\mathcal{A}^{2n} \subset \mathbb{P}^{2n}$. The Lagrangian Grassmannian has positive codimension in $Gr(n+1, 2n+1)$.

8.3 The construction of the symplectic structure

The construction of the symplectic structure on the relative Picard bundle is carried out in three steps:

In Step I we reduce it to the construction of the symplectic structure on the relative Pic^0-bundle.

In Step II we verify the cubic condition and thus construct the 2-form (or the 2-tensor in the Poisson case).

In Step III we prove the closedness of the 2-form.

Step I: Reduction to the Pic^0-Bundle Case:

The construction of a 2-form on the relative Picard bundle $\mathcal{M} \xrightarrow{h} B$ reduces to constructing it on its zero component $\mathcal{M}^0 \xrightarrow{h} B$, namely the Pic^0-bundle, by the following:

Proposition 8.8 *Any closed 2-form $\sigma_{\mathcal{M}^0}$ on $\mathcal{M}^0$, with respect to which the zero section of $\mathcal{M}^0$ is Lagrangian, extends to a closed 2-form $\sigma_{\mathcal{M}}$ on the whole Picard bundle $h : \mathcal{M} \to B$. The extension $\sigma_{\mathcal{M}}$ depends canonically on $\sigma_{\mathcal{M}^0}$ and the polarization $\mathcal{O}_X(1)$ of X.*

Proof: The point is that Picard bundles are rationally split. For any polarized projective variety $(Z, \mathcal{O}_Z(1))$, we have the Lefschetz map

$$Lef : Pic\, Z \longrightarrow Alb\, Z$$

$$[D] \mapsto \left[D \cap [\mathcal{O}_Z(1)]^{n-1}\right]$$

inducing an isogeny

$$Lef^0 : Pic^0\, Z \longrightarrow Alb^0\, Z.$$

We can set

$$L_Z := \left\{s \in Pic\, Z \mid \exists\, \ell, m,\ \ell \neq 0,\ \text{such that}\ \ell \cdot Lef(s) = m \cdot Lef(\mathcal{O}(1))\right\}.$$

This is an extension of $H_Z^{1,1}(Z)$ by the torsion subgroup L_Z^{tor} of $Pic(Z)$. In a family $\mathcal{Z} \to B$, these groups form a subsheaf $\mathcal{L}$ of $\mathcal{M} := Pic(\mathcal{Z}/B)$, intersecting $\mathcal{M}^0 := Pic^0(\mathcal{Z}/B)$ in its torsion subsheaf $\mathcal{L}^0$. In our situation, the 2-form $\sigma_{\mathcal{M}^0}$ is $\mathcal{L}^0$-invariant, so it extends uniquely to an $\mathcal{L}$-invariant closed 2-form $\sigma_{\mathcal{M}}$ on $\mathcal{M}$.

$\square$

Step II: Verification of the Cubic Condition:

In this step we construct the 2-form (or 2-vector) on the relative Picard bundle. In the next step we will prove that it is closed (respectively, a Poisson structure).

Let (X, ψ) be a smooth projective variety, ψ a generically non-degenerate holomorphic Poisson structure. Denote by D_∞ the degeneracy divisor of ψ. We will assume throughout this step that $Z \subset X$ is a smooth subvariety, $Z \cap (X - D_\infty)$ is non empty and Lagrangian, and

Condition 8.9 i) $[Z]$ *is a smooth point of the Hilbert scheme, and*

ii) *all deformations of Z in X are Lagrangian.*

As we saw in the previous section, conditions i) and ii) hold in case ψ is everywhere non-degenerate $((X, \psi^{-1})$ is a symplectic projective algebraic variety), and also in case X is a surface. Such $[Z]$ vary in a smooth Zariski open subset B of the Hilbert scheme and we denote by $h : \mathcal{M} \to B$ the relative Pic^0-bundle. Condition 8.9 can be relaxed considerably (see [Ma2]).

Let

$$\phi : N^*_{Z/X} \hookrightarrow T_Z \qquad (37)$$

be the injective homomorphism induced by the Poisson structure ψ. Its dual $\phi^* : T_Z^* \to N_{Z/X}$ induces an injective homomorphism.

$$i : H^0\left(Z, T_Z^*\right) \hookrightarrow H^0\left(Z, \mathcal{N}_{Z/X}\right). \qquad (38)$$

The vertical tangent bundle $V := h_* \mathcal{T}_{\mathcal{M}/B}$ is isomorphic to the Hodge bundle $\mathcal{H}^{0,1}(\mathcal{Z}/B)$. The polarization induces an isomorphism $V^* \simeq \mathcal{H}^{1,0}$. We get a global injective homomorphism $i : V^* \hookrightarrow T_B$.

Proposition 8.10 *The homomorphism i is induced by a canonical 2-vector $\psi_\mathcal{M} \in H^0(\mathcal{M}, \overset{2}{\Lambda} T_\mathcal{M})$ with respect to which $h : \mathcal{M} \to B$ is a Lagrangian fibration. (We do not assert yet that $\psi_\mathcal{M}$ is a Poisson structure).*

Proof: It suffices to show that i satisfies the (weak) cubic condition, namely, that $dp \circ i$ comes from a cubic. The derivative of the period map

$$dp : H^0\left(Z, N_{Z/X}\right) \longrightarrow \mathrm{Hom}\left(H^{1,0}(Z), H^{0,1}(Z)\right) \simeq \left[H^{1,0}(Z)^*\right]^{\otimes 2} \qquad (39)$$

is identified by the composition

$$H^0\left(Z, N_{Z/X}\right) \overset{\text{K-S}}{\longrightarrow} H^1(Z, T_Z) \overset{VHS}{\longrightarrow} \mathrm{Sym}^2 H^{1,0}(Z)^*$$

where K-S is the Kodaira-Spencer map given by cup product with the extension class of $T_{X|_Z}$:

$$\tau \in \mathrm{Ext}^1\left(N_{Z/X}, T_Z\right) \simeq H^1\left(Z, N^*_{Z/X} \otimes T_Z\right), \qquad (40)$$

and the variation of Hodge structure map VHS is given by cup product and contraction

$$H^1(Z, T_Z) \otimes H^0(Z, T_Z^*) \longrightarrow H^1(Z, \mathcal{O}_Z).$$

The composition $(K\text{--}S) \circ i : H^0(Z, T_Z^*) \to H^1(Z, T_Z)$ is then given by cup product with the class $(\phi \otimes \mathrm{id})(\tau) \in H^1(Z, T_Z \otimes T_Z)$. We will show that $(\phi \otimes \mathrm{id})(\tau)$ is symmetric, that is, an element of $H^1(Z, \mathrm{Sym}^2 T_Z)$. This would imply that dp, regarded as a section of $H^0\big(Z, N_{Z/X}\big)^* \otimes \mathrm{Sym}^2 H^{1,0}(Z)^* \overset{i}{\cong} H^{1,0}(Z)^* \otimes \mathrm{Sym}^2 H^{1,0}(Z)^*$, is symmetric also with respect to the first two factors. The cubic condition will follow.

Lemma 8.11 below implies that $(\phi \otimes \mathrm{id})(\tau)$ is in $H^1(Z, \mathrm{Sym}^2 T_Z)$ if and only if ϕ is induced by a section ψ in $H^0\big(Z, \overset{2}{\wedge} T_{X_{|Z}}\big)$ with respect to which Z is Lagrangian (i.e., $N^*_{Z/X}$ is isotropic). This is indeed the way ϕ was defined.

$\square$

Lemma 8.11 *Let T be an extension*

$$0 \to Z \to T \to N \to 0 \tag{41}$$

of a vector bundle N by a vector bundle Z. Then the following are equivalent for any homomorphism $\phi : N^* \to Z$.

i) The homomorphism ϕ is induced by a section $\psi \in H^0(\overset{2}{\wedge} T)$ with respect to which N^* is isotropic.

ii) The homomorphism $\phi_* := H^1(\phi \otimes 1) : H^1(N^* \otimes Z) \to H^1(Z \otimes Z)$ maps the extension class $\tau \in H^1(N^* \otimes Z)$ of T to a symmetric class $\phi_*(\tau) \in H^1(\mathrm{Sym}^2 Z) \subset H^1(Z \otimes Z)$.

Proof: We argue as in the proof of the cubic condition (lemma 7.1). The extension (41) induces an extension

$$0 \to \overset{2}{\wedge} Z \to F \to Z \otimes N \to 0,$$

where F is the subsheaf of $\overset{2}{\wedge} T$ of sections with respect to which N^* is isotropic. The homomorphism ϕ, regarded as a section of $Z \otimes N$, lifts to a section ψ of F if and only if it is in the kernel of the connecting homomorphism

$$\delta : H^0(Z \otimes N) \to H^1(\overset{2}{\wedge} Z).$$

The latter is given by a) pairing with the extension class τ

$$(\cdot)_* \tau : H^0(Z \otimes N) \to H^1(Z \otimes Z),$$

followed by b) wedge product

$$H^1(Z \otimes Z) \overset{\wedge}{\to} H^1(\overset{2}{\wedge} Z).$$

Thus, $\delta(\phi)$ vanishes if and only if $\phi_*(\tau)$ is in the kernel of $\wedge$, i.e., in $H^1(\mathrm{Sym}^2 Z)$.

$\square$

The identification of the cubic is particularly simple in the case of a curve Z on a surface X. In that case Serre's duality identifies VHS with the dual of the multiplication map

$$\mathrm{Sym}^2 H^0(Z, \omega_Z) \overset{\mathrm{VHS}^*}{\longrightarrow} H^0(Z, \omega_Z^{\otimes 2}).$$

The cubic $c \in \mathrm{Sym}^3 H^0(Z, \omega_Z)^*$ is given by composing the multiplication

$$\mathrm{Sym}^3 H^0(Z, \omega_Z) \longrightarrow H^0(Z, \omega_Z^{\otimes 3})$$

with the linear functional

$$(\phi \otimes \mathrm{id})(\tau) \in H^1(Z, T_Z^{\otimes 2}) \simeq H^0(Z, \omega_Z^{\otimes 3})^*$$

corresponding to the extension class τ.

In higher dimension (say n), the cubic depends on the choice of a polarization $\alpha \in H^{1,1}(X)$:

$$\mathrm{Sym}^3 H^0(Z, \Omega_Z^1) \to H^0(Z, \mathrm{Sym}^3 \Omega_Z^1) \xrightarrow{\phi_*(\tau)} H^1(Z, \Omega_Z^1) \xrightarrow{\alpha^{n-1}_{|z}} H^{n,n}(Z) \cong \mathbb{C}.$$

The choice of α is implicitly made in the proof of proposition 8.10 when we identify $H^{0,1}(Z)$ with $H^{1,0}(Z)^*$ via the Lefschetz isomorphism (see (39)) .

Step III: Closedness:

In this step we prove that the canonical 2-vector $\psi_{\mathcal{M}}$ constructed in the previous step is a Poisson structure. We first prove it in the symplectic case and later indicate the modifications needed for the Poisson case (assuming condition 8.9 of the previous step). This completes the proof of Theorems 8.1 and 8.3 stated in the introduction to this chapter.

Symplectic Case:

We assume, for simplicity of exposition, that (X, σ) is a smooth projective symplectic algebraic variety. The arguments apply verbatim to the more general setup involving a smooth projective algebraic variety X, a closed generically non-degenerate meromorphic 2-form σ on X with degeneracy divisor D_0 and polar divisor D_∞, and Lagrangian smooth projective subvarieties which do not intersect $D_0 \cup D_\infty$.

We then have a non-degenerate 2-tensor $\psi_{\mathcal{M}}$ on $h : \mathcal{M} \to B$ and hence a 2-form $\sigma_{\mathcal{M}}$. The closedness of $\sigma_{\mathcal{M}}$ follows from that of σ_X as we now show. A polarization of X induces a relative polarization on the universal Lagrangian subvariety

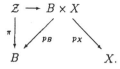

The relative polarization induces an isogeny

between the relative Pic^0-bundle and the relative Albanese $h : \mathcal{A} \to B$. Hence, a 2-form $\sigma_{\mathcal{A}}$ on $\mathcal{A}$. Clearly, closedness of $\sigma_{\mathcal{M}}$ is equivalent to that of $\sigma_{\mathcal{A}}$. Since the question is local, we may assume that we have a section $\xi : B \to \mathcal{Z}$. We then get for each positive integer t a relative Albanese map

$$
\begin{array}{ccc}
\mathcal{Z}^t & \xrightarrow{\;a_t\;} & \mathcal{A} \\
& \searrow_{\pi} \quad {}_{h}\swarrow & \\
& B &
\end{array}
$$

from the fiber product over B of t copies of the universal Lagrangian subvariety $\mathcal{Z} \to B$. For a fixed subvariety Z_b and points $(z_1, \ldots, z_t) \in Z_b^t$, a_t is given by integration

$$
\sum_{i=1}^{t} \int_{\xi(b)}^{z_i} (\cdot) \quad (\text{modulo } H_1(Z_b, \mathbb{Z})) \in H^{1,0}(Z)^* \big/ H_1(Z_b, \mathbb{Z}).
$$

We may assume, by choosing t large enough, that a_t is surjective. Thus, closedness of $\sigma_{\mathcal{A}}$ is equivalent to closedness of $a_t^*(\sigma_{\mathcal{A}})$. The closedness of $a_t^*(\sigma_{\mathcal{A}})$ now follows from that of σ_X by lemma 8.12.

Lemma 8.12 *Let $\ell : \mathcal{Z}^t \to X^t$ be the natural morphism; σ_{X^t} the product symplectic structure on X^t. Then,*

$$
a_t^*(\sigma_{\mathcal{A}}) = \ell^*(\sigma_{X^t}) - \pi^*(\xi^t)^* \, \ell^*(\sigma_{X^t}). \tag{42}
$$

Proof: The fibers of $\pi : \mathcal{Z}^t \to B$ are isotropic with respect to the 2-forms on both sides of the equation (42). Hence, these 2-forms induce (by contraction) homomorphisms

$$
f_{\mathcal{A}}, f_X : T_{\mathcal{Z}_b^t} \longrightarrow N_{\mathcal{Z}_b^t/\mathcal{Z}^t}^*.
$$

The section $\xi^t(B) \subset \mathcal{Z}^t$ is also isotropic with respect to the 2-forms on both sides of equation (42). Thus, equality in (42) will follow from equality of the induced homomorphisms $f_{\mathcal{A}}, f_X$. Proving the equality $f_{\mathcal{A}} = f_X$ is a straightforward, though lengthy, unwinding of cohomological identifications.

The relative normal bundle is identified as the pullback of the tangent bundle of the Hilbert scheme

$$
N_{\mathcal{Z}_b^t/\mathcal{Z}^t} \simeq \mathcal{O}_{\mathcal{Z}_b^t} \otimes (T_b B) \simeq \mathcal{O}_{\mathcal{Z}_b^t} \otimes H^0\big(Z_b, N_{Z_b/X}\big).
$$

We will show that the duals of both $f_{\mathcal{A}}$ and f_X

$$
f_{\mathcal{A}}^*, f_X^* : \mathcal{O}_{\mathcal{Z}_b^t} \otimes H^0\big(Z_b, N_{Z_b/X}\big) \to T_{\mathcal{Z}_b^t}^*
$$

are identified as the composition of
i) the diagonal homomorphism

$$
\mathcal{O}_{\mathcal{Z}_b^t} \otimes H^0(Z_b, N_{Z_b/X}) \xrightarrow{\Delta} \mathcal{O}_{\mathcal{Z}_b^t} \otimes \big[H^0(Z_b, N_{Z_b/X})\big]^t \quad \text{followed by}
$$

ii) the evaluation map

$$
e_t : \mathcal{O}_{\mathcal{Z}_b^t} \otimes \big[H^0(Z_b, N_{Z_b/X})\big]^t \simeq \mathcal{O}_{\mathcal{Z}_b^t} \otimes H^0\big(Z_b^t, N_{Z_b^t/X^t}\big) \longrightarrow N_{\mathcal{Z}_b^t/X^t} \quad \text{followed by}
$$

iii) contraction with the 2-form σ_{X^t}

$$
(\phi^{-1^*})^t : N_{\mathcal{Z}_b^t/X^t} \xrightarrow{\sim} T_{\mathcal{Z}_b^t}^*.
$$

(ϕ is given by contraction with the Poisson structure (37)).

Identification of f_A: (for simplicity assume t=1).

The 2-form σ_A is characterized as the unique 2-form with respect to which the three conditions of lemma 7.1 hold, i.e., i) $A \to B$ is a Lagrangian fibration, ii) the zero section is Lagrangian, and iii) σ_A induces the homomorphism

$$H^0(\phi^{-1^*}) = i^{-1} : H^0(Z, N_{Z/X}) \xrightarrow{\sim} H^0(Z, T_Z^*).$$

Thus, $a^*(\sigma_A)$ induces

$$f_A^* = \left(\mathcal{O}_{Z_b} \otimes H^0\left(Z_b, N_{Z_b/X}\right) \xrightarrow{(i^{-1})} \mathcal{O}_{Z_b} \otimes H^0\left(Z_b, T_{Z_b}^*\right) \xrightarrow{da^*} T_{Z_b}^* \right)$$

and the codifferential da^* of the Albanese map is the evaluation map.

Identification of f_X: $(t = 1)$

Both 2-forms $\ell^*(\sigma_X)$ and $\ell^*(\sigma_X) - \pi^*\xi^*\ell^*(\sigma_X)$ induce the same homomorphism f_X^* : $\mathcal{O}_{Z_b} \otimes T_b B \to T_{Z_b}^*$. This homomorphism is the composition $\phi^{-1^*} \circ \overline{(d\ell)}$, where $\overline{d\ell}$ is the homomorphism $N_{Z_b/Z} \to N_{Z_b/X}$ induced by the differential of $\ell : Z \to X$:

$$
\begin{array}{ccccccccc}
0 & \longrightarrow & T_{Z_b} & \longrightarrow & (T_Z)_{|Z_b} & \to & H^0(Z_b, N_{Z_b/X}) \otimes \mathcal{O}_{Z_b} & \longrightarrow & 0 \\
& & {\scriptstyle =}\downarrow & & {\scriptstyle d\ell}\downarrow & & {\scriptstyle \overline{d\ell}}\downarrow & & \\
0 & \longrightarrow & T_{Z_b} & \longrightarrow & (\ell^* TX)_{|Z_b} & \longrightarrow & N_{Z_b/X} & \longrightarrow & 0.
\end{array}
$$

Clearly $\overline{d\ell}$ is given by evaluation. This completes the proof of lemma 8.12.

$\square$

As a simple corollary of lemma 8.12 we have:

Corollary 8.13 *There exists a canonical symplectic structure σ_{A^t} on the relative Albanese of degree $t \in \mathbf{Z}$, depending canonically on the symplectic structure σ_X (independent of the polarization $\mathcal{O}_X(1)$!) and satisfying, for $t \geq 1$,*

$$a_t^*(\sigma_{A^t}) = \ell^*(\sigma_{X^t}).$$

(the pullback to the fiber product $Z^t := \times_B^t Z$ via the Albanese map coincides with the pullback of the symplectic structure σ_{X^t} on X^t).

Proof: The $t = 0$ case is proven. We sketch the proof of the $t \geq 1$ case. The $t \leq -1$ case is similar. Let ξ be a local section of $Z^t \to B$. Translation by the section $-a_t(\xi)$ of A^{-t} defines a local isomorphism

$$\tau_\xi : A^t \longrightarrow A^0.$$

Let

$$\sigma_{A^t} := \tau_\xi^*(\sigma_{A^0}) + h^*\xi^*\ell^*(\sigma_{X^t}).$$

We claim that σ_{A^t} is independent of ξ. This amounts to the identity

$$\tau_{(\xi_1 - \xi_2)}^*(\sigma_{A^0}) = \sigma_{A^0} - h^*[a_0(\xi_1 - \xi_2)]^*\sigma_{A^0}$$

for any two sections ξ_1, ξ_2 of $Z^t \to B$.

Poisson Case: (assuming condition 8.9)

Showing that the 2-vector $\psi_{\mathcal{M}}$ constructed in step II is a Poisson structure, amounts to showing that

Lemma 8.14 $\psi_{\mathcal{M}}(T_{\mathcal{M}}^*) \subset T_{\mathcal{M}}$ *is an involutive distribution,*.

and

Lemma 8.15 *the induced 2-form on each symplectic leaf is closed.*

Sketch of Proof of Lemma 8.14: Since $h : \mathcal{M} \to B$ is a Lagrangian fibration with respect to $\psi_{\mathcal{M}}$ (by proposition 8.10), the distribution is the pullback of the distribution on the base B. The latter is induced by the image of the injective homomorphism $i : V^* \hookrightarrow T_B$ identified by (38)

$$i = H^0(\phi^*) : H^0(Z, T_Z^*) \hookrightarrow H^0(Z, N_{Z/X}).$$

Recall (37) that ϕ, in turn, is induced by the Poisson structure ψ_X on X. The involutivity now follows from that of $\psi_X(T_X^*) \subset T_X$ by a deformation theoretic argument. The details are omitted.

□

In case X is a surface, the degeneracy divisor D_∞ of ψ_X is a curve and $iH^0(Z, T_Z^*) \subset H^0(Z, N_{Z/X})$ is the subspace of all infinitesimal deformations of Z which *fix* the divisor $Z \cap D_\infty$. Thus, the distribution $i(V^*)$ on the Hilbert scheme B (as in the proof of lemma 8.14) corresponds to the foliation by level sets of the algebraic morphism

$$R : B \longrightarrow \text{Hilb}(D_\infty)$$
$$Z \longmapsto Z \cap D_\infty.$$

The higher dimensional case is analogous. The degeneracy divisor D_∞ has an algebraic rank stratification

$$D_\infty = \bigcup_{r=0}^{n-1} D_\infty[2r] \qquad (\dim X = 2n).$$

Each rank stratum is foliated, local analytically, by symplectic leaves. The subspace

$$i : H^0(Z, T_Z^*) \subset H^0(Z, N_{Z/X})$$

is characterized as the subspace of all infinitesimal deformations of Z which deform the subscheme $Z \cap D_\infty[2r]$ fixing the image $f(Z \cap D_\infty[2r])$ with respect to any Casimir function f on $D_\infty[2r]$. As an illustration, consider the case where X is the logarithmic cotangent bundle $T_M^*(log(D))$ and Z is a 1-form with logarithmic poles along a divisor D with normal crossing. In this case the residues induce the symplectic leaves foliation.

Sketch of Proof of Lemma 8.15: The proof is essentially the same as in the symplectic case. We consider an open (analytic) subset B_1 of a leaf in B, the universal Lagrangian

$$Z_1 \xrightarrow{\ell} X$$

subvariety $\downarrow$, and the relative Albanese $\begin{matrix} \mathcal{A} \\ \downarrow \\ B_1 \end{matrix}$. One has to choose the section

B_1

$\xi : B_1 \to Z_1$ outside $\ell^{-1}(D_\infty)$ and notice that the identity (42) implies that the pullback $\ell^*(\sigma_X^t)$ of the *meromorphic* closed 2-form σ_X^t (inverse of the generically non-degenerate Poisson structure on the product of t copies of X) is a *holomorphic* 2-form on Z_1^t (because $a_t^*(\sigma_{\mathcal{A}})$ is) and that $a_t^*(\sigma_{\mathcal{A}})$ is *closed* (because $\ell^*(\sigma_{X^t})$ is).

□

8.4 Partial compactifications: a symplectic structure on the moduli space of Lagrangian sheaves

We describe briefly in this section the extension of the symplectic structure on the relative Picard $\mathcal{M}$ to an algebraic 2-form on the smooth locus of a partial compactification. For details see [Ma2]. For simplicity, we assume that (X, σ) is a smooth $2n$-dimensional projective symplectic variety. We note that with obvious modifications, the extension of the 2-form will hold in the setup $(X, \sigma, D_0, D_\infty)$ as in remark 8.2 allowing σ to degenerate and have poles away from the support of the sheaves.

When X is a symplectic surface, some of these extensions give rise to smooth projective symplectic compactifications [Mu1]. These projective symplectic compactifications appear also in the higher dimensional case:

Example 8.16 A somewhat trivial reincarnation of a relative Picard of a linear system on a K3 surface S as a birational model of a relative Picard of a Lagrangian Hilbert scheme over a higher dimensional symplectic variety X is realized as follows. Let X be the Beauville variety $S^{[n]}$ which is the resolution of the n-th symmetric product of S provided by the Hilbert scheme of zero cycles of length n [B2]. The symmetric powers $C^{[n]}$ of smooth curves on S are smooth Lagrangian subvarieties of $S^{[n]}$. Components of the relative Picard over the smooth locus in the linear system $|C|$ are isomorphic to Zariski open subsets of components of the relative Picard over the Lagrangian Hilbert scheme of $S^{[n]}$.

This leads us to speculate that genuinely new examples of smooth symplectic *projective* varieties will arise as birational models of moduli spaces of Lagrangian line bundles. (see section 8.5.2 for new *quasiprojective* examples).

We worked so far with a component $\mathcal{M} \to B$ of the relative Picard of the universal smooth Lagrangian subvariety $\mathcal{Z} \to B$ which dominates the corresponding component $\bar{B}$ of the Lagrangian Hilbert scheme (i.e., if L is supported on Z, $c_1(L) \in H_Z^{1,1}(Z)$ remains of type $(1,1)$ over B). Let $p(n) := \chi\big(L \otimes_{\mathcal{O}_X} \mathcal{O}_X(n)\big)$ be the Hilbert polynomial of a Lagrangian line bundle L parametrized by $\mathcal{M}$. A construction of C. Simpson enables us to compactify $\mathcal{M}$ as an open subset of a component $\mathcal{M}^{ss}$ of the moduli space of equivalence classes of coherent semistable sheaves on X with Hilbert polynomial p [Sim1]. Denote by $\mathcal{M}^s$ the open subset of $\mathcal{M}^{ss}$ parametrizing isomorphism classes of stable sheaves, $\mathcal{M}^{s,sm}$ the smooth locus of $\mathcal{M}^s$. Then $\mathcal{M} \subseteq \mathcal{M}^{s,sm} \subseteq \mathcal{M}^s \subseteq \mathcal{M}^{ss}$. In addition, the moduli space $\mathcal{M}^s$ embeds as a Zariski open subset of the moduli space of simple sheaves [AK2]. The

Zariski tangent space $T_{[L]}\mathcal{M}^s$ at a stable sheaf L is thus canonically isomorphic to the Zariski tangent space of the moduli space of simple sheaves. The latter is identified as the group $\mathrm{Ext}^1_{\mathcal{O}_X}(L,L)$ of extensions $0 \to L \to E \to L \to 0$ of L by L as an $\mathcal{O}_X$-module. When X is a $K3$ or abelian surface, Mukai's symplectic structure is given by the pairing

$$\mathrm{Ext}^1_{\mathcal{O}_X}(L,L) \otimes \mathrm{Ext}^1_{\mathcal{O}_X}(L,L) \xrightarrow{\text{Yoneda}} \mathrm{Ext}^2_{\mathcal{O}_X}(L,L) \xrightarrow{\text{S.D.}} \mathrm{Hom}_X(L, L \otimes \omega_X)^* \xrightarrow{id \otimes \sigma} \mathbb{C}$$

(Composition of the Yoneda pairing, Serre Duality, and evaluation at $id \otimes \sigma \in \mathrm{Hom}_X(L, L \otimes \omega_X)$).

The generalization of Mukai's pairing requires the construction of a homomorphism, depending linearly on the Poisson structure ψ,

$$y : H^{1,1}(X) \to \mathrm{Ext}^2_{\mathcal{O}_X}(L,L). \tag{43}$$

It sends the Kahler class $\alpha := c_1(\mathcal{O}_X(1)) \in H^{1,1}(X)$ to a 2-extension class $y(\alpha) \in \mathrm{Ext}^2_{\mathcal{O}_X}(L,L)$. Once this is achieved, the 2-form $\sigma_\mathcal{M}$ will become:

$$\mathrm{Ext}^1_{\mathcal{O}_X}(L,L) \otimes \mathrm{Ext}^1_{\mathcal{O}_X}(L,L) \xrightarrow{\text{Yoneda}} \mathrm{Ext}^2_{\mathcal{O}_X}(L,L) \xrightarrow{y(\alpha)^{n-1}} \mathrm{Ext}^{2n}_{\mathcal{O}_X}(L,L) \xrightarrow{\text{S.D.}}$$
$$\mathrm{Hom}_X(L, L \otimes \omega_X)^* \xrightarrow{id \otimes \sigma^n} \mathbb{C}. \tag{44}$$

Remark 8.17 When L is a line bundle on a smooth Lagrangian subvariety Z the construction involves a surprisingly rich polarized Hodge-like structure on the algebra

$$\mathrm{Ext}^*_{\mathcal{O}_X}(L,L) := \bigoplus_{k=0}^{2n} \mathrm{Ext}^k_{\mathcal{O}_X}(L,L).$$

Since $\mathrm{Ext}^k_{\mathcal{O}_X}(L,\cdot)$ is the right derived functor of the composition $\Gamma \circ \mathcal{H}om_{\mathcal{O}_X}(L,\cdot)$ of the Sheaf Hom and the global sections functors, there is a spectral sequence converging to $\mathrm{Ext}^k_{\mathcal{O}_X}(L,L)$ with

$$E_2^{p,q} = H^p\big(Z, \mathcal{E}xt^q_{\mathcal{O}_X}(L,L)\big)$$

(see [HS]). The sheaf of q-extensions $\mathcal{E}xt^q_{\mathcal{O}_X}(L,L)$ is canonically isomorphic to $\overset{q}{\bigwedge} N_{Z/X}$ and thus, via the symplectic structure, to Ω^q_Z. We obtain a canonical isomorphism $E_2^{p,q} \simeq H^{q,p}(Z)$ with the Dolbeault groups of Z. Notice however, that the Dolbeault groups appear in *reversed order* compared to their order in the graded pieces of the Hodge filtration on the cohomology ring $H^*(Z, \mathbb{C})$.

The construction of the 2-extension class $y(\alpha)$ and hence of the generalized Mukai pairing (44) can be carried out for all coherent sheaves parametrized by $\mathcal{M}^{s,sm}$. We obtain:

Theorem 8.18 *[Ma2] The symplectic structure $\sigma_\mathcal{M}$ on the relative Picard $\mathcal{M}$ extends to an algebraic 2-form over the smooth locus $\mathcal{M}^{s,sm}$ of the closure of $\mathcal{M}$ in the moduli space of stable sheaves on X. It is identified by the pairing (44).*

The non-degeneracy of $\sigma_\mathcal{M}$ at a point $[L] \in \mathcal{M}$ parametrizing a line bundle on a smooth Lagrangian subvariety Z follows from the Hard Lefschetz theorem. We expect $\sigma_\mathcal{M}$ to be non degenerate everywhere on $\mathcal{M}^{s,sm}$.

Finally we remark that the pairing (44) can be used to define a 2-form on other components of the moduli space of stable sheaves on X. This 2-form will, in general, be degenerate. In fact, some components are odd dimensional:

96

Example 8.19 Consider an odd dimensional complete linear system $|Z|$ whose generic element is a smooth ample divisor on an abelian variety X of even dimension ≥ 4, with a symplectic structure σ. The dimension of the component of the Hilbert scheme parameterizing deformations of Z is $\dim(\text{Pic } X) + \dim|Z| = \dim X + \dim|Z|$. Since $h^{1,0}(Z) = h^{1,0}(X)$, the component of the moduli space of sheaves parameterizing deformations of the structure sheaf $\mathcal{O}_Z$, as an $\mathcal{O}_X$-module, is of dimension $2 \cdot \dim X + \dim|Z|$ which is odd.

□

It is the Hodge theoretic interpretation of the graded pieces of the spectral sequence of $\text{Ext}^k_{\mathcal{O}_X}(L, L)$ for *Lagrangian* line bundles which assures the non degeneracy of $\sigma_{\mathcal{M}}$.

8.5 Examples

8.5.1 Higgs Pairs

In chapter 9 we define the notion of a 1-form valued Higgs pair (E, φ) over a smooth n-dimensional projective algebraic variety X. It consists of a torsion free sheaf E over X and a homomorphism $\varphi : E \to E \otimes \Omega^1_X$ satisfying the symmetry condition $\varphi \wedge \varphi = 0$.

The moduli space Higgs_X of semistable Higgs pairs of rank r with vanishing first and second Chern classes may be viewed as the Dolbeault non-abelian first $GL_r(\mathbb{C})$-cohomology group of X (cf. [Sim2] and theorem 4.7 when X is a curve):

Non-abelian Hodge theory introduces a hyperkahler structure on the smooth locus of the space $\mathcal{M}_{\text{Betti}}$ of isomorphism classes of semisimple $GL_r(\mathbb{C})$-representations of the fundamental group $\pi_1(X)$ of X [De2, H1, Sim3]. The hyperkahler structure consists of a Riemannian metric and an action of the quaternion algebra $\mathbb{H}$ on the real tangent bundle with respect to which

(i) the (purely imaginary) unit vectors $\{a|a\bar{a} = 1\}$ in $\mathbb{H}$ correspond to a (holomorphic) $\mathbb{P}^1$-family of integrable complex structures,

(ii) the metric is Kahler with respect to these complex structures.

All but two of the complex structures are isomorphic to that of $\mathcal{M}_{\text{Betti}}$, the two special ones are that of Higgs_X and its conjugate ($\mathcal{M}_{\text{Betti}}$ and Higgs_X are diffeomorphic).

The hyperkahler structure introduces a holomorphic symplectic structure σ on the smooth locus of Higgs_X. In case X is a Riemann surface, that symplectic structure is the one giving rise to the Hitchin integrable system of spectral Jacobians.

Our aim is to interpret the symplectic structure on Higgs_X as an example of a Lagrangian structure over the relative Picard of a Lagrangian component of the Hilbert scheme of the cotangent bundle T^*_X of X. This interpretation will apply to the Hitchin system (where $\dim X = 1$). For higher dimensional base varieties X it will apply only to certain particularly nice cases. See also [Bi] for a deformation theoretic study of the holomorphic symplectic structure.

The spectral construction (proposition 4.5) can be carried out also for Higgs pairs over a higher dimensional smooth projective variety X (cf. [Sim1]). We have a one to one correspondence between

(i) (Stable) Higgs pairs (E, φ) on X (allowing E to be a rank r torsion free sheaf) and

(ii) (Stable) sheaves F on the cotangent bundle T_X^* which are supported on (pure) n-dimensional projective subschemes of T_X^* which are finite, degree r, branched coverings (in a scheme theoretic sense) of X.

Projective subvarieties of T_X^* which are finite over X are called *spectral coverings*. Spectral coverings $\tilde{X}$ are necessarily Lagrangian since the symplectic form σ on T_X^*, which restricts to a global exact 2-form on $\tilde{X}$, must vanish on $\tilde{X}$.

Let B be the open subset of a component of the Hilbert scheme of $\mathbb{P}(T_X^* \oplus \mathcal{O}_X)$ parametrizing degree r smooth spectral coverings (closed subvarieties of $\mathbb{P}(T_X^* \oplus \mathcal{O}_X)$ which are contained in T_X^*). The above correspondence embeds components of the relative Picard $\mathcal{M} \to B$ as open subsets of components of the moduli spaces of stable rank r Higgs pairs over X. Theorem 8.1 of this chapter implies

Corollary 8.20 (i) *The open subset $\mathcal{M}$ of the moduli Higgs$_X$ of Higgs pairs over X which, under the spectral construction, parametrizes line bundles on smooth spectral covers, has a canonical symplectic structure $\sigma_{\mathcal{M}}$ (we do not require the Chern classes of the Higgs pairs to vanish).*

(ii) *The support morphism $h : \mathcal{M} \to B$ is a Lagrangian fibration.*

Remark 8.21 In general, when $\dim X > 1$, there could be components of the moduli spaces of Higgs pairs for which the open set $\mathcal{M}$ above is empty, i.e.,

1. the spectral coverings of all Higgs pairs in this component are singular, or

2. the corresponding sheaves on the spectral coverings are torsion free but not locally free.

8.5.2 Fano Varieties of Lines on Cubic Fourfolds

We will use theorem 8.1 to prove:

Example 8.22 *Let Y be a smooth cubic hypersurface in $\mathbb{P}^5$. The relative intermediate Jacobian $\mathcal{J} \to B$ over the family $B \subset |\mathcal{O}_{\mathbb{P}}(1)|$ of smooth cubic hyperplane sections of Y is an algebraically completely integrable Hamiltonian system.*

The statement follows from a description of the family $\mathcal{J} \to B$ as an open subset of the moduli space of Lagrangian sheaves on the Fano variety X of lines on Y.

A. Beauville and R. Donagi proved [BD] that X is symplectic (fourfold). Clemens and Griffiths proved in [CG] that the intermediate jacobian J_b of a smooth hyperplane section $Y \cap H_b$ is isomorphic to the Picard $Pic^0 Z_b$ of the 2-dimensional Fano variety Z_b of lines on the cubic 3-fold $Y \cap H_b$. C. Voisin observed that Z_b is a Lagrangian subvariety of X [V]. Since $h^{1,0}(Y \cap H_b) = 5$, B is isomorphic to a dense open subset of a component of the Hilbert scheme. In fact, using results of Altman and Kleiman, one can show that the corresponding component is isomorphic to $|\mathcal{O}_{\mathbb{P}}(1)|$ (see [AK1] Theorem 3.3 (iv)). Theorem 8.1 implies that the relative Picard $\mathcal{M} \to B$ has a completely integrable Hamiltonian system structure.

The symplectic structure $\sigma_{\mathcal{M}}$ is defined also at the fiber of the relative Picard corresponding to a Fano variety Z_b of lines on a hyperplane section $Y \cap H_b$ with an ordinary double point $x_b \in Y \cap H_b$ (Theorem 8.18). In that case, we have a genus 4 curve C_b in Z_b parametrizing lines through x_b. Z_b is isomorphic to the quotient $S^2 C_b/(C_1 \sim C_2)$ of the second symmetric product of C_b modulo the identification of two disjoint copies of C_b [CG]. It is not difficult to check that $\sigma_{\mathcal{M}}$ is *non-degenerate* also on the fiber $Pic^0(Z_b)$ of the relative Picard which is a $\mathbb{C}^\times$-extension of the Jacobian of genus 4. The non-degeneracy of the symplectic structure implies that we get an *induced boundary integrable system* on the relative Picard of the family of genus 4 curves

$$\mathcal{P}ic(\mathcal{C}) \to (Y^* - \Delta) \tag{45}$$

over the complement of the singular locus Δ of the dual variety of the cubic fourfold.

It is interesting to note that the boundary integrable system (45) can not be realized as the relative Picard of a family of curves on a symplectic surface. If this were the case, the generic rank of the pullback $a^*(\sigma_{\mathcal{P}ic^1(\mathcal{C})})$ of the symplectic structure from $\mathcal{P}ic^1(\mathcal{C}) \to (Y^* - \Delta)$ to $\mathcal{C} \to (Y^* - \Delta)$ via the Abel-Jacobi map would be 2. On the other hand, $a^*(\sigma_{\mathcal{P}ic^1(\mathcal{C})})$ is equal to the pullback of the symplectic structure σ_X on X via the natural dominant map $\mathcal{C} \to X$ (corollary 8.13). Thus, its generic rank is 4. The importance of this rank as an invariant of integrable systems supported by families of Jacobians is illustrated in an interesting recent study of J. Hurtubise [Hur].

More examples of nonrigid Lagrangian subvarieties can be found in [V, Ye].

9 Spectral covers

9.1 Algebraic extensions

We have seen that Hitchin's system, the geodesic flow on an ellipsoid, the polynomial matrices system of Chapter 4 , the elliptic solitons, and so on, all fit as special cases of the spectral system on a curve. In this final chapter, we consider some algebraic properties of the general spectral system. We are still considering families of Higgs pairs $(E \, , \, \varphi : E \longrightarrow E \otimes K)$, but we generalize in three separate directions:

1. The base curve C is replaced by an arbitrary complex algebraic variety S. The spectral curve $\tilde{C}$ then becomes a spectral cover $\tilde{S} \longrightarrow S$.

2. The line bundle K in which the endomorphism φ takes its values is replaced by a vector bundle, which we still denote by K. (this requires an integrability condition on φ.) Equivalently, $\tilde{S}$ is now contained in the total space $\mathbb{K}$ of a vector bundle over S.

3. Instead of the vector bundle E we consider a principal G-bundle $\mathcal{G}$, for an arbitrary complex reductive group G. The G-vector bundle E is then recovered as $E := \mathcal{G} \times^G V$, given the choice of a representation $\rho : G \longrightarrow Aut(V)$. The twisted endomorphism φ is replaced by a section of $K \otimes ad(\mathcal{G})$. Even in the original case of $G = GL(n)$ one encounters interesting phenomena in studying the dependence of $\tilde{S} := \tilde{S}_V$, for a given $(\mathcal{G}, \varphi)$, on the representation V of G.

We will see that essentially all *algebraic* properties (but not the *symplectic* structure) of the Hitchin system, or of the (line-bundle valued, $G = GL(n)$) spectral system on a curve, survive in this new context. In fact, the added generality forces the discovery of some symmetries which were not apparent in the original:

- Spectral curves are replaced by spectral covers. These come in several flavors: $\tilde{S}_V, \tilde{S}_\lambda, \tilde{S}_P$, indexed by representations of G, weights, and parabolic subgroups. The most basic object is clearly the *cameral* cover $\tilde{S}$; all the others can be considered as associated objects. In case $G = GL(n)$, the cameral cover specializes not to our previous spectral cover, which has degree n over S, but roughly to its Galois closure, of degree $n!$ over S.

- The spectral Picards, $Pic(\tilde{S}_V)$ etc., can all be written directly in terms of the decomposition of $Pic(\tilde{S})$ into Prym-type components under the action of the Weyl group W. In particular, there is a distinguished Prym component common to all the nontrivial $Pic(\tilde{S}_V)$. The identification of this component combines and unifies many interesting constructions in Prym theory.

- The Higgs bundle too can be relieved of its excess baggage. Stripping away the representation V as well as the values bundle K, one arrives (in subsection 9.3.1) at the notion of abstract, principal Higgs bundle. The abelianization procedure assigns to this a spectral datum, consisting of a cameral cover with an equivariant bundle on it.

- There is a Hitchin map (59) which is algebraically completely integrable in the sense that its fibers can be naturally identified, up to a "shift" and a "twist", with the distinguished Pryms (Theorem 9.10).

- The "shift" is a property of the group G, and is often nonzero even when $\tilde{S}$ is etale over S, cf. Proposition 9.8. The "twist", on the other hand, arises from the ramification of $\tilde{S}$ over S, cf. formula (57).

- The resulting abelianization procedure is local in the base S, and does not require particularly nice behavior near the ramification, cf. example 9.11. It does require that φ be regular (this means that its centralizer has the smallest possible dimension; for $GL(n)$, this means that each eigenvalue may have arbitrary multiplicity, but the eigen-*space* must be 1-dimensional), at least over the generic point of S. At present we can only guess at the situation for irregular Higgs bundles.

The consideration of general spectral systems is motivated in part by work of Hitchin [H2] and Simpson [Sim1]. In the remainder of this section we briefly recall those works. Our exposition in the following sections closely follows that of [D4], which in turn is based on [D2], for the group-theoretic approach to spectral decomposition used in section (9.2), and on [D3] for the Abelianization procedure, or equivalence of Higgs and spectral data, in section(9.3). Some of these results, especially in the case of a base curve, can also be found in [AvM, BK, F1, K, Me, MS, Sc].

Reductive groups

Principal G-bundles G for arbitrary reductive G were considered already in Hitchin's original paper [H2]. Fix a curve C and a line bundle K. There is a moduli space $\mathcal{M}_{G,K}$ parametrizing equivalence classes of semistable K-valued G-Higgs bundles, i.e. pairs $(\mathcal{G}, \varphi)$ with $\varphi \in K \otimes \mathfrak{ad}(\mathcal{G})$. The Hitchin map goes to

$$B := \oplus_i H^0(K^{\otimes d_i}),$$

where the d_i are the degrees of the f_i, a basis for the G-invariant polynomials on the Lie algebra $\mathfrak{g}$. It is:

$$h : (\mathcal{G}, \varphi) \longrightarrow (f_i(\varphi))_i.$$

When $K = \omega_C$, Hitchin showed [H2] that one still gets a completely integrable system, and that this system is algebraically completely integrable for the classical groups $GL(n), SL(n), SP(n), SO(n)$. The generic fibers are in each case (not quite canonically; one must choose various square roots! cf. sections 9.3.2 and 9.3.3) isomorphic to abelian varieties given in terms of the spectral curves $\tilde{C}$:

$GL(n)$ $\tilde{C}$ has degree n over C, the AV is $Jac(\tilde{C})$.

$SL(n)$ $\tilde{C}$ has degree n over C, the AV is $Prym(\tilde{C}/C)$.

$SP(n)$ $\tilde{C}$ has degree 2n over C and an involution $x \mapsto -x$.
        The map factors through the quotient $\overline{C}$.
        The AV is $Prym(\tilde{C}/\overline{C})$.

$SO(n)$ $\tilde{C}$ has degree n and an involution , with: (46)
        • a fixed component, when n is odd.
        • some fixed double points, when n is even.
        One must desingularize $\tilde{C}$ and the quotient $\overline{C}$,
        and ends up with the Prym of the
        desingularized double cover.

For the exceptional group G_2, the algebraic complete integrability was verified in [KP1]. A sketch of the argument for any reductive G is in [BK], and a complete proof was given in [F1]. We will outline a proof in section 9.3 below.

Higher dimensions

A sweeping extension of the notion of Higgs bundle is suggested by the work of Simpson [Sim1], which was already discussed in Chapter 8. To him, a Higgs bundle on a projective variety S is a vector bundle (or principal G-bundle ...) E with a *symmetric*, Ω_S^1-valued endomorphism

$$\varphi : E \longrightarrow E \otimes \Omega_S^1.$$

Here *symmetric* means the vanishing of:

$$\varphi \wedge \varphi : E \longrightarrow E \otimes \Omega_S^2,$$

a condition which is obviously vacuous on curves. Simpson constructs a moduli space for such Higgs bundles (satisfying appropriate stability conditions), and establishes diffeomorphisms to corresponding moduli spaces of connections and of representations of $\pi_1(S)$

In our approach, the Ω^1-valued Higgs bundle will be considered as a particular realization of an abstract Higgs bundle, given by a subalgebra of $ad(\mathcal{G})$. The symmetry condition will be expressed in the definition 9.5 of an abstract Higgs bundle by requiring the abelian subalgebras of $ad(\mathcal{G})$ to be abelian.

9.2 Decomposition of spectral Picards

9.2.1 The question

Throughout this section we fix a vector bundle K on a complex variety S, and a pair $(\mathcal{G}, \varphi)$ where $\mathcal{G}$ is a principal G-bundle on S and φ is a regular section of $K \otimes ad(\mathcal{G})$. (This data is equivalent to the regular case of what we call in section 9.3.1 a K-valued principal Higgs bundle.) Each representation

$$\rho : G \longrightarrow Aut(V)$$

determines an associated K-valued Higgs (vector) bundle

$$(\mathcal{V} := \mathcal{G} \times^G V, \qquad \rho(\varphi) \ ),$$

which in turn determines a spectral cover $\tilde{S}_V \longrightarrow S$.

The question, raised first in [AvM] when $S = \mathbf{P}^1$, is to relate the Picard varieties of the $\tilde{S}_V$ as V varies, and in particular to find pieces common to all of them. For Adler and van Moerbeke, the motivation was that many evolution DEs (of Lax type) can be *linearized* on the Jacobians of spectral curves. This means that the "Liouville tori", which live naturally in the complexified domain of the DE (and hence are independent of the representation V) are mapped isogenously to their image in $\text{Pic}(\tilde{S}_V)$ for each nontrivial V ; so one should be able to locate these tori among the pieces which occur in an isogeny decomposition of each of the $\text{Pic}(\tilde{S}_V)$. There are many specific examples where a pair of abelian varieties constructed from related covers of curves are known to be isomorphic or isogenous, and some of these lead to important identities among theta functions.

Example 9.1 Take $G = SL(4)$. The standard representation V gives a branched cover $\tilde{S}_V \longrightarrow S$ of degree 4. On the other hand, the 6-dimensional representation $\wedge^2 V$ (=the standard representation of the isogenous group $SO(6)$) gives a cover $\tilde{\tilde{S}} \longrightarrow S$ of degree 6, which factors through an involution:

$$\tilde{\tilde{S}} \longrightarrow \overline{S} \longrightarrow S.$$

One has the isogeny decompositions:

$$Pic\,(\tilde{S}) \sim Prym(\tilde{S}/S) \oplus Pic\,(S)$$

$$Pic\,(\tilde{\tilde{S}}) \sim Prym(\tilde{\tilde{S}}\,/\overline{S}) \oplus Prym(\overline{S}/S) \oplus Pic\,(S).$$

It turns out that

$$Prym(\tilde{S}/S) \sim Prym(\tilde{\tilde{S}}\,/\overline{S}).$$

For $S = \mathbf{P}^1$, this is Recillas' *trigonal construction* [Rec]. It says that every Jacobian of a trigonal curve is the Prym of a double cover of a tetragonal curve, and vice versa.

Example 9.2 Take $G = SO(8)$ with its standard 8-dimensional representation V. The spectral cover has degree 8 and factors through an involution, $\tilde{\tilde{S}} \longrightarrow \overline{S} \longrightarrow S$. The two half-spin representations V_1, V_2 yield similar covers

$$\tilde{\tilde{S}}_i \longrightarrow \overline{S}_i \longrightarrow S, \qquad i = 1, 2.$$

The *tetragonal construction* [D1] says that the three Pryms of the double covers are isomorphic. (These examples, as well as Pantazis' *bigonal construction* and constructions based on some exceptional groups, are discussed in the context of spectral covers in [K] and [D2].)

It turns out in general that there is indeed a distinguished, Prym-like isogeny component common to all the spectral Picards, on which the solutions to Lax-type DEs evolve linearly. This was noticed in some cases already in [AvM], and was greatly extended by Kanev's construction of Prym-Tyurin varieties. (He still needs S to be $\mathbf{P}^1$ and the spectral cover to have generic ramification; some of his results apply only to *minuscule representations*.) Various parts of the general story have been worked out recently by a number of authors, based on either of two approaches: one, pursued in [D2, Me, MS], is to decompose everything according to the action of the Weyl group W and to look for common pieces; the other, used in [BK, D3, F1, Sc], relies on the correspondence of spectral data and Higgs bundles . The group-theoretic approach is described in the rest of this section. We take up the second method, known as *abelianization*, in section 9.3.

9.2.2 Decomposition of spectral covers

The decomposition of spectral Picards arises from three sources. First, the spectral cover for a sum of representations is the union of the individual covers $\tilde{S}_V$. Next, the cover $\tilde{S}_V$ for an irreducible representation is still the union of subcovers $\tilde{S}_\lambda$ indexed by weight orbits. And finally, the Picard of $\tilde{S}_\lambda$ decomposes into Pryms. We start with a few observations about the dependence of the covers themselves on the representation. The decomposition of the Picards is taken up in the next subsection.

Spectral covers
Whenever a representation space V of G decomposes,

$$V = \oplus V_i,$$

there is a corresponding decomposition

$$\tilde{S}_V = \cup \tilde{S}_{V_i},$$

so we may restrict attention to irreducible representations V. There is an *infinite* collection (of irreducible representations $V := V_\mu$, hence) of spectral covers $\tilde{S}_V$, which can be parametrized by their highest weights μ in the dominant Weyl chamber $\overline{C}$, or equivalently by the W-orbit of extremal weights, in Λ/W. Here T is a maximal torus in G, $\Lambda := Hom(T, \mathbf{C}^*)$ is the *weight lattice* (also called *character lattice*) for G, and W is the Weyl group. Now V_μ decomposes as the sum of its weight subspaces V_μ^λ, indexed by certain weights λ in the convex hull in Λ of the W-orbit of μ. We conclude that each $\tilde{S}_{V_\mu}$ itself decomposes as the union of its subcovers $\tilde{S}_\lambda$, each of which involves eigenvalues in a given W-orbit $W\lambda$. (λ runs over the weight-orbits in V_μ.)

Parabolic covers
There is a *finite* collection of covers $\tilde{S}_P$, parametrized by the conjugacy classes in G of parabolic subgroups (or equivalently by arbitrary dimensional faces F_P of the chamber $\overline{C}$) such that (for general S) each eigenvalue cover $\tilde{S}_\lambda$ is birational to some parabolic cover $\tilde{S}_P$, the one whose open face F_P contains λ.

The cameral cover

There is a W-Galois cover $\tilde{S} \longrightarrow S$ such that each $\tilde{S}_P$ is isomorphic to $\tilde{S}/W_P$, where W_P is the Weyl subgroup of W which stabilizes F_P. We call $\tilde{S}$ the *cameral cover*, since, at least generically, it parametrizes the chambers determined by φ (in the duals of the Cartans). Informally, we think of $\tilde{S} \longrightarrow S$ as the cover which associates to a point $s \in S$ the set of Borel subalgebras of $ad(\mathcal{G})_s$ containing $\phi(s)$. More carefully, this is constructed as follows: There is a morphism $\mathfrak{g} \longrightarrow \mathfrak{t}/W$ sending $g \in \mathfrak{g}$ to the conjugacy class of its semisimple part g_{ss}. (More precisely, this is $Spec$ of the composed ring homomorphism $\mathbb{C}[\mathfrak{t}]^W \tilde{\leftarrow} \mathbb{C}[\mathfrak{g}]^G \hookrightarrow \mathbb{C}[\mathfrak{g}]$.) Taking fiber product with the quotient map $\mathfrak{t} \longrightarrow \mathfrak{t}/W$, we get the cameral cover $\tilde{\mathfrak{g}}$ of $\mathfrak{g}$. The cameral cover $\tilde{S} \longrightarrow S$ of a K-valued principal Higgs bundle on S is glued from covers of open subsets in S (on which K and $\mathcal{G}$ are trivialized) which in turn are pullbacks by φ of $\tilde{\mathfrak{g}} \longrightarrow \mathfrak{g}$.

9.2.3 Decomposition of spectral Picards

The decomposition of the Picard varieties of spectral covers can be described as follows:

The cameral Picard

From each isomorphism class of irreducible W-representations, choose an integral representative Λ_i. (This can always be done, for Weyl groups.) The group ring $\mathbb{Z}[W]$ which acts on $Pic(\tilde{S})$ has an isogeny decomposition:

$$\mathbb{Z}[W] \sim \oplus_i \Lambda_i \otimes_{\mathbb{Z}} \Lambda_i^*, \tag{47}$$

which is just the decomposition of the regular representation. There is a corresponding isotypic decomposition:

$$Pic(\tilde{S}) \sim \oplus_i \Lambda_i \otimes_{\mathbb{Z}} Prym_{\Lambda_i}(\tilde{S}), \tag{48}$$

where

$$Prym_{\Lambda_i}(\tilde{S}) := Hom_W(\Lambda_i, Pic(\tilde{S})). \tag{49}$$

Parabolic Picards

There are at least three reasonable ways of obtaining an isogeny decomposition of $Pic(\tilde{S}_P)$, for a parabolic subgroup $P \subset G$:

- The 'Hecke' ring $Corr_P$ of correspondences on $\tilde{S}_P$ over S acts on $Pic(\tilde{S}_P)$, so every irreducible integral representation M of $Corr_P$ determines a generalized Prym

$$Hom_{Corr_P}(M, Pic(\tilde{S}_P)),$$

and we obtain an isotypic decomposition of $Pic(\tilde{S}_P)$ as before.

- $Pic(\tilde{S}_P)$ maps, with torsion kernel, to $Pic(\tilde{S})$, so we obtain a decomposition of the former by intersecting its image with the isotypic components $\Lambda_i \otimes_{\mathbb{Z}} Prym_{\Lambda_i}(\tilde{S})$ of the latter.

- Since $\tilde{S}_P$ is the cover of S *associated* to the W-cover $\tilde{S}$ via the permutation representation $Z[W_P \backslash W]$ of W, we get an isogeny decomposition of $Pic(\tilde{S}_P)$ indexed by the irreducible representations in $Z[W_P \backslash W]$.

It turns out ([D2],section 6) that all three decompositions agree and can be given explicitly as

$$\oplus_i M_i \otimes Prym_{\Lambda_i}(\tilde{S}) \subset \oplus_i \Lambda_i \otimes Prym_{\Lambda_i}(\tilde{S}), \qquad M_i := (\Lambda_i)^{W_P}. \qquad (50)$$

Spectral Picards

To obtain the decomposition of the Picards of the original covers $\tilde{S}_V$ or $\tilde{S}_\lambda$, we need, in addition to the decomposition of $Pic(\tilde{S}_P)$, some information on the singularities. These can arise from two separate sources:

Accidental singularities of the $\tilde{S}_\lambda$. For a sufficiently general Higgs bundle, and for a weight λ in the interior of the face F_P of the Weyl chamber $\overline{C}$, the natural map:

$$i_\lambda : \tilde{S}_P \longrightarrow \tilde{S}_\lambda$$

is birational. For the *standard* representations of the classical groups of types A_n, B_n or C_n, this *is* an isomorphism. But for general λ it is *not*: In order for i_λ to be an isomorphism, λ must be a multiple of a fundamental weight, cf. [D2], lemma 4.2. In fact, the list of fundamental weights for which this happens is quite short; for the classical groups we have only: ω_1 for A_n, B_n and C_n, ω_n (the dual representation) for A_n, and ω_2 for B_2. Note that for D_n the list is *empty*. In particular, the covers produced by the standard representation of $SO(2n)$ are singular; this fact, noticed by Hitchin In [H2], explains the need for desingularization in his result (46).

Gluing the $\tilde{S}_V$. In addition to the singularities of each i_λ, there are the singularities created by the gluing map $\amalg_\lambda \tilde{S}_\lambda \longrightarrow \tilde{S}_V$. This makes explicit formulas somewhat simpler in the case, studied by Kanev [K], of *minuscule* representations, i.e. representations whose weights form a single W-orbit. These singularities account, for instance, for the desingularization required in the $SO(2n+1)$ case in (46).

9.2.4 The distinguished Prym

Combining much of the above, the Adler–van Moerbeke problem of finding a component common to the $Pic(\tilde{S}_V)$ for all non-trivial V translates into:

Find the non trivial irreducible representations Λ_i of W which occur in $Z[W_P \backslash W]$ with positive multiplicity for all proper Weyl subgroups $W_P \subsetneq W$.

It is easy to see that for arbitrary finite groups W, or even for Weyl groups W if we allow arbitrary rather than Weyl subgroups W_P, there may be no common factors [D2]. For example, when W is the symmetric group S_3 (=the Weyl group of $GL(3)$) and W_P is S_2 or A_3, the representations $Z[W_P \backslash W]$ are 3 or 2 dimensional, respectively, and have only the trivial representation as common component. In any case, our problem is equivalent (by Frobenius reciprocity or (50)) to

Find the irreducible representations Λ_i *of* W *such that for every proper Weyl subgroup* $W_P \subsetneq W$, *the space of invariants* $M_i := (\Lambda_i)^{W_P}$ *is non-zero.*

One solution is now obvious: the *reflection representation* of W acting on the weight lattice Λ has this property. In fact, Λ^{W_P} in this case is just the face F_P of $\overline{C}$. The corresponding component $Prym_\Lambda(\tilde{S})$, is called *the distinguished Prym*. We will see in section 9.3 that its points correspond, modulo some corrections, to Higgs bundles.

For the classical groups, this turns out to be the only common component. For G_2 and E_6 it turns out ([D2], section 6) that a second common component exists. The geometric significance of points in these extra components is not known. As far as we know, the only component other than the distinguished Prym which has arisen 'in nature' is the one associated to the 1-dimensional sign representation of W, cf. [KP2].

9.3 Abelianization

9.3.1 Abstract vs. K-valued objects

We want to describe the abelianization procedure in a somewhat abstract setting, as an equivalence between *principal Higgs bundles* and certain *spectral data*. Once we fix a *values* vector bundle K, we obtain an equivalence between K-*valued principal Higgs bundles* and K-*valued spectral data*. Similarly, the choice of a representation V of G will determine an equivalence of K-*valued Higgs bundles* (of a given representation type) with K-valued spectral data.

As our model of a W-cover we take the natural quotient map

$$G/T \longrightarrow G/N$$

and its partial compactification

$$\overline{G/T} \longrightarrow \overline{G/N}. \tag{51}$$

Here $T \subset G$ is a maximal torus, and N is its normalizer in G. The quotient G/N parametrizes maximal tori (=Cartan subalgebras) $\mathfrak{t}$ in $\mathfrak{g}$, while G/T parametrizes pairs $\mathfrak{t} \subset \mathfrak{b}$ with $\mathfrak{b} \subset \mathfrak{g}$ a Borel subalgebra. An element $x \in \mathfrak{g}$ is *regular* if the dimension of its centralizer $\mathfrak{c} \subset \mathfrak{g}$ equals $\dim T$ (=the rank of $\mathfrak{g}$). The partial compactifications $\overline{G/N}$ and $\overline{G/T}$ parametrize regular centralizers $\mathfrak{c}$ and pairs $\mathfrak{c} \subset \mathfrak{b}$, respectively.

In constructing the cameral cover in section 9.2.2, we used the W-cover $\mathfrak{t} \longrightarrow \mathfrak{t}/W$ and its pullback cover $\tilde{\mathfrak{g}} \longrightarrow \mathfrak{g}$. Over the open subset $\mathfrak{g}_{reg}$ of regular elements, the same cover is obtained by pulling back (51) via the map $\alpha : \mathfrak{g}_{reg} \longrightarrow \overline{G/N}$ sending an element to its centralizer:

$$
\begin{array}{ccccc}
\mathfrak{t} & \longleftarrow & \tilde{\mathfrak{g}}_{reg} & \longrightarrow & \overline{G/T} \\
\downarrow & & \downarrow & & \downarrow \\
\mathfrak{t}/W & \longleftarrow & \mathfrak{g}_{reg} & \stackrel{\alpha}{\longrightarrow} & \overline{G/N} \ .
\end{array}
\tag{52}
$$

When working with K-valued objects, it is usually more convenient to work with the left hand side of (52), i.e. with eigen*values*. When working with the abstract objects, this is unavailable, so we are forced to work with the eigen*vectors*, or the right hand side of (52). Thus:

Definition 9.3 *An abstract* cameral cover *of S is a finite morphism $\tilde{S} \longrightarrow S$ with W- action, which locally (etale) in S is a pullback of (51).*

Definition 9.4 *A K-valued cameral cover (K is a vector bundle on S) consists of a cameral cover $\pi : \tilde{S} \longrightarrow S$ together with an S-morphism*

$$\tilde{S} \times \Lambda \longrightarrow \mathbb{K} \tag{53}$$

which is W-invariant (W acts on $\tilde{S}, \Lambda$, hence diagonally on $\tilde{S} \times \Lambda$) and linear in Λ.

We note that a cameral cover $\tilde{S}$ determines quotients $\tilde{S}_P$ for parabolic subgroups $P \subset G$. A K-valued cameral cover determines additionally the $\tilde{S}_\lambda$ for $\lambda \in \Lambda$, as images in $\mathbb{K}$ of $\tilde{S} \times \{\lambda\}$. The data of (53) is equivalent to a W-equivariant map $\tilde{S} \longrightarrow \mathfrak{t} \otimes_{\mathbb{C}} K$.

Definition 9.5 *A G-principal Higgs bundle on S is a pair $(\mathcal{G}, c)$ with $\mathcal{G}$ a principal G- bundle and $c \subset ad(\mathcal{G})$ a subbundle of regular centralizers.*

Definition 9.6 *A K-valued G-principal Higgs bundle consists of $(\mathcal{G}, c)$ as above together with a section φ of $c \otimes K$.*

A principal Higgs bundle $(\mathcal{G}, c)$ determines a cameral cover $\tilde{S} \longrightarrow S$ and a homomorphism $\Lambda \longrightarrow \mathrm{Pic}(\tilde{S})$. Let F be a parameter space for Higgs bundles with a given $\tilde{S}$. Each non-zero $\lambda \in \Lambda$ gives a non-trivial map $F \longrightarrow \mathrm{Pic}(\tilde{S})$. For λ in a face F_P of $\overline{C}$, this factors through $\mathrm{Pic}(\tilde{S}_P)$. The discussion in section 9.2.4 now suggests that F itself should be given roughly by the distinguished Prym,

$$Hom_W(\Lambda, \mathrm{Pic}(\tilde{S})).$$

It turns out that this guess needs two corrections. The first correction involves restricting to a coset of a subgroup; the need for this is visible even in the simplest case where $\tilde{S}$ is etale over S, so $(\mathcal{G}, c)$ is everywhere regular and semisimple (i.e. c is a bundle of Cartans.) The second correction involves a twist along the ramification of $\tilde{S}$ over S. We explain these in the next two subsections.

9.3.2 The regular semisimple case: the shift

Example 9.7 Fix a smooth projective curve C and a line bundle $K \in \mathrm{Pic}(C)$ such that $K^{\otimes 2} \approx \mathcal{O}_C$. This determines an etale double cover $\pi : \tilde{C} \longrightarrow C$ with involution i, and homomorphisms

$$
\begin{aligned}
\pi^* &: \mathrm{Pic}(C) &\longrightarrow& \mathrm{Pic}(\tilde{C}) \ , \\
\mathrm{Nm} &: \mathrm{Pic}(\tilde{C}) &\longrightarrow& \mathrm{Pic}(C) \ , \\
i^* &: \mathrm{Pic}(\tilde{C}) &\longrightarrow& \mathrm{Pic}(\tilde{C}) \ ,
\end{aligned}
$$

satisfying

$$1 + i^* = \pi^* \circ \mathrm{Nm}.$$

- For $G = GL(2)$ we have $\Lambda = \mathbf{Z} \oplus \mathbf{Z}$, and $W = S_2$ permutes the summands, so

$$Hom_W(\Lambda, \mathrm{Pic}(\widetilde{C})) \approx \mathrm{Pic}(\widetilde{C}).$$

And indeed, the Higgs bundles corresponding to $\widetilde{C}$ are parametrized by $\mathrm{Pic}(\widetilde{C})$: send $L \in \mathrm{Pic}(\widetilde{C})$ to $(\mathcal{G}, \mathrm{c})$, where $\mathcal{G}$ has associated rank-2 vector bundle $\mathcal{V} := \pi_* L$, and $\mathrm{c} \subset \underline{End}(\mathcal{V})$ is $\pi_* \mathcal{O}_{\widetilde{C}}$.

- On the other hand, for $G = SL(2)$ we have $\Lambda = \mathbf{Z}$ and $W = S_2$ acts by ± 1, so

$$Hom_W(\Lambda, \mathrm{Pic}(\widetilde{S})) \approx \{L \in \mathrm{Pic}(\widetilde{C}) \mid i^* L \approx L^{-1}\} = \ker(1 + i^*).$$

This group has 4 connected components. The subgroup $\ker(\mathrm{Nm})$ consists of 2 of these. The connected component of 0 is the classical Prym variety, cf. [Mum3]. Now the Higgs bundles correspond, via the above bijection $L \mapsto \pi_* L$, to

$$\{L \in \mathrm{Pic}(\widetilde{C}) \mid \det(\pi_* L) \approx \mathcal{O}_C\} = \mathrm{Nm}^{-1}(K).$$

Thus they form the *non-zero* coset of the subgroup $\ker(\mathrm{Nm})$. (If we return to a higher dimensional S, there is no change in the $GL(2)$ story, but it is possible for K not to be in the image of Nm, so there may be *no* $SL(2)$-Higgs bundles corresponding to such a cover.)

This example generalizes to all G, as follows. The equivalence classes of extensions

$$1 \longrightarrow T \longrightarrow N' \longrightarrow W \longrightarrow 1$$

(in which the action of W on T is the standard one) are parametrized by the group cohomology $H^2(W, T)$. Here the 0 element corresponds to the semidirect product . The class $[N] \in H^2(W, T)$ of the normalizer N of T in G may be 0, as it is for $G = GL(n), \mathbf{P}GL(n), SL(2n+1)$; or not, as for $G = SL(2n)$.

Assume first, for simplicity, that $S, \widetilde{S}$ are connected and projective. There is then a natural group homomorphism

$$c: Hom_W(\Lambda, \mathrm{Pic}(\widetilde{S})) \longrightarrow H^2(W, T). \tag{54}$$

Algebraically, this is an edge homomorphism for the Grothendieck spectral sequence of equivariant cohomology, which gives the exact sequence

$$0 \longrightarrow H^1(W, T) \longrightarrow H^1(S, \mathcal{C}) \longrightarrow Hom_W(\Lambda, \mathrm{Pic}(\widetilde{S})) \xrightarrow{c} H^2(W, T). \tag{55}$$

where $\mathcal{C} := \widetilde{S} \times_W T$. Geometrically, this expresses a *Mumford group* construction: giving $\mathcal{L} \in Hom(\Lambda, \mathrm{Pic}(\widetilde{S}))$ is equivalent to giving a principal T-bundle $\mathcal{T}$ over $\widetilde{S}$; for $\mathcal{L} \in Hom_W(\Lambda, \mathrm{Pic}(\widetilde{S}))$, $c(\mathcal{L})$ is the class in $H^2(W, T)$ of the group N' of automorphisms of $\mathcal{T}$ which commute with the action on $\widetilde{S}$ of some $w \in W$.

To remove the restriction on $S, \widetilde{S}$, we need to replace each occurrence of T in (54,55) by $\Gamma(\widetilde{S}, T)$, the global sections of the trivial bundle on $\widetilde{S}$ with fiber T. The natural map $H^2(W, T) \longrightarrow H^2(W, \Gamma(\widetilde{S}, T))$ allows us to think of $[N]$ as an element of $H^2(W, \Gamma(\widetilde{S}, T))$.

Proposition 9.8 *[D3] Fix an etale W-cover $\pi : \tilde{S} \longrightarrow S$. The following data are equivalent:*

1. *Principal G-Higgs bundles $(\mathcal{G}, c)$ with cameral cover $\tilde{S}$.*

2. *Principal N-bundles $\mathcal{N}$ over S whose quotient by T is $\tilde{S}$.*

3. *W-equivariant homomorphisms $\mathcal{L} : \Lambda \longrightarrow Pic(\tilde{S})$ with $c(\mathcal{L}) = [N] \in H^2(W, \Gamma(\tilde{S}, T))$.*

We observe that while the shifted objects correspond to Higgs bundles, the unshifted objects

$$\mathcal{L} \in \mathrm{Hom}_W(\Lambda, \mathrm{Pic}(\tilde{S})), \qquad c(\mathcal{L}) = 0$$

come from the C-torsers in $H^1(S, C)$.

9.3.3 The regular case: the twist along the ramification

Example 9.9 Modify example 9.7 by letting $K \in \mathrm{Pic}(C)$ be arbitrary, and choose a section b of $K^{\otimes 2}$ which vanishes on a simple divisor $B \subset C$. We get a double cover $\pi : \tilde{C} \longrightarrow C$ branched along B, ramified along a divisor

$$R \subset \tilde{C}, \quad \pi(R) = B.$$

Via $L \mapsto \pi_* L$, the $SL(2)$-Higgs bundles still correspond to

$$\{L \in \mathrm{Pic}(\tilde{C}) \mid \det(\pi_* L) \approx \mathcal{O}_C\} = \mathrm{Nm}^{-1}(K).$$

But this is no longer in $Hom_W(\Lambda, \mathrm{Pic}(\tilde{S}))$; rather, the line bundles in question satisfy

$$i^* L \approx L^{-1}(R). \tag{56}$$

For arbitrary G, let Φ denote the root system and Φ^+ the set of positive roots. There is a decomposition

$$\overline{G/T} \smallsetminus G/T = \bigcup_{\alpha \in \Phi^+} R_\alpha$$

of the boundary into components, with R_α the fixed locus of the reflection σ_α in α. (Via (52), these correspond to the complexified walls in $\mathfrak{t}$.) Thus each cameral cover $\tilde{S} \longrightarrow S$ comes with a natural set of (Cartier) *ramification divisors*, which we still denote R_α, $\alpha \in \Phi^+$.

For $w \in W$, set

$$F_w := \left\{ \alpha \in \Phi^+ \mid w^{-1}\alpha \in \Phi^- \right\} = \Phi^+ \cap w\Phi^-,$$

and choose a W-invariant form $\langle, \rangle$ on Λ. We consider the variety

$$Hom_{W,R}(\Lambda, \mathrm{Pic}(\tilde{S}))$$

of R-twisted W-equivariant homomorphisms, i.e. homomorphisms $\mathcal{L}$ satisfying

$$w^*\mathcal{L}(\lambda) \approx \mathcal{L}(w\lambda)\left(\sum_{\alpha \in F_w} \frac{\langle -2\alpha, w\lambda\rangle}{\langle \alpha, \alpha\rangle} R_\alpha\right), \qquad \lambda \in \Lambda, \quad w \in W. \qquad (57)$$

This turns out to be the correct analogue of (56). (E.g. for a reflection $w = \sigma_\alpha$, F_w is $\{\alpha\}$, so this gives $w^*\mathcal{L}(\lambda) \approx \mathcal{L}(w\lambda)\left(\frac{\langle \alpha, 2\lambda\rangle}{\langle \alpha, \alpha\rangle} R_\alpha\right)$, which specializes to (56).) As before, there is a class map

$$c : Hom_{W,R}(\Lambda, \text{Pic}(\widetilde{S})) \longrightarrow H^2(W, \, \Gamma(\widetilde{S}, T)) \qquad (58)$$

which can be described via a Mumford-group construction.

To understand this twist, consider the formal object

$$\begin{aligned} \tfrac{1}{2}\text{Ram} : \Lambda &\longrightarrow \mathbf{Q} \otimes \text{Pic}\widetilde{S}, \\ \lambda &\longmapsto \textstyle\sum_{(\alpha \in \Phi+)} \frac{\langle \alpha, \lambda\rangle}{\langle \alpha, \alpha\rangle} R_\alpha. \end{aligned}$$

In an obvious sense, a principal T-bundle $\mathcal{T}$ on $\widetilde{S}$ (or a homomorphism $\mathcal{L} : \Lambda \longrightarrow \text{Pic}(\widetilde{S})$) is R-twisted W-equivariant if and only if $\mathcal{T}(-\tfrac{1}{2}\text{Ram})$ is W-equivariant, i.e. if $\mathcal{T}$ and $\tfrac{1}{2}\text{Ram}$ transform the same way under W. The problem with this is that $\tfrac{1}{2}\text{Ram}$ itself does not make sense as a T-bundle, because the coefficients $\frac{\langle \alpha, \lambda\rangle}{\langle \alpha, \alpha\rangle}$ are not integers. (This argument shows that if $Hom_{W,R}(\Lambda, \text{Pic}(\widetilde{S}))$ is non-empty, it is a torser over the untwisted $Hom_W(\Lambda, \text{Pic}(\widetilde{S}))$.)

Theorem 9.10 *[D3] For a cameral cover $\widetilde{S} \longrightarrow S$, the following data are equivalent:*
(1) G-principal Higgs bundles with cameral cover $\widetilde{S}$.
(2) R-twisted W-equivariant homomorphisms $\mathcal{L} \in c^{-1}([N])$.

The theorem has an essentially local nature, as there is no requirement that S be, say, projective. We also do not need the condition of generic behavior near the ramification, which appears in [F1, Me, Sc]. Thus we may consider an extreme case, where $\widetilde{S}$ is 'everywhere ramified':

Example 9.11 In example 9.9, take the section $b = 0$. The resulting cover $\widetilde{C}$ is a 'ribbon', or length-2 non-reduced structure on C: it is the length-2 neighborhood of C in $\mathbb{K}$. The SL(2)-Higgs bundles $(\mathcal{G}, c)$ for this $\widetilde{C}$ have an everywhere nilpotent c, so the vector bundle $\mathcal{V} := \mathcal{G} \times^{SL(2)} V \approx \pi_* L$ (where V is the standard 2-dimensional representation) fits in an exact sequence

$$0 \longrightarrow \mathcal{S} \longrightarrow \mathcal{V} \longrightarrow \mathcal{Q} \longrightarrow 0$$

with $\mathcal{S} \otimes K \approx \mathcal{Q}$. Such data are specified by the line bundle $\mathcal{Q}$, satisfying $\mathcal{Q}^{\otimes 2} \approx K$, and an extension class in $\text{Ext}^1(\mathcal{Q}, \mathcal{S}) \approx H^1(K^{-1})$. The kernel of the restriction map $\text{Pic}(\widetilde{C}) \longrightarrow \text{Pic}(C)$ is also given by $H^1(K^{-1})$ (use the exact sequence $0 \longrightarrow K^{-1} \longrightarrow \pi_*\mathcal{O}_{\widetilde{C}}^\times \longrightarrow \mathcal{O}_C^\times \longrightarrow 0$), and the R-twist produces the required square roots of K. (For more details on the nilpotent locus, cf. [L] and [DEL].)

9.3.4 Adding values and representations

Fix a vector bundle K, and consider the moduli space $\mathcal{M}_{S,G,K}$ of K-valued G-principal Higgs bundles on S. (It can be constructed as in Simpson's [Sim1], even though the objects we need to parametrize are slightly different than his. In this subsection we sketch a direct construction.) It comes with a Hitchin map:

$$h : \mathcal{M}_{S,G,K} \longrightarrow B_K \qquad (59)$$

where $B := B_K$ parametrizes all possible Hitchin data. Theorem 9.10 gives a precise description of the fibers of this map, independent of the values bundle K. This leaves us with the relatively minor task of describing, for each K, the corresponding base, i.e. the closed subvariety B_s of B parametrizing *split* Hitchin data, or K-valued cameral covers. The point is that Higgs bundles satisfy a symmetry condition, which in Simpson's setup is

$$\varphi \wedge \varphi = 0,$$

and is built into our definition 9.5 through the assumption that c is a bundle of regular centralizers, hence is abelian. Since commuting operators have common eigenvectors, this is translated into a splitness condition on the Hitchin data, which we describe below. (When K is a line bundle, the condition is vacuous, $B_s = B$.) The upshot is:

Lemma 9.12 *The following data are equivalent:*
(a) A K-valued cameral cover of S.
(b) A split, graded homomorphism $R^{\cdot} \longrightarrow Sym^{\cdot}K$.
(c) A split Hitchin datum $b \in B_s$.

Here $R^{\cdot}$ is the graded ring of W-invariant polynomials on $\mathfrak{t}$:

$$R^{\cdot} := (Sym^{\cdot}\mathfrak{t}^{*})^{W} \approx \mathbf{C}[\sigma_1, \ldots, \sigma_l], \qquad \deg(\sigma_i) = d_i \qquad (60)$$

where $l := \text{Rank}(\mathfrak{g})$ and the σ_i form a basis for the W-invariant polynomials. The Hitchin base is the vector space

$$B := B_K := \oplus_{i=1}^{l} H^0(S, Sym^{d_i}K) \approx \text{Hom}(R^{\cdot}, Sym^{\cdot}K).$$

For each $\lambda \in \Lambda$ (or $\lambda \in \mathfrak{t}^{*}$, for that matter), the expression in an indeterminate x:

$$q_\lambda(x,t) := \prod_{w \in W} (x - w\lambda(t)), \qquad t \in \mathfrak{t}, \qquad (61)$$

is W-invariant (as a function of t), so it defines an element $q_\lambda(x) \in R^{\cdot}[x]$. A Hitchin datum $b \in B \approx \text{Hom}(R^{\cdot}, Sym^{\cdot}K)$ sends this to

$$q_{\lambda,b}(x) \in Sym^{\cdot}(K)[x].$$

We say that b is *split* if, at each point of S and for each λ, the polynomial $q_{\lambda,b}(x)$ factors completely, into terms linear in x.

We note that, for λ in the interior of C (the positive Weyl chamber), $q_{\lambda,b}$ gives the equation in $\mathbf{K}$ of the spectral cover $\tilde{S}_\lambda$ of section (9.2.2): $q_{\lambda,b}$ gives a morphism $\mathbf{K} \longrightarrow$

$\mathrm{Sym}^N \mathbb{K}$, where $N := \#W$, and $\widetilde{S}_\lambda$ is the inverse image of the zero-section. (When λ is in a face F_P of $\overline{C}$, we define analogous polynomials $q_\lambda^P(x,t)$ and $q_{\lambda,b}^P(x)$ by taking the product in (61) to be over $w \in W_P \backslash W$. These give the reduced equations in this case, and q_λ is an appropriate power.)

Over B_s there is a universal K-valued cameral cover

$$\widetilde{S} \longrightarrow B_s$$

with ramification divisor $R \subset \widetilde{S}$. From the relative Picard,

$$\mathrm{Pic}(\widetilde{S}/B_s)$$

we concoct the relative N-shifted, R-twisted Prym

$$\mathrm{Prym}_{\Lambda,R}(\widetilde{S}/B_s).$$

By Theorem (9.10), this can then be considered as a parameter space $\mathcal{M}_{S,G,K}$ for all K-valued G-principal Higgs bundles on S. (Recall that our objects are assumed to be everywhere *regular*!) It comes with a 'Hitchin map', namely the projection to B_s, and the fibers corresponding to smooth projective $\widetilde{S}$ are abelian varieties. When S is a smooth, projective curve, we recover this way the algebraic complete integrability of Hitchin's system and its generalizations. More generally, for any S, one obtains an ACIHS (with symplectic, respectively Poisson structures) when the values bundle has the same (symplectic, respectively Poisson) structure, by a slight modification of the construction in Chapter 8. One considers only Lagrangian supports which retain a W-action, and only equivariant sheaves on them (with the numerical invariants of a line bundle). These two restrictions are symplecticly dual, so the moduli space of Lagrangian sheaves with these invariance properties is a symplectic (respectively, Poisson) subspace of the total moduli space, and the fibers of the Hitchin map are Lagrangian as expected.

9.3.5 Irregulars?

The Higgs bundles we consider in this survey are assumed to be everywhere regular. This is a reasonable assumption for line-bundle valued Higgs bundles on a curve or surface, but *not* in dim ≥ 3. This is because the complement of $\mathfrak{g}_{reg}$ has codimension 3 in $\mathfrak{g}$. The source of the difficulty is that the analogue of (52) fails over $\mathfrak{g}$. There are two candidates for the universal cameral cover: $\widetilde{\mathfrak{g}}$, defined by the left hand side of (52), is finite over $\mathfrak{g}$ with W action, but does not have a family of line bundles parametrized by Λ. These live instead on $\widetilde{\widetilde{\mathfrak{g}}}$, the object defined by the right hand side, which parametrizes pairs $(x,\mathfrak{b})$, $x \in \mathfrak{b} \subset \mathfrak{g}$. This suggests that the right way to analyze irregular Higgs bundles may involve spectral data consisting of a tower

$$\widetilde{\widetilde{S}} \overset{\sigma}{\longrightarrow} \widetilde{S} \longrightarrow S$$

together with a homomorphism $\mathcal{L} : \Lambda \longrightarrow \mathrm{Pic}(\widetilde{\widetilde{S}})$ such that the collection of sheaves

$$\sigma_*(\mathcal{L}(\lambda)), \qquad \lambda \in \Lambda$$

on $\tilde{S}$ is R-twisted W-equivariant in an appropriate sense. As a first step, one may wish to understand the direct images $R^i\sigma_*(\mathcal{L}(\lambda))$ and in particular the cohomologies $H^i(F, \mathcal{L}(\lambda))$ where F, usually called a *Springer fiber*, is a fiber of σ. For regular x, this fiber is a single point. For $x = 0$, the fiber is all of G/B, so the fiber cohomology is given by the Borel-Weil-Bott theorem. The question may thus be considered as a desired extension of BWB to general Springer fibers.

References

[AB]   Adams, M. R., Bergvelt, M. J.: *The Krichever map, vector bundles over algebraic curves, and Heisenberg algebras*, Comm. Math. Phys. 154 (1993), 265-305.

[AdC]   Arbarello, E., De Concini, C.: *Another proof of a conjecture of S. P. Novikov on periods and abelian integrals on Riemann surfaces*, Duke Math. J. 54 (1987) 163-178.

[AG]   Arnol'd, V. I., Givental', A. B.: *Symplectic Geometry*. In: Arnol'd, V.I., Novikov, S.P. (eds.) Dynamical systems IV, (EMS, vol. 4, pp. 1-136) Berlin Heidelberg New York: Springer 1988

[AHH]   Adams, M. R., Harnad, J., Hurtubise J.: *Isospectral Hamiltonian flows in finite and infinite dimensions, II. Integration of flows.* Commun. Math. Phys. 134, 555-585 (1990)

[AIK]   Altman, A., Iarrobino, A., Kleiman, S.: *Irreducibility of the Compactified Jacobian.* Proc. Nordic Summer School, 1-12 (1976)

[AK1]   Altman, A. B., Kleiman S. L.: *Foundation of the theory of Fano schemes*, Comp. Math. Vol 34, Fasc. 1, 1977, pp 3-47

[AK2]   Altman, A. B., Kleiman S. L.: *Compactifying the Picard Scheme*, Advances in Math. 35, 50-112 (1980)

[At]   Atiyah, M.: *Vector bundles over an elliptic curve.* Proc. Lond. Math. Soc. 7, 414-452 (1957)

[AvM]   Adler, M., van Moerbeke, P.: *Completely integrable systems, Euclidean Lie algebras, and curves*, Advances in Math. 38, (1980), 267-379.

[B1]   Beauville, A.: Beauville, A.: *Jacobiennes des courbes spectrales et systèmes hamiltoniens complètement intégrables.* Acta Math. 164, 211-235 (1990)

[B2]   Beauville, A.: *Varietes Kähleriennes dont la premiere classe de Chern est nulle.* J. Diff. Geom. 18, p. 755-782 (1983).

[Bi]   Biswas, I.: *A remark on a deformation theory of Green and Lazarsfeld*, J. reine angew. Math. 449 (1994) 103-124.

[Bo]   Bogomolov, F.: *Hamiltonian Kähler manifolds*, Soviet Math Dokl. 19 (1978), 1462-1465.

[Bn]   Bottacin, F.: Thesis, Orsay, 1992.

[BD]   Beauville, A., Donagi, R.: *La variété des droites d'une hypersurface cubique de dimension 4.* C. R. Acad. Sci. Paris Ser. I t. 301, 703-706 (1985)

[BG]   Bryant, R., Griffiths, P.: *Some observations on the infinitesimal period relations for regular threefolds with trivial canonical bundle*, in Arithmetic and Geometry, Papers dedicated to I.R. Shafarevich, vol. II, Birkhäuser, Boston (1983), 77-102.

115

[BK]   Beilinson, A., Kazhdan, D.: *Flat Projective Connections*, unpublished(1990).

[BL]   Beauville, A., Laszlo, Y.:   *Conformal blocks and generalized theta functions*, Comm. Math. Phys., to appear.

[BNR]   Beauville, A., Narasimhan, M. S., Ramanan, S.: *Spectral curves and the generalized theta divisor*. J. Reine Angew. Math. 398, 169-179 (1989)

[CdOGP] Candelas, P., de la Ossa, X. C., Green, P. S., and Parkes L.: *A pair of Calabi-Yau manifolds as an exactly soluble superconformal theory*, Phys. Lett. **B 258** (1991), 118-126; Nuclear Phys. **B 359** (1991), 21-74.

[CG]   Clemens, H., Griffiths, P.: *The intermediate Jacobian of the cubic threefold.* Annals of Math. 95 281-356 (1972)

[De1]   Deligne P.: *Theoreme de Lefschetz et criteres de degenerescence de suites spectrales*, Publ. Math. IHES, 35(1968) 107-126.

[De2]   Deligne, P.: letter to C. Simpson, March 20, 1989

[D1]   Donagi, R.: *The tetragonal construction*, Bull. Amer. Math. Soc. (N.S.) **4** (1981), 181-185.

[D2]   Donagi, R.: *Decomposition of spectral covers*, in: Journees de Geometrie Algebrique D'Orsay,Asterisque 218 (1993),145-175.

[D3]   Donagi, R.:   *Abelianization of Higgs bundles*, preprint.

[D4]   Donagi, R.:   *Spectral covers*, To appear in: Proceedings of special year in algebraic geometry 1992–1993, Mathematical Sciences Research Insitute.

[D5]   Donagi, R.: *Group law on the intersection of two quadrics.* Ann. Scuola Norm. Sup. Pisa Cl. Sci. (4) 7 (1980), no. 2, 217-239.

[DEL]   Donagi, R., Ein, L. and Lazarsfeld, R.:   *A non-linear deformation of the Hitchin dynamical system*, preprint, (Alg. Geom. eprint no. 9504017).

[DM]   Donagi, R. and Markman, E.: *Cubics, integrable systems, and Calabi-Yau threefolds*, to appear in the proceedings of the Algebraic Geometry Workshop on the occasion of the 65th birthday of F. Hirzebruch, May 1993. (Alg. Geom. eprint no. 9408004).

[DR]   Desale, U. V., Ramanan, S.: *Classification of Vector Bundles of Rank 2 on Hyperelliptic Curves.* Invent. Math. 38, 161-185 (1976)

[EZ]   El Zein F. and Zucker S.: *Extendability of normal functions associated to algebraic cycles.* In *Topics in transcendental Algebraic Geometry*, ed. by P. Griffiths, Annals of Mathematics

[F1]   Faltings, G.:   *Stable G-bundles and Projective Connections*, Jour. Alg. Geo 2 (1993), 507-568.

[F2]   Faltings, G.: *A proof of the Verlinde formula*, to appear in Jour. Alg. Geo.

[Gr]   Griffiths, P. A.: *Infinitesimal variations of Hodge structure III: determinantal varieties and the infinitesimal invariant of normal functions.* Compositio Math. **50** (1983), 267-324.

[GH]   Griffiths, P. A., Harris, J.: *Principles of algebraic geometry.* Pure and Applied Mathematics. Wiley-Interscience 1978

[Gro]   Grothendieck, A.: *Techniques de construction et théorèmes d'existence en géométrie algébrique, IV : Les schémas de Hilbert.* Séminaire Bourbaki 13e année (1960/61) Exposés 221

[HS]   Hilton P., Stammbach U.: *A course in homological algebra,* Graduate Texts in Mathematics, Vol. 4, Springer-Verlag, New York-Berlin,

[H1]   Hitchin, N.J.: *The self-duality equations on a Riemann surface,* Proc. Lond. Math. Soc. **55** (1987) 59-126.

[H2]   Hitchin, N. J.: *Stable bundles and integrable systems.* Duke Math. J. **54**, No 1 91-114 (1987)

[H3]   Hitchin, N. J.: *Flat Connections and Geometric Quantization.* Comm. Math. Phys. **131**, 347-380 (1990)

[Hur]   Hurtubise J.: *Integrable systems and Algebraic Surfaces,* Preprint 1994

[IVHS]   Carlson, J., Green, M., Griffiths, P., and Harris, J.: *Infinitesimal variations of Hodge structure (I).* Compositio Math. **50** (1983), 109-205.

[K]   Kanev, V.: *Spectral curves, simple Lie algebras and Prym-Tjurin varieties,* Proc. Symp. Pure Math. **49** (1989), Part I, 627-645.

[Ka]   Kawamata, Y.: *Unobstructed deformations - a remark on a paper of Z. Ran* J. Alg. Geom. 1 (1992) 183-190

[Kn]   Knörrer, H.: *Geodesics on the Ellipsoid.* Invent. Math. **59**, 119-143 (1980)

[Ko]   Kobayashi, S.: *Simple vector bundles over symplectic Kähler manifolds.* Proc. Japan Acad. Ser. A Math. Sci. **62**, 21-24 (1986)

[Ko-P]   Kouvidakis, A., Pantev, T.: *automorphisms of the moduli spaces of stable bundles on a curve.*, Math. Ann., to appear.

[KP1]   Katzarkov, L., Pantev, T.: *Stable G_2 bundles and algebraically completely integrable systems,* Comp. Math.**92** (1994), 43-60.

[KP2]   Katzarkov, L., Pantev, T.: *Representations of Fundamental groups whose Higgs bundles are pullbacks,* J. Diff. Geo. **39** (1994), 103-121.

[L]   Laumon, G.: *Un analogue global du cône nilpotent,* Duke Math. J. **57**(1988), 647-671.

[LM1]   Li, Y., Mulase, M.: *Category of morphisms of algebraic curves and a characterization of Prym varieties.* Preprint

[LM2]   Li, Y., Mulase, M.: *Hitchin systems and KP equations.* Preprint

[Ma1]   Markman, E.: *Spectral curves and integrable systems*, Compositio Math. 93, 255-290, (1994).

[Ma2]   Markman, E.: In preparation.

[Me]   Merindol, J. Y.: *Varietes de Prym d'un Revetement Galoisien*, prep.(1993).

[Mor]   Morrison, D.: *Mirror Symmetry and Rational Curves on Quintic Threefolds: A Guide for Mathematicians*, J. Am. Math. Soc. 6 (1993), 223.

[Mo]   Moser, J.: *Various aspects of integrable Hamiltonian systems*, in: Dynamical Systems 1978, Prog. Math. Vol. 8, 233-290.

[MS]   McDaniel, A., Smolinsky, L.: *A Lie theoretic Galois Theory for the spectral curves of an integrable system II*, prep. (1994).

[Mu1]   Mukai S.: *Symplectic structure of the moduli space of sheaves on an abelian or K3 surface*, Invent. math. 77,101-116 (1984)

[Mu2]   Mukai S.: *On the moduli space of bundles on K3 surfaces I*, Vector bundles on algebraic varieties, Proc. Bombay Conference, 1984, Tata Institute of Fundamental Research Studies, no. 11, Oxford University Press, 1987, pp. 341-413.

[Mul]   Mulase, M.: *Cohomological structure in soliton equations and Jacobian varieties*, J. Diff. Geom. 19 (1984), 403-430.

[Mum1]   Mumford, D.: *Lectures on curves on an algebraic surface*, Princeton University Press, 1966.

[Mum2]   Mumford, D.: *Tata Lectures on Theta II*, Birkhaeuser-Verlag, Basel, Switzerland, and Cambridge, MA 1984

[Mum3]   Mumford, D.: *Prym varieties I*, in:Contribution to analysis, Acad. Press (1974), 325-350.

[Nit]   Nitsure, N.: *Moduli space of semistable pairs on a curve.* Proc. London Math. Soc. (3) 62 (1991) 275-300

[NR]   Narasimhan, M. S., Ramanan, M. S.: *2θ-linear systems on abelian varieties*, in Vector bundles and algebraic varieties, Oxford University Press, 415-427 (1987)

[Ra]   Ran, Ziv: *Lifting of cohomology and unobstructedness of certain holomorphic maps.* Bull. Amer. Math. Soc. (N.S.) 26 (1992), no. 1, 113-117.

[Re]   Reid, M.: *The complete intersection of two or more quadrics*, Thesis, Cambridge (GB) 1972.

[Rec]   Recillas, S.: *Jacobians of curves with g_4^1's are Pryms of trigonal curves*, Bol. Soc. Mat. Mexicana **19** (1974) no.1.

[Se]   Seshadri, C. S.: *Fibrés vectoriels sur les courbes algébriques*. asterisque 96 1982

[Sc]   Scognamillo, R.: *Prym-Tjurin Varieties and the Hitchin Map*, preprint (1993).

[Sch]   Schmid, W.: *Variation of Hodge structure: the singularities of the period mapping*. Invent. Math. **22** (1973), 211-319.

[Sh]   Shiota, T.: *Characterization of Jacobian varieties in terms of soliton equations*, Inv. Math. **83** (1986) 333-382.

[Sim1]   Simpson, C.: *Moduli of representations of the fundamental group of a smooth projective variety I, II.* Preprint

[Sim2]   Simpson, C.: *Higgs bundles and local systems*. Inst. Hautes Etudes Sci. Publ. Math. (1992), No. 75 5-95.

[Sim3]   Simpson, C.: *Nonabelian Hodge Theory*, Proceedings of the International Congress of Mathematics, Kyoto, Japan, 1990.

[SW]   Segal, G., Wilson, G.,: *Loop groups and equations of KdV type*. Inst. Hautes Etudes Sci. Publ. Math. (1985), No. 61 5-65.

[Ti]   Tian, G.: *Smoothness of the universal deformation space of compact Calabi-Yau manifolds and its Petersson-Weil metric*, in Mathematical Aspects of String Theory (S.T. Yau, ed.), World Scientific, Singapore (1987), 629-646.

[To]   Todorov, A.: *The Weil-Petersson geometry of the moduli space of $SU(n \geq 3)$ (Calabi-Yau) manifolds, I*, Comm. Math. Phys. **126** (1989), 325-346.

[TV1]   Treibich, A., J.-L. Verdier *Solitons elliptiques*. The Grothendieck Festschrift, Vol. III, 437-480, Birkhauser Boston, 1990.

[TV2]   Treibich, A., J.-L. Verdier *Varietes de Kritchever des solitons elliptiques de KP*. Proceedings of the Indo-French Conference on Geometry (Bombay, 1989), 187-232, 1993.

[Ty1]   Tyurin, A. N.: *Symplectic structures on the varieties of moduli of vector bundles on algebraic surfaces with $p_g > 0$.* Math. USSR Izvestiya Vol. 33(1989), No. 1

[Ty2]   Tyurin, A. N.: *The geometry of moduli of vector bundles*. Russian Math. Surveys 29:6 57-88 (1974)

[V]   Voisin, C.: *Sur la stabilite des sous-varietes lagrangiennes des varietes symplectiques holomorphes*. Complex projective geometry (Trieste, 1989/Bergen, 1989), 294-303, Cambridge Univ. Press, Cambridge, 1992.

[We]   Weinstein, A.: *The Local Structure of Poisson Manifolds*. J. Diff. Geom. 18, 523-557 (1983)

[Ye]   Ye, Yun-Gang: *Lagrangian subvarieties of the moduli space of stable vector bundles on a regular algebraic surface with $p_g > 0$. Math. Ann.* 295, 411-425 (1993)

GEOMETRY OF 2D TOPOLOGICAL FIELD THEORIES

Boris DUBROVIN

SISSA, Trieste

Contents.

Introduction.   121
Lecture 1.
WDVV equations and Frobenius manifolds.   123
Appendix A.
Polynomial solutions of WDVV. Towards classification of Frobenius manifolds
with good analytic properties.   139
Appendix B.
Symmetries of WDVV. Twisted Frobenius manifolds.   144
Appendix C.
WDVV and Chazy equation. Affine connections on curves
with projective structure.   149
Lecture 2.
Topological conformal field theories and their moduli.   163
Lecture 3.
Spaces of isomonodromy deformations as Frobenius manifolds.   188
Appendix D.
Geometry of flat pencils of metrics.   226
Appendix E.
WDVV and Painlevé-VI.   229
Appendix F.
Analytic continuation of solutions of WDVV and braid group.   238
Appendix G.
Monodromy group of a Frobenius manifold.   246
Appendix H.
Generalized hypergeometric equation associated with a Frobenius manifold and its
monodromy.   254
Appendix I.
Determination of a superpotential of a Frobenius manifold.   265
Lecture 4.
Frobenius structure on the space of orbits of a Coxeter group.   268
Appendix J.
Extended complex crystallographic groups and twisted Frobenius manifolds.   289
Lecture 5.
Differential geometry of Hurwitz spaces.   297
Lecture 6.
Frobenius manifolds and integrable hierarchies. Coupling to topological
gravity.   318
References.   339

Introduction.

In these lecture notes I consider one remarkable system of differential equations that appeared in the papers of physicists on two-dimensional topological field theory (TFT) in the beginning of '90 [156, 42]. Roughly speaking, the problem is to find a quasihomogeneous function $F = F(t)$ of the variables $t = (t^1, \ldots, t^n)$ such that the third derivatives of it

$$c_{\alpha\beta\gamma}(t) := \frac{\partial^3 F(t)}{\partial t^\alpha \partial t^\beta \partial t^\gamma}$$

for any t are structure constants of an associative algebra A_t with a t-independent unity (the algebra will be automatically commutative) (see Lecture 1 for the precise formulation of the problem). For the function $F(t)$ one obtains a very complicated overdetermined system of PDEs. I call it WDVV equations. In the physical setting the solutions of WDVV describe moduli space of topological conformal field theories. One of the projects of these lectures is to try to reconstruct the building of 2D TFT on the base of WDVV equation.

From the point of view of a mathematician particular solutions of WDVV with certain "good" analytic properties are generating functions for the Gromov - Witten invariants of Kähler (and, more generally, of symplectic) manifolds [157]. They play the crucial role in the formulation (and, may be, in the future explanation) of the phenomenon of mirror symmetry of Calabi - Yau 3-folds [159]. Probably, they also play a central role in understanding of relations between matrix integrals, integrable hierarchies, and topology of moduli spaces of algebraic curves [157, 76, 96-108, 61].

We discuss briefly the "physical" and "topological" motivations of WDVV equations in Lecture 2. The other lectures are based mainly on the papers [48-50, 52-54] of the author. The material of Appendices was not published before besides Appendix D (a part of the preprint [54]).

In the abbreviated form our contribution to the theory of WDVV can be encoded by the following key words:

WDVV as Painlevé equations

Discrete groups and their invariants and particular solutions of WDVV

Symmetries of WDVV

To glue all these together we employ an amazingly rich (and nonstandard) differential geometry of WDVV. The geometric reformulation of the equations is given in Lecture 1. We observe then (on some simple but important examples) that certain analyticity conditions together with semisimplicity of the algebras A_t work as a very strong rule for selection of solutions (WDVV does almost not constrain the function $F(t)$ if the algebras A_t are nilpotent for any t). Probably, solutions with good analytic properties (in a sense to be formulated in a more precise way) are isolated points in the sea of all solutions of WDVV.

The main geometrical playing characters - the deformed affine connection and the deformed Euclidean metric - are introduced in Lecture 3. In this Lecture we find the general solution of WDVV for which the algebra A_t is semisimple for generic t. This is expressed via certain transcendental functions of the Painlevé-VI type and their higher order generalizations. The theory of linear differential operators with rational coefficients and of their monodromy preserving deformations plays an important role in these considerations.

In Appendix G we introduce another very important object: the monodromy group that can be constructed for any solution of WDVV. This is the monodromy group of a holonomic system of differential equations describing the deformation of Euclidean structure on the space of parameters t.

Our conjecture is that, for a solution of WDVV with good analytic properties the monodromy group is a discrete group. Some general properties of the monodromy group are obtained in Appendices G and H. We give many examples where the group is a finite Coxeter group, an extension of an affine Weyl group (Lecture 4) or of a complex crystallographic group (Appendix J). The solutions of WDVV for these monodromy groups are given by simple formulae in terms of the invariants of the groups. To apply this technique to the topological problems, like mirror symmetry, we need to find some natural groups related to, say, Calabi - Yau 3-folds (there are interesting results in this direction in the recent preprints [143]).

Bäcklund-type symmetries of the equations of associativity play an important technical role in our constructions. It turns out that the group of symmetries of WDVV is rich enough: for example, it contains elements that transform, in the physical notations, a solution with the given $d = \hat{c} = \frac{c}{3}$ to a solution with $d' = 2 - d$ (I recall that for topological sigma-models d is the complex dimension of the target space). Better understanding of the structure of the group of symmetries of WDVV could be useful in the mirror problem.

In the last Lecture we briefly discuss relation of the equations of associativity to integrable hierarchies and their semiclassical limits. Some of these relations were discussed also in [49, 88, 94-95, 104-105, 141] for the dispersionless limits of various integrable hierarchies of KdV type (the dispersionless limit corresponds to the tree-level approximation in TFT). Some of these observations were known also in the theory of Gauss - Manin equations [118]. Our approach is principally different: we construct an integrable hierarchy (in a semi-classical approximation) for any solution of WDVV. We obtain also for our hierarchies the semi-classical analogue of Lax representation. The problem of reconstruction of all the hierarchy (in all orders in the small dispersion expansion) is still open (see the recent papers [97, 61] where this problem was under investigation).

Acknowledgement. I wish to thank D.Abramovich, V.I.Arnol'd, V.V.Batyrev, A.B.Givental, C.Itzykson, Yu.I.Manin, D.R. Morrison, A.N.Todorov, C.Vafa, A.N.Varchenko, M.Verbitski, J.-B.Zuber for helpful discussions. I acknowledge hospitality of Harvard University and of MSRI where part of the work for this paper was carried out.

123

Lecture 1.
WDVV equations and Frobenius manifolds.

I start with formulation of the main subject of these lectures: a system of differential equations arising originally in the physical papers on two-dimensional field theory (see below Lecture 2). We look for a function $F = F(t)$, $t = (t^1, \ldots, t^n)$ such that the third derivatives of it

$$c_{\alpha\beta\gamma}(t) := \frac{\partial^3 F(t)}{\partial t^\alpha \partial t^\beta \partial t^\gamma}$$

obey the following equations
1) Normalization:

$$\eta_{\alpha\beta} := c_{1\alpha\beta}(t) \tag{1.1}$$

is a constant nondegenerate matrix. Let

$$(\eta^{\alpha\beta}) := (\eta_{\alpha\beta})^{-1}.$$

We will use the matrices $(\eta^{\alpha\beta})$ and $(\eta_{\alpha\beta})$ for raising and lowering indices.
2) Associativity: the functions

$$c^\gamma_{\alpha\beta}(t) := \eta^{\gamma\epsilon} c_{\epsilon\alpha\beta}(t) \tag{1.2}$$

(summation over repeated indices will be assumed in these lecture notes) for any t must define in the n-dimensional space with a basis $e_1, \ldots, e_n$ a structure of an assosciative algebra A_t

$$e_\alpha \cdot e_\beta = c^\gamma_{\alpha\beta}(t) e_\gamma. \tag{1.3}$$

Note that the vector e_1 will be the unity for all the algebras A_t:

$$c^\beta_{1\alpha}(t) = \delta^\beta_\alpha. \tag{1.4}$$

3) $F(t)$ must be quasihomogeneous function of its variables:

$$F(c^{d_1} t^1, \ldots, c^{d_n} t^n) = c^{d_F} F(t^1, \ldots, t^n) \tag{1.5}$$

for any nonzero c and for some numbers $d_1, \ldots, d_n, d_F$.
It will be convenient to rewrite the quasihomogeneity condition (1.5) in the infinitesimal form introducing the *Euler vector field*

$$E = E^\alpha(t)\partial_\alpha$$

as

$$\mathcal{L}_E F(t) := E^\alpha(t)\partial_\alpha F(t) = d_F \cdot F(t). \tag{1.6}$$

For the quasihomogeneity (1.5) $E(t)$ is a linear vector field

$$E = \sum_\alpha d_\alpha t^\alpha \partial_\alpha \tag{1.7}$$

generating the scaling transformations (1.5). Note that for the Lie derivative $\mathcal{L}_E$ of the unity vector field $e = \partial_1$ we must have

$$\mathcal{L}_E e = -d_1\, e. \tag{1.8}$$

Two generalizations of the quasihomogeneity condition will be important in our considerations:

1. We will consider the functions $F(t^1, \ldots, t^n)$ up to adding of a (nonhomogeneous) quadratic function in $t^1, \ldots, t^n$. Such an addition does not change the third derivatives. So the algebras A_t will remain unchanged. Thus the quasihomogeneity condition (1.6) could be modified as follows

$$\mathcal{L}_E F(t) = d_F F(t) + A_{\alpha\beta} t^\alpha t^\beta + B_\alpha t^\alpha + C. \tag{1.9}$$

This still provides quasihomogeneity of the functions $c_{\alpha\beta\gamma}(t)$. Moreover, if

$$d_F \neq 0, \ d_F - d_\alpha \neq 0, \ d_F - d_\alpha - d_\beta \neq 0 \text{ for any } \alpha, \ \beta \tag{1.10}$$

then the extra terms in (1.9) can be killed by adding of a quadratic form to $F(t)$.

2. We will consider more general linear nonhomogeneous Euler vector fields

$$E(t) = \left(q_\beta^\alpha t^\beta + r^\alpha \right) \partial_\alpha. \tag{1.11}$$

If the roots of $E(t)$ (i.e., the eigenvalues of the matrix $Q = (q_\beta^\alpha)$) are simple and nonzero then $E(t)$ can be reduced to the form (1.7) by a linear change of the variables t. If some of the roots of $E(t)$ vanish then, in general the linear nonhomogeneous terms in (1.11) cannot be killed by linear transformations of t. In this case for a diagonalizable matrix Q the Euler vector field can be reduced to the form

$$E(t) = \sum_\alpha d_\alpha t^\alpha \partial_\alpha + \sum_{\alpha | d_\alpha = 0} r^\alpha \partial_\alpha \tag{1.12}$$

(here d_α are the eigenvalues of the matrix Q). The numbers r^α can be changed by linear transformations in the kernel $\mathrm{Ker} Q$. However, in important examples the function $F(t)$ will be periodic (modulo quadratic terms) w.r.t. some lattice of periods in $\mathrm{Ker} Q$ (note that periodicity of $F(t)$ can happen only along the directions with zero scaling dimensions). In this case the vector (r^α) is defined modulo the group of automorphisms of the lattice of periods. Particularly, in topological sigma models with non-vanishing first Chern class of the target space the vector (r^α) is always nonzero (see below Lecture 2).

The degrees $d_1, \ldots, d_n, d_F$ are well-defined up to a nonzero factor. We will consider only the case

$$d_1 \neq 0$$

(the variable t^1 is marked due to (1.1)). It is convenient in this case to normalize the degrees $d_1, \ldots, d_n, d_F$ in such a way that

$$d_1 = 1. \tag{1.13a}$$

In the physical literature the normalized degrees usually are parametrized by some numbers $q_1 = 0, q_2, ..., q_n$ and d such that

$$d_\alpha = 1 - q_\alpha, \quad d_F = 3 - d. \tag{1.13b}$$

If the coordinates are normalized as in (1.18) then

$$q_n = d, \quad q_\alpha + q_{n-\alpha+1} = d. \tag{1.13c}$$

Mainly the case of real function $F(t)$ will be under consideration (although in the classification theorems we will work also with the complexified situation). All the numbers q_α, r_α, d in the real case are also to be real.

Associativity imposes the following system of nonlinear PDE for the function $F(t)$

$$\frac{\partial^3 F(t)}{\partial t^\alpha \partial t^\beta \partial t^\lambda} \eta^{\lambda\mu} \frac{\partial^3 F(t)}{\partial t^\gamma \partial t^\delta \partial t^\mu} = \frac{\partial^3 F(t)}{\partial t^\gamma \partial t^\beta \partial t^\lambda} \eta^{\lambda\mu} \frac{\partial^3 F(t)}{\partial t^\alpha \partial t^\delta \partial t^\mu} \tag{1.14}$$

for any $\alpha, \beta, \gamma, \delta$ from 1 to n. The quasihomogeneity (1.9) determines the scaling reduction of the system. The normalization (1.1) completely specifies the dependence of the function $F(t)$ on the marked variable t^1. The resulting system of equations will be called *Witten - Dijkgraaf - E. Verlinde - H. Verlinde (WDVV)* system: it was first found in the papers [156, 42] in topological field theory (see Lecture 2 below). A solution of the WDVV equations will be called *(primary) free energy*. The system (1.14) (without the scaling (1.9)) will be called *associativity equations*.

Remark 1.1. More general reduction of (1.14) is given by *conformal* transformations of the metric $\eta_{\alpha\beta}$. The generator E of the correspondent one-parameter group of diffeomorphisms of the t-space for $n \geq 3$ due to Liouville theorem [59] must have the form

$$E = a \left\{ t_1 t^\gamma \partial_\gamma - \frac{1}{2} t_\sigma t^\sigma \partial_1 \right\} + \sum_{\epsilon=1}^{n} d_\epsilon t^\epsilon \partial_\epsilon \tag{1.15a}$$

for some constants $a, d_1, ..., d_n$ with

$$d_\alpha = 1 - q_\alpha, \quad q_1 = 0, \quad q_\alpha + q_{n-\alpha+1} = d$$

(in the normalization (1.17)). The function $F(t)$ must obey the equation

$$\sum_{\epsilon=1}^{n} (at_1 + d_\epsilon) t^\epsilon \partial_\epsilon F = (3 - d + 2at_1) F + \frac{a}{8} \left(t_\sigma t^\sigma \right)^2 \tag{1.15b}$$

modulo quadratic terms. The equations of associativity (1.14) for F satisfying (1.15) also can be reduced to a system of ODE. I will not consider this system here. However, a *discrete* group of conformal symmetries of WDVV plays an important role in our considerations (see below Appendix B).

126

Observe that the system is invariant w.r.t. linear changes of the coordinates $t^1, \ldots, t^n$. To write down WDVV system more explicitly we use the following

Lemma 1.1 *The scaling transformations generated by the Euler vector field E (1.9) act as linear conformal transformations of the metric $\eta_{\alpha\beta}$*

$$\mathcal{L}_E \eta_{\alpha\beta} = (d_F - d_1)\eta_{\alpha\beta} \tag{1.16}$$

where the numbers d_F and d_1 are defined in (1.6) and (1.8).

Proof. Differentiating the equation (1.9) w.r.t. t^1, t^α and t^β and using $\partial_1 E^\rho = d_1 \delta_1^\rho$ (this follows from (1.8)) we obtain

$$q_\alpha^\rho \eta_{\rho\beta} + q_\beta^\rho \eta_{\rho\alpha} = (d_F - d_1)\eta_{\alpha\beta}.$$

The l.h.s. of this equality coincides with the Lie derivative $\mathcal{L}_E \eta_{\alpha\beta}$ of the metric $< , >$. Lemma is proved.

Corollary 1.1. *If $\eta_{11} = 0$ and all the roots of $E(t)$ are simple then by a linear change (possibly, with complex coefficients for odd n) of coordinates t^α the matrix $\eta_{\alpha\beta}$ can be reduced to the antidiagonal form*

$$\eta_{\alpha\beta} = \delta_{\alpha+\beta,n+1}. \tag{1.17}$$

In these coordinates

$$F(t) = \frac{1}{2}(t^1)^2 t^n + \frac{1}{2}t^1 \sum_{\alpha=2}^{n-1} t^\alpha t^{n-\alpha+1} + f(t^2, \ldots, t^n) \tag{1.18}$$

for some function $f(t^2, \ldots, t^n)$, the sum

$$d_\alpha + d_{n-\alpha+1} \tag{1.29}$$

does not depend on α and

$$d_F = 2d_1 + d_n. \tag{1.20}$$

If the degrees are normalized in such a way that $d_1 = 1$ then they can be represented in the form

$$d_\alpha = 1 - q_\alpha, \quad d_F = 3 - d \tag{1.21a}$$

where the numbers $q_1, \ldots, d_n, d$ satisfy

$$q_1 = 0, \quad q_n = d, \quad q_\alpha + q_{n-\alpha+1} = d. \tag{1.21b}$$

Exercise 1.1. Show that if $\eta_{11} \neq 0$ (this can happen only for $d_F = 3d_1$) the function F can be reduced by a linear change of t^α to the form

$$F = \frac{c}{6}(t^1)^3 + \frac{1}{2}t^1 \sum_{\alpha=1}^{n-1} t^\alpha t^{n-\alpha+1} + f(t^2, \ldots, t^n) \tag{1.22}$$

for a nonzero constant c where the degrees satisfy

$$d_\alpha + d_{n-\alpha+1} = 2d_1. \tag{1.23}$$

Proof of Corollary. If $< e_1, e_1 >= 0$ then one can chose the basic vector e_n such that $< e_1, e_n >= 1$ and e_n is still an eigenvector of Q. On the orthogonal complement of the span of e_1 and e_n we can reduce the bilinear form $< , >$ to the antidiagonal form using only eigenvectors of the scaling transformations. In these coordinates (1.18) follows from (1.1). Independence of the sum $d_\alpha + d_{n-\alpha+1}$ of α follows from (1.16). The formula for d_F follows from (1.16). Corollary is proved.

Remark 1.2. Over real numbers WDVV equations have an additional integer invariant: the signature of the quadratic form $\eta_{\alpha\beta} t^\alpha t^\beta$ restricted onto the subspace spanned by those basic vectors e_α with

$$q_\alpha = \frac{d}{2}.$$

I will mainly consider solutions of WDVV of the type (1.18). I do not know physical examples of the solutions of the second type (1.22). However, we are to take into account also the solutions with $\eta_{11} \neq 0$ for completeness of the mathematical theory of WDVV (see, e.g., Appendix B and Lecture 5).

Example 1.1. $n = 2$. Equations of associativity are empty. The other conditions specify the following general solution of WDVV

$$F(t_1, t_2) = \frac{1}{2} t_1^2 t_2 + t_2^k, \quad k = \frac{3-d}{1-d}, \ d \neq -1, 1, 3 \tag{1.24a}$$

$$F(t_1, t_2) = \frac{1}{2} t_1^2 t_2 + t_2^2 \log t_2, \quad d = -1 \tag{1.24b}$$

$$F(t_1, t_2) = \frac{1}{2} t_1^2 t_2 + \log t_2, \quad d = 3 \tag{1.24c}$$

$$F(t_1, t_2) = \frac{1}{2} t_1^2 t_2 + e^{\frac{2}{r} t_2}, \quad d = 1, \ r \neq 0 \tag{1.24d}$$

$$F(t_1, t_2) = \frac{1}{2} t_1^2 t_2, \quad d = 1, \ r = 0 \tag{1.24e}$$

(in concrete formulae I will often label the coordinates t^α by subscripts for the sake of graphical simplicity). In the last two cases $d = 1$ the Euler vector field is $E = t_1 \partial_1 + r \partial_2$.

Example 1.2. $n = 3$. In the three-dimensional algebra A_t with the basis $e_1 = 1$, e_2, e_3 the law of multiplication is determined by the following table

$$e_2^2 = f_{xxy} e_1 + f_{xxx} e_2 + e_3$$
$$e_2 e_3 = f_{xyy} e_1 + f_{xxy} e_2 \tag{1.25}$$
$$e_3^2 = f_{yyy} e_1 + f_{xyy} e_2$$

where the function F has the form

$$F(t) = \frac{1}{2}t_1^2 t_3 + \frac{1}{2}t_1 t_2^2 + f(t_2, t_3) \qquad (1.26)$$

for a function $f = f(x, y)$ (the subscripts denote the correspondent partial derivatives). The associativity condition

$$(e_2^2)e_3 = e_2(e_2 e_3) \qquad (1.27)$$

implies the following PDE for the function $f = f(x, y)$

$$f_{xxy}^2 = f_{yyy} + f_{xxx} f_{xyy}. \qquad (1.28)$$

It is easy to see that this is the only one equation of associativity for $n = 3$. The function f must satisfy also the following scaling condition

$$\left(1 - \frac{d}{2}\right) x f_x + (1 - d)y f_y = (3 - d)f, \quad d \neq 1, 2, 3 \qquad (1.29a)$$

$$\frac{1}{2}x f_x + r f_y = 2f, \quad d = 1 \qquad (1.29b)$$

$$r f_x - y f_y = f, \quad d = 2 \qquad (1.29c)$$

$$\frac{1}{2}x f_x + 2y f_y = c, \quad d = 3 \qquad (1.29d)$$

for some constants c, r. The correspondent scaling reductions of the equation (1.28)

$$f(x, y) = \frac{x^4}{y}\phi\left(\log(yx^q)\right), \quad q = -\frac{1 - d}{1 - \frac{1}{2}d}, \quad d \neq 1, 2, 3 \qquad (1.30a)$$

$$f(x, y) = x^4 \phi(y - 2r \log x), \quad d = 1 \qquad (1.30b)$$

$$f(x, y) = y^{-1}\phi(x + r \log y), \quad d = 2 \qquad (1.30c)$$

$$f(x, y) = 2c \log x + \phi(yx^{-4}), \quad d = 3 \qquad (1.30d)$$

are the following third order ODEs for the function $\phi = \phi(z)$,

$$-6\phi + 48\phi^2 + 11\phi' + 88q\phi\phi' - (144 + 144q - 3q^2)\phi'^2 - 6\phi'' + 48(2 + 2q + q^2)\phi\phi''$$

$$-4q(16 + 16q + q^2)\phi'\phi'' - (13q^2 + 13q^3 + q^4)\phi''^2$$

$$+\phi''' + 8q(3 + 3q + q^2)\phi\phi''' + 2q^2(1 + q + q^2)\phi'\phi''' - q^3(1 + q)\phi''\phi''' = 0 \qquad (1.31a)$$

for $d \neq 1, 2, 3$,

$$-144\phi'^2 + 96\phi\phi'' + 128r\phi'\phi'' - 52r^2\phi''^2 + \phi''' - 48r\phi\phi''' + 8r^2\phi'\phi''' + 8r^3\phi''\phi''' = 0 \qquad (1.31b)$$

for $d = 1$,

$$\phi'''[r^3 + 2\phi' - r\phi''] - (\phi'')^2 - 6r^2\phi'' + 11r\phi' - 6\phi = 0 \qquad (1.31c)$$

for $d = 2$,

$$\phi''' = 400\,\phi'^2 + 32\,c\,\phi'' + 1120\,z\,\phi'\,\phi'' + 784\,z^2\,\phi''^2 + 16\,c\,z\,\phi''' + 160\,z^2\,\phi'\,\phi''' + 192\,z^3\,\phi''\,\phi''' \tag{1.31d}$$

for $d = 3$.

Any solution of WDVV for $n = 3$ can be obtained from a solution of these ODEs. Later these will be shown to be reduceable to a particular case of the Painlevé-VI equation.

Remark 1.3. Let us compare (1.28) with the WDVV equations for the prepotential F of the second type (1.22). Here we look for a solutions of (1.14) in the form

$$F = \frac{1}{6}t_1^3 + t_1 t_2 t_3 + f(t_2, t_3). \tag{1.32}$$

The three-dimensional algebra with a basis $e_1 = e$, e_2, e_3 has the form

$$\begin{aligned} e_2^2 &= f_{xxy}e_2 + f_{xxx}e_3 \\ e_2 e_3 &= e_1 + f_{xyy}e_2 + f_{xxy}e_3 \\ e_3^2 &= f_{yyy}e_2 + f_{xyy}e_3. \end{aligned} \tag{1.33}$$

The equation of associativity has the form

$$f_{xxx}f_{yyy} - f_{xxy}f_{xyy} = 1. \tag{1.34}$$

It is interesting that this is the condition of unimodularity of the Jacobi matrix

$$\det \begin{pmatrix} \frac{\partial P}{\partial x} & \frac{\partial P}{\partial y} \\ \frac{\partial Q}{\partial x} & \frac{\partial Q}{\partial y} \end{pmatrix} = 1$$

for

$$P = f_{xx}(x, y), \quad Q = f_{yy}(x, y).$$

The function f must satisfy the quasihomogeneity condition

$$f(c^{1-a}x, c^{1+a}y) = c^3 f(x, y). \tag{1.35}$$

Example 1.3. $n = 4$. Here we have a system of 6 equations for the function $f = f(x, y, z)$ where

$$F(t_1, t_2, t_3, t_4) = \frac{1}{2}t_1^2 t_4 + t_1 t_2 t_3 + f(t_2, t_3, t_4) \tag{1.36}$$

$$\begin{aligned} -2f_{xyz} - f_{xyy}f_{xxy} + f_{yyy}f_{xxx} &= 0 \\ -f_{xzz} - f_{xyy}f_{xxz} + f_{yyz}f_{xxx} &= 0 \\ -2f_{xyz}f_{xxz} + f_{xzz}f_{xxy} + f_{yzz}f_{xxx} &= 0 \\ -f_{yyy}f_{xxz} + f_{yzz} + f_{yyz}f_{xxy} &= 0 \\ f_{zzz} - f_{xyz}{}^2 + f_{xzz}f_{xyy} - f_{yyz}f_{xxz} + f_{yzz}f_{xxy} &= 0 \\ f_{yyy}f_{xzz} - 2f_{yyz}f_{xyz} + f_{yzz}f_{xyy} &= 0. \end{aligned} \tag{1.37}$$

It is a nontrivial exercise even to verify compatibility of this overdetermined system of equations.
Particularly, any function f depending only on t_2 gives a solution of these equations. The multiplication in the correspondent algebras A_t have then the form

$$e_1 \cdot e_\alpha = e_\alpha, \; e_2 \cdot e_3 = e_4, \; e_2^2 = f'''(t_2)e_3, \qquad (1.38)$$

other products vanish. Thus A_t are nilpotent algebras for any t. These solutions of WDVV are compatible with the scaling (1.9) if the degree of t_2 equals 0.

I am going now to give a coordinate-free formulation of WDVV. Let me give first more details about the algebras A_t.

Definition 1.1. An algebra A over $\mathbf{C}$ is called (commutative) *Frobenius algebra* if:
1) It is a commutative associative $\mathbf{C}$-algebra with a unity e.
2) It is supplied with a $\mathbf{C}$-bilinear symmetric nondegenerate inner product

$$A \times A \to \mathbf{C}, \; a, b \mapsto < a, b >$$

being invariant in the following sense:

$$< ab, c > = < a, bc > . \qquad (1.39)$$

Remark 1.4. Let $\omega \in A^*$ be the linear functional

$$\omega(a) := < e, a > .$$

Then

$$< a, b > = \omega(ab). \qquad (1.40)$$

This formula determines a bilinear symmetric invariant inner product for arbitrary linear functional ω. It will be nondegenerate (for finite-dimensional Frobenius algebras) for generic $\omega \in A^*$. Note that we consider a Frobenius algebra with a *marked* invariant inner product.

Example 1.4. A is the direct sum of n copies of one-dimensional algebras. This means that a basis $e_1, \ldots, e_n$ can be chosen in the algebra with the multiplication law

$$e_i e_j = \delta_{ij} e_i, \; i, j = 1, \ldots, n. \qquad (1.41)$$

Then

$$< e_i, e_j > = 0 \text{ for } i \neq j, \qquad (1.42)$$

the nonzero numbers $< e_i, e_i >$, $i = 1, \ldots, n$ are the parameters of the Frobenius algebras of this type. This algebra is semisimple (it has no nilpotents).

Exercise 1.2. Prove that any Frobenius algebra over $\mathbf{C}$ without nilpotents is of the above form. Prove this statement not assuming *a priori* existence of a unity in the algebra.

I recal that a nonzero element $a \in A$ is called nilpotent if $a^m = 0$ for some m. The algebra A is semisimple if it contains no nilpotents. A basis of idempotents $\pi_1, \ldots, \pi_n$ exists in a semisimple n-dimensional Frobenius algebra

$$\pi_i \pi_j = \delta_{ij}\pi_i, \quad i, j = 1, \ldots, n.$$

The invariaant inner product is diagonal in the basis $\pi_1, \ldots, \pi_n$

$$< \pi_1, \pi_j >=< e, \pi_i \pi_j >= \delta_{ij}\eta_{ii}$$

where

$$\eta_{ii} :=< e, \pi_i > .$$

Introducing the orthonormalized basis

$$f_i = \frac{\pi_i}{\sqrt{\eta_{ii}}}$$

we define the transition matrix

$$e_\alpha = \sum_{i=1}^{n} \psi_{i\alpha} f_i.$$

It is clear that the matrix $(\psi_{i\alpha})$ can be computed in a pure algebraic way in terms of the structure constants $c_{\alpha\beta}^{\gamma}$ and of the invariant inner product $\eta_{\alpha\beta}$. We have

$$\eta_{ii} = \psi_{i1}^2$$

(due to the normalization $e_1 = e$)

$$\eta_{\alpha\beta} = \sum_{i=1}^{n} \psi_{i\alpha} \psi_{i\beta}$$

and

$$c_{\alpha\beta\gamma} =< e_\alpha e_\beta, e_\gamma >= \sum_{i=1}^{n} \frac{\psi_{i\alpha} \psi_{i\beta} \psi_{i\gamma}}{\psi_{i1}}.$$

For a solution of WDVV satisfying the semisimplicity condition for a generic t the matrix $\psi_{i\alpha}$ depends in a nice way on the parameters t. This will be used in Lecture 3 for local classification of these solutions.

An operation of *rescaling* is defined for an algebra with a unity e: we modify the multiplication law and the unity as folllows

$$a \cdot b \mapsto k\, a \cdot b, \quad e \mapsto k^{-1} e \tag{1.43}$$

for a given nonzero constant k. The rescalings preserve Frobenius property of the algebra.

Back to the main problem: we have a family of Frobenius algebras depending on the parameters $t = (t^1, \ldots, t^n)$. Let us denote by M the space of the parameters. We have thus a fiber bundle

$$\begin{array}{c} \downarrow \\ t \in M \end{array} A_t \qquad (1.44)$$

The basic idea is to identify this fiber bundle with the tangent bundle TM of the manifold M.

We come thus to our main definition. Let M be a n-dimensional manifold.

Definition 1.2. M is *Frobenius manifold* if a structure of Frobenius algebra is specified on any tangent plane $T_t M$ at any point $t \in M$ smoothly depending on the point such that

1. The invariant inner product $< , >$ is a flat metric on M.
2. The unity vector field e is covariantly constant w.r.t. the Levi-Cività connection ∇ for the metric $< , >$

$$\nabla e = 0. \qquad (1.45)$$

3. Let

$$c(u, v, w) := < u \cdot v, w > \qquad (1.46)$$

(a symmetric 3-tensor). We require the 4-tensor

$$(\nabla_z c)(u, v, w) \qquad (1.47)$$

to be symmetric in the four vector fields u, v, w, z.

4. A vector field E must be determined on M such that

$$\nabla(\nabla E) = 0 \qquad (1.48)$$

and that the correspondent one-parameter group of diffeomorphisms acts by conformal transformations of the metric $< , >$ and by rescalings on the Frobenius algebras $T_t M$.

In these lectures the word 'metric' stands for a C-bilinear quadratic form on M.

Note that the requirement 4 makes sense since we can locally identify the spaces of the algebras $T_t M$ using the Euclidean parallel transport on M. We will call E *Euler vector field* (see formula (1.9) above) of the Frobenius manifold. The covariantly constant operator

$$Q = \nabla E(t) \qquad (1.49)$$

on the tangent spaces $T_t M$ will be called *the grading operator* of the Frobenius manifold. The eigenvalues of the operator Q are constant functions on M. The eigenvalues q_α of $(\mathrm{id} - Q)$ will be called *scaling dimensions* of M. Particularly, as it follows from (1.45), the unity vector field e is an eigenvector of Q with the eigenvalue 1.

The infinitesimal form of the requirement 4 reads

$$\nabla_\gamma (\nabla_\beta E^\alpha) = 0 \qquad (1.50a)$$
$$\mathcal{L}_E c^\gamma_{\alpha\beta} = c^\gamma_{\alpha\beta} \qquad (1.50b)$$
$$\mathcal{L}_E e = -e \qquad (1.50c)$$
$$\mathcal{L}_E \eta_{\alpha\beta} = D\eta_{\alpha\beta} \qquad (1.50d)$$

for some constant $D = 2 - d$. Here $\mathcal{L}_E$ is the Lie derivative along the Euler vector field. In a coordinate-free way (1.50b) and (1.50d) read

$$\mathcal{L}_E(u \cdot v) - \mathcal{L}_E u \cdot v - u \cdot \mathcal{L}_E v = u \cdot v \qquad (1.50b')$$
$$\mathcal{L}_E < u, v > - < \mathcal{L}_E u, v > - < u, \mathcal{L}_E v > = D < u, v > \qquad (1.50d')$$

for arbitrary vector fields u and v.

Exercise 1.3. Show that the operator

$$\hat{V} := \nabla E - \frac{1}{2}(2 - d)\mathrm{id} \qquad (1.51)$$

is skew-symmetric w.r.t. $< , >$

$$< \hat{V}x, y > = - < x, \hat{V}y > . \qquad (1.52)$$

Lemma 1.2. *Any solution of WDVV equations with $d_1 \neq 0$ defined in a domain $t \in M$ determines in this domain the structure of a Frobenius manifold by the formulae*

$$\partial_\alpha \cdot \partial_\beta := c_{\alpha\beta}^\gamma(t)\partial_\gamma \qquad (1.53a)$$

$$< \partial_\alpha, \partial_\beta > := \eta_{\alpha\beta} \qquad (1.53b)$$

where

$$\partial_\alpha := \frac{\partial}{\partial t^\alpha} \qquad (1.53c)$$

etc.,

$$e := \partial_1 \qquad (1.53d)$$

and the Euler vector field has the form (1.9).

Conversely, locally any Frobenius manifold has the structure (1.53), (1.11) for some solution of WDVV equations.

Proof. The metric (1.53b) is manifestly flat being constant in the coordinates t^α. In these coordinates the covariant derivative of a tensor coincides with the correspondent partial derivatives. So the vector field (1.53d) is covariantly constant. For the covariant derivatives of the tensor (1.46)

$$c_{\alpha\beta\gamma}(t) = \frac{\partial^3 F(t)}{\partial t^\alpha \partial t^\beta \partial t^\gamma} \qquad (1.54)$$

we have a completely symmetric expression

$$\partial_\delta c_{\alpha\beta\gamma}(t) = \frac{\partial^4 F(t)}{\partial t^\alpha \partial t^\beta \partial t^\gamma \partial t^\delta}. \qquad (1.55)$$

This proves the property 3 of our definition. The property 4 is obvious since the one-parameter group of diffeomorphisms for the vector field (1.9) acts by rescalings (1.43).

134

Conversely, on a Frobenius manifold locally one can chose flat coordinates t^1, ..., t^n such that the invariant metric $< , >$ is constant in these coordinates. The symmetry condition (1.47) for the vector fields $u = \partial_\alpha$, $v = \partial_\beta$, $w = \partial_\gamma$ and $z = \partial_\delta$ reads

$$\partial_\delta c_{\alpha\beta\gamma}(t) \text{ is symmetric in } \alpha,\ \beta,\ \gamma,\ \delta$$

for

$$c_{\alpha\beta\gamma}(t) = < \partial_\alpha \cdot \partial_\beta, \partial_\gamma > .$$

Together with the symmetry of the tensor $c_{\alpha\beta\gamma}(t)$ this implies local existence of a function $F(t)$ such that

$$c_{\alpha\beta\gamma}(t) = \frac{\partial^3 F(t)}{\partial t^\alpha \partial t^\beta \partial t^\gamma}.$$

Due to covariant constancy of the unity vector field e we can do a linear change of coordinates in such a way that $e = \partial_1$. This gives (1.53d).

We are to prove now that the function $F(t)$ satisfies (1.9). Due to (1.48) in the flat coordinates $E(t)$ is a linear vector field of a form (1.11). From the definition of rescalings we have in the coordinates t^α

$$[\partial_1, E] = \partial_1. \tag{1.56}$$

Hence ∂_1 is an eigenvector of the operator $Q = \nabla E$ with the eigenvalue 1. I.e. $d_1 = 1$. The constant matrix $\left(Q_\beta^\alpha\right)$ must obey the equation

$$Q_{\alpha\beta} = D\,\eta_{\alpha\beta} \tag{1.57}$$

(this follows from (1.50d)) for some constant D. The last step is to use the condition (1.50b) (the definition of rescalings). From (1.50b) and (1.50d) we obtain

$$\mathcal{L}_E c_{\alpha\beta\gamma} = (1 + D)c_{\alpha\beta\gamma}.$$

Due to (1.54) this can be rewritten as

$$\partial_\alpha \partial_\beta \partial_\gamma \left[E^\epsilon \partial_\epsilon F - (1 + D)F \right] = 0.$$

This gives (1.9). Lemma is proved.

The function $F(t)$ determined by (1.54) will be called *prepotential* or *free energy* of the Frobenius manifold. In general it is well-defined only locally.

Exercise 1.4. In the case $d_1 = 0$ show that a two-dimensional commutative group of diffeomorphisms acts locally on the space of parameters t preserving the multiplication (1.53a) and the metric (1.53b).

This symmetry provides integrability in quadratures of the equation (1.31a) for $d_1 = 0$ (observation of [32]). Indeed, (1.31a) for $d_1 = 0$ ($q = -2$) reads

$$12z^2(\phi'')^2 + 8z^3\phi''\phi''' - \phi''' = 0.$$

The integral of this equation is obtained in elliptic quadratures from

$$\phi'' = \frac{1}{8z^3} \pm \frac{\sqrt{cz^3 + \frac{1}{64}}}{z^3}.$$

Definition 1.3. Two Frobenius manifolds M and $\tilde{M}$ are *equivalent* if there exists a diffeomorphism

$$\phi : M \to \tilde{M} \qquad (1.58a)$$

being a linear conformal transformation of the correspondent invariant metrics ds^2 and $d\tilde{s}^2$

$$\phi^* d\tilde{s}^2 = c^2 \, ds^2 \qquad (1.58b)$$

(c is a nonzero constant) with the differential ϕ_* acting as an isomorphism on the tangent algebras

$$\phi_* : T_t M \to T_{\phi(t)} \tilde{M}. \qquad (1.58c)$$

If ϕ is a local diffeomorphism with the above properties then it will be called *local equivalence*.

Note that for an equivalence ϕ not necessarily $F = \phi^* \tilde{F}$. For example, if the coordinates $(t^1, \ldots, t^n)$ and $(\tilde{t}^1, \ldots, \tilde{t}^n)$ on M and $\tilde{M}$ resp. are normalized as in (1.18) then the map

$$\tilde{t}^1 = t^1, \quad \tilde{t}^n = c^2 t^n, \quad \tilde{t}^\alpha = c t^\alpha \text{ for } \alpha \neq 1, n \qquad (1.59a)$$

$$\tilde{F}(\tilde{t}^1, \ldots, \tilde{t}^n) = c^2 F(t^1, \ldots, t^n). \qquad (1.59b)$$

for a constant $c \neq 0$ is an equivalence. Any equivalence is a superposition of (1.59) and of a linear η-orthogonal transformation of the coordinates $t^1, \ldots, t^n$.

Examples of Frobenius manifolds.

Example 1.5. Trivial Frobenius manifold. Let A be a graded Frobenius algebra. That means that some weights $q_1, \ldots, q_n$ are assigned to the basic vectors $e_1, \ldots, e_n$ such that

$$c^\gamma_{\alpha\beta} = 0 \text{ for } q_\alpha + q_\beta \neq q_\gamma \qquad (1.60a)$$

and also

$$\eta_{\alpha\beta} = 0 \text{ for } q_\alpha + q_\beta \neq d \qquad (1.60b)$$

for some d. Here

$$e_\alpha e_\beta = c^\gamma_{\alpha\beta} e_\gamma$$

$$\eta_{\alpha\beta} = <e_\alpha, e_\beta>$$

in the algebra A. Particularly, any Frobenius algebra can be considered as a graded one w.r.t. the trivial grading $q_\alpha = d = 0$.

These formulae define a structure of Frobenius manifold on $M = A$. The correspondent free energy F is a cubic function

$$F(t) = \frac{1}{6} c_{\alpha\beta\gamma} t^\alpha t^\beta t^\gamma = \frac{1}{6} <t^3, e> \qquad (1.61)$$

for

$$c_{\alpha\beta\gamma} = \eta_{\alpha\epsilon} c^{\epsilon}_{\alpha\beta},$$

$$\mathbf{t} = t^{\alpha} e_{\alpha}, \quad e = e_1 \text{ is the unity.}$$

Here the degrees of the coordinates t^{α} and of the function F are

$$d_{\alpha} = 1 - q_{\alpha}, \quad d_F = 3 - d.$$

For example, the cohomology ring $A = H^{\bullet}(X)$ of a $2d$-dimensional oriented closed manifold X satisfying

$$H^{2i+1}(X) = 0 \text{ for any } i \tag{1.62}$$

is a graded Frobenius algebra w.r.t. the cup product and the Poincaré duality pairing. The degree of an element $x \in H^{2q}(X)$ equals q.

Remark 1.5. To get rid of the restriction (1.62) one is to generalize the notion of Frobenius manifold to supermanifolds, i.e. to admit anticommuting coordinates t^{α}. Such a generalization was done by Kontsevich and Manin [92].

Example 1.6. The direct product $M' \times M''$ of two Frobenius manifolds of the dimensions n and m resp. carries a natural structure of a Frobenius manifold if the scaling dimensions satisfy the constraint

$$\frac{d'_1}{d''_1} = \frac{d'_F}{d''_F}. \tag{1.63}$$

If the flat coordinates $t^{1'}, \ldots, t^{n'}, t^{1''}, \ldots, t^{m''}$ are normalized as in (1.18), and $\deg t^{1'} = \deg t^{1''} = 1$, then (1.63) reads $\deg t^{n'} = \deg t^{m''}$. Thus only the case $m > 1$, $n > 1$ is of interest. The prepotential F for the direct product has the form

$$F\left(t^1, \hat{t}^1, t^{2'}, \ldots, t^{n-1'}, t^{2''}, \ldots, t^{m-1''}, \hat{t}^N, t^N\right) =$$

$$= \frac{1}{2} t^{1\,2} t^N + t^1 \hat{t}^1 \hat{t}^N + \frac{1}{2} t^1 \sum_{\alpha=2}^{n-1} t^{\alpha'} t^{n-\alpha+1'} + \frac{1}{2} t^1 \sum_{\beta=2}^{m-1} t^{\beta''} t^{m-\beta+1''} +$$

$$+ f'\left(t^{2'}, \ldots, t^{n-1'}, \frac{1}{2}(t^N + \hat{t}^N)\right) + f''\left(t^{2''}, \ldots, t^{m-1''}, \frac{1}{2}(t^N - \hat{t}^N)\right) \tag{1.64a}$$

where the functions $f'(t^{2'}, \ldots, t^{n'})$ and $f''(t^{2''}, \ldots, t^{m''})$ determine the prepotentials of M' and M'' in the form (1.18). Here $N = n + m$,

$$t^1 = \frac{t^{1'} + t^{1''}}{2}, \quad \hat{t}^1 = \frac{t^{1'} - t^{2''}}{2}$$
$$t^N = t^{n'} + t^{m''}, \quad \hat{t}^N = t^{n'} - t^{m''}. \tag{1.64b}$$

Observe that only trivial Frobenius manifolds can be multiplied by a one-dimensional Frobenius manifold.

Example 1.7. [42] M is the space of all polynomials of the form

$$M = \{\lambda(p) = p^{n+1} + a_n p^{n-1} + \ldots + a_1 | a_1, \ldots, a_n \in \mathbf{C}\} \tag{1.65}$$

with a nonstandard affine structure. We identify the tangent plane to M with the space of all polynomials of the degree less than n. The algebra A_λ on $T_\lambda M$ by definition coincides with the algebra of truncated polynomials

$$A_\lambda = \mathbf{C}[p]/(\lambda'(p)) \tag{1.66a}$$

(the prime denotes d/dp). The invariant inner product is defined by the formula

$$< f, g >_\lambda = \operatorname*{res}_{p=\infty} \frac{f(p)g(p)}{\lambda'(p)}. \tag{1.66b}$$

The unity vector field e and the Euler vector field E read

$$e := \frac{\partial}{\partial a_1}, \quad E := \frac{1}{n+1} \sum_i (n - i + 1) a_i \frac{\partial}{\partial a_i}. \tag{1.66c}$$

We will see that this is an example of a Frobenius manifold in Lecture 4.

Remark 1.6. The notion of Frobenius manifolfd admits algebraic formalization in terms of the ring of functions on a manifold. More precisely, let R be a commutative associative algebra with a unity over a field k of characteristics $\neq 2$. We are interested in structures of Frobenius algebra over R in the R-module of k-derivations $Der(R)$ (i.e. $u(\kappa) = 0$ for $\kappa \in k, u \in Der(R)$) satisfying

$$\tilde{\nabla}_u(\lambda)\tilde{\nabla}_v(\lambda) - \tilde{\nabla}_v(\lambda)\tilde{\nabla}_u(\lambda) = \tilde{\nabla}_{[u,v]}(\lambda) \quad \text{identicaly in } \lambda \tag{1.67a}$$

$$\text{for } \tilde{\nabla}_u(\lambda)v = \nabla_u v + \lambda u \cdot v, \tag{1.67b}$$

(see below Lemma 3.1)

$$\nabla_u e = 0 \quad \text{for all } u \in Der(R) \tag{1.67c}$$

where e is the unity of the Frobenius algebra $Der(R)$. Non-degenerateness of the symmetric inner product

$$< , >: Der(R) \times Der(R) \to R$$

means that it provides an isomorphism $\operatorname{Hom}_R(Der(R), R) \to Der(R)$. I recall that the covariant derivative is a derivation $\nabla_u v \in Der(R)$ defined for any u, $v \in Der(R)$ being determined from the equation

$$< \nabla_u v, w >=$$

$$\frac{1}{2}[u < v, w > +v < w, u > -w < u, v > + < [u,v], w > + < [w,u], v > + < [w,v], u >] \tag{1.68}$$

for any $w \in Der(R)$ (here $[,]$ denotes the commutator of derivations).

To reformulate algebraically the scaling invariance (1.50) we need to introduce gradings to the algebras R and $Der(R)$. In the case of algebras of functions the gradings are determined by the assumptions

$$\deg t^\alpha = 1 - q_\alpha, \text{ for } q_\alpha \neq 1, \text{ or } \deg e^{t^\alpha} = r_\alpha \text{ for } q_\alpha = 1, \qquad (1.69a)$$

$$\deg \partial_\alpha = q_\alpha \qquad (1.69b)$$

where the numbers q_α, r_α are defined by the Euler vector field

$$E = \sum_\alpha [(1 - q_\alpha)t^\alpha + r_\alpha]\partial_\alpha. \qquad (1.69c)$$

Remark 1.7. From the definition it follows that the topology of a Frobenius manifold is very simple: it is a domain in $\mathbf{C}^k \times \mathbf{C}^{*l}$ for some k, l, $k + l = n$ due to existence of locally Euclidean metric on M. Below (in Appendix B) we will modify the definition introducing twisted Frobenius manifolds. They are glued from Frobenius manifolds as from the building blocks. The topology of twisted Frobenius manifolds can be nontrivial (see also Lecture 5).

Appendix A.
Polynomial solutions of WDVV. Towards classification of Frobenius manifolds with good analytic properties.

Let all the structure constants of a Frobenius manifold be analytic in the point $t = 0$. Then the germ of the Frobenius manifold near the point $t = 0$ can be considered as a deformation of the Frobenius algebra $A_0 := T_{t=0}M$. This is a graded Frobenius algebra with a basis $e_1, \ldots, e_n$ and with the structure constants $c_{\alpha\beta}^\gamma(0)$. The degrees of the basic vectors are

$$\deg e_\alpha = q_\alpha$$

where the numbers q_α are defined in (1.21). The algebras $T_t M$ for $t \neq 0$ can be considered thus as *deformations* of the graded Frobenius algebra $T_0 M$. In the physical setting (see Lecture 2 below) $T_0 M$ is the primary chiral algebra of the correspondent topological conformal field theory. The algebras $T_t M$ are operator algebras of the perturbed topological field theory. So the problem of classification of analytic deformations of graded Frobenius algebras looks to be also physically motivated. (Probably, analytic deformability in the sense that the graded Frobenius algebra can be the tangent algebra at the origin of an analytic Frobenius manifold imposes a strong constraint on the algebra A_0.)

We consider here the case where all the degrees $\deg t^\alpha$ are *real positive* numbers and not all of them are equal. In the normalization (1.18) that means that $0 < d < 1$.

Problem. To find all the solutions of WDVV being analytic in the origin $t = 0$ with real positive degrees of the flat coordinates.

Notice that for the positive degrees analyticity in the origin and the quasihomogeneity (1.5) implies that the function $F(t)$ is a polynomial in $t^1, \ldots, t^n$. So the problem coincides with the problem of finding of the *polynomial solutions* of WDVV.

For $n = 2$ all the noncubic polynomial solutions have the form (1.24a) where k is an integer and $k \geq 4$. Let us consider here the next case $n = 3$. Here we have a function F of the form (1.26) and

$$\deg t^1 = 1, \ \deg t^2 = 1 - \frac{d}{2}, \ \deg t^3 = 1 - d, \ \deg f = 3 - d. \tag{A.1}$$

The function f must satisfy the equation (1.28). If

$$f(x,y) = \sum a_{pq} x^p y^q$$

then the condition of quasihomogeneity reads

$$a_{pq} \neq 0 \text{ only for } p + q - 3 = (\tfrac{1}{2}p + q - 1)d. \tag{A.2}$$

Hence d must be a rational number. Solving the quasihomogeneity equation (A.2) we obtain the following two possibilities for the function f: 1.

$$f = \sum_k a_k x^{4-2km} y^{kn-1} \tag{A.3a}$$

$$d = \frac{n - 2m}{n - m} \qquad (A.3b)$$

for some natural numbers n, m, n is odd, and 2.

$$f = \sum_k a_k x^{4-km} y^{kn-1} \qquad (A.4a)$$

$$d = \frac{2(n - m)}{2n - m} \qquad (A.4b)$$

for some natural numbers n, m, m is odd. Since the powers in the expansions of f must be nonnegative, we obtain the following three possibilities for f:

$$f = ax^2 y^{n-1} + by^{2n-1}, \quad n \geq 3 \qquad (A.5a)$$

$$f = ay^{n-1}, \quad n \geq 5 \qquad (A.5b)$$

$$f = ax^3 y^{n-1} + bx^2 y^{2n-1} + cxy^{3n-1} + dy^{4n-1}, \quad n \geq 2. \qquad (A.5c)$$

The inequalities for n in these formulae can be assumed since the case when f is at most cubic polynomial is not of interest. (It is easy to see that for a cubic solution with $n = 3$ necessary $f = 0$.)

Substituting (A.5) to (1.28) we obtain resp.

$$\begin{aligned} a(n-1)(n-2)(n-3) &= 0 \\ -4a^2(n-1)^2 + b(2n-1)(2n-2)(2n-3) &= 0 \end{aligned} \qquad (A.6a)$$

$$a(n-1)(n-2)(n-3) = 0 \qquad (A.6b)$$

$$\begin{aligned} a(n-1)(n-2)(n-3) &= 0 \\ 18a^2\left((n-1)(n-2) - 2(n-1)^2\right) + b(2n-1)(2n-2)(2n-3) &= 0 \\ c(3n-1)(3n-2)(3n-3) &= 0 \\ -4b^2(2n-1)^2 + 6ac(3n-1)(3n-2) + d(4n-1)(4n-2)(4n-3) &= 0. \end{aligned} \qquad (A.6c)$$

It is clear that a must not vanish for a nonzero f. So n must equal 3 in the first case (A.6a), 2 or 3 in the third one, and in the second case (A.6b) there is no non-zero solutions. Solving the system (A.6a) for $n = 3$ and the system (A.6c) for $n = 2$ and $n = 3$ we obtain the following 3 polynomial solutions of WDVV:

$$F = \frac{t_1^2 t_3 + t_1 t_2^2}{2} + \frac{t_2^2 t_3^2}{4} + \frac{t_3^5}{60} \qquad (A.7)$$

$$F = \frac{t_1^2 t_3 + t_1 t_2^2}{2} + \frac{t_2^3 t_3}{6} + \frac{t_2^2 t_3^3}{6} + \frac{t_3^7}{210} \qquad (A.8)$$

$$F = \frac{t_1^2 t_3 + t_1 t_2^2}{2} + \frac{t_2^3 t_3^2}{6} + \frac{t_2^2 t_3^5}{20} + \frac{t_3^{11}}{3960}. \qquad (A.9)$$

These are unique up to the equivalence noncubic polynomial solutions of WDVV with $n = 3$ with positive degrees of t^α.

For the case $d = 1$ one can easily find a polynomial solution

$$F = \frac{t_1^2 t_3 + t_1 t_2^2}{2} + t_2^4. \qquad (A.10)$$

This admits a one-parameter group of equivalences

$$t_3 \mapsto t_3 + const.$$

The corresponding Frobenius manifold is essentially reduced to a two-dimensional one.

Refining above arguments we prove the following

Theorem A.1. *1). All noncubic polynomial solutions of WDVV for $n = 3$ are equivalent to one of (A.7), (A.8), (A.9), (A.10).*

2). Besides these for $n = 3$ there exist solutions analytic at $t = 0$ only for $d = 2\frac{m+1}{m+2}$ (one-parameter family of solutions), $d = 2\frac{m+2}{m+4}$ (one solution) and $d = 2\frac{m+3}{m+6}$ (one solution) for an arbitrary integer m.

Proof. From the ansatz (1.30a) we obtain for the function $f(x, y)$ an expansion

$$f(x, y) = \sum a_k y^{k-1} x^{4+qk}$$

where, we recall that

$$q = -2\frac{1 - d}{2 - d}.$$

So q must be a rational number. For $q = 0$ we obtain the solution (A.10). For negative

$$q = -\frac{m}{n}$$

we rewrite the expansion for f in the form

$$f = \sum_{k \geq 1} a_k x^{4-mk} y^{kn-1}.$$

So $m \leq 4$. This gives the solutions (A.7) - (A.9). For positive

$$q = \frac{m}{n}$$

substitution of the series

$$f = \sum_{k \geq 1} a_k x^{4+mk} y^{kl-1}$$

into (1.31a) gives

$$(n - 1)(n - 2)(n - 3)a_1 = 0$$

and the recursion relations of the form

$$(kn-1)(kn-2)(kn-3)a_k = \text{polynomial in } a_1, \ldots, a_{k-1}.$$

This proves the second part of the theorem.

The polynomial (A.7) coincides with the prepotential of Example 1.7 with $n = 3$. The crucial observation to understand the nature of the other polynomials (A.8) and (A.9) was done by V.I.Arnold. He observed that the degrees $5 = 4+1$, $7 = 6+1$, and $11 = 10+1$ of the polynomials have a simple relation to the Coxeter numbers of the groups of symmetries of the Platonic solids (4, 6, 10 for the tetrahedron, cube and icosahedron resp.). In Lecture 4 I will show how to explain this observation using a hidden symmetry of WDVV (see also Appendix G). I do not know the sense of the solutions in the form of infinite power series described in the second part of the theorem.

We will consider now solutions of WDVV with $n = 3$ and $d = 1$. The Euler vector field must have the form

$$E = t_1\partial_1 + \frac{1}{2}t_2\partial_2 + r\partial_3. \qquad (A.11)$$

We will assume that $r > 0$ (the case $r = 0$ will be considered in Appendix C). We will be interested in solutions of the form

$$F = \frac{1}{2}t_1^2 t_3 + \frac{1}{2}t_1 t_2^2 + \sum_{k,l \geq 0} a_{kl} t_2^k e^{lt_3}. \qquad (A.12)$$

Theorem A.2. *For $d = 1$ and $r > 0$ there exist only 3 solutions of WDVV of the form (A.12):*

$$F = \frac{1}{2}t_1^2 t_3 + \frac{1}{2}t_1 t_2^2 - \frac{1}{24}t_2^4 + t_2 e^{t_3}, \quad r = \frac{3}{2} \qquad (A.13)$$

$$F = \frac{1}{2}t_1^2 t_3 + \frac{1}{2}t_1 t_2^2 + \frac{1}{2}t_2^4 + t_2^2 e^{t_3} - \frac{1}{48}e^{2t_3}, \quad r = 1 \qquad (A.14)$$

$$F = \frac{1}{2}t_1^2 t_3 + \frac{1}{2}t_1 t_2^2 - \frac{1}{72}t_2^4 + \frac{2}{3}t_2^3 e^{t_3} + \frac{2}{3}t_2^2 e^{2t_3} + \frac{9}{16}e^{4t_3}, \quad r = \frac{1}{2}. \qquad (A.15)$$

The proof is obtained by a direct substitution of the ansatz (1.30b) to the equation (1.31b)

In Lecture 4 it will be shown that the three solutions correspond to certain extensions of the affine Weyl groups of the type A_2, B_2 and G_2 resp.

More generally, we will be interested in solutions of WDVV of the form

$$F(t) = \text{cubic} + \sum_{k_1,k_2,\ldots \geq 0} a_{k_1 k_2 \ldots}(t'') \exp\left(k_1 t_1' + k_2 t_2' + \ldots\right) \qquad (A.16)$$

where the coordinates t are subdivided in two parts $t = (t', t'')$ in such a way that $\deg t' = 0$, $\deg t'' \neq 0$, assuming convergence of the series in a neibourghood of $t' = -\infty$, $t'' = 0$. These will be called *solutions with good analytic properties*. From the first experiments of this Appendix it follows that solutions with good analytic properties are isolated points in the space of all solutions of WDVV (at least, under some additional assumptions about the solution, like the semi-simplicity assumption below). The results of this paper suggest a conjectural correspondence between Frobenius manifolds with good analytic properties and certain reflection groups and their extensions.

Appendix B.
Symmetries of WDVV. Twisted Frobenius manifolds.

By definition *symmetries* of WDVV are the transformations

$$t^\alpha \mapsto \hat{t}^\alpha,$$
$$\eta_{\alpha\beta} \mapsto \hat{\eta}_{\alpha\beta}, \qquad\qquad (B.1)$$
$$F \mapsto \hat{F}$$

preserving the equations. First examples of the symmetries have been introduced above: they are equivalencies of Frobenius manifolds and shifts along vectors belonging to the kernel of the grading operator Q.

Here we describe two types of less trivial symmetries for which the map $t^\alpha \mapsto \hat{t}^\alpha$ preserves the multiplication of the vector fields.

Type 1. Legendre-type transformation S_κ for a given $\kappa = 1, \ldots, n$

$$\hat{t}_\alpha = \partial_\alpha \partial_\kappa F(t) \qquad\qquad (B.2a)$$

$$\frac{\partial^2 \hat{F}}{\partial \hat{t}^\alpha \partial \hat{t}^\beta} = \frac{\partial^2 F}{\partial t^\alpha \partial t^\beta} \qquad\qquad (B.2b)$$

$$\hat{\eta}_{\alpha\beta} = \eta_{\alpha\beta}. \qquad\qquad (B.2c)$$

We have

$$\partial_\alpha = \partial_\kappa \cdot \hat{\partial}_\alpha. \qquad\qquad (B.3)$$

So the transformation S_κ is invertible where ∂_κ is an invertible element of the Frobenius algebra of vector fields. Note that the unity vector field

$$e = \frac{\partial}{\partial \hat{t}^\kappa}. \qquad\qquad (B.2d)$$

The transformation S_1 is the identity; the transformations S_κ commute for different κ.

To describe what happens with the scaling degrees (assuming diagonalizability of the degree operator Q) we shift the degrees putting

$$\mu_\alpha := q_\alpha - \frac{d}{2}, \quad \alpha = 1, \ldots, n. \qquad\qquad (B.4)$$

Observe that the spectrum consists of the eigenvalues of the operator $-\hat{V}$ (see (1.51)). The shifted degrees are symmetric w.r.t. zero

$$\mu_\alpha + \mu_{n-\alpha+1} = 0. \qquad\qquad (B.5)$$

We will call the numbers $\mu_1, \ldots, \mu_n$ *spectrum* of the Frobenius manifold. Knowing the spectrum we can uniquely reconstruct the degrees putting

$$q_\alpha = \mu_\alpha - \mu_1, \quad d = -2\mu_1. \qquad\qquad (B.6)$$

It is easy to see that the transformations S_κ preserve the spectrum up to permutation of the numbers $\mu_1, \ldots, \mu_n$: for $\kappa \neq \frac{n}{2}$ it interchanges the pair (μ_1, μ_n) with the pair $(\mu_\kappa, \mu_{n-\kappa+1})$. For $\kappa = \frac{n}{2}$ the transformed Frobenius manifold is of the second type (1.22).

To prove that (B.2) determines a symmetry of WDVV we introduce on the Frobenius manifold M a new metric $<\,,\,>_\kappa$ putting

$$< a, b >_\kappa := < \partial_\kappa^2, a \cdot b > . \qquad (B.7)$$

Exercise B.1. Prove that the variables $\hat{t}^\alpha$ (B.2a)) are the flat coordinates of the metric (B.7). Prove that

$$< \hat{\partial}_\alpha \cdot \hat{\partial}_\beta, \hat{\partial}_\gamma >_\kappa = \hat{\partial}_\alpha \hat{\partial}_\beta \hat{\partial}_\gamma \hat{F}(\hat{t}). \qquad (B.8)$$

Example B.1. For

$$F = \frac{1}{2} t^1{}^2 t^2 + e^{t^2}$$

$(d = 1)$ the transformation S_2 gives

$$\hat{t}^1 = e^{t^2}$$
$$\hat{t}^2 = t^1.$$

Renumbering $\hat{t}^1 \leftrightarrow \hat{t}^2$ (due to (B.2d)) we obtain

$$\hat{F} = \frac{1}{2}(\hat{t}^1)^2 \hat{t}^2 + \frac{1}{2}(\hat{t}^2)^2 \left(\log t^2 - \frac{3}{2} \right). \qquad (B.9)$$

This coincides with (1.24b) (now $d = -1$). See also Example 5.5 below.

If there are coincidences between the degrees

$$q_{\kappa_1} = \ldots = q_{\kappa_s} \qquad (B.10)$$

then we can construct more general transformation S_c putting

$$\hat{t}_\alpha = \sum_{i=1}^{s} c^i \partial_\alpha \partial_{\kappa_i} F(t) \qquad (B.11)$$

for arbitrary constants $(c^1, \ldots, c^s) =: c$. This is invertible when the vector field

$$\sum c^i \partial_{\kappa_i}$$

is invertible. The transformed metric on M depends quadratically on c^i

$$< a, b >_c := < \left(\sum c^i \partial_{\kappa_i} \right)^2, a \cdot b > . \qquad (B.12)$$

Type 2. The inversion I:

$$\hat{t}^1 = \frac{1}{2}\frac{t_\sigma t^\sigma}{t^n}$$

$$\hat{t}^\alpha = \frac{t^\alpha}{t^n} \text{ for } \alpha \neq 1, n \qquad (B.13a)$$

$$\hat{t}^n = -\frac{1}{t^n}$$

(the coordinates are normalized as in (1.18)),

$$\hat{F}(\hat{t}) = (t^n)^{-2}\left[F(t) - \frac{1}{2}t^1 t_\sigma t^\sigma\right] = (\hat{t}^n)^2 F + \frac{1}{2}\hat{t}^1 \hat{t}_\sigma \hat{t}^\sigma, \qquad (B.13b)$$

$$\hat{\eta}_{\alpha\beta} = \eta_{\alpha\beta}.$$

Note that the inversion acts as a conformal transformation of the metric $<,>$

$$\eta_{\alpha\beta}d\hat{t}^\alpha d\hat{t}^\beta = (t^n)^{-2}\eta_{\alpha\beta}dt^\alpha dt^\beta. \qquad (B.14)$$

The inversion *changes* the spectrum $\mu_1, \ldots, \mu_n$.

Lemma B.1. *If*

$$E(t) = \sum(1 - q_\alpha)t^\alpha \partial_\alpha \qquad (B.15)$$

then after the transform one obtains

$$\hat{E}(\hat{t}) = \sum(1 - \hat{q}_\alpha)\hat{t}^\alpha \hat{\partial}_\alpha \qquad (B.16a)$$

where

$$\hat{\mu}_1 = -1 + \mu_n, \ \hat{\mu}_n = 1 + \mu_1, \ \hat{\mu}_\alpha = \mu_\alpha \text{ for } \alpha \neq 1, n. \qquad (B.16b)$$

Particularly,

$$\hat{d} = 2 - d. \qquad (B.16d)$$

If

$$E(t) = \sum(1 - q_\alpha)t^\alpha \partial_\alpha + \sum_{q_\sigma=1} r^\sigma \partial_\sigma$$

and $d \neq 1$, or $d = 1$ but $r^n = 0$, then

$$\hat{E}(\hat{t}) = \hat{E}^\alpha(\hat{t})\hat{\partial}_\alpha \qquad (B.17a)$$

where

$$\hat{E}^1 = \hat{t}^1 + \sum_{q_\sigma=1} r^\sigma \hat{t}^{n-\sigma+1}$$

$$\hat{E}^\alpha = (d - q_\alpha)\hat{t}^\alpha \text{ for any } \alpha \text{ s.t. } q_\alpha \neq 1 \qquad (B.17b)$$

$$\hat{E}^\sigma = (d - 1)\hat{t}^\sigma - r^\sigma \hat{t}^n \text{ for any } \sigma \text{ s.t. } q_\sigma = 1$$

$$\hat{E}^n = (d - 1)\hat{t}^n.$$

Proof is straightforward.

The transformation of the type 2 looks more misterious (we will see in Lecture 3 that this is a Schlesinger transformation of WDVV in the sense of [135]). We leave to the reader to verify that the inversion preserves the multiplication of vector fields. Hint: use the formulae

$$\hat{c}_{\alpha\beta\gamma} = t_1 c_{\alpha\beta\gamma} - t_\alpha \eta_{\beta\gamma} - t_\beta \eta_{\alpha\gamma} - t_\gamma \eta_{\alpha\beta}$$

$$\hat{c}_{\alpha\beta n} = t_1 c_{\alpha\beta\sigma} t^\sigma - \frac{1}{2}\eta_{\alpha\beta} t_\sigma t^\sigma - 2 t_\alpha t_\beta$$

$$\hat{c}_{\alpha n n} = t_1 c_{\alpha\lambda\mu} t^\lambda t^\mu - 2 t_\alpha t_\sigma t^\sigma$$

$$\hat{c}_{nnn} = t_1 c_{\lambda\mu\nu} t^\lambda t^\mu t^\nu - \frac{3}{2}(t_\sigma t^\sigma)^2$$

(B.18a)

(here $\alpha, \beta, \gamma \neq 1, n$) together with (B.14) to prove that

$$\hat{c}_{\alpha\beta\gamma} = (t^n)^{-2} \frac{\partial t^\lambda}{\partial \hat{t}^\alpha} \frac{\partial t^\mu}{\partial \hat{t}^\beta} \frac{\partial t^\nu}{\partial \hat{t}^\gamma} c_{\lambda\mu\nu}.$$

(B.18b)

Exercise B.2. Show that the solution (1.24c) is the I-transform of the solution (1.24b).

Exercise B.3. Prove that the group $SL(2, \mathbf{C})$ acts on the space of solutions of WDVV with $d = 1$ by

$$t^1 \mapsto t^1 + \frac{1}{2}\frac{c}{ct^n + d}\sum_{\sigma \neq 1} t_\sigma t^\sigma$$

$$t^\alpha \mapsto \frac{t^\alpha}{ct^n + d}$$

(B.19)

$$t^n \mapsto \frac{at^n + b}{ct^n + d},$$

$$ad - bc = 1.$$

[Hint: consider superpositions of I with the shifts along t^n.]

The inversion is an involutive transformation up to an equivalence

$$I^2 : (t^1, t^2, \dots, t^{n-1}, t^n) \mapsto (t^1, -t^2, \dots, -t^{n-1}, t^n)$$

$$F \mapsto F.$$

(B.20)

Proposition B.1. *Assuming invertibility of the transformations S_κ, I we can reduce by these transformations any solution of WDVV to a solution with*

$$0 \leq q_\alpha \leq d \leq 1.$$

(B.21)

Definition B.1. A Frobenius manifold will be called *reduced* if it satisfies the inequalities (B.21).

Particularly, the transformations S_κ are invertible near those points t of a Frobenius manifold M where the algebra $T_t M$ has no nilpotents. In the next Lecture we will obtain complete local classification of such Frobenius manifolds.

Using the above transformations I, S we can glue together a few Frobenius manifolds to obtain a more complicated geometrical object that will be called *twisted Frobenius manifold*. The multiplication of tangent vector fields is globally well-defined on a twisted Frobenius manifold. But the invariant inner product (and therefore, the function F) is defined only locally. On the intersections of the coordinate charts these are to transform according to the formula (B.12) or (B.14). We will construct examples of twisted Frobenius manifolds in Appendices C, J. Twisted Frobenius manifolds could also appear as the moduli spaces of the topological sigma models of the B-type [151] where the flat metric is well-defined only locally.

149

Appendix C.
WDVV and Chazy equation.
Affine connections on curves with projective structure.

Here we consider three-dimensional Frobenius manifolds with $d = 1$. The degrees of the flat variables are

$$\deg t^1 = 1, \ \deg t^2 = 1/2, \ \deg t^3 = 0, \tag{C.1a}$$

and the Euler vector field

$$E = t^1 \partial_1 + \frac{1}{2} t^2 \partial_2. \tag{C.1b}$$

Let us look for a solution of WDVV being periodic in t^3 with the period 1 (modulo quadratic terms) and analytic in the point $t^1 = t^2 = 0$, $t^3 = i\infty$. The function F must have the form

$$F = \frac{1}{2}(t^1)^2 t^3 + \frac{1}{2} t^1 (t^2)^2 - \frac{(t^2)^4}{16} \gamma(t^3) \tag{C.2}$$

for some unknown 1-periodic function $\gamma = \gamma(\tau)$ analytic at $\tau = i\infty$

$$\gamma(\tau) = \sum_{n \geq 0} a_n q^n, \quad q = e^{2\pi i \tau}. \tag{C.3}$$

The coefficients a_n are determined up to a shift along τ,

$$\tau \mapsto \tau + \tau_0, \ a_n \mapsto a_n e^{2\pi i n \tau_0}. \tag{C.4}$$

For the function γ we obtain from (1.28)

$$\gamma''' = 6\gamma\gamma'' - 9\gamma'^2. \tag{C.5}$$

Exercise C.1. Prove that the equation (C.5) has a unique (modulo the ambiguity (C.4)) nonconstant solution*) of the form (C.3),

$$\gamma(\tau) = \frac{\pi i}{3} \left[1 - 24q - 72q^2 - 96q^3 - 168q^4 - \ldots \right], \quad q = e^{2\pi i \tau}. \tag{C.6}$$

The equation (C.5) was considered by J.Chazy [31] as an example of ODE with the general solution having moving natural boundary. It arose as a reduction of the self-dual Yang - Mills equation in [1]. Following [1, 142] I will call (C.5) *Chazy equation*.

---

*) The solution can be reduced to a real one by the change

$$t_3 \mapsto it_3, \ t_1 \mapsto i^{-1} t_1, \ F \mapsto i^{-1} F.$$

Exercise C.2. Show that the roots $\omega_1(\tau)$, $\omega_2(\tau)$, $\omega_3(\tau)$ of the cubic equation

$$\omega^3 + \frac{3}{2}\gamma(\tau)\omega^2 + \frac{3}{2}\gamma'(\tau)\omega + \frac{1}{4}\gamma''(\tau) = 0 \qquad (C.7)$$

satisfy the system

$$\dot{\omega}_1 = -\omega_1(\omega_2 + \omega_3) + \omega_2\omega_3$$
$$\dot{\omega}_2 = -\omega_2(\omega_1 + \omega_3) + \omega_1\omega_3$$
$$\dot{\omega}_3 = -\omega_3(\omega_1 + \omega_2) + \omega_1\omega_2. \qquad (C.8)$$

The system (C.8) was integrated by Halphen [75]. It was rediscovered in the context of the self-dual Einstein equations by Atiyah and Hitchin [11].

The main property [31] of Chazy equation is the invariance w.r.t. the group $SL(2,\mathbf{C})$

$$\tau \mapsto \tilde{\tau} = \frac{a\tau + b}{c\tau + d}, \quad ad - bc = 1 \qquad (C.9a)$$

$$\gamma(\tau) \mapsto \tilde{\gamma}(\tilde{\tau}) = (c\tau + d)^2\gamma(\tau) + 2c(c\tau + d). \qquad (C.9b)$$

The invariance (C.9) follows immediately from the invariance of WDVV w.r.t. the transformation (3.19). Observe that (C.9b) coincides with the transformation law of one-dimensional affine connection w.r.t. the Möbius transformations (C.9a) (cf. [142]).

We make here a digression about one-dimensional affine connections. One-dimensional real or complex manifolds will be considered; in the complex case only holomorphic connections will be of interest. The connection is determined by a function

$$\gamma(\tau) := \Gamma^1_{11}(\tau)$$

(holomorphic in the complex case, and $\Gamma^{\bar{1}}_{\bar{1}\bar{1}} = \overline{\gamma(\tau)}$) for any given local coordinate τ. The covariant derivative of a k-tensor $f(\tau)\,d\tau^k$ by definition is a $(k+1)$-tensor of the form

$$\nabla f(\tau)\,d\tau^{k+1} := \left(\frac{df}{d\tau} - k\gamma(\tau)f(\tau)\right)d\tau^{k+1}. \qquad (C.10)$$

This implies that under a change of coordinate

$$\tau \mapsto \tilde{\tau} = \tilde{\tau}(\tau) \qquad (C.11a)$$

(holomorphic in the complex case) the connection must transform as follows

$$\tilde{\gamma}(\tilde{\tau}) = \frac{1}{d\tilde{\tau}/d\tau}\gamma(\tau) - \frac{d^2\tilde{\tau}/d\tau^2}{d\tilde{\tau}/d\tau}. \qquad (C.11b)$$

One-dimensional affine connection has no local invariants: it can be reduced to zero by an appropriate change of coordinate. To find the *flat* local parameter x one is to look for a 1-form $\omega = \phi(\tau)\,d\tau$ such that $\nabla\omega = 0$, i.e.

$$\frac{d\phi}{d\tau} - \gamma\phi = 0 \qquad (C.12a)$$

and then put

$$\omega = dx. \qquad (C.12b)$$

The covariant derivative of a k-tensor $f\,dx^k$ coincides with the usual derivative w.r.t. the flat coordinate x

$$\nabla f\,dx^{k+1} \equiv \frac{df}{dx}\,dx^{k+1}. \qquad (C.13)$$

In arbitrary coordinate τ the covariant derivative can be written in the form

$$\nabla f\,d\tau^{k+1} = \phi^k \frac{d}{d\tau}\left(f\phi^{-k}\right)d\tau^{k+1}. \qquad (C.14)$$

Let us assume now that there is fixed a projective structure on the one-dimensional manifold. That means that the transition functions (C.11a) now are not arbitrary but they are the Möbius transformations (C.9a). Then the transformation law (C.11b) coincides with (C.9b).

When is it possible to reduce the connection to zero by a Möbius transformation? What is the complete list of differential-geometric invariants of an affine connection on one-dimensional manifold with a projective structure?

The following simple construction gives the answer to the questions.

Proposition C.1.

1. *For a one-dimensional connection γ the quadratic differential*

$$\Omega\,d\tau^2, \quad \Omega := \frac{d\gamma}{d\tau} - \frac{1}{2}\gamma^2 \qquad (C.15)$$

is invariant w.r.t. the Möbius transformations.

2. *The connection γ can be reduced to zero by a Möbius transformation iff $\Omega = 0$.*

Proof. The verification of the invariance of $\Omega\,d\tau^2$ is straightforward. From this it follows that $\Omega = 0$ when γ is reducible to 0 by a Möbius transformation. Conversely, solving the equation $\Omega = 0$ we obtain

$$\gamma = -\frac{2}{\tau - \tau_0}.$$

After the inversion

$$\tilde{\tau} = \frac{1}{\tau - \tau_0}$$

we obtain $\tilde{\gamma}(\tilde{\tau}) = 0$. Proposition is proved.

Remark C.1. For arbitrary change of coordinate $\tilde{\tau} = \tilde{\tau}(\tau)$ the "curvature" Ω transforms like projective connection

$$\tilde{\Omega} = \left(\frac{d\tau}{d\tilde{\tau}}\right)^2 \Omega + S_{\tilde{\tau}}(\tau) \qquad (C.16)$$

where $S_\tau(\tau)$ stands for the Schwartzian derivative

$$S_z(w) := \frac{d^3w/dz^3}{dw/dz} - \frac{3}{2}\left(\frac{d^2w/dz^2}{dw/dz}\right)^2. \qquad (C.17)$$

We obtain a map

$$\text{affine connections} \longrightarrow \text{projective connections.}$$

This is the appropriate differential-geometric interpretation of the well-known Miura transformation.

Exercise C.3. Let $P = P(\gamma, d\gamma/d\tau, d^2\gamma/d\tau^2, \ldots)$ be a polynomial such that for any affine connection γ the tensor $P\,d\tau^k$ for some k is invariant w.r.t. Möbius transformations. Prove that P can be represented as

$$P = Q(\Omega, \nabla\Omega, \nabla^2\Omega, \ldots) \qquad (C.18)$$

where Q is a graded homogeneous polynomial of the degree k assuming that $\deg \nabla^l\Omega = l+2$.

In other words, the "curvature" Ω and the covariant derivatives of it provide the complete set of differential-geometric invariants of an affine connection on a one-dimensional manifold with a projective structure.

Example C.1. Consider a Sturm - Liouville operator

$$L = -\frac{d^2}{dx^2} + u(x), \quad x \in D, \quad D \subset S^1 \text{ or } D \subset \mathbf{CP}^1. \qquad (C.19)$$

It determines a projective structure on D in the following standard way. Let $y_1(x)$, $y_2(x)$ be two linearly independent solutions of the differential equation

$$L\,y = 0. \qquad (C.20)$$

We introduce a new local coordinate τ in D putting

$$\tau = \frac{y_2(x)}{y_1(x)}. \qquad (C.21)$$

This specifies a projective structure in D. If D is not simply connected then a continuation of $y_1(x)$, $y_2(x)$ along a closed curve gives a linear substitution

$$y_1(x) \mapsto cy_2(x) + dy_1(x)$$

$$y_2(x) \mapsto ay_2(x) + by_1(x) \qquad (C.22)$$

for some constants a, b, c, d, $ad - bc = 1$ (conservation of the Wronskian $y_1y_2' - y_2y_1'$). This is a Möbius transformation of the local parameter τ. Another choice of the basis $y_1(x)$, $y_2(x)$ produces an equivalent projective structure in D.

153

We have also a natural affine connection in D. It is uniquely specified by saying that x is the flat coordinate for the connection. What is the "curvature" Ω of this affine connection w.r.t. the projective structure (C.21)? The answer is

$$\Omega d\tau^2 = 2u(x)dx^2 \qquad (C.23)$$

(verify it!).

Let us come back to the Chazy equation. It is natural to consider the general class of equations of the form

$$P(\gamma, d\gamma/d\tau, \ldots, d^{k+1}\gamma/d\tau^{k+1}) = 0 \qquad (C.24)$$

for a polynomial P invariant w.r.t. the transformations (C.9). Due to (C.18) these can be rewritten in the form

$$Q(\Omega, \nabla\Omega, \ldots, \nabla^k\Omega) = 0 \qquad (C.25a)$$

for

$$\Omega = \frac{d\gamma}{d\tau} - \frac{1}{2}\gamma^2, \qquad (C.25b)$$

Q is a graded homogeneous polynomial with $\deg \nabla^l\Omega = l + 2$. Putting

$$u := \frac{1}{2}\frac{\Omega d\tau^2}{\omega^2} \qquad (C.26)$$

(cf. (C.23)) where $\nabla\omega = 0$, $\omega =: dx$, we can represent (C.25) as

$$Q(2u, 2u', \ldots, 2u^{(k)}) = 0 \qquad (C.27)$$

for $u' = du/dx$ etc. Solving (C.27) we can reconstruct $\gamma(\tau)$ via two independent solutions $y_1(x)$, $y_2(x)$ of the Sturm - Liouville equation (C.20) normalized by $y_2'y_1 - y_1'y_2 = 1$

$$\tau = \frac{y_2(x)}{y_1(x)}, \quad \gamma = \frac{d(y_1^2)}{dx}. \qquad (C.28)$$

Let us consider examples of the equations of the form (C.25). I will consider only the equations linear in the highest derivative $\nabla^k\Omega$.

For $k = 0$ we have only the conditions of flatness. For $k = 1$ there exists only one invariant differential equation $\nabla\Omega = 0$ or

$$\gamma'' - 3\gamma\gamma' + \gamma^3 = 0. \qquad (C.29)$$

We have $u(x) = c^2$ (a constant); a particular solution of (C.29) is

$$\gamma = -\frac{2}{\tau}. \qquad (C.30)$$

The general solution can be obtained from (C.30) using the invariance (C.9).

For $k = 2$ the equations of our class must have the form

$$\nabla^2 \Omega + c\Omega^2 = 0 \qquad (C.31a)$$

for a constant c or more explicitly

$$\gamma''' - 6\gamma\gamma'' + 9\gamma'^2 + (c - 12)\left(\gamma' - \frac{1}{2}\gamma^2\right)^2 = 0. \qquad (C.31b)$$

For $c = 12$ this coincides with the Chazy equation (C.5). The corresponding equation (C.27)

$$u'' + 2cu^2 = 0 \qquad (C.32)$$

for $c \neq 0$ can be integrated in elliptic functions

$$u(x) = -\frac{3}{c}\wp_0(x) \qquad (C.33)$$

where $\wp_0(x)$ is the equianharmonic Weierstrass elliptic function, i.e. the inverse to the elliptic integral

$$x = \int_\infty^{\wp_0} \frac{dz}{2\sqrt{z^3 - 1}}. \qquad (C.34)$$

[All the solutions of (C.32) can be obtained from (C.33) by shifts and dilations along x. There is also a particular solution $u = -3/cx^2$ and the orbit of this w.r.t. (C.9).] So the solutions of (C.31) can be expressed as in (C.28) via the solutions of the Lamé equation with the equianharmonic potential

$$y'' + \frac{3}{c}\wp_0(x)y = 0 \qquad (C.35)$$

(for $c = 0$ via Airy functions). It can be reduced to the hypergeometric equation

$$t(t - 1)\frac{d^2y}{dt^2} + \left(\frac{7}{6}t - \frac{1}{2}\right)\frac{dy}{dt} + \frac{1}{12c}y = 0 \qquad (C.36)$$

by the substitution

$$t = 1 - \wp_0^3(x). \qquad (C.37)$$

From (C.28) we express the solution of (C.31) in the form

$$\tau = \frac{y_2(t)}{y_1(t)}, \quad \gamma = \frac{d\log y_1^2}{d\tau} \qquad (C.38)$$

for two linearly independent solutions $y_1(t)$, $y_2(t)$ of the hypergeometric equation.

Particularly, for the Chazy equation one obtains [31] the hypergeometric equation

$$t(t - 1)\frac{d^2y}{dt^2} + \left(\frac{7}{6}t - \frac{1}{2}\right)\frac{dy}{dt} + \frac{1}{144}y = 0. \qquad (C.39)$$

Note that the function $t = t(\tau)$ is the Schwartz triangle function $S(0, \pi/2, \pi/3; \tau)$. So the (projective) monodromy group of (C.39) coincides with the modular group.

Remark C.2. In the theory of the Lamé equation (C.35) the values

$$\frac{3}{c} = -m(m+1) \qquad (C.40)$$

for an integer m are of particular interest [56]. These look not to be discussed from the point of view of the theory of projective structures.

Chazy considered also the equation

$$\gamma''' - 6\gamma\gamma'' + 9\gamma'^2 + \frac{432}{n^2 - 36}\left(\gamma' - \frac{1}{2}\gamma^2\right)^2 = 0 \qquad (C.41)$$

for an integer $n > 6$. The correspondent Lamé equation

$$y'' - m(m+1)\wp_0(x)y = 0 \qquad (C.42)$$

has

$$m = \frac{3}{n} - \frac{1}{2}. \qquad (C.43)$$

Particularly, for the equation (C.5) $m = -\frac{1}{2}$. The solutions of (C.41), according to Chazy, can be expressed via the Schwartz triangle function $S(\pi/n, \pi/2, \pi/3; \tau)$. This can be seen from (C.36).

Exercise C.4.
1. Show that the equation of the class (C.25) of the order $k = 3$ can be integrated via solutions of the Lamé equation $y'' + A\wp(x)y = 0$ with arbitrary Weierstrass elliptic potential.
2. Show that for $k = 4$ γ can be expressed via the solutions of the equation (C.20) with the potential $u(x)$ satisfying

$$u^{IV} + au^3 + buu'' + cu'^4 = 0 \qquad (C.44)$$

for arbitrary constants a, b, c. Observe that for $a = -b = 10$, $c = -5$ the equation (C.44) is a particular case of the equation determining the genus two algebraic-geometrical (i.e. "two gap") potentials of the Sturm - Liouville operator [56].

Let me explain now the geometrical meaning of the solution (C.6) of the Chazy equation (C.5). The underlined complex one-dimensional manifold M here will be the modular curve

$$M := \{Im\tau > 0\}/SL(2, \mathbf{Z}). \qquad (C.45)$$

(This is not a manifold but an orbifold. So I will drop away the "bad" points $\tau = i\infty$, $\tau = e^{2\pi i/3}$, $\tau = i$ and the $SL(2, \mathbf{Z})$-images of them.) A construction of a natural affine connection on M essentially can be found in the paper [67] of Frobenius and Stickelberger. They described an elegant approach to the problem of differentiating of elliptic functions w.r.t.

their periods. I recall here the basic idea of this not very wellknown paper because of its very close relations to the subject of the present lectures.

Let us consider a lattice on the complex plane

$$L = \{2m\omega + 2n\omega' | m, n \in \mathbf{Z}\} \tag{C.46}$$

with the basis 2ω, $2\omega'$ such that

$$Im \left(\tau = \frac{\omega'}{\omega} \right) > 0. \tag{C.47}$$

Another basis

$$\omega', \ \omega \mapsto \tilde{\omega}' = a\omega' + b\omega, \ \tilde{\omega} = c\omega' + d\omega, \tag{C.48}$$

$$\begin{pmatrix} a & b \\ c & d \end{pmatrix} \in SL(2, \mathbf{Z}) \tag{C.49}$$

determines the same lattice.

Let $\mathcal{L}$ be the set of all lattices. I will drop away (as above) the orbifold points of $\mathcal{L}$ corresponding to the lattices with additional symmetry. So $\mathcal{L}$ is a two-dimensional manifold.

By E_L we denote the complex torus (elliptic curve)

$$E_L := \mathbf{C}/L. \tag{C.50}$$

We obtain a natural fiber bundle

$$\mathcal{M} = \begin{matrix} \downarrow & E_L \\ \mathcal{L} & \end{matrix} . \tag{C.51}$$

The space of this fiber bundle will be called *universal torus*. (Avoid confusion with universal elliptic curve: the latter is two-dimensional while our universal torus is three-dimensional. The points of the universal torus corresponding to proportional lattices give isomorphic elliptic curves.) Meromorphic functions on $\mathcal{M}$ will be called *invariant elliptic functions*. They can be represented as

$$f = f(z; \omega, \omega') \tag{C.52a}$$

with f satisfying the properties

$$f(z + 2m\omega + 2n\omega'; \omega, \omega') = f(z; \omega, \omega'), \tag{C.52b}$$

$$f(z; c\omega' + d\omega, \ a\omega' + b\omega) = f(z; \omega, \omega') \tag{C.52c}$$

for

$$\begin{pmatrix} a & b \\ c & d \end{pmatrix} \in SL(2, \mathbf{Z}). \tag{C.52d}$$

An example is the Weierstrass elliptic function

$$\wp \equiv \wp(z;\omega,\omega') = \frac{1}{z^2} + \sum_{m^2+n^2\neq 0}\left(\frac{1}{(z-2m\omega-2n\omega')^2} - \frac{1}{(2m\omega+2n\omega')^2}\right). \quad (C.53)$$

It satisfies the differential equation

$$(\wp')^2 = 4\wp^3 - g_2\wp - g_3 \quad (C.54)$$

with

$$g_2 \equiv g_2(\omega,\omega') = 60\sum_{m^2+n^2\neq 0}\frac{1}{(2m\omega+2n\omega')^4}, \quad (C.55)$$

$$g_3 \equiv g_3(\omega,\omega') = 140\sum_{m^2+n^2\neq 0}\frac{1}{(2m\omega+2n\omega')^6}. \quad (C.56)$$

Frobenius and Stickelberger found two vector fields on the universal torus $\mathcal{M}$. The first one is the obvious Euler vector field

$$\omega\frac{\partial}{\partial\omega} + \omega'\frac{\partial}{\partial\omega'} + z\frac{\partial}{\partial z}. \quad (C.57)$$

In other words, if f is an invariant elliptic function then so is

$$\omega\frac{\partial f}{\partial\omega} + \omega'\frac{\partial f}{\partial\omega'} + z\frac{\partial f}{\partial z}.$$

(There is even more simple example of a vector field on the universal torus: $\partial/\partial z$.) To construct the second vector field we need the Weierstrass ζ-function

$$\zeta \equiv \zeta(z;\omega,\omega') = \frac{1}{z} + \sum_{m^2+n^2\neq 0}\left(\frac{1}{z-2m\omega-2n\omega'} + \frac{1}{2m\omega+2n\omega'} + \frac{z}{(2m\omega+2n\omega')^2}\right),$$
$$(C.58)$$

$$\frac{d\zeta}{dz} = -\wp. \quad (C.59)$$

The ζ-function depends on the lattice L (but not on the particular choice of the basis ω, ω') but it is not an invariant elliptic function in the above sense since

$$\zeta(z+2m\omega+2n\omega';\omega,\omega') = \zeta(z;\omega,\omega') + 2m\eta + 2n\eta' \quad (C.60a)$$

where

$$\eta \equiv \eta(\omega,\omega') := \zeta(\omega;\omega,\omega'), \quad (C.60b)$$
$$\eta' \equiv \eta'(\omega,\omega') := \zeta(\omega';\omega,\omega'). \quad (C.60c)$$

The change (C.48) of the basis in the lattice acts on η, η' as

$$\tilde{\eta}' = a\eta' + b\eta, \quad \tilde{\eta} = c\eta' + d\eta. \quad (C.61)$$

Lemma C.1. [67] *If f is an invariant elliptic function then so is*

$$\eta\frac{\partial f}{\partial\omega} + \eta'\frac{\partial f}{\partial\omega'} + \zeta\frac{\partial f}{\partial z}. \qquad (C.62)$$

Proof is in a simple calculation using (C.60) and (C.61).

Exercise.

1). For $f = \wp(z;\omega,\omega')$ obtain [67, formula 11.]

$$\eta\frac{\partial\wp}{\partial\omega} + \eta'\frac{\partial\wp}{\partial\omega'} + \zeta\frac{\partial\wp}{\partial z} = -2\wp^2 + \frac{1}{3}g_2. \qquad (C.63)$$

2). For $f = \zeta(z;\omega,\omega')$ (warning: this is not an elliptic function!) obtain [67, formula 29.]

$$\eta\frac{\partial\zeta}{\partial\omega} + \eta'\frac{\partial\zeta}{\partial\omega'} + \zeta\frac{\partial\zeta}{\partial z} = \frac{1}{2}\wp' - \frac{1}{12}g_2 z. \qquad (C.64)$$

Consider now the particular class of invariant elliptic functions not depending on z.

Corollary C.1. *If $f = f(\omega,\omega')$ is a homogeneous function on the lattice of the weight* $(-2k)$,

$$f(c\omega, c\omega') = c^{-2k}f(\omega,\omega') \qquad (C.65)$$

then

$$\eta\frac{\partial f}{\partial\omega} + \eta'\frac{\partial f}{\partial\omega'} \qquad (C.66)$$

is a homogeneous function of the lattice of the degree $(-2k-2)$.

Exercise C.5. Using (C.63) and (C.64) prove that

$$\eta\frac{\partial g_2}{\partial\omega} + \eta'\frac{\partial g_2}{\partial\omega'} = -6g_3 \qquad (C.67)$$

$$\eta\frac{\partial g_3}{\partial\omega} + \eta'\frac{\partial g_3}{\partial\omega'} = -\frac{g_2^2}{3} \qquad (C.68)$$

[67, formula 12.] and

$$\eta\frac{\partial\eta}{\partial\omega} + \eta'\frac{\partial\eta}{\partial\omega'} = -\frac{1}{12}g_2\omega \qquad (C.69)$$

[67, formula 31.].

Any homogeneous function $f(\omega,\omega')$ on $\mathcal{L}$ of the weight $(-2k)$ determines a k-tensor

$$\hat{f}(\tau)d\tau^k \qquad (C.70a)$$

on the modular curve M where

$$f(\omega,\omega') = \omega^{-2k}\hat{f}(\tau), \quad \tau = \frac{\omega'}{\omega}. \qquad (C.70b)$$

In the terminology of the theory of automorphic functions $\hat{f}$ is an automorphic form of the modular group of the weight $2k$. [Also some assumptions about behaviour of $\hat{f}$ in the orbifold points are needed in the definition of an automorphic form; we refer the reader to a textbook in automorphic functions (e.g., [87]) for the details.] Due to Corollary we obtain a map

$$k - \text{tensors on } M \rightarrow (k+1) - \text{tensors on } M, \qquad (C.71a)$$

$$\hat{f}(\tau) \mapsto \nabla \hat{f}(\tau) := -\frac{2}{\pi i} \omega^{2k+2} \left(\eta \frac{\partial f}{\partial \omega} + \eta' \frac{\partial f}{\partial \omega'} \right). \qquad (C.71b)$$

(Equivalently: an automorphic form of the weight $2k$ maps to an automorphic form of the weight $2k + 2$.) This is the affine connection on M we need. We call it FS-connection.
Explicitly:

$$\nabla \hat{f} = -\frac{2}{\pi i} \left(\eta \frac{\partial}{\partial \omega} + \eta' \frac{\partial}{\partial \omega'} \right) \left[\omega^{-2k} \hat{f} \left(\frac{\omega'}{\omega} \right) \right]$$

$$= -\frac{2}{\pi i} \left[(-\eta \omega' + \eta' \omega) \frac{d\hat{f}}{d\tau} - 2k \omega \eta \hat{f} \right] = \frac{d\hat{f}}{d\tau} - k\gamma \hat{f} \qquad (C.72a)$$

(I have used the Legendre identity $\eta \omega' - \eta' \omega = \pi i/2$) where

$$\gamma \equiv \gamma(\tau) := -\frac{4}{\pi i} \omega \eta(\omega, \omega'). \qquad (C.72b)$$

The FS-connection was rediscovered in the theory of automorphic forms by Rankin [126] (see also [87, page 123]). From [107, page 389] we obtain

$$\gamma(\tau) = \frac{1}{3\pi i} \frac{\theta_1'''(0; \tau)}{\theta_1'(0; \tau)}. \qquad (C.72c)$$

From (C.60) it follows the representation of $\gamma(\tau)$ via the normalized Eisenstein series $E_2(\tau)$ (this is not an automorphic form!)

$$\gamma(\tau) = \frac{i\pi}{3} E_2(\tau) \qquad (C.72d)$$

$$E_2(\tau) = 1 + \frac{3}{\pi^2} \sum_{m \neq 0} \sum_{n=-\infty}^{\infty} \frac{1}{(m\tau + n)^2} = 1 - 24 \sum_{n=1}^{\infty} \sigma(n) q^n. \qquad (C.73)$$

Here $\sigma(n)$ stands for the sum of all the divisors of n.

Proposition C.2. *The FS-connection on the modular curve satisfies the Chazy equation (C.5).*

Proof (cf. [142]). Put

$$\hat{g}_2 \equiv \hat{g}_2(\tau) = \omega^4 g_2(\omega, \omega'), \quad \hat{g}_3 = \hat{g}_3(\tau) = \omega^6 g_3(\omega, \omega'). \qquad (C.74)$$

From (C.69) we obtain that

$$\Omega \equiv \gamma' - \frac{1}{2}\gamma^2 = \frac{2}{3(\pi i)^2}\hat{g}_2. \qquad (C.75)$$

Substituting to (C.67), (C.68) we obtain

$$\nabla^2\Omega + 12\Omega^2 = 0.$$

Proposition is proved.

From (C.72d), (C.73) we conclude that the solution (C.6) specified by the analyticity at $\tau = i\infty$ coincides with the FS-connection.

Exercise C.6. Derive from (C.5) the following recursion relation for the sums of divisors of natural numbers

$$\sigma(n) = \frac{12}{n^2(n-1)} \sum_{k=1}^{n-1} k(3n - 5k)\sigma(k)\sigma(n-k). \qquad (C.76)$$

To construct the flat coordinate for the FS-connection we observe that [67] for the discriminant

$$\Delta \equiv \Delta(\omega,\omega') = g_2^3 - 27g_3^2 \qquad (C.77)$$

we have from (C.67), (C.68)

$$\eta\frac{\partial\Delta}{\partial\omega} + \eta'\frac{\partial\Delta}{\partial\omega'} = 0. \qquad (C.78)$$

So

$$\nabla\hat{\Delta}(\tau) = 0 \qquad (C.79)$$

where $\hat{\Delta}(\tau)$ is a 6-tensor

$$\hat{\Delta}(\tau) = (2\pi)^{12}q\prod_{n=1}^{\infty}(1 - q^n)^{24}. \qquad (C.80)$$

The sixth root of $(2\pi)^{-12}\hat{\Delta}(\tau)d\tau^6$ gives the covariantly constant 1-form dx

$$dx := \eta^4(\tau)d\tau \qquad (C.81)$$

where $\eta(\tau)$ is the Dedekind eta-function

$$\eta(\tau) = q^{\frac{1}{24}}\prod_{n\geq 1}(1 - q^n) \qquad (C.82)$$

(avoid confusions with the function $\eta = \zeta(\omega;\omega,\omega')$!). We obtain particularly that the FS covariant derivative of a k-tensor $\hat{f}(\tau)$ can be written as

$$\nabla\hat{f} = \eta^{4k}(\tau)\frac{d}{d\tau}\left[\frac{\hat{f}}{\eta^{4k}(\tau)}\right]. \qquad (C.82)$$

Another consequence is the following formula for the FS-connection

$$\gamma(\tau) = \frac{1}{6}\frac{d}{d\tau}\log\hat{\Delta}(\tau) = 4\frac{d}{d\tau}\log\eta(\tau) = 8\pi i\left(\frac{1}{24} - \sum_{n=1}^{\infty}\frac{nq^n}{1-q^n}\right). \tag{C.83}$$

Remark C.3. Substituting (C.84) in the Chazy equation we obtain a 4-th order differential equation for the modular discriminant. It is a consequence of the third order equation of Jacobi [83, S. 103]

$$\left[12\psi^3\frac{d^2\psi}{dz^2}\right]^3 - 27\left[\frac{1}{8}\psi^4\frac{d^3(\psi^2)}{dz^2}\right]^2 = 1, \tag{C.85a}$$

$$\psi = \eta^{-2}(\tau), \quad z = 2\pi i\tau. \tag{C.85b}$$

Notice also the paper [80] of Hurwitz where it is shown that any holomorphic automorphic form or a meromorphic automorphic function associated with a group arising from Riemann surfaces of algebraic functions satisfies certain algebraic differential equation of the third order.

Consider now the Frobenius structure on the space

$$\hat{\mathcal{M}} := \{t^1, t^2, t^3 | Imt^3 > 0\} \tag{C.86}$$

specified by the FS solution (C.72) of the Chazy equation. So

$$\begin{aligned}F &= \frac{1}{2}(t^1)^2 t^3 + \frac{1}{2}t^1(t^2)^2 - \frac{\pi i}{2}(t^2)^4\left(\frac{1}{24} - \sum_{n=1}^{\infty}\frac{nq^n}{1-q^n}\right)\\ &= \frac{1}{2}(t^1)^2 t^3 + \frac{1}{2}t^1(t^2)^2 - \frac{\pi i}{2}(t^2)^4\left(\frac{1}{24} - \sum_{n=1}^{\infty}\sigma(n)q^n\right)\end{aligned} \tag{C.87}$$

where $q = \exp 2\pi i t^3$. Here we have $\tilde{\gamma} = \gamma$, i.e. the solution $\gamma(\tau)$ obeys the transformation rule

$$\gamma\left(\frac{a\tau+b}{c\tau+d}\right) = (c\tau+d)^2\gamma(\tau) + 2c(c\tau+d), \quad \begin{pmatrix} a & b \\ c & d \end{pmatrix} \in SL(2,\mathbf{Z}). \tag{C.88}$$

The formulae (B.19) for integer a, b, c, d determine a realisation of the group $SL(2,\mathbf{Z})$ as a group of symmetries of the Frobenius manifold (C.87). Factorizing $\hat{\mathcal{M}}$ over the transformations (B.19) we obtain a first example of twisted Frobenius manifold in the sense of Appendix B.

The invariant metric is a section of a line bundle over the manifold. This is the pull-back of the tangent bundle of the modular curve under the natural projection

$$(t^1, t^2, t^3) \mapsto t^3.$$

162

Indeed, the object

$$\left((dt^2)^2 + 2dt^1 dt^3\right) \otimes \frac{\partial}{\partial \tau} \qquad (C.89)$$

is invariant w.r.t. the transformations (B.19) (this follows from (B.14)).

Exercise C.7. Show that the formulae

$$t^1 = -\frac{1}{2\pi i} \left[\wp(z; \omega, \omega') + \omega^{-1} \eta(\omega; \omega, \omega')\right]$$

$$t^2 = \frac{\sqrt{2}}{\omega} \qquad (C.90)$$

$$t^3 = \tau = \omega'/\omega$$

establish an isomorphism of the twisted Frobenius manifold (C.87) with the universal torus $\mathcal{M}$.

In Appendix J I will explain the relation of this example to geometry of complex crystallographic group.

Remark C.4. The triple correlators $c_{\alpha\beta\gamma}(t)$ can be represented like a "sum over instanton corrections" [157] in topological sigma models (see the next lecture). For example,

$$c_{333} = 4\pi^4 (t^2)^4 \sum_{n \geq 1} n^3 A(n) \frac{q^n}{1 - q^n} \qquad (C.91)$$

where

$$A(n) = n^{-3} \prod \left[p_i^{k_i}(p_i^3 + p_i^2 + p_i + 1) - (p_i^2 + p_i + 1)\right] \qquad (C.92a)$$

for

$$n = \prod_i p_i^{k_i} \qquad (C.92b)$$

being the factorization of n in the product of powers of different primes $p_1, p_2, \ldots$.

Lecture 2.
Topological conformal field theories
and their moduli.

A quantum field theory (QFT) on a D-dimensional manifold Σ consists of:

1). a family of local fields $\phi_\alpha(x)$, $x \in \Sigma$ (functions or sections of a fiber bundle over Σ). A metric $g_{ij}(x)$ on Σ usualy is one of the fields (the gravity field).

2). A Lagrangian $L = L(\phi, \phi_x, ...)$. Classical field theory is determined by the Euler – Lagrange equations

$$\frac{\delta S}{\delta \phi_\alpha(x)} = 0, \quad S[\phi] = \int_\Sigma L(\phi, \phi_x, ...). \tag{2.1}$$

As a rule, the metric $g_{ij}(x)$ on Σ is involved explicitly in the Lagrangian even if it is not a dynamical variable.

3). Procedure of quantization usualy is based on construction of an appropriate path integration measure $[d\phi]$. The partition function is a result of the path integration over the space of all fields $\phi(x)$

$$Z_\Sigma = \int [d\phi] e^{-S[\phi]}. \tag{2.2}$$

Correlation functions (non normalized) are defined by a similar path integral

$$< \phi_\alpha(x)\phi_\beta(y) ... >_\Sigma = \int [d\phi] \phi_\alpha(x)\phi_\beta(y) ... e^{-S[\phi]}. \tag{2.3}$$

Since the path integration measure is almost never well-defined (and also taking into account that different Lagrangians could give equivalent QFTs) an old idea of QFT is to construct a self-consistent QFT by solving a system of differential equations for correlation functions. These equations were scrutinized in 2D conformal field theories where D=2 and Lagrangians are invariant with respect to conformal transformations

$$\delta g_{ij}(x) = \epsilon g_{ij}(x), \quad \delta S = 0.$$

This theory is still far from being completed.

Here I will consider another class of solvable 2-dimensional QFT: *topological field theories*. These theories admit *topological invariance*: they are invariant with respect to arbitrary change of the metric $g_{ij}(x)$ on the 2-dimensional surface Σ

$$\delta g_{ij}(x) = \text{arbitrary}, \quad \delta S = 0. \tag{2.4}$$

On the quantum level that means that the partition function Z_Σ depends only on topology of Σ. All the correlation functions also are topological creatures: they depend only on the labels of operators and on topology of Σ but not on the positions of the operators

$$< \phi_\alpha(x)\phi_\beta(y) ... >_\Sigma \equiv < \phi_\alpha\phi_\beta \cdots >_g \tag{2.5}$$

where g is the genus of Σ. The simplest example is 2D gravity with the Hilbert – Einstein action

$$S = \int R\sqrt{g}d^2x = \text{Euler characteristic of } \Sigma. \tag{2.6}$$

There are two ways of quantization of this functional. The first one is based on an appropriate discrete version of the model ($\Sigma \to$ polihedron). This way leads to considering matrix integrals of the form [23]

$$Z_N(t) = \int_{X^*=X} \exp\{-\text{tr}(X^2 + t_1 X^4 + t_2 X^6 + \ldots)\}dX \tag{2.7}$$

where the integral should be taken over the space of all $N \times N$ Hermitean matrices X. Here t_1, t_2 ... are called coupling constants. A solution of 2D gravity is based on the observation that after an appropriate limiting procedure $N \to \infty$ (and a renormalization of t) the limiting partition function coincides with τ-function of a particular solution of the KdV-hierarchy [24, 46, 73].

Another approach called *topological 2D gravity* is based on an appropriate supersymmetric extension of the Hilbert – Einstein Lagrangian [155 - 157]. This reduces the path integral over the space of all metrics $g_{ij}(x)$ on a surface Σ of the given genus g to an integral over the finite-dimensional space of conformal classes of these metrics, i.e. over the moduli space $\mathcal{M}_g$ of Riemann surfaces of genus g. Correlation functions of the model are expressed via intersection numbers of certain cycles on the moduli space [157, 41]

$$\sigma_p \leftrightarrow c_p \in H_*(\mathcal{M}_g), \quad p = 0, 1, \ldots \tag{2.8a}$$

$$< \sigma_{p_1} \sigma_{p_2} \ldots >_g = \#(c_{p_1} \cap c_{p_2} \cap \ldots) \tag{2.8b}$$

(here the subscript g means correlators on a surface of genus g). This approach is often called *cohomological field theory.*

More explicitly, let g, s be integers satisfying the conditions

$$g \geq 0, \quad s > 0, \quad 2 - 2g - s < 0. \tag{2.9}$$

Let

$$\mathcal{M}_{g,s} = \{(\Sigma, x_1, \ldots, x_s)\} \tag{2.10}$$

be the moduli space of smooth algebraic curves Σ of genus g with s ordered distinct marked points $x_1, \ldots, x_s$ (the inequalities (2.9) provide that the curve with the marked points is *stable*, i.e. it admits no infinitesimal automorphisms). By $\overline{\mathcal{M}}_{g,s}$ we will denote the Deligne – Mumford compactification of $\mathcal{M}_{g,s}$. Singular curves with double points obtained by a degeneration of Σ keeping the marked points off the singularities are to be added to compactify $\mathcal{M}_{g,s}$. Any of the components of $\Sigma \setminus$ (singularities) with the marked and the singular points on it is required to be stable. Natural line bundles $L_1, \ldots, L_s$ over $\overline{\mathcal{M}}_{g,s}$ are defined. By definition,

$$\text{fiber of } L_i|_{(\Sigma, x_1, \ldots, x_s)} = T_{x_i}^* \Sigma. \tag{2.11}$$

The Chern classes $c_1(L_i) \in H^*(\overline{\mathcal{M}}_{g,s})$ of the line bundles and their products are *Mumford - Morita - Miller classes* of the moduli space [112]. The genus g correlators of the topological gravity are defined via the intersection numbers of these cycles

$$< \sigma_{p_1} \ldots \sigma_{p_s} >_g := \prod_{i=1}^{s}(2p_i + 1)!! \int_{\overline{\mathcal{M}}_{g,s}} c_1^{p_1}(L_1) \wedge \ldots \wedge c_1^{p_s}(L_s). \qquad (2.12)$$

These numbers could be nonzero only if

$$\sum(p_i - 1) = 3g - 3. \qquad (2.13)$$

These are nonnegative rational numbers but not integers since $\overline{\mathcal{M}}_{g,s}$ is not a manifold but an orbifold.

It was conjectured by Witten that the both approaches to 2D quantum gravity should give the same results. This conjecture was proved by Kontsevich [89 - 90] (another proof was obtained by Witten [160]). He showed that the generating function

$$F(t) = \sum_{g,n} \sum_{p_1 < \ldots < p_n} \sum_{k_1,\ldots,k_n=0}^{\infty} \frac{T_{p_1}^{k_1} \ldots T_{p_n}^{k_n}}{k_1! \ldots k_n!} < \sigma_{p_1}^{k_1} \ldots \sigma_{p_n}^{k_n} >_g$$

$$= \sum_{g=0}^{\infty} \left\langle \exp \sum_{p=0}^{\infty} T_p \sigma_p \right\rangle_g \qquad (2.14)$$

(the free energy of 2D gravity) is logarythm of τ-function of a solution of the KdV hierarchy where $T_0 = x$ is the spatial variable of the hierarchy, $T_1, T_2, \ldots$ are the times (this was the original form of the Witten's conjecture). The τ-function is specified by the string equation (see eq. (6.54b) below for genus zero). Warning: the matrix gravity and the topological one correspond to two different τ-functions of KdV (in the terminology of Witten these are *different phases* of 2D gravity).

Other examples of 2D TFT's (see below) proved out to have important mathematical applications, probably being the best tool for treating sophisticated topological objects. For some of these 2D TFT's a description in terms of integrable hierarchies was conjectured. This gives rise to the following

Problem. To find a rigorous mathematical foundation of 2D topological field theory. More concretely, to elaborate a system of axioms providing the description (if any) of 2D TFT's in terms of integrable hierarchies of KdV type.

A first step on the way to the solution of the problem was done by Atiyah [10] (for any dimension D) in the spirit of G.Segal's axiomatization of conformal field theory. He proposed simple axioms specifying properties of correlators of the fields in the *matter sector* of a 2D topological field theory. In the matter sector the set of local fields $\phi_1(x),\ldots,\phi_n(x)$ (the so-called *primary fields* of the model) does not contain the metric on Σ. (Afterwards one should integrate over the space of metrics. This gives rise to a procedure of *coupling to topological gravity* that will be described below in Lecture 6. In the above example of topological gravity the matter sector consists only of the identity operator.) Then the correlators of the fields $\phi_1(x)$, ..., $\phi_n(x)$ obey very simple algebraic axioms. According to these axioms the matter sector of a 2D TFT is specified by:

1. The space of the local physical states A. I will consider only finite-dimensional spaces of the states

$$\dim A = n < \infty.$$

2. An assignment

$$(\Sigma, \partial\Sigma) \mapsto v_{(\Sigma,\partial\Sigma)} \in A_{(\Sigma,\partial\Sigma)} \tag{2.15}$$

for any oriented 2-surface Σ with an oriented boundary $\partial\Sigma$ that depends only on the topology of the pair $(\Sigma, \partial\Sigma)$ *). Here the linear space $A_{(\Sigma,\partial\Sigma)}$ is defined as follows:

$$\begin{aligned} A_{(\Sigma,\partial\Sigma)} &= \mathbf{C} \text{ if } \partial\Sigma = \emptyset \\ &= A_1 \otimes \ldots \otimes A_k \end{aligned} \tag{2.16}$$

if the boundary $\partial\Sigma$ consists of k components $C_1, \ldots, C_k$ (oriented cycles) and

$$A_i := \begin{cases} A & \text{if the orientation of } C_i \text{ is coherent to the orientation of } \Sigma \\ A^* & \text{(the dual space) otherwise.} \end{cases} \tag{2.17}$$

Drawing the pictures I will assume that the surfaces are oriented via the external normal vector; so only the orientation of the boundary will be shown explicitly.

The assignment (2.15) is assumed to satisfy the following three axioms.

---

*) We can modify this axiom assuming that the assignment (2.15) is covariant w.r.t. some representation in $A_{(\Sigma,\partial\Sigma)}$ of the mapping class group $(\Sigma, \partial\Sigma) \to (\Sigma, \partial\Sigma)$. The simplest generalisation of such a type is that, where the space of physical states is $\mathbf{Z}_2$-graded

$$A = A_{even} \oplus A_{odd}.$$

A homeomorphism $(\Sigma, \partial\Sigma) \to (\Sigma, \partial\Sigma)$ permuting the co-oriented components $C_1, \ldots, C_k$ of $\partial\Sigma$

$$C_1, \ldots, C_k \to C_{i_1}, \ldots, C_{i_k}$$

acts trivially on A_{even} but it multiplies the vectors of A_{odd} by the sign of the permutation $(i_1, \ldots, i_k)$. In this lectures we will not consider such a generalisation.

167

1. *Normalization:*

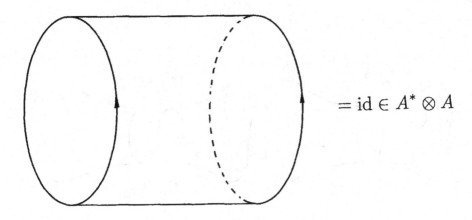

$$= \mathrm{id} \in A^* \otimes A$$

Figure 1

2. *Multiplicativity:* if

$$(\Sigma, \partial\Sigma) = (\Sigma_1, \partial\Sigma_1) \cup (\Sigma_2, \partial\Sigma_2) \tag{2.18a}$$

(disjoint union) then

$$v_{(\Sigma,\partial\Sigma)} = v_{(\Sigma_1,\partial\Sigma_1)} \otimes v_{(\Sigma_2,\partial\Sigma_2)} \in A_{(\Sigma,\partial\Sigma)} = A_{(\Sigma_1,\partial\Sigma_1)} \otimes A_{(\Sigma_2,\partial\Sigma_2)}. \tag{2.18b}$$

3. *Factorization.* To formulate this axiom I recall the operation of contraction defined in tensor products like (2.16), (2.17). By definition, ij-contraction

$$A_1 \otimes \ldots \otimes A_k \to A_1 \otimes \ldots \otimes \hat{A}_i \otimes \ldots \otimes \hat{A}_j \otimes \ldots \otimes A_k \tag{2.19}$$

(the i-th and the j-th factors are omitted in the r.h.s.) is defined when A_i and A_j are dual one to another using the standard pairing

$$A^* \otimes A \to \mathbf{C}$$

of the i-th and j-th factors and the identity on the other factors.

Let $(\Sigma, \partial\Sigma)$ and $(\Sigma', \partial\Sigma')$ coincide outside of a ball; inside the ball the two have the form

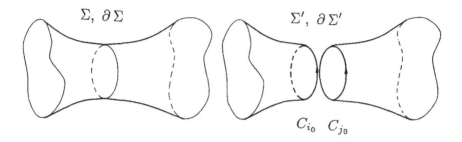

Figure 2

(I draw a cycle on the neck of Σ to emphasize that it is obtained from Σ' by gluing together the cycles C_{i_0} and C_{j_0}.) Then we require that

$$v_{(\Sigma, \partial\Sigma)} = i_0 j_0 - \text{contraction of } v_{(\Sigma', \partial\Sigma')}. \tag{2.20}$$

Particularly let us redenote by $v_{g,s}$ the vector

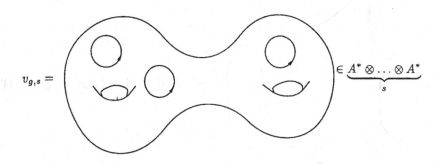

Figure 3. On the picture $g = 2$, $s = 3$.

This is a symmetric polylinear function on the space of the states. Choosing a basis

$$\phi_1, \ldots, \phi_n \in A \tag{2.21}$$

we obtain the components of the polylinear function

$$v_{g,s} \left(\phi_{\alpha_1} \otimes \ldots \phi_{\alpha_s} \right) =: < \phi_{\alpha_1} \ldots \phi_{\alpha_s} >_g \tag{2.22}$$

that by definition are called the genus g correlators of the fields $\phi_{\alpha_1}, \ldots, \phi_{\alpha_s}$.

We will prove, following [40], that the space of the states A carries a natural structure of a Frobenius algebra. All the correlators can be expressed in a pure algebraic way in terms of this algebra.

Let

c = 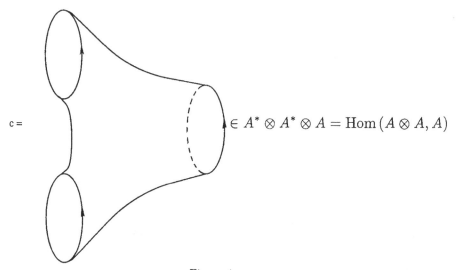 $\in A^* \otimes A^* \otimes A = \mathrm{Hom}\,(A \otimes A, A)$

Figure 4

$\eta =$ 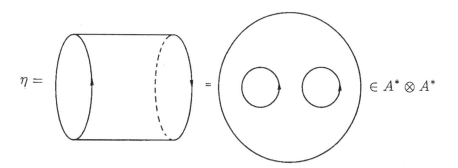 $\in A^* \otimes A^*$

Figure 5. Bilinear form $<\,,\,>$ on A.

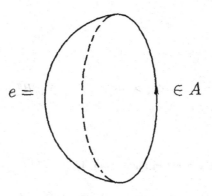

$$e = \qquad \in A$$

Figure 6

Theorem 2.1.

1. *The tensors c, η specify on A a structure of a Frobenius algebra with the unity e.*
2. *Let*

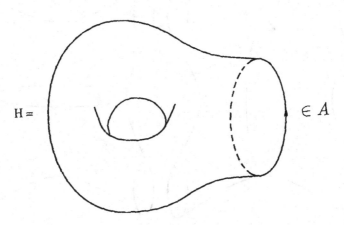

$$H = \qquad \in A$$

Figure 7

Then

$$< \phi_{\alpha_1} \cdots \phi_{\alpha_k} >_g \, = \, < \phi_{\alpha_1} \cdot \ldots \cdot \phi_{\alpha_k}, H^g > \qquad (2.23)$$

(in the r.h.s. · means the product in the algebra A).

Proof. Commutativity of the multiplication is obvious since we can interchange the legs of the pants on Fig. 4 by a homeomorphism. Similarly, we obtain the symmetry of the inner product $< \, , \, >$. Associativity follows from Fig. 8

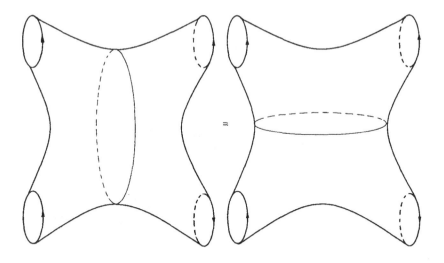

Figure 8

Particularly, the k-product is determined by the k-leg pants

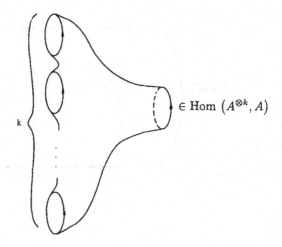

$\in \mathrm{Hom}\,\left(A^{\otimes k}, A\right)$

Figure 9

Unity:

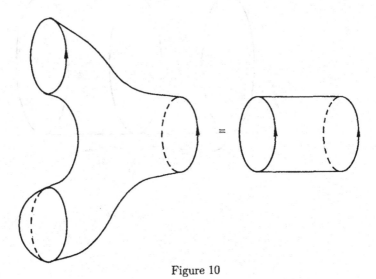

Figure 10

Nondegencrateness of η. We put

$$\tilde{\eta} = \quad 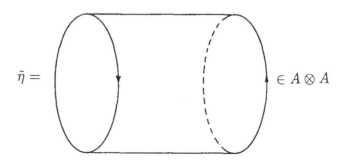 \quad \in A \otimes A$$

Figure 11

and prove that $\tilde{\eta} = \eta^{-1}$. This follows from Fig. 12

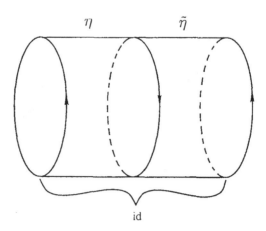

Figure 12

Compatibility of the multiplication with the inner product is proved on the next picture:

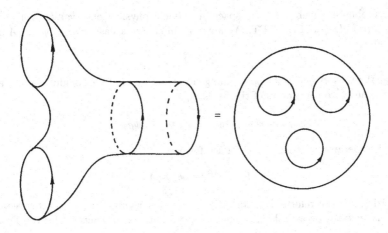

Figure 13

The first part of the theorem is proved.

The proof of the second part is given on the following picture:

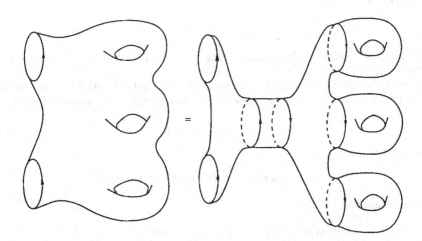

Figure 14

Theorem is proved.

Remark 2.1. If the space of physical states is $\mathbf{Z_2}$-graded (see the footnote on page 48) then we obtain a $\mathbf{Z_2}$-graded Frobenius algebra. Such a generalization was considered by Kontsevich and Manin in [92].

The Frobenius algebra on the space A of local physical observables will be called *primary chiral algebra* of the TFT. We always will choose a basis $\phi_1, \ldots, \phi_n$ of A in such a way that

$$\phi_1 = 1. \tag{2.24}$$

Note that the tensors $\eta_{\alpha\beta}$ and $c_{\alpha\beta\gamma}$ defining the structure of the Frobenius algebra are the following genus zero correlators of the fields ϕ_α

$$\eta_{\alpha\beta} = < \phi_\alpha\phi_\beta >_0, \quad c_{\alpha\beta\gamma} = < \phi_\alpha\phi_\beta\phi_\gamma >_0 . \tag{2.25}$$

The handle operator H is the vector of the form

$$H = \eta^{\alpha\beta}\phi_\alpha \cdot \phi_\beta \in A. \tag{2.26}$$

Summarizing, we can reformulate the Atiyah's axioms saying that the matter sector of a 2D TFT is encoded by a Frobenius algebra. No additional restrictions for the Frobenius algebra can be read out of the axioms.

On this way

Topologicaly invariant Lagrangian $\to$ correlators of local physical fields

we lose too much relevant information. To capture more information on a topological Lagrangian we will consider a topological field theory together with its deformations preserving topological invariance

$$L \to L + \sum t^\alpha L_\alpha^{(pert)} \tag{2.27}$$

(t^α are coupling constants). To construct these moduli of a TFT we are to say more words about the construction of a TFT.

A realization of the topological invariance is provided by QFT with a nilpotent symmetry. We have a Hilbert space $\mathcal{H}$ where the operators of the QFT act and an endomorphism (symmetry)

$$Q : \mathcal{H} \to \mathcal{H}, \quad Q^2 = 0. \tag{2.28}$$

In the classical theory

{physical observables} = {invariants of symmetry}.

In the quantum theory

{physical observables} = {operators commuting with Q}.

I will denote by $\{Q, \phi\}$ the commutator/anticommutator of Q with the operator ϕ (depending on the statistics of ϕ).

Lemma 2.1. $\{Q, \{Q, \cdot\}\} = 0.$

Proof follows from $Q^2 = 0$ and from the Jacobi identity.

Hence the operators of the form

$$\phi = \{Q, \psi\} \tag{2.29}$$

are always physical. However, they do not contribute to the correlators

$$< \{Q, \psi\}\phi_1\phi_2 \ldots >= 0 \tag{2.30}$$

if ϕ_1, ϕ_2 ... are physical fields. So the space of physical states can be identified with the *cohomology* of the operator Q

$$A = \mathrm{Ker}Q/\mathrm{Im}Q. \tag{2.31}$$

The operators in A are called *primary states*.

The topological symmetry will follow if we succeed to construct operators $\phi_\alpha^{(1)}$, $\phi_\alpha^{(2)}$ for any primary field $\phi_\alpha = \phi_\alpha(x)$ such that

$$d\phi_\alpha(x) = \{Q, \phi_\alpha^{(1)}\}, \ \ d\phi_\alpha^{(1)}(x) = \{Q, \phi_\alpha^{(2)}\}. \tag{2.32}$$

(We assume here that the fields $\phi_\alpha(x)$ are scalar functions of $x \in \Sigma$. So $\phi_\alpha^{(1)}(x)$ and $\phi_\alpha^{(2)}(x)$ will be 1-forms and 2-forms on Σ resp.) Indeed,

$$d_x < \phi_\alpha(x)\phi_\beta(y) \ldots >=< \{Q, \phi_\alpha^{(1)}(x)\}\phi_\beta(y) \ldots >= 0. \tag{2.33}$$

The operators $\phi_\alpha^{(1)}$ and $\phi_\alpha^{(2)}$ can be constructed for a wide class of QFT obtained by a procedure of *twisting* [100] from a $N = 2$ supersymmetric quantum field theory (see [41]). Particularly, in this case the primary chiral algebra is a *graded* Frobenius algebra in the sense of Lecture 1. The degrees q_α of the fields ϕ_α are the correspondent eigenvalues of the $U(1)$-charge of the $N = 2$ algebra; d is just the label of the $N = 2$ algebra (it is called *dimension**) since in the case of topological sigma-models it coincides with the complex dimension of the target space). The class of TFT's obtained by the twisting procedure from $N = 2$ superconformal QFT is called *topological conformal field theories* (TCFT).

From

$$\{Q, \oint_C \phi_\alpha^{(1)}\} = \oint_C d\phi_\alpha = 0 \tag{2.34}$$

we see that $\oint_C \phi_\alpha^{(1)}$ is a physical observable for any closed cycle C on Σ. Due to (2.30) this operator depends only on the homology class of the cycle.

Similarly, we obtain that

$$\int\int_\Sigma \phi_\alpha^{(2)} \tag{2.35}$$

is also a physical observable. (Both the new types of observables are non-local!)

---

*) In the physical literature it is sometimes denoted by $d = \hat{c} = c/3$.

Using the operators (2.35) we can construct a very important class of perturbations of the TCFT modifying the action as follows

$$S \mapsto \tilde{S}(t) := S - \sum_{\alpha=1}^{n} t^{\alpha} \int \int_{\Sigma} \phi_{\alpha}^{(2)} \qquad (2.36)$$

where the parameters $t = (t^1, \ldots t^n)$ are called *coupling constants*. The perturbed correlators will be functions of t

$$< \phi_{\alpha}(x)\phi_{\beta}(y) \ldots > (t) := \int [d\phi]\phi_{\alpha}(x)\phi_{\beta}(y) \ldots e^{-\tilde{S}(t)}. \qquad (2.37)$$

Theorem 2.2. [42]
1. *The perturbation (2.36) preserves the topological invariance.*
2. *The perturbed primary chiral algebra A_t satisfies the WDVV equations.*

Due to this theorem the construction (2.36) determines a *canonical moduli space* of dimension n of a TCFT with n primaries. And this moduli space carries a structure of Frobenius manifold.

I will not reproduce here the proof of this (physical) theorem (it looks like the statement holds true under more general assumptions than those were used in the proof [42] - see below). It would be interesting to derive the theorem directly from Segal's-type axioms (see in [150]) of TCFT.

The basic idea of my further considerations is to add the statement of this theorem as *a new axiom* of TFT. In other words, we will axiomatize not an isolated TCFT but the TCFT together with its canonical moduli space (2.36). Let me repeat that the axioms of TCFT now read:

The canonical moduli space of a TCFT is a Frobenius manifold.

The results of Appendices A, C above show that the axiom together with certain analiticity assumptions of the primary free energy could give rise to a reasonable classification of TCFT. Particularly, the formula (A.7) gives the free energy of the A_3 topological minimal model (see below). More general relation of Frobenius manifolds to discrete groups will be established in Appendix G. In lecture 6 we will show that the axioms of coupling (at tree-level) to topological gravity of Dijkgraaf and Witten can be derived from geometry of Frobenius manifolds. The description of Zamolodchikov metric (the $t\,t^*$ equations of Cecotti and Vafa [28]) is an additional differential-geometric structure on the Frobenius manifold [51].

For the above example of topological gravity the matter sector is rather trivial: it consists only of the unity operator. The correspondent Frobenius manifold (the moduli space) is one-dimensional,

$$F = \frac{1}{6}(t^1)^3.$$

All the nontrivial fields σ_p for $p > 0$ in the topological gravity come from the integration over the space of metrics (coupling to topological gravity that we will discuss in Lecture 6).

We construct now other examples of TCFT describing their matter sectors. I will skip to describe the Lagrangians of these TCFT giving only the "answer": the description of the primary correlators in topological terms.

Example 2.1. Witten's algebraic-geometrical description [158] of the A_n topological minimal models [42] (due to K.Li [100] this is just the topological counterpart of the n-matrix model). We will construct some coverings over the moduli spaces $\mathcal{M}_{g,s}$. Let us fix numbers $\alpha = (\alpha_1, \ldots, \alpha_s)$ from 1 to n such that

$$n(2g - 2) + \sum_{i=1}^{s}(\alpha_i - 1) = (n + 1)l \tag{2.38}$$

for some integer l. Consider a line bundle

$$\mathcal{L} \to \Sigma \tag{2.39}$$

of the degree l such that

$$\mathcal{L}^{\otimes(n+1)} = K_\Sigma^n \bigotimes_i O(x_i)^{\alpha_i - 1}. \tag{2.40}$$

Here K_Σ is the canonical line bundle of the curve Σ (of the genus g). The sections of the line bundle in the r.h.s. are n-tensors on Σ having poles only at the marked points x_i of the orders less than α_i.

We have $(n + 1)^{2g}$ choices of the line bundle $\mathcal{L}$. Put

$$\mathcal{M}'_{g,s}(\alpha) := \{(\Sigma, x_1, \ldots, x_s, \mathcal{L})\}. \tag{2.41}$$

This is a $(n + 1)^{2g}$-sheeted covering over the moduli space of stable algebraic curves. An important point [158] is that the covering can be extended onto the compactification $\overline{\mathcal{M}}_{g,s}$. Riemann – Roch implies that, generically

$$\dim H^0(\Sigma, \mathcal{L}) = l + 1 - g = d(g - 1) + \sum_{i=1}^{s} q_{\alpha_i} =: N(\alpha) \tag{2.42}$$

where we have introduced the notations

$$d := \frac{n - 1}{n + 1}, \quad q_\alpha := \frac{\alpha - 1}{n + 1}. \tag{2.43}$$

Let us consider now the vector bundle

$$\begin{array}{c} \downarrow \quad V(\alpha) \\ \mathcal{M}'_{g,s}(\alpha) \end{array} \tag{2.44}$$

where

$$V(\alpha) := H^0(\Sigma, \mathcal{L}). \tag{2.45}$$

180

Strictly speaking, this is not a vector bundle since the dimension (2.42) can jump on the curves where $H^1(\Sigma, \mathcal{L}) \neq 0$. However, the top Chern class $c_N(V(\alpha))$, $N = N(\alpha)$ of the bundle is well-defined (see [158] for more detail explanation). We define the primary correlators by the formula

$$< \phi_{\alpha_1} \ldots \phi_{\alpha_s} >_g := (n+1)^{-g} \int_{\overline{\mathcal{M}'_{g,s}}(\alpha)} c_N(V(\alpha)) \tag{2.46}$$

(the $\overline{\mathcal{M}'_{g,s}}$ is an appropriate compactification of $\mathcal{M}'_{g,s}$). These are nonzero only if

$$\sum_{i=1}^{s}(q_{\alpha_i} - 1) = (3 - d)(g - 1). \tag{2.47}$$

The generating function of the genus zero primary correlators

$$F(t) := < \exp \sum_{\alpha=1}^{n} t^\alpha \phi_\alpha >_0 \tag{2.48}$$

due to (2.47) is a quasihomogeneous polynomial of the degree $3-d$ where the degrees of the coupling constants t^α equal $1-q_\alpha$. One can verify directly that $F(t)$ satisfies WDVV [158]. It turns out that the Frobenius manifold (2.48) coincides with the Frobenius manifold of polynomials of Example 1.7!

One can describe in algebraic-geometrical terms also the result of "coupling to topological gravity" (whatever it means) of the above matter sector. One should take into consideration an analogue of Mumford – Morita – Miller classes (see above in the construction of topological gravity) $c_1(L_i)$ (I recall that the fiber of the line bundle L_i is the cotangent line $T^*_{x_i}\Sigma$). After coupling to topological gravity of the matter sector (2.46) we will obtain an infinite number of fields $\sigma_p(\phi_\alpha)$, $p = 0, 1, \ldots$ (also denoted by $\phi_{\alpha,p}$ in Lecture 6). The fields $\sigma_0(\phi_\alpha)$ can be identified with the primaries ϕ_α. For $p > 0$ the fields $\sigma_p(\phi_\alpha)$ are called gravitational descendants of ϕ_α. Their correlators are defined by the following intersection number

$$< \sigma_{p_1}(\phi_{\alpha_1}) \ldots \sigma_{p_s}(\phi_{\alpha_s}) >_g :=$$

$$= \frac{\left(\frac{\alpha_1}{n+1}\right)_{p_1} \ldots \left(\frac{\alpha_s}{n+1}\right)_{p_s}}{(n+1)^{g-p_1-\ldots-p_s}} \int_{\overline{\mathcal{M}'_{g,s}}(\alpha)} c_1^{p_1}(L_1) \wedge \ldots \wedge c_1^{p_s}(L_s) \wedge c_N(V(\alpha)). \tag{2.49}$$

Here we introduce the notation

$$(r)_p := r(r+1)\ldots(r+p-1), \quad (r)_0 := 1. \tag{2.50}$$

The correlator is nonzero only if

$$\sum_{i=1}^{s}(p_i + q_{\alpha_i} - 1) = (3 - d)(g - 1). \tag{2.51}$$

The generating function of the correlators

$$\mathcal{F}(T) := \sum_g \left\langle \exp \sum_{p=0}^{\infty} \sum_{\alpha=1}^{n} T_{\alpha,p} \sigma_p(\phi_\alpha) \right\rangle_g \qquad (2.52)$$

is conjectured by Witten [158] to satisfy the n-th generalized KdV (or Gelfand - Dickey) hierarchy. Here T is the infinite vector of the indeterminates $T_{\alpha,p}$, $\alpha = 1, \ldots, n$, $p = 0, 1, \ldots$. We will come back to the discussion of the conjecture in Lecture 6.

Example 2.2. Topological sigma-models [155] (A-models in the terminology of Witten [151]). Let X be a compact Kähler manifold of (complex) dimension d with non-positive canonical class. The fields in the matter sector of the TFT will be in 1-1-correspondence with the cohomologies $H^*(X, \mathbf{C})$. I will describe only the genus zero correlators of the fields. For simplicity I will consider only the case when the odd-dimensional cohomologies of X vanish (otherwise one should consider $\mathbf{Z}_2$-graded Frobenius manifolds [92]).

The two-point correlator of two cocycles ϕ_α, $\phi_\beta \in H^*(X, \mathbf{C})$ coincides with the intersection number

$$< \phi_\alpha, \phi_\beta >= \int_X \phi_\alpha \wedge \phi_\beta. \qquad (2.53)$$

The definition of multipoint correlators is given [151] in terms of the intersection theory on the moduli spaces of "instantons", i.e. holomorphic maps of the Riemann sphere CP^1 to X.

Let

$$\psi : CP^1 \to X \qquad (2.54)$$

be a holomorphic map of a given homotopical type $[\psi]$. Also some points x, y, $\ldots$ are to be fixed on CP^1. Let $\mathcal{M}[\psi]$ be the moduli space of all such maps for the given homotopical type $[\psi]$. We consider the "universal instanton": the natural map

$$\Psi : CP^1 \times \mathcal{M}[\psi] \to X. \qquad (2.55)$$

We define the primary correlators putting

$$< \phi_\alpha \phi_\beta \ldots >_0 := \int_{\mathcal{M}[\psi]} \Psi^*(\phi_\alpha)|_{x \times \mathcal{M}[\psi]} \wedge \Psi^*(\phi_\beta)|_{y \times \mathcal{M}[\psi]} \wedge \cdots \qquad (2.56)$$

This TFT can be obtained by twisting of a N=2 superconformal theory if X is a Calabi - Yau manifold, i..e. if the canonical class of X vanishes. So one could expect to describe this class of TFT's by Frobenius manifolds only for Calabi - Yau X. However, we still obtain a Frobenius manifold for more general Kähler manifolds X (though it has been proved rigorously only for some particular classes of Kähler manifolds [111, 127]). But the scaling invariance (1.7) must be modified to (1.12).

To define a generating function we are to be more careful in choosing of a basis in $H^{1,1}(X, \mathbf{C})$. We choose this basis $\phi_{\alpha_1}, \ldots, \phi_{\alpha_k}$ of integer Kähler forms $\in H^{1,1}(X, \mathbf{C}) \cap H^2(X, \mathbf{Z})$. By definition, integrals of the form

$$d_i \equiv d_i[\psi] := \int_{CP^1} \psi^*(\phi_{\alpha_i}), \quad i = 1, \ldots, k \qquad (2.57)$$

are all nonnegative. They are homotopy invariants of the map (2.54). The generating function

$$F(t) := \Big\langle \sum_{[\psi]} \sum_{\phi_\alpha \in H^*(X)} t^\alpha \phi_\alpha \Big\rangle_0 \qquad (2.58)$$

will be a formal series in $e^{-t^{\alpha_i}}$, $i = 1, \ldots, k$ and a formal power series in other coupling constants. So the Frobenius manifold coincides with the cohomology space

$$M = \oplus H^{2i}(X, \mathbf{C}). \qquad (2.59)$$

The free energy $F(t)$ is $2\pi i$-periodic in t^{α_i} for $\phi_{\alpha_i} \in H^{1,1}(X, \mathbf{C}) \cap H^2(X, \mathbf{Z})$. In the limit

$$t^{\alpha_i} \to +\infty \text{ for } \phi_{\alpha_i} \in H^{1,1}(X, \mathbf{C}) \cap H^2(X, \mathbf{Z}), \ t^\alpha \to 0 \text{ for other } \phi_\alpha \qquad (2.60)$$

the multiplication on the Frobenius manifold coincides with the multiplication in the cohomologies. In other words, the cubic part of the corresponding free energy is determined by the graded Frobenius algebra $H^*(X)$ in the form (1.61). Explicitly, the free energy has the structure of (formal) Fourier series

$$F(t, \tilde{t}) = \text{cubic part} + \sum_{k_i, [\psi]} N_{[\psi]}(k_1, \ldots, k_m) \frac{(\tilde{t}^{\beta_1})^{k_1} \ldots (\tilde{t}^{\beta_m})^{k_m}}{k_1! \ldots k_m!} e^{-t^{\alpha_1} d_1 - \ldots - t^{\alpha_k} d_k}. \qquad (2.61)$$

Here the coupling constants $t^{\alpha_1}, \ldots t^{\alpha_k}$ correspond to a basis of Kähler forms as above, the coupling constants $\tilde{t}^{\beta_j}$ correspond to a basis in the rest part of cohomologies

$$\tilde{t}^{\beta_j} \leftrightarrow \tilde{\phi}_{\beta_j} \in \oplus_{k>1} H^{2k}(X, \mathbf{Z}) (\text{modulo torsion}). \qquad (2.62)$$

The coefficients $N_{[\psi]}(k_1, \ldots, k_m)$ are defined as the Gromov - Witten invariants [72, 157] of X^*). By definition, they count the numbers of rational maps (2.54) of the homotopy type $[\psi] = (d_1, \ldots, d_k)$ intersecting with the Poincaré-dual cycles

$$\# \left\{ \psi(CP^1) \cap \mathcal{D}(\tilde{\phi}_{\beta_j}) \right\} = k_j, \ j = 1, \ldots, m. \qquad (2.63)$$

(We put $N_{[\psi]}(k_1, \ldots, k_m) = 0$ if the set of maps satisfying (2.63) is not discrete.)
The Euler vector field E has the form

$$E = \sum (1 - q_\alpha) t^\alpha \partial_\alpha - \sum_{i=1}^k r_{\alpha_i} \partial_{\alpha_i} \qquad (2.64)$$

---

*) One needs to make some technical genericity assumptions about the maps ψ in order to prove WDVV for the free energy (2.61). One possibility is to perturb the complex structure on X and to consider pseudoholomorphic curves ψ. This was done (under some restrictions on X) in [111, 127]. Another scheme was proposed recently by Kontsevich [91]. It based on considerations of *stable maps*, i.e. of such maps that the restrictions (2.63) allow no infinitesimal deformations of them.

where q_α is the complex dimension of the cocycle ϕ_α and the integers r_{α_i} are defined by the formula

$$c_1(X) = \sum_{i=1}^{k} r_{\alpha_i} \phi_{\alpha_i}. \tag{2.65}$$

Particularly, suppressing all the couplings but those corresponding to the Kähler forms $\in H^{1,1}(X, \mathbf{C}) \cap H^2(X, \mathbf{Z})$ we reduce the perturbed primary chiral ring (2.61) to the *quantum cohomology ring* of C.Vafa [146]. For a Calabi - Yau manifold X (where $c_1(X) = 0$) this is the only noncubic part of the Frobenius structure. The quantum multiplication on Calabi - Yau manifolds is still defined on $H^*(X)$. It has a structure of a graded algebra with the same gradings as in usual cohomology algebra, but this structure depends on the parameters $t^{\alpha_1}, \ldots, t^{\alpha_k}$.

An equivalent reformulation of the multiplication (depending on the coupling constants $t^{\alpha_1}, \ldots, t^{\alpha_k}$) in the quantum cohomology ring reads as follows [146]

$$< \phi_\alpha \cdot \phi_\beta, \phi_\gamma >:= \sum_{[\psi]} e^{-t^{\alpha_1} d_1 - \ldots - t^{\alpha_k} d_k} \tag{2.66}$$

where the summation is taken over all the homotopy classes of the maps

$$\psi : CP^1 \to X \text{ such that } \psi(0) \in \mathcal{D}(\phi_\alpha), \ \psi(1) \in \mathcal{D}(\phi_\beta), \ \psi(\infty) \in \mathcal{D}(\phi_\gamma) \tag{2.67}$$

(here $\mathcal{D}$ is the Poincaré duality operator $\mathcal{D} : H^i(X) \to H_{d-i}(X)$). By definition we put zero in the r.h.s. of (2.66) for those classes $[\psi]$ when the set of maps satisfying (2.67) is not discrete. In the limit

$$t^{\alpha_i} \to +\infty \tag{2.68}$$

the quantum cohomology ring coincides with $H^*(X)$.

The most elementary example is the quantum cohomology ring of CP^d

$$\mathbf{C}[x]/(x^{d+1} = e^{-t}). \tag{2.69}$$

Here $t = t_2$ corresponds to the Kähler class of the standard metric on CP^d.

The trivial case is that $d = 1$ (complex projective line). We obtain two coupling parameters $t_1 \leftrightarrow e_1 \in H^0(CP^1, \mathbf{Z})$, $t_2 \leftrightarrow -e_2 \in H^2(CP^1, \mathbf{Z})$ (I change the sign for convenience). The Frobenius structure on $H^*(CP^1)$ is completely determined by the quantum multiplication (2.66). So

$$F(t_1, t_2) = \frac{1}{2} t_1^2 t_2 + e^{t_2}. \tag{2.70}$$

The Euler vector field has the form, according to (2.64)

$$E = t_1 \partial_1 + 2\partial_2. \tag{2.71}$$

In the case of the projective plane CP^2 $(d = 2)$ we choose again the basic elements e_1, e_2, e_3 in H^0, H^2, H^4 resp. The classical cohomology ring has the form

$$e_2^2 = e_3 \tag{2.72a}$$

$$e_2 e_3 = 0. \tag{2.72b}$$

In the quantum cohomology ring instead of (2.72b) we have

$$e_2 e_3 = e^{t_2} e_1. \tag{2.73}$$

So we must have

$$\partial_2 \partial_3^2 F|_{t_1 = t_3 = 0} = e^{t_2}. \tag{2.74}$$

Hence the Frobenius structure on $H^*(CP^2)$ must have the free energy of the form

$$F(t_1, t_2, t_3) = \frac{1}{2} t_1^2 t_3 + \frac{1}{2} t_1 t_2^2 + \frac{1}{2} t_3^2 e^{t_2} + o(t_3^2 e^{t_2}). \tag{2.75}$$

The Euler vector field for the Frobenius manifold reads

$$E = t_1 \partial_1 + 3 \partial_2 - t_3 \partial_3. \tag{2.76}$$

So we obtain finally the following structure of the free energy

$$F = \frac{1}{2} t_1^2 t_3 + \frac{1}{2} t_1 t_2^2 + t_3^{-1} \phi(t_2 + 3 \log t_3) \tag{2.77}$$

where the function $\phi = \phi(x)$ is a solution of the differential equation

$$9\phi''' - 18\phi'' + 11\phi' - 2\phi = \phi'' \phi''' - \frac{2}{3} \phi' \phi''' + \frac{1}{3} \phi''^2 \tag{2.78}$$

of the form

$$\phi = \sum_{k \geq 1} A_k e^{kx}. \tag{2.79}$$

Due to (2.74) one must have

$$A_1 = \frac{1}{2}. \tag{2.80}$$

For the coefficients of the Fourier series (2.79) we obtain from (2.78) the recursion relations for $n > 1$

$$A_n = \frac{1}{(n-1)(9n^2 - 9n + 2)} \sum_{k+l=n,\ 0<k,l} kl \left[\left(n + \frac{2}{3} \right) kl - \frac{2}{3}(k^2 + l^2) \right] A_k A_l. \tag{2.81}$$

So the normalization (2.80) uniquely specifies the solution (2.79) (observation of [92]). The coefficients A_k are identified by Kontsevich and Manin [92] as

$$A_k = \frac{N_k}{(3k-1)!}$$

where N_k is the number of rational curves of the degree k on CP^2 passing through generic $3k - 1$ points.

Let us prove convergence of the series (2.79)[*].

Lemma 2.2. A_k *are positive numbers satisfying*

$$A_k \leq \frac{3}{5k^4} \left(\frac{5}{6}\right)^k \tag{2.83}$$

for any k.

Proof. Positivity of A_n follows from positivity of the coefficients in (2.81). The inequality (2.83) holds true for $k = 1$, 2 ($A_2 = 1/120$). We continue the proof by induction in k. For $k \geq 3$ we have

$$(n-1)(9n^2 - 9n + 2) \geq 3n^3. \tag{2.84}$$

Assuming the inequality (2.83) to be valid for any $k < n$, we obtain from (2.83), (2.84)

$$A_n \leq \frac{3}{50n^3} \cdot \left(\frac{5}{6}\right)^n \sum_{k+l=n} \left[\frac{\left(n + \frac{2}{3}\right)}{k^2 \, l^2} + \frac{2}{3}\left(\frac{1}{k \, l^3} + \frac{1}{k^3 \, l}\right)\right]$$

$$= \frac{3}{50n^3}\left(\frac{5}{6}\right)^n \left[\frac{4(n+1)}{n^3}\sum_1^{n-1}\frac{1}{k} + \frac{2n + \frac{8}{3}}{n^2}\sum_1^{n-1}\frac{1}{k^2} + \frac{4}{3n}\sum_1^{n-1}\frac{1}{k^3}\right] < \frac{3}{5n^4}\left(\frac{5}{6}\right)^n$$

where we use the following elementary inequalities

$$\frac{4(n+1)}{n^3}\sum_1^{n-1}\frac{1}{k} < \frac{3}{n}$$

$$\frac{2n + \frac{8}{3}}{n^2}\sum_1^{n-1}\frac{1}{k^2} < \frac{5}{n}$$

$$\frac{4}{3n}\sum_1^{n-1}\frac{1}{k^3} < \frac{2}{n}.$$

Lemma is proved.

Corollary 2.1. *The function $\phi(x)$ is analytic in the domain*

$$\operatorname{Re} x < \log\frac{6}{5}. \tag{2.85}$$

It would be interesting to give more neat description of the analytic properties of the function $\phi(x)$. An analytic expression for this function in a closed form is still not available,

---

[*] As I learned very recently from D.Morrison, a general scheme of proving convergence of the series for the free energy in topological sigma-models was recently proposed by J.Kollár.

although there are some very interesting formulae in the recent preprint of Kontsevich [91]. The asymptotics of the numbers N_k for big k was found in a very recent preprint [39] of Di Francesco and Itzykson.

More complicated example is the quantum cohomology ring of the quintic in CP^4 (the simplest example of a Calabi - Yau three-fold). Here $\dim H^{1,1}(X) = 1$. We denote by ϕ the basic element in $H^{1,1}(X)$ (the Kähler class) and by t the correspondent coupling constant. The only nontrivial term in the quantum cohomology ring is

$$< \phi \cdot \phi, \phi >= 5 + \sum_{n=1}^{\infty} A(n)n^3 \frac{q^n}{1 - q^n}, \quad q = e^{-t} \tag{2.86}$$

where $A(n)$ is the number of rational curves in X of the degree n. The function (2.86) has been found in [26] in the setting of mirror conjecture. Quantum cohomologies for other manifolds were calculated in [9, 15, 30, 71, 86]. In [71] quantum cohomologies of flag varieties were found. A remarkable description of them in terms of the ring of functions on a particular Lagrangian manifold of the Toda system was discovered. Also a relation of quantum cohomologies and Floer cohomologies [129, 124] was elucidated in these papers. A general approach to calculating numbers of higher genera curves in Calabi - Yau varieties was found in [17].

Example 2.3. Topological sigma-models of B-type [151]. Let X be a compact Calabi - Yau manifold (we consider only 3-folds, i.e. $d = \dim_C X = 3$). The correlation functions in the model are expressed in terms of periods of some differential forms on X. The structure of the Frobenius manifold $M = M(X)$ is decribed only on the sublocus $M_0(X)$ of complex structures on X. I recall that $\dim M_0(X) = \dim H^{2,1}(X)$ while the dimension of M is equal to the dimension of the full cohomology space of X. The bilinear form $\eta_{\alpha\beta}$ coincides with the intersection number

$$< \phi', \phi'' >= \int_X \phi' \wedge \phi''. \tag{2.87}$$

To define the trilinear form on the tangent space to $M_0(X)$ we fix a holomorphic 3-form

$$\Omega \in H^{3,0}(X) \tag{2.88}$$

and normalize it by the condition

$$\oint_{\gamma_0} \Omega = 1 \tag{2.89}$$

(I recall that $\dim H^{3,0}(X) = 1$ for a Calabi - Yau 3-fold) for an appropriate cycle $\gamma_0 \in H_3(X, \mathbf{Z})$. Then the trilinear form reads

$$< \partial, \partial', \partial'' >:= \int_X \partial\partial'\partial''\Omega \wedge \Omega \tag{2.90}$$

for three tangent vector fields ∂, ∂', ∂'' on $M_0(X)$ (I refer the reader to [114] regarding the details of the construction of holomorphic vector fields on $M_0(X)$ using technique of variations of Hodge structures). From the definition it follows that the Frobenius structure (2.90) is well-defined only locally. Globally we would expect to obtain a twisted Frobenius manifold in the sense of Appendix B because of the ambiguity in the choice of the normalizing cycle γ_0.

The mirror conjecture claims that for any Calabi - Yau manifold X there exists another Calabi - Yau manifold $\hat{X}$ such that the Frobenius structure of A-type determined by the quantum multiplication on the cohomologies of X is locally isomorphic to the Frobenius structure of the B-type defined by the periods of $\hat{X}$. See [146, 151] concerning motivations of the conjecture and [8, 26, 30, 79, 115] for consideration of the particular examples of mirror dual manifolds X and $\hat{X}$.

Example 2.4. Topological Landau - Ginsburg (LG) models [145]. The bosonic part of the LG-action S has the form

$$S = \int d^2 z \left(|\frac{\partial p}{\partial z}|^2 + |\lambda'(p)|^2 \right)$$
(2.91)

where the holomorphic function $\lambda(p)$ is called *superpotential* and S is considered as a functional of the holomorphic *superfield* $p = p(z)$. The classical states are thus in the one-to-one correspondence with the critical points of $\lambda(p)$

$$p(z) \equiv p_i, \quad \lambda'(p_i) = 0, \quad i = 1, \ldots, n.$$
(2.92)

Quantum correlations can be computed [145, 29] in terms of solitons propagating between the classical vacua (2.92). If the critical points of the superpotential are non-degenerate and the critical values are pairwise distinct then masses of the solitons are proportional to the differences of the critical values. In this case we obtain a *massive* TFT.

The moduli space of a LG theory can be realized as a family of LG models with an appropriately deformed superpotential

$$\lambda = \lambda(p; t^1, \ldots, t^n).$$
(2.93)

The Frobenius structure on the space of parameters is given by the following formulae [145, 29]

$$< \partial, \partial' >_\lambda = \sum \operatorname*{res}_{\lambda'=0} \frac{\partial(\lambda dp)\partial'(\lambda dp)}{d\lambda(p)}$$
(2.94a)

$$< \partial, \partial', \partial'' >_\lambda = \sum \operatorname*{res}_{\lambda'=0} \frac{\partial(\lambda dp)\partial'(\lambda dp)\partial''(\lambda dp)}{d\lambda(p)dp.}$$
(2.94b)

By definition, in these formulae the vector fields ∂, ∂', ∂'' on the space of parameters act trivially on p.

Particularly, for the superpotential

$$\lambda(p) = p^{n+1}$$

the deformed superpotential coincides with the generic polynomial of the degree $n + 1$ of the form (1.65). The Frobenius structure (2.94) coincides with the one of Example 1.7 (see Lecture 4 below).

We will not discuss here interesting relations between topological sigma-models and topological LG models (see [161]).

Lecture 3.
Spaces of isomonodromy deformations as Frobenius manifolds.

Let us consider a linear differential operator of the form

$$\Lambda = \frac{d}{dz} - U - \frac{1}{z}V \tag{3.1}$$

where U and V are two complex $n \times n$ z-independent matrices, U is a diagonal matrix with pairwise distinct diagonal entries, and the matrix V is skew-symmetric. The solutions of the differential equation with rational coefficients

$$\Lambda\psi = 0 \tag{3.2}$$

are analytic multivalued vector functions in $z \in \mathbf{C} \setminus 0$. The monodromy of these solutions will be called monodromy of the differential operator Λ (we will give below the precise definition of the monodromy).

Let $\mathcal{M}(\Lambda)$ be the space of all operators of the form (3.1) with a given monodromy. As it follows from the general theory of isomonodromy deformations [82, 109, 113, 135] the diagonal entries $u_1, \ldots, u_n$ of the matrix U can serve as local coordinates near generic point of $\mathcal{M}(\Lambda)$ (thus a function $V = V(u)$ locally is well defined). We define a Frobenius structure on $\mathcal{M}(\Lambda)$ by the multiplication

$$\partial_i \cdot \partial_j = \delta_{ij}\partial_i, \quad \partial_i := \frac{\partial}{\partial u_i}$$

the quadratic form

$$< , >:= \sum_{i=1}^{n} \psi_i^2(u)du_i^2$$

where $\psi = (\psi_1(u), \ldots, \psi_n(u))^T$ is an eigenvector of $V(u)$ (it can be normalized in such a way that the metric $< , >$ is flat), the unity

$$e := \sum_{i=1}^{n} \partial_i$$

and the Euler vector field

$$E := \sum_{i=1}^{n} u_i\partial_i.$$

Observe that the Frobenius algebra on the tangent planes to $\mathcal{M}(\Lambda)$ is semisimple in any point u (we will say that such a Frobenius manifold satisfies semisimplicity condition).

The main result of this Lecture is the following

Main Theorem. *The above formulae determine a Frobenius structure on the space $\mathcal{M}(\Lambda)$ of all the operators of the form (3.1) with a given monodromy. Choice of another* *eigenvector ψ changes the Frobenius structure by a transformation of the form (B.2). Conversely, any Frobenius manifold satisfying semisimplisity assumption locally can be obtained by such a construction.*

The following two creatures are the principal playing characters on a Frobenius manifold M.

1. Deformed Euclidean connection

$$\tilde{\nabla}_u(z)v := \nabla_u v + zu \cdot v. \tag{3.3}$$

It is a symmetric connection depending on the parameter z. Here u, v are two vector fields on M, $\nabla_u v$ is the Levi-Cività connection for the invariant metric $< \,,\, >$.

Lemma 3.1. *The connection $\tilde{\nabla}(z)$ is flat identically in z iff the algebra $c_{\alpha\beta}^\gamma(t)$ is associative and a function $F(t)$ locally exists such that*

$$c_{\alpha\beta\gamma}(t) = \partial_\alpha \partial_\beta \partial_\gamma F(t). \tag{3.4}$$

Proof. In the flat coordinates vanishing of the curvature of the deformed connection reads

$$[\tilde{\nabla}_\alpha(z), \tilde{\nabla}_\beta(z)]^\epsilon = [z(\partial_\beta c_{\alpha\gamma}^\epsilon - \partial_\alpha c_{\beta\gamma}^\epsilon) + z^2(c_{\alpha\gamma}^\sigma c_{\beta\sigma}^\epsilon - c_{\beta\gamma}^\sigma c_{\alpha\sigma}^\epsilon)] = 0.$$

Vanishing of the coefficients of the quadratic polynomial in z together with the symmetry of the tensor $c_{\alpha\beta\gamma} := \eta_{\gamma\epsilon} c_{\alpha\beta}^\epsilon$ is equivalent to the statements of the lemma.

Remark 3.1. Vanishing of the curvature is equivalent to the compatibility of the linear system

$$\partial_\alpha \xi_\beta = z c_{\alpha\beta}^\gamma(t) \xi_\gamma. \tag{3.5a}$$

This gives a "Lax pair" for the associativity equations (1.14) (z plays the role of the spectral parameter). Note that

$$\partial_1 \xi_\beta = z \xi_\beta. \tag{3.5b}$$

This coincides with the normalization (1.1). For any given z the system has n-dimensional space of solutions. The solutions are closely related to the flat coordinates of the deformed connection $\tilde{\nabla}(z)$, i.e to the independent functions $\tilde{t}^1(t, z)$, ..., $\tilde{t}^n(t, z)$ such that in these new coordinates the deformed covariant derivatives coincide with partial derivatives

$$\tilde{\nabla}_\alpha(z) = \frac{\partial}{\partial \tilde{t}^\alpha}, \quad \tilde{t}^\alpha = \tilde{t}^\alpha(t, z). \tag{3.6}$$

Exercise 3.1. Prove that 1) any solution of the system (3.5) is the gradient of some function

$$\xi_\alpha = \partial_\alpha \tilde{t}; \tag{3.7}$$

2) if ξ_α^1, ..., ξ_α^n is a fundamental system of solutions of the system (3.5) for a given z then the correspondent functions $\tilde{t}^1$, ..., $\tilde{t}^n$ are flat coordinates for the deformed connection $\tilde{\nabla}(z)$.

Exercise 3.2. Derive from (3.5) the following Lax pair for the Chazy equation (C.5)

$$[\partial_z + U, \partial_x + V] = 0 \tag{3.8}$$

where the matrices U and V have the form

$$U = \begin{pmatrix} 0 & -1 & 0 \\ \frac{3}{4}z^2\gamma' & \frac{3}{2}z\gamma & -1 \\ \frac{1}{4}z^3\gamma'' & \frac{3}{4}z^2\gamma' & 0 \end{pmatrix}, \quad V = \begin{pmatrix} 0 & 0 & -1 \\ \frac{1}{4}z^3\gamma'' & \frac{3}{4}z^2\gamma' & 0 \\ \frac{1}{16}z^4\gamma''' & \frac{1}{4}z^3\gamma'' & 0 \end{pmatrix}. \tag{3.9}$$

Observe that the more strong commutativity condition

$$[U, V] = 0 \tag{3.10}$$

holds true for the matrices U, V. It is easy to see that one can put arbitrary three unknown functions in (3.9) instead of γ', γ'', γ'''. Then the commutativity (3.8) together with (3.10) is still equivalent to the Chazy equation.

The commutation representation (3.8), (3.10) looks to be intermediate one between Lax pairs with a derivative w.r.t. the spectral parameter z (those being typical in the theory of isomonodromic deformations) and "integrable algebraic systems" of [55], i.e. the equations of commutativity of matrices depending on the spectral parameter.

A Lax pair for the Chazy equation was obtained also in [1]. But instead of finite dimensional matrices some differential operators with partial derivatives are involved. This gives no possibility to apply the machinery of the theory of integrable systems.

I learned recently from S.Chakravarty that he has found another finite-dimensional Lax pair for Chazy equation. This looks similar to the Lax pair of the Painlevé-VI equation with a nontrivial dependence of the poles on the both dependent and independent variables.

The further step is to consider WDVV as the scaling reduction of the equations of associativity (1.14) as of an integrable system. The standard machinery of integration of scaling reductions of integrable systems [64, 82, 135] suggests to add to (3.5) a differential equation in the spectral parameter z for the auxiliary function $\xi = \xi(t, z)$.

Proposition 3.1. *WDVV is equivalent to compatibility of the system of equations (3.5) together with the equation*

$$z\partial_z\xi_\alpha = zE^\gamma(t)c_{\gamma\alpha}^\beta(t)\xi_\beta + Q_\alpha^\gamma\xi_\gamma \tag{3.11}$$

where $Q_\alpha^\gamma = \nabla_\alpha E^\gamma$.

Proof. Due to (1.50b) the system (3.5) is invariant w.r.t. the group of rescalings (1.43) together with the transformations $z \mapsto kz$. So the system (3.5) for the covector ξ is compatible with the equation

$$x\partial_z\xi = \mathcal{L}_E\xi.$$

On the solutions of (3.5) the last equation can be rewritten in the form (3.11). Proposition is proved.

The compatibility of (3.5) and (3.11) can be reformulated [92] as vanishing of the curvature of the connection on $M \times CP^1$ (the coordinates are (t, z)) given by the operators (3.5) and (3.11). For tangent vectors v on $M \times CP^1$ parallel to M the z-component $\tilde{\nabla}_z$ of the connection acts by the formula

$$\tilde{\nabla}_z v = \partial_z v + E \cdot v + \frac{1}{z} Q v. \tag{3.12}$$

Along CP^1 the connection acts as $\tilde{\nabla}_z = \partial_z$, $\tilde{\nabla}_\alpha(z) = \partial_\alpha$.

The further step of the theory (also being standard [64, 82, 135]) is to parametrize the solutions of WDVV by the monodromy data of the operator (3.11) with rational coefficients. We are not able to do this in general. The difficulty is to describe quantitavely the monodromy of the operator. The main problem is to choose a trivialization in the space of solutions of the equation (3.11) for big z. This problem looks not to be purely technical: one can see that WDVV does not satisfy the Painlevé property of absence of movable critical points in the t-coordinates (see, e.g., the discussion of the analytic properties of solutions of Chazy equation in Appendix C above). Another point is that, the monodromy of the equation (3.11) can be trivial for a nontrivial solution of WDVV (e.g., for the nilpotent solutions (1.38)).

Our main strategy will be to find an appropriate coordinate system on M and to do a gauge transform of the operator (3.11) providing applicability of the isomonodromy deformations technique to WDVV. This can be done under semi-simplicity assumptions imposed onto M (see below).

Another important playing character is a new metric on a Frobenius manifold. It is convenient to define it as a metric on the cotangent bundle T^*M i.e. as an inner product of 1-forms. For two 1-forms ω_1 and ω_2 we put

$$(\omega_1, \omega_2)^* := i_E(\omega_1 \cdot \omega_2) \tag{3.13}$$

(I label the metric by $*$ to stress that this is an inner product on T^*M). Here i_E is the operator of contraction of a 1-form with the vector field E; we multiply two 1-forms using the operation of multiplication of tangent vectors on the Frobenius manifold and the duality between tangent and cotangent spaces established by the invariant inner product.

Exercise 3.3. Prove that the inner product (u, v) of two vector fields w.r.t. the new metric is related to the old inner product $< u, v >$ by the equation

$$(E \cdot u, v) = < u, v > . \tag{3.14}$$

Thus the new metric on the *tangent* bundle is welldefined in the points t of M where $E(t)$ is an invertible element of the algebra $T_t M$.

In the flat coordinates t^α the metric $(\ ,\ )^*$ has the components

$$g^{\alpha\beta}(t) := (dt^\alpha, dt^\beta)^* = E^\epsilon(t) c_\epsilon^{\alpha\beta}(t) \tag{3.15}$$

where
$$c_\epsilon^{\alpha\beta}(t) := \eta^{\alpha\sigma}c_{\sigma\epsilon}^\beta(t). \tag{3.16}$$

If the degree operator is diagonalizable then

$$g^{\alpha\beta}(t) = (d + 1 - q_\alpha - q_\beta)F^{\alpha\beta}(t) + A^{\alpha\beta} \tag{3.17}$$

where

$$F^{\alpha\beta}(t) := \eta^{\alpha\lambda}\eta^{\beta\mu}\frac{\partial^2 F(t)}{\partial t^\lambda \partial t^\mu} \tag{3.18}$$

(I recall that we normalise the degrees d_α in such a way that $d_1 = 1$) and the matrix $A_{\alpha\beta} = \eta_{\alpha\alpha'}\eta_{\beta\beta'}A^{\alpha'\beta'}$ is defined in (1.9).

Lemma 3.2. *The metric (3.13) does not degenerate identically near the t^1-axis for sufficiently small $t^1 \neq 0$.*

Proof. We have
$$c_1^{\alpha\beta}(t) \equiv \eta^{\alpha\beta}.$$

So for small $t^2, ..., t^n$
$$g^{\alpha\beta}(t) \simeq t^1 c_1^{\alpha\beta} + A^{\alpha\beta} = t^1\eta^{\alpha\beta} + A^{\alpha\beta}.$$

This cannot be degenerate identically in t^1. Lemma is proved.

It turns out that the new metric also is flat. In fact I will prove a more strong statement, that any linear combination of the metrics (,)* and < , >* is a flat metric (everywhere when being nondegenerate). To formulate the precise statement I recall some formulae of Riemannian geometry.

Let (,)* be a symmetric nondegenerate bilinear form on the cotangent bundle T^*M to a manifold M. In a local coordinate system $x^1, ..., x^n$ the metric is given by its components

$$g^{ij}(x) := (dx^i, dx^j)^* \tag{3.19}$$

where (g^{ij}) is an invertible symmetric matrix. The inverse matrix $(g_{ij}) := (g^{ij})^{-1}$ specifies a metric on the manifold i.e. a nondegenerate inner product on the tangent bundle TM

$$(\partial_i, \partial_j) := g_{ij}(x) \tag{3.20}$$

$$\partial_i := \frac{\partial}{\partial x^i}. \tag{3.21}$$

The *Levi-Cività connection* ∇_k for the metric is uniquely specified by the conditions

$$\nabla_k g_{ij} := \partial_k g_{ij} - \Gamma_{ki}^s g_{sj} - \Gamma_{kj}^s g_{is} = 0 \tag{3.22a}$$

or, equivalently,

$$\nabla_k g^{ij} := \partial_k g^{ij} + \Gamma_{ks}^i g^{sj} + \Gamma_{ks}^j g^{is} = 0 \tag{3.22b}$$

and

$$\Gamma_{ij}^k = \Gamma_{ji}^k. \tag{3.23}$$

193

(I recall that summation over twice repeated indices here and below is assumed. We will keep the symbol of summation over more than twice repeated indices.) Here the coefficients Γ_{ij}^k of the connection (the Christoffel symbols) can be expressed via the metric and its derivatives as

$$\Gamma_{ij}^k = \frac{1}{2} g^{ks} \left(\partial_i g_{sj} + \partial_j g_{is} - \partial_s g_{ij} \right). \tag{3.24}$$

For us it will be more convenient to work with the *contravariant components* of the connection

$$\Gamma_k^{ij} := (dx^i, \nabla_k dx^j)^* = -g^{is}\Gamma_{sk}^j. \tag{3.25}$$

The equations (3.22) and (3.23) for the contravariant components read

$$\partial_k g^{ij} = \Gamma_k^{ij} + \Gamma_k^{ji} \tag{3.26}$$

$$g^{is}\Gamma_s^{jk} = g^{js}\Gamma_s^{ik}. \tag{3.27}$$

It is also convenient to introduce operators

$$\nabla^i = g^{is}\nabla_s \tag{3.28a}$$

$$\nabla^i \xi_k = g^{is}\partial_s \xi_k + \Gamma_k^{is}\xi_s. \tag{3.28b}$$

For brevity we will call the operators ∇^i and the correspondent coefficients Γ_k^{ij} *contravariant connection*.

The *curvature tensor* R_{slt}^k of the metric measures noncommutativity of the operators ∇_i or, equivalently ∇^i

$$(\nabla_s \nabla_l - \nabla_l \nabla_s)\xi_t = -R_{slt}^k \xi_k \tag{3.29a}$$

where

$$R_{slt}^k = \partial_s \Gamma_{lt}^k - \partial_l \Gamma_{st}^k + \Gamma_{sr}^k \Gamma_{lt}^r - \Gamma_{lr}^k \Gamma_{st}^r. \tag{3.29b}$$

We say that the metric is *flat* if the curvature of it vanishes. For a flat metric local *flat coordinates* $p^1, ..., p^n$ exist such that in these coordinates the metric is constant and the components of the Levi-Città connection vanish. Conversely, if a system of flat coordinates for a metric exists then the metric is flat. The flat coordinates are determined uniquely up to an affine transformation with constant coefficients. They can be found from the following system

$$\nabla^i \partial_j p = g^{is}\partial_s \partial_j p + \Gamma_j^{is}\partial_s p = 0, \ i,j = 1,...,n. \tag{3.30}$$

If we choose the flat coordinates orthonormalized

$$(dp^a, dp^b)^* = \delta^{ab} \tag{3.31}$$

then for the components of the metric and of the Levi-Città connection the following formulae hold

$$g^{ij} = \frac{\partial x^i}{\partial p^a}\frac{\partial x^j}{\partial p^a} \tag{3.32a}$$

$$\Gamma_k^{ij} dx^k = \frac{\partial x^i}{\partial p^a} \frac{\partial^2 x^j}{\partial p^a \partial p^b} dp^b. \tag{3.32b}$$

All these facts are standard in geometry (see, e.g., [59]). We need to represent the formula (3.29b) for the curvature tensor in a slightly modified form (cf. [57, formula (2.18)]).

Lemma 3.3. *For the curvature of a metric the following formula holds*

$$R_l^{ijk} := g^{is} g^{jt} R_{slt}^k = g^{is} \left(\partial_s \Gamma_l^{jk} - \partial_l \Gamma_s^{jk} \right) + \Gamma_s^{ij} \Gamma_l^{sk} - \Gamma_s^{ik} \Gamma_l^{sj}. \tag{3.33}$$

Proof. Multiplying the formula (3.29b) by $g^{is} g^{jt}$ and using (3.25) and (3.26) we obtain (3.33). The lemma is proved.

Let us consider now a manifold supplied with two nonproportional metrics $(\ ,\ )_1^*$ and $(\ ,\ )_2^*$. In a coordinate system they are given by their components g_1^{ij} and g_2^{ij} resp. I will denote by Γ_{1k}^{ij} and Γ_{2k}^{ij} the correspondent Levi-Cività connections ∇_1^i and ∇_2^i. Note that the difference

$$\Delta^{ijk} = g_2^{is} \Gamma_{1s}^{jk} - g_1^{is} \Gamma_{2s}^{jk} \tag{3.34}$$

is a tensor on the manifold.

Definition 3.1. We say that the two metrics form a *flat pencil* if:
1. The metric

$$g^{ij} = g_1^{ij} + \lambda g_2^{ij} \tag{3.35a}$$

is flat for arbitrary λ and
2. The Levi-Cività connection for the metric (3.35a) has the form

$$\Gamma_k^{ij} = \Gamma_{1k}^{ij} + \lambda \Gamma_{2k}^{ij}. \tag{3.35b}$$

I will describe in more details the conditions for two metrics to form a flat pencil in Appendix D below (it turns out that these conditions are very close to the axioms of Frobenius manifolds).

Let us consider the metrics $(\ ,\ )^*$ and $<\ ,\ >^*$ on a Frobenius manifold M (the second metric is induced on T^*M by the invariant metric $<\ ,\ >$). We will assume that the Euler vector field E in the flat coordinates has the form

$$E = \sum_\alpha [(1 - q_\alpha) t^\alpha + r_\alpha] \partial_\alpha.$$

Proposition 3.2. *The metrics $(\ ,\ )^*$ and $<\ ,\ >^*$ on a Frobenius manifold form a flat pencil.*

Lemma 3.4. *In the flat coordinates the contravariant components of the Levi-Cività connection for the metric $(\ ,\ )^*$ have the form*

$$\Gamma_\gamma^{\alpha\beta} = \left(\frac{d+1}{2} - q_\beta \right) c_\gamma^{\alpha\beta}. \tag{3.36}$$

Proof. Substituting (3.36) to (3.26), (3.27) and (3.33) we obtain identities. Lemma is proved.

Proof of proposition. We repeat the calculation of the lemma for the same connection and for the metric $g^{\alpha\beta} + \lambda\eta^{\alpha\beta}$. The equations (3.26) and (3.27) hold true identically in λ. Now substitute the connection into the formula (3.33) for the curvature of the metric $g^{\alpha\beta} + \lambda\eta^{\alpha\beta}$. We again obtain identity. Proposition is proved.

Definition 3.2. The metric $(\ ,\ )^*$ of the form (3.13) will be called *intersection form of the Frobenius manifold.*

We borrow this name from the singularity theory [5, 6]. The motivation becomes clear from the consideration of the Example 1.7. In this example the Frobenius manifold coincides with the universal unfolding of the simple singularity of the A_n-type [5]. The metric (3.13) for the example coincides with the intersection form of (even-dimensional) vanishing cycles of the singularity [3, 6, 70] as an inner product on the cotangent bundle to the universal unfolding space (we identify [6] the tangent bundle to the universal unfolding with the middle homology fibering using the differential of the period mapping). It turns out that the intersection form of odd-dimensional vanishing cycles (assuming that the base of the bundle is even-dimensional) coincides with the skew-symmetric form $< \hat{V} .\ ,\ . >^*$ where the operator $\hat{V}$ was defined in (1.51). This can be derived from the results of Givental [69] (see below (3.46)).

Example 3.1. For the trivial Frobenius manifold corresponding to a graded Frobenius algebra $A = \{c_{\alpha\beta}^{\gamma}, \eta_{\alpha\beta}, q_{\alpha}, d\}$ (see Example 1 of Lecture 1) the intersection form is a linear metric on the dual space A^*

$$g^{\alpha\beta} = \sum (1 - q_\epsilon)t^\epsilon c_\epsilon^{\alpha\beta}, \tag{3.37}$$

for

$$c_\epsilon^{\alpha\beta} = \eta^{\alpha\sigma}c_{\sigma\epsilon}^{\beta}, \quad (\eta^{\alpha\beta}) := (\eta_{\alpha\beta})^{-1}.$$

From the above considerations it follows that the Christoffel coefficients for this flat metric are

$$\Gamma_\gamma^{\alpha\beta} = \left(\frac{d+1}{2} - q_\beta\right)c_\gamma^{\alpha\beta}. \tag{3.38}$$

Flat linear metrics with constant contravariant Christoffel coefficients were first studied by S.Novikov and A.Balinsky [12] due to their close relations to vector analogues of the Virasoro algebra. We will come back to this example in Lecture 6 (see also Appendix G below).

Remark 3.2. Knowing the intersection form of a Frobenius manifold and the Euler and the unity vector fields E and e resp. we can uniquely reconstruct the Frobenius structure if $d + 1 \neq q_\alpha + q_\beta$ for any $1 \leq \alpha, \beta \leq n$. Indeed, we can put

$$< ,\ >^* := \mathcal{L}_e(,\ )^* \tag{3.39}$$

(the Lie derivative along e). Then we can choose the coordinates t^α taking the flat coordinates for the metric $< , >^*$ and choosing them homogeneous for E. Putting

$$g^{\alpha\beta} := (dt^\alpha, dt^\beta)^* \tag{3.40}$$

$$\deg g^{\alpha\beta} := \frac{\mathcal{L}_E g^{\alpha\beta}}{g^{\alpha\beta}} \tag{3.41}$$

we can find the function F from the equations

$$g^{\alpha\beta} = \deg g^{\alpha\beta} F^{\alpha\beta} \tag{3.42}$$

($F^{\alpha\beta}$ are the contravariant components of the Hessian of F, see formula (3.17)). This observation will be very important in the constructions of the next lecture.

Exercise 3.4. Let ω will be the 1-form on a Frobenius manifold defined by

$$\omega(\, . \,) = < e, \, . \, > . \tag{3.43}$$

Show that the formula

$$\{x^k, x^l\} := \frac{1}{2}\mathcal{L}_e\left[(dx^l, dx^i)\,\partial_i\left(dx^k, \omega\right) - \left(dx^k, dx^i\right)\partial_i\left(dx^l, \omega\right)\right] \tag{3.44}$$

defines a Poisson bracket on the Frobenius manifold. [Hint: prove that the bracket is constant in the flat coordinates for the metric $< , >$,

$$\{t^\alpha, t^\beta\} = -\frac{1}{2}(q_\alpha - q_\beta)\eta^{\alpha\beta}.] \tag{3.45}$$

Observe that the tensor of the Poisson bracket has the form

$$\{ . \, , . \} = -\frac{1}{2} < \hat{V}., \, . >^* . \tag{3.46}$$

For the case of the Frobenius manifold of Example 1.7 the Poisson structure coincides with the skew-symmetric intersection form on the universal unfolding of the A_n singularity (see [65]). The formula (3.44) for this case was obtained by Givental [65, Corollary 3].

We add now the following *assumption of semisimplicity* on the Frobenius manifold M. We say that a point $t \in M$ is *semisimple* if the Frobenius algebra $T_t M$ is semisimple (i.e. it has no nilpotents). It is clear that semisimplicity is an open property of the point. The assumption of semisimplicity for a Frobenius manifold M means that a generic point of M is semisimple. In physical context this corresponds to massive perturbations of TCFT [29]. So we will also call M satisfying the semisimplicity assumption *massive* Frobenius manifold.

Main lemma. *In a neighborhood of a semisimple point local coordinates u^1, ..., u^n exist such that*

$$\partial_i \cdot \partial_j = \delta_{ij}\,\partial_i, \quad \partial_i = \frac{\partial}{\partial u^i}. \tag{3.47}$$

Proof. In a neighborhood of a semisimple point t *vector fields* ∂_1, ..., ∂_n exist such that $\partial_i \cdot \partial_j = \delta_{ij} \partial_i$ (idempotents of the algebra $T_t M$). We need to prove that these vector fields commute pairwise. Let

$$[\partial_i, \partial_j] =: f_{ij}^k \partial_k. \tag{3.48}$$

We rewrite the condition of flatness of the deformed connection $\tilde{\nabla}(z)$ in the basis ∂_1, ..., ∂_n. I recall that the curvature operator for a connection ∇ is defined by

$$R(X,Y)Z := [\nabla_X, \nabla_Y]Z - \nabla_{[X,Y]}Z \tag{3.49}$$

for any three vector fields X, Y, Z. We define the coefficients of the Euclidean connection on the Frobenius manifold in the basis ∂_1, ..., ∂_n by the formula (see [59], sect. 30.1)

$$\nabla_{\partial_i}\partial_j =: \Gamma_{ij}^k \partial_k. \tag{3.50}$$

Vanishing of the curvature of $\tilde{\nabla}(z)$ in the terms linear in z reads

$$\Gamma_{kj}^l \delta_i^l + \Gamma_{ki}^l \delta_{kj} - \Gamma_{ki}^l \delta_j^l - \Gamma_{kj}^l \delta_{ki} = f_{ij}^l \delta_k^l \tag{3.51}$$

(no summation over the repeated indices in this formula!). For $l = k$ this gives $f_{ij}^k = 0$. Lemma is proved.

Observe that neither the scaling invariance (1.9) nor constancy of the unity $e = \partial/\partial t^1$ have been used in the proof of Main Lemma.

Remark 3.3. The main lemma can be reformulated in terms of *algebraic symmetries* of a massive Frobenius manifold. We say that a diffeomorphism $f : M \to M$ of a Frobenius manifold is algebraic symmetry if it preserves the multiplication law of vector fields:

$$f_*(u \cdot v) = f_*(u) \cdot f_*(v) \tag{3.52}$$

(here f_* is the induced linear map $f_* : T_x M \to T_{f(x)} M$).

It is easy to see that algebraic symmetries of a Frobenius manifold form a finite-dimensional Lie group $G(M)$. The generators of action of $G(M)$ on M (i.e. the representation of the Lie algebra of $G(M)$ in the Lie algebra of vector fields on M) are the vector fields w such that

$$[w, u \cdot v] = [w, u] \cdot v + [w, v] \cdot u \tag{3.53}$$

for any vector fields u, v.

Note that the group $G(M)$ always is nontrivial: it contains the one-parameter subgroup of shifts along the coordinate t^1. The generator of this subgroup coincides with the unity vector field e.

Main lemma'. *The connect component of the identity in the group $G(M)$ of algebraic symmetries of a n-dimensional massive Frobenius manifold is a n-dimensional commutative Lie group that acts localy transitively on M.*

I will call the local coordinates u^1, ..., u^n on a massive Frobenius manifold *canonical coordinates*. They can be found as independent solutions of the system of PDE

$$\partial_\gamma u\, c^\gamma_{\alpha\beta}(t) = \partial_\alpha u\, \partial_\beta u \qquad (3.54)$$

or, equivalently, the 1-form du must be a homomorphism of the algebras

$$du : T_t M \to \mathbb{C}. \qquad (3.55)$$

The canonical coordinates are determined uniquely up to shifts and permutations. We solve now explicitly this system of PDE using the scaling invariance (1.9).

Proposition 3.3. *In a neighborhood of a semisimple point all the roots $u^1(t)$, ..., $u^n(t)$ of the characteristic equation*

$$\det(g^{\alpha\beta}(t) - u\eta^{\alpha\beta}) = 0 \qquad (3.56)$$

are simple. They are canonical coordinates in this neighborhood. Conversely, if the roots of the characteristic equation are simple in a point t then t is a semisimple point on the Frobenius manifold and $u^1(t)$, ..., $u^n(t)$ are canonical coordinates in the neighbourhood of the point.

Here $g^{\alpha\beta}(t)$ is the intersection form (3.15) of the Frobenius manifold.

Lemma 3.5. *Canonical coordinates in a neighborhood of a semisimple point can be chosen in such a way that the Euler vector field E have the form*

$$E = \sum_i u^i \partial_i. \qquad (3.57)$$

Proof. Rescalings generated by E act on the idempotents ∂_i as $\partial_i \mapsto k^{-1}\partial_i$. So an appropriate shift of u^i provides $u^i \mapsto ku^i$. Lemma is proved.

Lemma 3.6. *The invariant inner product $<\ ,\ >$ is diagonal in the canonical coordinates*

$$< \partial_i, \partial_j >= \eta_{ii}(u)\, \delta_{ij} \qquad (3.58)$$

for some nonzero functions $\eta_{11}(u)$, ..., $\eta_{nn}(u)$. The unity vector field e in the canonical coordinates has the form

$$e = \sum_i \partial_i. \qquad (3.59)$$

The proof is obvious (cf. (1.41), (1.42)).

Proof of proposition. In the canonical coordinates of Main Lemma we have

$$du^i \cdot du^j = \eta_{ii}^{-1} du^i\, \delta_{ij}. \qquad (3.60)$$

So the intersection form reads

$$g^{ij}(u) = u^i \eta_{ii}^{-1} \delta_{ij}. \qquad (3.61)$$

The characteristic equation (3.56) reads

$$\prod_i (u - u^i) = 0.$$

This proves the first part of the proposition.

To prove the second part we consider the linear operators $U = (U_\beta^\alpha(t))$ on $T_t M$ where

$$U_\beta^\alpha(t) := g^{\alpha\epsilon}(t)\eta_{\epsilon\beta}. \tag{3.62}$$

From (3.14) it folows that U is the operator of multiplication by the Euler vector field E. The characteristic equation for this operator coincides with (3.56). So under the assumptions of the proposition the operator of multiplication by E in the point t is a semisimple one. This implies the semisimplicity of all the algebra $T_t M$ because of the commutativity of the algebra.

Proposition is proved.

Using canonical coordinates we reduce the problem of local classification of massive Frobenius manifolds to an integrable system of ODE. To obtain such a system we will study the properties of the invariant metric in the canonical coordinates. I recall that this metric has diagonal form in the canonical coordinates. In other words, u^1, ..., u^n are curvilinear orthogonal coordinates in the (locally) Euclidean space with the (complex) Euclidean metric $< , >$. The familiar object in the geometry of curvilinear orthogonal coordinates is the *rotation coefficients*

$$\gamma_{ij}(u) := \frac{\partial_j \sqrt{\eta_{ii}(u)}}{\sqrt{\eta_{jj}(u)}}, \quad i \neq j \tag{3.63}$$

(locally we can fix some branches of $\sqrt{\eta_{ii}(u)}$). They determine the law of rotation with transport along the u-axes of the natural orthonormal frame related to the orthogonal system of coordinates (see [35]).

Lemma 3.7. *The coefficients $\eta_{ii}(u)$ of the invariant metric have the form*

$$\eta_{ii}(u) = \partial_i t_1(u), \quad i = 1, \ldots, n. \tag{3.64}$$

Proof. According to (1.39) the invariant inner product $< , >$ has the form

$$< a, b >=< e, a \cdot b >\equiv \omega(a \cdot b) \tag{3.65}$$

for any two vector fields a, b where the 1-form ω is

$$\omega(.) :=< e, . >, \quad e = \frac{\partial}{\partial t^1}. \tag{3.66}$$

Hence $\omega = dt_1$. Lemma is proved.

We summarize the properties of the invariant metric in the canonical coordinates in the following

Proposition 3.4. *The rotation coefficients (3.63) of the invariant metric are symmetric*

$$\gamma_{ij}(u) = \gamma_{ji}(u). \tag{3.67}$$

The metric is invariant w.r.t. the diagonal translations

$$\sum_k \partial_k \eta_{ii}(u) = 0, \quad i = 1, \ldots, n. \tag{3.68}$$

The functions $\eta_{ii}(u)$ and $\gamma_{ij}(u)$ are homogeneous functions of the canonical coordinates of the degrees $-d$ and -1 resp.

Proof. The symmetry (3.67) follows from (3.64):

$$\gamma_{ij}(u) = \frac{1}{2} \frac{\partial_i \partial_j t_1(u)}{\sqrt{\partial_i t_1(u) \partial_j t_1(u)}}. \tag{3.69}$$

To prove (3.68) we use (3.64) and covariant constancy of the vector field

$$e = \sum_i \partial_i.$$

This reads

$$\sum_{k=1}^{n} \Gamma_{ik}^j = 0.$$

For $i = j$ using the Christoffel formulae (3.24) we obtain (3.68). The homogeneity follows from (1.50). Proposition is proved.

Corollary 3.1. *The rotation coefficients (3.63) satisfy the following system of equations*

$$\partial_k \gamma_{ij} = \gamma_{ik} \gamma_{kj}, \quad i, j, k \text{ are distinct} \tag{3.70a}$$

$$\sum_{k=1}^{n} \partial_k \gamma_{ij} = 0 \tag{3.70b}$$

$$\sum_{k=1}^{n} u^k \partial_k \gamma_{ij} = -\gamma_{ij}. \tag{3.70c}$$

Proof. The equations (3.70a) and (3.70b) for a symmetric off-diagonal matrix $\gamma_{ij}(u)$ coincide with the equations of flatness of the diagonal metric obtained for the metrics of the form (3.64) by Darboux and Egoroff [35]. The equation (3.70c) follows from homogeneity.

We have shown that any massive Frobenius manifold determines a scaling invariant (3.70c) solution of the *Darboux - Egoroff system* (3.70a,b). We show now that, conversely,

any solution of the system (3.70) under some genericity assumptions determines locally a massive Frobenius manifold.
Let
$$\Gamma(u) := (\gamma_{ij}(u))$$
be a solution of (3.70).

Lemma 3.8. *The linear system*

$$\partial_k \psi_i = \gamma_{ik}\psi_k, \quad i \neq k \tag{3.71a}$$

$$\sum_{k=1}^{n} \partial_k \psi_i = 0, \quad i = 1, \dots, n \tag{3.71b}$$

for an auxiliary vector-function $\psi = (\psi_1(u), \dots, \psi_n(u))^T$ *has n-dimensional space of solutions.*

Proof. Compatibility of the system (3.71) follows from the Darboux - Egoroff system (3.70). Lemma is proved.

Let us show that, under certain genericity assumptions a basis of homogeneous in u solutions of (3.71) can be chosen. We introduce the $n \times n$-matrix

$$V(u) := [\Gamma(u), U] \tag{3.72}$$

where
$$U := \mathrm{diag}(u^1, \dots, u^n), \tag{3.73}$$

[,] stands for matrix commutator.

Lemma 3.9. *The matrix* $V(u)$ *satisfies the following system of equations*

$$\partial_k V(u) = [V(u), [E_k, \Gamma]], \quad k = 1, \dots, n$$
$$V(u) = [\Gamma(u), U] \tag{3.74}$$

where E_k *are the matrix unities*
$$(E_k)_{ij} = \delta_{ik}\delta_{kj}. \tag{3.75}$$

Conversely, all the differential equations (3.70) follow from (3.74).

Proof. From (3.70) we obtain

$$\partial_i \gamma_{ij} = \frac{1}{u^i - u^j} \left(\sum_{k \neq i,j} (u^j - u^k)\gamma_{ik}\gamma_{kj} - \gamma_{ij} \right). \tag{3.76}$$

The equation (3.74) follows from (3.70a) and (3.76). Lemma is proved.

Corollary 3.2.
1). *The matrix* $V(u)$ *acts on the space of solutions of the linear system (3.71).*
2). *Eigenvalues of* $V(u)$ *do not depend on u.*

3). A solution $\psi(u)$ of the system (3.71) is a homogeneous function of u

$$\psi(cu) = c^{\mu}\psi(u) \tag{3.77}$$

iff $\psi(u)$ is an eigenvector of the matrix $V(u)$

$$V(u)\psi(u) = \mu\psi(u). \tag{3.78}$$

Proof. We rewrite first the linear system (3.71) in the matrix form. This reads

$$\partial_k\psi = -[E_k, \Gamma]\psi. \tag{3.79}$$

From (3.74) and (3.79) it follows immediately the first statement of the lemma. Indeed, if ψ is a solution of (3.79) then

$$\partial_k(V\psi) = (V[E_k, \Gamma] - [E_k, \Gamma]V)\psi - V[E_k, \Gamma]\psi = -[E_k, \Gamma]V\psi.$$

The second statement of the lemma also follows from (3.74). The third statement is obvious since

$$\sum u^i \partial_i \psi = V\psi$$

(this follows from (3.79)). Lemma is proved.

Remark 3.4. We will show below that a spectral parameter can be inserted in the linear system (3.71). This will give a way to integrate the system (3.70) using the isomonodromy deformations technique.

We denote by μ_1, ..., μ_n the eigenvalues of the matrix $V(u)$. Due to skew-symmetry of $V(u)$ these can be ordered in such a way that

$$\mu_\alpha + \mu_{n-\alpha+1} = 0. \tag{3.80}$$

Proposition 3.5. *For a massive Frobenius manifold corresponding to a scaling invariant solution of WDVV the matrix $V(u)$ is diagonalizable. Its eigenvectors $\psi_\alpha = (\psi_{1\alpha}(u), \ldots, \psi_{n\alpha}(u))^T$ satisfying (3.71) are*

$$\psi_{i\alpha}(u) = \frac{\partial_i t_\alpha(u)}{\sqrt{\eta_{ii}(u)}}, \quad i, \alpha = 1, \ldots, n. \tag{3.81}$$

The correspondent eigenvalues are

$$\mu_\alpha = q_\alpha - \frac{d}{2} \tag{3.82}$$

(the spectrum of the Frobenius manifold in the sense of Appendix C). Conversely, let $V(u)$ be any diagonalizable solution of the system (3.74) and $\psi_\alpha = (\psi_{i\alpha}(u))$ be the solutions of (3.71) satisfying

$$V\psi_\alpha = \mu_\alpha\psi_\alpha. \tag{3.83a}$$

If the eigenvalues μ_α of the matrix V are simple then the functions $\psi_\alpha = (\psi_{i\alpha})$ can be found in quadratures of some rational differential forms on the V-space

$$\psi_\alpha = e^{-\sigma_\alpha}\phi_\alpha \qquad (3.83b)$$

$$\sigma_\alpha = \int \sum_i \phi^T_{n-\alpha+1}\left(\partial_i + [E_i, \Gamma]\right)\phi_\alpha \, du^i \qquad (3.83c)$$

where ϕ_α are eigenvectors of V with the eigenvalues μ_α normalized by the condition

$$\phi^T_\alpha \phi_\beta = \delta_{\alpha+\beta, n+1}. \qquad (3.83d)$$

Then the formulae

$$\eta_{\alpha\beta} = \sum_i \psi_{i\alpha}\psi_{i\beta} \qquad (3.84a)$$

$$\partial_i t_\alpha = \psi_{i1}\psi_{i\alpha} \qquad (3.84b)$$

$$c_{\alpha\beta\gamma} = \sum_i \frac{\psi_{i\alpha}\psi_{i\beta}\psi_\gamma}{\psi_{i1}} \qquad (3.84c)$$

determine locally a massive Frobenius manifold with the scaling dimensions

$$q_\alpha = \mu_\alpha - \mu_1, \quad d = -2\mu_1. \qquad (3.84d)$$

Proof is straightforward.

Note that we obtain a Frobenius manifold of the second type (1.22) if the marked vector ψ_{i1} belongs to the kernel of V.

Remark 3.5. The construction of Proposition works also for Frobenius manifolds with nondiagonalizable matrices $\nabla_\alpha E^\beta$. They correspond to nondiagonalizable matrices $V(u)$. The reason of appearing of linear nonhomogeneous terms in the Euler vector field (when some of q_α is equal to 1) is more subtle. We will discuss it in terms of monodromy data below.

Remark 3.6. The change of the coordinates $(u^1, \ldots, u^n) \mapsto (t^1, \ldots, t^n)$ is not invertible in the points where one of the components of the vector-function $\psi_{i1}(u)$ vanishes.

Exercise 3.5. Prove the formula

$$V_{ij}(u) = \sum_{\alpha,\beta} \eta^{\alpha\beta}\mu_\alpha\psi_{i\alpha}(u)\psi_{j\beta}(u) \qquad (3.85)$$

for the matrix $V(u)$ where $\psi_{i\alpha}(u)$ are given by the formula (3.81).

Due to (3.85) one can represent (3.71) as a closed system of differential equations for the vector functions $\psi_{i1}(u), \ldots, \psi_{in}(u)$ putting

$$\gamma_{ij}(u) = \frac{V_{ij}(u)}{u_j - u_i}.$$

This is an alternative representation of the equations (3.70)

$$\partial_i \psi_{j\beta}(u) = \sum_{1 \le \alpha \le \frac{n}{2}} \mu_\alpha \frac{\psi_{i\alpha}\psi_{j\,n-\alpha+1} - \psi_{j\alpha}\psi_{i\,n-\alpha+1}}{u_j - u_i}\psi_{i\beta}(u), \ i \ne j$$

$$\sum_i \partial_i \psi_{i\beta} = 0$$

for any $\beta = 1, \ldots, n$.

Remark 3.7. Forgetting the scaling invariance (1.9) one can still reduce, in the semisimple case the associativity equations (1.14) with the normalization (1.1) to the Darboux - Egoroff system (3.70a,b). The reduction is done by the formula (3.69). This is still an integrable system [47] (the linear system (3.118a,b) below gives a commutation representation for (3.70a,b)). General solution of the system (3.70a,b) depends on $n(n-1)/2$ arbitrary functions of one variable. Conversely, any solution of the Darboux - Egoroff system generates an n-dimensional family of solutions of the equations (1.14), (1.1) by the formulae (3.84a-c). In these formulae $\psi_{i\alpha}(u)$, $\alpha = 1, \ldots, n$ must be arbitrary linearly independent solutions of the system (3.71).

In a similar way one can treat (still in the semisimple case) the associativity equation (1.14) *without* the normalization (1.1). In this case $(\eta^{\alpha\beta})$ in (1.14) is a given symmetric nondegenerate matrix (not related to the derivatives of $F(t)$). To construct solutions of (1.14) starting from a solution $\gamma_{ij}(u) = \gamma_{ji}(u)$ of the Darboux - Egoroff system (3.70a,b) one needs to fix a solution $\psi_{i0}(u)$ of the linear subsystem (3.71a) and a basis $\psi_{i1}(u), \ldots, \psi_{in}(u)$ of solutions of the full linear system (3.71). Then the formulae (3.84a) and

$$t^\alpha = \eta^{\alpha\beta}t_\beta, \quad \partial_i t_\alpha = \psi_{i\alpha}\psi_{i0}, \quad \partial_\alpha\partial_\beta\partial_\gamma F = \sum_i \frac{\psi_{i\alpha}\psi_{i\beta}\psi_{i\gamma}}{\psi_{i0}}$$

determine a solution $F = F(t)$ of the equation (1.14).

Observe that the choice of the solution $\psi_{i0}(u)$ of (3.71a) depends on n arbitrary functions of one variable. The solution satisfies the semisimplicity condition. We leave as an exercise for the reader to prove that, conversely any solution of the associativity equation (1.14) satisfying the semisimplicity condition can be constructed in such a way. [Hint: due to Exercise 1.2 a semisimple algebra with the structure constants

$$c_{\alpha\beta}^\gamma(t) = \eta^{\gamma\epsilon}\partial_\epsilon\partial_\alpha\partial_\beta F(t)$$

has a unity element $e = (e^\alpha(t))$. Prove that the 1-form

$$\omega_\alpha(t) = \eta_{\alpha\beta}e^\beta(t)$$

is closed. So locally

$$\omega_\alpha(t) = \partial_\alpha v(t)$$

for some function $v(t)$. Put

$$\psi_{i0} := \sqrt{\partial_i v}$$

as a function of the canonical coordinates u to obtain the needed solution $\psi_{i0}(u)$ of (3.71a).]

From the construction it follows that a solution $\gamma_{ij}(u)$ of the system (3.70) determines n essentially different (up to an equivalence) solutions of WDVV. This comes from the freedom in the choice of the solution ψ_{i1} in the formulae (3.84). We will see now that these ambiguity is described by the transformations (B.2) (or (B.11), in the case of coincidences between the eigenvalues of V).

Definition 3.3. A 1-form σ on a massive Frobenius manifold M is called *admissible* if the new invariant metric

$$< a, b >_\sigma := \sigma(a \cdot b) \tag{3.86}$$

together with the old multiplication law of tangent vectors and with the old unity e and the old Euler vector field E determines on M a structure of Frobenius manifold with the same rotation coefficients $\gamma_{ij}(u)$.

For example, the 1-form

$$\sigma = dt_1$$

is an admissible one: it determines on M the given Frobenius structure.

Proposition 3.6. *All the admissible forms on a massive Frobenius manifold are*

$$\sigma_c(\,\cdot\,) := \left\langle \left(\sum_k c^k \partial_{\kappa_k} \right)^2, \cdot \right\rangle \tag{3.87a}$$

for arbitary constants c^k and

$$\deg t^{\kappa_1} = \deg t^{\kappa_2} = \ldots \tag{3.87b}$$

The form σ_c can be written also as follows

$$\sigma_c = \sum_{i,j} c^i c^j F_{\kappa_i \kappa_j \alpha} dt^\alpha. \tag{3.87c}$$

Proof. Flat coordinates $t^{\alpha'}$ for a Egoroff metric

$$< \,,\, >' = \sum_i \eta'_{ii}(u) du^i \tag{3.88}$$

with the given rotation coefficients $\gamma_{ij}(u)$ are determined by the system

$$\partial_i \psi'_{j\alpha} = \gamma_{ij}(u) \psi'_{i\alpha}, \quad i \neq j$$

$$\sum_{i=1}^n \partial_i \psi'_{j\alpha} = 0 \tag{3.89}$$

$$\partial_i t'_\alpha = \sqrt{\eta'_{ii}(u)} \psi'_{i\alpha}.$$

Particularly,

$$\psi'_{i1}(u) = \sqrt{\eta'_{ii}(u)}.$$

Also $\psi'_{i\alpha}(u)$ must be homogeneous functions of u. From (3.89) we conclude, as in Corollary 3.2 that they must be eigenvectors of the matrix $V(u)$. So we must have

$$\sqrt{\eta'_{ii}(u)} = \sum_k c^k \psi_{i\,\kappa_k}(u).$$

This gives (3.87). Reversing the calculations we obtain that the metric (3.87) is admissible. Proposition is proved.

From Propositions 3.5 and 3.6 we obtain

Corollary 3.3. *There exists a one-to-one correspondence*

$$\left\{ \begin{matrix} \text{Massive Frobenius manifolds} \\ \text{modulo transformations (B.11)} \end{matrix} \right\} \leftrightarrow \left\{ \begin{matrix} \text{solutions of the system (3.74)} \\ \text{with diagonalizable } V(u) \end{matrix} \right\} \qquad (3.90)$$

Remark 3.8. Solutions of WDVV equations without semisimplicity assumption depend on functional parameters. Indeed, for nilpotent algebras the associativity conditions are very weak (eventually empty, see for example the solution (1.38)). However, it is possible to describe a closure of the class of massive Frobenius manifolds as the set of all Frobenius manifolds with n-dimensional commutative group of algebraic symmetries. Let A be a fixed n-dimensional Frobenius algebra with structure constants c^k_{ij} and an invariant inner nondegenerate inner product $\epsilon = (\epsilon_{ij})$. Let us introduce matrices

$$C_i = (c^k_{ij}). \qquad (3.91)$$

An analogue of the Darboux – Egoroff system (3.70) for an operator-valued function

$$\gamma(u) : A \to A, \ \gamma = (\gamma^j_i(u)), \ u = (u^1, \ldots, u^n) \qquad (3.92)$$

(an analogue of the rotation coefficients) where the operator γ is symmetric with respect to ϵ,

$$\epsilon\gamma = \gamma^{\mathrm{T}}\epsilon \qquad (3.93)$$

has the form

$$[C_i, \partial_j\gamma] - [C_j, \partial_i\gamma] + [[C_i, \gamma], [C_j, \gamma]] = 0, \ i,j = 1, \ldots, n, \qquad (3.94)$$

$\partial_i = \partial/\partial u^i$. This is an integrable system with the Lax representation

$$\partial_i\Psi = \Psi(zC_i + [C_i, \gamma]), \ i = 1, \ldots, n. \qquad (3.95)$$

It is convenient to consider $\Psi = (\psi_1(u), \ldots, \psi_n(u))$ as a function with values in the dual space A^*. Note that A^* also is a Frobenius algebra with the structure constants $c^{ij}_k = c^i_{ks}\epsilon^{sj}$ and the invariant inner product $< , >_*$ determined by $(\epsilon^{ij}) = (\epsilon_{ij})^{-1}$.

Let $\Psi_\alpha(u)$, $\alpha = 1, \ldots ,n$ be a basis of solutions of (3.95) for $z = 0$

$$\partial_i \Psi_\alpha = \Psi_\alpha[C_i, \gamma], \quad \alpha = 1, ..., n \tag{3.96a}$$

such that the vector $\Psi_1(u)$ is invertible in A^*. We put

$$\eta_{\alpha\beta} = <\Psi_\alpha(u), \Psi_\beta(u)>_* \tag{3.96b}$$

$$\text{grad}_u t_\alpha = \Psi_\alpha(u) \cdot \Psi_1(u) \tag{3.96c}$$

$$c_{\alpha\beta\gamma}(t(u)) = \frac{\Psi_\alpha(u) \cdot \Psi_\beta(u) \cdot \Psi_\gamma(u)}{\Psi_1(u)}. \tag{3.96d}$$

Theorem 3.1. *Formulae (3.96) for arbitrary Frobenius algebra A localy parametrize all Frobenius manifolds with n-dimensional commutative group of algebraic symmetries.*

Considering u as a vector in A and $\Psi_1^2 = \Psi_1 \cdot \Psi_1$ as a linear function on A one obtains the following analogue of Egoroff metrics (on A)

$$ds^2 = \Psi_1^2(du \cdot du). \tag{3.97}$$

Examples. We start with the simplest example $n = 2$. The equations (3.70) are linear in this case. They can be solved as

$$\gamma_{12}(u) = \gamma_{21}(u) = \frac{i\mu}{u^1 - u^2} \tag{3.98}$$

where $\pm\mu$ are the eigenvalues of the matrix $V(u)$ being constant in this case. The basis $\psi_{i\alpha}(u)$ of solutions of the system (3.71) has the form

$$\psi_1 = \frac{1}{\sqrt{2}}\begin{pmatrix} r^\mu \\ ir^\mu \end{pmatrix}, \quad \psi_2 = \frac{1}{\sqrt{2}}\begin{pmatrix} r^{-\mu} \\ -ir^{-\mu} \end{pmatrix}, \quad r = u^1 - u^2 \tag{3.99}$$

(we omit the inessential normalization constant). For $\mu \neq -1/2$ the flat coordinates are

$$t^1 = \frac{u^1 + u^2}{2}, \quad t^2 = \frac{r^{2\mu+1}}{2(2\mu + 1)}. \tag{3.100}$$

We have

$$q_1 = 0, \quad q_2 = -2\mu = d.$$

For $\mu \neq \pm 1/2$, $-3/2$ the function F has the form

$$F = \frac{1}{2}(t^1)^2 t^2 + c(t^2)^k$$

for

$$k = (2\mu + 3)/(2\mu + 1), \quad c = \frac{(1 + 2\mu)^3}{2(1 - 2\mu)(2\mu + 3)}[2(2\mu + 1)]^{-4\mu/(2\mu+1)}.$$

For $\mu = 1/2$ the function F has the form

$$F = \frac{1}{2}(t^1)^2 t^2 + \frac{(t^2)^2}{8}\left(\log t^2 - \frac{3}{2}\right).$$

For $\mu = -3/2$ we have

$$F = \frac{1}{2}(t^1)^2 t^2 + \frac{1}{3 \cdot 2^7}\log t^2.$$

For $\mu = -1/2$ the flat coordinates are

$$t^1 = \frac{u^1 + u^2}{2}, \quad t^2 = \frac{1}{2}\log r. \tag{3.101}$$

The function F is

$$F = \frac{1}{2}(t^1)^2 t^2 + 2^{-6} e^{4t^2}.$$

For $n \geq 3$ the system (3.70) is non-linear. I will rewrite the first part of it, i.e. the Darboux - Egoroff equations (3.70a,b) in a more recognizable (for the theory of integrable systems) shape. Let us restrict the functions $\gamma_{ij}(u)$ onto the plane

$$u^i = a_i x + b_i t, \quad i = 1, \ldots, n$$

where the vectors $a = (a_1, \ldots, a_n)$, $b = (b_1, \ldots, b_n)$, and $(1, 1, \ldots, 1)$ are linearly independent. After the substitution we obtain the following matrix form of the system (3.70a,b)

$$\partial_t[A, \Gamma] - \partial_x[B, \Gamma] + [[A, \Gamma], [B, \Gamma]] = 0 \tag{3.102a}$$

where

$$A = \mathrm{diag}(a_1, \ldots, a_n), \quad B = \mathrm{diag}(b_1, \ldots, b_n), \quad \Gamma = (\gamma_{ij}) \tag{3.102b}$$

$[\,,\,]$ stands for the commutator of $n \times n$ matrices. This is a particular reduction of the wellknown n-wave system [119] (let us forget at the moment that all the matrices in (3.102) are complex but not real or hermitean etc.). 'Reduction' means that the matrices $[A, \Gamma]$, $[B, \Gamma]$ involved in (3.102) are skew-symmetric but not generic. I recall [56] that particularly, when the x-dependence drops down from (3.102), the system reduces to the equations of free rotations of a n-dimensional solid (the so-called *Euler - Arnold top* on the Lie algebra $so(n)$)

$$V_t = [\Omega, V], \tag{3.103a}$$

$$V = [A, \Gamma], \quad \Omega = [B, \Gamma], \tag{3.103b}$$

$$\Omega, \ V \in so(n), \quad \Omega = ad_B ad_A^{-1} V. \tag{3.104}$$

This is a hamiltonian system on $so(n)$ with the standard Lie - Poisson bracket [56] and with the quadratic Hamiltonian

$$H = -\frac{1}{2}\operatorname{tr}\Omega V = \frac{1}{2}\sum_{i<j}\frac{b_i - b_j}{a_i - a_j}V_{ij}^2 \tag{3.105}$$

(we identify $so(n)$ with the dual space using the Killing foorm).

The equation (3.70c) determines a scaling reduction of the n-wave system (3.102). It turns out that this has still the form of the Euler – Arnold top on $so(n)$ but the Hamiltonian depends explicitly on time.

Proposition 3.7. *The dependence(3.74) of the matrix $V(u)$ on u^i is determined by a hamiltonian system on $so(n)$ with a time-dependent quadratic Hamiltonian*

$$H_i = \frac{1}{2} \sum_{j \neq i} \frac{V_{ij}^2}{u^i - u^j},$$ (3.106)

$$\partial_i V = \{V, H_i\} \equiv \left[V, ad_{E_i} ad_U^{-1} V\right]$$ (3.107)

for any $i = 1, \ldots, n$.

Proof follows from (3.74), (3.103).

The variables $u^1, \ldots, u^n$ play the role of the times for the pairwise commuting hamiltonian systems (3.106). From (3.107) one obtains

$$\sum_{i=1}^{n} \partial_i V = 0$$
$$\sum_{i=1}^{n} u^i \partial_i V = 0.$$ (3.108)

So there are only $n - 2$ independent parameters among the "times" $u^1, \ldots, u^n$.

The systems (3.74) have "geometrical integrals" (i.e. the Casimirs of the Poisson bracket on $so(n)$). These are the Ad-invariant polynomials on $so(n)$. They are the symmetric combinations of the eigenvalues of the matrix $V(u)$. I recall that, due to Corollary 3.2 the values of these geometrical integrals are expressed via the scaling dimensions of the Frobenius manifold.

It turns out that the non-autonomic tops (3.106) are still integrable. But the integrability of them is more complicated than those for the equations (3.103): they can be integrated by the method of isomonodromic deformations (see below).

Example 3.2. $n = 3$. Take the Hamiltonian

$$H := (u_2 - u_1)H_3 = \frac{1}{2} \left[\frac{\Omega_1^2}{s - 1} + \frac{\Omega_2^2}{s}\right]$$ (3.109)

where we put

$$\Omega_k(s) := -V_{ij}(u)$$ (3.110)

$$s = \frac{u^3 - u^1}{u^2 - u^1},$$ (3.111)

and (i, j, k) is an even permutation of $(1, 2, 3)$. The Poisson bracket on $so(3)$ has the form

$$\{\Omega_1, \Omega_2\} = \Omega_3, \ \{\Omega_2, \Omega_3\} = \Omega_1, \ \{\Omega_3, \Omega_1\} = \Omega_2.$$ (3.112)

The correspondent hamiltonian system reads

$$\frac{d\Omega_1}{ds} = \frac{1}{s}\Omega_2\Omega_3$$

$$\frac{d\Omega_2}{ds} = -\frac{1}{s-1}\Omega_1\Omega_3 \qquad (3.113)$$

$$\frac{d\Omega_3}{ds} = \frac{1}{s(s-1)}\Omega_1\Omega_2.$$

The system has an obvious first integral (the Casimir of (3.112))

$$\Omega_1^2 + \Omega_2^2 + \Omega_3^2 = -\mu^2. \qquad (3.114)$$

The value of the Casimir can be expressed via the scaling dimension d. Indeed, the matrix $\Omega(u)$ (3.110) has the form

$$\Omega(u) = \begin{pmatrix} 0 & \Omega_3 & -\Omega_2 \\ -\Omega_3 & 0 & \Omega_1 \\ \Omega_2 & -\Omega_1 & 0 \end{pmatrix}. \qquad (3.115)$$

The eigenvalues of this matrix are 0 and $\pm R$ where R is defined in (3.114). From (3.84) we know that the eigenvalues are related to the scaling dimensions. For $n = 3$ the scaling dimensions in the problem can be expressed via one parameter d as

$$q_1 = 0, \quad q_2 = \frac{d}{2}, \quad q_3 = d. \qquad (3.116)$$

From this we obtain that

$$\mu = -\frac{d}{2}$$

(minus is chosen for convenience).

Proposition 3.8. *WDVV equations for $n = 3$ with the scaling dimensions (3.116) are equivalent to the following Painlevé-VI equation*

$$y'' = \frac{1}{2}\left(\frac{1}{y} + \frac{1}{y-1} + \frac{1}{y-s}\right)(y')^2 - \left(\frac{1}{s} + \frac{1}{s-1} + \frac{1}{y-s}\right)y'$$

$$+ \frac{y(y-1)(y-s)}{s^2(s-1)^2}\left(\frac{(1+d)^2}{2} + \frac{s(s-1)}{2(y-s)^2}\right). \qquad (3.117)$$

Proof can be obtained using [65, 66]. In Appendix E we give another proof that will enable us to construct solutions of the Painlevé-VI starting from particular solutions of WDVV

For $n > 3$ the system (3.74) can be considered as a high-order analogue of the Painlevé-VI. To show this we introduce a commutation representation [47] of the Darboux - Egoroff system (3.70).

Lemma 3.10. *The equations (3.70a,b) are equivalent to the compatibility conditions of the following linear system of differential equations depending on the spectral parameter z for an auxiliary function $\psi = (\psi_1(u,z), \ldots, \psi_n(u,z))$*

$$\partial_k \psi_i = \gamma_{ik} \psi_k, \quad i \neq k \qquad (3.118a)$$

$$\sum_{k=1}^{n} \partial_k \psi_i = z \psi_i, \quad i = 1, \ldots, n. \qquad (3.118b)$$

Proof is in a straightforward calculation.

Exercise. Show that the formula

$$z^{\frac{d}{2}-1} \frac{\partial_i \tilde{t}(t(u), z)}{\sqrt{\eta_{ii}(u)}} = \psi_i(u, z) \qquad (3.119)$$

establishes one-to-one correspondence between solutions of the linear systems (3.118) and (3.5).

We can say thus that (3.118) gives a gauge transformation of the "Lax pair" of WDVV to the "Lax pair" of the Darboux - Egoroff system (3.70a,b).

The equation (3.70c) specifies the similarity reduction of the system (3.70a,b). It is clear that this is equivalent to a system of ODE of the order $n(n-1)/2$. Indeed, for a given Cauchy data $\gamma_{ij}(u_0)$ we can uniquely find the solution $\gamma_{ij}(u)$ of the system (3.70). We will consider now this ODE system for arbitrary n in more details.

We introduce first a very useful differential operator in z with rational coefficients

$$\Lambda = \partial_z - U - \frac{1}{z} V(u). \qquad (3.120)$$

Here

$$U = \text{diag}(u^1, \ldots, u^n) \qquad (3.121)$$

the matrix $V(u)$ is defined by (3.72).

Proposition 3.9. *Equations (3.70) (or, equivalently, equations (3.74)) are equivalent to the compatibility conditions of the linear problem (3.118) with the differential equation in z*

$$\Lambda \psi = 0. \qquad (3.122)$$

Proof is straightforward (cf. the proof of Corollary 3.2).

Observe that the equations (3.118), (3.122) are obtained by transformation (3.119) from the equations for the flat coordinates $\tilde{t}(t, z)$ of the perturbed connection $\left(\tilde{\nabla}_\alpha(z), \tilde{\nabla}_z \right)$ of the formm (3.3), (3.12).

Solutions $\psi(u, z) = (\psi_1(u, z), \ldots, \psi_n(u, z))^T$ of the equation (3.122) for a fixed u are multivalued analytic functions in $\mathbf{C} \setminus z = 0 \cup z = \infty$. The multivaluedness is described by *monodromy* of the operator Λ. It turns out the parameters of monodromy of the operator Λ with the coefficients depending on u as on the parameters *do not depend on* u. So they are first integrals of the system (3.74). We will see that a part of the first integrals are the eigenvalues of the matrix $V(u)$ (I recall that these are expressed via the degrees of the variables t^α). But for $n \geq 3$ other integrals are not algebraic in $V(u)$.

Let us describe the monodromy of the operator in more details. We will fix some vector $u = (u^1, \ldots, u^n)$ with the only condition $u^i - u^j \neq 0$ for $i \neq j$. For the moment we will concentrate ourselves on the z-dependence of the solution of the differential equation (3.122) taking aside the dependence of it on u. The equation (3.122) has two singularities in the z-sphere $C \cup \infty$: the regular singularity at $z = 0$ and the irregular one at $z = \infty$. The monodromy around the origin is defined as an invertible matrix M_0 such that

$$\Psi\left(z\, e^{2\pi i}\right) = \Psi\left(z\right) M_0.$$

Here $\Psi(z)$ is an invertible matrix solution of the equation (3.122) (the fundamental matrix). The monodromy matrix M_0 depends on the choice of the fundamental matrix $\Psi(z)$. Thus it is determined up to a similarity transformation

$$M_0 \mapsto T^{-1} M_0 T$$

with a nondegenerate T. The eigenvalues of M_0 are determined uniquelly by the eigenvalues $\mu_1, \ldots, \mu_n$ of the matrix V

$$\text{eigen}\,(M_0) = e^{2\pi i \mu_1}, \ldots, e^{2\pi i \mu_n}. \tag{3.123}$$

Particularly, if the matrix V is diagonalizable and none of the differences $\mu_\alpha - \mu_\beta$ for $\alpha \neq \beta$ is an integer then a fundamental matrix $\Psi_0(z) = \left(\psi_{j\alpha}^0(z)\right)$ of (3.122) can be constructed such that

$$\psi_{j\alpha}^0(z) = z^{\mu_\alpha}\psi_{j\alpha}(1 + o(z)), \quad z \to 0. \tag{3.124}$$

Here α is the label of the solution of (3.122); the vectors $\psi_{j\alpha}$ for any $\alpha = 1, \ldots, n$ are the eigenvectors of the matrix V with the eigenvalues μ_α resp. The monodromy of the matrix around $z = 0$ is a diagonal matrix

$$\Psi^0(ze^{2\pi i}) = \Psi^0(z)\text{diag}(e^{2\pi i \mu_1}, \ldots, e^{2\pi i \mu_n}). \tag{3.125}$$

Monodromy at $z = \infty$ (irregular singularity of the operator Λ) is specified by a $n \times n$ *Stokes matrix*. I recall here the definition of Stokes matrices adapted to the operators of the form (3.120).

One of the definitions of the Stokes matrix of the operator (3.120) is based on the theory of reduction of ODEs to a canonical form by analytic gauge transformations. Let us consider an operator

$$L = \frac{d}{dz} - A(z)$$

with an analytic at $z = \infty$ matrix valued function $A(z)$ satisfying

$$A(z) = U + O\left(\frac{1}{z}\right)$$
$$A^T(-z) = A(z).$$

The gauge transformations of L have the form

$$L \mapsto g^{-1}(z)Lg(z) \qquad (3.126a)$$

where the matrix valued function $g(z)$ is analytic near $z = \infty$ with

$$g(z) = 1 + O(z^{-1}) \qquad (3.126b)$$

satisfying

$$g(z)g^T(-z) = 1. \qquad (3.126c)$$

According to the idea of Birkhoff [19] any operator L can be reduced by transformations (3.126) to an operator of the form (3.120). The orbits of the gauge transformations (3.126) form a finite-dimensional family. The local coordinates in this family are determined by the Stokes matrix of the operator Λ.

To give a constructive definition of the Stokes matrix one should use the asymptotic analysis of solutions of the equation (3.122) near $z = \infty$.

Let us fix a vector of the parameters $(u^1, \ldots, u^n)$ with $u^i \neq u^j$ for $i \neq j$. We define first *Stokes rays* in the complex z-plane. These are the rays R_{ij} defined for $i \neq j$ of the form

$$R_{ij} := \left\{ z \,|\, Re[z(u^i - u^j)] = 0, \ Re[e^{i\epsilon}z(u^i - u^j)] > 0 \text{ for a small } \epsilon > 0. \right\}. \qquad (3.127)$$

The ray R_{ji} is opposite to R_{ij}.

Let l be an arbitrary oriented line in the complex z-plane passing through the origin containing no Stokes rays. It divides $\mathbf{C}$ into two half-planes $\mathbf{C}_{\text{right}}$ and $\mathbf{C}_{\text{left}}$. There exist [13, 152] two matrix-valued solutions $\Psi^{\text{right}}(z)$ and $\Psi^{\text{left}}(z)$ of (3.122) analytic in the half-planes $\mathbf{C}_{\text{right/left}}$ resp. with the asymptotic

$$\Psi^{\text{right/left}}(z) = \left(1 + O\left(\frac{1}{z}\right)\right)e^{zU} \text{ for } z \to \infty, \ z \in \mathbf{C}_{\text{right/left}}. \qquad (3.128)$$

These functions can be analytically continued (preserving the asymptotics) into some sectorial neighbourhoods of the half-planes. On the intersections of the neighbourhoods of $\mathbf{C}_{\text{right/left}}$ the solutions $\Psi^{\text{right}}(z)$ and $\Psi^{\text{left}}(z)$ must be related by a linear transformation. To be more specific let the line l have the form

$$l = \left\{ z = \rho e^{i\phi_0} | \rho \in \mathbf{R}, \ \phi_0 \text{ is fixed} \right\} \qquad (3.129)$$

with the natural orientation of the real ρ-line. The half-planes $\mathbf{C}_{\text{right/left}}$ will be labelled in such a way that the vectors

$$\pm i e^{i\phi_0} \tag{3.130}$$

belong to $\mathbf{C}_{\text{left/right}}$ resp. For the matrices $\Psi^{\text{right}}(z)$ and $\Psi^{\text{left}}(z)$ analytic in the sectors

$$-\phi_0 - \epsilon < \arg z < \phi_0 + \epsilon$$

and

$$\phi_0 - \epsilon < \arg z < -\phi_0 + \epsilon$$

resp. for a sufficiently small positive ϵ we obtain

$$\begin{aligned}\Psi^{\text{left}}(\rho e^{i\phi_0}) &= \Psi^{\text{right}}(\rho e^{i\phi_0})S_+ \\ \Psi^{\text{left}}(-\rho e^{i\phi_0}) &= \Psi^{\text{right}}(-\rho e^{i\phi_0})S_-\end{aligned} \tag{3.131}$$

$\rho > 0$ for some constant nondegenerate matrices S_+ and S_-. The boundary-value problem (3.131) together with the asymptotic (3.128) is a particular case of *Riemann – Hilbert* b.v.p.

Proposition 3.10.

1. *The matrices $S_\pm$ must have the form*

$$S_+ = S, \quad S_- = S^T \tag{3.132a}$$

$$S \equiv (s_{ij}), \quad s_{ii} = 1, \quad s_{ij} = 0 \text{ if } i \neq j \text{ and } R_{ij} \subset \mathbf{C}_{\text{right}}. \tag{3.132b}$$

2. *The monodromy at the origin is similar to the matrix*

$$M := S^T S^{-1} \tag{3.133a}$$

$$M_0 = C^{-1} S^T S^{-1} C \tag{3.133b}$$

for a nondegenerate matrix C. Particularly, the eigenvalues of the matrix $S^T S^{-1}$ are

$$(e^{2\pi i \mu_1}, \ldots, e^{2\pi i \mu_n}) \tag{3.134}$$

where $\mu_1, \ldots, \mu_n$ is the spectrum of the matrix V. The solution $\Psi^0(z)$ (3.125) has the form

$$\Psi^0(z) = \Psi^{\text{right}}(z) \, C. \tag{3.135}$$

Proof. We will show first that the restriction for the matrices S_+ and S_- follows from the skew-symmetry of the matrix V. Indeed, the skew-symmetry is equivalent to constancy of the natural inner product in the space of solutions of (3.122)

$$\Psi^T(-z)\Psi(z) = const. \tag{3.136}$$

Let us proof that for the piecewise-analytic function

$$\Psi(z) := \Psi^{\text{right/left}}(z) \tag{3.137}$$

the identity

$$\Psi(z)\Psi^T(-z) = 1 \tag{3.138}$$

follows from the restriction (3.132a). Indeed, for $\rho > 0$, $z = \rho e^{i(\phi_0+0)}$, the l.h.s. of (3.138) reads

$$\Psi^{\text{right}}(z)\Psi^{\text{left}\,T}(-z) = \Psi^{\text{right}}(z)S\Psi^{\text{right}\,T}(-z). \tag{3.139}$$

For $z = \rho^{i(\phi_0-0)}$ we obtain the same expression for the l.h.s. So the piecewise-analytic function $\Psi(z)\Psi^T(-z)$ has no jump on the semiaxis $z = \rho^{i\phi_0}$, $\rho > 0$. Neither it has a jump on the opposite semiaxis (it can be verified similarly). So the matrix-valued function is analytic in the whole complex z-plane. As $z \to \infty$ we have from the asymptotic (3.128) that this function tends to the unity matrix. The Liouville theorem implies (3.138). So the condition (3.132a) is sufficient for the skew-symmetry of the matrix V. Inverting the considerations we obtain also the necessity of the condition.

To prove the restrictions (3.132b) for the Stokes matrix we put

$$\Psi^{\text{right/left}}(z) =: \Phi^{\text{right/left}}(z)\exp zU. \tag{3.140}$$

For the boundary values of the matrix-valued functions $\Phi^{\text{right/left}}(z)$ on the line l we have the following relations

$$\Phi^{\text{left}}(\rho e^{i\phi_0}) = \Phi^{\text{right}}(\rho e^{i\phi_0})\,\tilde{S} \tag{3.141a}$$

$$\Phi^{\text{left}}(-\rho e^{i\phi_0}) = \Phi^{\text{right}}(-\rho e^{i\phi_0})\,\tilde{S}^T \tag{3.141b}$$

$$\Phi^{\text{right/left}}(z) = 1 + O\left(\frac{1}{z}\right), \quad z \to \infty. \tag{3.141c}$$

Here

$$\tilde{S} := e^{zU}Se^{-zU}. \tag{3.141d}$$

From the asymptotic (3.128) we conclude that the matrix $\tilde{S}$ must tend to 1 when $z = \rho e^{i\phi_0}$, $\rho \to +\infty$. This gives (3.132b).

To prove the second statement of the proposition it is enough to compare the monodromy around $z = 0$ of the solution Ψ^{right} analytically continued through the half-line l_+ and of Ψ^0. Proposition is proved.

Definition 3.4. The matrix S in (3.132) is called *Stokes matrix* of the operator (3.120). The matrix M_0 is called *monodromy about the origin* of the operator. The matrix C in (3.135) is called *(central) connection matrix* of the operator. The set

$$S,\ M_0,\ C,\ \mu_1,\dots,\mu_n$$

with the S matrix of the form (3.132) satisfying the constraints

$$C^{-1}S^TS^{-1}C = M_0 \tag{3.142a}$$

$$\text{eigen } M_0 = (\mu_1, \dots, \mu_n) \tag{3.142b}$$

is called *monodromy data* of the operator (3.120).

Observe that precisely $n(n-1)/2$ off-diagonal entries of the Stokes matrix can be nonzero due to (3.132).

The Stokes matrix changes when the line l passes through a separating ray R_{ij}. We will describe these changes in Appendix F.

The connection matrix C is defined up to transformations

$$C \mapsto CQ, \quad \det Q \neq 0, \quad QM_0 = M_0 Q. \tag{3.143}$$

These transformations change the solution Ψ_0 preserving unchanged the matrix V. Note that if the matrix $M = S^T S^{-1}$ has simple spectrum then the solution Ψ^0 is completely determined (up to a normalization of the columns) by the Stokes matrix S. Non-semisimplicity of the matrix M can happen when some of the differences between the eigenvalues of V is equal to an integer (see below).

The eigenvalues of the skew-symmetric complex matrix V in general are complex numbers. We will obtain below sufficient conditions for them to be real.

To reconstruct the operator (3.120) from the monodromy data one is to solve the following *Riemann - Hilbert* boundary value problem: to construct functions $\Phi_{\text{right}}(z)$ analytic for $z \in C_{\text{right}}$ for $|z| \geq 1$, $\Phi_{\text{left}}(z)$ analytic for $z \in C_{\text{left}}$ for $|z| \geq 1$ having

$$\Phi_{\text{right/left}}(z) = 1 + O\left(\frac{1}{z}\right) \text{ for } |z| \to \infty \tag{3.144}$$

and $\Phi_0(z)$ analytic in the unit circle $|z| \leq 1$ such that the boundary values of these functions are related by the following equations

$$\Phi_0(z) = \Phi_{\text{right}}(z) e^{zU} C z^{-L_0} \text{ for } z \in C_{\text{right}}, \ |z| = 1 \tag{3.145a}$$

$$\Phi_0(z) = \Phi_{\text{left}}(z) e^{zU} S^{-1} C z^{-L_0} \text{ for } z \in C_{\text{left}}, \ |z| = 1 \tag{3.145b}$$

$$\Phi_{\text{left}}(z) = \Phi_{\text{right}}(z) e^{zU} S e^{-zU} \text{ for } z \in l_+, \ |z| > 1 \tag{3.145c}$$

$$\Phi_{\text{left}}(z) = \Phi_{\text{right}}(z) e^{zU} S^T e^{-zU} \text{ for } z \in l_-, \ |z| > 1 \tag{3.145d}$$

where the matrix L_0 must satisfy the conditions

$$e^{2\pi i L_0} = M_0 \tag{3.145e}$$

$$\text{eigen } L_0 = (\mu_1, \dots, \mu_n). \tag{3.145f}$$

Exercise 3.7. Prove that the asymptotic expansion of the solution of the Riemann - Hilbert problem (3.145) for $z \to \infty$ has the form

$$\Phi^{\text{right/left}}(z) = \left(1 + \frac{\Gamma}{z} + O\left(\frac{1}{z^2}\right)\right) \tag{3.146a}$$

where
$$V = [\Gamma, U] = \Phi_0(0)L_0\Phi_0^{-1}(0) \tag{3.146b}$$

Proposition 3.11. *The Riemann - Hilbert problem (3.145) together with (3.146) determines a meromorphic function*

$$V = V(\hat{u}; S, M_0, C, \mu_1, \ldots, \mu_n). \tag{3.147a}$$

Here $\hat{u}$ is a point of the universal covering of the space

$$\mathbf{C}^n \setminus \operatorname{diag} = \{(u_1, \ldots, u_n) | u_i \neq u_j \text{ for } i \neq j\} \tag{3.147b}$$

The regular points of the function (3.147) correspond to the monodromy data and the vectors u for which the Riemann - Hilbert problem (3.145) has a unique solution. For any such a regular point the matrices $\Phi_{\text{right/left}}$ and Φ_0 are invertible everywhere.

This is a consequence of the general theory of Riemann - Hilbert problem (see [109, 113]).

Let us assume that the Riemann - Hilbert problem (3.145) for given monodromy data has a unique solution for a given u. Since the solvability of the problem is an open property, we obtain for the given (S, M_0, C, μ) locally a well-defined skew-symmetric matrix-valued analytic function $V(u)$. We show now that this is a solution of the system (3.74). More precisely,

Proposition 3.12. *If the dependence on u of the matrix $V(u)$ of the coefficients of the system (3.122) is specified by the system (3.74) then the monodromy data do not depend on u. Conversely, if the u-dependence of $V(u)$ preserves the matrices S, M_0, C and the numbers μ_α unchanged then $V(u)$ satisfies the system (3.74).*

Proof. We prove first the second part of the proposition. For the piecewise-analytic function

$$\Psi(u, z) = \begin{cases} \Phi^{\text{right/left}}(u, z)e^{zU} & \text{for } z \in \mathbf{C}_{\text{right/left}}, \; |z| > 1 \\ \Phi_0(u, z)z^{L_0} & \text{for } |z| < 1 \end{cases}$$

determined by the Riemann - Hilbert problem (3.145) the combination

$$\partial_i \Psi(u, z) \cdot \Psi^{-1}(u, z) \tag{3.148}$$

for any $i = 1, \ldots, n$ has no jumps on the line l neither on the unit circle. So it is analytic in the whole complex z-plane. From (3.144) we obtain that for $z \to \infty$

$$\partial_i \Psi(u, z) \cdot \Psi^{-1}(u, z) = zE_i - [E_i, \Gamma] + O(1/z)$$

and from the constancy of M_0, C it follows that (3.148) is analytic also at the origin. Applying the Liouville theorem we conclude that the solution of the Riemann - Hilbert problem satisfies the linear system

$$\partial_i \Psi(u, z) = (zE_i - [E_i, \Gamma(u)])\Psi(u, z), \quad i = 1, \ldots, n. \tag{3.149a}$$

Furthermore, due to z-independence of the monodromy data the function $\Psi(cu, c^{-1}z)$ is a solution of the same Riemann - Hilbert problem. From the uniqueness we obtain that the solution satisfies also the condition

$$\left(z\frac{d}{dz} - \sum u^i \partial_i\right)\Psi(u, z) = 0. \qquad (3.149b)$$

Compatibility of the equations (3.149a) reads

$$0 = (\partial_i\partial_j - \partial_j\partial_i)\Psi(u, z) \equiv ([E_j, \partial_i\Gamma] - [E_i, \partial_j\Gamma] + [[E_i, \Gamma], [E_j, \Gamma]])\,\Psi(u, z).$$

Since the matrix

$$([E_j, \partial_i\Gamma] - [E_i, \partial_j\Gamma] + [[E_i, \Gamma], [E_j, \Gamma]])$$

does not depend on z, we conclude that

$$([E_j, \partial_i\Gamma] - [E_i, \partial_j\Gamma] + [[E_i, \Gamma], [E_j, \Gamma]]) = 0.$$

This coincides with the equations (3.70a,b). From (3.149b) we obtain the scaling condition

$$\Gamma(cu) = c^{-1}\Gamma(u).$$

This gives the last equation (3.70c).

Conversely, if the matrix $\Gamma(u)$ satisfies the system (3.70) (or, equivalently, $V = [\Gamma, U]$ satisfies the system (3.74)) then the equations (3.149a) are compatible with the equation (3.149b). Hence for a solution of the Riemann - Hilbert problem the matrices

$$(\partial_i - zE_i + [E_i, \Gamma])\Psi$$

and

$$\left(z\frac{d}{dz} - \sum u^i \partial_i\right)\Psi$$

satisfy the same differential equation (3.122). Hence

$$(\partial_i - zE_i + [E_i, \Gamma])\Psi(u, z) = \Psi(u, z)\,T_i \qquad (3.150a)$$

and

$$\left(z\frac{d}{dz} - \sum u^i \partial_i\right)\Psi(u, z) = \Psi(u, z)\,T \qquad (3.150b)$$

for some matrices $T_i = T_i(u)$, $T = T(u)$. Comparing the expansions of the both sides of (3.150) at $z \to \infty$ we obtain that $T_i = T = 0$. So the solution of the Riemann - Hilbert problem satisfies the equations (3.149). From this immediately follows that the monodromy data do not depend on u. Proposition is proved.

According to the proposition the monodromy data of the operators (3.120) locally parametrize the solutions of the system (3.74). For generic Stokes matrices S all the monodromy data are locally uniquelly determined by the Stokes matrix due to the relations (3.142). Note that, due to the conditions (3.132) there are precisely $n(n-1)/2$ independent complex parameters in the Stokes matrix of the operator (3.120). These can be considered as the local coordinates on the space of solutions of the system (3.74).

Local meromorphicity of the function (3.147) in u claimed by Proposition 3.11 is usually referred to as to the *Painlevé property* of the system (3.74). We summarize it as the following

Corollary 3.4. *Any solution $V(u)$ of the system (3.74) is a single-valued meromorphic function on the universal covering of domain $u^i \neq u^j$ for all $i \neq j$ i.e., on $CP^{n-1} \setminus$* diagonals.

Remark 3.9. One can obtain the equations (3.74) also as the equations of isomonodromy deformations of an operator with regular singularities

$$\frac{d\phi}{d\lambda} + \sum_{i=1}^{n} \frac{A_i}{\lambda - u_i} \phi = 0. \tag{3.151a}$$

The points $\lambda = u_1, \ldots, \lambda = u_n$ are the regular singularities of the coefficients. If $A_1 + \ldots + A_n \neq 0$ then $\lambda = \infty$ is also a regular singularity. The monodromy preserving deformations of (3.151a) were described by Schlesinger [138]. They can be represented in the form of compatibility conditions of (3.151a) with linear system

$$\partial_i \phi = \left(\frac{A_i}{\lambda - u_i} + B_i \right) \phi, \quad i = 1, \ldots, n \tag{3.151b}$$

for some matrices B_i. To represent (3.74) as a reduction of the Schlesinger equations one put

$$A_i = E_i V, \quad B_i = -ad_{E_i} ad_U^{-1} V. \tag{3.151c}$$

Observe that the hamiltonian structure (3.107) of the equations (3.74) is obtained by the reduction (3.151c) of the hamiltonian structure of general Schlesinger equations found in [84].

Doing the substitution

$$\phi = \Psi(u)\chi \tag{3.152}$$

where $\Psi(u) = (\psi_{i\alpha}(u))$ is the matrix of eigenvectors of $V(u)$ normalized in such a way that

$$\partial_i \Psi = B_i \Psi$$

(this coincides with (3.71)) we obtain an equivalent form of the equations (3.151)

$$\frac{d\chi}{d\lambda} = -\eta \Psi^T (\lambda - U)^{-1} \Psi \hat{\mu}\chi \tag{3.153a}$$

$$\partial_i \chi = \frac{\eta \Psi^T E_i \Psi}{\lambda - u_i} \hat{\mu}\chi. \tag{3.153b}$$

To obtain (3.151) from (3.120) and (3.122) we apply the following trick (essentially due to Poincaré and Birkhoff, see the textbook [81, Section 19.4]). Do (just formally) the inverse Laplace transform

$$\psi(z) = z \oint e^{\lambda z} \phi(\lambda) \, d\lambda. \qquad (3.154)$$

Substituting to (3.122), (3.144a) and integrating by parts we obtain (3.151) and (3.152).

We will show now that non-semisimplicity of the matrix $M = S^T S^{-1}$ is just the reason of appearing of linear nonhomogeneous terms in the Euler vector field $E(t)$. We consider here only the simplest case of Frobenius manifolds with the pairwise distinct scaling dimensions satisfying the inequalities

$$0 \le q_\alpha < q_n = d \le 1. \qquad (3.155)$$

Such Frobenius manifolds were called reduced in Appendix B. We showed that any massive Frobenius manifold can be reduced to the form (3.155) by the transformations of Appendix B. Later we will show that these transformations essentially do not change the Stokes matrix.

Proposition 3.13. *1). For the case $d < 1$ the matrix M is diagonalizable.*
2). For $d = 1$ the matrix M is diagonalizable iff $E(t)$ is a linear homogeneous vector field.

Proof. If $d < 1$ then all the numbers $e^{2\pi i \mu_1}, \ldots, e^{2\pi i \mu_n}$ are pairwise distinct (I recall that the scaling dimensions are assumed to be pairwise distinct). This gives diagonalizability of the matrix M.

If $d = 1$ then $\mu_1 = -\frac{1}{2}$, $\mu_n = \frac{1}{2}$. So $e^{2\pi i \mu_1} = e^{2\pi i \mu_n} = -1$, and the characteristic roots of the matrix M are not simple. To prove diagonalizability of the matrix M we are to construct a fundamental system of solutions of the equation $\Lambda \psi = 0$ of the form

$$\psi(u, z) = (\psi(u) + O(z)) z^\mu.$$

We will show that the linear nonhomogeneous terms in the Euler vector field give just the obstruction to construct such a fundamental system.

Lemma 3.11. *If the Euler vector field of a Frobenius manifold with $d = 1$ is*

$$E = \sum_{\alpha=1}^{n-1} (1 - q_\alpha) t^\alpha \partial_\alpha + r \partial_n$$

then a fundamental system of solutions $\psi_{i\alpha}(u, z)$ of the equation $\Lambda \psi = 0$ exists such that

$$\psi_{i\alpha}(u, z) = (\psi_{i\alpha}(u) + O(z)) z^{\mu_\alpha}, \quad \alpha \ne 1 \qquad (3.156a)$$

$$\psi_{i1}(u, z) = \frac{1}{\sqrt{z}} [\psi_{i1}(u) + rz \log z \, \psi_{in}(u) + O(z)] \qquad (3.156b)$$

where $\psi_{i\alpha}(u)$ are defined in (3.81).

Proof. Existence of the solutions of the form (3.156a) is a standard fact (see, e.g., [81]). Let us look for a solution $\psi_{i1}(u,z)$ in the form

$$\psi_{i1}(u,z) = \psi^{(0)} + \psi^{(1)} + \dots$$

where

$$\psi^{(0)} = \frac{1}{\sqrt{z}}\psi_{i1}(u)$$

and the successive approximations are determined by the recursion relations

$$z\partial_z\psi^{(k+1)} = zU\psi^{(k)} + V\psi^{(k+1)}, \quad k = 1, 2, \dots.$$

Again, existence and convergence of the expansion is a standard fact of the theory of ODEs with a regular singularity in the presence of a resonance. Using the identities

$$u_i\psi_{i1}(u) = \sum_{\alpha=1}^{n} g^{n\alpha}(t)\psi_{i\alpha}(u) \tag{3.157}$$

and

$$g^{n\alpha} = \begin{cases} (1 - q_\alpha)t^\alpha & \text{if } \alpha \neq n \\ r & \text{for } \alpha = n \end{cases}$$

we obtain the first correction in the form

$$\psi^{(1)} = \sqrt{z}\left(r \log z\, \psi_{in}(u) + \sum_{\alpha=1}^{n-1} t^\alpha \psi_{i\alpha}(u)\right).$$

The subsequent corrections are at least of order $\sqrt{z}$. Lemma is proved.

We conclude that the basis of solutions of (3.122) being also eigenvectors of the monodromy around $z = 0$ can be constructed *iff* $r = 0$. Proposition is proved.

Corollary 3.5. *Under the assumptions of Lemma 3.11 the monodromy matrix M_0 in the origin $z = 0$ w.r.t. the basis (3.156) has the following form*

$$(M_0)_{\alpha\alpha} = e^{2\pi i\mu_\alpha}, \tag{3.158a}$$

$$(M_0)_{n1} = -2\pi i r, \tag{3.158b}$$

other entries vanish.

We conclude that the linear nonhomogeneous terms in the Euler vector field possibly existing when some of the charges $q_\alpha = 1$ are not determined by the Stokes matrix but by the monodromy at the origin M_0.

Similar arguments allow us to compute the monodromy at the origin also in the case of more complicated resonances.

Example 3.3. *) Consider the Frobenius manifold M^n constructed from the quantum cohomologies of a compact Kähler manifold X of the complex dimension d (see Lecture 2 above). I recall that $n = \dim H^*(X)$ and the flat coordinates t^α are in 1-to-1 correspondence with the cocycles

$$\phi_\alpha \in H^{2q_\alpha}(X).$$

Due to the isomonodromicity property it is enough to compute the monodromy at the origin in the classicall limit

$$t^\alpha \to 0 \text{ when } q_\alpha \neq 1$$
$$t^\alpha \to -\infty \text{ for } q_\alpha = 1 \, .$$

In the limit the operator of multiplication by the Euler vector field coincides with the operator

$$R : H^*(X) \to H^*(X)$$

of cohomological multiplication by the first Chern class of X. It shifts gradings in $H^*(X)$ by 1. From this we easily derive the identity

$$z^\mu R = R z^{\mu+1}$$

where, as above $\mu = \mathrm{diag}(\mu_1, \ldots, \mu_n)$, $\mu_\alpha = q_\alpha - d/2$. Using this identity we obtain the classical limit of the fundamental matrix of the system (3.122) in the form

$$\Psi_0 = z^R z^{\mu+1}. \tag{3.160}$$

So the monodromy of (3.122) around $z = 0$ is given by the matrix

$$M_0 = (-1)^d r^{2\pi i R}. \tag{3.161}$$

We are close now to formulate the precise statement about parametrization of solutions of WDVV by Stokes matrices of the operators (3.120).

Lemma 3.13. *For two equivalent Frobenius manifolds satisfying the semisimplicity condition the corresponding solutions $V(u) = (V_{ij}(u))$ of the system (3.74) are related by a permutation of coordinates*

$$(u^1, \ldots, u^n)^T \mapsto P(u^1, \ldots, u^n)^T, \tag{3.162a}$$

$$V(u) \mapsto \epsilon P^{-1} V(u) P \epsilon, \tag{3.162b}$$

P is the matrix of the permutation, ϵ is an arbitrary diagonal matrix with ± 1 diagonal entries.

---

*) This example was inspired by the observation of Di Francesco and Itzykson (after Lemma 2 in Section 2.5 of [39]) of a relation between the monodromy at the origin for the CP^2 sigma-model and the classical cohomologies of CP^2.

Observe that the permutations act on the differential operators Λ as

$$\Lambda \mapsto \epsilon P^{-1}\Lambda P\epsilon. \qquad (3.163)$$

The Stokes matrix S of the operator Λ changes as

$$S \mapsto P^{-1}\epsilon S \epsilon P. \qquad (3.164)$$

Note that the Legendre-type transformations (B.2) change the operator Λ only as in (3.163). So the correspondent transformations of the Stokes matrix have the form (3.164). Summarizing the considerations of this section, we obtain

Theorem 3.2. *There exists a local one-to-one correspondence*

$$\left\{ \begin{array}{l} \text{Massive Frobenius manifolds} \\ \text{modulo transformations (B.2)} \end{array} \right\} \leftrightarrow \left\{ \begin{array}{l} \text{Stokes matrices of differential} \\ \text{operators } \Lambda \text{ modulo transformations (3.164)} \end{array} \right\}.$$

Definition 3.5. The Stokes matrix S of the operator (3.120) considered modulo the transformations (3.164) will be called *Stokes matrix of the Frobenius manifold*.

Remark 3.10. In the paper [29] Cecotti and Vafa found a physical interpretation of the matrix entries S_{ij} for a Landau - Ginsburg TFT as the algebraic numbers of solitons propagating between classical vacua. In this interpretation S always is an integer-valued matrix. Due to (3.134) they arrive thus at the problem of classification of integral matrices S such that all the eigenvalues of $S^T S^{-1}$ are unimodular. This is the main starting point in the programme of classification of $N = 2$ superconformal theories proposed in [29].

It is interesting that *the same* Stokes matrix appears, according to [29], in the Riemann - Hilbert problem of [51] specifying the Zamolodchikov (or $t\,t^*$) hermitean metric on these Frobenius manifolds.

At the end of this section we explain the sense of the transformations (B.13) of WDVV from the point of view of the operators (3.120).

Proposition 3.14. *The rotation coefficients $\gamma_{ij}(u)$ and $\hat{\gamma}_{ij}(u)$ of two Frobenius manifolds related by the inversion (B.11) are related by the formula*

$$\hat{\gamma}_{ij} = \gamma_{ij} - A_{ij} \qquad (3.165a)$$

where

$$A_{ij} := \frac{\sqrt{\partial_i t_1 \partial_j t_1}}{t_1}. \qquad (3.165b)$$

The solutions $\psi(u, z)$ and $\hat{\psi}(u, z)$ of the correspondent systems (3.122) are related by the gauge transformation

$$\psi = \left(1 + \frac{A}{z}\right)\hat{\psi} \qquad (3.166)$$

for $A = (A_{ij})$.

We leave the proof of this statement as an exercise for the reader.

The gauge transformations of the form

$$\psi(z) \mapsto g(z)\psi(z)$$

with rational invertible matrix valued function $g(z)$ preserving the form of the operator Λ are called *Schlesinger transformations* of the operator [84]. They preserve unchanged the monodromy property of the operator. However, they change some of the eigenvalues of the matrix V by an integer (see (B.16)).

It can be proved that all *elementary* Schlesinger transformations of the system (3.122) where $g(z) = (1 + A z^{-1})$ are superpositions of the transformation (B.13) and of the Legendre-type transformations (B.2). These generate all the group of Schlesinger transformations of (3.120). This group is a group of symmetries of WDVV according to Appendix B.

To conclude this long lecture we will discuss briefly the reality conditions of the solutions of WDVV. We say that the Frobenius manifold is real if it admits an antiholomorphic automorphism $\tau : M \to M$. This means that in some coordinates on M the structure functions $c_{\alpha\beta}^{\gamma}(t)$ all are real. The scaling dimensions q_α also are to be real.

The antiinvolution τ could either preserve or permute the canonical coordinates $u^1(t)$, $\ldots$, $u^n(t)$. We consider here only the case when the canonical coordinates are τ-invariant near some real point $t \in M$, $\tau^* u_i = \bar{u}_i$, $i = 1, \ldots, n$.

Exercise 3.8. Prove that the canonical coordinates are real near a point $t \in M$ where the intersection form is definite positive. Prove that in this case for even n half of the canonical coordinates are positive and half of them are negative, while for odd n one obtains $(n+1)/2$ negative and $(n-1)/2$ positive canonical coordinates.

For real canonical coordinates the diagonal metric $\eta_{ii}(u)$ is real as well. We put

$$J_i := sign\, \eta_{ii}(u), \quad i = 1, \ldots, n \tag{3.167}$$

near the point u under consideration. The matrix $\Gamma(u)$ of the rotation coefficients and, hence, the matrix $V(u)$ obeys the symmetry

$$\Gamma^\dagger = J\Gamma J, \quad V^\dagger = -JVJ \tag{3.168a}$$

where

$$J = \mathrm{diag}(J_1, \ldots, J_n). \tag{3.168b}$$

Here dagger stands for the hermitean conjugation.

Proposition 3.15. *If the coefficients of the operator Λ for real u satisfy the symmetry (3.168) then the Stokes matrix w.r.t. the line $l = \{\mathrm{Im}\, z = 0\}$ satisfies the equation*

$$\bar{S} J S J = 1. \tag{3.169}$$

Conversely, if the Stokes matrix satisfies the equation (3.169) and the Riemann - Hilbert problem (3.145) has a unique solution for a given real u then the corresponding solution of the system (3.74) satisfies (3.168).

Here the bar denotes the complex conjugation of all the entries of S.

Proof. Let l be the real line on the z-plane. As in the proof of Proposition 3.10 we obtain that the equation (3.169) is equivalent to the equation

$$\Psi_{\text{right/left}}(u, z) = \left(\Psi_{\text{left/right}}(u, \bar{z})\right)^{\dagger}. \tag{3.170}$$

Proposition is proved.

To derive from (3.168) the reality of the Frobenius manifold we are to provide also reality of the Euler vector field. For this we need the eigenvalues of the matrix $M = S^T S^{-1}$ to be unimodular.

Lemma 3.14. *The eigenvalues λ of a matrix $M = S^T S^{-1}$ with the matrix S satisfying (3.169) are invariant w.r.t. the transformations*

$$\lambda \mapsto \lambda^{-1}, \quad \lambda \mapsto \bar{\lambda}. \tag{3.171}$$

Proof is obvious.

We conclude that for a generic matrix S satisfying (3.169) the collection of the eigenvalues must consist of:

1). Quadruples $\lambda, \lambda^{-1}, \bar{\lambda}, \bar{\lambda}^{-1}$ for a nonreal λ with $|\lambda| \neq 1$.
2). Pairs $\lambda, \bar{\lambda}$ for a nonreal λ with $|\lambda| = 1$.
3). Pairs λ, λ^{-1} for a real λ distinct from ± 1.
4). The point $\lambda = 1$ for the matrices of odd dimension.

All these types of configurations of eigenvalues are stable under small perturbations of S. Absence of the eigenvalues of the types 1 and 3 specifies an open domain in the space of all complex S-matrices.

Example 3.4. For $n = 3$ and $J = \text{diag}(-1, -1, 1)$ the matrices satisfying (3.169) are parametrized by 3 real numbers a, b, c as

$$S = \begin{pmatrix} 1 & ia & b + \frac{i}{2}ac \\ 0 & 1 & c \\ 0 & 0 & 1 \end{pmatrix}. \tag{3.172}$$

The eigenvalues of $S^T S^{-1}$ are unimodular *iff*

$$0 \leq b^2 + c^2 + \frac{1}{4}a^2c^2 - a^2 \leq 4. \tag{3.173}$$

Appendix D.
Geometry of flat pencils of metrics.

Proposition D.1. *For a flat pencil of metrics a vector field* $f = f^i \partial_i$ *exists such that the difference tensor (3.34) and the metric* g_1^{ij} *have the form*

$$\Delta^{ijk} = \nabla_2^i \nabla_2^j f^k \tag{D.1a}$$

$$g_1^{ij} = \nabla_2^i f^j + \nabla_2^j f^i + c g_2^{ij} \tag{D.1b}$$

for a constant c. *The vector field satisfies the equations*

$$\Delta_s^{ij} \Delta_l^{sk} = \Delta_s^{ik} \Delta_l^{sj} \tag{D.2}$$

where

$$\Delta_k^{ij} := g_{2ks} \Delta^{sij} = \nabla_{2k} \nabla_2^i f^j,$$

and

$$(g_1^{is} g_2^{jt} - g_2^{is} g_1^{jt}) \nabla_{2s} \nabla_{2t} f^k = 0. \tag{D.3}$$

Conversely, for a flat metric g_2^{ij} *and for a solution* f *of the system (D.2), (D.3) the metrics* g_2^{ij} *and (D.1b) form a flat pencil.*

Proof. Let us assume that x^1, ..., x^n is the flat coordinate system for the metric g_2^{ij}. In these coordinates we have

$$\Gamma_{2k}^{ij} = 0, \ \Delta_k^{ij} := g_{2ks} \Delta^{sij} = \Gamma_{1k}^{ij}. \tag{D.4}$$

The equation $R_l^{ijk} = 0$ in these coordinates reads

$$\left(g_1^{is} + \lambda g_2^{is}\right) \left(\partial_s \Delta_l^{jk} - \partial_l \Delta_s^{jk}\right) + \Delta_s^{ij} \Delta_l^{sk} - \Delta_s^{ik} \Delta_l^{sj} = 0. \tag{D.5}$$

Vanishing of the linear in λ term provides existence of a tensor f^{ij} such that

$$\Delta_k^{ij} = \partial_k f^{ij}.$$

The rest part of (D.5) gives (D.2). Let us use now the condition of symmetry (3.27) of the connection $\Gamma 1^{ij}_k + \lambda \Gamma_{2k}^{ij}$. In the coordinate system this reads

$$\left(g_1^{is} + \lambda g_2^{is}\right) \partial_s f^{jk} = \left(g_1^{js} + \lambda g_2^{js}\right) \partial_s f^{ik}. \tag{D.6}$$

Vanishing of the terms in (D.6) linear in λ provides existence of a vector field f such that

$$f^{ij} = g_2^{is} \partial_s f^j.$$

This implies (D.1a). The rest part of the equation (D.6) gives (D.3). The last equation (3.26) gives (D.1b). The first part of the proposition is proved. The converse statement follows from the same equations.

Remark D.1. The theory of S.P.Novikov and the author establishes a one-to-one correspondence between flat contravariant metrics on a manifold M and Poisson brackets of hydrodynamic type on the loop space

$$L(M) := \{\text{smooth maps } S^1 \to M\}$$

with certain nondegeneracy conditions [57, 58]. For a flat metric $g^{ij}(x)$ and the correspondent contravariant connection ∇^i the Poisson bracket of two functionals of the form

$$I = I[x] = \frac{1}{2\pi} \int_0^{2\pi} P(s, x(s))\, ds, \quad J = J[x] = \frac{1}{2\pi} \int_0^{2\pi} Q(s, x(s))\, ds,$$

$x = (x^i(s))$, $x(s + 2\pi) = x(s)$ is defined by the formula

$$\{I, J\} := \frac{1}{2\pi} \int_0^{2\pi} \frac{\delta I}{\delta x^i(s)} \nabla^i \frac{\delta J}{\delta x^j(s)}\, dx^j(s) + \frac{1}{2\pi} \int_0^{2\pi} \frac{\delta I}{\delta x^i(s)} g^{ij}(x) d_s \frac{\delta J}{\delta x^j(s)}. \tag{D.7}$$

Here the variational derivative $\delta I / \delta x^i(s) \in T_* M|_{x = x(s)}$ is defined by the equality

$$I[x + \delta x] - I[x] = \frac{1}{2\pi} \int_0^{2\pi} \frac{\delta I}{\delta x^i(s)} \delta x^i(s)\, ds + o(|\delta x|); \tag{D.8}$$

$\delta J / \delta x^j(s)$ is defined by the same formula, $d_s := ds \frac{\partial}{\partial s}$. The Poisson bracket can be uniquely extended to all "good" functionals on the loop space by Leibnitz rule [57, 58]. Flat pencils of metrics correspond to compatible pairs of Poisson brackets of hydrodynamic type. By the definition, Poisson brackets $\{\ ,\ \}_1$ and $\{\ ,\ \}_2$ are called compatible if an arbitrary linear combination with constant coefficients

$$a\{\ ,\ \}_1 + b\{\ ,\ \}_2$$

again is a Poisson bracket. Compatible pairs of Poisson brackets are important in the theory of integrable systems [108].

The main source of flat pencils is provided by the following statement.

Lemma D.1. *If for a flat metric in some coordinate system x^1, ..., x^n both the components $g^{ij}(x)$ of the metric and $\Gamma^{ij}_k(x)$ of the correspondent Levi-Cività connection depend linearly on the coordinate x^1 then the metrics*

$$g_1^{ij} := g^{ij} \text{ and } g_2^{ij} := \partial_1 g^{ij} \tag{D.9}$$

form a flat pencil assuming that $\det(g_2^{ij}) \neq 0$. The correspondent Levi-Cività connections have the form

$$\Gamma^{ij}_{1k} := \Gamma^{ij}_k, \quad \Gamma^{ij}_{2k} := \partial_1 \Gamma^{ij}_k. \tag{D.10}$$

Proof. The equations (3.26), (3.27) and the equation of vanishing of the curvature have constant coefficients. Hence the transformation

$$g^{ij}(x^1,...,x^n) \mapsto g^{ij}(x^1 + \lambda,...,x^n), \ \Gamma_k^{ij}(x^1,...,x^n) \mapsto \Gamma_k^{ij}(x^1 + \lambda,...,x^n)$$

for an arbitrary λ maps the solutions of these equations to the solutions. By the assumption we have

$$g^{ij}(x^1 + \lambda,...,x^n) = g_1^{ij}(x) + \lambda g_2^{ij}(x), \ \Gamma_k^{ij}(x^1 + \lambda,...,x^n) = \Gamma_{1k}^{ij}(x) + \lambda \Gamma_{2k}^{ij}(x).$$

The lemma is proved.

All the above considerations can be applied also to complex (analytic) manifolds where the metrics are nondegenerate quadratic forms analyticaly depending on the point of M.

Appendix E.
WDVV and Painlevé-VI.

To reduce WDVV for $n = 3$ to a particular form of the Painlevé-VI equation we will use the commutation representation (3.153). For simplicity of the derivation we will assume the matrix V to be diagonalizable (although the result holds true without this assumption).

Since the matrix $\hat{\mu} = \mathrm{diag}(\mu, 0, -\mu)$ where $\mu = -d/2$ has a zero eigenvalue, the component χ_2 of the vector-function $\chi = (\chi_1, \chi_2, \chi_3)^T$ drops from the r.h.s. of the system (3.153). For the vector $\tilde{\chi} := (\chi_1, \chi_3)^T$ we obtain a closed system

$$\frac{d\tilde{\chi}}{d\lambda} = -\mu \left[\frac{A_1}{\lambda - u_1} + \frac{A_2}{\lambda - u_2} + \frac{A_3}{\lambda - u_3} \right] \tilde{\chi} \equiv A(\lambda)\tilde{\chi} \qquad (E.1a)$$

$$\partial_i \tilde{\chi} = \mu \frac{A_i}{\lambda - u_i} \tilde{\chi}, \quad i = 1, 2, 3 \qquad (E.1b)$$

where the matrices A_i have the form

$$A_i = \begin{pmatrix} \psi_{i1}\psi_{i3} & -\psi_{i3}^2 \\ \psi_{i1}^2 & -\psi_{i1}\psi_{i3} \end{pmatrix}. \qquad (E.2)$$

From the definition of the intersection form (3.13) we obtain for the matrix $A(\lambda)$ the formula

$$A(\lambda) = \mu \begin{pmatrix} g_{13}(t, \lambda) & -g_{33}(t, \lambda) \\ g_{11}(t, \lambda) & -g_{13}(t, \lambda) \end{pmatrix} \qquad (E.3)$$

where the matrix $g_{\alpha\beta}(t, \lambda)$ is the inverse to the matrix $\left(g^{\alpha\beta}(t) - \lambda \eta^{\alpha\beta} \right)$ and $g^{\alpha\beta}(t)$ has the form (3.15). Note that the component χ_2 can be found by quadratures

$$\frac{d\chi_2}{d\lambda} = \mu \left[g_{12}\chi_1 - g_{23}\chi_3 \right]$$

$$\partial_i \chi_2 = \frac{\mu}{\lambda - u_i} \left[\psi_{i1}\psi_{i2}\chi_1 - \psi_{i2}\psi_{i3}\chi_3 \right].$$

The matrices A_i satisfy the equation

$$\sum_i A_i = \begin{pmatrix} 1 & 0 \\ 0 & -1 \end{pmatrix} \qquad (E.4)$$

following from the normalization $\eta_{\alpha\beta} = \delta_{\alpha+\beta,4}$ of the t-coordinates. η-orthogonal changes of the t-coordinates determine simultaneous conjugations of the matrices A_i

$$A_i \mapsto K^{-1} A_i K, \quad K = \mathrm{diag}(k_1, k_2). \qquad (E.5)$$

Let us introduce (essentially following [84]) coordinates p, q on the space of matrices A_i satisfying (E.4) modulo transformations (E.5) in the following way: q is the root λ of the equation

$$[A(\lambda)]_{12} = 0, \qquad (E.6)$$

and

$$p = [A(q)]_{11} \, . \tag{E.7}$$

Explicitly

$$q = \left(g^{11}g^{22} - g^{12^2} \right) / g^{11}$$

$$p = \mu \frac{g^{11}g^{22}}{g^{12^3} + g^{11}g^{12}g^{13} - g^{11}g^{12}g^{22} - g^{11^2}g^{23}} \tag{E.8}$$

The entries of the matrices A_i can be expressed via the coordinates p, q and an auxilliary parameter k as follows

$$\psi_{i1}\psi_{i3} = -\frac{q - u_i}{2\mu^2 P'(u_i)} \left[P(q)p^2 + 2\mu\frac{P(q)}{q - u_i}p + \mu^2(q + 2u_i - \sum u_j) \right] \tag{E.9a}$$

$$\psi_{i3}^2 = -k\frac{q - u_i}{P'(u_i)} \tag{E.9b}$$

$$\psi_{i1}^2 = -k^{-1}\frac{q - u_i}{4\mu^4 P'(u_i)} \left[P(q)p^2 + 2\mu\frac{P(q)}{q - u_i}p + \mu^2(q + 2u_i - \sum u_j) \right]^2 \tag{E.9c}$$

where the polynomial $P(\lambda)$ has the form

$$P(\lambda) := (\lambda - u_1)(\lambda - u_2)(\lambda - u_3). \tag{E.10}$$

Substituting these formulae to the equations

$$\partial_i A(\lambda) = \frac{A_i}{(\lambda - u_i)^2} + \frac{[A(\lambda), A_i]}{\lambda - u_i} \tag{E.11}$$

of compatibility of the system (E.1) we obtain a closed system of equations for the functions p and q

$$\partial_i q = \frac{P(q)}{P'(u_i)} \left[2p + \frac{1}{q - u_i} \right]$$

$$\partial_i p = -\frac{P'(q)p^2 + (2q + u_i - \sum u_j)\,p + \mu(1 - \mu)}{P'(u_i)} \tag{E.12}$$

and a quadrature for the function $\log k$

$$\partial_i \log k = (2\mu - 1)\frac{q - u_i}{P'(u_i)}. \tag{E.13}$$

Eliminating p from the system we obtain a second order differential equation for the function $q = q(u_1, u_2, u_3)$

$$\partial_i^2 q = \frac{1}{2}\frac{P'(q)}{P(q)}\left(\partial_i q\right)^2 - \left[\frac{1}{2}\frac{P''(u_i)}{P'(u_i)} + \frac{1}{q - u_i}\right]\partial_i q$$

$$+\frac{1}{2}\frac{P(q)}{\left(P'(u_i)\right)^2}\left[(2\mu-1)^2+\frac{P'(u_i)}{(q-u_i)^2}\right],\quad i=1,2,3.\qquad(E.14)$$

The system (E.14) is invariant w.r.t. transformations of the form

$$u_i \mapsto au_i + b$$
$$q \mapsto au_i + b.$$

Introducing the invariant variables

$$x=\frac{u_3-u_1}{u_2-u_1}$$
$$y=\frac{q}{u_2-u_1}-\frac{u_1}{u_2-u_1}$$

we obtain for the function $y=y(x)$ the following particular Painlevé-VI equation

$$y''=\frac{1}{2}\left[\frac{1}{y}+\frac{1}{y-1}+\frac{1}{y-x}\right](y')^2-\left[\frac{1}{x}+\frac{1}{x-1}+\frac{1}{y-x}\right]y'$$
$$+\frac{1}{2}\frac{y(y-1)(y-x)}{x^2(x-1)^2}\left[(2\mu-1)^2+\frac{x(x-1)}{(y-x)^2}\right].\qquad(E.15)$$

Conversely, for a solution $y(x)$ of the equation (E.15) we construct functions $q=q(u_1,u_2,u_3)$ and $p=p(u_1,u_2,u_3)$ putting

$$q=(u_2-u_1)y\left(\frac{u_3-u_1}{u_2-u_1}\right)+u_1$$
$$p=\frac{1}{2}\frac{P'(u_3)}{P(q)}y'\left(\frac{u_3-u_1}{u_2-u_1}\right)-\frac{1}{2}\frac{1}{q-u_3}.\qquad(E.16)$$

Then we compute the quadrature (E.13) determining the function k (this provides us with one more arbitrary integration constant). After this we are able to compute the matrix $(\psi_{i\alpha}(u))$ from the equations (E.9) and

$$(\psi_{12},\psi_{22},\psi_{32})=\pm i\,(\psi_{21}\psi_{33}-\psi_{23}\psi_{31},\psi_{13}\psi_{31}-\psi_{11}\psi_{33},\psi_{11}\psi_{23}-\psi_{13}\psi_{21}).\qquad(E.17)$$

The last step is in reconstructing the flat coordinates $t=t(u)$ and the tensor $c_{\alpha\beta\gamma}$ using the formulae (3.84).

We will show now that the hamiltonian structure of the Painlevé-VI is inherited from the $so(3)$-hamiltonian structure of the time-dependent Euler equations (3.113). We observe first that the squares of the matrix elements $V_{ij}=\mu(\psi_{i1}\psi_{j3}-\psi_{i3}\psi_{j1})$ can be expressed via p and q using identity

$$V_{ij}^2=-\mu^2\mathrm{tr}A_iA_j.\qquad(E.18)$$

This gives

$$V_{ij}^2=-\frac{(p(q-u_k)+\mu)^2P(q)}{(q-u_k)P'(u_k)}\qquad(E.19)$$

where i, j, k are three distinct indices from 1 to 3. The matrix elements V_{ij} can be uniquelly reconstructed from (E.19) and from the following requirements: 1) the vector (V_{23}, V_{31}, V_{12}) must be proportional to (E.17) and 2) the identity

$$V_{12}V_{23}V_{31} = p^3 \frac{P(q)P\left(q + \frac{\mu}{p}\right)}{(u_1 - u_2)(u_2 - u_3)(u_3 - u_1)} \qquad (E.20)$$

holds true (this specifies the common sign of the entries V_{ij}). It is easy to verify now that the transformation

$$V_{23}, \; V_{31}, \; V_{12} \mapsto q, \, p, \, \mu \qquad (E.21)$$

transforms the standard $so(3)$-Lie-Poisson brackets for the entries V_{ij} to the canonical brackets for q and p commuting with the Casimir $\mu = \sqrt{-(V_{23}^2 + V_{31}^2 + V_{12})}$. Observe that the change of coordinates (E.8) depends explicitly on the times u_1, u_2, u_3. Due to this dependence the hamiltonians of the systems (E.12) *are not* obtained by the reduction of the hamiltonians (3.106) on the spheres (3.114).

To obtain the hamiltonians of the equations (E.12) from the quadratic hamiltonians of the Euler equations (3.113) we use the following elementary statement.

Lemma E.1. *Let*

$$\frac{dy^i}{dt} = \left\{ y^i, H(y, t) \right\} \qquad (E.22)$$

be a hamiltonian system on a Poisson manifold with Poisson brackets { , } *and*

$$y = \varphi(x, t) \qquad (E.23)$$

a local diffeomorphism depending explicitly on time t Let the vector field $\partial_t \varphi$ be a hamiltonian one with the hamiltonian $-\delta H$. Then (E.22) is a hamiltonian system also in the x-coordinnates

$$\frac{dx^i}{dt} = \left\{ x^i, \hat{H}(x, t) \right\} \qquad (E.24a)$$

with the hamiltonian

$$\hat{H}(x, t) = H(\varphi(x, t), t) + \delta H(\varphi(x, t), t), \qquad (E.24b)$$

Proof is obvious.

Applying this procedure to the transformation (E.8) we obtain, after some calculations

$$\delta H_i = \frac{p \frac{P(q)}{q - u_i} + \mu(q - u_i)}{P'(u_i)} \qquad (E.25)$$

and the restriction of the hamiltonians (3.106) on the spheres (3.114) we obtain easyly using (E.19)

$$H_i = \frac{p^2 P(q) + \mu^2(q - u_i)}{P'(u_i)}. \qquad (E.26)$$

Using the reduced hamiltonians $H_i + \delta H_i$ we obtain the hamiltonian of the equation (E.15)

$$\hat{H} = \frac{y(y-1)(y-x)p^2 + y(y-1)p + \mu(1-\mu)(y-x)}{y(y-1)}. \qquad (E.27)$$

Applying this formalism to the polynomial solutions of Appendix A we obtain three algebraic solutions of Painlevé-VI as follows.

Intersection form of the type A_3 (the solution (A.7))

$$\begin{pmatrix} \frac{3}{4}t_2{}^2 + \frac{1}{2}t_3{}^3 & \frac{5}{4}t_2 t_3 & t_1 \\[2mm] \frac{5}{4}t_2 t_3 & t_1 + \frac{1}{2}t_3{}^2 & \frac{3}{4}t_2 \\[2mm] t_1 & \frac{3}{4}t_2 & \frac{1}{2}t_3 \end{pmatrix}. \qquad (E.28)$$

Characteristic polynomial

$$\det\left(g^{\alpha\beta}(t) - u\eta^{\alpha\beta}\right) = u^3 + u^2\left(-3t_1 - \frac{1}{2}t_3{}^2\right)$$

$$+ u\left(3t_1{}^2 - \frac{9}{4}t_2{}^2 t_3 + t_1 t_3{}^2 - \frac{1}{4}t_3{}^4\right)$$

$$-t_1{}^3 - \frac{27}{64}t_2{}^4 + \frac{9}{4}t_1 t_2{}^2 t_3 - \frac{1}{2}t_1{}^2 t_3{}^2 - \frac{7}{8}t_2{}^2 t_3{}^3 + \frac{1}{4}t_1 t_3{}^4 + \frac{1}{8}t_3{}^6.$$

We obtain the following solution of (E.15) with $\mu = -1/4$ represented in the parametric form

$$y = \frac{1}{\omega_2 - \omega_1}\left[\frac{-19t+4}{6t+4} - \omega_1\right]$$

$$x = \frac{\omega_3 - \omega_1}{\omega_2 - \omega_1} \qquad (E.29a)$$

where $\omega_1, \omega_2, \omega_3$ are the roots of the cubic equation

$$\omega^3 - \omega^2 - (9t+1)\omega - \frac{27}{8}t^2 - 7t + 1 = 0. \qquad (E.29b)$$

Intersection form of the type B_3 (the solution (A.8))

$$\begin{pmatrix} \frac{5}{3}t_2{}^2 t_3 + \frac{1}{3}t_3{}^5 & \frac{2}{3}t_2{}^2 + \frac{4}{3}t_2 t_3{}^2 & t_1 \\[2mm] \frac{2}{3}t_2{}^2 + \frac{4}{3}t_2 t_3{}^2 & t_1 + t_2 t_3 + \frac{1}{3}t_3{}^3 & \frac{2}{3}t_2 \\[2mm] t_1 & \frac{2}{3}t_2 & \frac{1}{3}t_3 \end{pmatrix}. \qquad (E.30)$$

Characteristic polynomial

$$\det\left(g^{\alpha\beta}(t) - u\eta^{\alpha\beta}\right) = u^3 + u^2\left(-3\,t_1 - t_2\,t_3 - \frac{1}{3}t_3{}^3\right)$$

$$+u\left(3\,t_1{}^2 - \frac{8}{9}t_2{}^3 + 2\,t_1\,t_2\,t_3 - \frac{7}{3}t_2{}^2\,t_3{}^2 + \frac{2}{3}t_1\,t_3{}^3 - \frac{1}{9}t_3{}^6\right)$$

$$-t_1{}^3 + \frac{8}{9}t_1\,t_2{}^3 - t_1{}^2\,t_2\,t_3 - \frac{8}{9}t_2{}^4\,t_3 + \frac{7}{3}t_1\,t_2{}^2\,t_3{}^2 - \frac{1}{3}t_1{}^2\,t_3{}^3$$

$$-\frac{1}{27}t_2{}^3\,t_3{}^3 - \frac{5}{9}t_2{}^2\,t_3{}^5 + \frac{1}{9}t_1\,t_3{}^6 + \frac{1}{9}t_2\,t_3{}^7 + \frac{1}{27}t_3{}^9.$$

This gives a solution of (E.15) with $\mu = -1/3$ of the form

$$y = \frac{(2-s)(s+1)(s^2-3)^2}{(2+s)(5s^4-10s^2+9)}$$

$$x = \frac{(2-s)^2(1+s)}{(2+s)^2(1-s)}.$$

(E.31)

Intersection form of the type H_3 (the solution (A.9))

$$\begin{pmatrix} \frac{3}{5}t_2{}^3 + \frac{9}{5}t_2{}^2\,t_3{}^3 + \frac{1}{20}t_3{}^9 & \frac{7}{5}t_2{}^2\,t_3 + \frac{7}{10}t_2\,t_3{}^4 & t_1 \\[2mm] \frac{7}{5}t_2{}^2\,t_3 + \frac{7}{10}t_2\,t_3{}^4 & t_1 + t_2\,t_3{}^2 + \frac{1}{10}t_3{}^5 & \frac{3}{5}t_2 \\[2mm] t_1 & \frac{3}{5}t_2 & \frac{1}{5}t_3 \end{pmatrix}.$$

(E.32)

Characteristic polynomial

$$\det\left(g^{\alpha\beta}(t) - u\eta^{\alpha\beta}\right) = u^3 + u^2\left(-3\,t_1 - t_2\,t_3{}^2 - \frac{1}{10}t_3{}^5\right)$$

$$+u\left(3\,t_1{}^2 - \frac{9}{5}t_2{}^3\,t_3 + 2\,t_1\,t_2\,t_3{}^2 - \frac{6}{5}t_2{}^2\,t_3{}^4 + \frac{1}{5}t_1\,t_3{}^5 - \frac{1}{100}t_3{}^{10}\right)$$

$$-t_1{}^3 - \frac{27}{125}t_2{}^5 + \frac{9}{5}t_1\,t_2{}^3\,t_3 - t_1{}^2\,t_2\,t_3{}^2 - \frac{23}{25}t_2{}^4\,t_3{}^3 + \frac{6}{5}t_1\,t_2{}^2\,t_3{}^4$$

$$-\frac{1}{10}t_1{}^2\,t_3{}^5 - \frac{1}{50}t_2{}^3\,t_3{}^6 - \frac{2}{25}t_2{}^2\,t_3{}^9 + \frac{1}{100}t_1\,t_3{}^{10} + \frac{1}{100}t_2\,t_3{}^{12} + \frac{1}{1000}t_3{}^{15}.$$

We obtain the following solution of (E.15) with $\mu = -2/5$

$$y = \frac{1}{\omega_2 - \omega_1}\left[\frac{-272t^4 - 20t^3 - 62t^2 + 10t + 1}{12t^3 + 36t^2 + 1} - \omega_1\right]$$

$$x = \frac{\omega_3 - \omega_1}{\omega_2 - \omega_1}$$

(E.33a)

235

$$\omega^3 - (10t+1)\omega^2 - (180t^3 + 120t^2 + 1)\omega - 216t^5 - 920t^4 - 20t^3 - 80t^2 + 10t + 1 = 0. \quad (E.33b)$$

Intersection form of the type $\tilde{W}(A_2)$ (the solution (A.13)) is given in the formula (5.85). Although this is not an algebraic solution of WDVV the correspondent solution of the Painlevé-VI (for $\mu = -\frac{1}{2}$) is algebraic. It has the form

$$y = \frac{1}{\omega_1 - \omega_2}\left[\frac{9 + 4t^3}{8t} + \omega_1\right]$$
$$x = \frac{\omega_3 - \omega_1}{\omega_2 - \omega_1} \quad\quad\quad (E.34a)$$

where ω_i are the roots of the equation

$$\omega^3 + \frac{1}{2}t^2\omega^2 - \frac{9}{2}t\omega - \frac{1}{8}(27 + 16t^3) = 0. \quad (E.34b)$$

Intersection form of the type $\tilde{W}(B_2)$ (the solution (A.14)) reads as follows

$$\left(g^{\alpha\beta}(t)\right) = \begin{pmatrix} \frac{1}{4}e^{2t_3} + \frac{1}{2}t_2^2 e^{t_3} & \frac{3}{4}t_2 e^{t_3} & t_1 \\ \frac{3}{4}t_2 e^{t_3} & t_1 - \frac{1}{4}t_2^2 + \frac{1}{2}e^{t_3} & \frac{1}{2}t_2 \\ t_1 & \frac{1}{2}t_2 & 1 \end{pmatrix}. \quad (E.35)$$

It gives the following algebraic solution of the Painlevé-VI (again for $\mu = -\frac{1}{2}$)

$$y = \frac{(1+s)^2(s+2)}{4(1+2s^2)}$$
$$x = \frac{(s+2)^2}{8s}. \quad\quad\quad (E.36)$$

Intersection form of the type $\tilde{W}(G_2)$ (the solution (A.15)) is given by the matrix

$$\begin{pmatrix} 2\left(9e^{4t_3} + 6e^{2t_3}t_2^2 + \frac{2}{3}e^{t_3}t_2^3\right) & \frac{3}{2}\left(6e^{2t_3}t_2 + 2e^{t_3}t_2^2\right) & t_1 \\ \frac{3}{2}\left(6e^{2t_3}t_2 + 2e^{t_3}t_2^2\right) & 3e^{2t_3} + t_1 + 4e^{t_3}t_2 - \frac{1}{6}t_2^2 & \frac{1}{2}t_2 \\ t_1 & \frac{1}{2}t_2 & \frac{1}{2} \end{pmatrix}$$

The solution of Painlevé-VI equation with $\mu = -1/2$ has the form

$$y = \frac{3(3-t)(t+1)(t^2-3)^2}{(3+t)^2(t^6 + 3t^4 - 9t^2 + 9)}$$
$$x = \frac{(3-t)^3(1+t)}{(1-t)(3+t)^3}. \quad\quad\quad (E.37)$$

An approach to classification of particular solutions of Painlevé-VI equation that can be expressed via classical transcendental functions was proposed by Okamoto [121]. This is based on the theory of canonical transformations of Painlevé-VI as of a time-dependent Hamiltonian system. For our particular Painlevé-VI for a non-integer d the solutions of Okamoto have the form

$$q \equiv u_i$$

for some i, p can be found from the equations (E.12). Particularly, for $q \equiv u_3$ the solution can be expressed via hypergeometric functions

$$p = \frac{1}{u_2 - u_1} \frac{d}{dx} \log f(\mu, 1 - \mu, 1; x) \qquad (E.38a)$$

where $f(\mu, 1 - \mu, 1; x)$ stands for the general solution of the hypergeometric equation

$$x(1 - x)f'' + (1 - 2x)f' - \mu(1 - \mu)f = 0 \qquad (E.38b)$$

Also for $\mu = 0$ or $\mu = 1$ there is a particular solution of [121] of the form $p \equiv 0$. This gives a rational solution of (3.74) of the form

$$V_{ij} = \frac{a_i a_j (u_i - u_j)}{\sum a_i^2 u_i} \qquad (E.39a)$$

when the constants a_i must satisfy the constraint

$$\sum_{i=1}^{n} a_i^2 = 0 \qquad (E.39b)$$

(this is a solution of (3.74) for an arbitrary n).

It is easy to check that none of our algebraic solutions of the Painlevé-VI is of the above form (it is sufficient to check that $\det\left(g^{\alpha\beta} - q\eta^{\alpha\beta}\right) \neq 0$). The Legendre-type transformations of Appendix B correspond to the transpositions of the matrix $A(\lambda)$. This generates solutions of the Painlevé-VI equation of the form (3.15) with the change $\mu \mapsto -\mu$. Particularly, for the function q this gives a transformation of the form

$$q \mapsto \tilde{q} = \left(g^{22}g^{33} - g^{23^2}\right)/g^{33}. \qquad (E.40)$$

The inversion (B.13) maps a solution of (E.15) to a solution of the same equation with $\mu \mapsto -(1 + \mu)$.

These are the only canonical transformations of the Okamoto type of our Painlevé-VI equation.

We conclude that the constructed algebraic solutions of the Painlevé-VI equations cannot be reduced to previously known solutions of these equations. *)

---

*) I learned recently from N.Hitchin [77] that he also found a particular solution of the Painlevé-VI equation in terms of algebraic functions using the Gauss - Manin connection of the type A_3. In a more recent preprint [78] Hitchin constructs an infinite family of algebraic solutions of certain equations of the Painlevé-VI type. It would be interesting to compare these equations with those of the form (E.15).

Exercise E.1. 1). Show that the canonical coordinates u_1, u_2, u_3 for a 3-dimensional Frobenius manifold with the free energy of the form (C.2) with $\gamma = \gamma(t_3)$ being an arbitrary solution of the Chazy equation (C.5) have the form

$$u_i = t_1 + \frac{1}{2}t_2^2\omega_i(t_3), \quad i = 1, 2, 3 \qquad (E.41)$$

where $\omega_i(\tau)$ are the roots of the cubic equation (C.7).

2). Show that the correspondent solutions of the Euler equations (3.113) has the form

$$\Omega_i = \frac{\omega_i}{2\sqrt{(\omega_j - \omega_i)(\omega_i - \omega_k)}} \qquad (E.42)$$

(here i, j, k are distinct indices) where the dependence $t_3 = t_3(s)$ is determined by the equation

$$\frac{\omega_3(t_3) - \omega_1(t_3)}{\omega_2(t_3) - \omega_1(t_3)} = s. \qquad (E.43)$$

Appendix F.
Analytic continuation of solutions of WDVV and braid group

Recall that the characteristic feature of ODEs of the Painlevé type is absence of movable critical singularities for a generic solution of the ODE. For example, generic solution of the system (3.74) is a meromorphic (matrix-valued) function of $(u^1, \ldots, u^n)$ outside the diagonals $u^i = u^j$ for some $i \neq j$. The diagonals form the locus of fixed critical singularities of the solutions. Particularly, for $n = 3$ the solutions of (3.74) have the critical singularities on the diagonals $u_1 = u_2$, $u_1 = u_3$, $u_2 = u_3$. The transformation of Appendix E translates these to the critical points $z = 0$, 1, ∞ of the PVI equation (E.1).

The behaviour of solutions of equations of the Painlevé type "in the large" is very complicated. Multivaluedness of the analytic continuation of the solutions around the critical locus can be called *nonlinear monodromy* of equations of the Painlevé type.

We propose here an approach to the problem of description of the nonlinear monodromy of equations of the Painlevé type in the setting of the isomonodromy integration method. This method is based on the representation of equations of the Painlevé type as monodromy preserving deformations of an auxilliary linear differential operator with rational coefficients. Solutions of the equation of the Painlevé type are parametrized by the monodromy data of the auxilliary linear differential operator. To avoid confusions we will call these *linear monodromy* of the auxilliary linear differential operator.

Our aim is to describe the nonlinear monodromy of solutions of equations of the Painlevé type in terms of linear monodromy of the associated linear differential operator with rational coefficients.

As an example we consider here isomonodromy deformations of the operator

$$\Lambda = \frac{d}{dz} - U - \frac{1}{z}V, \qquad (F.1)$$

i.e., the system (3.74) where, as above, $U = \operatorname{diag}(u^1, \ldots, u^n)$ is a constant diagonal matrix with $u^i \neq u^j$ for $i \neq j$ and V is a skew-symmetric matrix. In this Appendix we will explain how to relate the *nonlinear monodromy* of the isomonodromy deformations of the operator (F.1) around the fixed critical locus $u^i = u^j$ for $i \neq j$ with the linear monodromy of the operator.

We define first *Stokes factors* of the operator (see [13]).

For any ordered pair ij with $i \neq j$ we define the *Stokes ray* R_{ij}

$$R_{ij} = \left\{ z = -ir\left(\bar{u}^i - \bar{u}^j\right) \mid r \geq 0 \right\}. \qquad (F.2)$$

Note that the ray R_{ji} is the opposite to R_{ij}. The line $R_{ij} \cup R_{ji}$ divides $\mathbf{C}$ into two half-planes P_{ij} and P_{ji} where the half-plan P_{ij} is on the left of the ray R_{ij}. We have

$$|e^{zu^i}| > |e^{zu^j}| \text{ for } z \in P_{ij}. \qquad (F.3)$$

Separating rays are those who coincide with some of the Stokes rays.

Let l be an oriented line going through the origin not containing Stokes rays. It divides $\mathbf{C}$ into two half-planes P_{left} and P_{right}. We order the separating rays $R_1, \ldots, R_{2m}$ starting with the first one in P_{right}. In the formulae below the labels of the separating rays will be considered modulo $2m$.

Let Ψ_j be the matrix solution of the equation

$$\Lambda \Psi = 0 \qquad (F.4)$$

uniquely determined by the asymptotic

$$\Psi = \left(1 + O\left(\frac{1}{z}\right)\right) e^{zU} \qquad (F.5)$$

in the sector from $R_j e^{-i\epsilon/2}$ to $R_{m+j} e^{-i\epsilon}$. Here ϵ is a sufficiently small positive number. This can be extended analytically into the open sector from R_{j-1} to R_{m+j}. On the intersection of two such subsequent sectors we have

$$\Psi_{j+1} = \Psi_j K_{R_j} \qquad (F.6)$$

for some nondegenerate matrix K_{R_j}.

For a given choice of the oriented line l we obtain thus a matrix K_R for any separating ray R. These matrices will be called *Stokes factors* of the operator Λ.

Lemma F.1. *The matrices $K = K_R$ satisfy the conditions*

$$K_{ii} = 1, \ i = 1, \ldots, n, \ K_{ij} \neq 0 \text{ for } i \neq j \text{ only when } R_{ji} \subset R \qquad (F.7a)$$

$$K_{-R} = K_R^{-T}. \qquad (F.7b)$$

The solutions Ψ_{right} and Ψ_{left} of Lecture 3 have the form

$$\Psi_{\text{right}} = \Psi_1, \ \Psi_{\text{left}} = \Psi_{m+1}. \qquad (F.8)$$

The Stokes matrix S is expressed via the Stokes factors in the form

$$S = K_{R_1} K_{R_2} \ldots K_{R_m}. \qquad (F.9)$$

Conversely, for a given configuration of the line l and of the Stokes rays the Stokes factors of the form (F.7) are uniquelly determined from the equation (F.9).

Proofs of all of the statement of the lemma but (F.7b) can be found in [13]. The relation (F.7b) follows from the skew-symmetry of V as in Proposition 3.10.

Example F.1. For generic $u_1, \ldots, u_n$ all the Stokes rays are pairwise distinct. Then the Stokes factors have only one nonzero off-diagonal element, namely

$$\left(K_{R_{ij}}\right)_{ji} \neq 0. \qquad (F.10)$$

Let the matrix $V = V(u)$ depend now on $u = (u^1, \ldots, u^n)$ in such a way that small deformations of u are isomonodromic. After a large deformation the separating rays could pass through the line l. The correspondent change of the Stokes matrix is described by the following

Corollary F.1. *If a separating ray R passes through the positive half-line l_+ moving clockwise then the solutions Ψ_{right}, Ψ_{left} and the Stokes matrix S are transformed as follows*

$$\Psi_{\mathrm{right}} = \Psi'_{\mathrm{right}} K_R^T, \quad \Psi'_{\mathrm{left}} = \Psi_{\mathrm{left}} K_R, \quad S' = K_R^T S K_R. \qquad (F.11)$$

Remark F.1. A similar statement holds as well without skew-symmetry of the matrix V. Instead of the matrices K_R and K_R^T in the formulae there will be two independent matrices K_R and K_{-R}^{-1}.

Particularly, let us assume that the real parts of u^i are pairwise distinct. We order them in such a way that

$$\mathrm{Re}\, u^1 < \ldots < \mathrm{Re}\, u^n. \qquad (F.12)$$

The real line with the natural orientation will be chosen as the line l. The correspondent Stokes matrix will be upper triangular for such a choice. Any closed path in the space of pairwise distinct ordered parameters u determines a transformation of the Stokes matrix S that can be read of (F.11) (just permutation of the Stokes factors). We obtain an action of the pure braid group on the space of the Stokes matrices. We can extend it onto all the braid group adding permutations of $u_1, \ldots, u_n$. Note that the eigenvalues of the matrix $S^T S^{-1}$ (the monodromy of (F.1) in the origin) are preserved by the action (F.11).

Proposition F.1. *If the solution $V(u)$ of the equations of the isomonodromy deformations of the operator Λ with the given Stokes matrix S is an algebraic function with branching along the diagonals $u^i = u^j$ then S belongs to a finite orbit of the action of the pure braid group.*

Proof. We know from Proposition 3.11 that the matrix function $V(u)$ is meromorphic on the universal covering of $CP^{n-1} \setminus \cup \{u^i = u^j\}$. Closed paths in the deformation space will interchange the branches of this function. Due to the assumptions we will have only finite number of branches. Proposition is proved.

In the theory of Frobenius manifolds the parameters u^i (i.e. the canonical coordinates) are determined only up to a permutation. So we obtain the action of the braid group B_n on the space of Stokes matrices. Explicitly, the standard generator σ_i of B_n ($1 \le i \le n-1$) interchanging u^i and u^{i+1} moving u^i clockwise around u^{i+1} acts as follows

$$\Psi'_{\mathrm{right}} = P\Psi_{\mathrm{right}} K^{-1}, \quad \Psi'_{\mathrm{left}} = P\Psi_{\mathrm{left}} K, \qquad (F.13a)$$

$$S' = KSK \qquad (F.13b)$$

where the matrix $K = K_i(S)$ has the form

$$K_{ss} = 1, s = 1, \ldots, n, \ s \ne i, \ i+1,$$

$$K_{ii} = -s_{i\,i+1}, \quad K_{i\,i+1} = K_{i+1\,i} = 1, \quad K_{i+1\,i+1} = 0 \qquad (F.14)$$

other matrix entries of K vanish, $P = P_i$ is the matrix of permutation $i \leftrightarrow i + 1$.

It is clear that finite orbits of the full braid group B_n must consist of finite orbits of the subgroup of pure braids. So it is sufficient to find finite orbits of the action (F.13).

Exercise F.1. Verify that the braid

$$(\sigma_1 \ldots \sigma_{n-1})^n \tag{F.15}$$

acts trivially on the space of Stokes matrices.

The braid (F.15) is the generator of the center of the braid group B_n for $n \geq 3$ [20]. We obtain thus an action of the quotient

$$A_n^* = B_n/\text{center}$$

on the space of the Stokes matrices. Note that A_n^* coincides with the mapping class group of the plane with n marked points [20]. For $n = 3$ the group A_n^* is isomorphic to the modular group $PSL(2, \mathbf{Z})$.

In order to construct examples of finite orbits of the action (F.13b), (F.14) we represent it in a more geometric way (cf. [29]).

Let V be a n-dimensional space supplied with a symmetric bilinear form (,). For any vector $f \in V$ satisfying $(f, f) = 2$ the *reflection* $R_f : V \to V$ is defined by the formula

$$R_f(x) = x - (x, f)f.$$

It preserves the hyperplane orthogonal to f and it inverts f

$$R_f(f) = -f.$$

The hyperplane is called the *mirror* of the reflection. The linear map R_f preserves also the bilinear form (,). So it can be projected onto the quotient V/V_0 where $V_0 \subset V$ is the annihilator of the bilinear form.

We say that a basis $e_1, \ldots, e_n$ in $(V, (,))$ is an *admissible* one if

$$(e_i, e_i) = 2, \quad i = 1, \ldots, n$$

and the reflections $R_1 := R_{e_1}, \ldots, R_n := R_{e_n}$ generate a finite group of linear transformations of the space V/V_0. Equivalently, this means that, applying the reflections $R_1, \ldots, R_n$ and their products to the vectors $e_1, \ldots, e_n$ and projecting them onto V/V_0 we obtain a finite system of vectors.

An alternative definition of admissibility can be given in terms of (generalized) root systems. We say that a system of nonzero vectors f_α in a space V with a symmetric bilinear form (,) is a *generalized root system* if (i) the vectors span the space V, and their projections is a finite system of vectors spanning V/V_0, (ii) the square lengths (f_α, f_α) all equal 2 and (iii) the system is invariant w.r.t. the reflections R_{f_α}.

Exercise F.2. Prove that the basis $e_1, \ldots, e_n$ in $(V, (\ , \ ))$ is an admissible one *iff* it is a part of a generalized root system in $(V, (\ , \ ))$. [Hint: use the identity

$$R_{ij} = R_i R_j R_i$$

where R_{ij} is the reflection in the hyperplane orthogonal to $R_i(e_j)$.]

We introduce now an action of the braid group B_n on the set of admissible bases in $(V, (\ , \ ))$. The standard generator σ_i $(i = 1, \ldots, n-1)$ of the braid group (as above) acts as follows

$$
\begin{aligned}
\sigma_i(e_k) &= e_k, \quad k \neq i, i+1 \\
\sigma_i(e_i) &= R_i(e_{i+1}) \equiv e - i + 1 - (e_i, e_{i+1})e_i \\
\sigma_i(e_{i+1}) &= e_i.
\end{aligned}
\tag{F.16}
$$

Observe that the action of the braid (F.15) is not trivial.

Exercise F.3. Prove that the action of the braid $(\sigma_{n-1} \ldots \sigma_1)^n$ on an admissible basis $e_1, \ldots, e_n$ coincides with the transformation $-R_1 R_2 \ldots R_n$.

It is easy to relate the action (F.16) to the action (F.13). To an admissible basis we associate an upper triangular matrix $S = S(e_1, \ldots, e_n) = (s_{ij})$ taking a half of the Gram matrix

$$s_{ii} = 1, \quad s_{ij} = (e_i, e_j) \text{ for } i < j, \quad s_{ij} = 0 \text{ for } i > j. \tag{F.17}$$

Lemma F.2. *1). The transformation law (F.13a) of the columns of the matrix Ψ_{right} is dual to (F.16). 2). The map*

$$(e_1, \ldots, e_n) \mapsto S(e_1, \ldots, e_n)$$

intertwines the action (F.16) of the braid group on admissible bases with the action (F.13b) on the Stokes matrices

$$S\left(\sigma_i(e_1, \ldots, \sigma_n)\right) = \sigma_i\left(S(e_1, \ldots, e_n)\right).$$

Proof. The matrix K of the transformation (F.16) coincides with (F.14). This proves the lemma.

The computation of Appendix H of the intersection form of a Frobenius manifold in terms of the Stokes matrix will elucidate the correspondence (F.17).

Now we can prove

Proposition F.2. *Let $e_1, \ldots, e_n$ be an admissible basis in $(V, (\ , \ ))$. Then the orbit of the matrix $S = S(e_1, \ldots, e_n)$ under the action (F.13b) of the braid group B_n is finite.*

Proof. Let f_α, $\alpha = 1, \ldots, N$ for some $N < \infty$ be the generalized root system consisted of the vectors $e_1, \ldots, e_n$ and their images under the reflections $R_1, \ldots, R_n$ and their iterations. Applying any element of the braid group to the basis $(e_1, \ldots, e_n) \mapsto (e_1', \ldots, e_n')$ we will always obtain a vector of the root system. So there is only finite number of possibilities for the Gram matrix (e'_i, e'_j) depending only on the projections of e'_i onto V/V_0. Hence we obtain a finite orbit of $S = S(e_1, \ldots, e_n)$. Proposition is proved.

We recall that a finite Coxeter group is a finite group W of linear transformations of n-dimensional Euclidean space V generated by reflections (see details in Lecture 4 below). For any finite Coxeter group we will construct a finite orbit of the action (F.13b) of the braid group B_n (avoid confusions with the Coxeter group of the type B_n!).

Let $R_1, \ldots, R_n$ be a system of generated reflections of W. Let $e_1, \ldots, e_n$ be the vectors orthogonal to the mirrors of the reflections normalized by the condition $(e_i, e_i) = 2$ (here $(\,,\,)$ is the W-invariant Euclidean inner product on V). The Stokes matrix (F.17) in this case coincides with the upper half of the Coxeter matrix of the group with respect to the given system of generated reflections (probably, such a matrix S was considered first by Coxeter in [34]. From Proposition F.2 we obtain

Corollary F.2. *The B_n-orbit of the upper half of the Coxeter matrix of an arbitrary finite Coxeter group w.r.t. arbitrary system of generating reflections is finite.*

Example F.2. For $n = 3$ we put $s_{12} = x$, $s_{13} = y$, $s_{23} = z$. The transformations of the braid group act as follows:

$$\sigma_1 : \; (x, y, z) \mapsto (-x, z - xy, y), \qquad (F.18a)$$

$$\sigma_2 : \; (x, y, z) \mapsto (y - xz, x, -z). \qquad (F.18b)$$

These preserve the polynomial

$$x^2 + y^2 + z^2 - xyz. \qquad (F.19)$$

Indeed, the characteristic equation of the matrix $S^T S^{-1}$ has the form

$$(\lambda - 1)[\lambda^2 + (x^2 + y^2 + z^2 - xyz - 2)\lambda + 1] = 0. \qquad (F.20)$$

The action of the group B_3 (in fact, this can be reduced to the action of $PSL(2, \mathbf{Z})$) admits also an invariant Poisson bracket

$$\begin{aligned} \{x, y\} &= xy - 2z \\ \{y, z\} &= yz - 2x \,. \qquad (F.21) \\ \{z, x\} &= zx - 2y \end{aligned}$$

The polynomial (F.19) is the Casimir of the Poisson bracket. Thus an invariant symplectic structure is induced on the level surfaces

$$x^2 + y^2 + z^2 - xyz = \text{const}.$$

A B_n-invariant Poisson bracket exists also on the space of Stokes matrices of the order n. But it has more complicated structure.

For integer x, y, z this action on the invariant surface $x^2 + y^2 + z^2 = xyz$ was discussed first by Markoff in 1876 in the theory of Diophantine approximations [27]. The general

244

action (F.13b), (F.14) (still on integer valued matrices) appeared also in the theory of exceptional vector bundles over projective spaces [128]. Essentially it was also found from physical considerations in [29] (again for integer matrices S) describing "braiding of Landau - Ginsburg superpotential". The invariant Poisson structure (F.21) looks to be new.

There are 3 finite Coxeter groups in the three-dimensional space: the groups of symmetries of tetrahedron, cube and icosahedron. According to Corollary F.2 they give the finite B_3-orbits of the points

$$(0, -1, -1), \quad (0, -1, -\sqrt{2}), \quad (0, -1, -\frac{\sqrt{5}+1}{2}).$$

The orbits consist of 16, 36, 40 points resp. Two more finite orbits of the braid group can be obtained using another system of generating reflections in the icosahedron group. The first orbit of 40 points consists of the images of the point

$$\left(\frac{1-\sqrt{5}}{2}, \frac{1-\sqrt{5}}{2}, \frac{1-\sqrt{5}}{2} \right).$$

The corresponding mirrors are the planes passing through the origin and through the three edges of some face of icosahedron. The orbit of the point

$$\left(0, \frac{1-\sqrt{5}}{2}, -\frac{1+\sqrt{5}}{2} \right)$$

consists of 72 points. Two of the mirrors in these case pass through the origin and through two of the edges of some face of icosahedron; the third mirror passes through the origin and a median of the face (not between the first two mirrors). We see that in the icosahedron group not all the systems of generating reflections are equivalent w.r.t. the action of the braid group.

We will see below that the first three finite orbits are the Stokes matrices of the polynomial solutions (A.7), (A.8), (A.9) resp. I do not know the solutions of WDVV with the last two Stokes matrices.

We construct now 3-dimensional generalized root systems with a degenerate bilinear form. Let us take the Stokes matrix

$$S = \begin{pmatrix} 1 & -2\cos\phi & -2\sin\phi \\ 0 & 1 & 0 \\ 0 & 0 & 1 \end{pmatrix}.$$

The symmetrization $S + S^T$ for an arbitrary ϕ gives a bilinear form of rank 2 on the three-dimensional space with a marked basis e_1, e_2, e_3. The basis will be admissible iff ϕ is commensurable with π

$$\phi = \pi \frac{k}{n}.$$

Particularly, for $k = 1$, $n = 3$ we obtain the Stokes matrix of the solution (4.71) of WDVV, and for $k = 1$, $n = 4$ the Stokes matrix of the solution (4.75).

The last example of a 3-dimensional generalized root system is given by the Stokes matrix

$$S = \begin{pmatrix} 1 & 2 & 2 \\ 0 & 1 & 2 \\ 0 & 0 & 1 \end{pmatrix}.$$

The symmetrized bilinear form has rank 1. This is the Stokes matrix of the twisted Frobenius manifold (C.87).

It is an interesting problem to classify periodic orbits of the action (F.13b), (F.14) of the braid group, and to figure out what of them correspond to algebraic solutions of the Painlevé-type equations (3.74).

Remark F.1. Cecotti and Vafa in [29] conjectured that

$$S = \begin{pmatrix} 1 & 3 & 3 \\ 0 & 1 & 3 \\ 0 & 0 & 1 \end{pmatrix}$$

is the Stokes matrix of the CP^2 topological sigma model (see above Lecture 2). Points of the orbit of the Stokes matrix are in one-to-one correspondence with triples of Markoff numbers (see in [25]). So the orbit is not finite, and the function (2.77) is not algebraic.

Appendix G.
Monodromy group of a Frobenius manifold.

In Lecture 3 I have defined the intersection form of arbitrary Frobenius manifold M. This is another flat (contravariant) metric $(\,,\,)^*$ on M defined by the formula (3.13). In this Appendix we will study the Euclidean structure on M determined by the new metric. Let us assume that the Frobenius manifold M is *analytic*. This means that the structure functions $(c^\gamma_{\alpha\beta}(t))$ are analytic in t. As it follows from the resullts of Lecture 3 the assumption is not restrictive in the semisimple case.

The contravariant metric $(\,,\,)^*$ degenerates on the sublocus where the determinant

$$\Delta(t) := \det\left(g^{\alpha\beta}(t)\right). \tag{G.1}$$

vanishes. Let $\Sigma \subset M$ be specified by the equation

$$\Sigma := \left\{t|\Delta(t) := \det(g^{\alpha\beta}(t)) = 0\right\}. \tag{G.2}$$

This is a proper analytic subset in M. We will call it *the discriminant locus* of the Frobenius manifold. The analytic function $\Delta(t)$ will be called *the discriminant* of the manifold.

On $M \setminus \Sigma$ we have a locally Euclidean metric determined by the inverse of the intersection form. This specifies an isometry

$$\Phi : \Omega \to \widehat{M \setminus \Sigma} \tag{G.3}$$

of a domain Ω in the standard n-dimensional (complex) Euclidean space E^n to the universal covering of $M \setminus \Sigma$. Action of the fundamental group $\pi_1(M \setminus \Sigma)$ on the universal covering can be lifted to an action by isometries on E^n. We obtain a representation

$$\mu : \pi_1(M \setminus \Sigma) \to Isometries(E^n). \tag{G.4}$$

Definition G.1. The group

$$W(M) := \mu\left(\pi_1(M \setminus \Sigma)\right) \subset Isometries(E^n) \tag{G.5}$$

is called *the monodromy group of the Frobenius manifold.*

By the construction
$$M \setminus \Sigma = \Omega/W(M).$$

To construct explicitly the isometry (G.3) we are to fix a point $t \in M \setminus \Sigma$ and to find the flat coordinates of the intersection form in a neibourghood of the point. The flat coordinates $x = x(t^1, \ldots, t^n)$ are to be found from the system of differential equations

$$\hat\nabla^\alpha \hat\nabla_\beta x := g^{\alpha\epsilon}(t)\partial_\epsilon\partial_\beta x + \Gamma^{\alpha\epsilon}_\beta(t)\partial_\epsilon x = 0, \quad \alpha, \beta = 1, \ldots, n. \tag{G.6}$$

(Here $\hat\nabla$ is the Levi-Cività connection for $(\,,\,)^*$; the components of the metric and of the connection are given by the formulae (3.15), (3.36).) This is an overdetermined holonomic system. Indeed, vanishing of the curvature of the intersection form (Proposition 3.2) provides compatibility of the system. More precisely,

Proposition G.1. *Near a point $t_0 \in M$ where*

$$\det\left(g^{\alpha\beta}(t_0)\right) \neq 0$$

the space of solutions of the system (G.6) modulo constants has dimension n. Any linearly independent (modulo constants) solutions $x^1(t), \ldots, x^n(t)$ of (G.6) can serve as local coordinates near t_0. The metric $g^{\alpha\beta}(t)$ in these coordinates is constant

$$g^{ab} := \frac{\partial x^a}{\partial t^\alpha}\frac{\partial x^b}{\partial t^\beta}g^{\alpha\beta}(t) = const.$$

This is a reformulation of the standard statement about the flat coordinates of a zero-curvature metric.

Exercise G.1. For $d \neq 1$ prove that:
1. $x_a(t)$ are quasihomogeneous functions of t^1, ..., t^n of the degree

$$\deg x_a(t) = \frac{1-d}{2}; \qquad (G.7)$$

2. that

$$t_1 \equiv \eta_{1\alpha}t^\alpha = \frac{1-d}{4}g_{ab}x^a x^b \qquad (G.8)$$

where $(g_{ab}) = \left(g^{ab}\right)^{-1}$.

Example G.1. For the two-dimensional Frobenius manifold with the polynomial free energy

$$F(t_1, t_2) = \frac{1}{2}t_1^2 t_2 + t_2^{k+1}, \quad k \geq 2 \qquad (G.9)$$

the system (G.6) can be easily solved in elementary functions. The flat coordinates x and y can be introduced in such a way that

$$\begin{aligned} t_1 &= 4\sqrt{k(k^2-1)}\,\mathrm{Re}\,(x+iy)^k \\ t_2 &= x^2 + y^2. \end{aligned} \qquad (G.10)$$

Thus the monodromy group of the Frobenius manifold (G.9) is the group $I_2(k)$ of symmetries of the regular k-gon.

For the polynomial solutions (A.7), (A.8), (A.9) the calculation of the monodromy group is more involved. In the next Lecture we will see that the monodromy groups of these three polynomial solutions of WDVV coincide with the groups A_3, B_3, H_3 of symmetries of the regular tetrahedron, cube and icosahedron in the three-dimensional space.

More generally, for the Frobenius manifolds of Lecture 4 where $M = \mathbf{C}^n/W$ for a finite Coxeter group W, the solutions of the system (G.6) are the Euclidean coordinates in $\mathbf{C}^n$ as the functions on the space of orbits. If we identify the space of orbits with the universal unfolding of the correspondent simple singularity [5, 6, 25, 139] then the map

$$M \ni t \mapsto (x^1(t), \ldots, x^n(t)) \in \mathbf{C}^n \qquad (G.11)$$

coincides with the period mapping. The components $x^a(t)$ are sections of the bundle of vanishing cycles being locally horizontal w.r.t. the Gauss – Manin connection. Note that globally (G.11) is a multivalued mapping. The multivaluedness is just described by the action of the Coxeter group W coinciding with the monodromy group of the Frobenius manifolds.

Basing on this example we introduce

Definition G.2. The system (G.6) is called *Gauss – Manin equation* of the Frobenius manifold.

Note that the coefficients of Gauss – Manin equation on an analytic Frobenius manifold also are analytic in t. This follows from (3.15), (3.36). However, the solutions may not be analytic everywhere. Indeed, if we rewrite Gauss – Manin equations in the form solved for the second order derivatives

$$\partial_\alpha \partial_\beta x - \Gamma^\gamma_{\alpha\beta}(t)\partial_\gamma x = 0 \qquad (G.12)$$

$$\Gamma^\gamma_{\alpha\beta}(t) := -g_{\alpha\epsilon}\Gamma^{\epsilon\beta}_\gamma(t), \quad (g_{\alpha\epsilon}) := (g^{\alpha\epsilon})^{-1}$$

then the coefficients will have poles on the discriminant Σ.

So the solutions of Gauss – Manin equation (G.6) are analytic in t on $M \setminus \Sigma$. Continuation of some basic solution $x^1(t), \ldots, x^n(t)$ along a closed path γ on $M \setminus \Sigma$ can give new basis of solutions $\tilde{x}^1(t), \ldots, \tilde{x}^n(t)$. Due to Proposition G.1 it must have the form

$$\tilde{x}^a(t) = A^a_b(\gamma)x^b(t) + B^a(\gamma) \qquad (G.13)$$

for some constants $A^a_b(\gamma), B^a(\gamma)$. The matrix $A^a_b(\gamma)$ must be orthogonal w.r.t. the intersection form

$$A^a_b(\gamma)g^{bc}A^d_c(\gamma) = g^{ad}. \qquad (G.14)$$

The formula (G.13) determines the representation (G.4) of the fundamental group $\pi_1(M \setminus \Sigma, t) \ni \gamma$ to the group of isometries of the n-dimensional complex Euclidean space E^n. This is just the monodromy representation (G.5).

Proposition G.2. *For $d \neq 1$ the monodromy group is a subgroup in $O(n)$ (linear orthogonal transformations).*

Proof. Due to Exercise G.1 for $d \neq 1$ one can choose the coordinates $x^a(t)$ to be invariant w.r.t. the scaling transformations

$$x^a(c^{\deg t^1}t^1, \ldots, c^{\deg t^n}t^n) = c^{\frac{1-d}{2}}x^a(t^1, \ldots, t^n). \qquad (G.15)$$

The monodromy preserves such an invariance. Proposition is proved.

Example G.2. If some of the scaling dimensions $q_\alpha = 1$ then the Frobenius structure may admit a discrete group of translations along these variables. The Gauss - Manin equations then will be a system with periodic coefficients. The correspondent monodromy transformation (i.e. the shift of solutions of (G.6) along the periods) will contribute to the monodromy group of the Frobenius manifold.

To see what happens for $d = 1$ let us find the monodromy group for the two-dimensional Frobenius manifold with

$$F = \frac{1}{2}t^{1^2}t^2 + e^{t^2}. \qquad (G.16)$$

I recall that this describes the quantum cohomology of CP^1. The manifold M is the cylinder $\left(t^1, t^2(\mathrm{mod}\, 2\pi i)\right)$. The Euler vector field is

$$E = t^1 \partial_1 + 2\partial_2. \qquad (G.17)$$

The intersection form has the matrix

$$\left(g^{\alpha\beta}\right) = \begin{pmatrix} 2e^{t^2} & t^1 \\ t^1 & 2 \end{pmatrix}. \qquad (G.18)$$

The Gauss – Manin system reads

$$\begin{aligned}
2e^{t^2}\partial_1{}^2 x + t^1 \partial_1 \partial_2 x &= 0 \\
2e^{t^2}\partial_1\partial_2 x + t^1 \partial_2{}^2 x + e^{t^2}\partial_1 x &= 0 \\
t^1 \partial_1{}^2 x + 2\partial_1\partial_2 x + \partial_1 x &= 0 \\
t^1 \partial_1 \partial_2 x + 2\partial_2{}^2 x &= 0.
\end{aligned} \qquad (G.19)$$

The basic solutions are

$$\begin{aligned}
x^1 &= -it^2 \\
x^2 &= 2\arccos\frac{1}{2}t^1 e^{-\frac{t^2}{2}}.
\end{aligned} \qquad (G.20)$$

The intersection form in these coordinates is

$$\frac{1}{2}\left(-dx^{1^2} + dx^{2^2}\right). \qquad (G.21)$$

The Euler vector field reads

$$E = \frac{\partial}{\partial x^1}. \qquad (G.22)$$

The discriminant locus is specified by the equation

$$t^1 = \pm 2e^{\frac{t^2}{2}} \qquad (G.23a)$$

or, equivalently

$$x^2 = 0, \quad x^2 = 2\pi. \qquad (G.23b)$$

The monodromy group is generated by the transformations of the following two types.

The transformations of the first type are obtained by continuation of the solutions (G.20) along the loops around the discriminant locus. This gives the transformations

$$\left(x^1, x^2\right) \mapsto \left(x^1, \pm x^2 + 2\pi n\right), \quad n \in \mathbf{Z}. \tag{G.24a}$$

This is nothing but the action of the simplest affine Weyl group of the type $A_1^{(1)}$. The transformations of the second type

$$\left(x^1, x^2\right) \mapsto \left(x^1 + 2\pi n, (-1)^n x^2\right), \quad n \in \mathbf{Z} \tag{G.24b}$$

are generated by the closed loops $t^2 \mapsto t^2 + 2\pi i n$ on M. This gives an extension of the affine Weyl group (G.24a). So M is the quotient of $\mathbf{C}^2$ over the extended affine Weyl group (G.24) (although t^2 is not a globally single-valued function on the quotient). The coordinate

$$t^1 = 2e^{ix^1/2}\cos\frac{x^2}{2} \tag{G.25}$$

is the basic invariant of the group (G.24) homogeneous w.r.t. the Euler vector field (G.22). That means that any other invariant is a polynomial (or a power series) in t^1 with the coefficients being arbitrary $2\pi i$-periodic functions in t^2. Another flat coordinate

$$t^2 = ix^1$$

is invariant w.r.t. the affine Weyl group (G.24a) and it gets a shift w.r.t. the transformations (G.24b).

Example G.3. The monodromy group of a Frobenius manifold can be in principle computed even if we do not know the structure of it using the isomonodromicity property of the solutions of the Gauss - Manin system. For the example of the CP^2 model (see above Lecture 2) it is enough to compute the monodromy group on the sublocus $t^1 = t^3 = 0$. We obtain that the linear part of the monodromy of the CP^2-model is generated by the monodromy group of the operator with rational coefficients

$$\begin{pmatrix} 3q & 0 & \lambda \\ 0 & \lambda & 3 \\ \lambda & 3 & 0 \end{pmatrix} \frac{\partial \xi}{\partial \lambda} + \begin{pmatrix} 0 & 0 & -1/2 \\ 0 & 1/2 & 0 \\ 3/2 & 0 & 0 \end{pmatrix} \xi = 0 \tag{G.26a}$$

for

$$q = e^t, \quad t = t_2$$

and of the operator with $2\pi i$-periodic coefficients

$$\begin{pmatrix} 3q & 0 & \lambda \\ 0 & \lambda & 3 \\ \lambda & 3 & 0 \end{pmatrix} \frac{\partial \xi}{\partial t} + \begin{pmatrix} \frac{3}{2}q & 0 & 0 \\ 0 & 0 & -1/2 \\ 0 & 1/2 & 0 \end{pmatrix} \xi = 0. \tag{G.26b}$$

The last one can be also reduced to an operator with rational coefficients (with irregular singularity at $q = \infty$) by the substitution $t \to q$. It is still an open problem to solve these equations and to compute the monodromy.

The monodromy group can be defined also for twisted Frobenius manifolds. Particularly, the inversion (B.11) acts as a conformal transformation of the intersection form. We will see in Appendix J an example of such a situation.

Another important sublocus in a Frobenius manifold satisfying the semisimplicity condition is *the nilpotent locus* $\Sigma_{\rm nil}$ consisting of all the points t of M where the algebra on $T_t M$ is not semisimple. According to Proposition 3.3 the nilpotent discriminant is contained in the discriminant locus of the polynomial

$$\det\left(g^{\alpha\beta}(t) - \lambda\eta^{\alpha\beta}\right) = \Delta(t^1 - \lambda, t^2, \ldots, t^n). \qquad (G.27)$$

The discriminant $\Delta_{\rm nil}(t)$ of (G.27) as of the polynomial on λ will be called *nilpotent discriminant*.

Example G.4. For the Frobenius manifold of Example 1.7 the discriminant locus

$$\Sigma = \{\text{set of polynomials } \lambda(p) \text{ having a critical value } \lambda = 0\}. \qquad (G.28)$$

The discriminant $\Delta(t)$ coincides with the discriminant of the polynomial $\lambda(p)$ (I recall that the coefficients of the polynomial $\lambda(p)$ are certain functions on t). The nilpotent locus of M is the *caustic* (see [6])

$$\Sigma_{\rm nil} = \{\text{set of polynomials } \lambda(p) \text{ with multiple critical points } p_0,$$

$$\lambda^{(k)}(p_0) = 0 \text{ for } k = 1, 2, \ldots, k_0 \geq 2\}. \qquad (G.29)$$

The nilpotent discriminant is a divisor of the discriminant of the polynomial $\mathrm{discr}_p(\lambda(p) - \lambda)$ as of the polynomial in λ.

On the complement $M \setminus \Sigma$ a metric

$$ds^2 := g_{\alpha\beta}(t)dt^\alpha dt^\beta \qquad (G.30)$$

is well-defined. Here the matrix $g_{\alpha\beta}(t)$ is the inverse to the matrix $g^{\alpha\beta}(t)$ (3.15). The metric has a pole on the discriminant locus. We will show that the singularity of the metric can be eliminated after lifting to a covering of M.

Let $\hat{M}$ be the two-sheet covering of the Frobenius manifold M ramifying along the discriminant locus

$$\hat{M} := \left\{(w, t), \ w \in \mathbf{C}, \ t \in M | w^2 = \Delta(t)\right\}. \qquad (G.31)$$

We have a natural projection

$$\pi : \hat{M} \to M. \qquad (G.32)$$

Lemma G.1. *The pullback $\pi^* ds^2$ of the metric (G.30) onto $\hat{M}$ is analytic on $\hat{M} \setminus (\Sigma \cap \Sigma_{\rm nil})$.*

Proof. Outside of Σ on M we can use the canonical coordinates. In these the metric has the form

$$ds^2 = \sum_{i=1}^{n} \frac{\eta_{ii}(u)}{u^i}(du^i)^2. \qquad (G.33)$$

252

The canonical coordinates serve near a point $t_0 \in M \setminus (\Sigma \cap \Sigma_{\text{nil}})$ as well, but some of them vanish. The vanishing is determined by a splitting $(i_1, \ldots, i_p) \cup (j_1, \ldots, j_q) = (1, 2, \ldots, n)$, $p + q = n$ such that

$$u^{i_1}(t_0) = 0, \ldots, u^{i_p}(t_0) = 0, \quad u^{j_s}(t_0) \neq 0, \ldots, u^{j_q} \neq 0. \quad (G.34)$$

The local coordinates near the correspondent point $\pi^{-1}(t_0) \in \hat{M}$ are

$$\sqrt{u^{i_1}}, \ldots, \sqrt{u^{i_p}}, u^{j_1}, \ldots, u^{i_q}. \quad (G.35)$$

Rewriting (G.33) near the point t_0 as

$$ds^2 = 4 \sum_{s=1}^{p} \eta_{i_s i_s}(u)(d\sqrt{u^{i_s}})^2 + \sum_{s=1}^{q} \eta_{j_s j_s}(u)(du^{j_s})^2 \quad (G.36)$$

we obtain analyticity of ds^2 on $\hat{M}$. Lemma is proved.

Due to Lemma G.1 the flat coordinates $x^a(t)$ as the functions on $\hat{M}$ can be extended to any component of $\Sigma \setminus \Sigma_{\text{nil}}$. The image

$$\left(x^1(t), \ldots, x^n(t)\right)_{t \in \Sigma \setminus \Sigma_{\text{nil}}} \quad (G.37)$$

is the discriminant locus written in the flat coordinates $x^1, \ldots, x^n$.

Lemma G.2. *Any component of $\Sigma \setminus \Sigma_{\text{nil}}$ in the coordinates $x^1(t), \ldots, x^n(t)$ is a hyperplane.*

Proof. We will show that the second fundamental form of the hypersurface $\Delta(t) = 0$ in $\hat{M}$ w.r.t. the metric (G.30) vanishes. Near $\Sigma \setminus \Sigma_{\text{nil}}$ we can use the canonical coordinates $u^1, \ldots, u^n$. Let $\Sigma \setminus \Sigma_{\text{nil}}$ be specified locally by the equation $u^n = 0$. Let us first calculate the second fundamental form of the hypersurface

$$u^n = u_0^n \neq 0. \quad (G.38)$$

The unit normal vector to the hypersurface is

$$N = \sqrt{\frac{u_0^n}{\eta_{nn}}} \partial_n. \quad (G.39)$$

The vectors ∂_i, $i \neq n$ span the tangent plane to the hypersurface. The second fundamental form is

$$b_{ij} := \left(\hat{\nabla}_i \partial_j, N\right) = \sqrt{\frac{\eta_{nn}}{u_0^n}} \Gamma_{ij}^n, \quad 1 \leq i, j \leq n-1$$

$$= -\frac{\delta_{ij}}{2}\sqrt{\frac{u_0^n}{\eta_{nn}}} \partial_n \left(\frac{\eta_{ii}}{u^i}\right) = -\delta_{ij}\sqrt{\frac{u_0^n}{\eta_{nn}}} \frac{\psi_{i1}\psi_{n1}\gamma_{in}}{u^i} \quad (G.40)$$

in the notations of Lecture 3 (but Γ^k_{ij} here are the Christoffel coefficients for the intersection form). This vanishes when $u^n_0 \to 0$. Lemma is proved.

I recall that a linear orthogonal transformation $A : \mathbf{C}^n \to \mathbf{C}^n$ of a complex Euclidean space is called *reflection* if $A^2 = 1$ and A preserves the points of a hyperplane in $\mathbf{C}^n$.

Theorem G.1. *For $d \neq 1$ the monodromy along a small loop around the discriminant on an analytic Frobenius manifold satisfying semisimplicity condition is a reflection.*

Proof. Since the flat coordinates are analytic and single valued on the two-sheet covering $\hat{M}$ the monodromy transformations A along loops around Σ are involutions, $A^2 = 1$.They preserves the hyperplanes (G.28). Note that the hyperplanes necessary pass through the origin for $d \neq 1$. Theorem is proved.

In Lecture 4 we will show that any finite reflection group arises as the monodromy group of a Frobenius manifold. This gives a very simple constructionn of polynomial Frobenius manifolds with $d < 1$. Similarly, the construction of Example G.2 can be generalized to arbitrary affine Weyl groups (properly extended). This gives a Frobenius structure with $d = 1$ and linear nonhomogeneous Euler vector field E on their orbit spaces (i.e. $\mathcal{L}_E t^n \neq 0$). We will consider the details of this construction in a separate publication. Finally, in Appendix J we will construct a twisted Frobenius manifold whose monodromy is a simplest extended complex crystallographic group. This will give Froobenius manifolds again with $d = 1$ but with linear homogeneous Euler vector field (i.e. $\mathcal{L}_E t^n = 0$).

Appendix H.
Generalized hypergeometric equation
associated with a Frobenius manifold and its monodromy.

The main instrument to calculate the monodromy of a Frobenius manifold coming from the loops winding around the discriminant is a differential equation with rational coefficients wich we are going to define now. This will be the equation for the flat coordinates of the linear pencil of the metrics

$$(\, , \,)^*_\lambda := (\, , \,)^* - \lambda < \, , \, >^* = \left(g^{\alpha\beta} - \lambda \eta^{\alpha\beta} \right) \qquad (H.1)$$

as the functions of the parameter λ. We will obtain also an integral transform relating the flat coordinates of the deformed connection (3.3) and the flat coordinnates of the deformed metric (H.1).

Let $x_1 = x_1(t)$, ..., $x_n = x_n(t)$ be flat coordinates for the metric $g^{\alpha\beta}(t)$. The flat coordinates for the flat pencil H.1 can be construct easily.

Lemma H.1. *The functions*

$$\tilde{x}_a(t, \lambda) := x_a(t^1 - \lambda, t^2, \ldots, t^n), \quad a = 1, \ldots, n \qquad (H.2)$$

are flat coordinates for the deformed metric (H.1).

Proof. The linear combination (H.1) can be written in the form

$$g^{\alpha\beta}(t) - \lambda \eta^{\alpha\beta} = g^{\alpha\beta}(t^1 - \lambda, t^2, \ldots, t^n). \qquad (H.3)$$

Lemma is proved.

Corollary H.1. *The gradient* $\xi_\epsilon := \partial_\epsilon x(\lambda, t)$ *of the flat coordinates of the pencil* $g^{\alpha\beta}(t) - \lambda\eta^{\alpha\beta}$ *satisfies the following system of linear differential equations in* λ

$$(\lambda\eta^{\alpha\epsilon} - g^{\alpha\epsilon}(t)) \frac{d}{d\lambda}\xi_\epsilon = \eta^{\alpha\epsilon}\left(-\frac{1}{2} + \mu_\epsilon\right)\xi_\epsilon. \qquad (H.4)$$

Proof. We have from (3.38)

$$\Gamma_1^{\alpha\epsilon} = \left(\frac{d+1}{2} - q_\epsilon\right)c_1^{\alpha\epsilon} = \left(\frac{1}{2} - \mu_\epsilon\right)\eta^{\alpha\epsilon}. \qquad (H.5)$$

So the equation (G.6) for $\beta = 1$ reads

$$(g^{\alpha\epsilon}(t) - \lambda\eta^{\alpha\epsilon})\,\partial_1\partial_\epsilon x + \eta^{\alpha\epsilon}\left(\frac{1}{2} - \mu_\epsilon\right)\partial_\epsilon x = 0. \qquad (H.6)$$

Due to Lemma H.1 we have

$$\partial_1 = -\frac{d}{d\lambda}. \qquad (H.7)$$

Corollary is proved.

The equation (H.4) is a system of linear ordinary differential equations with rational coefficients depending on the parameters $t^1, \ldots, t^n$. The coefficients have poles on the shifted discriminant locus

$$\Sigma_\lambda := \left\{ t \mid \Delta(t^1 - \lambda, t^2, \ldots, t^n) = 0 \right\}. \qquad (H.8)$$

Lemma H.2. *Monodromy of the system (H.4) of differential equations with rational coefficients around Σ_λ coincides with the monodromy of the Frobenius manifold around the discriminant Σ. The monodromy does not depend on the parameters t.*

Proof. The first statement is obvious. The second one follows from the compatibility of the equations (G.6) with the equations in λ (H.4).

Definition H.1. The differential equation (H.4) with rational coefficients will be called *generalized hypergeometric equation associated with the Frobenius manifold.*

In this definition we are motivated also by [122] where it was shown that the differential equations for the functions $_nF_{n-1}$ are particular cases of the system (H.15) (this is an equivalent form of (H.4), see below) however, in general without the skew-symmetry of the matrix V. In the semisimple case the equation (H.4) has only regular singularities on (H.8) and in the infinite point $\lambda = \infty$.

We construct now an integral transform relating the flat coordinates $\tilde{t}(t, z)$ of the deformed connection (3.3) and the flat coordinates $x(t, \lambda)$ of the pencil of metrics (H.1). We recall that the coordinates $x(t, \lambda)$ are the solutions of the differential equations

$$(g^{\alpha\epsilon} - \lambda\eta^{\alpha\epsilon}) \partial_\epsilon \partial_\beta x + \Gamma_\beta^{\alpha\epsilon} \partial_\epsilon x = 0. \qquad (H.9)$$

Proposition H.1. *Let $\tilde{t}(t, z)$ be a flat coordinate of the deformed connection (3.3) normalized by the condition*

$$z\partial_z \tilde{t} = \mathcal{L}_E \tilde{t}. \qquad (H.10)$$

Then the function

$$x(t, \lambda) := \oint z^{\frac{d-3}{2}} e^{-\lambda z} \tilde{t}(t, z) \, dz \qquad (H.11)$$

is a flat coordinate for the pencil (H.1).

Here the integral is considered along any closed loop in the extended complex plane $z \in \mathbb{C} \cup \infty$. We will specify later how to choose the contour of the integration to obtain a well-defined integral.

Proof is based on the following

Lemma H.3. *The following identity holds true*

$$\left(dt^\alpha, \hat{\nabla}_\gamma d\tilde{t} \right)^* dt^\gamma = z^{-1} d\left[< dt^\alpha, \left(z\partial_z + \frac{d-3}{2} \right) d\tilde{t} >^* \right]. \qquad (H.12)$$

Here $\tilde{t} = \tilde{t}(t, z)$ is a flat coordinate of the deformed affine connection (3.3), $\hat{\nabla}$ is the Levi-Cività connection for the intersection form, $d = dt^\gamma \partial_\gamma$.

Proof. The l.h.s. of (H.12) reads

$$\left(g^{\alpha\sigma}\partial_\sigma\partial_\gamma\tilde{t} + \Gamma^{\alpha\sigma}_\gamma\partial_\sigma\tilde{t}\right)dt^\gamma =$$

$$= \left(z\sum_\rho E^\rho c^{\alpha\sigma}_\rho c^\nu_{\sigma\gamma}\partial_\nu\tilde{t} + \sum_\sigma \left(\frac{d+1}{2} - q_\sigma\right)c^{\alpha\sigma}_\gamma\partial_\sigma\tilde{t}\right)dt^\gamma.$$

On the other side, using the equation (H.10) we obtain

$$\partial_\gamma\left(z\partial_z + \frac{d-3}{2}\right)\partial_\epsilon\tilde{t} = z^2 c^\sigma_{\epsilon\gamma}U^\nu_\sigma\partial_\nu\tilde{t} + zc^\sigma_{\epsilon\gamma}\partial_\sigma\tilde{t} + z\sum_\sigma c^\sigma_{\epsilon\gamma}(1 - q_\sigma)\partial_\sigma\tilde{t} + z\frac{d-3}{2}c^\sigma_{\epsilon\gamma}\partial_\sigma\tilde{t} =$$

$$= z^2 E^\rho c^\nu_{\rho\sigma}c^\sigma_{\epsilon\gamma}\partial_\nu\tilde{t} + z\sum_\sigma c^\sigma_{\epsilon\gamma}\left(\frac{d+1}{2} - q_\sigma\right)\partial_\sigma\tilde{t}.$$

Multiplying by $\eta^{\alpha\epsilon}$ and using the associativity condition

$$c^\sigma_{\epsilon\gamma}c^\nu_{\rho\sigma} = c^\sigma_{\rho\epsilon}c^\nu_{\sigma\gamma}$$

we obtain, after multiplication by dt^γ and division over z, the expression (H.12). Lemma is proved.

Proof of Proposition. For the function $x = x(t,\lambda)$ of the form (H.11) we obtain, using Lemma and integrating by parts

$$\left(dt^\alpha, \hat{\nabla}_\gamma dx\right)^* dt^\gamma = dt^\gamma \oint z^{\frac{d-3}{2}}e^{-\lambda z}\left(dt^\alpha, \hat{\nabla}_\gamma d\tilde{t}\right)^* dz$$

$$= d_t \oint z^{\frac{d-3}{2}}e^{-\lambda z} < dt^\alpha, \partial_z d\tilde{t} + \frac{d-3}{2z}d\tilde{t} >^* dz$$

$$= d_t \left\{\oint\left(\lambda z^{\frac{d-3}{2}}e^{-\lambda z} - \frac{d-3}{2}z^{\frac{d-5}{2}}e^{-\lambda z}\right) < dt^\alpha, d\tilde{t} >^* dz\right.$$

$$\left.+\frac{d-3}{2}\oint z^{\frac{d-5}{2}}e^{-\lambda z} < dt^\alpha, d\tilde{t} >^* dz\right\}$$

$$= \lambda d_t \oint z^{\frac{d-3}{2}}e^{-\lambda z} < dt^\alpha, d\tilde{t} >^* dz = \lambda\eta^{\alpha\epsilon}\partial_\epsilon\partial_\gamma x\,dt^\gamma.$$

So x satisfies the differential equation (H.9). Proposition is proved.

We study now the monodromy of our generalized hypergeometric equation (G.6) in a neiborghood of a semisimple point $t \in M$. First we rewrite the differential equations (3.5) and (H.9) and the integral transform (H.11) in the canonical coordinates u^i.

Proposition H.2. *Let $x = x(t,\lambda)$ be a flat coordinate of the metric (H.1). Put*

$$\phi_i(u,\lambda) := \partial_i x(t,\lambda)/\sqrt{\eta_{ii}(u)}, \quad t = t(u). \tag{H.13}$$

The vector-function $\phi = (\phi_1, \ldots, \phi_n)^T$, $\phi_i = \phi_i(u, \lambda)$ *satisfies the system*

$$\partial_j \phi_i = \gamma_{ij} \phi_j, \quad i \neq j \qquad (H.14a)$$

$$\sum_{k=1}^{n} (u^k - \lambda) \partial_k \phi_i = -\frac{1}{2} \phi_i \qquad (H.14b)$$

and also the following differential equation in λ

$$(\lambda \cdot 1 - U) \frac{d}{d\lambda} \phi = -\left(\frac{1}{2} \cdot 1 + V(u) \right) \phi \qquad (H.15)$$

where $U = \operatorname{diag}(u^1, \ldots, u^n)$ *and* $V(u) = \left((u^j - u^i) \gamma_{ij}(u) \right)$ *(cf. (3.156) above).*
If $\psi = (\psi_1, \ldots, \psi_n)^T$ *is a solution to the linear system (3.118), (3.122) then*

$$\phi(u, \lambda) = \oint e^{-\lambda z} \psi(u, z) \frac{dz}{\sqrt{z}}$$

satisfies the system (H.9).
   The proof is omitted.

   In the semisimple case also the system (H.15) will be called generalized hypergeometric equation associated to the Frobenius manifold.
   For the example $n = 2$ the substitution

$$\phi_1(u_1, u_2, \lambda) = \frac{1}{\sqrt{\lambda - u_1}} \psi_1(t), \quad \phi_2(u_1, u_2, \lambda) = \frac{1}{\sqrt{\lambda - u_2}} \psi_2(t)$$

$$\lambda = u_1 + t(u_2 - u_1)$$

reduces the system (H.15) to a very elementary particular case of the Gauss equation

$$t(t - 1) \frac{d^2 \psi_1}{dt^2} + \left(t - \frac{1}{2} \right) \frac{d\psi_1}{dt} - \mu^2 \psi_1 = 0,$$

$\mu = -d/2$. The solutions are expressed via elementary functions

$$\psi_1 = a \cos(\mu \arccos t) + b \sin(\mu \arccos t)$$

for arbitrary constants a, b.

   We consider now the reduced case (see above Appendix B) where $0 \leq q_\alpha \leq d < 1$. In this case instead of the loop integrals (H.11) (or (H.16)) it's better to use more convenient Laplace integrals. We will use these Laplace integrals to express the monodromy of our generalized hypergeometric equation in terms of the Stokes matrix of the Frobenius manifold. This will establish a relation between the Stokes matrix and the monodromy group of Frobenius manifold with $d < 1$.

Let $\Psi(u,z) = (\psi_{ia}(u,z))$, $i,a = 1,\ldots,n$ be a solution of the equation (3.122) analytic in a half-plane. Let us assume that

$$0 \leq q_\alpha \leq d < 1. \qquad (H.17)$$

We construct the functions $x_a(u,\lambda)$ taking the Laplace transform of these solutions:

$$\partial_i x_a(u,\lambda) = \sqrt{\overline{\eta_{ii}(u)}}\hat{\psi}_{ia}(u,\lambda) \qquad (H.18a)$$

where

$$\hat{\psi}_{ia}(u,\lambda) := \frac{1}{\sqrt{-2\pi}} \int_0^\infty e^{-\lambda z}\psi_{ia}(u,z)\frac{dz}{\sqrt{z}}. \qquad (H.18b)$$

We can normalize them uniquelly by the homogeneity requirement

$$\left(\lambda\frac{d}{d\lambda} - \mathcal{L}_E\right) x_a(u,\lambda) = \frac{1-d}{2}x_a(u,\lambda). \qquad (H.19)$$

Theorem H.1. *Functions $x_a(u,\lambda)$ are flat coordinates of the deformed metric (H.1).*
Proof coincides with the proof of Proposition H.2 (due to the inequalities (H.17) the boundary terms at $z = 0$ vanish).

Corollary H.2. *The intersection form*

$$g^{ab} := (dx_a(u,\lambda), dx_b(u,\lambda))_\lambda^* = \sum_{i=1}^n (u^i - \lambda)\hat{\psi}_{ia}(u,\lambda)\hat{\psi}_{ib}(u,\lambda) \qquad (H.20)$$

does not depend on λ neither on u.

The coordinates $x_a(u,\lambda)$ are multivalued analytic functions of λ. They have also singularities at the points $\lambda = u^i$. The monodromy of these functions coincide with the monodromy of the differential operator (H.15) with regular singular points at $\lambda = u^i$, $i = 1,\ldots,n$ and $\lambda = \infty$. We will calculate now this monodromy in terms of the Stokes matrix of the original operator.

To calculate the monodromy I will use the following elementary way of analytic continuation of Laplace transforms of a function analytic in a halfplane.

Lemma H.4. *Let the function $\psi(z)$ be analytic in the right halfplane and*

$$\begin{aligned} |\psi(z)| &\to 1 \text{ for } z \to \infty \\ z|\psi(z)| &\to 0 \text{ for } |z| \to 0 \end{aligned} \qquad (H.21)$$

uniformly in the sector $-\frac{\pi}{2} + \epsilon \leq \arg z \leq \frac{\pi}{2} - \epsilon$ for arbitrary small $\epsilon > 0$. Then the Laplace transform

$$\hat{\psi}(\lambda) := \int_0^\infty e^{-\lambda z}\psi(z)\,dz \qquad (H.22)$$

can be analytically continued in the complex λ-plane with a cut along the negative real half-line.

Proof. $\hat{\psi}(\lambda)$ is an analytic function in the right half-plane $\operatorname{Re}\lambda > 0$. Let us show that for these λ the equality

$$\hat{\psi}(\lambda) = \int_0^\infty e^{-\lambda z e^{i\alpha}} \psi(ze^{i\alpha})\, d(ze^{i\alpha}) \qquad (H.23)$$

holds true for any α such that

$$-\frac{\pi}{2} - \arg\lambda < \alpha < \frac{\pi}{2} - \arg\lambda.$$

Indeed, let us consider the contour integral

$$\oint_C e^{-\lambda z}\psi(z)\, dz.$$

Integrals along the arcs tend to zero when $r \to 0$, $R \to \infty$ (see Fig.15). In the limit we obtain (H.23).

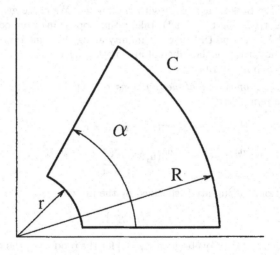

Fig.15

Now observe that the r.h.s. of (H.23) is analytic in the halfplane

$$-\frac{\pi}{2} - \alpha < \arg\lambda < \frac{\pi}{2} - \alpha.$$

Varying α from $-\frac{\pi}{2}+0$ to $\frac{\pi}{2}-0$ we obtain the needed analytic continuation.

Let us fix some oriented line $l = l_+ \cup l_-$ not containing the separating rays of the operator (3.120). Let $\Psi^{\text{right}/\text{left}}(u,z) = \left(\psi_{ia}^{\text{right}/\text{left}}(u,z)\right)$ be the canonical solutions (3.128) of (3.122) in the correspondent half-planes. Their Laplace transforms will be defined by the integrals

$$\hat{\psi}_{ia}^{\text{right}}(u,\lambda) = \frac{1}{\sqrt{-2\pi}} \int_{il_-} e^{-\lambda z} \psi_{ia}^{\text{right}}(u,z) \frac{dz}{\sqrt{z}} \qquad (H.24)$$

(analytic function outside the cut $u_a + i\bar{l}_+$) where we chose the branch of $\sqrt{z}$ with the cut along l_-, and

$$\hat{\psi}_{ia}^{\text{left}}(u,\lambda) = \frac{1}{\sqrt{-2\pi}} \int_{il_+} e^{-\lambda z} \psi_{ia}^{\text{left}}(u,z) \frac{dz}{\sqrt{z}} \qquad (H.25)$$

(analytic in λ outside the cut $u_a + i\bar{l}_+$). By $x_a^{\text{right}}(u,\lambda)$ and $x_a^{\text{left}}(u,\lambda)$ we denote the correspondent coordinates (H.18). By

$$A := S + S^T \qquad (H.26)$$

we denote the symmetrized Stokes matrix. This matrix does not degenerate due to inequalities (H.17).

Note that the rays $u_1 + i\bar{l}_+, \ldots, u_n + i\bar{l}_+$ are pairwise distinct (this is equivalent to nonintersecting of the Stokes rays (F.2) with the line l). We chose generators g_a in the fundamental group $\pi_1(\mathbf{C} \setminus (u^1 \cup \ldots \cup u^n))$ taking the loops going from ∞ along these rays to u^a then around $\lambda = u^a$ and then back to infinity along the same ray.

The monodromy group of the differential equation (H.15) w.r.t. the chosen basis of the fundamental group is described by

Theorem H.2. *Monodromy of the functions* $x_1^{\text{right}}(u,\lambda), \ldots, x_n^{\text{right}}(u,\lambda)$ *around the point* $\lambda = u_b$ *is the reflection*

$$x_a^{\text{right}}(u,\lambda) \mapsto x_a^{\text{right}}(u,\lambda) \text{ for } a \neq b$$
$$x_b^{\text{right}}(u,\lambda) \mapsto x_b^{\text{right}}(u,\lambda) - \sum_{a=1}^{n} A_{ba} x_a^{\text{right}}(u,\lambda). \qquad (H.27)$$

Remark. Monodromy at infinity is specified by the matrix

$$-T = -S^T S^{-1}. \qquad (H.28)$$

This is in agreement with the Coxeter identity [34] for the product of the reflections (H.27):

$$R_1 \ldots R_n = -S^T S^{-1} \qquad (H.29)$$

where R_b is the matrix of the reflection (H.27).

Proof (cf. [14]). When λ comes clockwise/counter-clockwise to the cut $u_a + i\bar{l}_-$ the ray of integration in the Laplace integral (H.24) for $b = a$ (only!) rotates counter-clockwise/clockwise to l_+/l_-. To continue the integrals through the cut we express them via $x_b^{\text{left}}(u,\lambda)$ using the formula (3.131). Since the functions $x_b^{\text{left}}(u,\lambda)$ and the functions $x_a^{\text{right}}(u,\lambda)$ for $a \neq b$ have no jump on the cut $u_a + i\bar{l}_-$ we obtain the monodromy transformation

$$x_b^{\text{right}} + \sum_{R_{ab} \subset P_{\text{right}}} s_{ab} x_a^{\text{right}} \mapsto - \left(x_b^{\text{right}} + \sum_{R_{ab} \subset P_{\text{left}}} s_{ba} x_a^{\text{right}} \right)$$

the sign "$-$" is due to the change of the branch of $\sqrt{z}$ when the ray of the integration is moving through l_-. This coincides with (H.27). Theorem is proved.

We can construct another system of flat coordinates using the Laplace transform of the columns of the matrix $\Psi_0(u,z) = \Psi_{\text{right}}(u,z)C$. I recall that the matrix C consists of the eigenvectors of $S^T S^{-1}$

$$S^T S^{-1} C = C e^{2\pi i \mu}. \tag{H.30}$$

Here we introduce a diagonal matrix

$$\mu = \text{diag}\,(\mu_1, \ldots, \mu_n). \tag{H.31}$$

The numbers $\mu_\alpha = q_\alpha - \frac{d}{2}$ are ordered in such a way that

$$\mu_\alpha + \mu_{n-\alpha+1} = 0. \tag{H.32}$$

This gives a useful identity

$$\mu\eta + \eta\mu = 0. \tag{H.33}$$

Note that the case $0 \leq q_\alpha \leq d < 1$ corresponds to

$$-\frac{1}{2} < \mu_\alpha < \frac{1}{2}. \tag{H.34}$$

We normalize the eigenvectors (H.30) in such a way that the entries of the matrix $\Psi_0(u,z) = (\psi_{0i\alpha}(u,z))$ have the following expansions near the origin

$$\psi_{0i\alpha}(u,z) = z^{\mu_\alpha}\,(\psi_{i\alpha}(u) + O(z)) \text{ when } z \to 0,\ z \in P_{\text{right}}. \tag{H.35}$$

We continue analytically the matrix $\Psi_0(u,z)$ in the left half-plane $z \in P_{\text{left}}$ with a cut along the ray il_+.

Lemma H.5. *Under the assumption (H.34) and the normalization (H.35) the matrix C satisfies the relations*

$$C\eta e^{\pi i \mu}C^T = S, \quad C\eta e^{-\pi i \mu}C^T = S^T. \tag{H.36}$$

Proof. We use the identity

$$\Psi_{\text{right}}(u,z)\Psi_{\text{left}}^T(u,-z) = 1 \tag{H.37}$$

(see above). Let z belong to the sector from il_- to l_+. Then $-z = ze^{-\pi i}$ belongs to the sector from il_+ to l_-. For such z we have

$$\Psi_0(u, ze^{-\pi i}) = \Psi_{\text{left}}(u, -z)S^{-T}C. \qquad (H.38)$$

Substituting (H.38) and (H.33) to (H.37), we obtain

$$\Psi_0(u, z)C^{-1}SC^{-T}\Psi_0^T(u, ze^{-\pi i}) = 1,$$

or

$$C^{-1}SC^{-T} = \Psi_0^{-1}(u, z)\Psi_0^{-T}(u, ze^{-\pi i}). \qquad (H.39)$$

Let z tend to zero keeping it within the sector from il_- to l_+. Using the identity

$$\Psi^{-1}(u)\Psi^{-T}(u) \equiv \eta,$$

where

$$\Psi(u) := (\psi_{i\alpha}(u))$$

in the leading term in z we obtain from (H.39) and (H.35)

$$C^{-1}SC^{-T} = z^{-\mu}\eta z^{-\mu}e^{\pi i\mu} = \eta e^{\pi i\mu}$$

and we can truncate the terms $O(z)$ off the expansion due to (H.34). This proves the first of the equations (H.36). Transposing this equation and applying again the identity (H.33) we obtain the second equation (H.36). Lemma is proved.

We introduce the coordinates $y_\alpha(u, \lambda)$ such that

$$\partial_i y_\alpha(u, \lambda) = \sqrt{\eta_{ii}(u)} \int_{il_-} e^{-\lambda z}\psi_{0i\alpha}(u, z)\frac{dz}{\sqrt{z}} \qquad (H.40)$$

normalizing them as in (H.19). They are still flat coordinates of the metric $(\ ,\ )^* - \lambda < \ ,\ >^*$. We calculate now the matrix of the correspondent covariant metric in these coordinates.

Lemma H.6. *In the coordinates $y_\alpha(u, \lambda)$ the metric $g_{\alpha\beta}(u, \lambda) := \left(g^{\alpha\beta}(u) - \lambda\eta^{\alpha\beta}\right)^{-1}$ is a constant matrix of the form*

$$-\frac{1}{\pi}\eta \cos\pi\mu. \qquad (H.41)$$

Proof. The contravariant metric $g^{\alpha\beta} - \lambda\eta^{\alpha\beta}$ in the coordinates (H.40) has the matrix

$$\hat{g}^{\alpha\beta} = \sum_{i=1}^{n}(u^i - \lambda)\int_{il_-} e^{-\lambda z}\psi_{0i\alpha}(u, z)\frac{dz}{\sqrt{z}}\int_{il_-} e^{-\lambda z}\psi_{0i\beta}(u, z)\frac{dz}{\sqrt{z}}. \qquad (H.42)$$

The matrix is λ-independent. So we can calculate it taking the asymptotic with $\lambda \to \infty$. From (H.35) we obtain asymptotically

$$\int_{il_-} e^{-\lambda z}\psi_{0i\alpha}(u, z)\frac{dz}{\sqrt{z}} \simeq \Gamma\left(\frac{1}{2} + \mu_\alpha\right)\lambda^{-\left(\frac{1}{2} + \mu_\alpha\right)}\psi_{i\alpha}(u). \qquad (H.43)$$

So

$$\hat{g}^{\alpha\beta} = -\lambda\Gamma\left(\frac{1}{2} + \mu_\alpha\right)\Gamma\left(\frac{1}{2} + \mu_\beta\right)\lambda^{-(1+\mu_\alpha+\mu_\beta)}\eta_{\alpha\beta} =$$

$$-\delta_{\alpha+\beta,n+1}\Gamma\left(\frac{1}{2} + \mu_\alpha\right)\Gamma\left(\frac{1}{2} - \mu_\alpha\right) = -\frac{\pi\delta_{\alpha+\beta,n+1}}{\cos\pi\mu_\alpha}. \tag{H.44}$$

Taking the inverse matrix, we obtain (H.41). Lemma is proved.

Corollary H.3. *In the coordinates x_a^{right} the intersection form has the matrix*

$$G = A = S + S^T. \tag{H.45}$$

Proof. From (H.30) we obtain the transformation of the coordinates

$$(y_1, \ldots, y_n) = \sqrt{-2\pi}(x_1, \ldots, x_n)C. \tag{H.46}$$

Here we denote $x_a = x_a^{\text{right}}$. So the matrix G has the form, due to Lemma H.6

$$G = 2C\eta\cos\pi\mu C^T = S + S^T.$$

Corollary is proved.

The corollary shows that $x_1^{\text{right}}, \ldots, x_n^{\text{right}}$ are the coordinates w.r.t. the basis of the root vectors of the system of generating reflections (H.27). This establishes a relation of the monodromy of the differential operator (3.120) to the monodromy group of the Frobenius manifold.

I recall that the root vectors $e_1, \ldots, e_n$ of the system of reflections $R_1, \ldots, R_n$ are defined by the following two conditions

$$R_i e_i = -e_i, \quad i = 1, \ldots, n$$

$$(e_i, e_i) = 2.$$

The reflection R_i in the basis of the root vectors acts as

$$R_i(e_j) = e_j - (e_i, e_j)e_i.$$

The matrix

$$A_{ij} := (e_i, e_j)$$

is called *Coxeter matrix* of the system of reflections. For the monodromy group of our generalized hypergeometric equation (H.15) the Coxeter matrix coincides with the symmetrized Stokes matrix (H.45). The assumption $d < 1$ is equivalent to nondegenerateness of the symmetrized Stokes matrix.

For the finite Coxeter groups (see Lecture 4 below) the system of generating reflections can be chosen in such a way that all A_{ij} are nonpositive. For the particular subclass of

Weyl groups of simple Lie algebras the Coxeter matrix coincides with the symmetrized Cartan matrix of the Lie algebra.

The coordinates $y_1, \ldots, y_n$ are dual to the basis of eigenvectors of the *Coxeter transformation* $R_1 \ldots R_n$ due to the formula (H.29). The basis of the eigenvectors $f_1, \ldots, f_n$ of the Coxeter transform due to (H.41) is normalized as

$$(f_\alpha, f_\beta) = -\frac{1}{\pi} \cos \pi \mu_\beta \delta_{\alpha+\beta, n+1}. \tag{H.47}$$

At the end of this Appendix we consider an application of the integral formula (H.11) to computation of the flat coordinates of the intersection form on a trivial Frobenius manifold. In this case we have a linear contravariant metric (3.37) parametrized by a graded Frobenius algebra A. The gradings q_α of the basic vectors e_α of the algebra are determined up to a common nonzero factor

$$q_\alpha \mapsto \kappa q_\alpha, \quad d \mapsto \kappa d.$$

The normalized flat coordinates $\tilde{t}_\alpha(t, z)$ of the deformed connection can be found easily ((3.5) is an equation with constant coefficients)

$$\tilde{t}_\alpha(t, z) = z^{q_\alpha - d} < e_\alpha, e^{zt} - 1 >, \quad \alpha = 1, \ldots, n \tag{H.48}$$

for

$$t = t^\alpha e_\alpha \in A.$$

So the integral formula (H.11) for the flat coordinates of the pencil (H.1) reads

$$x_\alpha(t, \lambda) = < e_\alpha, \int z^{-\frac{3}{2} + \mu_\alpha} \left(e^{z(t-\lambda)} - e^{-\lambda z} \right) dz > =$$

$$= \Gamma \left(-\frac{1}{2} + \mu_\alpha \right) < e_\alpha, (\lambda - t)^{\frac{1}{2} - \mu_\alpha} > \tag{H.49}$$

for $\mu_\alpha = q_\alpha - \frac{d}{2}$ if $\mu_\alpha \neq \frac{1}{2}$. For $\lambda = 0$ renormalizing (H.49) we obtain the flat coordinates $x_\alpha(t)$ of the intersection form (3.37)

$$x_\alpha(t) = < e_\alpha, t^{\frac{1}{2} - \mu_\alpha} >, \quad \alpha = 1, \ldots, n. \tag{H.50}$$

The r.h.s. is a polynomial in $t^2, \ldots, t^n$ but it ramifies as a function of t^1.

Particularly, for an arbitrary Frobenius algebra A with the trivial grading $q_\alpha = d = 0$ the intersection form (3.37) coincides with the metric on the dual space A^* introduced by Balinski and Novikov [12]. The formula (H.50) in this case gives the quadratic transformation of ref. [12] to the flat coordinates $x_1, \ldots, x_n$ of the metric

$$t = x^2, \quad x = x^\alpha e_\alpha \in A. \tag{H.51}$$

Remark H.1. Inverting the Laplace-type integrals (H.11) and integrating by parts we arrive to an integral representation of the deformed coordinates $\tilde{t}(t, z)$ via the flat coordinates of the intersection form (therefore, via solutions of our generalized hypergeometric equation). This gives "oscillating integrals" for the solutions of (3.5), (3.6):

$$\tilde{t}(t, z) = -z^{\frac{1-d}{2}} \oint e^{z\lambda(x,t)} dx \tag{H.52}$$

where $\lambda = \lambda(x, t)$ is a function inverse to $x = x(t, \lambda) = x(t^1 - \lambda, t^2, \ldots, t^n)$ for a flat coordinate $x(t)$ of the intersection form.

265

Appendix I.
Determination of a superpotential of a Frobenius manifold.

In this Appendix we will show that *any* irreducible massive Frobenius manifold with $d < 1$ can be described by the formulae (2.94) for some LG superpotential $\lambda(p; t)$. The superpotential will always be a function of *one* variable p (may be, a multivalued one) depending on the parameters $t = (t^1, \ldots, t^n)$.

We first construct a function $\lambda(p; t)$ for any massive Frobenius manifold such that the critical values of it are precisely the canonical coordinates on the Frobenius manifold

$$u^i(t) = \lambda(q^i(t); t), \quad \frac{d\lambda}{dp}\big|_{p=q^i(t)} = 0, \quad i = 1, \ldots, n. \qquad (I.1)$$

For the construction we will use the flat coordinates of the flat pencil (H.1) of metrics on M. I recall that these can be represented as

$$x_a(\lambda; t^1, \ldots, t^n) = x_a(t^1 - \lambda, t^2, \ldots, t^n) \qquad (I.2)$$

where $x_a(t)$ are flat coordinates of the intersection form.

Due to Lemma G.1 the coordinates $x_a(t^1 - \lambda, t^2, \ldots, t^n)$ are analytic in t and λ outside of the locus

$$\Delta(t^1 - \lambda, t^2, \ldots, t^n) = 0. \qquad (I.3)$$

On the semisimple part of the locus we have

$$\lambda = u^i(t) \qquad (I.4)$$

for some i. Near such a point $x_a(t^1 - \lambda, t^2, \ldots, t^n)$ is analytic in $\sqrt{\lambda - u^i(t)}$.

Let us fix some $t_0 \in M \setminus (\Sigma \cup \Sigma_{\text{nil}})$ and some a between 1 and n such that

$$\partial_1 x_a(t_0^1 - u^i(t_0^1, t_0^2, \ldots, t_0^n), t_0^2, \ldots, t_0^n) \neq 0 \qquad (I.5)$$

for any $i = 1, \ldots, n$ (such a exists since x_1, ..., x_n are local coordinates) and put

$$p = p(\lambda, t) := x_a(t^1 - \lambda, t^2, \ldots, t^n). \qquad (I.6)$$

By $\lambda = \lambda(p, t)$ we denote the inverse function.

Proposition I.1. *For t close to t_0 the critical points of the function $\lambda(p, t)$ are*

$$q^i = p(u^i(t), t), \quad i = 1, \ldots, n. \qquad (I.7)$$

The correspondent critical values equal $u^i(t)$.

Proof. Near $\lambda = u^i(t_0)$ we have

$$x_a = q^i + x_a^1 \sqrt{\lambda - u^i} + O(\lambda - u^i)$$

and $x_a^1 \neq 0$ by the assumption. For the inverse function we have locally

$$\lambda = u^i(t) + (x_a^1)^{-1}(p - q^i)^2 + o\left((p - q^i)^2\right). \tag{I.8}$$

This proves that λ has the prescribed critical values.

Let us assume now that $d < 1$. In this case we will construct a particular flat coordinate $p = p(t)$ of the intersection form such that the function inverse to $p = p(t, \lambda) = p(t^1 - \lambda, t^2, \ldots, t^n)$ is the LG superpotential of the Frobenius manifold. I will use the flat coordinates $x_a^{\text{right}}(u, \lambda)$ constructed in the previous Appendix.

Lemma I.1. *For* $\lambda \to u_i$

$$x_a^{\text{right}}(u, \lambda) = q_{ai}(u) + \delta_{ai}\sqrt{2\eta_{ii}(u)}\sqrt{u_i - \lambda} + O(u_i - \lambda) \tag{I.9}$$

for some functions $q_{ai}(u)$.

Proof. Using the asymptotic

$$\psi_{ia}^{\text{right}}(u, z) = \left[\delta_{ai} + O\left(\frac{1}{z}\right)\right] e^{zu_i}$$

for $z \to \infty$ we obtain (I.9). Lemma is proved.

We consider now the following particular flat coordinate

$$p = p(t; \lambda) := \sum_{a=1}^n x_a^{\text{right}}(u, \lambda). \tag{I.10}$$

By $\lambda = \lambda(p; t)$ we denote, as above, the inverse function.

Theorem I.1. *For the metrics* $< \, , \, >$, $(\, , \,)$ *and for the trilinear form (1.46) the following formulae hold true*

$$< \partial', \partial'' >_t = -\sum_{i=1}^n \mathop{\text{res}}_{p=q^i} \frac{\partial'(\lambda(p, t)dp)\, \partial''(\lambda(p, t)dp)}{d\lambda(p, t)} \tag{I.11}$$

$$(\partial', \partial'')_t = -\sum_{i=1}^n \mathop{\text{res}}_{p=q^i} \frac{\partial'(\log \lambda(p, t)dp)\, \partial''(\log \lambda(p, t)dp)}{d\log \lambda(p, t)} \tag{I.12}$$

$$c(\partial', \partial'', \partial''')_t = -\sum_{i=1}^n \mathop{\text{res}}_{p=q^i} \frac{\partial'(\lambda(p, t)dp)\, \partial''(\lambda(p, t)dp)\, \partial'''(\lambda(p, t)dp)}{dp\, d\lambda(p, t)}. \tag{I.13}$$

In these formulae

$$d\lambda := \frac{\partial \lambda(p, t)}{\partial p} dp, \quad d\log \lambda := \frac{\partial \log \lambda(p, t)}{\partial p} dp.$$

Proof. From (I.9) we obtain

$$p(\lambda, t) = \sum_a x_a^{\text{right}}(u, \lambda) = q_i(u) + \sqrt{2\eta_{ii}(u)}\sqrt{u_i - \lambda} + O(u_i - \lambda) \qquad (I.14)$$

near $\lambda = u_i$ where we put

$$q_i(u) := \sum_a q_{ai}(u). \qquad (I.15)$$

So $q_i(u)$ is a critical point of λ with the critical value u_i. Near this point

$$\lambda = u_i - \frac{(p - q_i)^2}{2\eta_{ii}(u)} + O(p - q_i)^3. \qquad (I.16)$$

From this formula we immediately obtain that for $\partial' = \partial_i$, $\partial'' = \partial_j$ the r.h.s. of the formula (I.11) is equal to

$$\eta_{ii}(u)\delta_{ij} = <\partial_i, \partial_j>_u .$$

This proves the equality (I.11). The other equalities are proved in a similar way. Theorem is proved.

Example I.1. Using the flat coordinates from Example G.2 we obtain the LG superpotential of the CP^1-model (see Lecture 2)

$$\lambda(p; t_1, t_2) = t_1 - 2e^{\frac{t_2}{2}} \cos p. \qquad (I.17)$$

The cosine is considered as an analytic function on the cylinder $p \simeq p + 2\pi$, so it has only 2 critical points $p = 0$ and $p = \pi$.

Other examples will be considered in Lectures 4 and 5.

268

Lecture 4.

Frobenius structure on the space of orbits of a Coxeter group.

Let W be a *Coxeter group*, i.e. a finite group of linear transformations of real n-dimensional space V generated by reflections. In this Lecture we construct Frobenius manifolds whose monodromy is a given Coxeter group W. All of these Frobenius structures will be polynomial. The results of Appendix A suggest that the construction of this Lecture gives all the polynomial solution of WDVV with $d < 1$ satisfying the semisimplicity assumption, although this is still to be proved.

We always can assume the transformations of the group W to be orthogonal w.r.t. a Euclidean structure on V. The complete classification of irreducible Coxeter groups was obtained in [33]; see also [22]. The complete list consists of the groups (dimension of the space V equals the subscript in the name of the group) A_n, B_n, D_n, E_6, E_7, E_8, F_4, G_2 (the Weyl groups of the correspondent simple Lie algebras), the groups H_3 and H_4 of symmetries of the regular icosahedron and of the regular 600-cell in the 4-dimensional space resp. and the groups $I_2(k)$ of symmetries of the regular k-gone on the plane. The group W also acts on the symmetric algebra $S(V)$ (polynomials of the coordinates of V) and on the $S(V)$-module $\Omega(V)$ of differential forms on V with polynomial coefficients. The subring $R = S(V)^W$ of W-invariant polynomials is generated by n algebraicaly independent homogeneous polynomials y^1, ..., y^n [22]. The submodule $\Omega(V)^W$ of the W-invariant differential forms with polynomial coefficients is a free R-module with the basis $dy^{i_1} \wedge \ldots \wedge dy^{i_k}$ [22]. Degrees of the basic invariant polynomials are uniquely determined by the Coxeter group. They can be expressed via the *exponents* m_1, ..., m_n of the group, i.e. via the eigenvalues of a Coxeter element C in W [22]

$$d_i := \deg y^i = m_i + 1, \tag{4.1a}$$

$$\{\text{eigen } C\} = \{\exp \frac{2\pi i(d_1 - 1)}{h}, ..., \exp \frac{2\pi i(d_n - 1)}{h}\}. \tag{4.1b}$$

The maximal degree h is called *Coxeter number* of W. I will use the reversed ordering of the invariant polynomials

$$d_1 = h > d_2 \geq \ldots \geq d_{n-1} > d_2 = 2. \tag{4.2}$$

The degrees satisfy the *duality condition*

$$d_i + d_{n-i+1} = h + 2, \; i = 1, ..., n. \tag{4.3}$$

The list of the degrees for all the Coxeter groups is given in Table 1.

W	$d_1, \ldots, d_n$
A_n	$d_i = n + 2 - i$
B_n	$d_i = 2(n - i + 1)$
$D_n,\ n = 2k$	$d_i = 2(n - i),\ i \le k,$
	$d_i = 2(n - i + 1),\ k + 1 \le i$
$D_n,\ n = 2k + 1$	$d_i = 2(n - i),\ i \le k,$
	$d_{k+1} = 2k + 1,$
	$d_i = 2(n - i + 1),\ k + 2 \le i$
E_6	12, 9, 8, 6, 5, 2
E_7	18, 14, 12, 10, 8, 6, 2
E_8	30, 24, 20, 18, 14, 12, 8, 2
F_4	12, 8, 6, 2
G_2	6, 2
H_3	10, 6, 2
H_4	30, 20, 12, 2
$I_2(k)$	k, 2

Table 1.

I will extend the action of the group W to the complexified space $V \otimes \mathbf{C}$. The space of orbits

$$M = V \otimes \mathbf{C}/W$$

has a natural structure of an affine algebraic variety: the coordinate ring of M is the (complexified) algebra R of invariant polynomials of the group W. The coordinates $y^1, \ldots, y^n$ on M are defined up to an invertible transformation

$$y^i \mapsto y^{i'}(y^1, \ldots, y^n), \tag{4.4}$$

where $y^{i'}(y^1, \ldots, y^n)$ is a graded homogeneous polynomial of the same degree d_i in the variables $y^1, \ldots, y^n$, $\deg y^k = d_k$. Note that the Jacobian $\det(\partial y^{i'}/\partial y^j)$ is a constant (it should not be zero). The transformations (4.4) leave invariant the vector field $\partial_1 := \partial/\partial y^1$ (up to a constant factor) due to the strict inequality $d_1 > d_2$. The coordinate y^n is determined uniquely within a factor. Also the vector field

$$E = \frac{1}{h}\left(d_1 y^1 \partial_1 + \ldots + d_n y^n \partial_n\right) = \frac{1}{h} x^a \frac{\partial}{\partial x^a} \tag{4.5}$$

(the generator of scaling transformations) is well-defined on M. Here we denote by x^a the coordinates in the linear space V.

Let $(\ ,\ )$ denotes the W-invariant Euclidean metric in the space V. I will use the orthonormal coordinates $x^1, \ldots, x^n$ in V with respect to this metric. The invariant y^n can be chosen as

$$y^n = \frac{1}{2h}((x^1)^2 + \ldots + (x^n)^2). \tag{4.6}$$

We extend (,) onto $V \otimes \mathbf{C}$ as a complex quadratic form.

The factorization map $V \otimes \mathbf{C} \to M$ is a local diffeomorphism on an open subset of $V \otimes \mathbf{C}$. The image of this subset in M consists of *regular orbits* (i.e. the number of points of the orbit equals $\# W$). The complement is the *discriminant locus* $\operatorname{Discr} W$. By definition it consists of all irregular orbits. Note that the linear coordinates in V can serve also as local coordinates in small domains in $M \setminus \operatorname{Discr} W$. It defines a metric (,) (and (,)*) on $M \setminus \operatorname{Discr} W$. The contravariant metric can be extended onto M according to the following statement (cf. [134, Sections 5 and 6]).

Lemma 4.1. *The Euclidean metric of V induces polynomial contravariant metric* (,)* *on the space of orbits*

$$g^{ij}(y) = (dy^i, dy^j)^* := \frac{\partial y^i}{\partial x^a} \frac{\partial y^j}{\partial x^a} \tag{4.7}$$

and the correspondent contravariant Levi-Cività connection

$$\Gamma^{ij}_k(y) dy^k = \frac{\partial y^i}{\partial x^a} \frac{\partial^2 y^j}{\partial x^a \partial x^b} dx^b \tag{4.8}$$

also is a polynomial one.

Proof. The right-hand sides in (4.7)/(4.8) are W-invariant polynomials/differential forms with polynomial coefficients. Hence $g^{ij}(y)/\Gamma^{ij}_k(y)$ are polynomials in y^1, ..., y^n. Lemma is proved.

Remark 4.1. The matrix $g^{ij}(y)$ does not degenerate on $M \setminus \operatorname{Discr} W$ where the factorization $V \otimes \mathbf{C} \to M$ is a local diffeomorphism. So the polynomial (also called *discriminant* of W)

$$D(y) := \det(g^{ij}(y)) \tag{4.9}$$

vanishes precisely on the discriminant locus $\operatorname{Discr} W$ where the variables x^1, ..., x^n fail to be local coordinates. Due to this fact the matrix $g^{ij}(y)$ is called *discriminant matrix* of W. The contravariant metric (4.7) was introduced by V.I.Arnold [3] in the form of operation of convolution of invariants $f(x)$, $g(x)$ of a reflection group

$$f, \ g \mapsto (df, \, dg)^* = \sum \frac{\partial f}{\partial x^a} \frac{\partial g}{\partial x^a}.$$

Note that the image of V in the real part of M is specified by the condition of positive semidefiniteness of the matrix $(g^{ij}(y))$ (cf. [125]). The Euclidean connection (4.8) on the space of orbits is called *Gauss - Manin connection*.

The main result of this lecture is

Theorem 4.1. *There exists a unique, up to an equivalence, Frobenius structure on the space of orbits of a finite Coxeter group with the intersection form (4.7), the Euler vector field (4.5) and the unity vector field $e := \partial / \partial y^1$.*

We start the proof of the theorem with the following statement.

Proposition 4.1. *The functions $g^{ij}(y)$ and $\Gamma^{ij}_k(y)$ depend linearly on y^1.*

Proof. From the definition one has that $g^{ij}(y)$ and $\Gamma^{ij}_k(y)$ are graded homogeneous polynomials of the degrees

$$\deg g^{ij}(y) = d_i + d_j - 2 \tag{4.10}$$

$$\deg \Gamma^{ij}_k(y) = d_i + d_j - d_k - 2. \tag{4.11}$$

Since $d_i + d_j \leq 2h = 2d_1$ these polynomials can be at most linear in y^1. Proposition is proved.

Corollary 4.1 (K.Saito) *The matrix*

$$\eta^{ij}(y) := \partial_1 g^{ij}(y) \tag{4.12}$$

has a triangular form

$$\eta^{ij}(y) = 0 \text{ for } i + j > n + 1, \tag{4.13}$$

and the antidiagonal elements

$$\eta^{i(n-i+1)} =: c_i \tag{4.14}$$

are nonzero constants. Particularly,

$$c := \det(\eta^{ij}) = (-1)^{\frac{n(n-1)}{2}} c_1 ... c_n \neq 0. \tag{4.15}$$

Proof. One has

$$\deg \eta^{ij}(y) = d_i + d_j - 2 - h.$$

Hence $\deg \eta^{i(n-i+1)} = 0$ (see (4.3)) and $\deg \eta^{ij} < 0$ for $i+j > n+1$. This proves triangularity of the matrix and constancy of the antidiagonal entries c_i. To prove nondegenerateness of $(\eta^{ij}(y))$ we consider, following Saito, the discriminant (4.9) as a polynomial in y^1

$$D(y) = c(y^1)^n + a_1(y^1)^{n-1} + ... + a_n$$

where the coefficients $a_1, ..., a_n$ are quasihomogeneous polynomials in $y^2, ..., y^n$ of the degrees $h, 2h, ..., nh$ resp. and the leading coefficient c is given in (4.15). Let γ be the eigenvector of a Coxeter transformation C with the eigenvalue $\exp(2\pi i/h)$. Then

$$y^k(\gamma) = y^k(C\gamma) = y^k(\exp(2\pi i/h)\gamma) = \exp(2\pi i d_k/h)y^k(\gamma).$$

For $k > 1$ we obtain

$$y^k(\gamma) = 0, \ k = 2, ..., n.$$

But $D(\gamma) \neq 0$ [22]. Hence the leading coefficient $c \neq 0$. Corollary is proved.

Corollary 4.2. *The space M of orbits of a finite Coxeter group carries a flat pencil of metrics $g^{ij}(y)$ (4.7) and $\eta^{ij}(y)$ (4.12) where the matrix $\eta^{ij}(y)$ is polynomially invertible globally on M.*

We will call (4.12) *Saito metric* on the space of orbits. This was introduced by Saito, Sekiguchi and Yano in [131] using the classification of Coxeter groups for all of them but E_7 and E_8. The general proof of flatness (not using the classification) was obtained in [130].

This metric will be denoted by $< , >^*$ (and by $< , >$ if considered on the tangent bundle TM). Let us denote by

$$\gamma_k^{ij}(y) := \partial_1 \Gamma_k^{ij}(y) \qquad (4.16)$$

the components of the Levi-Cività connection for the metric $\eta^{ij}(y)$. These are quasihomogeneous polynomials of the degrees

$$\deg \gamma_k^{ij}(y) = d_i + d_j - d_k - h - 2. \qquad (4.17)$$

Corollary 4.3 (K.Saito). *There exist homogeneous polynomials* $t^1(x)$, ..., $t^n(x)$ *of degrees* d_1, ..., d_n *resp. such that the matrix*

$$\eta^{\alpha\beta} := \partial_1 (dt^\alpha, dt^\beta)^* \qquad (4.18)$$

is constant.

The coordinates t^1, ..., t^n on the orbit space will be called *Saito flat coordinates*. They can be chosen in such a way that the matrix (4.18) is antidiagonal

$$\eta^{\alpha\beta} = \delta^{\alpha+\beta,n+1}.$$

Then the Saito flat coordinates are defined uniquely up to an η-orthogonal transformation

$$t^\alpha \mapsto a_\beta^\alpha t^\beta,$$

$$\sum_{\lambda+\mu=n+1} a_\lambda^\alpha a_\mu^\beta = \delta^{\alpha+\beta,n+1}.$$

Proof. From flatness of the metric $\eta^{ij}(y)$ it follows that the flat coordinates $t^\alpha(y)$, $\alpha = 1$, ..., n exist at least localy. They are to be determined from the following system

$$\eta^{is}\partial_s\partial_j t + \gamma_j^{is}\partial_s t = 0 \qquad (4.19)$$

(see (3.30)). The inverse matrix $(\eta_{ij}(y)) = (\eta^{ij}(y))^{-1}$ also is polynomial in y^1, ..., y^n. So rewriting the system (4.19) in the form

$$\partial_k\partial_l t + \eta_{il}\gamma_k^{is}\partial_s t = 0 \qquad (4.20)$$

we again obtain a system with polynomial coefficients. It can be written as a first-order system for the entries $\xi_l = \partial_l t$,

$$\partial_k\xi_l + \eta_{il}\gamma_k^{is}\xi_s = 0, \ k,l = 1,...,n \qquad (4.21)$$

(the integrability condition $\partial_k \xi_l = \partial_l \xi_k$ follows from vanishing of the curvature). This is an overdetermined holonomic system. So the space of solutions has dimension n. We can choose a fundamental system of solutions $\xi_l^\alpha(y)$ such that $\xi_l^\alpha(0) = \delta_l^\alpha$. These functions are analytic in y for sufficiently small y. We put $\xi_l^\alpha(y) =: \partial_l t^\alpha(y)$, $t^\alpha(0) = 0$. The system of solutions is invariant w.r.t. the scaling transformations

$$y^i \mapsto c^{d_i} y^i, \; i = 1, ..., n.$$

So the functions $t^\alpha(y)$ are quasihomogeneous in y of the same degrees d_1, ..., d_n. Since all the degrees are positive the power series $t^\alpha(y)$ should be polynomials in y^1, ..., y^n. Because of the invertibility of the transformation $y^i \mapsto t^\alpha$ we conclude that $t^\alpha(y(x))$ are polynomials in x^1, ..., x^n. Corollary is proved.

We need to calculate particular components of the metric $g^{\alpha\beta}$ and of the correspondent Levi-Cività connection in the coordinates t^1, ..., t^n (in fact, in arbitrary homogeneous coordinates y^1, ..., y^n).

Lemma 4.2. *Let the coordinate t^n be normalized as in (4.6). Then the following formulae hold:*

$$g^{n\alpha} = \frac{d_\alpha}{h} t^\alpha \tag{4.22}$$

$$\Gamma_\beta^{n\alpha} = \frac{(d_\alpha - 1)}{h} \delta_\beta^\alpha. \tag{4.23}$$

(In the formulae there is no summation over the repeated indices!)
Proof. We have

$$g^{n\alpha} = \frac{\partial t^n}{\partial x^a} \frac{\partial t^\alpha}{\partial x^a} = \frac{1}{h} x^a \frac{\partial t^\alpha}{\partial x^a} = \frac{d_\alpha}{h} t^\alpha$$

due to the Euler identity for the homogeneous functions $t^\alpha(x)$. Furthermore,

$$\Gamma_\beta^{n\alpha} dt^\beta = \frac{\partial t^n}{\partial x^a} \frac{\partial^2 t^\alpha}{\partial x^a \partial x^b} dx^b = \frac{1}{h} x^a \frac{\partial^2 t^\alpha}{\partial x^a \partial x^b} dx^b = \frac{1}{h} x^a d \left(\frac{\partial t^\alpha}{\partial x^a} \right)$$

$$= \frac{1}{h} d \left(x^a \frac{\partial t^\alpha}{x^a} \right) - \frac{1}{h} \frac{\partial t^\alpha}{\partial x^a} dx^a = \frac{(d_\alpha - 1)}{h} dt^\alpha.$$

Lemma is proved.

We can formulate now

Main lemma. *Let t^1, ..., t^n be the Saito flat coordinates on the space of orbits of a finite Coxeter group and*

$$\eta^{\alpha\beta} = \partial_1 (dt^\alpha, dt^\beta)^* \tag{4.24}$$

be the correspondent constant Saito metric. Then there exists a quasihomogeneous polynomial $F(t)$ of the degree $2h + 2$ such that

$$(dt^\alpha, dt^\beta)^* = \frac{(d_\alpha + d_\beta - 2)}{h} \eta^{\alpha\lambda} \eta^{\beta\mu} \partial_\lambda \partial_\mu F(t). \tag{4.25}$$

The polynomial $F(t)$ determines on the space of orbits a polynomial Frobenius structure with the structure constants

$$c_{\alpha\beta}^{\gamma}(t) = \eta^{\gamma\epsilon}\partial_\alpha\partial_\beta\partial_\epsilon F(t) \qquad (4.26a)$$

the unity

$$e = \partial_1 \qquad (4.26b)$$

the Euler vector field

$$E = \sum \left(1 - \frac{\deg t^\alpha}{h}\right) t^\alpha \partial_\alpha$$

and the invariant inner product η.

Proof. Because of Corollary 4.3 in the flat coordinates the tensor $\Delta_\gamma^{\alpha\beta} = \Gamma_\gamma^{\alpha\beta}$ should satisfy the equations (D.1) - (D.3) where $g_1^{\alpha\beta} = g^{\alpha\beta}(t)$, $g_2^{\alpha\beta} = \eta^{\alpha\beta}$. First of all according to (D.1a) we can represent the tensor $\Gamma_\gamma^{\alpha\beta}(t)$ in the form

$$\Gamma_\gamma^{\alpha\beta}(t) = \eta^{\alpha\epsilon}\partial_\epsilon\partial_\gamma f^\beta(t) \qquad (4.27)$$

for a vector field $f^\beta(t)$. The equation (3.27) (or, equivalently, (D.3)) for the metric $g^{\alpha\beta}(t)$ and the connection (4.27) reads

$$g^{\alpha\sigma}\Gamma_\sigma^{\beta\gamma} = g^{\beta\sigma}\Gamma_\sigma^{\alpha\gamma}.$$

For $\alpha = n$ because of Lemma 4.2 this gives

$$\sum_\sigma d_\sigma t^\sigma \eta^{\beta\epsilon}\partial_\sigma\partial_\epsilon f^\gamma = (d_\gamma - 1)g^{\beta\gamma}.$$

Applying to the l.h.s. the Euler identity (here deg $\partial_\epsilon f^\gamma = d_\gamma - d_\epsilon + h$) we obtain

$$(d_\gamma - 1)g^{\beta\gamma} = \sum_\epsilon \eta^{\beta\epsilon}(d_\gamma - d_\epsilon + h)\partial_\epsilon f^\gamma = (d_\gamma + d_\beta - 2)\eta^{\beta\epsilon}\partial_\epsilon f^\gamma. \qquad (4.28a)$$

From this one obtains the symmetry

$$\frac{\eta^{\beta\epsilon}\partial_\epsilon f^\gamma}{d_\gamma - 1} = \frac{\eta^{\gamma\epsilon}\partial_\epsilon f^\beta}{d_\beta - 1}.$$

Let us denote

$$\frac{f^\gamma}{d_\gamma - 1} =: \frac{F^\gamma}{h}. \qquad (4.28b)$$

We obtain

$$\eta^{\beta\epsilon}\partial_\epsilon F^\gamma = \eta^{\gamma\epsilon}\partial_\epsilon F^\beta.$$

Hence a function $F(t)$ exists such that

$$F^\alpha = \eta^{\alpha\epsilon}\partial_\epsilon F. \qquad (4.28c)$$

It is clear that $F(t)$ is a quasihomogeneous polynomial of the degree $2h + 2$. From the formula (4.28) one immediately obtains (4.25).

Let us prove now that the coefficients (4.26a) satisfy the associativity condition. It is more convenient to work with the dual structure constants

$$c_\gamma^{\alpha\beta}(t) = \eta^{\alpha\lambda}\eta^{\beta\mu}\partial_\lambda\partial_\mu\partial_\gamma F.$$

Because of (4.27), (4.28) one has

$$\Gamma_\gamma^{\alpha\beta} = \frac{d_\beta - 1}{h}c_\gamma^{\alpha\beta}.$$

Substituting this in (D.2) we obtain associativity. Finaly, for $\alpha = n$ the formulae (4.22), (4.23) imply

$$c_\beta^{n\alpha} = \delta_\beta^\alpha.$$

Since $\eta^{1n} = 1$, the vector (4.26b) is the unity of the algebra. Lemma is proved.

Proof of Theorem.

Existence of a Frobenius structure on the space of orbits satisfying the conditions of Theorem 4.1 follows from Main lemma. We are now to prove uniqueness. Let us consider a polynomial Frobenius structure on M with the Euler vector field (4.5) and with the Saito invariant metric. In the Saito flat coordinates we have

$$dt^\alpha \cdot dt^\beta = \eta^{\alpha\lambda}\eta^{\beta\mu}\partial_\lambda\partial_\mu\partial_\gamma F(t)dt^\gamma.$$

The r.h.s. of (3.13) reads

$$i_E(dt^\alpha \cdot dt^\beta) = \frac{1}{h}\sum_\gamma d_\gamma t^\gamma \eta_{\alpha\lambda}\eta^{\beta\mu}\partial_\lambda\partial_\mu\partial_\gamma F(t) = \frac{1}{h}(d_\alpha + d_\beta - 2)\eta_{\alpha\lambda}\eta^{\beta\mu}\partial_\lambda\partial_\mu F(t).$$

This should be equal to $(dt^\alpha, dt^\beta)^*$. So the function $F(t)$ must satisfy (4.25). It is determined uniquely by this equation up to terms quadratic in t^α. Such an ambiguity does not affect the Frobenius structure. Theorem is proved.

We will show know that the Frobenius manifolds we have constructed satisfy the semisimplicity condition. This will follow from the following construction.

Let $R = \mathbf{C}[y_1, ..., y_n]$ be the coordinate ring of the orbit space M. The Frobenius algebra structure on the tangent planes T_yM for any $y \in M$ provides the R-module $Der\, R$ of invariant vector fields with a structure of Frobenius algebra over R. To describe this structure let us consider a homogeneous basis of invariant polynomials $y_1, ..., y_n$ of the Coxeter group. Let $D(y_1, ..., y_n)$ be the discriminant of the group. We introduce a polynomial of degree n in an auxiliary variable u putting

$$P(u; y_1, ..., y_n) := D(y_1 - u, y_2, ..., y_n). \tag{4.29}$$

Let $D_0(y_1, ..., y_n)$ be the discriminant of this polynomial in u. It does not vanish identicaly on the space of orbits.

Theorem 4.2. *The map*

$$1 \mapsto e, \quad u \mapsto E \tag{4.30a}$$

can be extended uniquely to an isomorphism of R-algebras

$$R[u]/(P(u;y)) \to Der\, R. \tag{4.30b}$$

Corollary 4.4. *The algebra on $T_y M$ has no nilpotents outside the zeroes of the polynomial $D_0(y_1, ..., y_n)$.*

We start the proof with an algebraic remark: let T be a n-dimensional space and $U : T \to T$ an endomorphism (linear operator). Let

$$P_U(u) := \det(U - u \cdot 1)$$

be the characteristic polynomial of U. We say that the endomorphism U is semisimple if all the n roots of the characteristic polynomial are simple. For a semisimple endomorphism there exists a cyclic vector $e \in T$ such that

$$T = \mathrm{span}\,(e, Ue, ..., U^{n-1}e).$$

The map

$$\mathbf{C}[u]/(P_U(u)) \to T, \quad u^k \mapsto U^k e, \ k = 0, 1, ..., n - 1 \tag{4.31}$$

is an isomorphism of linear spaces.

Let us fix a point $y \in M$. We define a linear operator

$$U = (U_j^i(y)) : T_y M \to T_y M \tag{4.32}$$

(being also an operator on the cotangent bundle of the space of orbits) taking the ratio of the quadratic forms g^{ij} and η^{ij}

$$< U\omega_1, \omega_2 >^* = (\omega_1, \omega_2)^* \tag{4.33}$$

or, equivalently,

$$U_j^i(y) := \eta_{js}(y)g^{si}(y). \tag{4.34}$$

Lemma 4.3. *The characteristic polynomial of the operator $U(y)$ is given up to a nonzero factor c^{-1} (4.15) by the formula (4.29).*

Proof. We have

$$P(u; y^1, \dots, y^n) := \det(U - u \cdot 1) = \det(\eta_{js})\det(g^{si} - u\eta^{si}) =$$

$$c^{-1}\det(g^{si}(y^1 - u, y^2, \dots, y^n) = c^{-1}D(y^1 - u, y^2, \dots, y^n).$$

Lemma is proved.

277

Corollary 4.5. *The operator $U(y)$ is semisimple at a generic point $y \in M$.*

Proof. Let us prove that the discriminant $D_0(y^1, \ldots, y^n)$ of the characteristic polynomial $P(u; y^1, \ldots, y^n)$ does not vanish identicaly on M. Let us fix a Weyl chamber $V_0 \subset V$ of the group W. On the inner part of V_0 the factorization map

$$V_0 \to M_{Re}$$

is a diffeomorphism. On the image of V_0 the discriminant $D(y)$ is positive. It vanishes on the images of the n walls of the Weyl chamber:

$$D(y)_{i-\text{th wall}} = 0, \quad i = 1, \ldots, n. \tag{4.35}$$

On the inner part of the i-th wall (where the surface (4.35) is regular) the equation (4.35) can be solved for y^1:

$$y^1 = y_i^1(y^2, \ldots, y^n). \tag{4.36}$$

Indeed, on the inner part

$$(\partial_1 D(y))_{i-\text{th wall}} \neq 0.$$

This holds since the polynomial $D(y)$ has simple zeroes at the generic point of the discriminant of W (see, e.g., [4]) .

Note that the functions (4.36) are the roots of the equation $D(y) = 0$ as the equation in the unknown y^1. It follows from above that this equation has simple roots for generic y^2, ..., y^n. The roots of the characteristic equation

$$D(y^1 - u, y^2, \ldots, y^n) = 0 \tag{4.37a}$$

are therefore

$$u_i = y^1 - y_i^1(y^2, \ldots, y^n), \quad i = 1, \ldots, n. \tag{4.37b}$$

Genericaly these are distinct. Lemma is proved.

Lemma 4.4. *The operator U on the tangent planes $T_y M$ coincides with the operator of multiplication by the Euler vector field E.*

Proof. We check the statement of the lemma in the Saito flat coordinates:

$$\sum_\sigma \frac{d_\sigma}{h} t^\sigma c_{\sigma\beta}^\alpha = \frac{h - d_\beta + d_\alpha}{h} \eta^{\alpha\epsilon} \partial_\epsilon \partial_\beta F =$$

$$\sum_\lambda \frac{d_\lambda + d_\alpha - 2}{h} \eta_{\beta\lambda} \eta^{\alpha\epsilon} \eta^{\lambda\mu} \partial_\epsilon \partial_\mu F = \eta_{\beta\lambda} g^{\alpha\lambda} = U_\beta^\alpha.$$

Lemma is proved.

Proof of Theorem 4.2.

Because of Lemmas 4.3, 4.4 the vector fields

$$e, \; E, \; E^2, \ldots, E^{n-1} \tag{4.38}$$

genericaly are linear independent on M. It is easy to see that these are polynomial vector fields on M. Hence e is a cyclic vector for the endomorphism U acting on $Der\,R$. So in generic point $y \in M$ the map (4.30a) is an isomorphism of Frobenius algebras

$$\mathbf{C}[u]/(P(u;x)) \to T_xM.$$

This proves Theorem 2.

Remark 4.2. The Euclidean metric (4.7) also defines an invariant inner product for the Frobenius algebras (on the cotangent planes T_*M). It can be shown also that the trilinear form

$$(\omega_1 \cdot \omega_2, \omega_3)^*$$

can be represented (localy, outside the discriminant locus $Discr\,W$) in the form

$$(\hat{\nabla}^i\hat{\nabla}^j\hat{\nabla}^k\hat{F}(x))\partial_i \otimes \partial_j \otimes \partial_k$$

for some function $\hat{F}(x)$. Here $\hat{\nabla}$ is the Gauss-Manin connection (i.e. the Levi-Cività connection for the metric (4.7)). The unity dt^n/h of the Frobenius algebra on T_*M is not covariantly constant w.r.t. the Gauss-Manin connection.

Remark 4.3. The vector fields

$$l^i := g^{is}(y)\partial_s, \quad i = 1,\dots,n \tag{4.39}$$

form a basis of the R-module $Der_R(-\log(D(y)))$ of the vector fields on M tangent to the discriminant locus [4]. By the definition, a vector field $u \in Der_R(-\log(D(y)))$ iff

$$uD(y) = p(y)D(y)$$

for a polynomial $p(y) \in R$. The basis (4.39) of $Der_R(-\log(D(y)))$ depends on the choice of coordinates on M. In the Saito flat coordinates commutators of the basic vector fields can be calculated via the structure constants of the Frobenius algebra on T_*M. The following formula holds:

$$[l^\alpha, l^\beta] = \frac{d_\beta - d_\alpha}{h}c_\epsilon^{\alpha\beta}l^\epsilon. \tag{4.40}$$

This can be proved using (4.25).

Example 4.1. $W = I_2(k)$, $k \geq 0$. The action of the group on the complex z-plane is generated by the transformations

$$z \mapsto e^{\frac{2\pi i}{k}}z, \quad z \mapsto \bar{z}.$$

The invariant metric on $\mathbf{R}^2 = \mathbf{C}$ is

$$ds^2 = dzd\bar{z},$$

the basic invariant polynomials are

$$t^1 = z^k + \bar{z}^k, \quad \deg t^1 = k,$$

$$t^2 = \frac{1}{2k} z\bar{z}, \quad \deg t^2 = 2.$$

We have

$$g^{11}(t) = (dt^1, dt^1) = 4\frac{\partial t^1}{\partial z}\frac{\partial t^1}{\partial \bar{z}} = 4k^2(z\bar{z})^{k-1} = (2k)^{k+1}(t^2)^{k-1}$$

$$g^{12}(t) = (dt^1, dt^2) = 2\left(\frac{\partial t^1}{\partial z}\frac{\partial t^2}{\partial \bar{z}} + \frac{\partial t^1}{\partial \bar{z}}\frac{\partial t^2}{\partial z}\right) = (z^k + \bar{z}^k) = t^1$$

$$g^{22}(t) = 4\frac{\partial t^2}{\partial z}\frac{\partial t^2}{\partial \bar{z}} = \frac{2}{k}t^2.$$

The Saito metric (4.12) is constant in these coordinates. The formula (4.25) gives

$$F(t^1, t^2) = \frac{1}{2}(t^1)^2 t^2 + \frac{(2k)^{k+1}}{2(k^2-1)}(t^2)^{k+1}.$$

This coincides with (1.24a) (up to an equivalence) for $\mu = \frac{1}{2}(k-1)/(k+1)$. Particularly, for $k = 3$ this gives the Frobenius structure on $\mathbf{C}^2/A_2$, for $k = 4$ on $\mathbf{C}^2/B_2$, for $k = 6$ on $\mathbf{C}^2/G_2$.

Example 4.2. $W = A_n$. The group acts on the $(n+1)$-dimensional space $\mathbf{R}^{n+1} = \{(\xi_0, \xi_1, \ldots, \xi_n)\}$ by the permutations

$$(\xi_0, \xi_1, \ldots, \xi_n) \mapsto (\xi_{i_0}, \xi_{i_1}, \ldots, \xi_{i_n}).$$

Restricting the action onto the hyperplane

$$\xi_0 + \xi_1 + \ldots, +\xi_n = 0 \tag{4.41}$$

we obtain the desired action of A_n on the n-dimensional space (4.41). The invariant metric on (4.41) is obtained from the standard Euclidean metric on $\mathbf{R}^{n+1}$ by the restriction.

The invariant polynomials on (4.41) are symmetric polynomials on $\xi_0, \xi_1, \ldots, \xi_n$. The elementary symmetric polynomials

$$a_k = (-1)^{n-k+1}(\xi_0\xi_1 \ldots \xi_k + \ldots), \quad k = 1, \ldots, n \tag{4.42}$$

can be taken as a homogeneous basis in the graded ring of the W-invariant polynomials on (4.41). So the complexified space of orbits $M = \mathbf{C}^n/A_n$ can be identified with the space of polynomials $\lambda(p)$ of an auxiliary variable p of the form (1.65).

Let us show that the Frobenius structure (4.25) on M coincides with the structure (1.66) (this will give us the simplest proof of that the formulae (1.66) give an example of Frobenius manifold). It will be convenient first to rewrite the formulae (1.66) in a slightly modified way (cf. (2.94))

Lemma 4.5.

1. For the example 3 of Lecture 1 the inner product $<\,,\,>_\lambda$ and the 3-d rank tensor $c(.,.,.) =<\,\cdot\,\cdot\,.,\,.\,>_\lambda$ have the form

$$< \partial', \partial'' >_\lambda = -\sum_{|\lambda|<\infty} \operatorname*{res}_{d\lambda=0} \frac{\partial'(\lambda(p)dp)\,\partial''(\lambda(p)dp)}{d\lambda(p)} \tag{4.43}$$

$$c(\partial', \partial'', \partial''') = -\sum_{|\lambda|<\infty} \operatorname*{res}_{d\lambda=0} \frac{\partial'(\lambda(p)dp)\,\partial''(\lambda(p)dp)\,\partial'''(\lambda(p)dp)}{dp\,d\lambda(p)}. \tag{4.44}$$

2. Let $q^1,\ldots,\,q^n$ be the critical points of the polynomial $\lambda(p)$,

$$\lambda'(q^i) = 0, \quad i = 1,\ldots,n$$

and

$$u^i = \lambda(q^i), \quad i = 1,\ldots,n \tag{4.45}$$

be the correspondent critical values. The variables $u^1,\ldots,\,u^n$ are local coordinates on M near the points λ where the polynomial $\lambda(p)$ has no multiple roots. These are canonical coordinates for the multiplication (1.66a). The metric (1.66b) in these coordinates has the diagonal form

$$< \,,\, >|_\lambda = \sum_{i=1}^n \eta_{ii}(u)(du^i)^2, \quad \eta_{ii}(u) = \frac{1}{\lambda''(q^i)}. \tag{4.46}$$

3. The metric on M induced by the invariant Euclidean metric in a point λ where the polynomial $\lambda(p)$ has simple roots has the form

$$(\partial', \partial'')_\lambda = -\sum_{|\lambda|<\infty} \operatorname*{res}_{d\lambda=0} \frac{\partial'(\log\lambda(p)dp)\,\partial''(\log\lambda(p)dp)}{d\log\lambda(p)}. \tag{4.47}$$

Here ∂', ∂'', ∂''' are arbitrary tangent vectors on M in the point λ, the derivatives $\partial'(\lambda(p)dp)$ etc. are taken keeping $p = const$; $\lambda'(p)$ and $\lambda''(p)$ are the first and the second derivatives of the polynomial $\lambda(p)$ w.r.t. p. In other words, the formulae (4.43) - (4.44) mean that (1.65) is the LG superpotential for the Frobenius manifold (1.66) [42].

Proof. The first formula follows immediately from (1.66b) since the sum of residues of a meromorphic differential ω on the Riemann p-sphere vanishes:

$$\operatorname*{res}_{p=\infty}\omega + \sum \operatorname*{res}_{|\lambda|<\infty}\omega = 0. \tag{4.48}$$

Here we apply the residue theorem to the meromorphic differential

$$\omega = \frac{\partial'(\lambda(p)dp)\,\partial''(\lambda(p)dp)}{d\lambda(p)}.$$

From (4.48) it also follows that the formula (1.66a) can be rewritten as

$$c(\partial', \partial'', \partial''') = \operatorname*{res}_{p=\infty} \frac{\partial'(\lambda(p)dp)\, \partial''(\lambda(p)dp)\, \partial'''(\lambda(p)dp)}{dp\, d\lambda(p)}. \qquad (4.49)$$

Let

$$f(p) = \partial'(\lambda(p)), \quad g(p) = \partial''(\lambda(p)), \quad h(p) = \partial'''(\lambda(p)),$$

$$f(p)g(p) = q(p) + r(p)\lambda'(p)$$

for polynomials $q(p)$, $r(p)$, $\deg q(p) < n$. In the algebra $\mathbf{C}[p]/(\lambda'(p))$ we have then

$$f \cdot g = q.$$

On the other side, for the residue (4.49) we obtain

$$\operatorname*{res}_{p=\infty} \frac{\partial'(\lambda(p)dp)\, \partial''(\lambda(p)dp)\, \partial'''(\lambda(p)dp)}{dp\, d\lambda(p)} = \operatorname*{res}_{p=\infty} \frac{q(p)h(p)dp}{d\lambda(p)} + \operatorname*{res}_{p=\infty} r(p)h(p)dp.$$

The second residue in the r.h.s. of the formula equals zero while the first one coincides with the inner product $< q, h >_\lambda = < f \cdot g, h >_\lambda$.

Let us prove the second statement of Lemma. Let $\lambda(p)$ be a polynomial without multiple roots. Independence of the critical values $u^1, ..., u^n$ as functions of the polynomial is a standard fact (it also follows from the explicit formula (4.55) for the Jacobi matrix). Let us choose $\xi_1, ..., \xi_n$ as the coordinates on the hyperplane (4.41). These are not orthonormal: the matrix of the (contravariant) W-invariant metric in these coordinates has the form

$$g^{ab} = \delta^{ab} - \frac{1}{n+1}. \qquad (4.50)$$

We have

$$\lambda(p) = (p + \xi_1 + ... + \xi_n)\prod_{a=1}^{n}(p - \xi_a), \quad \lambda'(p) = \prod_{i=1}^{n}(p - q^i), \qquad (4.51)$$

$$\partial_i \lambda(p) = \frac{1}{p - q^i}\frac{\lambda'(p)}{\lambda''(q^i)}. \qquad (4.52)$$

The last one is the Lagrange interpolation formula since

$$\partial_i \lambda(p)|_{p=q^j} = \delta_{ij}. \qquad (4.53)$$

Substituting $p = \xi_a$ to the identity

$$(\partial_i \xi_1 + ... + \partial_i \xi_n)\prod_{b=1}^{n}(p - \xi_b) - \sum_{a=1}^{n}\frac{\lambda(p)}{p - \xi_a}\partial_i \xi_a = \partial_i \lambda(p) \qquad (4.54)$$

we obtain the formula for the Jacobi matrix

$$\partial_i \xi_a = -\frac{1}{(\xi_a - q^i)\lambda''(q^i)}, \quad i, a = 1, ..., n. \qquad (4.55)$$

Note that for a polynomial $\lambda(p)$ without multiple roots we have $\xi_a \neq q^i$, $\lambda''(q^i) \neq 0$. For the metric (4.43) from (4.53) we obtain

$$< \partial_i, \partial_j > = -\delta_{ij} \frac{1}{\lambda''(q^i)}. \tag{4.56}$$

For the tensor (4.44) for the same reasons only $c(\partial_i, \partial_i, \partial_i)$ could be nonzero and

$$c(\partial_i, \partial_i, \partial_i) \equiv < \partial_i \cdot \partial_i, \partial_i > = -\frac{1}{\lambda''(q^i)}. \tag{4.57}$$

Hence

$$\partial_i \cdot \partial_j = \delta_{ij} \partial_i \tag{4.58}$$

in the algebra (1.66).

To prove the last statement of the lemma we observe that the metric (4.47) also is diagonal in the coordinates u^1, ..., u^n with

$$g_{ii}(u) := (\partial_i, \partial_i) = -\frac{1}{u^i \lambda''(q^i)}. \tag{4.59}$$

The inner product of the gradients $(d\xi_a, d\xi_b)$ w.r.t. the metric (4.59) is

$$\sum_{i=1}^{n} \frac{1}{g_{ii}(u)} \frac{\partial \xi_a}{\partial u^i} \frac{\partial \xi_b}{\partial u^i} = -\sum_{i=1}^{n} \frac{u^i}{(\xi_a - q^i)(\xi_b - q^i)\lambda''(q^i)}$$

$$= -\sum_{i=1}^{n} \operatorname*{res}_{d\lambda=0} \frac{\lambda(p)}{(p - \xi_a)(p - \xi_b)\lambda'(p)} = \left[\operatorname*{res}_{p=\infty} + \operatorname*{res}_{p=\xi_a} + \operatorname*{res}_{p=\xi_b} \right] \frac{\lambda(p)}{(p - \xi_a)(p - \xi_b)\lambda'(p)}$$

$$= \delta^{ab} - \frac{1}{n+1}.$$

So the metric (4.47) coincides with the W-invariant Euclidean metric (4.7). Lemma is proved.

Exercise 4.1. Prove that the function

$$V(u) := -\frac{1}{2(n+1)} \left[\xi_0^2 + \ldots + \xi_n^2 \right] |_{\xi_0 + \ldots + \xi_n = 0} \tag{4.60}$$

is the potential for the metric (4.59):

$$\partial_i V(u) = \eta_{ii}(u).$$

Let us check that the curvature of the metric (1.66b) vanishes. I will construct explicitly the flat coordinates for the metric (cf. [42, 131]). Let us consider the function $p = p(\lambda)$ inverse to the polynomial $\lambda = \lambda(p)$. It can be expanded in a Puiseaux series as $\lambda \to \infty$

$$p = p(k) = k + \frac{1}{n+1} \left(\frac{t^n}{k} + \frac{t^{n-1}}{k^2} + \ldots + \frac{t^1}{k^n} \right) + O\left(\frac{1}{k^{n+1}} \right) \tag{4.61}$$

where $k := \lambda^{\frac{1}{n+1}}$, the coefficients

$$t^1 = t^1(a_1, \ldots, a_n), \ldots, t^n = t^n(a_1, \ldots, a_n) \tag{4.62}$$

are determined by this expansion. The inverse functions can be found from the identity

$$(p(k))^{n+1} + a_n(p(k))^{n-1} + \ldots + a_1 = k^{n+1}. \tag{4.63}$$

This gives a triangular change of coordinates of the form

$$a_i = -t^i + f_i(t^{i+1}, \ldots, t^n), \quad i = 1, \ldots, n. \tag{4.64}$$

So the coefficients t^1, ..., t^n can serve as global coordinates on the orbit space M (they give a distinguished basis of symmetric polynomials of $(n + 1)$ variables).

Exercise 4.2. Show the following formula [42, 131] for the coordinates t^α

$$t^\alpha = -\frac{n+1}{n-\alpha+1} \operatorname*{res}_{p=\infty} \left(\lambda^{\frac{n-\alpha+1}{n+1}} (p) \, dp \right). \tag{4.65}$$

Let us prove that the variables t^α are the flat coordinates for the metric (1.66b),

$$< \partial_\alpha, \partial_\beta >= \delta_{\alpha+\beta,n+1}. \tag{4.66}$$

To do this (and also in other proofs) we will use the following "thermodynamical identity".

Lemma 4.6. *Let $\lambda = \lambda(p, t^1, \ldots, t^n)$ and $p = p(\lambda, t^1, \ldots, t^n)$ be two mutually inverse functions depending on the parameters $t^1, \ldots, t^n$. Then*

$$\partial_\alpha(\lambda \, dp)_{p=const} = -\partial_\alpha(p \, d\lambda)_{\lambda=const}, \tag{4.67}$$

$\partial_\alpha = \partial/\partial t^\alpha$.

Proof. Differentiating the identity

$$\lambda(p(\lambda, t), t) \equiv \lambda$$

w.r.t. t^α we obtain

$$\frac{d\lambda}{dp} \partial_\alpha p(\lambda, t)_{\lambda=const} + \partial_\alpha \lambda(p(\lambda, t), t)_{p=const} = 0.$$

Lemma is proved.

Observe that $k = \lambda^{\frac{1}{n+1}}$ can be expanded as a Laurent series in $1/p$

$$k = p + O\left(\frac{1}{p}\right).$$

By $[\ \ ]_+$ I will denote the polynomial part of a Laurent series in $1/p$. For example, $[k]_+ = p$. Similarly, for a differential $f\,dk$ where f is a Laurent series in $1/p$ we put $[f\,dk]_+ := [f\,dk/dp]_+dp$.

Lemma 4.7. *The following formula holds true*

$$\partial_\alpha(\lambda\,dp)_{p=const} = -[k^{\alpha-1}dk]_+, \quad \alpha = 1,\ldots,n. \tag{4.68}$$

Proof. We have

$$-\partial_\alpha(\lambda\,dp)_{p=const} = \partial_\alpha(p\,d\lambda)_{\lambda=const} = \left(\frac{1}{n+1}\frac{1}{k^{n-\alpha+1}} + O\left(\frac{1}{k^{n+1}}\right)\right)dk^{n+1}$$

$$= k^{\alpha-1}dk + O\left(\frac{1}{k}\right)dk$$

since $k = const$ while $\lambda = const$. The very l.h.s. of this chain of equalities is a polynomial differential in p. And $[O(1/k)dk]_+ = 0$. Lemma is proved.

Corollary 4.6. *The variables t^1, ..., t^n are the flat coordinates for the metric (1.66b). The coefficients of the metric in these coordinates are*

$$\eta_{\alpha\beta} = \delta_{\alpha+\beta,n+1}. \tag{4.69}$$

Proof. From the previous lemma we have

$$\partial_\alpha(\lambda\,dp)_{p=const} = -k^{\alpha-1}dk + O(1/k)dk.$$

Substituting to the formula (4.43) we obtain

$$< \partial_\alpha, \partial_\beta >_\lambda = \operatorname*{res}_{p=\infty} \frac{k^{\alpha-1}dk\,k^{\beta-1}dk}{dk^{n+1}} = \frac{1}{n+1}\delta_{\alpha+\beta,n+1}$$

(the terms of the form $O(1/k)dk$ do not affect the residue). Corollary is proved.

Now we can easyly prove that the formulae of Example 1.7 describe a Frobenius structure on the space M of polynomials $\lambda(p)$. Indeed, the critical values of $\lambda(p)$ are the canonical coordinates u^i for the multiplication in the algebra of truncated polynomials $C/(\lambda'(p)) = T_\lambda M$. The metric (1.66b) is flat on M and it is diagonal in the canonical coordinates. From the flatness and from (4.60) it follows that this is a Darboux - Egoroff metric on M. From Lemma 4.5 we conclude that M with the structure (1.66) is a Frobenius manifold. It also follows that the correspondent intersection form coincides with the A_n-invariant metric on C^n. From the uniqueness part of Theorem 4.1 we conclude that the Frobenius structure (1.66) coincides (up to an equivalence) with the Frobenius structure of Theorem 4.1.

Remark 4.4. For the derivatives of the correspondent polynomial $F(t)$ in [42] the following formula was obtained

$$\partial_\alpha F = \frac{1}{(\alpha + 1)(n + \alpha + 2)} \operatorname*{res}_{p=\infty} \lambda^{\frac{n+\alpha+2}{n+1}} dp.$$

(4.70)

Other examples of polynomial solutions of WDVV associated with finite Coxeter groups can be found in [38, 165].

The constructions of this Lecture can be generalized for the case when the monodromy group is an extension of affine Weyl groups. The simplest solution of this type is given by the quantum multiplication on CP^1 (see Example G.2 above). We will not describe here the general construction (to be published elsewhere) but we will give two examples of it. In these examples one obtains three-dimensional Frobenius manifolds.

Exercise 4.3. Prove that

$$F = \frac{1}{2}t_1^2 t_3 + \frac{1}{2}t_1 t_2^2 - \frac{1}{24}t_2^4 + t_2 e^{t_3}$$

(4.71)

is a solution of WDVV with the Euler vector field

$$E = t_1\partial_1 + \frac{1}{2}t_2\partial_2 + \frac{3}{2}\partial_3.$$

(4.72)

Prove that the monodromy group of the Frobenius manifold coinsides with an extension of the affine Weyl group $\tilde{W}(A_2)$. Hint: Prove that the flat coordinates x, y, z of the intersection form are given by

$$t_1 = 2^{-\frac{1}{3}}e^{\frac{2}{3}z}\left[e^{x+y} + e^{-x} + e^{-y}\right]$$
$$t_2 = 2^{-\frac{2}{3}}e^{\frac{1}{3}z}\left[e^{-x-y} + e^{x} + e^{y}\right]$$
$$t_3 = z.$$

(4.73)

The intersection form in these coordinates is

$$ds^2 = -2(dx^2 + dxdy + dy^2) + \frac{2}{3}dz^2.$$

(4.74)

Exercise 4.4. Prove that

$$F = \frac{1}{2}t_1^2 t_3 + \frac{1}{2}t_1 t_2^2 - \frac{1}{48}t_2^4 + \frac{1}{4}t_2^2 e^{t_3} + \frac{1}{32}e^{2t_3}$$

(4.75)

is a solution of WDVV with the Euler vector field

$$E = t_1\partial_1 + \frac{1}{2}t_2\partial_2 + \partial_3.$$

(4.76)

Prove that the monodromy group of the Frobenius manifold is an extension of the affine Weyl group $\tilde{W}(B_2)$. Hint: Show that the flat coordinates of the intersection form are given by

$$t_1 = e^{2\pi i z}\left[\cos 2\pi x \cos 2\pi y + \frac{1}{2}\right]$$

$$t_2 = e^{\pi i z}\left[\cos 2\pi x + \cos 2\pi y\right] \qquad (4.77)$$

$$t_3 = 2\pi z.$$

The intersection form in these coordinates is proportional to

$$ds^2 = dx^2 + dy^2 - \frac{1}{2}dz^2.$$

Remark 4.5. The correspondent extension of the dual affine Weyl group $\tilde{W}(C_2)$ gives an equivalent Frobenius 3-manifold.

In the appendix to this Lecture we outline a generalization of our constructions to the case of extended complex crystallographic groups.

We obtain now an integral representation of the solution of the Riemann - Hilbert b.v.p. of Lecture 3 for the polynomial Frobenius manifolds on the space of orbits of a finite Coxeter group W.

Let us fix a system of n reflections $T_1, \ldots, T_n$ generating the group W (the order of the reflections also will be fixed). Via $e_1, \ldots, e_n$ I denote the normal vectors to the mirrors of the reflections normalized as

$$T_i(e_i) = -e_i, \qquad (4.78a)$$

$$(e_i, e_i) = 2, \quad i = 1, \ldots, n. \qquad (4.78b)$$

Let $x_1, \ldots, x_n$ be the coordinates in $\mathbf{R}^n$ w.r.t. the basis (4.78).

Let us consider the system of equations for the unknowns $x_1, \ldots, x_n$

$$y_1(x_1, \ldots, x_n) = y_1 - \lambda$$

$$y_2(x_1, \ldots, x_n) = y_2$$

$$\ldots \qquad (4.79)$$

$$y_n(x_1, \ldots, x_n) = y_n$$

where $y_i(x)$ are basic homogeneous W-invariant polynomials for the group W, deg $y_1 = h = \max$ (see above). Let

$$x_1 = x_1(y, \lambda) \qquad (4.80_1)$$

$$\ldots$$

$$x_n = x_n(y, \lambda) \qquad (4.80_n)$$

be the solution of this system (these are algebraic functions), $y = (y_1, \ldots, y_n) \in M = \mathbf{C}/W$. (Note that these are the basic solutions of the Gauss - Manin equations on the

Frobenius manifold.) By $\lambda = \lambda_1(x_1, y), \ldots, \lambda = \lambda_n(x_n, y)$ we denote the inverse functions to $(4.80_1), \ldots, (4.80_n)$ resp.

Proposition 4.2. *The functions*

$$h_a(y, z) := -z^{\frac{1}{k}} \int e^{z\lambda(x,y)} dx, \quad a = 1, \ldots, n \tag{4.81}$$

are flat coordinates of the deformed connection (3.3) on the Frobenius manifold $\mathbf{C}/W$. *Taking*

$$\psi_{ia}(y, z) := \frac{\partial_i h_a(u, z)}{\sqrt{\eta_{ii}(y)}}, \quad i = 1, \ldots, n \tag{4.82}$$

where $\partial_i = \partial/\partial u^i$, u_i *are the roots of (4.29), we obtain the solution of the Riemann - Hilbert problem of Lecture 3 for the Frobenius manifold.*

Proof. The formula (4.81) follows from (H.51). The second statement is the inversion of Theorem H.2. Proposition is proved.

Corollary 4.7. *The nonzero off-diagonal entries of the Stokes matrix of the Frobenius manifold* $\mathbf{C}/W$ *for a finite Coxeter group* W *coincide with the entries of the Coxeter matrix of* W.

So the Stokes matrix of the Frobenius manifolds is "a half" of the correspondent Coxeter matrix. For the simply-laced groups (i.e., the A - D - E series) this was obtained from physical considerations in [29].

To obtain the LG superpotential for the Frobenius manifold $\mathbf{C}/W$ we are to find the inverse function $\lambda = \lambda(p, y)$ to

$$p = x_1(y, \lambda) + \ldots + x_n(y, \lambda) \tag{4.83}$$

according to (I.10).

Example 4.3. We consider again the group A_n acting on the hyperplane (4.41) of the Euclidean space $\mathbf{R}^{n+1}$ with a standard basis $f_0, f_1, \ldots, f_n$. We chose the permutations

$$T_1 : \xi_0 \leftrightarrow \xi_1, \ldots, T_n : \xi_0 \leftrightarrow \xi_n \tag{4.84}$$

as the generators of the reflection group. The correspondent root basis (in the hyperplane (4.41)) is

$$e_1 = f_1 - f_0, \ldots, e_n = f_n - f_0. \tag{4.85}$$

The coordinates of a vector $\xi_0 f_0 + \xi_1 f_1 + \ldots + \xi_n f_n$ are

$$x_1 = \xi_1, \ldots, x_n = \xi_n.$$

Note that the sum

$$p = x_1 + \ldots + x_n = -\xi_0 \tag{4.86}$$

is one of the roots (up to a sign) of the equation

$$p^{n+1} + a_1 p^{n-1} + \ldots + a_n = 0.$$

The invariant polynomial of the highest degree is a_n. So to construct the LG superpotential we are to solve the equation

$$p^{n+1} + a_1 p^{n-1} + \ldots + a_n - \lambda = 0. \tag{4.87}$$

and then to invert it. It's clear that we obtain

$$\lambda = \lambda(p, a_1, \ldots, a_n) = p^{n+1} + a_1 p^{n-1} + \ldots + a_n. \tag{4.88}$$

We obtain a new proof of Lemma 4.5.

For other Coxeter groups W the above algorithm gives a universal construction of an analogue of the versal deformation of the correspondent simple singularity. But the calculations are more involved, and λ becomes an algebraic function of p.

Appendix J.
Extended complex crystallographic Coxeter groups
and twisted Frobenius manifolds.

Complex crystallographic groups were introduced by Bernstein and Schwarzman in [16] (implicitly they had been already used by Looijenga in [102]). These are the groups of affine transformations of a complex affine n-dimensional space V with the linear part generated by reflections. The very important subclass is *complex crystallographic Coxeter groups* (CCC groups briefly). In this case by definition V is the complexification of a real space $V_{\mathbf{R}}$; it is required that the linear parts of the transformations of a CCC group form a Coxeter group acting in the real linear space of translations of $V_{\mathbf{R}}$.

CCC groups are labelled by Weyl groups of simple Lie algebras. For any fixed Weyl group W the correspondent CCC group $\tilde{W}$ depends on a complex number τ in the upper half-plane as on the parameter. Certain factorization w.r.t. a discrete group of Möbius transformations of the upper half-plane that we denote by Γ_W must be done to identify equivalent CCC groups with the given Weyl group W. Bernstein and Schvarzman found also an analogue of the Chevalley theorem for CCC groups. They proved that for a fixed τ the space of orbits of a CCC group $\tilde{W}$ is a weighted projective space. The weights coincide with the markings on the extended Dynkin graph of W. Note that the discriminant locus (i.e. the set of nonregular orbits) depends on τ.

We have a natural fiber bundle over the quotient $\{Im\tau > 0\}/\Gamma_W$ with the fiber $V/\tilde{W}$. It turns out that the space of this bundle (after adding of one more coordinate, see below the precise construction) carries a natural structure of a twisted Frobenius manifold in the sense of Apendix B (above).

For the simply-laced case $W = A_l$, D_l, E_l the construction [16] of CCC groups is of special simplicity. The Weyl group W acts by integer linear transformations in the space $\mathbf{C}^l$ of the complexified root lattice $\mathbf{Z}^l$. This action preserves the lattice $\mathbf{Z}^l \oplus \tau \mathbf{Z}^l$. The CCC group $\tilde{W} = \tilde{W}(\tau)$ is the semidirect product of W and of the lattice $\mathbf{Z}^l \oplus \tau \mathbf{Z}^l$. The groups $\tilde{W}(\tau)$ and $\tilde{W}(\tau')$ are equivalent *iff*

$$\tau' = \frac{a\tau + b}{c\tau + d}, \quad \begin{pmatrix} a & b \\ c & d \end{pmatrix} \in SL(2, \mathbf{Z}).$$

The space of orbits $\mathbf{C}^l/\tilde{W}(\tau)$ can be obtained in two steps. First we can factorize over the translations $\mathbf{Z}^l \oplus \tau \mathbf{Z}^l$. We obtain the direct product of l copies of identical elliptic curves $E_\tau = \mathbf{C}/\{\mathbf{Z} \oplus \tau \mathbf{Z}\}$. The Weyl group W acts on E_τ^l. After factorization E_τ^l/W we obtain the space of orbits.

To construct the Frobenius structure we need a $\tilde{W}$-invariant metric on the space of the fiber bundle over the modular curve $M = \{Im\tau > 0\}/SL(2, \mathbf{Z})$ with the fiber $\mathbf{C}^l/\tilde{W}(\tau)$. Unfortunately, such a metric does not exist.

To resolve the problem we will consider a certain central extension of the group $\tilde{W}(\tau)$ acting in an extended space $\mathbf{C}^l \oplus \mathbf{C}$. The invariant metric we need lives on the extended space. The modular group acts by conformal transformations of the metric. Factorizing over the action of all these groups we obtain the twisted Frobenius manifold that corresponds to the given CCC group.

I will explain here the basic ideas of the construction for the simplest example of the CCC group $\tilde{A}_1$ leaving more general considerations for a separate publication. The Weyl group A_1 acts on the complex line $\mathbf{C}$ by reflections

$$v \mapsto -v. \qquad (J.1a)$$

The group $\tilde{A}_1(\tau)$ is the semidirect product of the reflections and of the translations

$$v \mapsto v + m + n\tau, \quad m, n \in \mathbf{Z}. \qquad (J.1b)$$

The quotient

$$\mathbf{C}/\tilde{A}_1(\tau) = E_\tau/\{\pm 1\} \qquad (J.2)$$

is the projective line. Indeed, the invariants of the group (J.1) are even elliptic functions on E_τ. It is well-known that any even elliptic function is a rational function of the Weierstrass $\wp$. This proves the Bernstein - Schwarzman's analogue of the Chevalley theorem for this very simple case.

Let us try to invent a metric on the space (v, τ) being invariant w.r.t. the transformations (J.1). We immediately see that the candidate dv^2 invariant w.r.t. the reflections does not help since under the transformations (J.1b)

$$dv^2 \mapsto (dv + n d\tau)^2 \neq dv^2.$$

The problem of constructing of an invariant metric can be solved by adding of one more auxiliary coordinate to the (v, τ)-space adjusting the transformation law of the new coordinate in order to preserve invariance of the metric. The following statement gives the solution of the problem.

Lemma J.1.

1. The metric

$$ds^2 := dv^2 + 2d\phi \, d\tau \qquad (J.3)$$

remains invariant under the transformations $v \mapsto -v$ and

$$\phi \mapsto \phi - nv - \frac{1}{2}n^2\tau + k$$
$$v \mapsto v + m + n\tau \qquad (J.4a)$$
$$\tau \mapsto \tau$$

for arbitrary m, n, k.

2. The formulae

$$\phi \mapsto \tilde{\phi} = \phi + \frac{1}{2}\frac{cv^2}{c\tau + d}$$
$$v \mapsto \tilde{v} = \frac{v}{c\tau + d} \qquad (J.4b)$$
$$\tau \mapsto \tilde{\tau} = \frac{a\tau + b}{c\tau + d}$$

with $ad - bc = 1$ *determine a conformal transformation of the metric* ds^2

$$d\tilde{v}^2 + 2d\tilde{\phi}\,d\tilde{\tau} = \frac{dv^2 + 2d\phi\,d\tau}{(c\tau + d)^2}. \tag{J.5}$$

Proof is in a simple calculation.

We denote by $\hat{A}_1$ the group generated by the reflection $v \mapsto -v$ and by the transformations (J.4) with integer m, n, k, a, b, c, d.

Observe that the subgroup of the translations (J.1b) in $\tilde{A}_1$ becomes non-commutative as the subgroup in $\hat{A}_1$.

The group generated by the transformations (J.4) with integer parameters is called *Jacobi group*. This name was proposed by Eichler and Zagier [62]. The automorphic forms of subgroups of a finite index of Jacobi group are called *Jacobi forms* [*ibid.*]. They were systematically studied in [62]. An analogue $\hat{W}$ of the group $\hat{A}_1$ (the transformations (J.4) together with the Weyl group $v \mapsto -v$) can be constructed for any CCC group $\tilde{W}$ taking the Killing form of W instead of the squares v^2, n^2 and nv. Invariants of these Jacobi groups were studied by Saito [134] and Wirthmüller [154] (see also the relevant papers [103, 85, 7]). Particularly, Saito constructed flat coordinates for the so-called "codimension 1" case. (This means that there exists a unique maximum among the markings of the extended Dynkin graph. On the list of our examples only the E-groups are of codimension one.) Explicit formulae of the Jacobi forms for $\hat{E}_6$ that are the flat coordinates in the sense of [134] have been obtained recently in [136]. Theory of Jacobi forms for all the Jacobi groups but $\hat{E}_8$ was constructed in [154].

Let $\mathbf{C}^3 = \{(\phi, v, \tau), Im\,\tau > 0\}$. I will show that the space of orbits

$$\mathcal{M}_{\hat{A}_1} := \mathbf{C}^3 / \hat{A}_1 \tag{J.6}$$

carries a natural structure of a twisted Frobenius manifold with the intersection form proportional to (J.3). It turns out that this coincides with the twisted Frobenius manifold of Appendix C. Furthermore, it will be shown that this twisted Frobenius manifold can be described by the LG superpotential

$$\lambda(p; \omega, \omega', c) := \wp(2\omega p; \omega, \omega') + c. \tag{J.7}$$

We will construct below the flat coordinates t^1, t^2, t^3 as functions of ω, ω', c.

The factorization over $\hat{A}_1$ will be done in two steps. First we construct a map

$$\mathbf{C}_0^3 = \{(\phi, v, \tau),\ Im\,\tau > 0,\ v \neq m + n\tau\} \to \mathbf{C}_0^3 = \{(z, \omega, \omega'),\ Im\frac{\omega'}{\omega} > 0,\ z \neq 2m\omega + 2n\omega'\} \tag{J.8}$$

such that the action of the group $\hat{A}_1$ transforms to the action of the group of translations

$$z \mapsto z + 2m\omega + 2n\omega' \tag{J.9a}$$

reflections

$$z \mapsto -z \qquad (J.9b)$$

and changes of the basis of the lattice

$$\begin{array}{l} \omega' \mapsto a\omega' + b\omega \\ \omega \mapsto c\omega' + d\omega \end{array}, \quad \begin{pmatrix} a & b \\ c & d \end{pmatrix} \in SL(2, \mathbf{Z}). \qquad (J.9c)$$

Note that the subgroup generated by (J.9) look very similar to the CCC group $\tilde{A}_1$ but the non-normalized lattice $\{2m\omega + 2n\omega'\}$ is involved in the construction of this subgroup.
We will use the Weierstrass σ-function

$$\sigma(z; \omega, \omega') = z \prod_{m^2 + n^2 \neq 0} \left\{ \left(1 - \frac{z}{w}\right) \exp\left[\frac{z}{w} + \frac{1}{2}\left(\frac{z}{w}\right)^2\right] \right\}, \qquad (J.10a)$$

$$w := 2m\omega + 2n\omega'$$

$$\frac{d}{dz} \log \sigma(z; \omega, \omega') = \zeta(z; \omega, \omega'). \qquad (J.10b)$$

It is not changed when changing the basis 2ω, $2\omega'$ of the lattice while for the translations (J.9a)

$$\sigma(z + 2m\omega + 2n\omega'; \omega, \omega')$$

$$= (-1)^{m+n+mn}\sigma(z; \omega, \omega') \exp\left[(z + m\omega + n\omega')(2m\eta + 2n\eta')\right] \qquad (J.11)$$

where $\eta = \eta(\omega, \omega')$, $\eta' = \eta'(\omega, \omega')$ are defined in (C.60)

Lemma J.2. *The equations*

$$\phi = \frac{1}{2\pi i}\left[\log \sigma(z; \omega, \omega') - \frac{\eta}{2\omega}z^2\right]$$

$$v = \frac{z}{2\omega} \qquad (J.12)$$

$$\tau = \frac{\omega'}{\omega}$$

determine a map (J.8). This map locally is a bi-holomorphic equivalence.
Proof. Expressing σ-function via the Jacobi theta-functions we obtain

$$\phi = \frac{1}{2\pi i}\left[\log 2\omega + \log \frac{\theta_1(v; \tau)}{\theta_1'(0; \tau)}\right].$$

This can be uniquely solved for ω

$$\omega = \frac{1}{2}\frac{\theta_1'(0; \tau)}{\theta_1(v; \tau)}e^{2\pi i\phi}. \qquad (J.13a)$$

Substituting to (J.12b) we obtain

$$z = v \frac{\theta_1'(0;\tau)}{\theta_1(v;\tau)} e^{2\pi i \phi} \qquad (J.13b)$$

$$\omega' = \frac{\tau}{2} \frac{\theta_1'(0;\tau)}{\theta_1(v;\tau)} e^{2\pi i \phi}. \qquad (J.13c)$$

Using the transformation law of σ-function and (J.4) we complete the proof of the lemma.

Corollary J.1. *The space of orbits $\mathbf{C}^3/\hat{A}_1$ coincides with the universal torus (C.51) factorized over the involution $z \mapsto -z$.*

Let us calculate the metric induced by (J.3) on the universal torus. To simplify this calculation I will use the following basis of vector fields on the universal torus (cf. Appendix C above)

$$D_1 = \omega \frac{\partial}{\partial \omega} + \omega' \frac{\partial}{\partial \omega'} + z \frac{\partial}{\partial z}$$

$$D_2 = \frac{\partial}{\partial z} \qquad (J.14)$$

$$D_3 = \eta \frac{\partial}{\partial \omega} + \eta' \frac{\partial}{\partial \omega'} + \zeta(z;\omega,\omega') \frac{\partial}{\partial z}.$$

As it follows from Lemma the inner products of these vector fields are functions on the universal torus (invariant elliptic functions in the terminology of Appendix C).

Proposition J.1. *In the basis (J.14) the metric $(D_a, D_b) =: g_{ab}$ has the form*

$$(g_{ab}) = \frac{1}{4\omega^2} \begin{pmatrix} 0 & 0 & -1 \\ 0 & 1 & 0 \\ -1 & 0 & -\wp(z;\omega,\omega') \end{pmatrix}. \qquad (J.15)$$

Proof. By the definition we have

$$(D_a, D_b) = D_a v D_b v + D_a \phi D_b \tau + D_a \tau D_b \phi. \qquad (J.16)$$

We have

$$D_1 \phi = \frac{1}{2\pi i}, \quad D_1 v = D_1 \tau = 0,$$

$$D_2 \phi = \frac{1}{2\pi i} \left[\zeta(z;\omega,\omega') - \frac{\eta}{\omega} z \right], \quad D_2 v = \frac{1}{2\omega}, \quad D_2 \tau = 0,$$

$$D_3 \phi = \frac{1}{4\pi i} \left[\left(\zeta(z;\omega,\omega') - \frac{\eta}{\omega} z \right)^2 + \wp(z;\omega,\omega') \right],$$

$$D_3 v = \frac{1}{2\omega} \left[\zeta(z;\omega,\omega') - \frac{\eta}{\omega} z \right], D_3 \tau = -\frac{\pi i}{2\omega^2}. \qquad (J.17)$$

In the derivation of these formulae we used the formulae of Appendix C and also one more formula of Frobenius and Stickelberger

$$\eta\frac{\partial\log\sigma}{\partial\omega} + \eta'\frac{\partial\log\sigma}{\partial\omega'} = -\frac{1}{2}\zeta(z)^2 + \frac{1}{2}\wp(z) - \frac{1}{24}g_2 z \qquad (J.18)$$

[67, formula 30.]. From (J.17) we easily obtain (J.15). Proposition is proved.

Remark J.1. The metric (J.15) on the universal torus has still its values in the line bundle ℓ. I recall (see above Appendix C) that ℓ is the pull-back of the tangent bundle of the modular curve w.r.t. the projection $(z,\omega,\omega') \mapsto \omega'/\omega$.

Remark J.2. The Euler vector field D_1 in the coordinates (ϕ, v, τ) has the form

$$D_1 = \frac{1}{2\pi i}\frac{\partial}{\partial\phi}. \qquad (J.19)$$

Let

$$\hat{\mathcal{M}} := \left\{ (z,\omega,\omega'),\ Im\frac{\omega'}{\omega} > 0,\ z \neq 0 \right\} /\{z \mapsto \pm z + 2m\omega + 2n\omega'\}. \qquad (J.20)$$

The metric (J.15) is well-defined om $\hat{\mathcal{M}}$. The next step is to construct a Frobenius structure on $\hat{\mathcal{M}}$.

The functions on $\hat{\mathcal{M}}$ are rational combinations of ω, ω' and $\wp(z;\omega,\omega')$. We define a grading in the ring of these functions with only a pole at $z = 0$ allowed putting

$$\deg\omega = \deg\omega' = -\frac{1}{2}, \quad \deg\wp = 1. \qquad (J.21a)$$

The correspondent Euler vector field for this grading is

$$E = -\frac{1}{2}D_1. \qquad (J.21b)$$

Note the relation of the grading to that defined by the vector field $\partial/\partial\phi$.

Proposition J.2. *There exists a unique Frobenius structure on the manifold $\hat{\mathcal{M}}$ with the intersection form $-4\pi i\,ds^2$ (where ds^2 is defined in (J.3)), the Euler vector field (J.21) and the unity vector field $e = \partial/\partial\varphi$. This Frobenius structure coincides with (C.87).*

Proof. I introduce coordinates t^1, t^2, t^3 on $\hat{\mathcal{M}}$ putting

$$t^1 = -\frac{1}{\pi i}\left[\wp(z;\omega,\omega') + \frac{\eta}{\omega}\right] = -\frac{1}{\pi i}\left[\frac{\theta_1''(v;\tau)\theta_1(v;\tau) - {\theta_1'}^2(v;\tau)}{{\theta_1'}^2(0;\tau)}\right]e^{-4\pi i\phi}$$

$$t^2 = \frac{1}{\omega} = 2\frac{\theta_1(v;\tau)}{\theta_1'(0;\tau)}e^{-2\pi i\phi} \qquad (J.22)$$

$$t^3 = \frac{\omega'}{\omega} = \tau.$$

Let us calculate the contravariant metric in these coordinates. To do this I use again the vector fields (J.14):

$$(dt^\alpha, dt^\beta) = g^{ab} D_a t^\alpha D_b t^\beta \tag{J.23}$$

where $g^{ab} = g^{ab}(z; \omega, \omega')$ is the inverse matrix to $-4\pi i g_{ab}$ (J.15)

$$(g^{ab}) = -\frac{\omega^2}{\pi i} \begin{pmatrix} \wp & 0 & -1 \\ 0 & 1 & 0 \\ -1 & 0 & 0 \end{pmatrix}. \tag{J.24}$$

Using the formulae of Apppendix C we obtain

$$D_1 t^1 = \frac{2}{\pi i}\left[\wp + \frac{\eta}{\omega}\right], \quad D_1 t^2 = -\frac{1}{\omega}, \quad D_1 t^3 = 0$$

$$D_2 t^1 = -\frac{1}{\pi i}\wp', \quad D_2 t^2 = D_2 t^3 = 0$$

$$D_3 t^1 = \frac{1}{\pi i}\left[2\wp^2 - \frac{1}{4}g_2 + \frac{\eta^2}{\omega^2}\right], \quad D_3 t^2 = -\frac{\eta}{\omega^2}, \quad D_3 t^3 = -\frac{\pi i}{2\omega^2}. \tag{J.25}$$

From this it follows that the matrix $g^{\alpha\beta} := (dt^\alpha, dt^\beta)$ has the form

$$g^{11} = \frac{i}{\pi^3 \omega^4}\left[\omega^6 g_3 - \eta\omega\,\omega^4 g_2 + 4(\eta\omega)^3\right],$$

$$g^{12} = \frac{3}{\pi^2 \omega^3}\left[(\eta\omega)^2 - \frac{1}{12}\omega^4 g_2\right], \quad g^{13} = \frac{i}{\pi}\left[\wp + \frac{\eta}{\omega}\right],$$

$$g^{22} = \frac{i}{\pi}\left[\wp - 2\frac{\eta}{\omega}\right], \quad g^{23} = \frac{1}{2\omega}, \quad g^{33} = 0. \tag{J.26}$$

Differentiating $g^{\alpha\beta}$ w.r.t. e we obtain a constant matrix

$$(\eta^{\alpha\beta}) = \left(-\frac{1}{\pi i}\frac{\partial g^{\alpha\beta}}{\partial \wp}\right) \equiv \left(\frac{\partial g^{\alpha\beta}}{\partial t^1}\right) = \begin{pmatrix} 0 & 0 & 1 \\ 0 & 1 & 0 \\ 1 & 0 & 0 \end{pmatrix}.$$

So t^1, t^2, t^3 are the flat coordinates.

The next step is to calculate the matrix

$$F^{\alpha\beta} := \frac{g^{\alpha\beta}}{\deg g^{\alpha\beta}} \tag{J.27}$$

and then to find the function F from the condition

$$\frac{\partial^2 F}{\partial t^\alpha \partial t^\beta} = \eta_{\alpha\alpha'}\eta_{\beta\beta'}F^{\alpha'\beta'}.$$

(We cannot do this for the F^{33} entry. But we know that F^{33} must be equal to $\partial^2 F/\partial t^{1^2}$.) It is straightforward to verify that the function F of the form (C.87) satisfies (J.27). Proposition is proved.

Remark J.3. Another way to prove the proposition is close to the proof of the main theorem of Lecture 4. We verify that the entries of the contravariant metric on the space of orbits $\hat{\mathcal{M}}$ (J.6) and the contravariant Christoffel symbols (3.25) of it are elliptic functions with at most second order pole in the point $z = 0$. From this it immediately follows that the metrics $g^{\alpha\beta}$ and $\partial g^{\alpha\beta}/\partial\wp$ form a flat pencil. This gives a Frobenius structure on $\hat{\mathcal{M}}$. The last step is to prove analyticity of the components of $g^{\alpha\beta}$ in the point $\omega'/\omega = i\infty$. This completes the alternative proof of the proposition.

Factorizing the Frobenius manifold we have obtained over the action of the modular group we obtain the twisted Frobenius structure on the space of orbits of the extended CCC group $\hat{A}_1$ coinciding with this of Appendix C. Observe that the flat coordinates t^α are not globally single-valued functions on the orbit space due to the transformation law (B.19) determining the structure of the twisted Frobenius manifold.

We will consider briefly the twisted Frobenius manifolds for the CCC groups $\hat{A}_n$ in Lecture 5. The twisted Frobenius structures for the groups $\hat{E}_6$ and $\hat{E}_7$ can be obtained using the results of [149] and [106].

Lecture 5.

Differential geometry of Hurwitz spaces.

Hurwitz spaces are the moduli spaces of Riemann surfaces of a given genus g with a given number of sheets $n + 1$. In other words, these are the moduli spaces of pairs (C, λ) where C is a smooth algebraic curve of the genus g and λ is a meromorphic function on C of the degree $n + 1$. Just this function realizes C as a $n + 1$-sheet covering over CP^1 (i.e. as a $n + 1$-sheet Riemann surface). I will consider Hurwitz spaces with an additional assumption that the type of ramification of C over the infinite point is fixed.

Our aim is to construct a structure of twisted Frobenius manifold on a Hurwitz space for any g and n. The idea is to take the function $\lambda = \lambda(p)$ on the Riemann surface $p \in C$ depending on the moduli of the Riemann surface as the superpotential in the sense of Appendix I. But we are to be more precise to specify what is the argument of the superpotential to be kept unchanged whith the differentiation along the moduli (see (I.11) - (I.13)).

I will construct first a Frobenius structure on a covering of these Hurwitz spaces corresponding to fixation of a symplectic basis of cycles in the homologies $H_1(C, \mathbf{Z})$. Factorization over the group $Sp(2g, \mathbf{Z})$ of changes of the basis gives us a twisted Frobenius structure on the Hurwitz space. It will carry a g-dimensional family of metrics being sections of certain bundle over the Hurwitz space.

We start with explicit description of the Hurwitz spaces.

Let $M = M_{g;n_0,\dots,n_m}$ be a moduli space of dimension

$$n = 2g + n_0 + \dots + n_m + 2m \tag{5.1}$$

of pairs

$$(C; \lambda) \in M_{g;n_0,\dots,n_m} \tag{5.2}$$

where C is a Riemann surface with marked meromorphic function

$$\lambda : C \to CP^1, \quad \lambda^{-1}(\infty) = \infty_0 \cup \dots \cup \infty_m \tag{5.3}$$

where $\infty_0, \dots, \infty_m$ are distinct points of the curve C. We require that the function λ has the degree $n_i + 1$ near the point ∞_i. (This is a connected manifold as it follows from [117].) We need the critical values of λ

$$u^j = \lambda(P_j), \ d\lambda|_{P_j} = 0, \ j = 1, \dots, n \tag{5.4}$$

(i.e. the ramification points of the Riemann surface (5.3)) to be local coordinates in open domains in $\hat{M}$ where

$$u^i \neq u^j \text{ for } i \neq j \tag{5.5}$$

(due to the Riemann existence theorem). Note that P_j in (5.4) are the branch points of the Riemann surface. Another assumption is that the one-dimensional affine group acts on $\hat{M}$ as

$$(C; \lambda) \mapsto (C; a\lambda + b) \tag{5.6a}$$

$$u^i \mapsto au^i + b, \ i = 1, \ldots, n. \tag{5.6b}$$

Example 5.1. For the case $g = 0$, $m = 0$, $n_0 = n$ the Hurwitz space consists of all the polynomials of the form

$$\lambda(p) = p^{n+1} + a_n p^{n-1} + \ldots + a_1, \ a_1, \ldots, a_n \in \mathbf{C}. \tag{5.7}$$

We are to supress the term with p^n in the polynomial $\lambda(p)$ to provide a possibility to coordinatize (5.7) by the n critical values of λ. The affine transformations $\lambda \mapsto a\lambda + b$ act on (5.7) as

$$p \mapsto a^{\frac{1}{n+1}} p, \ a_i \mapsto a_i a^{\frac{n-i+1}{n+2}} \text{ for } i > 1, \ a_1 \mapsto a a_1 + b. \tag{5.8}$$

Example 5.2. For $g = 0$, $m = n$, $n_0 = \ldots = n_m = 0$ the Hurwitz space consists of all rational functions of the form

$$\lambda(p) = p + \sum_{i=1}^{n} \frac{q_i}{p - p_i}. \tag{5.9}$$

The affine group $\lambda \mapsto a\lambda + b$ acts on (5.9) as

$$p \mapsto ap + b, \ p_i \mapsto p_i + \frac{b}{a}, \ q_i \mapsto aq_i. \tag{5.10}$$

Example 5.3. For a positive genus g, $m = 0$, $n_0 = 1$ the Hurwitz space consists of all hyperelliptic curves

$$\mu^2 = \prod_{j=1}^{2g+1} (\lambda - u^j).$$

The critical values $u^1, \ldots, u^{2g+1}$ of the projection $(\lambda, \mu) \mapsto \lambda$ are the local coordinates on the moduli space. Globally they are well-defined up to a permutation.

Example 5.4. $g > 0$, $m = 0$, $n := n_0 \geq q$. In this case the quotient of the Hurwitz space over the affine group (5.6) is obtained from the moduli space of *all* smooth algebraic curves of the genus g by fixation of a non-Weierstrass point $\infty_0 \in C$.

I describe first the basic idea of introducing of a Frobenius structure in the covering of a Hurwitz space. We define the multiplication of the tangent vector fields declaring that the ramification points $u^1, \ldots, u^n$ are the canonical coordinates for the multiplication, i.e.

$$\partial_i \cdot \partial_j = \delta_{ij} \partial_i \tag{5.11}$$

for

$$\partial_i := \frac{\partial}{\partial u^i}.$$

This definition works for Riemann surfaces with pairwise distinct ramification points of the minimal order two. Below we will extend the multiplication onto all the Hurwitz space.

The unity vector field e and the Euler vector field E are the generators of the action (5.6) of the affine group

$$e = \sum_{i=1}^{n} \partial_i, \quad E = \sum_{i=1}^{n} u^i \partial_i. \tag{5.12}$$

To complete the description of the Frobenius structure we are to describe admissible one-forms on the Hurwitz space. I recall that a one-form Ω on the manifold with a Frobenius algebra structure in the tangent planes is called admissible if the invariant inner product

$$< a, b >_\Omega := \Omega(a \cdot b) \tag{5.13}$$

(for any two vector-fields a and b) determines on the manifold a Frobenius structure (see Lecture 3).

Any quadratic differential Q on C holomorphic for $|\lambda| < \infty$ determines a one-form Ω_Q on the Hurwitz space by the formula

$$\Omega_Q := \sum_{i=1}^{n} du^i \operatorname*{res}_{P_i} \frac{Q}{d\lambda}. \tag{5.14}$$

A quadratic differential Q is called $d\lambda$-*divisible* if it has the form

$$Q = q d\lambda \tag{5.15}$$

where the differential q has no poles in the branch points of C. For a $d\lambda$-divisible quadratic differential Q the correspondent one-form $\Omega_Q = 0$. Using this observation we extend the construction of the one-form Ω_Q to *multivalued* quadratic differentials. By the definition this is a quadratic differential Q on the universal covering of the curve C such that the monodromy transformation along any cycle γ acts on Q by

$$Q \mapsto Q + q_\gamma d\lambda \tag{5.16}$$

for a differential q_λ. Multivalued quadratic differentials determine one-forms on the Hurwitz space by the same formula (5.14).

We go now to an appropriate covering of a Hurwitz space in order to describe multivalued quadratic differentials for which the one-forms (5.14) will define metrics on the Hurwitz space according to (5.13).

The covering $\hat{M} = \hat{M}_{g;n_0,\ldots,n_m}$ will consist of the sets

$$(C; \lambda; k_0, \ldots, k_m; a_1, \ldots, a_g, b_1, \ldots, b_g) \in \hat{M}_{g;n_0,\ldots,n_m} \tag{5.17}$$

with the same C, λ as above and with a marked symplectic basis $a_1, \ldots, a_g, b_1, \ldots, b_g \in H_1(C, \mathbf{Z})$, and marked branches $k_0, \ldots, k_m$ of roots of λ near $\infty_0, \ldots, \infty_m$ of the orders $n_0 + 1, \ldots, n_m + 1$ resp.,

$$k_i^{n_i+1}(P) = \lambda(P), \quad P \text{ near } \infty_i. \tag{5.18}$$

(This is still a connected manifold.)

The admissible quadratic differentials on the Hurwitz space will be constructed as squares $Q = \phi^2$ of certain differentials ϕ on C (or on a covering of C). I will call them *primary* differentials. I give now the list of primary differentials.

Type 1. One of the normalized Abelian differentials of the second kind on C with poles only at $\infty_0, \ldots, \infty_m$ of the orders less then the correspondent orders of the differential $d\lambda$. More explicitly,

$$\phi = \phi_{t^{i;\alpha}}, \quad i = 0, \ldots, m, \quad \alpha = 1, \ldots, n_i \tag{5.19a}$$

is the normalized Abelian differential of the second kind with a pole in ∞_i,

$$\phi_{t^{i;\alpha}} = -\frac{1}{\alpha} dk_i^\alpha + \text{regular terms} \quad \text{near } \infty_i, \tag{5.19b}$$

$$\oint_{a_j} \phi_{t^{i;\alpha}} = 0; \tag{5.19c}$$

Type 2.

$$\phi = \sum_{i=1}^m \delta_i \phi_{v^i} \text{ for } i = 1, \ldots, m \tag{5.20a}$$

with the coefficients $\delta_1, \ldots, \delta_i$ independent on the point of $\hat{M}$. Here ϕ_{v^i} is one of the normalized Abelian differentials of the second kind on C with a pole only at ∞_i with the principal part of the form

$$\phi_{v^i} = -d\lambda + \text{regular terms} \quad \text{near } \infty_i, \tag{5.20b}$$

$$\oint_{a_j} \phi_{v^i} = 0; \tag{5.20c}$$

Type 3.

$$\phi = \sum_{i=1}^m \alpha_i \phi_{w^i} \tag{5.21a}$$

$$\oint_{a_j} \phi = 0 \tag{5.21b}$$

with the coefficients $\alpha_1, \ldots, \alpha_m$ independent on the point of $\hat{M}$. Here ϕ_{w^i} is the normalized Abelian differential of the third kind with simple poles at ∞_0 and ∞_i with residues -1 and $+1$ resp.;

Type 4.

$$\phi = \sum_{i=1}^g \beta_i \phi_{r^i} \tag{5.22a}$$

with the coefficients $\beta_1, \ldots, \beta_g$ independent on the point of $\hat{M}$. Here ϕ_{r^i} is the normalized multivalued differential on C with increments along the cycles b_j of the form

$$\phi_{r^i}(P + b_j) - \phi_{r^i}(P) = -\delta_{ij} d\lambda, \tag{5.22b}$$

$$\oint_{a_j} \phi_{r^i} = 0 \tag{5.22c}$$

without singularities but those prescribed by (5.22b);

Type 5.

$$\phi = \sum_{i=1}^{g} \gamma_i \phi_{s^i} \tag{5.23a}$$

with the coefficients $\gamma_1, \ldots, \gamma_g$ independent on the point of $\hat{M}$. Here ϕ_{s^i} is the holomorphic differentials on C normalized by the condition

$$\oint_{a_j} \phi_{s^i} = \delta_{ij}. \tag{5.23b}$$

Exercise 5.1. Prove that the one-forms (5.14) where $Q =$ square of any of the primary differentials (5.19) - (5.23) span the cotangent space to $\hat{M}$ at any point with pairwise distinct ramification points.

Let ϕ be one of the primary differentials of the above list. We put

$$Q = \phi^2 \tag{5.24}$$

and we will show that the correspondent one-form Ω_Q on the Hurwitz space is admissible. This will give a Frobenius structure on the covering $\hat{M}$ of the Hurwitz space for any of the primary differentials. We recall that the metric correspondent to the one-form Ω_{ϕ^2} has by definition the form

$$< \partial', \partial'' >_\phi := \Omega_{\phi^2}(\partial' \cdot \partial'') \tag{5.25}$$

for any two tangent fields ∂', ∂'' on $\hat{M}$.

To construct the superpotential of this Frobenius structure we introduce a multivalued function p on C taking the integral of ϕ

$$p(P) := \text{v.p.} \int_{\infty_0}^{P} \phi. \tag{5.26}$$

The principal value is defined by the subtraction of the divergent part of the integral as the correspondent function on k_0. So

$$\phi = dp. \tag{5.27}$$

Now we can consider the function λ on C locally as the function $\lambda = \lambda(p)$ of the complex variable p. This function also depends on the point of the space $\hat{M}$ as on parameter.

302

Let $\hat{M}_\phi$ be the open domain in $\hat{M}$ specifying by the condition

$$\phi(P_i) \neq 0, \quad i = 1, \ldots, n. \tag{5.28}$$

Theorem 5.1.

1. *For any primary differential ϕ of the list (5.19) - (5.23) the multiplication (5.11), the unity and the Euler vector field (5.12), and the one-form Ω_{ϕ^2} determine on $\hat{M}_\phi$ a structure of Frobenius manifold. The correspondent flat coordinates t^A, $A = 1, \ldots, N$ consist of the five parts*

$$t^A = \left(t^{i;\alpha}, i = 0, \ldots, m, \alpha = 1, \ldots, n_i; \; p^i, q^i, i = 1, \ldots, m; \; r^i, s^i, i = 1, \ldots, g\right) \tag{5.29}$$

where

$$t^{i;\alpha} = \operatorname*{res}_{\infty_i} k_i^{-\alpha} p \, d\lambda, \quad i = 0, \ldots, m, \; \alpha = 1, \ldots, n_i; \tag{5.30a}$$

$$p^i = \text{v.p.} \int_{\infty_0}^{\infty_i} dp, \quad i = 1, \ldots, m; \tag{5.30b}$$

$$q^i = -\operatorname*{res}_{\infty_i} \lambda \, dp, \quad i = 1, \ldots, m; \tag{5.30c}$$

$$r^i = \oint_{b_i} dp, \tag{5.30d}$$

$$s^i = -\frac{1}{2\pi i} \oint_{a_i} \lambda \, dp, \quad i = 1, \ldots, g. \tag{5.30e}$$

The metric (5.22) in the coordinates has the following form

$$\eta_{t^{i;\alpha} t^{i;\beta}} = \frac{1}{n_i + 1} \delta_{ij} \delta_{\alpha+\beta, n_i+1} \tag{5.31a}$$

$$\eta_{v^i w^j} = \frac{1}{n_i + 1} \delta_{ij} \tag{5.31b}$$

$$\eta_{r^i s^j} = \frac{1}{2\pi i} \delta_{ij}, \tag{5.31c}$$

other components of the η vanish.

The function $\lambda = \lambda(p)$ is the superpotential of this Frobenius manifold in the sense of Appendix I.

2. *For any other primary differential φ the one-form Ω_φ is an admissible one-form on the Frobenius manifold.*

Proof. From (5.11) it follows that the metric (5.25) which we will denote also by ds_ϕ^2 is diagonal in the coordinates $u^1, \ldots, u^N$

$$ds_\phi^2 = \sum_{i=1}^N \eta_{ii}(u) du^{i^2} \tag{5.32a}$$

with

$$\eta_{ii}(u) = \operatorname*{res}_{P_i} \frac{\phi^2}{d\lambda}. \tag{5.32b}$$

We first prove that this is a Darboux - Egoroff metric for any primary differential.

Main lemma. *For any primary differential ϕ on the list (5.19) - (5.23) the metric (5.32) is a Darboux - Egoroff metric satisfying also the invariance conditions*

$$\mathcal{L}_e ds_\phi^2 = 0, \tag{5.33a}$$

$$\mathcal{L}_E ds_\phi^2 \text{ is proportional to } ds_\phi^2. \tag{5.33b}$$

The rotation coefficients of the metric do not depend on the choice of the primary differential ϕ.

Proof. We prove first some identity relating a bilinear combination of periods and principal parts at $\lambda = \infty$ of differentials on C as functions on the moduli to the residues in the branch points.

We introduce a local parameter z_a near the point ∞_a putting

$$z_a = k_a^{-1}. \tag{5.34}$$

Let $\omega^{(1)}$, $\omega^{(2)}$ be two differentials on the universal covering of $C \setminus (\infty_0 \cup \ldots \cup \infty_m)$ holomorphic outside of the infinity with the following properties at the infinite points

$$\omega^{(i)} = \sum_k c_{ka}^{(i)} z_a^k dz_a + d\sum_{k>0} r_{ka}^{(i)} \lambda^k \log \lambda, \quad P \to \infty_a \tag{5.35a}$$

$$\oint_{a_\alpha} \omega^{(i)} = A_\alpha^{(i)} \tag{5.35b}$$

$$\omega^{(i)}(P + a_\alpha) - \omega^{(i)}(P) = dp_\alpha^{(i)}(\lambda), \quad p_\alpha^{(i)}(\lambda) = \sum_{s>0} p_{s\alpha}^{(i)} \lambda^s \tag{5.35c}$$

$$\omega^{(i)}(P + b_\alpha) - \omega^{(i)}(P) = dq_\alpha^{(i)}(\lambda), \quad q_\alpha^{(i)}(\lambda) = \sum_{s>0} q_{s\alpha}^{(i)} \lambda^s, \tag{5.35d}$$

$i = 1, 2$, where $c_{ka}^{(i)}$, $r_{ka}^{(i)}$, $A_\alpha^{(i)}$, $p_{s\alpha}^{(i)}$, $q_{s\alpha}^{(i)}$ are some constants (i.e. independent on the curve). We introduce also a bilinear pairing of such differentials putting

$$< \omega^{(1)} \omega^{(2)} > :=$$

$$= -\sum_{a=0}^m \left[\sum_{k \geq 0} \frac{c_{-k-2,a}^{(1)}}{k+1} c_{k,a}^{(2)} + c_{-1,a} \text{v.p.} \int_{P_0}^{\infty_a} \omega^{(2)} + 2\pi i \text{v.p.} \int_{P_0}^{\infty_a} r_{k,a}^{(1)} \lambda^k \omega^{(2)} \right]$$

$$+ \frac{1}{2\pi i} \sum_{\alpha=1}^g \left[-\oint_{a_\alpha} q_\alpha^{(1)}(\lambda)\omega^{(2)} + \oint_{b_\alpha} p_\alpha^{(1)}(\lambda)\omega^{(2)} + A_\alpha^{(1)} \oint_{b_\alpha} \omega^{(2)} \right]. \tag{5.36}$$

The principal values, as above, are obtained by subtraction of the divergent parts of the integrals as the corresponding functions on $k_0, \ldots, k_m$; P_0 is a marked point of the curve C with

$$\lambda(P_0) = 0. \tag{5.37}$$

We will use also a natural connection in the tautological bundle

$$\begin{matrix} & C \\ \downarrow & \\ \hat{M} & \end{matrix}. \tag{5.38}$$

The connection is uniquely determined by the requirement that for the horizontal lifts of the vector fields ∂_i

$$\partial_i \lambda = 0. \tag{5.39}$$

Lemma 5.1. *The following identity holds*

$$\operatorname*{res}_{P_j} \frac{\omega^{(1)} \omega^{(2)}}{d\lambda} = \partial_j < \omega^{(1)} \omega^{(2)} >. \tag{5.40}$$

Proof. We realize the symplectic basis $a_1, \ldots, a_g$, $b_1, \ldots, b_g$ by oriented cycles passing through the point P_0. After cutting C along these cycles we obtain a $4g$-gon. We connect one of the vertices of the $4g$-gon (denoting this again by P_0) with the infinite points $\infty_0, \ldots, \infty_m$ by pairwise nonintersecting paths running inside of the $4g$-gon. Adding cuttings along these paths we obtain a domain $\tilde{C}$. We assume that the λ-images of all the cuttings do not depend on the moduli $u \in \hat{M}$. This can be done locally. Then we have an identity

$$\frac{1}{2\pi i} \oint_{\partial \tilde{C}} \left(\omega^{(1)}(P) \int_{P_0}^{P} \partial_j \omega^{(2)} \right) = - \operatorname*{res}_{P_j} \frac{\omega^{(i)} \omega^{(2)}}{d\lambda}. \tag{5.41}$$

After calculation of all the residues and of all contour integrals we obtain (5.40). Lemma is proved.

Corollary 5.1. *The pairing (5.36) of the differentials of the form (5.35) is symmetric up to an additive constant not depending on the moduli.*

Proof of Main Lemma. From Lemma 5.1 we obtain

$$\eta_{jj}(u) = \partial_j < \phi\phi >, \quad j = 1, \ldots, N. \tag{5.42}$$

From this we obtain the symmetry of the rotation coefficients of the metric (5.32). To prove the identity (3.70a) for the rotation coeffients let us consider the differential

$$\partial_i \partial_j \phi \int \partial_k \phi$$

for distinct i, j, k. It has poles only in the branch points P_i, P_j, P_k. The contour integral of the differential along $\partial \tilde{C}$ (see above the proof of Lemma 5.1) equals zero. Hence the sum of the residues vansish. This reads

$$\partial_j \sqrt{\eta_{ii}} \partial_k \sqrt{\eta_{ii}} + \partial_i \sqrt{\eta_{jj}} \partial_k \sqrt{\eta_{jj}} = \sqrt{\eta_{kk}} \partial_i \partial_j \sqrt{\eta_{kk}}.$$

This can be written in the form (3.70a) due to the symmetry $\gamma_{ji} = \gamma_{ij}$.

Let us prove now that the rotation coefficients do not depend on ϕ. Let φ be another primary differential. We consider the differential

$$\partial_i \phi \int \partial_j \varphi$$

for $i \neq j$. From vanishing of the sum of the residues we obtain

$$\sqrt{\eta_{jj}^\varphi} \partial_i \sqrt{\eta_{jj}^\phi} = \sqrt{\eta_{ii}^\phi} \partial_j \sqrt{\eta_{ii}^\varphi}.$$

Using the symmetry $\gamma_{ji} = \gamma_{ij}$ we immediately obtain that the rotation coefficients of the two metrics coincide.

Now we are to prove the identity (5.33a). Let us define an operator D_e on functions $f = f(P, u)$ by the formula

$$D_e f = \frac{\partial f}{\partial \lambda} + \partial_e f. \tag{5.43}$$

The operator D_e can be extended to differentials as the Lie derivative (i.e. requiring $dD_e = D_e d$). We have

$$D_e \phi = 0 \tag{5.44}$$

for any of the primary differentials ϕ. Indeed, from the definition of these differentials it folows that these do not change with the transformations

$$\lambda \mapsto \lambda + b, \quad u^j \mapsto u^j + b, \quad j = 1, \ldots, N.$$

Such invariance is equivalent to (3.70b). From (5.42) we immediately obtain

$$\partial_e \eta_{jj} = 0$$

for the metric (5.32). Note that this implies also (3.70b).

Doing in a similar way we prove also (5.33b). Introduce the operator

$$D_E := \lambda \frac{\partial}{\partial \lambda} + \partial_E \tag{5.45}$$

we have

$$D_E \phi = [\phi] \phi. \tag{5.46}$$

Here the numbers $[\phi]$ for the differentials (5.19) - (5.23) have the form

$$[\phi_{t^{i;\alpha}}] = \frac{\alpha}{n_i + 1}$$
$$[\phi_{v^i}] = 1$$
$$[\phi_{w^i}] = 0$$
$$[\phi_{r^i}] = 1$$
$$[\phi_{s^i}] = 0. \tag{5.47}$$

306

From this we obtain

$$\partial_E \eta_{ii}^\phi = (2[\phi] - 1)\eta_{ii}^\phi.$$ (5.48)

The equation (3.70c) also follows from (5.48). Lemma is proved.

So we have obtained for any primary differential ϕ a flat Darboux - Egoroff metric in an open domain of $\hat{M}$ being invariant w.r.t. the multiplication (5.11). It is easy to see that the unity vector field is covariantly constant w.r.t. the Levi-Cività connection for the metric. According to the results of Lecture 3, this determine a Frobenius structure on the domain in $\hat{M}$. Any other metric for another primary differential will be admissible for this Frobenius structure due to Lemma 5.1 and Proposition 3.6. So all the Frobenius structures are obtained one from another by the Legendre-type transformations (B.2).

The open domains $\hat{M}_\phi$ cover all the universal covering of the Hurwitz space under consideration.

We are to prove the formulae for the flat coordinates of the metric and to establish that $\lambda = \lambda(p)$ is the superpotential for this Frobenius manifold.

We start from the superpotential.

Lemma 5.2. *Derivatives of* $\lambda(p)dp$ *along the variables (5.19) - (5.23) have the form*

$$\partial_{t^{i;\alpha}} \lambda(p)dp = -\phi_{t^{i;\alpha}}$$
$$\partial_{v^i} \lambda(p)dp = -\phi_{v^i}$$
$$\partial_{w^i} \lambda(p)dp = -\phi_{w^i} \qquad , \qquad (5.49)$$
$$\partial_{r^i} \lambda(p)dp = -\phi_{r^i}$$
$$\partial_{s^i} \lambda(p)dp = -\phi_{s^i}$$

I recall that the differentiation in (5.49) is to be done with $p = const$.

Proof. Using the "thermodynamic identity" (4.67) we can rewrite any of the derivatives (5.49) as

$$\partial_{t^A} (\lambda(p)dp)_{p=const} = -\partial_{t^A} (p\,d\lambda)_{\lambda=const}.$$ (5.50)

The derivatives like

$$\partial_{t^A} p(\lambda)_{\lambda=const}$$

are holomorphic on the finite part of the curve C outside the branch points P_j. In the branch point these derivatives have simple poles. But the differential $d\lambda$ vanishes precisely at the branch points. So (5.50) is holomorphic everywhere.

Let us consider now behaviour of the derivatives (5.50) at the infinity. We have

$$p = \text{singular part} -$$

$$-(1 - \delta_{i0})v^i - \frac{1}{n_i + 1}\sum_{\alpha=1}^{n_i} t^{i;\alpha} k_i^{-(n_i-\alpha+1)} - \frac{1}{n_i+1}w^i k_i^{-(n_i+1)} + O\left(k_i^{-(n_i+2)}\right)$$ (5.51)

near the point ∞_i. The singular part by the construction does not depend on the moduli. Also we have

$$p(P + b_\alpha) - p(P) = r^\alpha, \quad \alpha = 1, \ldots, g.$$ (5.52)

For the differential $\omega := p\,d\lambda$ we will obtain from (5.51), (5.52) and from

$$d\lambda = (n_i + 1)k_i^{n_i}dk_i \ \text{ near } \infty_i \tag{5.53}$$

the following analytic properties

$$\omega = \text{singular part } - (1 - \delta_{i0})v^i d\lambda - \sum_{\alpha=1}^{n_i} t^{i;\alpha}k_i^{\alpha-1}dk_i - w^i\frac{dk_i}{k_i} + O\left(k_i^{-2}\right)dk_i \tag{5.54a}$$

$$\omega(P + b_\alpha) - \omega(P) = r^\alpha d\lambda \tag{5.54b}$$

$$\oint_{a_\alpha} \omega = -s^\alpha. \tag{5.54c}$$

Differentiating these formulae w.r.t. one of the variables (5.30) we obtain precisely one of the differentials with the analytic properties (5.19) - (5.23). This proves the lemma.

To complete the proof of Theorem we need to prove that (5.30) are the flat coordinates[*] for the metric $< \ , \ >_\phi$ and that the matrix of this metric in the coordinates (5.30) has the form (5.31).

Let t^A be one of the coordinates (5.30). We denote

$$\phi_A := -\partial_{t^A}\lambda\, dp. \tag{5.55}$$

By the definition and using Lemma 5.1 we obtain

$$< \partial_{t^A}, \partial_{t^B} >_\phi = \sum_{|\lambda|<\infty} \operatorname*{res}_{d\lambda=0} \frac{\phi_A\phi_B}{d\lambda} = \partial_e < \phi_A\,\phi_B >. \tag{5.56}$$

Note that in the formula (5.56) for $< \phi_A\,\phi_B >$ only the contribution of the second differential ϕ_B depends on the moduli. Let us define the coefficients $c_{ka}^{(A,\,B)}$, $A_\alpha^{(A,\,B)}$, $q_{1\alpha}^{(A,\,B)}$ for the differentials $\phi_{A,\,B}$ as in the formula (5.35). Note that by the construction of the primary differentials all the coefficients $r_{ka}^{(A,\,B)}$, $p_{s\alpha}^{(A,\,B)}$ vanish; the same is true for $q_{sa}^{(A,\,B)}$ for $s > 1$. From the equation $D_e\phi_B = 0$ (see (5.44) above) we obtain

$$\partial_e c_{ka}^{(B)} = \frac{k+1}{n_a+1}c_{k-n_a-1,a}^{(B)},$$

$$\partial_e \int_{P_0}^{\infty_a} \phi_B = \frac{c_{-n_a-2,a}^{(B)}}{n_a+1} + \left(\frac{\phi_B}{d\lambda}\right)_{P_0}$$

---

[*] The statement of my papers [48, 49] that these are global coordinates on the Hurwitz space is wrong. They are coordinates on $\hat{M}_\phi$ only. Changing ϕ we obtain a coordinate system of the type (5.30) in a neighbourhood of any point of the Hurwitz space.

$$\partial_e \oint_{a_\alpha} \lambda \phi_B = \oint_{a_\alpha} \phi_B,$$

$$\partial_e \oint_{b_\alpha} \phi_B = -\frac{\phi_B}{d\lambda}(P + b_\alpha) + \frac{\phi_B}{d\lambda}(P) \tag{5.57}$$

(this does not depend on the point P).

The proof of all of these formulae is essentially the same. I will prove for example the last one.

Let us represent locally the differential ϕ_B as

$$\phi_B = d\Phi(\lambda; u^1, \ldots, u^N).$$

We have for arbitrary a, b

$$\partial_e \int_a^b \phi_B = \partial_e \int_a^b d\Phi(\lambda; u^1, \ldots, u^N) = \frac{d}{d\epsilon} \int_a^b d\Phi(\lambda; u^1 + \epsilon, \ldots, u^N + \epsilon)|_{\epsilon=0}. \tag{5.58}$$

Doing the change of the variable $\lambda \mapsto \lambda - \epsilon$ we rewrite (5.58) as

$$\frac{d}{d\epsilon} \int_{a-\epsilon}^{b-\epsilon} d\Phi(\lambda + \epsilon; u^1 + \epsilon, \ldots, u^N + \epsilon)|_{\epsilon=0} = -\left(\frac{d\Phi(\lambda; u^1, \ldots, u^N)}{d\lambda} \right)_{\lambda=a}^{\lambda=b} \tag{5.59}$$

since

$$\frac{d}{d\epsilon} d\Phi(\lambda + \epsilon; u^1 + \epsilon, \ldots, u^N + \epsilon) \equiv D_e \phi_B = 0$$

(see (5.44) above). Using (5.59) for the contour integral we obtain (5.57).

From this and from (5.36) we obtain

$$\partial_e < \phi_A \, \phi_B > = -\sum_{a=0}^m \left(\frac{1}{n_a + 1} \sum_{k=0}^{n_a - 1} c^{(A)}_{-k-2,a} c^{(B)}_{k-n_a-1,a} \right) -$$

$$-\sum_{a=1}^m \frac{1}{n_a + 1} \left(c^{(A)}_{-1,a} c^{(B)}_{-n_a-2,a} + c^{(B)}_{-1,a} c^{(A)}_{-n_a-2,a} \right) - \frac{1}{2\pi i} \sum_{\alpha=1}^g \left(q^{(A)}_{1\alpha} A^{(B)}_\alpha + A^{(A)}_\alpha q^{(B)}_{1\alpha} \right). \tag{5.60}$$

The r.h.s. of this expression is a constant that can be easyly calculated using the explicit form of the differentials ϕ_A, ϕ_B. This gives (5.31).

Observe now that the structure functions $c_{ABC}(t)$ of the Frobenius manifold can be calculated as

$$c_{ABC}(t) = \sum_{i=1}^n \operatorname*{res}_{P_i} \frac{\phi_{t^A} \phi_{t^B} \phi_{t^C}}{d\lambda dp}. \tag{5.61}$$

Extension of the Frobenius structure on *all* the moduli space $\hat{M}$ is given by the condition that the differential

$$\frac{\phi_{t^A} \phi_{t^B} - c^C_{AB} \phi_{t^C} dp}{d\lambda} \tag{5.62}$$

is holomorphic for $|w| < \infty$. (The Frobenius algebra on $T_t M$ will be nilpotent for Riemann surfaces $\lambda : C \to CP^1$ with more than double branch points.) Theorem is proved.

Remark 5.1. Interesting algebraic-geometrical examples of solutions of equations of associativity (1.14) (not satisfying the scaling invariance) generalizing ours were constructed in [95]. In these examples M is a moduli space of Riemann surfaces of genus g with marked points, marked germs of local parameters near these points, and with a marked normalized Abelian differential of the second kind $d\lambda$ with poles at marked points and with fixed b-periods

$$\oint_{b_i} d\lambda = B_i, \quad i = 1, \ldots, g. \tag{5.63}$$

For $B_i = 0$ $d\lambda$ is a perfect differential of a function λ. So one obtains the above Frobenius structures on $M_{g;n_0,\ldots,n_m}$.

Exercise 5.2. Prove that the function F for the Frobenius structure constructed in Theorem has the form

$$F = -\frac{1}{2} < p\,d\lambda\, p\,d\lambda > \tag{5.64}$$

where the pairing $< \ >$ is defined in (5.36).

Exercise 5.3. Prove the formula

$$\partial_{t^A}\partial_{t^B} F = - < \phi_A\, \phi_B > . \tag{5.65}$$

Note, particularly, that for the t^A-variables of the fifth type (5.30e) the formula (5.65) reads

$$\partial_{s^\alpha}\partial_{s^\beta} F = -\tau_{\alpha\beta} \tag{5.66a}$$

where

$$\tau_{\alpha\beta} := \oint_{b_\beta} \phi_{s^\alpha} \tag{5.66b}$$

is the period matrix of holomorphic differentials on the curve C.

Remark 5.2. The formula (5.66) means that the Jacobians $J(C)$ of the curve C

$$J(C) := \mathbf{C}^g/\{m + \tau n\}, \quad m, n \in \mathbf{Z}^g$$

are Lagrangian manifolds for the symplectic structure

$$\sum_{\alpha=1}^{g} ds^\alpha \wedge dz_\alpha \tag{5.67}$$

where $z_1, \ldots, z_g$ are the natural coordinates on $J(C)$ (i.e., coming from the linear coordinates in $\mathbf{C}^g$). Indeed, the shifts

$$z \mapsto z + m + \tau n \tag{5.68}$$

preserve the symplectic form (5.67) due to (5.66). Conversely, if the shifts (5.68) preserve (5.67) then the matrix τ can be presented in the form (5.66a). So the representation

(5.66) is a manifestation of the phenomenon that the Jacobians are *complex Liouville tori* and the coordinates z_α, s^α are the *complex action-angle variables* on the tori. The origin of the symplectic structure (5.67) in the geometry of Hurwitz spaces is in realization of these spaces as the moduli spaces [93] of the algebraic-geometrical solutions of integrable hierarchies of the KdV type (see also the next Lecture).

In the recent paper [45] an interesting symplectic structure has been constructed on the fiber bundle of the intermediate Jacobians of a Calabi - Yau three-folds X over the moduli space of pairs (X, Ω), $\Omega \in H^{3,0}(X)$. The intermediate Jacobians are also complex Liouville tori (i.e., Lagrangian manifolds) although they are not Abelian varieties. Their period matrix thus also can be represented in the form (5.66a) for appropriate coordinates on the moduli space.

Remark 5.3. A part of the flat coordinates of the intersection form of the Frobenius manifold can be obtained by the formulae similar to (5.30b) - (5.30d) with the substitution $\lambda \mapsto \log \lambda$. Another part is given, instead of (5.30a), by the formula

$$\tilde{t}^a := p(Q_a), \quad a = 1, \ldots, n \tag{5.69a}$$

where

$$n + 1 := n_0 + 1 + n_1 + 1 + \ldots + n_m + 1$$

is the number of sheets of the Riemann surface $\lambda : C \to CP^1$, Q_0, Q_1, $\ldots$, Q_n are the zeroes of λ on C,

$$\lambda(Q_a) = 0. \tag{5.69b}$$

The last step in our construction is to factorize over the group $Sp(g, \mathbf{Z})$ of changes of the symplectic basis $a_1, \ldots, a_g$, $b_1, \ldots, b_g$

$$a_i \mapsto \sum_{j=1}^{g} (C_{ij} b_j + D_{ij} a_j)$$

$$b_i \mapsto \sum_{j=1}^{g} (A_{ij} b_j + B_{ij} a_j) \tag{5.70}$$

where the matrices $A = (A_{ij})$, $B = (B_{ij})$, $C = (C_{ij})$, $D = (D_{ij})$ are integer-valued matrices satisfying

$$\begin{pmatrix} A & B \\ C & D \end{pmatrix} \begin{pmatrix} 0 & -1 \\ 1 & 0 \end{pmatrix} \begin{pmatrix} A & B \\ C & D \end{pmatrix}^T = \begin{pmatrix} 0 & -1 \\ 1 & 0 \end{pmatrix} \tag{5.71}$$

and changes of the branches k_j of the roots of λ near ∞_j

$$k_j \mapsto e^{\frac{2\pi i l}{n_j + 1}} k_j, \quad l = 1, \ldots, n_j + 1. \tag{5.72}$$

We can calculate the transformation law of the primary differentials (5.19) - (5.23) with the transformations (5.70) - (5.72). This determines the transformation law of the metrics

(5.32). They transform like squares of the primary differentials. All this gives a messy picture on the quotient of the twisted Frobenius manifold coinciding with the Hurwitz space. An important simplification we have is that the multiplication law of tangent vectors to the Hurwitz space stays invariant w.r.t. the transformations.

The picture is simplified drastically if the genus of the curves equals zero. In this case the modular group (5.70) disappears and the action of the group of roots of unity (5.72) is very simple.

Exercise 5.4. Verify that the Frobenius structure of Theorem 5.1 on the Hurwitz space (5.7) for the primary differential $\phi = dp$ coincides with the structure of Example 1.7. Check also that the formulae (5.30a) for the flat coordinates in this case coincide with the formulae (4.61).

For the positive genus case the above transformation law of the invariant metrics splits into five blocks corresponding to the five types of the primary differentials (5.19) - (5.23). Let us consider only the invariant metrics and the correspondent Frobenius structures that correspond to the holomorphic primary differentials (the type five in the previous notations). Any of these structures for a given symplectic basis of cycles is parametrized by g constants $\gamma_1, \ldots, \gamma_g$ in such a way that the correspondent primary differential is

$$\phi = \sum_{i=1}^{g} \gamma_i \phi_{s^i}. \tag{5.73}$$

I recall that ϕ_{s^i} are the basic normalized holomorphic differentials on the curve C w.r.t. the given symplectic basis of cycles. The correspondent metric (5.32) we denote by $< \, , \, >_\gamma$. A change (5.70) of the symplectic basis of cycles determines the transformation law of the metric:

$$< \, , \, >_\gamma \mapsto < \, , \, >_{(C\tau + D)^{-1}\gamma}. \tag{5.74}$$

Here

$$\gamma = (\gamma_1, \ldots, \gamma_g)^T.$$

This is an example of twisted Frobenius manifold: g-parameter family of Frobenius structures on the Hurwitz space with the action (5.74) of the Siegel modular group (5.70), (5.71).

We stop now the general considerations postponing for further publications, and we consider examples.

Example 5.5. $M_{0;1,0}$. This is the space of rational functions of the form

$$\lambda = \frac{1}{2}p^2 + a + \frac{b}{p - c} \tag{5.75}$$

where a, b, c are arbitrary complex parameters. We take first dp as the basic primary differential. So $\lambda(p)$ is the LG superpotential (I.11). Using the definition (I.11) we immediately obtain for the metric

$$< \partial_a, \partial_a > = < \partial_b, \partial_c > = 1 \tag{5.76}$$

other components vanish, and for the trilinear tensor (I.12)

$$< \partial_a, \partial_a, \partial_a > = 1$$
$$< \partial_a, \partial_b, \partial_c > = 1$$
$$< \partial_b, \partial_b, \partial_b > = b^{-1} \qquad (5.77)$$
$$< \partial_b, \partial_c, \partial_c > = c$$
$$< \partial_c, \partial_c \partial_c > = b$$

otherwise zero. The free energy and the Euler operator read

$$F = \frac{1}{6}a^3 + abc + \frac{1}{2}b^2 \log b + \frac{1}{6}bc^3 - \frac{3}{4}b^2 \qquad (5.78)$$

(we add the quadratic term for a convenience later on)

$$E = a\partial_a + \frac{3}{2}b\partial_b + \frac{1}{2}c\partial_c. \qquad (5.79)$$

This is the solution of WDVV of the second type (i.e. $< e, e > \neq 0$).

To obtain a more interesting solution of WDVV let us do the Legendre transform S_b (in the notations of Appendix B). The new flat coordinates read

$$\hat{t}^a = c$$
$$\hat{t}^b = a + \frac{1}{2}c^2 \qquad (5.80)$$
$$\hat{t}^c = \log b.$$

The new free energy $\hat{F}$ is to be determined from the equations (B.2b). After simple calculations we obtain

$$\hat{F} = \frac{1}{2}t_1^2 t_3 + \frac{1}{2}t_1 t_2^2 - \frac{1}{24}t_2^4 + t_2 e^{t_3} \qquad (5.81a)$$

where

$$t_1 := \hat{t}^b$$
$$t_2 := \hat{t}^a \qquad (5.81b)$$
$$t_3 := \hat{t}^c.$$

The Euler vector field of the new solution is

$$E = t_1 \partial_1 + \frac{1}{2}t_2 \partial_2 + \frac{3}{2}\partial_3. \qquad (5.82)$$

So for the solution (5.81) of WDVV $d = 1$, $r = 3/2$. This is just the Frobenius manifold of Exercise 4.3. The correspondent primary differential is

$$d\hat{p} := \partial_b(\lambda dp) = \frac{dp}{p - c}.$$

From (5.26) we obtain

$$\hat{p} = \log \frac{p-c}{\sqrt{2}}.$$

Substituting to (5.75) we derive the superpotential for the Frobenius manifold (5.81)

$$\lambda = \lambda(\hat{p}) = e^{2\hat{p}} + t_2\sqrt{2}e^{\hat{p}} + t_1 + \frac{1}{\sqrt{2}}e^{t_3-\hat{p}}. \tag{5.84}$$

The intersection form of the Frobenius manifold (5.81) is given by the matrix

$$\left(g^{\alpha\beta}(t)\right) = \begin{pmatrix} 2t_2e^{t_3} & \frac{3}{2}e^{t_3} & t_1 \\ \frac{3}{2}e^{t_3} & t_1 - \frac{1}{2}t_2^2 & \frac{1}{2}t_2 \\ t_1 & \frac{1}{2}t_2 & \frac{3}{2} \end{pmatrix}. \tag{5.85}$$

The flat coordinates x, y, z of the intersection form are obtained using Remark 5.3 in the form (4.73) where

$$\hat{p} = -\frac{1}{6}\log 2 + \frac{z}{3} + x + (2m+1)\pi i \text{ and } \hat{p} = -\frac{1}{6}\log 2 + \frac{z}{3} + y + (2n+1)\pi i \tag{5.86}$$

are two of the roots of the equation $\lambda(\hat{p}) = 0$ (m, n are arbitrary integers). As in Exercise 4.3 the monodromy around the discriminant of the Frobenius manifold is an affine Weyl group (this time of the type $A_3^{(1)}$). It is generated, say, by the transformations

$$\begin{aligned} x &\mapsto y \\ y &\mapsto x \\ z &\mapsto z \end{aligned} \tag{5.87a}$$

and

$$\begin{aligned} x &\mapsto x \\ y &\mapsto -x - y \\ z &\mapsto z \end{aligned} \tag{5.87b}$$

and by translations

$$\begin{aligned} x &\mapsto x + 2m\pi i \\ y &\mapsto y + 2n\pi i \\ z &\mapsto z. \end{aligned} \tag{5.87c}$$

The monodromy around the obvious closed loop along t_3 gives an extension of the affine Weyl group by means of the transformation

$$\begin{aligned} x &\mapsto y - \frac{2\pi i}{3} \\ y &\mapsto -x - y - \frac{2\pi i}{3} \\ z &\mapsto z + 2\pi i. \end{aligned} \tag{5.87d}$$

314

This is the analogue of the gliding reflection (G.24b) since the cube of it is just a translation

$$x \mapsto x$$
$$y \mapsto y \qquad (5.87e)$$
$$z \mapsto z + 6\pi i.$$

We conclude that the monodromy group of the Frobenius manifold (5.81) coincides with the extension of the affine Weyl group of the type $A_2^{(1)}$ by means of the group of cubic roots of the translation (5.87e) (cf. Exercise 4.3 above).

Exercise 5.5. Using an appropriate generalization of the above example to the Hurwitz space $M_{0;n-1,0}$ show, that the monodromy group of the Frobenius manifold with the superpotential

$$\lambda(p) = e^{np} + a_1 e^{(n-1)p} + \ldots + a_n + a_{n+1} e^{-p} \qquad (5.88a)$$

$(a_1, \ldots, a_{n+1}$ are arbitrary complex numbers) is an extension of the affine Weyl group of the type $A_n^{(1)}$ by means of the group of roots of the order $n + 1$ of the translation

$$\log a_{n+1} \mapsto \log a_{n+1} + 2(n+1)\pi i. \qquad (5.88b)$$

Example 5.6. $M_{1;1}$. By the definition this is the space of elliptic curves of the form

$$\mu^2 = 4\lambda^3 + a_1 \lambda^2 + a_2 \lambda + a_3 \qquad (5.89)$$

with arbitrary coefficients a_1, a_2, a_3 providing that the polynomial in the r.h.s. of (5.89) has no multiple roots. I will show that in this case the twisted Frobenius structure of Theorem 5.1 coincides with the twisted Frobenius structure on the space of orbits of the extended CCC group $\hat{A}_1$ (see above Appendix C and Appendix J).

It is convenient to use elliptic uniformization of the curves. I will use the Weierstrass uniformization. For this I rewrite the curve in the form

$$\mu^2 = 4(\lambda - c)^2 - g_2(\lambda - c) - g_3 = 4(\lambda - c - e_1)(\lambda - c - e_2)(\lambda - c - e_3) \qquad (5.90)$$

where c and the parameters g_2, g_3 of the Weierstrass normal form are uniquely specified by a_1, a_2, a_3. The Weierstrass uniformization of (5.90) reads

$$\lambda = \wp(z) + c$$
$$\mu = \wp'(z) \qquad (5.91)$$

where $\wp = \wp(z; g_2, g_3)$ is the Weierstrass function. The infinite point is $z = 0$. Fixation of a basis of cycles on the curve (5.90) corresponds to fixation of an ordering of the roots e_1, e_2, e_3. The correspondent basis of the lattice of periods is 2ω, $2\omega'$ where

$$\wp(\omega) = e_1, \quad \wp(\omega') = e_3. \qquad (5.92)$$

Let us use the holomorphic primary differential

$$dp = \frac{dz}{2\omega}$$

to construct a Frobenius structure on M. The corresponding superpotential is

$$\lambda(p) := \wp(2\omega p; \omega, \omega') + c. \tag{5.93}$$

$$p \simeq p + m + n\tau. \tag{5.94}$$

The flat coordinates (5.30) read

$$t^1 := \frac{1}{\pi i} \oint_a \lambda \, dp = \frac{1}{2\pi i \omega} \int_0^{2\omega} [\wp(z; \omega, \omega') + c] \, dz = \frac{1}{\pi i} \left[-c + \frac{\eta}{\omega} \right] \tag{5.95a}$$

$$t^2 := - \operatorname*{res}_{z=0} \lambda^{-1/2} p \, d\lambda = 1/\omega \tag{5.95b}$$

$$t^3 := \oint_b dp = \tau \quad \text{where } \tau = \omega'/\omega. \tag{5.95c}$$

(I have changed slightly the normalization of the flat coordinates (5.30).) The metric $<\,,\,>_{dp}$ in these coordinates has the form

$$ds^2 = dt^{2^2} + 2dt^1 dt^3. \tag{5.96}$$

Changes of the basis of cycles on the elliptic curve determined by the action of $SL(2, \mathbf{Z})$

$$\begin{array}{l} \omega' \mapsto a\omega' + b\omega \\ \omega \mapsto c\omega' + d\omega \end{array} \quad \begin{pmatrix} a & b \\ c & d \end{pmatrix} \in SL(2, \mathbf{Z}) \tag{5.97}$$

gives rise to the following transformation of the metric

$$ds^2 \mapsto \frac{ds^2}{(c\tau + d)^2}. \tag{5.98}$$

These formulae determine a twisted Frobenius structure on the moduli space $M_{1;1}$ of elliptic curves (5.89).

The coincidence of the formulae for the flat coordinates and for the action (5.97) of the modular group with the above formulae in the constructions in the theory of the extended CCC group $\hat{A}_1$ is not accidental. We will prove now the following statement.

Theorem 5.2. *The space of orbits of $\hat{A}_1$ and the manifold $M_{1;1}$ are isomorphic as twisted Frobenius manifolds.*

Proof. Let us consider the intersection form of the Frobenius manifold $M_{1;1}$. Substituting

$$c = -\wp(z; \omega, \omega') \tag{5.99}$$

we obtain a flat metric on the $(z; \omega, \omega')$-space. Calculating this metric in the basis of the vector fields D_1, D_2, D_3 (J.14) we obtain the metric (J.24). It is clear that the unity and the Euler vector fields of $M_{1;1}$ coincide with $e = \partial/\partial\wp$ and (J.21) resp. So Theorem 5.1 implies the isomorphism of the Frobenius manifolds. The action of the modular group on these manifolds is given by the same formulae. Theorem is proved.

Example 5.7. We consider very briefly the Hurwitz space $M_{1;n}$ coinciding, as we will see, with the space of orbits of $\hat{A}_n$.

This is the moduli space of all elliptic curves

$$E_L := \mathbf{C}/L, \quad L = \{2m\omega + 2n\omega'\} \tag{5.100}$$

with a marked meromorphic function of degree $n + 1$ with only one pole. We can assume this pole to coincide with $z = 0$.

To realize the space of orbits of the extended CCC group $\hat{A}_n$ we consider the family of the Abelian manifolds E_L^{n+1} fibered over the space $\mathcal{L}$ of all lattices L in $\mathbf{C}$. The symmetric group acts on the space of the fiber bundle by permutations $(z_0, z_1, \ldots, z_n) \mapsto (z_{i_0}, z_{i_1}, \ldots, z_{i_n})$ (we denote by z_k the coordinate on the k-th copy of the elliptic curve). We must restrict this onto the hyperplane

$$z_0 + z_1 + \ldots + z_n = 0 \tag{5.101}$$

After factorization over the permuations we obtain the space of orbits of $\hat{A}_n$.

Note that the conformal invariant metric for $\hat{A}_n$ on the space $(z_0, z_1, \ldots, z_n; \omega, \omega')$ can be obtained as the restriction of the direct sum of the invariant metrics (J.15) onto the hyperplane (5.101).

We assign now to any point of the space of orbits of $\hat{A}_n$ a point (E_L, λ) in the Hurwitz space $M_{1;n}$ where E_L is the same elliptic curve and the function λ is defined by the formula

$$\lambda := (-1)^{n-1} \frac{\prod_{k=0}^{n} \sigma(z - z_k)}{\sigma^{n+1}(z) \prod_{k=0}^{n} \sigma(z_k)} = \frac{1}{n!} \frac{\det \begin{pmatrix} 1 & \wp(z) & \wp'(z) & \ldots & \wp^{(n-1)}(z) \\ 1 & \wp(z_1) & \wp'(z_1) & \ldots & \wp^{(n-1)}(z_1) \\ \vdots & \vdots & \vdots & & \vdots \\ 1 & \wp(z_n) & \wp'(z_n) & \ldots & \wp^{(n-1)}(z_n) \end{pmatrix}}{\det \begin{pmatrix} 1 & \wp(z_1) & \wp'(z_1) & \ldots & \wp^{(n-2)}(z_1) \\ 1 & \wp(z_2) & \wp'(z_2) & \ldots & \wp^{(n-2)}(z_2) \\ \vdots & \vdots & \vdots & & \vdots \\ 1 & \wp(z_n) & \wp'(z_n) & \ldots & \wp^{(n-2)}(z_n) \end{pmatrix}}$$

$$= \frac{(-1)^{n-1}}{n!} \wp^{(n-1)}(z) + c_n(z_0, z_1, \ldots z_n)\wp^{(n-2)}(z) + \ldots + c_1(z_0, z_1, \ldots, z_n) \tag{5.102}$$

where the coefficients $c_1, \ldots, c_n$ are defined by this equation (I have used the classical addition formula [153, p.458] for sigma-functions). Taking, as above, the normalized holomorphic differential

$$dp := \frac{dz}{2\omega}$$

we obtain the twisted Frobenius structure on the space of functions (5.102). By the construction, the superpotential of this Frobenius manifold has the form (5.102) where one must substitute

$$z \mapsto 2\omega p.$$

It is easy to check that the intersection form of this Frobenius structure coincides with the conformal invariant metric of $\hat{A}_n$. Indeed, the flat coordinates (5.69) of the intersection form are the zeroes of $\lambda(p)$

$$\tilde{t}^0 := \frac{z_0}{2\omega}, \dots, \tilde{t}^n := \frac{z_n}{2\omega} \tag{5.103a}$$

related by the linear constraint

$$\tilde{t}^0 + \dots + \tilde{t}^n = 0. \tag{5.103b}$$

These coordinates are not well-defined. The ambiguity in their definition comes from two origins. The first one is the monodromy group of the Riemann surface $\lambda : C \to CP^1$. It acts by permutations on the set of the zeroes of the function λ. Another origin of the ambiguity is in the multivaluedness of the function p (Abelian integral) on the Riemann surface. So the translations of the flat coordinates

$$\tilde{t}^a \mapsto m + n\tau, \quad m, n \in \mathbf{Z} \tag{5.104}$$

give the same point of the Hurwitz space. The permutations and the translations just give the action of the CCC group $\tilde{A}_n$. To see the extended CCC group $\hat{A}_n$ one should consider all the flat coordinates of the intersection form. They are obtained by adding two more coordinates (5.30d) and (5.30e) (with the substitution $\lambda \mapsto \log \lambda$) to (5.103)

$$\tau = \frac{\omega'}{\omega} = \oint_b dp, \quad \phi := -\frac{1}{2\pi i} \oint_a \log \lambda \, dp \tag{5.105}$$

(cf. (J.12)).

For the general case $g > 1$ on the twisted Fobenius manifold $M_{g;n_0,\dots,n_m}$ with the g-dimensional family $< \,,\, >_\gamma$ (see (5.73) above) of the metrics corresponding to the holomorphic primary differentials the ambiguity in the definition of the flat coordinates (5.69) can be described by an action of the semidirect product of the monodromy group of the Riemann surfaces by the lattice of periods of holomorphic differentials. This can be considered as a higher genus generalization of CCC groups.

Our construction even for $g = 0$ does not cover the Frobenius structures of Lecture 4 on the spaces of orbits of finite Coxeter groups besides A_n. To include this class of examples into the general scheme of geometry on Hurwitz spaces and also to cover the orbit spaces of the extended CCC groups but $\hat{A}_n$ we are to consider equivariant Hurwitz spaces. These by definition consist of the pairs (C, λ) as above where a finite group acts on C preserving invariant the function λ. We are going to do it in further publications.

Lecture 6.

Frobenius manifolds and integrable hierarchies.

Coupling to topological gravity.

We start with explanation of the following observation: all the examples of Frobenius manifolds constructed in Lecture 5 are finite dimensional invariant manifolds of integrable hierarchies of the KdV type. To explain this relation we are to explain briefly the notion of semi-classical limit of an integrable hierarchy. In physical language the semi-classical (more particularly, the dispersionless) limit will correspond to the description of the quantum field theory after coupling to topological gravity considered in the genus zero (i.e., the tree-level) approximation.

Let

$$\partial_{t_k} y^a = f_k^a(y, \partial_x y, \partial_x^2 y, \ldots), \quad a = 1, \ldots, l, \quad k = 0, 1, \ldots \tag{6.1}$$

be a commutative hierarchy of Hamiltonian integrable systems of the KdV type. "Hierarchy" means that the systems are ordered, say, by action of a recursion operator. Number of recursions determine a level of a system in the hierarchy. Systems of the level zero form a primary part of the hierarchy (these correspond to the primary operators in TFT); others can be obtained from the primaries by recursions.

Example 6.1. The $nKdV$ (or Gelfand - Dickey) hierarchy is an infinite system of commuting evolutionary PDEs for functions $a_1(x), \ldots, a_n(x)$. To construct the equations of the hierarchy we consider the operator

$$L = \partial^{n+1} + a_1(x)\partial^{n-1} + \ldots + a_n(x), \tag{6.2}$$

$$\partial = d/dx.$$

For any pair

$$(\alpha, p), \quad \alpha = 1, \ldots, n, \quad p = 0, 1, \ldots \tag{6.3}$$

we consider the evolutionary system of the Lax form

$$\partial_{t^{\alpha,p}} L = \left[L, \left[L^{\frac{\alpha}{n+1}+p} \right]_+ \right]. \tag{6.4}$$

The brackets $[\,,\,]$ stand for commutator of the operators, $\left[L^{\frac{\alpha}{n+1}+p} \right]_+$ denotes the differential part of the pseudodifferential operator $L^{\frac{\alpha}{n+1}+p}$. The commutator in the r.h.s. is an ordinary differential operator (in x) of the order at most $n-1$. So (6.4) is a system of PDE for the functions $a_1(x, t), \ldots, a_n(x, t)$, the time variable is $t = t^{\alpha,p}$. For example, for $\alpha = 1$, $p = 0$ we have $\left[L^{\frac{\alpha}{n+1}+p} \right]_+ = \partial$, so the correspondent PDE is the x-translations

$$\partial_{t^{1,0}} a_i(x) + \partial_x a_i(x) = 0. \tag{6.5}$$

The equations (6.4) have a bihamiltonian structure: they can be represented in the form

$$\partial_{t^{\alpha,p}} a_i(x) = \{a_i(x), H_{\alpha,p}\} = \{a_i(x), H_{\alpha,p-1}\}_1 \tag{6.6}$$

for some family of local Hamiltonians $H_{\alpha,p} = H_{\alpha,p}[a(x)]$ and w.r.t. a pair of Poisson brackets $\{\ ,\ \}$ and $\{\ ,\ \}_1$ on an appropriate space of functionals of $a_i(x)$. There is no symmetry between the Poisson brackets: the Casimirs (i.e., generators of the annihilator) of the *first* Poisson bracket $\{\ ,\ \}$ only are local functionals $H_{\alpha,-1}[a(x)]$. For the example of the KdV hierarchy $(n = 1)$ the two Poisson brackets are

$$\{u(x), u(y)\} = \delta'(x - y) \tag{6.7a}$$

$$\{u(x), u(y)\}_1 = -\frac{1}{2}\delta'''(x - y) + 2u(x)\delta'(x - y) + u'(x)\delta(x - y). \tag{6.7b}$$

Here $u(x) = -a_1(x)$ in the previous notations. The annihilator of the first Poisson bracket is generated by the local Casimir

$$H_{-1} = \int u(x)\,dx. \tag{6.8}$$

The Poisson brackets satisfy a very important property of *compatibility*: any linear combination

$$\{\ ,\ \}_1 - \lambda\{\ ,\ \} \tag{6.9}$$

for an arbitrary parameter λ is again a Poisson bracket. This gives a possibility to construct an infinite sequence of the commuting Hamiltonians $H_{\alpha,p}$ starting from the Casimirs $H_{\alpha,-1}$, $\alpha = 1, \ldots, n$ of the first Poisson bracket $\{\ ,\ \}$ using (6.6) as the recursion relations (see [108] for details). For any *primary* Hamiltonian $H_{\alpha,-1}$ (a Casimir of the first Poisson structure) we obtain an infinite chain of its *descendants* $H_{\alpha,p}$, $p \geq 0$ determined from the recursion relations

$$\{a_i(x), H_{\alpha,p+1}\} = \{a_i(x), H_{\alpha,p}\}_1, \quad p = -1, 0, 1, 2, \ldots. \tag{6.10}$$

They all are local functionals (i.e. integrals of polynomials of $a_i(x)$ and of their derivatives).

The hierarchy posesses a rich family of finite-dimensional invariant manifolds. Some of them can be found in a straightforward way; one needs to apply algebraic geometry methods [56] to construct more wide class of invariant manifolds. Any of these manifolds after an extension to complex domain turns out to be fibered over some base M (a complex manifold of some dimension n) with m-dimensional tori as the fibers (common invariant tori of the hierarchy). For $m = 0$ M is nothing but the family of common stationary points of the hierarchy. For any $m \geq 0$ M is a moduli space of Riemann surfaces of some genus g with certain additional structures: marked points and a marked meromorphic function with poles of a prescribed order in these points [93]. This is just a Hurwitz space considered in Lecture 5. Therefore they are the families of parameters of the finite-gap ("g-gap") solutions of the hierarchy. Our main observation is that any such M carries a natural structure of a Frobenius manifold.

Example 6.2. For $nKdV$ the family of stationary solutions of the hierarchy consists of all operators L with constant coefficients (any two such operators commute pairwise). This coincides with Frobenius manifold of Example 1.7.

Example 6.3. For the same $nKdV$ the family of "g-gap" solutions is the Hurwitz space $M_{g,n}$ in the notations of Lecture 5.

To give an idea how an integrable Hamiltonian hierarchy of the above form induces tensors $c_{\alpha\beta}^{\gamma}$, $\eta_{\alpha\beta}$ on a finite dimensional invariant manifold M I need to introduce the notion of semiclassical limit of a hierarchy near a family M of invariant tori (sometimes it is called also a *dispersionless limit* or *Whitham averaging* of the hierarchy; see details in [57, 58]. In the simplest case of the family of stationary solutions the semiclassical limit is defined as follows: one should substitute in the equations of the hierarchy

$$x \mapsto \epsilon x = X, \ t_k \mapsto \epsilon t_k = T_k \tag{6.11}$$

and tend ϵ to zero. For more general M (family of invariant tori) one should add averaging over the tori. As a result one obtains a new integrable Hamiltonian hierarchy where the dependent variables are coordinates $v^1, ..., v^n$ on M and the independent variables are the slow variables X and $T_0, T_1, ...$. This new hierarchy always has a form of a quasilinear system of PDE of the first order

$$\partial_{T_k} v^p = c_{kq}^p(v)\partial_X v^q, \ k = 0,1,... \tag{6.12}$$

for some matrices of coefficients $c_{kq}^p(v)$. One can keep in mind the simplest example of a semiclassical limit (just the dispersionless limit) of the KdV hierarchy. Here M is the one-dimensional family of constant solutions of the KdV hierarchy. For example, rescaling the KdV one obtains

$$u_T = uu_X + \epsilon^2 u_{XXX} \tag{6.13}$$

(KdV with small dispersion). After $\epsilon \to 0$ one obtains

$$u_T = uu_X. \tag{6.14}$$

The semiclassical limit of all the KdV hierarchy has the form

$$\partial_{T_k} u = \frac{u^k}{k!}\partial_X u, \ k = 0,1,.... \tag{6.15}$$

A semiclassical limit of spatialy discretized hierarchies (like Toda system) is obtained by a similar way. It is still a system of quasilinear PDE of the first order.

It is important to note that the commutation representation (6.4) in the semiclassical limit takes the form

$$\partial_{T^{\alpha,p}} \lambda(X,p) = \{\lambda(X,p), \rho_{\alpha,p}(X,p)\}. \tag{6.16}$$

In the r.h.s. $\{\ ,\ \}$ stands for the standard Poisson bracket on the (X,p)-plane

$$\{\lambda(x,p), \rho(x,p)\} = \frac{\partial\lambda}{\partial p}\frac{\partial\rho}{\partial x} - \frac{\partial\rho}{\partial p}\frac{\partial\lambda}{\partial x}. \tag{6.17}$$

We will call (6.16), (6.17) *semiclassical Lax representation*. For the dispersionless limit the function $\lambda(x,p)$ is just the symbol of the L-operator obtained by the substitution

$d/dx \to p$. The function $\rho(x,p)$ can be computed using fractional powers as in (6.4). For the case of the semiclassical limits on a family of oscillating finite-gap solutions the construction of the functions λ and ρ is more complicated (roughly speaking, $\lambda(X,p)$ is the Bloch dispersion law, i.e. the dependence of the eigenvalue λ of the operator L with periodic or quasiperiodic coefficients on the quasimomentum p and on the slow spatial variable $X = \epsilon x$).

Let us come back to determination of tensors $\eta_{\alpha\beta}$, $c_{\alpha\beta}^\gamma$ on M. Let $v_1, \ldots, v^n$ be arbitrary coordinates on M. A semiclassical limit (or "averaging") of both the Hamiltonian structures in the sense of general construction of S.P.Novikov and the author induces a compatible pair of Hamiltonian structures of the semiclassical hierarchy: a pair of Poisson brackets of the form

$$\{v^p(X), v^q(Y)\}_{\text{semiclassical}} = \eta^{ps}(v(X))[\delta_s^q \partial_X \delta(X-Y) - \gamma_{sr}^q(v)v_X^r \delta(X-Y)] \quad (6.18a)$$

$$\{v^p(X), v^q(Y)\}_{\text{semiclassical}}^1 = g^{ps}(v(X))[\delta_s^q \partial_X \delta(X-Y) - \Gamma_{sr}^q(v)v_X^r \delta(X-Y)] \quad (6.18b)$$

where $\eta^{pq}(v)$ and $g^{pq}(v)$ are contravariant components of two metrics on M and $\gamma_{pr}^q(v)$ and $\Gamma_{pr}^q(v)$ are the Christoffel symbols of the corresponding Levi-Cività connections for the metrics $\eta^{pq}(v)$ and $g^{pq}(v)$ resp. (the so-called *Poisson brackets of hydrodynamic type*). Observe that the metric $\eta_{\alpha\beta}$ is obtained from the semiclassical limit of the first Hamiltonian structure of the original hierarchy. From the general theory of Poisson brackets of hydrodynamic type [57, 58] one concludes that both the metrics on M have zero curvature. [In fact, from the compatibility of the Poisson brackets it follows that the metrics $\eta^{pq}(v)$ and $g^{pq}(v)$ form a flat pencil in the sense of Lecture 3.] So local flat coordinates $t^1, \ldots, t^n$ on M exist such that the metric $\eta^{pq}(v)$ in this coordinates is constant

$$\frac{\partial t^\alpha}{\partial v^p} \frac{\partial t^\beta}{\partial v^q} \eta^{pq}(v) = \eta_{\alpha\beta} = \text{const}.$$

The Poisson bracket $\{\ ,\ \}_{\text{semiclassical}}$ in these coordinates has the form

$$\{t^\alpha(X), t^\beta(Y)\}_{\text{semiclassical}} = \eta^{\alpha\beta} \delta'(X-Y). \quad (6.19)$$

The tensor $(\eta_{\alpha\beta}) = (\eta^{\alpha\beta})^{-1}$ together with the flat coordinates t^α is the first part of a structure we want to construct. (The flat coordinates $t^1, \ldots, t^n$ can be expressed via Casimirs of the original Poisson bracket and action variables and wave numbers along the invariant tori - see details in [57, 58, 120].)

To define a tensor $c_{\alpha\beta}^\gamma(t)$ on M (or, equivalently, the "primary free energy" $F(t)$) we need to use a semiclassical limit of the τ-function of the original hierarchy [94, 95, 48, 49, 141]. For the dispersionless limit the definition of the semiclassical τ-function reads

$$\log \tau_{\text{semiclassical}}(T_0, T_1, \ldots) = \lim_{\epsilon \to 0} \epsilon^{-2} \log \tau(\epsilon t_0, \epsilon t_1, \ldots). \quad (6.20)$$

Then

$$F = \log \tau_{\text{semiclassical}} \quad (6.21)$$

for a particular τ-function of the hierarchy. Here $\tau_{\text{semiclassical}}$ should be considered as a function only of the n primary slow variables. The semiclassical τ-function as the function of all slow variables coincides with the tree-level partition function of the matter sector $\eta_{\alpha\beta}$, $c_{\alpha\beta}^{\gamma}$ coupled to topological gravity.

Summarizing, we can say that a structure of Frobenius manifold (i.e., a solution of WDVV) on an invariant manifold M of an integrable Hamiltonian hierarchy is induced by a semiclassical limit of the first Poisson bracket of the hierarchy and of a particular τ-function of the hierarchy.

Now we will try to solve the inverse problem: starting from a Frobenius manifold M to construct a bi-hamiltonian hierarchy and to realize M as an invariant manifold of the hierarchy in the sense of the previous construction. By now we are able only to construct the semiclassical limit of the unknown hierarchy corresponding to any Frobenius manifold M. The problem of recovering of the complete hierarchy looks to be more complicated. Probably, this can be done not for arbitrary Frobenius manifold - see an interesting discussion of this problem in the recent preprint [61].

We describe now briefly the corresponding construction. After this we explain why this is equivalent to coupling of the matter sector of TFT described by the Frobenius manifold to topological gravity in the tree-level approximation.

Let us fix a Frobenius manifold M. Considering this as the matter sector of a 2D TFT model, let us try to calculate the tree-level (i.e., the zero-genus) approximation of the complete model obtained by coupling of the matter sector to topological gravity. The idea to use hierarchies of Hamiltonian systems of hydrodynamic type for such a calculation was proposed by E.Witten [157] for the case of topological sigma-models. An advantage of my approach is in effective construction of these hierarchies for *any* solution of WDVV. The tree-level free energy of the model will be identified with τ-function of a particular solution of the hierarchy. The hierarchy carries a bihamiltonian structure under a non-resonance assumption for scaling dimensions of the model.

So let $c_{\alpha\beta}^{\gamma}(t)$, $\eta_{\alpha\beta}$ be a solution of WDVV, $t = (t^1, \ldots, t^n)$. I will construct a hierarchy of the first order PDE systems linear in derivatives (*systems of hydrodynamic type*) for functions $t^{\alpha}(T)$, T is an infinite vector of times

$$T = (T^{\alpha,p}), \quad \alpha = 1, \ldots, n, \quad p = 0, 1, \ldots; \quad T^{1,0} = X, \tag{6.22}$$

$$\partial_{T^{\alpha,p}} t^{\beta} = c_{(\alpha,p)\gamma}^{\beta}(t)\partial_X t^{\gamma} \tag{6.23}$$

for some matrices of coefficients $c_{(\alpha,p)\gamma}^{\beta}(t)$. The marked variable $X = T^{1,0}$ usualy is called *cosmological constant*.

I will consider the equations (6.23) as dynamical systems (for any (α, p)) on the loop space $\mathcal{L}(M)$ of functions $t = t(X)$ with values in the Frobenius manifold M.

A. Construction of the systems. I define a Poisson bracket on the space of functions $t = t(X)$ (i.e. on the loop space $\mathcal{L}(M)$) by the formula

$$\{t^{\alpha}(X), t^{\beta}(Y)\} = \eta^{\alpha\beta}\delta'(X - Y). \tag{6.24}$$

All the systems (6.23) have hamiltonian form

$$\partial_{T^{\alpha,p}} t^{\beta} = \{t^{\beta}(X), H_{\alpha,p}\} \tag{6.25}$$

with the Hamiltonians of the form

$$H_{\alpha,p} = \int h_{\alpha,p+1}(t(X))dX. \tag{6.26}$$

The generating functions of densities of the Hamiltonians

$$h_\alpha(t,z) = \sum_{p=0}^\infty h_{\alpha,p}(t)z^p, \quad \alpha = 1,\ldots,n \tag{6.27}$$

coincide with the flat coordinates $\tilde{t}(t,z)$ of the perturbed connection $\tilde{\nabla}(z)$ (see (3.5) - (3.7)). That means that they are determined by the system (cf. (3.5))

$$\partial_\beta\partial_\gamma h_\alpha(t,z) = zc_{\beta\gamma}^\epsilon(t)\partial_\epsilon h_\alpha(t,z). \tag{6.28}$$

This gives simple recurrence relations for the densities $h_{\alpha,p}$. Solutions of (6.28) can be normalized in such a way that

$$h_\alpha(t,0) = t_\alpha = \eta_{\alpha\beta}t^\beta, \tag{6.29a}$$

$$< \nabla h_\alpha(t,z), \nabla h_\beta(t,-z) >= \eta_{\alpha\beta} \tag{6.29b}$$

$$\partial_1 h_\alpha(t,z) = zh_\alpha(t,z) + \eta_{1\alpha}. \tag{6.29c}$$

Here ∇ is the gradient (in t) w.r.t. the metric $<\,,\,>$.

Example 6.4. For a massive Frobenius manifold with $d < 1$ the functions $h_\alpha(t,z)$ can be determined uniquely from the equation

$$\partial_i h_\alpha = z^{-\mu_\alpha}\sqrt{\eta_{ii}(u)}\psi_{i\alpha}^0(u,z) \tag{6.30}$$

where the solution $\psi_{i\alpha}^0(u,z)$ of the system (3.122) has the form (3.124). Thus these functions can be continued analytically in any sector of the complex z-plane.

Exercise 6.1. Prove the following identities for the gradients of the generating functions $h_\alpha(t,z)$:

$$\nabla < \nabla h_\alpha(t,z), \nabla h_\beta(t,w) >= (z+w)\nabla h_\alpha(t,z) \cdot \nabla h_\beta(t,w) \tag{6.31}$$

$$[\nabla h_\alpha(t,z), \nabla h_\beta(t,w)] = (w-z)\nabla h_\alpha(t,z) \cdot \nabla h_\beta(t,w). \tag{6.32}$$

There is the product of the vector fields on the Frobenius manifold in the r.h.s. of the formulae.

Let us show that the Hamiltonians (6.26) are in involution. So all the systems of the hierarchy (6.23) commute pairwise.

Lemma 6.1. *The Poisson brackets (6.24) of the functionals $h_\alpha(t(X),z)$ for any fixed z have the form*

$$\{h_\alpha(t(X),z_1), h_\beta(t(Y),z_2)\} = [q_{\alpha\beta}(t(Y),z_1,z_2) + q_{\beta\alpha}(t(X),z_2,z_1)]\delta'(X-Y) \tag{6.33a}$$

where

$$q_{\alpha\beta}(t, z_1, z_2) := \frac{z_2}{z_1 + z_2} < \nabla h_\alpha(t, z_1), \nabla h_\beta(t, z_2) > . \qquad (6.33b)$$

Proof. For the derivatives of $q_{\alpha\beta}(t, z_1, z_2)$ one has from (6.31)

$$\nabla q_{\alpha\beta}(t, z_1, z_2) = z_2 \nabla h_\alpha(t, z_1) \cdot \nabla h_\beta(t, z_2).$$

The l.h.s. of (6.33a) has the form

$$\begin{aligned}
\{h_\alpha(t(X), z_1), h_\beta(t(Y), z_2)\} &=< \nabla h_\alpha(t(X), z_1), \nabla h_\beta(t(Y), z_2) > \delta'(X - Y) \\
&=< \nabla h_\alpha(t(X), z_1), \nabla h_\beta(t(X), z_2) > \delta'(X - Y) \\
&+ z_2 < \nabla h_\alpha(t, z_1) \cdot \nabla h_\beta(t, z_2), \partial_X t > \delta(X - Y).
\end{aligned}$$

This completes the proof.

Exercise 6.2. For any solution $h(t, z)$ of the equation (6.28) prove that

$$\partial_{T^{\alpha,k}} h(t, z) = \underset{w=0}{\mathrm{res}} \frac{w^{-k-1}}{z + w} < \nabla h(t, z), \nabla h_\alpha(t, w) > . \qquad (6.34)$$

(The denominator $z + w$ was lost in the formula (3.53) of [50].)

Corollary 6.1. *The Hamiltonians (6.26) commute pairwise.*

Observe that the functionals

$$H_{\alpha,-1} = \int t_\alpha(X) \, dX$$

span the annihilator of the Poisson bracket (6.24).

Exercise 6.3. Show that the equations (6.25) for $p = 0$ (the primary part of the hierarchy) have the form

$$\partial_{T^{\alpha,0}} t^\gamma = c^\gamma_{\alpha\beta}(t) \partial_X t^\beta. \qquad (6.35)$$

For the equations with $p > 0$ obtain the recursion relation

$$\partial_{T^{\alpha,p}} = \nabla^\epsilon h_{\alpha,p} \partial_{T^{\epsilon,0}}. \qquad (6.36)$$

Exercise 6.4. Prove that for a massive Frobenius manifold all the systems of the hierarchy (6.25) are diagonal in the canonical coordinates $u^1, \ldots, u^n$.

In this case from the results of Tsarev [144] it follows completeness of the family of the conservation laws (6.26) for any of the systems in the hierarchy (6.25).

B. Specification of a solution $t = t(T)$. The hierarchy (6.25) admits an obvious scaling group

$$T^{\alpha,p} \mapsto cT^{\alpha,p}, \quad t \mapsto t. \qquad (6.37)$$

Let us take the nonconstant invariant solution for the symmetry

$$(\partial_{T^{1,1}} - \sum T^{\alpha,p}\partial_{T^{\alpha,p}})t(T) = 0 \qquad (6.38)$$

(I identify $T^{1,0}$ and X. So the variable X is supressed in the formulae.) This solution can be found without quadratures from a fixed point equation for the gradient map

$$t = \nabla\Phi_T(t), \qquad (6.39)$$

$$\Phi_T(t) = \sum_{\alpha,p} T^{\alpha,p}h_{\alpha,p}(t). \qquad (6.40)$$

Proposition 6.1. *The hierarchy (6.25) in the domain*

$$T^{1,1} = \epsilon, \quad X, \quad T^{\alpha,p} = o(\epsilon) \text{ for } (\alpha,p) \neq (1,1), \quad \epsilon \to 0 \qquad (6.41)$$

has a unique nonconstant solution $t^\beta = t^\beta(X,T)$ invariant w.r.t. the transformations (6.38). This can be found from the variational equations

$$\mathrm{grad}_t\left[\Phi_T(t) + Xt_1\right] = 0 \qquad (6.42)$$

or, equivalently, as the fixed point of the gradient map

$$t = \nabla\Phi_T(t) \qquad (6.43)$$

where

$$\Phi_T(t) = \sum_{\alpha,p} T^{\alpha,p}h_{\alpha,p}(t). \qquad (6.44)$$

Proof. For the invariant solutions of (6.23) one has

$$\left(X\delta_\gamma^\beta + \sum c_{(\alpha,p)\,y}^\beta(t)\right)\partial_X t^\gamma = 0, \quad \beta = 1,\ldots,n. \qquad (6.45)$$

Using the recursion (6.36) this system can be represented in the form (6.72). In the domain (6.41) one has

$$\partial_\mu\partial_\nu\left[\Phi_T(t) + Xt_1\right] = \epsilon\eta_{\mu\nu} + o(\epsilon).$$

Hence the solution is locally unique. Therefore it satisfies (6.25). Proposition is proved.

C. τ-function. Let us define functions $< \phi_{\alpha,p}\phi_{\beta,q} > (t)$ from the expansion

$$(z+w)^{-1}(< \nabla h_\alpha(t,z), \nabla h_\beta(t,w) > -\eta_{\alpha\beta})$$

$$= \sum_{p,q=0}^{\infty} < \phi_{\alpha,p}\phi_{\beta,q} > (t)z^p w^q =:< \phi_\alpha(z)\phi_\beta(w) > (t). \qquad (6.46)$$

326

The infinite matrix of the coefficients $< \phi_{\alpha,p}\phi_{\beta,q} > (t)$ has a simple meaning: it is the energy-momentum tensor of the commutative Hamiltonian hierarchy (6.25). That means that the matrix entry $< \phi_{\alpha,p}\phi_{\beta,q} > (t)$ is the density of flux of the Hamiltonian $H_{\alpha,p}$ along the flow $T^{\beta,q}$:

$$\partial_{T^{\beta,q}} h_{\alpha,p+1}(t) = \partial_X < \phi_{\alpha,p}\phi_{\beta,q} > (t). \tag{6.47}$$

Then we define

$$\log \tau(T) = \frac{1}{2} \sum < \phi_{\alpha,p}\phi_{\beta,q} > (t(T))T^{\alpha,p}T^{\beta,q}$$

$$+ \sum < \phi_{\alpha,p}\phi_{1,1} > (t(T))T^{\alpha,p} + \frac{1}{2} < \phi_{1,1}\phi_{1,1} > (t(T)). \tag{6.48}$$

Exercise 6.5. Prove that the τ-function satisfies the identity

$$\partial_{T^{\alpha,p}}\partial_{T^{\beta,q}} \log \tau = < \phi_{\alpha,p}\phi_{\beta,q} > . \tag{6.49}$$

The equations (6.47), (6.49) are the main motivation for the name "τ-function" (cf. [135, 84]).

Exercise 6.6. Derive the following formulae

$$\begin{aligned}
< \phi_{\alpha,0}\phi_{\beta,0} > &= \partial_\alpha\partial_\beta F \\
< \phi_{\alpha,p}\phi_{1,0} > &= h_{\alpha,p} \\
< \phi_{\alpha,p}\phi_{\beta,0} > &= \partial_\beta h_{\alpha,p+1} \\
< \phi_{\alpha,p}\phi_{1,1} > &= (t^\gamma\partial_\gamma - 1) h_{\alpha,p+1} \\
< \phi_{\alpha,0}\phi_{1,1} > &= t^\lambda\partial_\alpha\partial_\lambda F - \partial_\alpha F \\
< \phi_{1,1}\phi_{1,1} > &= t^\alpha t^\beta\partial_\alpha\partial_\beta F - 2t^\alpha\partial_\alpha F + 2F.
\end{aligned} \tag{6.50}$$

Exercise 6.7. Show that

$$< \phi_\alpha(z)\phi_1(w) > = h_\alpha(t,z) + O(w). \tag{6.51}$$

Remark 6.1. More general solutions of (6.25) has the form

$$\nabla[\Phi_T(t) - \Phi_{T_0}(t)] = 0 \tag{6.52}$$

for arbitrary constant infinite vector $T_0 = (T_0^{\alpha,p})$. For massive Frobenius manifolds these form a dense subset in the space of all solutions of (6.25) (see [144, 50]). Formally they can be obtained from the solution (6.42) by a shift of the arguments $T^{\alpha,p}$. τ-function of the solution (6.52) can be formaly obtained from (6.48) by the same shift. For the example of topological gravity [156, 157] such a shift is just the operation that relates the tree-level free energies of the topological phase of 2D gravity and of the matrix model. It should be taken into account that the operation of such a time shift in systems of hydrodynamic type is a subtle one: it can pass through a point of gradient catastrophe where derivatives become infinite. The correspondent solution of the KdV hierarchy has no gradient catastrophes but oscillating zones arise (see [74] for details).

Theorem 6.1. *Let*

$$\mathcal{F}(T) = \log \tau(T), \qquad (6.53a)$$

$$< \phi_{\alpha,p}\phi_{\beta,q} \ldots >_0 = \partial_{T^{\alpha,p}}\partial_{T^{\beta,q}} \ldots \mathcal{F}(T). \qquad (6.53b)$$

Then the following relations hold

$$\mathcal{F}(T)|_{T^{\alpha,p}=0 \text{ for } p>0, \ T^{\alpha,0}=t^\alpha} = F(t) \qquad (6.54a)$$

$$\partial_X \mathcal{F}(T) = \sum T^{\alpha,p}\partial_{T^{\alpha,p-1}}\mathcal{F}(T) + \frac{1}{2}\eta_{\alpha\beta}T^{\alpha,0}T^{\beta,0} \qquad (6.54b)$$

$$< \phi_{\alpha,p}\phi_{\beta,q}\phi_{\gamma,r} >_0 = < \phi_{\alpha,p-1}\phi_{\lambda,0} >_0 \eta^{\lambda\mu} < \phi_{\mu,0}\phi_{\beta,q}\phi_{\gamma,r} >_0 . \qquad (6.54c)$$

Proof can be obtained using the results of the Exercises (6.5) - (6.7) (see [50] for details).

Let me establish now a 1-1 correspondence between the statements of the theorem and the axioms of Dijkgraaf and Witten of coupling to topological gravity. In a complete model of 2D TFT (i.e. a matter sector coupled to topological gravity) there are infinite number of operators. They usualy are denoted by $\phi_{\alpha,p}$ or $\sigma_p(\phi_\alpha)$. The operators $\phi_{\alpha,0}$ can be identified with the primary operators ϕ_α; the operators $\phi_{\alpha,p}$ for $p > 0$ are called *gravitational descendants* of ϕ_α. Respectively one has infinite number of coupling constants $T^{\alpha,p}$. For example, for the A_n-topological minimal models (see Lecture 2 above) the gravitational descendants come from the Mumford - Morita - Miller classes of the moduli spaces $\mathcal{M}_{g,s}$. A similar description of gravitational descendants can be done also for topological sigma-models [157, 41].

The formula (6.53a) expresses the tree-level (i.e. genus zero) partition function of the model of 2D TFT via logarythm of the τ-function (6.48). Equation (6.53b) is the standard relation between the correlators (of genus zero) in the model and the free energy. Equation (6.54a) manifests that before coupling to gravity the partition function (6.53a) coincides with the primary partition function of the given matter sector. Equation (6.54b) is the string equation for the free energy [156, 157, 41]. And equations (6.54c) coincide with the genus zero recursion relations for correlators of a TFT [43, 157].

We conclude with another formulation of Theorem 6.1.

Theorem 6.2. *For any Frobenius manifold the formulae give a solution of the Dijk-graaf - Witten relations defining coupling to topological gravity at the tree level.*

I recall that the solution of these relations for a given matter sector of a TFT (i.e., for a given Frobenius manifold) is unique, according to [43].

Remark 6.2. According to Lecture 2, in topological sigma-models the genus zero correlators of the primary fields can be computed as the values on the fundamental class $[M_{0,n}]$ of the products $\phi_\alpha\phi_\beta \ldots$ of some cohomology classes ϕ_α, $\phi_\beta, \ldots \in H^*(M_{0,n})$ of the moduli space $M_{0,n}$ of rational curves with n punctures. Coupling to topological gravity is given in terms of the values of the same products on the cycles dual to Mumford - Morita - Miller classes. So, our Theorem gives an algorythm of reconstructing of the generating function of these values starting from a solution of the equations of associativity (note that the scaling (1.5) has nnot been used in the construction). Recently Kontsevich aand Manin proved that the cohomology classes ϕ_α, ϕ_β,... themselves can be uniquely reconstructed starting fromm a solution of the associativity equation (the second reconstruction theorem of [92]).

Particularly, from (6.53) one obtains

$$< \phi_{\alpha,p}\phi_{\beta,q} >_0 = < \phi_{\alpha,p}\phi_{\beta,q} > \tag{6.55a}$$

where the r.h.s. is defined in (6.46),

$$< \phi_{\alpha,p}\phi_{1,0} >_0 = h_{\alpha,p}(t(T)), \tag{6.55b}$$

$$< \phi_{\alpha,p}\phi_{\beta,q}\phi_{\gamma,r} >_0 = < \nabla h_{\alpha,p} \cdot \nabla h_{\beta,q} \cdot \nabla h_{\gamma,r}, [e - \sum T^{\alpha,p}\nabla h_{\alpha,p-1}]^{-1} > . \tag{6.55c}$$

The second factor of the inner product in the r.h.s. of (6.55c) is an invertible element (in the Frobenius algebra of vector fields on M) for sufficiently small $T^{\alpha,p}$, $p > 0$. From the last formula one obtains

Proposition 6.2. *The coefficients*

$$c_{p,\alpha\beta}^\gamma(T) = \eta^{\gamma\mu}\partial_{T^{\alpha,p}}\partial_{T^{\beta,p}}\partial_{T^{\mu,p}}\log\tau(T) \tag{6.56}$$

for any p and any T are structure constants of a commutative associative algebra with the invariant inner product $\eta_{\alpha\beta}$.

As a rule such an algebra has no unity.

In fact the Proposition holds also for a τ-function of an arbitrary solution of the form (6.52).

We see that the hierarchy (6.25) determines a family of Bäcklund transforms of the associativity equations (1.14)

$$F(t) \mapsto \tilde{F}(\tilde{t}),$$

$$\tilde{F} = \log\tau, \quad \tilde{t}^\alpha = T^{\alpha,p} \tag{6.57}$$

for a fixed p and for arbitrary τ-function of (6.25). So it is natural to consider equations of the hierarchy as Lie – Bäcklund symmetries of WDVV.

Up to now I even did not use the scaling invariance (1.9). It turns out that this gives rise to a bihamiltonian structure of the hierarchy (6.25) under certain nonresonancy conditions.

We say that a pair α, p is *resonant* if

$$\frac{d+1}{2} - q_\alpha + p = 0. \tag{6.58}$$

Here p is a nonnegative integer. The Frobenius manifold with the scaling dimensions q_α, d is *nonresonant* if all pairs α, p are nonresonant. For example, manifolds satisfying the inequalities

$$0 = q_1 \leq q_2 \leq \ldots \leq q_n = d < 1 \tag{6.59}$$

all are nonresonant.

Theorem 6.3. *1) For a Frobenius manifold with the scaling dimensions q_α and d the formula*

$$\{t^\alpha(X), t^\beta(Y)\}_1 = g^{\alpha\beta}(t(X))\delta'(X-Y) + \left(\frac{d+1}{2} - q_\beta\right) c_\gamma^{\alpha\beta}(t(X))\partial_X t^\gamma(X)\delta(X-Y) \tag{6.60}$$

where $g^{\alpha\beta}(t)$ is the intersection form of the Frobenius manifold determines a Poisson bracket compatible with the Poisson bracket (6.24). 2) For a nonresonant TCFT model all the equations of the hierarchy (6.25) are Hamiltonian equations also with respect to the Poisson bracket (6.60).

Proof follows from (H.12).

The nonresonancy condition is essential: equations (6.25) with resonant numbers (α, p) do not admit another Poisson structure.

Example 6.5. Trivial Frobenius manifold corresponding to a graded n-dimensional Frobenius algebra A. Let

$$\mathbf{t} = t^\alpha e_\alpha \in A. \tag{6.61}$$

The linear system (6.28) can be solved easily:

$$h_\alpha(t, z) = z^{-1} < e_\alpha, e^{zt} - 1 > . \tag{6.62}$$

This gives the following form of the hierarchy (6.25)

$$\partial_{T^{\alpha,p}} \mathbf{t} = \frac{1}{p!} e_\alpha t^p \partial_X \mathbf{t}. \tag{6.63}$$

The solution (6.43) is specified as the fixed point

$$G(\mathbf{t}) = \mathbf{t}, \tag{6.64a}$$

$$G(\mathbf{t}) = \sum_{p=0}^{\infty} \frac{\mathbf{T}_p}{p!} t^p. \tag{6.64b}$$

Here I introduce A-valued coupling constants

$$\mathbf{T}_p = T^{\alpha,p} e_\alpha \in A, \; p = 0, 1, \ldots. \tag{6.65}$$

The solution of (6.64a) has the well-known form

$$\mathbf{t} = G(G(G(\ldots))) \tag{6.66}$$

(infinite number of iterations). The τ-function of the solution (6.66) has the form

$$\log \tau = \frac{1}{6} < 1, \mathbf{t}^3 > - \sum_p \frac{< \mathbf{T}_p, \mathbf{t}^{p+2} >}{(p+2)p!} + \frac{1}{2} \sum_{p,q} \frac{< \mathbf{T}_p \mathbf{T}_q, \mathbf{t}^{p+q+1} >}{(p+q+1)p!q!}. \qquad (6.67)$$

For the tree-level correlation functions of a TFT-model with constant primary correlators one immediately obtains

$$< \phi_{\alpha,p} \phi_{\beta,q} >_0 = \frac{< e_\alpha e_\beta, \mathbf{t}^{p+q+1} >}{(p+q+1)p!q!}, \qquad (6.68a)$$

$$< \phi_{\alpha,p} \phi_{\beta,q} \phi_{\gamma,r} >_0 = \frac{1}{p!q!r!} < e_\alpha e_\beta e_\gamma, \frac{\mathbf{t}^{p+q+r}}{1 - \sum_{s\geq 1} \frac{\mathbf{T}_s \mathbf{t}^{s-1}}{(s-1)!}} >. \qquad (6.68b)$$

Let us consider now the second hamiltonian structure (6.60). I start with the most elementary case $n = 1$ (the pure gravity). Let me redenote the coupling constant

$$u = t^1.$$

The Poisson bracket (6.60) for this case reads

$$\{u(X), u(Y)\}_1 = \frac{1}{2}(u(X) + u(Y))\delta'(X - Y). \qquad (6.69)$$

This is nothing but the Lie – Poisson bracket on the dual space to the Lie algebra of one-dimensional vector fields.

For arbitrary graded Frobenius algebra A the Poisson bracket (6.60) also is linear in the coordinates t^α

$$\{t^\alpha(X), t^\beta(Y)\}_1 = [(\frac{d+1}{2} - q_\alpha)c_\gamma^{\alpha\beta} t^\gamma(X) + (\frac{d+1}{2} - q_\beta)c_\gamma^{\alpha\beta} t^\gamma(Y)]\delta'(X - Y). \qquad (6.70)$$

It determines therefore a structure of an infinite dimensional Lie algebra on the loop space $\mathcal{L}(A^*)$ where A^* is the dual space to the graded Frobenius algebra A. Theory of linear Poisson brackets of hydrodynamic type and of corresponding infinite dimensional Lie algebras was constructed in [12] (see also [58]). But the class of examples (6.70) is a new one.

Let us come back to a general (i.e. nontrivial) Frobenius manifold. I will assume that the scaling dimensions are ordered in such a way that

$$0 = q_1 < q_2 \leq \ldots \leq q_{n-1} < q_n = d. \qquad (6.71)$$

Then from (6.60) one obtains

$$\{t^n(X), t^n(Y)\}_1 = \frac{1-d}{2}(t^n(X) + t^n(Y))\delta'(X - Y). \qquad (6.72)$$

Since

$$\{t^\alpha(X), t^n(Y)\}_1 = [(\frac{d+1}{2} - q_\alpha)t^\alpha(X) + \frac{1-d}{2}t^\alpha(Y)]\delta'(X - Y), \qquad (6.73)$$

the functional

$$P = \frac{2}{1-d} \int t^n(X)dX \qquad (6.74)$$

generates spatial translations. We see that for $d \neq 1$ the Poisson bracket (6.60) can be considered as a nonlinear extension of the Lie algebra of one-dimensional vector fields. An interesting question is to find an analogue of the Gelfand – Fuchs cocycle for this bracket. I found such a cocycle for a more particular class of Frobenius manifolds. We say that a Frobenius manifold is *graded* if for any t the Frobenius algebra $c^\gamma_{\alpha\beta}(t)$, $\eta_{\alpha\beta}$ is graded.

Theorem 6.4. *For a graded Frobenius manifold the formula*

$$\{t^\alpha(X), t^\beta(Y)\}_1^{\hat{}} = \{t^\alpha(X), t^\beta(Y)\}_1 + \epsilon^2\eta^{1\alpha}\eta^{1\beta}\delta'''(X - Y) \qquad (6.75)$$

determines a Poisson bracket compatible with (6.24) and (6.60) for arbitrary ϵ^2 (the central charge). For a generic graded Frobenius manifold this is the only one deformation of the Poisson bracket (6.60) proportional to $\delta'''(X - Y)$.

For $n = 1$ (6.75) determines nothing but the Lie – Poisson bracket on the dual space to the Virasoro algebra

$$\{u(X), u(Y)\}_1^{\hat{}} = \frac{1}{2}[u(X) + u(Y)]\delta'(X - Y) + \epsilon^2\delta'''(X - Y) \qquad (6.76)$$

(the second Poisson structure of the KdV hierarchy). For $n > 1$ and constant primary correlators (i.e. for a constant graded Frobenius algebra A) the Poisson bracket (6.75) can be considered as a vector-valued extension (for $d \neq 1$) of the Virasoro.

The compatible pair of the Poisson brackets (6.24) and (6.75) generates an integrable hierarchy of PDE for a non-resonant graded Frobenius manifold using the standard machinery of the bihamiltonian formalism (see above)

$$\partial_{T^{\alpha,p}}t^\beta = \{t^\beta(X), \hat{H}_{\alpha,p}\} = \{t^\beta(X), \hat{H}_{\alpha,p-1}\}_1^{\hat{}}. \qquad (6.77)$$

Here the Hamiltonians have the form

$$\hat{H}_{\alpha,p} = \int \hat{h}_{\alpha,p+1}dX, \qquad (6.78a)$$

$$\hat{h}_{\alpha,p+1} = [\frac{d+1}{2} - q_\alpha + p]^{-1}h_{\alpha,p+1}(t) + \epsilon^2\Delta\hat{h}_{\alpha,p+1}(t, \partial_X t, \ldots, \partial_X^p t; \epsilon^2) \qquad (6.78b)$$

where $\Delta\hat{h}_{\alpha,p+1}$ are some polynomials determined by (6.77). They are graded-homogeneous of degree 2 where $\deg\partial_X^k t = k$, $\deg\epsilon = -1$. The small dispersion parameter ϵ also plays the role of the string coupling constant. It is clear that the hierarchy (6.25) is the zero-dispersion limit of this hierarchy. For $n = 1$ using the pair (6.24) and (6.76) one immediately recover the KdV hierarchy. Note that this describes the topological gravity. For a trivial manifold (i.e. for a graded Frobenius algebra A) the first nontrivial equations of the hierarchy are

$$\partial_{T^{\alpha,1}}t = e_\alpha t t_X + \frac{2\epsilon^2}{3-d}e_\alpha e_n t_{XXX}. \tag{6.79}$$

For non-graded Frobenius manifolds it could be of interest to find nonlinear analogues of the cocycle (6.75). These should be differential geometric Poisson brackets of the third order [58] of the form

$$\{t^\alpha(X), t^\beta(Y)\}_1^\cdot = \{t^\alpha(X), t^\beta(Y)\}_1 +$$

$$\epsilon^2\{g^{\alpha\beta}(t(X))\delta'''(X-Y) + b_\gamma^{\alpha\beta}(t(X))t_X^\gamma\delta''(X-Y) +$$

$$[f_{\gamma}^{\alpha\beta}(t(X))t_{XX}^\gamma + h_{\gamma\delta}^{\alpha\beta}(t(X))t_X^\gamma t_X^\delta]\delta'(X-Y) +$$

$$[p_\gamma^{\alpha\beta}(t)t_{XXX}^\gamma + q_{\gamma\delta}^{\alpha\beta}(t)t_{XX}^\gamma t_X^\delta + r_{\gamma\delta\lambda}^{\alpha\beta}(t)t_X^\gamma t_X^\delta t_X^\lambda]\delta(X-Y)\}. \tag{6.80}$$

I recall (see [58]) that the form (6.80) of the Poisson bracket should be invariant with respect to nonlinear changes of coordinates in the manifold M. This implies that the leading term $g^{\alpha\beta}(t)$ transforms like a metric (may be, degenerate) on the cotangent bundle T_*M, $b_\gamma^{\alpha\beta}(t)$ are contravariant components of a connection on M etc. The Poisson bracket (6.80) is assumed to be compatible with (6.24). Then the compatible pair (6.24), (6.80) of the Poisson brackets generates an integrable hierarchy of the same structure (6.77), (6.78). The hierarchy (6.25) will be the dispersionless limit of (6.77).

Example 6.6. Let me describe the hierarchies (6.25) for two-dimensional Frobenius manifolds. Let us redenote the coupling constants

$$t^1 = u, \quad t^2 = \rho. \tag{6.81}$$

For $d \neq -1, 1, 3$ the primary free energy F has the form

$$F = \frac{1}{2}\rho u^2 + \frac{g}{a(a+2)}\rho^{a+2}, \tag{6.82}$$

$$a = \frac{1+d}{1-d} \tag{6.83}$$

where we introduce an arbitrary constant g. Let me give an example of equations of the hierarchy (6.25) (the $T = T^{1,1}$-flow)

$$u_T + uu_X + g\rho^a\rho_X = 0 \tag{6.84a}$$

$$\rho_T + (\rho u)_X = 0. \tag{6.84b}$$

These are the equations of isentropic motion of one-dimensional fluid with the dependence of the pressure on the density of the form $p = \frac{g}{a+2}\rho^{a+2}$. The Poisson structure (6.24) for these equations was proposed in [123]. For $a = 0$ (equivalently $d = -1$) the system coincides with the equations of waves on shallow water (the dispersionless limit [164] of the nonlinear Schrödinger equation (NLS)).

For $d = 1$ the primary free energy has the form

$$F = \frac{1}{2}\rho u^2 + g e^\rho. \tag{6.85}$$

This coincides with the free energy of the topological sigma-model with CP^1 as the target space. The corresponding $T = T^{2,0}$-system of the hierarchy (6.25) reads

$$u_T = g(e^\rho)_X$$

$$\rho_T = u_X.$$

Eliminating u one obtains the long wave limit

$$\rho_{TT} = g(e^\rho)_{XX} \tag{6.86}$$

of the Toda system

$$\rho_{n\,tt} = e^{\rho_{n+1}} - 2e^{\rho_n} + e^{\rho_{n-1}}. \tag{6.87}$$

Example 6.7. For the Frobenius manifold of Example 1.7 the hierarchy (6.25) is just the dispersionless limit of the $nKdV$-hierarchy. This was essentially obtained in [42] and elucidated by Krichever in [94]. The two metrics on the Frobenius manifold (I recall that this coincides with the space of orbits of the group A_n) are just obtained from the two hamiltonian structures of the $nKdV$ hierarchy: the Saito metric is obtained by the semiclassical limit of from the first Gelfand-Dickey Poisson bracket of $nKdV$ and the Euclidean metric is obtained by the same semiclassical limit from the second Gelfand-Dickey Poisson bracket. The Saito and the Euclidean coordinates on the orbit space are the Casimirs for the corresponding Poisson brackets. The factorization map $V \to M = V/W$ is the semiclassical limit of the Miura transformation.

Example 6.8. The Hurwitz spaces $M_{g;n_0,\ldots,n_m}$ parametrize "g-gap" solutions of certain integrable hierarchies. The Lax operator L for these hierarchies must have a form of a $(m+1) \times (m+1)$-matrix. The equation $L\psi = 0$ for a $(m+1)$-component vector-function ψ must read as a system of ODE (in x) of the orders $n_0 + 1, \ldots, n_m + 1$ resp. Particularly, for $m = 0$ one obtains the scalar operator L of the order $n = n_0$. So the Hurwitz spaces $M_{g;n}$ parametrize algebraic-geometrical solutions [93] (of the genus g) of the $nKdV$ hierarchy.

To describe the hierarchy (6.25) we are to solve the recursion system (6.28).

Proposition 6.3. *The generating functions $h_{t^A}(t;z)$ (6.27) (where t^A is one of the flat coordinates (5.30) on the Hurwitz space) have the form*

$$h_{t^{i;\alpha}}(t;z) = -\frac{n_i+1}{\alpha} \operatorname*{res}_{p=\infty_i} k_i^\alpha \,_1F_1(1; 1 + \frac{\alpha}{n_i+1}; z\lambda(p))dp$$

$$h_{p^i} = \text{v.p.} \int_{\infty_0}^{\infty_i} e^{\lambda(p)z} dp$$

$$h_{q^i} = \operatorname*{res}_{\infty_i} \frac{e^{\lambda z} - 1}{z} dp \tag{6.88}$$

$$h_{r^i} = \oint_{b_i} e^{\lambda z} dp$$

$$h_{s^i} = \frac{1}{2\pi i} \oint_{a_i} p e^{\lambda z} d\lambda.$$

Here $_1F_1(a,b,z)$ is the Kummer confluent hypergeometric function (see [107]). We leave the proof as an exercise for the reader (see [48]).

Remark 6.3. Integrals of the form (6.88) seem to be interesting functions on the moduli space of the form $M = M_{g;n_0,\ldots,n_m}$. A simplest example of such an integral for a family of elliptic curves reads

$$\int_0^\omega e^{\lambda \wp(z)} dz \tag{6.89}$$

where $\wp(z)$ is the Weierstrass function with periods 2ω, $2\omega'$. For real negative λ a degeneration of the elliptic curve ($\omega \to \infty$) reduces (6.89) to the standard probability integral $\int_0^\infty e^{\lambda x^2} dx$. So the integral (6.89) is an analogue of the probability integral as a function on λ and on moduli of the elliptic curve. I recall that dependence on these parameters is specified by the equations (6.28).

Gradients of this functions on the Hurwitz space M have the form

$$\partial_{t^A} h_{t^{i;\alpha}} = \operatorname*{res}_{\infty_i} k_i^{\alpha - n_i - 1} {}_1F_1(1; \frac{\alpha}{n_i + 1}; z\lambda(p))\phi_{t^A}$$

$$\partial_{t^A} h_{p^i} = \eta_{t^A p^i} - z \text{v.p.} \int_{\infty_0}^{\infty_i} e^{\lambda z} \phi_{t^A}$$

$$\partial_{t^A} h_{q^i} = \operatorname*{res}_{\infty_i} e^{\lambda z} \phi_{t^A} \tag{6.90}$$

$$\partial_{t^A} h_{r^i} = \eta_{t^A r^i} - z \oint_{b_i} e^{\lambda z} \phi_{t^A}$$

$$\partial_{t^A} h_{s^i} = \frac{1}{2\pi i} \oint_{a_i} e^{\lambda z} \phi_{t^A}.$$

The pairing (6.46) $< \phi_\alpha(z)\phi_\beta(w) >$ involved in the definition of the τ-function (6.48) coincides with (5.36).

Remark 6.4. For any Hamiltonian $H_{A,k}$ of the form (6.26), (6.88) one can construct a differential $\Omega_{A,k}$ on C or on a covering $\tilde{C}$ with singulariries only at the marked infinite points such that

$$\frac{\partial}{\partial u^i} h_{t^A,k} = \operatorname*{res}_{P_i} \frac{\Omega_{A,k} dp}{d\lambda}, \quad i = 1, \ldots, n. \tag{6.91}$$

We give here, following [48], an explicit form of these differentials (for $m = 0$ also see [49]). All these will be normalized (i.e. with vanishing a-periods) differentials on C or on the universal covering of C with no other singularities or multivaluedness but those indicated in (6.92) - (6.97)

$$\Omega_{t^{i;\alpha},k} = -\frac{1}{n_i + 1} \left[\left(\frac{\alpha}{n_i + 1} \right)_{k+1} \right]^{-1} d\lambda^{\frac{\alpha}{n_i+1}+k} + \text{regular terms} \tag{6.92}$$

$$\Omega_{v^i,k} = -d\left(\frac{\lambda^{k+1}}{(k+1)!} \right) + \text{regular terms}, \quad i = 1, \ldots, m \tag{6.93}$$

$$\Omega_{w^i,k} = \begin{cases} -\frac{1}{n_i+1} d\psi_k(\lambda) + \text{reg. terms} & \text{near } \infty_i \\ \frac{1}{n_0+1} d\psi_k(\lambda) + \text{reg. terms} & \text{near } \infty_0 \end{cases} \qquad (6.94)$$

where

$$\psi_k(\lambda) := \frac{\lambda^k}{k!}\left[\log \lambda - \left(1 + \frac{1}{2} + \ldots + \frac{1}{k}\right)\right], \; k > 0 \qquad (6.95)$$

$$\Omega_{r^i,k}(P + b_j) - \Omega_{r^i,k}(P) = -\delta_{ij}\frac{\lambda^k}{k!}d\lambda \qquad (6.96)$$

$$\Omega_{s^i,k}(P + a_j) - \Omega_{s^i,k}(P) = \delta_{ij}\frac{\lambda^{k-1}}{(k-1)!}d\lambda. \qquad (6.97)$$

Using these differentials the hierarchy (6.25) can be written in the Flaschka – Forest –McLaughlin form [63]

$$\partial_{T^{A,p}}dp = \partial_X \Omega_{A,p} \qquad (6.98)$$

(derivatives of the differentials are to be calculated with $\lambda =$ const.).

Integrating (6.98) along the Riemann surface C one obtains for the the Abelian integrals

$$q_{A,k} := \int \Omega_{A,k}$$

a similar representation

$$\partial_{T^{A,p}}p(\lambda) = \partial_X q_{A,p}(\lambda).$$

Rewriting this for the inverse function $\lambda = \lambda(p)$ we obtain the semiclassical Lax representation of the hierarchy (6.25) for the Hurwitz space (see details below in a more general setting)

$$\partial_{T^{A,p}}\lambda(p) = \{\lambda, \rho_{A,k}\}$$

where

$$\rho_{A,k} := q_{A,k}(\lambda(p)).$$

The matrix $< \phi_{A,p}\phi_{B,q} >$ determines a pairing of these differentials with values in functions on the moduli space

$$< \Omega_{A,p}\,\Omega_{B,q} >=< \phi_{A,p}\phi_{B,q} > (t) \qquad (6.99)$$

This pairing coincides with the two-point correlators (6.55a). Particularly, the primary free energy F as a function on M can be written in the form

$$F = -\frac{1}{2} < pd\lambda\,pd\lambda > . \qquad (6.100)$$

Note that the differential $pd\lambda$ can be written in the form

$$pd\lambda = \sum \frac{n_i + 1}{n_i + 2}\Omega_{\infty_i}^{(n_i+2)} + \sum t^A \phi_{t^A} \qquad (6.101)$$

where $\Omega_{\infty_i}^{(n_i+2)}$ is the Abelian differential of the second kind with a pole at ∞_i of the form

$$\Omega_{\infty_i}^{(n_i+2)} = dk_i^{n_i+2} + \text{regular terms} \quad \text{near } \infty_i. \tag{6.102}$$

For the pairing (6.99) one can obtain from [95] the following formula

$$< f_1 d\lambda \, f_2 d\lambda > = \frac{1}{2} \int \int_C (\bar{\partial} f_1 \partial f_2 + \partial f_1 \bar{\partial} f_2) \tag{6.103}$$

where the differentials ∂ and $\bar{\partial}$ along the Riemann surface should be understood in the distribution sense. The meromorphic differentials $f_1 dw$ and $f_2 dw$ on the covering $\tilde{C}$ should be considered as piecewise meromorphic differentials on C with jumps on some cuts.

Exercise 6.8. Prove that the three-point correlators $\partial_{T^{A,p}} \partial_{T^{B,q}} \partial_{T^{C,r}} \mathcal{F}$ as functions on $T^{\alpha,0} = t^\alpha$ with $T^{\alpha,p} = 0$ for $p > 0$ can be written in the form

$$< \phi_{A,p} \phi_{B,q} \phi_{C,r} > = \sum \operatorname*{res}_{d\lambda=0} \frac{\Omega_{A,p} \Omega_{B,q} \Omega_{C,r}}{d\lambda dp}. \tag{6.104}$$

Here A, B, C denote the labels of one of the flat coordinates (5.29), the numbers p, q, r take values $0, 1, 2, \ldots$. [Hint: use (6.55c) and (6.91).]

For the space of polynomials $M_{0;n}$ (the Frobenius manifold of the A_n-topological minimal model) the formula (6.104) was obtained in [97, 60].

The corresponding hierarchy (6.25) is obtained by averaging along invariant tori of a family of g-gap solutions of a KdV-type hierarchy related to a matrix operator L of the matrix order $m + 1$ and of orders n_0, ..., n_m in $\partial/\partial x$. The example $m = 0$ (the averaged Gelfand – Dickey hierarchy) was considered in more details in [49].

Also for $g + m > 0$ one needs to extend the KdV-type hierarchy to obtain (6.25) (see [49]). To explain the nature of such an extension let us consider the simplest example of $m = 0$, $n_0 = 1$. The moduli space M consists of hyperelliptic curves of genus g with marked homology basis

$$y^2 = \prod_{i=1}^{2g+1} (\lambda - u_i). \tag{6.105}$$

This parametrizes the family of g-gap solutions of the KdV. The L operator has the well-known form

$$L = -\partial_x^2 + u. \tag{6.106}$$

In real smooth periodic case $u(x + T) = u(x)$ the quasimomentum $p(\lambda)$ is defined by the formula

$$\psi(x + T, \lambda) = e^{ip(\lambda)T} \psi(x, \lambda) \tag{6.107}$$

for a solution $\psi(x, \lambda)$ of the equation

$$L\psi = \lambda\psi \tag{6.108}$$

(the Bloch – Floquet eigenfunction). The differential dp can be extended onto the family of all (i.e. quasiperiodic complex meromorphic) g-gap operators (6.105) as a normalized

Abelian differential of the second kind with a double pole at the infinity $\lambda = \infty$. (So the superpotential $\lambda = \lambda(p)$ has the sense of the Bloch dispersion law, i.e. the dependence of the energy λ on the quasimomentum p.) The Hamiltonians of the KdV hierarchy can be obtained as coefficients of expansion of dp near the infinity. To obtain a complete family of conservation laws of the averaged hierarchy (6.25) one needs to extend the family of the KdV integrals by adding nonlocal functionals of u of the form

$$\oint_{a_i} \lambda^k dp, \quad \oint_{b_i} \lambda^{k-1} dp, \quad k = 1, 2, \ldots. \tag{6.109}$$

We will obtain now the semiclassical Lax representation for the equations of the hierarchy (6.25) for *arbitrary* Frobenius manifold.

Let $h(t, z)$ be any solution of the equation

$$\partial_\alpha \partial_\beta h(t, z) = z c^\gamma_{\alpha\beta}(t) \partial_\gamma h(t, z) \tag{6.110}$$

normalized by the homogeneity condition

$$z \partial_z h = \mathcal{L}_E h. \tag{6.111}$$

By $p(t, \lambda)$ I will denote the correspondent flat coordinate of the pencil (H.1) given by the integral (H.11)

$$p(t, \lambda) = \oint z^{\frac{d-3}{2}} e^{-\lambda z} h(t, z) \, dz. \tag{6.112}$$

We introduce the functions

$$q_{\alpha,k}(t, \lambda) := \operatorname*{res}_{w=0} \oint \frac{z^{\frac{d-3}{2}} w^{-k-1}}{z + w} < \nabla h(t, z), \nabla h_\alpha(t, w) > \, dz. \tag{6.113}$$

Lemma 6.2. *The following identity holds*

$$\partial_{T^{\alpha,k}} p(t, \lambda) = \partial_X q_{\alpha,k}(t, \lambda). \tag{6.114}$$

Proof. Integrating the formula (6.34) with the weight $z^{\frac{d-3}{2}} e^{-\lambda z}$ we obtain (6.114). Lemma is proved.

Exercise 6.9. Let $p(\lambda)$, $q(\lambda)$ be two functions of λ depending also on parameters x and t in such a way that

$$\partial_t p(\lambda)_{\lambda=const} = \partial_x q(\lambda)_{\lambda=const}. \tag{6.115}$$

Let $\lambda = \lambda(p)$ be the function inverse to $p = p(\lambda)$ and

$$\rho(p) := q(\lambda(p)). \tag{6.116}$$

Prove that

$$\partial_t \lambda(p)_{p=const} = \{\lambda, \rho\} := \frac{\partial \lambda}{\partial x}\frac{\partial \rho}{\partial p} - \frac{\partial \rho}{\partial x}\frac{\partial \lambda}{\partial p}. \tag{6.117}$$

Theorem 6.5. *The hierarchy (6.25) admits the semiclassical Lax representation*

$$\partial_{T^{\alpha,k}}\lambda = \{\lambda, \rho_{\alpha,k}\} \tag{6.118}$$

where

$$\rho_{\alpha,k} := q_{\alpha,k}(t, \lambda(p,t)) \tag{6.119}$$

and the functions $\lambda = \lambda(p,t)$ is the inverse to (6.112).
Proof follows from (6.114) and (6.117).

In fact we obtain many semiclassical Lax representations of the hierarchy (6.25): one can take any solution of (6.114) and the correspondent flat coordinate of the intersection form and apply the above procedure. The example of A_n Frobenius manifold suggests that for $d < 1$ one should take in (6.118) the flat coordinate $p = p(\lambda, t)$ (I.10) inverse to the LG superpotential constructed in Appendix I for any massive Frobenius manifold with $d < 1$.

The next chapter in our story about Frobenius manifolds could be a quantization of the dispersionless Lax pairs (6.118). We are to substitute back $p \to d/dx$ and to obtain a hierarchy of the KdV type. We hope to address the problem of the quantization in subsequent publications.

References.

1. Ablowitz M., Chakravarty S., and Takhtajan L., Integrable systems, self-dual Yang-Mills equations and connections with modular forms, Univ. Colorado Preprint PAM # 113 (December 1991).

2. Arnol'd V.I., Normal forms of functions close to degenerate critical points. The Weyl groups A_k, D_k, E_k, and Lagrangian singularities, *Functional Anal.* **6** (1972) 3 - 25.

3. Arnol'd V.I., Wave front evolution and equivariant Morse lemma, *Comm. Pure Appl. Math.* **29** (1976) 557 - 582.

4. Arnol'd V.I., Indices of singular points of 1-forms on a manifold with boundary, convolution of invariants of reflection groups, and singular projections of smooth surfaces, *Russ. Math. Surv.* **34** (1979) 1 - 42.

5. Arnol'd V.I., Gusein-Zade S.M., and Varchenko A.N., Singularities of Differentiable Maps, volumes I, II, Birkhäuser, Boston-Basel-Berlin, 1988.

6. Arnol'd V.I., Singularities of Caustics and Wave Fronts, Kluwer Acad. Publ., Dordrecht - Boston - London, 1990.

7. van Asch A., Modular forms and root systems, *Math. Anal.* **222** (1976), 145-170.

8. Aspinwall P.S., Morrison D.R., Topological field theory and rational curves, *Comm. Math. Phys.* **151** (1993) 245 - 262.

9. Astashkevich A., and Sadov V., Quantum cohomology of partial flag manifolds, hep-th/9401103.

10. Atiyah M.F., Topological quantum field theories, *Publ. Math. I.H.E.S.* **68** (1988) 175.

11. Atiyah M.F., and Hitchin N., The Geometry and Dynamics of Magnetic Monopoles, Princeton Univ. Press, Princeton, 1988.

12. Balinskii A., and Novikov S., *Sov. Math. Dokl.* **32** (1985) 228.

13. Balser W., Jurkat W.B., and Lutz D.A., Birkhoff invariants and Stokes multipliers for meromorphic linear differential equations, *J. Math. Anal. Appl.* **71** (1979), 48-94.

14. Balser W., Jurkat W.B., and Lutz D.A., On the reduction of connection problems for differential equations with an irregular singular point to ones with only regular singularities, *SIAM J. Math. Anal.* **12** (1981) 691 - 721.

15. Batyrev V.V., Quantum cohomology rings of toric manifolds, *Astérisque* **218** (1993), 9-34.

16. Bernstein J.N., and Shvartsman O.V., Chevalley theorem for complex crystallographic group, *Funct. Anal. Appl.* **12** (1978), 308-310;
Chevalley theorem for complex crystallographic groups, Complex crystallographic Coxeter groups and affine root systems, In: *Seminar on Supermanifolds*, 2, ed. D.Leites, Univ. Stockholm (1986).

17. Bershadsky M., Cecotti S., Ooguri H., and Vafa C., Kodaira - Spencer theory of gravity and exact results for quantum string amplitudes, Preprint HUTP-93/A025, RIMS-946, SISSA-142/93/EP.

18. Beukers F., and Heckman G., Monodromy for hypergeometric function $_nF_{n-1}$, Invent. Math. **95** (1989), 325-354.

19. Birkhoff G.D., Singular points of ordinary linear differential equations, Proc. Amer. Acad. **49** (1913), 436-470.

20. Birman J., Braids, Links, and Mapping Class Groups, Princeton Univ. Press, Princeton, 1974.

21. Blok B. and Varchenko A., Topological conformal field theories and the flat coordinates, Int. J. Mod. Phys. **A7** (1992) 1467.

22. Bourbaki N., Groupes et Algèbres de Lie, Chapitres 4, 5 et 6, Masson, Paris-New York-Barcelone-Milan-Mexico-Rio de Janeiro, 1981.

23. Brezin E., Itzykson C., Parisi G., and Zuber J.-B., Comm. Math. Phys. **59** (1978) 35.
Bessis D., Itzykson C., and Zuber J.-B., Adv. Appl. Math. **1** (1980) 109.
Mehta M.L., Comm. Math. Phys. **79** (1981) 327;
Chadha S., Mahoux G., and Mehta M.L., J.Phys. **A14** (1981) 579.

24. Brezin E., and Kazakov V, Exactly solvable field theories of closed strings, Phys. Lett. **B236** (1990) 144.

25. Brieskorn E. Singular elements of semisimple algebraic groups, In: Actes Congres Int. Math., **2**, Nice (1970), 279 - 284.

26. Candelas P., de la Ossa X.C., Green P.S., and Parkes L., A pair of Calabi - Yau manifolds as an exactly soluble superconformal theory, Nucl. Phys. **B359** (1991), 21-74.

27. Cassels J.W., An Introduction to the Geometry of Numbers, Springer, N.Y. etc., 1971.

28. Cecotti S. and Vafa C., Topological-antitopological fusion, Nucl. Phys. **B367** (1991) 359-461.

29. Cecotti S. and Vafa C., On classification of $N=2$ supersymmetric theories, Comm. Math. Phys. **158** (1993), 569-644.

30. Ceresole A., D'Auria R., and Regge T., Duality group for Calabi - Yau 2-modulii space, Preprint POLFIS-TH. 05/93, DFTT 34/93.

31. Chazy J., Sur les équations différentiellles dont l'intégrale générale possède un coupure essentielle mobile, C.R. Acad. Sc. Paris **150** (1910), 456-458.

32. Chakravarty S., Private communication, June 1993.

33. Coxeter H.S.M., Discrete groups generated by reflections, Ann. Math. **35** (1934) 588 - 621.

34. Coxeter H.S.M., The product of the generators of a finite group generated by reflections, *Duke Math. J.* **18** (1951) 765 - 782.

35. Darboux G., *Leçons sur les systèmes ortogonaux et les cordonnées curvilignes*, Paris, 1897.
    Egoroff D.Th., *Collected papers on differential geometry*, Nauka, Moscow (1970) (in Russian).

36. Date E., Kashiwara M, Jimbo M., and Miwa T., Transformation groups for soliton equations, In: *Nonlinear Integrable Systems - Classical Theory and Quantum Theory*, World Scientific, Singapore, 1983, pp. 39-119.

37. Deligne P., and Mumford D., The irreducibility of the space of curves of given genus, *Inst. Hautes Études Sci. Publ. Math.* **45** (1969), 75.

38. Di Francesco P., Lesage F., and Zuber J.B., Graph rings and integrable perturbations of N=2 superconformal theories, *Nucl. Phys.* **B408** (1993), 600-634.

39. Di Francesco P., and Itzykson C., Quantum intersection rings, Preprint SPhT 94/111, September 1994.

40. Dijkgraaf R., *A Geometrical Approach to Two-Dimensional Conformal Field Theory*, Ph.D. Thesis (Utrecht, 1989).

41. Dijkgraaf R., Intersection theory, integrable hierarchies and topological field theory, Preprint IASSNS-HEP-91/91, December 1991.

42. Dijkgraaf R., E.Verlinde, and H.Verlinde, *Nucl. Phys.* **B 352** (1991) 59;
    Notes on topological string theory and 2D quantum gravity, Preprint PUPT-1217, IASSNS-HEP-90/80, November 1990.

43. Dijkgraaf R., and Witten E., *Nucl. Phys.* **B 342** (1990) 486.

44. Drinfel'd V.G. and Sokolov V.V., *J. Sov. Math.* **30** (1985) 1975.

45. Donagi R., and Markman E., Cubics, Integrable systems, and Calabi - Yau threefolds, Preprint (1994).

46. Douglas M., and Shenker S., Strings in less than one dimension, *Nucl. Phys.* **B335** (1990) 635.

47. Dubrovin B., On differential geometry of strongly integrable systems of hydrodynamic type, *Funct. Anal. Appl.* **24** (1990).

48. Dubrovin B., Differential geometry of moduli spaces and its application to soliton equations and to topological field theory, Preprint No.117, Scuola Normale Superiore, Pisa (1991).

49. Dubrovin B., Hamiltonian formalism of Whitham-type hierarchies and topological Landau - Ginsburg models, *Comm. Math. Phys.* **145** (1992) 195 - 207.

50. Dubrovin B., Integrable systems in topological field theory, *Nucl. Phys.* **B 379** (1992) 627 - 689.

51. Dubrovin B., Geometry and integrability of topological-antitopological fusion, *Comm. Math. Phys.***152** (1993), 539-564.

52. Dubrovin B., Integrable systems and classification of 2-dimensional topological field theories, In "Integrable Systems", Proceedings of Luminy 1991 conference dedicated to the memory of J.-L. Verdier. Eds. O.Babelon, O.Cartier, Y.Kosmann-Schwarbach, Birkhäuser, 1993.

53. Dubrovin B., Topological conformal field theory from the point of view of integrable systems, In: Proceedings of 1992 Como workshop "Quantum Integrable Field Theories". Eds. L.Bonora, G.Mussardo, A.Schwimmer, L.Girardello, and M.Martellini, Plenum Press, 1993.

54. Dubrovin B., Differential geometry of the space of orbits of a Coxeter group, Preprint SISSA-29/93/FM (February 1993).

55. Dubrovin B., Fokas A.S., and Santini P.M., Integrable functional equations and algebraic geometry, *Duke Math. J.* (1994).

56. Dubrovin B., Krichever I., and Novikov S., *Integrable Systems*. I. Encyclopaedia of Mathematical Sciences, vol.4 (1985) 173, Springer-Verlag.

57. Dubrovin B. and Novikov S.P., The Hamiltonian formalism of one-dimensional systems of the hydrodynamic type and the Bogoliubov - Whitham averaging method,*Sov. Math. Doklady* **27** (1983) 665 - 669.

58. Dubrovin B. and Novikov S.P., Hydrodynamics of weakly deformed soliton lattices. Differential geometry and Hamiltonian theory, *Russ. Math. Surv.* **44:6** (1989) 35 - 124.

59. Dubrovin B., Novikov S.P., and Fomenko A.T., Modern Geometry, Parts 1 - 3, Springer Verlag.

60. Eguchi T., Kanno H., Yamada Y., and Yang S.-K., Topological strings, flat coordinates and gravitational descendants, *Phys. Lett.***B305** (1993), 235-241;
Eguchi T., Yamada Y., and Yang S.-K., Topological field theories and the period integrals, *Mod. Phys. Lett.* **A8** (1993), 1627-1638.

61. Eguchi T., Yamada Y., and Yang S.-K., On the genus expansion in the topological string theory, Preprint UTHEP-275 (May 1994).

62. Eichler M., and Zagier D., The Theory of Jacobi Forms, Birkhäuser, 1983.

63. Flaschka H., Forest M.G., and McLaughlin D.W., *Comm. Pure Appl. Math.* **33** (1980) 739.

64. Flaschka H., and Newell A.C., *Comm. Math. Phys.* **76** (1980) 65.

65. Fokas A.S., Ablowitz M.J., On a unified approach to transformations and elementary solutions of Painlevé equations, *J. Math. Phys.* **23** (1982), 2033-2042.

66. Fokas A.S., Leo R.A., Martina L., and Soliani G., *Phys. Lett.* **A115** (1986) 329.

343

67. Frobenius F.G., and Stickelberger L., Über die Differentiation der elliptischen Funktionen nach den Perioden und Invarianten, *J. Reine Angew. Math.* **92** (1882).

68. Gelfand I.M., and Dickey L.A., *A Family of Hamilton Structures Related to Integrable Systems*, preprint IPM/136 (1978) (in Russian).
Adler M., *Invent. Math.* **50** (1979) 219.
Gelfand I.M., and Dorfman I., *Funct. Anal. Appl.* **14** (1980) 223.

69. Givental A.B., Sturm theorem for hyperelliptic integrals, *Algebra and Analysis* **1** (1989), 95-102 (in Russian).

70. Givental A.B., Convolution of invariants of groups generated by reflections, and connections with simple singularities of functions, *Funct. Anal.* **14** (1980) 81 - 89.

71. Givental A., Kim B., Quantum cohomology of flag manifolds and Toda lattices, Preprint UC Berkeley (December 1993).

72. Gromov M., Pseudo-holomorphic curves in symplectic manifolds, *Invent. Math.* **82** (1985), 307.

73. Gross D.J., and Migdal A.A., A nonperturbative treatment of two dimensional quantum gravity, *Phys. Rev. Lett.* **64** (1990) 127.

74. Gurevich A.V., and Pitaevskii L.P., *Sov. Phys. JETP* **38** (1974) 291; *ibid.*, **93** (1987) 871; *JETP Letters* **17** (1973) 193.
Avilov V., and Novikov S., *Sov. Phys. Dokl.* **32** (1987) 366.
Avilov V., Krichever I., and Novikov S., *Sov. Phys. Dokl.* **32** (1987) 564.

75. Halphen G.-H., Sur un systèmes d'équations différentielles, *C.R. Acad. Sc. Paris* **92** (1881), 1001-1003, 1004-1007.

76. Hanany A., Oz Y., and Plesser M.R., Topological Landau - Ginsburg formulation and integrable structures of 2d string theory, IASSNS-HEP 94/1.

77. Hitchin N., Talk at ICTP (Trieste), April, 1993.

78. Hitchin N., Poncelet polygons and the Painlevé equations, Preprint, February, 1994.

79. Hosono S., Klemm A., Theisen S., and Yau S.-T., Mirror symmetry, mirror map and applications to complete intersection Calabi - Yau spaces, Preprint HUTMP-94/02, CERN-TH.7303/94, LMU-TPW-94-03 (June 1994).

80. Hurwitz A., Über die Differentsialglechungen dritter Ordnung, welchen die Formen mit linearen Transformationen in sich genügen, *Math. Ann.* **33** (1889), 345-352.

81. Ince E.L., Ordinary Differential Equations, London - New York etc., Longmans, Green and Co., 1927.

82. Its A.R., and Novokshenov V.Yu., *The Isomonodromic Deformation Method in the Theory of Painlevé Equations*, Lecture Notes in Mathematics 1191, Springer-Verlag, Berlin 1986.

83. Jacobi C.G.J., *Crelle J.* **36** (1848), 97-112.

84. Jimbo M., and Miwa T., Monodromy preserving deformations of linear ordinary differential equations with rational coefficients. II. *Physica* **2D** (1981) 407 - 448.

85. Kac V.G., and Peterson D., Infinite-dimensional Lie algebras, theta functions and modular forms, *Adv. in Math.* **53** (1984), 125-264.

86. Kim B., Quantum cohomology of partial flag manifolds and a residue formula for their intersection pairing, Preprint UC Berkeley (May 1994).

87. Koblitz N., Introduction to Elliptic Curves and Modular Forms, Springer, New York etc., 1984.

88. Kodama Y. A method for solving the dispersionless KP equation and its exact solutions, *Phys. Lett.* **129A** (1988), 223-226;
Kodama Y. and Gibbons J., A method for solving the dispersionless KP hierarchy and its exact solutions, II, *Phys. Lett.* **135A** (1989) 167-170;
Kodama Y., Solutions of the dispersionless Toda equation, *Phys. Lett.* **147A** (1990), 477-482.

89. Kontsevich M., *Funct. Anal.* **25** (1991) 50.

90. Kontsevich M., Intersection theory on the moduli space of curves and the matrix Airy function, *Comm. Math. Phys.* **147** (1992) 1-23.

91. Kontsevich M., Enumeration of rational curves via torus action, *Comm. Math. Phys.* **164** (1994) 525 - 562.

92. Kontsevich M., Manin Yu.I., Gromov - Witten classes, quantum cohomology and enumerative geometry, MPI preprint (1994).

93. Krichever I.M., Integration of nonlinear equations by the methods of algebraic geometry, *Funct. Anal. Appl.* **11** (1977), 12-26.

94. Krichever I.M., The dispersionless Lax equations and topological minimal models, *Commun. Math. Phys.* **143** (1991), 415-426.

95. Krichever I.M., The τ-function of the universal Whitham hierarchy, matrix models and topological field theories, *Comm. Pure Appl. Math.* **47** (1994) 437 - 475.

96. Lerche W., Generalized Drinfeld - Sokolov hierarchies, quantum rings, and W-gravity, Preprint CERN-TH.6988/93.

97. Lerche W., Chiral rings and integrable systems for models of topological gravity, Preprint CERN-TH.7128/93.

98. Lerche W., On the Landau - Ginsburg realization of topological gravities, Preprint CERN-TH.7210/94 (March 1994).

99. Lerche W., Vafa C., and Warner N.P., Chiral rings in N=2 superconformal theories, *Nucl. Phys.* **B324** (1989), 427-474.

100. Li K., *Topological gravity and minimal matter*, CALT-68-1662, August 1990; *Recursion relations in topological gravity with minimal matter*, CALT-68-1670, September 1990.

345

101. Looijenga E., A period mapping for certain semiuniversal deformations, *Compos. Math.* **30** (1975) 299 - 316.

102. Looijenga E., Root systems and elliptic curves, *Invent. Math.* **38** (1976), 17-32.

103. Looijenga E., Invariant theory for generalized root systems, *Invent. Math.* **61** (1980), 1-32.

104. Losev A., Descendants constructed from matter field in topological Landau - Ginsburg theories coupled to topological gravity, Preprint ITEP (September 1992).

105. Losev A.S., and Polyubin I., On connection between topological Landau - Ginsburg gravity and integrable systems, hep-th/9305079 (May, 1993).

106. Maassarani Z., *Phys. Lett.* **273B** (1992) 457.

107. Magnus W., Oberhettinger F., and Soni R.P., *Formulas and Theorems for the Special Functions of Mathematical Physics*, Springer-Verlag, Berlin - Heidelberg - New York, 1966.

108. Magri F., *J. Math. Phys.* **19** (1978) 1156.

109. Malgrange B., Équations Différentielles à Coefficients Polynomiaux, Birkhäuser, 1991.

110. McCoy B.M., Tracy C.A., Wu T.T., *J. Math. Phys.* **18** (1977) 1058.

111. McDuff D., and Salamon D., J-Holomorphic curves and quantum cohomology, Preprint SUNY and Math. Inst. Warwick (April 1994).

112. Miller E., The homology of the mapping class group, *J. Diff. Geom.* **24** (1986), 1. Morita S., Characteristic classes of surface bundles, *Invent. Math.* **90** (1987), 551. Mumford D., Towards enumerative geometry of the moduli space of curves, In: *Arithmetic and Geometry*, eds. M.Artin and J.Tate, Birkhäuser, Basel, 1983.

113. Miwa T., Painlevé property of monodromy presereving equations and the analyticity of τ-functions, *Publ. RIMS* **17** (1981), 703-721.

114. Morrison D.R., Mirror symmetry and rational curves on quintic threefolds: a guide for mathematicians, *J. Amer. Math. Soc.* **6** (1993), 223-247.

115. Nagura M., and Sugiyama K., Mirror symmetry of K3 and torus, Preprint UT-663.

116. Nakatsu T., Kato A., Noumi M., and Takebe T., Topological strings, matrix integrals, and singularity theory, *Phys. Lett.* **B322** (1994), 192-197.

117. Natanzon S., *Sov. Math. Dokl.* **30** (1984) 724.

118. Noumi M., Expansions of the solution of a Gauss - Manin system at a point of infinity, *Tokyo J. Math.* **7:1** (1984), 1-60.

119. Novikov S.P. (Ed.), *The Theory of Solitons: the Inverse Problem Method*, Nauka, Moscow, 1980. Translation: Plenum Press, N.Y., 1984.

120. Novikov S.P., and Veselov A.P., *Sov. Math. Doklady* (1981);

Proceedings of Steklov Institute (1984).

121. Okamoto K., Studies on the Painlevé equations, *Annali Mat. Pura Appl.* **146** (1987), 337-381.

122. Okubo K., Takano K., and Yoshida S., A connection problem for the generalized hypergeometric equation, *Fukcial. Ekvac.* **31** (1988), 483-495.

123. Olver P.J., *Math. Proc. Cambridge Philos. Soc.* **88** (1980) 71.

124. Piunikhin S., Quantum and Floer cohomology have the same ring structure, Preprint MIT (March 1994).

125. Procesi C. and Schwarz G., Inequalities defining orbit spaces, *Invent. Math.* **81** (1985) 539 - 554.

126. Rankin R.A., The construction of automorphic forms from the derivatives of a given form, *J. Indian Math. Soc.* **20** (1956), 103-116.

127. Ruan Y., Tian G., A mathematical theory of quantum cohomology, *Math. Res. Lett.* **1** (1994), 269-278.

128. Rudakov A.N., Integer valued bilinear forms and vector bundles, *Math. USSR Sbornik* **180** (1989), 187-194.

129. Sadov V., On equivalence of Floer's and quantum cohomology, Preprint HUTP-93/A027.

130. Saito K., On a linear structure of a quotient variety by a finite reflection group, Preprint RIMS-288 (1979).

131. Saito K., Yano T., and Sekeguchi J., On a certain generator system of the ring of invariants of a finite reflection group, *Comm. in Algebra* **8(4)** (1980) 373 - 408.

132. Saito K., On the periods of primitive integrals. I., Harvard preprint, 1980.

133. Saito K., Period mapping associated to a primitive form, *Publ. RIMS* **19** (1983) 1231 - 1264.

134. Saito K., Extended affine root systems II (flat invariants), *Publ. RIMS* **26** (1990) 15 - 78.

135. Sato M., Miwa T., and Jimbo A., *Publ. RIMS* **14** (1978) 223; **15** (1979) 201, 577, 871; **16** (1980) 531.
     Jimbo A., Miwa T., Mori Y., Sato M., *Physica* **1D** (1980) 80.

136. Satake I., Flat structure of the simple elliptic singularity of type $\tilde{E}_6$ and Jacobi form, hep-th/9307009.

137. Shcherbak O.P., Wavefronts and reflection groups, *Russ. Math. Surv.* **43:3** (1988) 149 - 194.

138. Schlesinger L., Über eine Klasse von Differentsialsystemen beliebliger Ordnung mit festen kritischer Punkten, *J. für Math.* **141** (1912), 96-145.

139. Slodowy P., Einfache Singularitaten und Einfache Algebraische Gruppen, Preprint, Regensburger Mathematische Schriften **2**, Univ. Regensburg (1978).

140. Spiegelglass M., Setting fusion rings in topological Landau - Ginsburg, *Phys. Lett.* **B274** (1992), 21-26.

141. Takasaki K. and Takebe T., SDiff(2) Toda equation - hierarchy, tau function and symmetries, *Lett. Math. Phys.* **23** (1991), 205-214;
Integrable hierarchies and dispersionless limi, Preprint UTMS 94-35.

142. Takhtajan L., Modular forms as tau-functions for certain integrable reductions of the Yang-Mills equations, Preprint Univ. Colorado (March 1992).

143. Todorov A.N., Global properties of the moduli of Calabi - Yau manifolds-II (Teichmüller theory), Preprint UC Santa Cruz (December 1993);
Some ideas from mirror geometry applied to the moduli space of K3, Preprint UC Santa Cruz (April 1994).

144. Tsarev S., *Math. USSR Izvestija* **36** (1991); *Sov. Math. Dokl.* **34** (1985) 534.

145. Vafa C., *Mod. Phys. Let.* **A4** (1989) 1169.

146. Vafa C., Topological mirrors and quantum rings, in [162].

147. Varchenko A.N. and Chmutov S.V., Finite irreducible groups, generated by reflections, are monodromy groups of suitable singularities, *Func. Anal.* **18** (1984) 171 - 183.

148. Varchenko A.N. and Givental A.B., Mapping of periods and intersection form, *Funct. Anal.* **16** (1982) 83 - 93.

149. Verlinde E. and Warner N., *Phys. Lett.* **269B** (1991) 96.

150. Voronov A.A., Topological field theories, string backgrounds, and homotopy algebras, Preprint University of Pennsylvania (December 1993).

151. Wall C.T.C., A note on symmetry of singularities, *Bull. London Math. Soc.* **12** (1980) 169 - 175;
A second note on symmetry of singularities, *ibid.*, 347 - 354.

152. Wasow W., Asymptotic expansions for ordinary differential equations, Wiley, New York, 1965.

153. Whittaker E.T., and Watson G.N., A Course of Modern Analysis, N.Y. AMS Press, 1979.

154. Wirthmüller K., Root systems and Jacobi forms, *Compositio Math.* **82** (1992), 293-354.

155. Witten E., *Comm. Math. Phys.* **117** (1988) 353;
ibid., **118** (1988) 411.

156. Witten E., On the structure of the topological phase of two-dimensional gravity, *Nucl. Phys.* **B 340** (1990) 281-332.

157. Witten E., Two-dimensional gravity and intersection theory on moduli space, *Surv. Diff. Geom.* **1** (1991) 243-210.

158. Witten E., Algebraic geometry associated with matrix models of two-dimensional gravity, Preprint IASSNS-HEP-91/74.

159. Witten E., Lectures on mirror symmetry, In: [162].

160. Witten E., On the Kontsevich model and other models of two dimensional gravity, Preprint IASSNS-HEP-91/24.

161. Witten E., Phases of N=2 theories in two dimensions, *Nucl. Phys.* **B403** (1993), 159-222.

162. Yau S.-T., ed., Essays on Mirror Manifolds, International Press Co., Hong Kong, 1992.

163. Yano T., Free deformation for isolated singularity, *Sci. Rep. Saitama Univ.* **A9** (1980), 61-70.

164. Zakharov V.E., On the Benney's equations, *Physica* **3D** (1981), 193-202.

165. Zuber J.-B., On Dubrovin's topological field theories, Preprint SPhT 93/147.

INTEGRALS OF MOTION AND QUANTUM GROUPS

BORIS FEIGIN AND EDWARD FRENKEL

CONTENTS

1. Introduction
2. Classical Toda field theories associated to finite-dimensional simple Lie algebras
   2.1. The case of $\mathfrak{sl}_2$ – classical Liouville theory
   2.2. General case
   2.3. BGG resolution
   2.4. Extended complex and its cohomology
3. Classical affine Toda field theories
   3.1. The case of $\widehat{\mathfrak{sl}}_2$ – classical sine-Gordon theory
   3.2. General case
4. Quantum Toda field theories
   4.1. Vertex operator algebras
   4.2. The VOA of the Heisenberg algebra
   4.3. Quantum integrals of motion
   4.4. Liouville theory
   4.5. Quantum groups and quantum BGG resolutions
   4.6. Toda field theories associated to finite-dimensional simple Lie algebras
   4.7. Affine Toda field theories
   4.8. Concluding remarks

References

1. INTRODUCTION.

1.1. In these lectures we propose a new approach to the study of local integrals of motion in the classical and quantum Toda field theories. Such a theory is associated to a Lie algebra $\mathfrak{g}$, which is either a finite-dimensional simple Lie algebra or an affine Kac-Moody algebra.

The classical Toda field theory, associated to $\mathfrak{g}$, revolves around the system of equations

$$(1.1.1) \qquad \partial_\tau \partial_t \phi_i(t,\tau) = \frac{1}{2} \sum_{j \in S} (\alpha_i, \alpha_j) \exp[\phi_j(t,\tau)], \qquad i \in S$$

where each $\phi_i(t,\tau)$ is a family of functions in t, depending on the time variable τ, S is the set of simple roots of $\mathfrak{g}$, and (α_i, α_j) is the scalar product of the ith and jth simple roots [81].

The simplest examples of the Toda equations are the Liouville equation

$$(1.1.2) \qquad \partial_\tau \partial_t \phi(t,\tau) = e^{\phi(t,\tau)},$$

corresponding to $\mathfrak{g} = \mathfrak{sl}_2$, and the sine-Gordon equation

(1.1.3) $$\partial_\tau \partial_t \phi(t, \tau) = e^{\phi(t,\tau)} - e^{-\phi(t,\tau)},$$

corresponding to $\mathfrak{g} = \widehat{\mathfrak{sl}}_2$.

Many aspects of the classical Toda field theories have been studied by both physicists and mathematicians (cf., e.g., [1, 113, 96, 97, 102, 103, 32, 33, 86, 116, 90, 38, 7, 3, 4, ?, 104] and references therein): realization as a zero-curvature equation, complete integrability, soliton solutions, dressing transformations, connection with generalized KdV hierarchies, lattice analogues, etc. There are also many interesting works devoted to the quantum Toda field theory, cf., e.g., [118, 112, 74, 13, 77, 35, 106, 23, 115, 39, 100] and references therein.

1.2. In this paper we will study a Hamiltonian formalism for the Toda field theories. By that we mean constructing a Hamiltonian space M and a hamiltonian H, such that the system of equations (1.1.1) can be rewritten in the Hamiltonian form:

(1.2.1) $$\partial_\tau U = \{U, H\}.$$

Here $\{\cdot, \cdot\}$ stands for a Poisson bracket on the space $F(M)$ of functions on M. We will be primarily interested in the integrals of motion for the equation (1.2.1). An integral of motion is an element X of the space $F(M)$, which satisfies the equation

$$\{X, H\} = 0.$$

It is conserved with respect to the evolution of the Hamiltonian system, defined by the equation (1.2.1).

Note that this definition does not require H to be an element of $F(M)$, it merely requires the Poisson bracket with H to be a well-defined linear operator, acting from $F(M)$ to some other vector space. Given H, we can define the space of integrals of motion of the system (1.2.1) as the kernel of this linear operator. If this operator preserves the Poisson bracket, then the space of integrals of motion is itself a Poisson algebra.

1.3. For the Toda equation (1.1.1) we choose as the Hamiltonian space, the space $L\mathfrak{h}$ of polynomial functions on the circle with values in the Cartan subalgebra $\mathfrak{h}$ of $\mathfrak{g}$ and as the space of functions, the space $\mathcal{F}_0$ of local functionals on $L\mathfrak{h}$. Such a functional can be presented in the form of the residue

$$F[\mathbf{u}(t)] = \int P(\mathbf{u}, \partial_t \mathbf{u}, \dots)dt,$$

where P is a polynomial in the coordinates $u^i(t)$ of $\mathbf{u}(t) \in L\mathfrak{h}$ with respect to the basis of the simple roots, and their derivatives (cf. § 2 for the precise definition). The Poisson structure on $L\mathfrak{h}$ has an interpretation as a Kirillov-Kostant structure, because $L\mathfrak{h}$ can be viewed as a hyperplane in the dual space to the Heisenberg Lie algebra $\widehat{\mathfrak{h}}$ – the central extension of $L\mathfrak{h}$. This defines a Poisson bracket on $\mathcal{F}_0$.

The space $\mathcal{F}_0$ was one of the first examples of Poisson algebras of functions on infinite-dimensional hamiltonian spaces. It has been studied since the discovery of integrability of the KdV equation and its generalizations, and exhaustive literature is devoted to it, cf., e.g. [68, 67, 117, 88, 69, 70, 2, 95, 33, 116, 37, 27]. We essentially follow the approach of Gelfand and Dickey and treat the space $\mathcal{F}_0$ in a purely algebraic way. In addition, we also consider the spaces $\mathcal{F}_{\alpha_i}$, consisting of functionals of the form

$$\int P(\mathbf{u}, \partial_t \mathbf{u}, \dots) e^{\phi_i(t)} dt,$$

where $\phi_i(t)$ is such that $\partial_t \phi_i(t) = u^i(t)$. It is possible to extend the Poisson bracket $\mathcal{F}_0 \times \mathcal{F}_0 \to \mathcal{F}_0$ to a bilinear map $\mathcal{F}_0 \times \mathcal{F}_{\alpha_i} \to \mathcal{F}_{\alpha_i}$, cf. [86, 116, 33], and this allows to write the Toda equation (1.1.1) in the Hamiltonian form

$$\partial_\tau \mathbf{u}(t) = \{\mathbf{u}(t), H\}.$$

Here the hamiltonian H is given by

$$H = \frac{1}{2} \sum_i \int e^{\phi_i(t)} dt.$$

It is an element of $\oplus_i \mathcal{F}_{\alpha_i}$, and the Poisson bracket with H is a well-defined linear operator, acting from $\mathcal{F}_0$ to $\oplus_i \mathcal{F}_{\alpha_i}$.

So we can define the space of local integrals of motion of the Toda equation (1.1.1) as the kernel of the operator $\{\cdot, H\}$, or, in other words, as the intersection of the kernels of the operators $\bar{Q}_i = \{\cdot, \int e^{\phi_i(t)} dt\} : \mathcal{F}_0 \to \mathcal{F}_{\alpha_i}$. These operators preserve the Poisson structure, and hence the space of integrals of motion is a Poisson subalgebra of $\mathcal{F}_0$.

1.4. The crucial observation, which will enable us to compute this space, is that, roughly speaking, *the operators $\bar{Q}_i$ satisfy the Serre relations of the Lie algebra $\mathfrak{g}$, or, in other words, they generate the nilpotent subalgebra $\mathfrak{n}_+$ of $\mathfrak{g}$*. Using this fact, we will be able to interpret the space of local integrals of motion as a cohomology space of a certain complex $F^*(\mathfrak{g})$. To construct this complex, we will use the so-called Bernstein–Gelfand–Gelfand (BGG) resolution, which is the resolution of the trivial representation of $\mathfrak{n}_+$ by Verma modules. The cohomologies of this complex coincide with the cohomologies of $\mathfrak{n}_+$ with coefficients in some $\mathfrak{n}_+$–module.

More precisely, we can lift the operators $\bar{Q}_i$ to certain linear operators Q_i, acting on the space π_0 of differential polynomials in $u^i(t)$. These operators give us an action of the Lie algebra $\mathfrak{n}_+$ on π_0.

In the case, when $\mathfrak{g}$ is a finite-dimensional simple Lie algebra, the 0th cohomology of $\mathfrak{n}_+$ with coefficients in π_0 can be identified with the space of differential polynomials in $W^{(1)}, \dots, W^{(l)} \in \pi_0$ of degrees $d_1 + 1, \dots, d_l + 1$, respectively. Here the d_i's are the exponents of $\mathfrak{g}$, and the grading is defined on π_0 in such a way that the degree of $\partial_t^n u^i(t)$ is equal to $n + 1$. We prove that the integrals of motion of the Toda field theory, corresponding to $\mathfrak{g}$, coincide with all residues of differential polynomials in the $W^{(i)}$'s. They form a Poisson subalgebra of $\mathcal{F}_0$, which is called the Adler-Gelfand-Dickey algebra, or the classical $\mathcal{W}$–algebra, associated with the Lie algebra $\mathfrak{g}$.

The space of integrals of motion of the affine Toda field theory, associated to an affine algebra $\mathfrak{g}$, can be identified with the first cohomology of the nilpotent subalgebra $\mathfrak{n}_+$ of $\mathfrak{g}$ with coefficients in π_0. This space is naturally embedded into the space of integrals of motion of the Toda theory, associated to the finite-dimensional Lie algebra $\bar{\mathfrak{g}}$, whose Dynkin diagram is obtained by deleting the 0th nod of the Dynkin diagram of $\mathfrak{g}$ (or any other nod).

We can compute the latter cohomology and obtain the well-known result that the integrals of motion of the affine Toda theory have degrees equal to the exponents of $\mathfrak{g}$ modulo the Coxeter number. These integrals of motion commute with each other. This is especially easy to see in the case when all the exponents of the corresponding affine algebra are odd, and the Coxeter number is even (this excludes $A_n^{(1)}, n > 1, D_{2n}^{(1)}, E_6^{(1)}$ and $E_7^{(1)}$). In such a case the degrees of all integrals of motion are odd, so that the Poisson bracket of any two of them should be an integral of motion of an even degree, and hence should vanish. The set of local integrals of motion of the affine Toda field theory, associated to $\mathfrak{g}$, coincides with the set of hamiltonians of the corresponding generalized KdV system. These integrals of motion generate a maximal abelian subalgebra in the Poisson algebra of integrals of motion of the corresponding finite-dimensional Toda field theory.

In our next paper [51] we explain further the geometric meaning of higher KdV hamiltonians. Namely, we will identify the vector space with the coordinates $\partial^n u^i, i = 1, \ldots, l, n \geq 0$, with a homogeneous space of the nilpotent subgroup N_+ of the corresponding affine group. This homogeneous space is the quotient of the group N_+ by its principal commutative subgroup – the Lie group of $\mathfrak{a}$. The vector fields on this space, which, by Gelfand-Dickey formalism correspond to the KdV hamiltonians, coincide with the vector fields of the infinitesimal action of the opposite principal abelian subalgebra $\mathfrak{a}_- \subset \mathfrak{n}_-$ on this homogeneous space. In particular, this identification enables us to prove the mutual commutativity of hamiltonians in general case.

1.5. Thus we obtain an interpretation of the integrals of motion of the classical Toda field theories as cohomologies of certain complexes. This formulation not only allows us to describe the spaces of classical integrals of motion, but also to prove the existence of their quantum deformations.

The quantum integrals of motion are defined as elements of the quantum Heisenberg algebra, which is a quantization of the Poisson algebra $\mathcal{F}_0$. More precisely, we define a Lie algebra $\mathcal{F}_0^\beta$ of all Fourier components of vertex operators from the vertex operator algebra of the Heisenberg algebra, cf. § 4.2. For completeness, we include in § 4.1 a survey of vertex operator algebras, which closely follows § 3 of [58].

The Lie bracket in $\mathcal{F}_0^\beta$ is polynomial in the deformation parameter β^2 with zero constant term, and the linear term coincides with the Poisson bracket in $\mathcal{F}_0$, so that $\mathcal{F}_0^\beta$ degenerates into $\mathcal{F}_0$ when $\beta \to 0$. We can also interpret the operators $\bar{Q}_i$ as classical limits of integrals of bosonic vertex operators, $\bar{Q}_i^\beta$. Therefore it is natural to define the space of quantum integrals of motion of the affine Toda field theory as the intersection of kernels of the operators $\bar{Q}_i^\beta$.

Thus, if $x = x^{(0)} + \beta^2 x^{(1)} + \cdots \in \mathcal{F}_0^\beta$ is a quantum integral of motion, then $x^{(0)}$ is a classical integral of motion. It remains to be seen however, whether for each classical integral of motion there exists its quantum deformation, and whether such deformed integrals of motion commute with each other.

Such quantum deformations do not necessarily exist in general. Indeed, if we have a family of linear operators acting between two vector spaces, then the dimension of the kernel may increase for a special value of parameter. Thus, the space of classical integrals of motion, which is defined as the kernel of the operator $\sum_i \bar{Q}_i^\beta$ for the *special* value $\beta = 0$, may well be larger than the space of quantum integrals of motion, which is defined as the kernel of the operator $\sum_i \bar{Q}_i^\beta$ for *generic* values of β.

1.6. In order to prove the existence of quantum integrals of motion we will use higher cohomologies. The usefulness of higher cohomologies can be illustrated by the following toy example.

Suppose, we have two finite-dimensional vector spaces, A and B, and a family of linear operators ϕ_β depending on a parameter β. Assume that for $\beta = 0$ the 1st cohomology of the complex $A \longrightarrow B$ (= the cokernel of the operator ϕ_0) is equal to 0. One can show then that the 0th cohomology of this complex (= the kernel of ϕ_0) can be deformed.

Indeed, vanishing of the 1st cohomology of the complex $A \longrightarrow B$ for $\beta = 0$ entails vanishing of the 1st cohomology for generic β, because the dimension of cohomology stays the same for generic values of parameter, and it may only *increase* for special values. But the Euler characteristics of our complex, i.e. the difference between the dimension of the kernel and the dimension of the cokernel, is also equal to $\dim A - \dim B$ and hence does not depend on β. Since the 1st cohomology vanishes for generic β and $\beta = 0$, we see that the dimension of the 0th cohomology for generic β is the same as for $\beta = 0$.

We can apply this idea in our situation. Although our spaces are infinite-dimensional, they are $\mathbb{Z}$–graded with finite-dimensional homogeneous components, and our operator H preserves the grading. So, our infinite-dimensional linear problem splits into a set of finite-dimensional linear problems. In the simplest case of $\mathfrak{g} = \mathfrak{sl}_2$, these finite-dimensional problems can be solved in the same way as in the example above. By proving vanishing of the cokernel of our operator $\bar{Q}_1$, we can prove that all classical local integrals of motion in the Liouville theory can be quantized. The quantum algebra of integrals of motion in this case is the quantum Virasoro algebra, which is a well-known fact.

In general, the cokernel of our operator is not equal to 0 for $\beta = 0$, so that this simple trick does not work. However, a deformation of our extended complex $F^*(\mathfrak{g})$ will do the job. Roughly speaking, it turns out that the operators Q_i^β generate the quantized universal enveloping algebra $U_q(\mathfrak{n}_+)$ of the nilpotent subalgebra $\mathfrak{n}_+$ with $q = \exp(\pi i \beta^2)$. A quantum analogue of the BGG resolution will allow us to construct such a deformed complex, $F_\beta^*(\mathfrak{g})$, for generic β in the same way as for $\beta = 0$.

1.7. In the case when $\mathfrak{g}$ is finite-dimensional, we prove that all higher cohomologies of our complex vanish when $\beta = 0$, so they also vanish for generic β. Using Euler characteristics we then prove that all classical integrals of motion of the corresponding Toda field theory can be quantized. These quantum integrals of motion are Fourier components of vertex operators from a certain vertex operator algebra – the so-called $\mathcal{W}$–algebra. A comprehensive review of the theory of $\mathcal{W}$–algebras and references can be found in [20].

The $\mathcal{W}$–algebra, corresponding to $\mathfrak{g} = \mathfrak{sl}_2$, is the vertex operator algebra of the Virasoro algebra. The $\mathcal{W}$–algebra, corresponding to $\mathfrak{g} = \mathfrak{sl}_3$, was constructed by Zamolodchikov [119]. For A and D series of simple Lie algebras the $\mathcal{W}$–algebras were constructed by Fateev and Lukyanov [40, 41, 42]. We give a general proof of the existence of $\mathcal{W}$–algebras, associated to arbitrary finite-dimensional simple Lie algebras. A similar construction also appeared in [99] for $\mathfrak{g} = \mathfrak{sl}_n$.

The $\mathcal{W}$–algebra can also be defined by means of quantum Drinfeld-Sokolov reduction [45, 48, 59] as the 0th cohomology of a certain BRST complex. The definition via quantum Drinfeld-Sokolov reduction is related to the definition in this paper, because the first term of a spectral sequence associated to the BRST complex coincides with our quantum complex $F_\beta^*(\mathfrak{g})$ for generic values of β [48, 57]. Therefore the cohomologies of the BRST complex and the complex $F_\beta^*(\mathfrak{g})$ coincide for generic β. Recently these cohomologies were computed using the opposite spectral sequence of the BRST complex [25]. This gives an alternative proof of existence of $\mathcal{W}$–algebras.

If $\mathfrak{g}$ is affine, our extended complex has non-trivial higher cohomologies when $\beta = 0$.

354

However, in the case when all exponents of $\mathfrak{g}$ are odd and the Coxeter number is even, we can still derive that all cohomology classes can be deformed, from the Euler character argument. We then immediately see that the corresponding quantum integrals of motion commute with each other. In the remaining cases we can prove these results by a more refined argument. When $\mathfrak{g} = \widehat{\mathfrak{sl}}_2$, we obtain a proof of the existence of quantum KdV hamiltonians. In this setting, it was conjectured in [72] and some partial results were obtained in [120, 35, 85, 29, 30, 108].

1.8. The $\mathcal{W}-$algebra is the chiral algebra of a certain two-dimensional conformal field theory; this explains its importance for quantum field theory. Our construction gives a realization of this chiral algebra in terms of the simplest chiral algebra, the chiral algebra of the free fields. Such a realization, which is usually referred to as a free field realization, is very important for computation of the correlation functions in the corresponding models of quantum field theory. The free field realization is an attempt to immerse a complicated structure into a simpler one and then give the precise description of the image of the complicated structure by finding the constraints, to which it satisfies inside the simple one. Our construction resembles the construction of the Harish-Chandra homomorphism, which identifies the center of the universal enveloping algebra of a simple Lie algebra $\mathfrak{g}$ (complicated object) with the polynomials on the Cartan subalgebra $\mathfrak{h}$ – an abelian Lie algebra (simple object), which are invariant with respect to the action of the Weyl group. In other words, the center can be described as the subalgebra in the algebra of polynomials on $\mathfrak{h}$, which are invariant with respect to the simple reflections s_i. Likewise, we have been able to embed the $\mathcal{W}-$algebra of $\mathfrak{g}$ into the vertex operator algebra of the Heisenberg algebra $\widehat{\mathfrak{h}}$, which is something like the algebra of polynomials on $L\mathfrak{h}$. The image of this embedding coincides with the kernel of the operators $\bar{Q}_i^\beta$, which play the role of simple reflections from the Weyl group.

The quantum integrals of motion of the affine Toda field theory, associated to an affine algebra $\mathfrak{g}$, constitute an infinite-dimensional abelian subalgebra in the $\mathcal{W}-$algebra, associated to the finite-dimensional Lie algebra $\bar{\mathfrak{g}}$. This abelian subalgebra consists of the local integrals of motion of a deformation of this conformal field theory. The knowledge of the existence of infinitely many integrals of motion and of their degrees (or spins) is very important for understanding this non-conformal field theory, and in many cases it allows to construct the S-matrix of this theory explicitly [120].

Some of the results of this paper have previously appeared in our papers [45, 48, 57, 50]. An earlier version of this paper appeared in September of 1993 as a preprint YITP/K-1036 of Yukawa Institute of Kyoto University, and also as hep-th/9310022 on hep-th computer net.

2. CLASSICAL TODA FIELD THEORIES ASSOCIATED TO FINITE-DIMENSIONAL SIMPLE LIE ALGEBRAS.

2.1. The case of $\mathfrak{sl}_2$ – classical Liouville theory.

2.1.1. Hamiltonian space. Denote by $\mathfrak{h}$ the Cartan subalgebra of $\mathfrak{sl}_2$ – the one-dimensional abelian Lie algebra, and by $L\mathfrak{h}$ the abelian Lie algebra of polynomial functions on the circle with values in $\mathfrak{h}$. This will be our hamiltonian space. It is isomorphic to the space of Laurent polynomials $\mathbb{C}[t, t^{-1}]$. We would like to introduce a suitable space of functions on $L\mathfrak{h}$, which we will denote by $\mathcal{F}_0$, together with a Poisson bracket.

First let us introduce the space π_0 of differential polynomials, i.e. the space of polynomials in variables $u, \partial u, \partial^2 u, \ldots$. It is equipped with an action of derivative ∂, which sends $\partial^n u$ to $\partial^{n+1} u$ and satisfies the Leibnitz rule.

We define $\mathcal{F}_0$ as the space of local functionals on $L\mathfrak{h}$. A local functional F is a functional, whose value at a point $u(t) \in L\mathfrak{h}$ can be represented as the formal residue

$$F[u(t)] = \int P(u(t), \partial_t u(t), \ldots) dt,$$

where $P \in \pi_0$ is a differential polynomial, and $\partial_t = \partial/\partial t$. In words: we insert $u(t), \partial_t u(t), \ldots$ into P; this gives us a Laurent polynomial, and we take its residue, i.e. the (-1)st Fourier component.

We can represent local functionals as series of the form

$$(2.1.1) \qquad \sum_{i_1 + \cdots + i_m = -m+1} c_{i_1 \ldots i_m} \cdot u_{i_1} \ldots u_{i_m},$$

where the coefficients $c_{i_1 \ldots i_m}$ are polynomials in $i_1, \ldots, i_m$. Here u_i's are the Fourier components of $u(t) : u(t) = \sum_{i \in \mathbb{Z}} u_i t^{-i-1}$. For example, $\int u(t) dt = u_0$, $\int u(t)^2 dt = \sum_{i+j=-1} u_i u_j$. Note that since we deal with Laurent *polynomials*, only finitely many summands of the series $(2.1.1)$ can be non-zero for a given $u(t)$.

We have a map $\int : \pi_0 \to \mathcal{F}_0$, which sends $P \in \pi_0$ to $\int P dt \in \mathcal{F}_0$. The following Lemma, for the proof of which cf. [88, 69], shows that the kernel of the residue map consists of total derivatives and constants.

2.1.2. *Lemma. The sequence*

$$0 \longrightarrow \pi_0/\mathbb{C} \stackrel{\partial}{\longrightarrow} \pi_0/\mathbb{C} \stackrel{\int}{\longrightarrow} \mathcal{F}_0 \longrightarrow 0$$

is exact.

2.1.3. *Poisson bracket.* We can now define the Poisson bracket of two local functionals F and G, corresponding to two differential polynomials, P and R, as follows:

$$(2.1.2) \qquad \{F, G\}[u(t)] = - \int \frac{\delta P}{\delta u} \partial_t \frac{\delta R}{\delta u} dt,$$

where

$$\frac{\delta P}{\delta u} = \frac{\partial P}{\partial u} - \partial \frac{\partial P}{\partial(\partial u)} + \partial^2 \frac{\partial P}{\partial(\partial^2 u)} - \cdots$$

denotes the variational derivative. Note that the variational derivative of a differential polynomial, which is a total derivative, is equal to 0, and so formula $(2.1.2)$ defines a well-defined bracket map $\mathcal{F}_0 \times \mathcal{F}_0 \to \mathcal{F}_0$.

2.1.4. This bracket satisfies all axioms of the Lie bracket and so it defines a structure of Lie algebra on $\mathcal{F}_0$. One can prove this in terms of differential polynomials [88, 69], or in terms of Fourier components [67]. We will recall the latter proof following [95], §§7.21-7.23, since we will use it later in § 4.

Note that we can extend our Poisson algebra of local functionals $\mathcal{F}_0$ by adjoining all Fourier components of differential polynomials, not only the (-1)st ones. Let $\widehat{\mathcal{F}}_0$ be the space of functionals on $L\mathfrak{h}$, which can be represented as residues of differential polynomials with explicit dependence on t:

$$F[u(t)] = \int P(\partial^n u(t); t) dt.$$

We can define a Poisson bracket on $\widehat{\mathcal{F}}_0$ by the same formula (2.1.2). We will prove now that this bracket makes $\widehat{\mathcal{F}}_0$ into a Lie algebra. This will imply that $\mathcal{F}_0$ is a Lie algebra as well, because the bracket of two elements of $\mathcal{F}_0$ is again an element of $\mathcal{F}_0$.

Any element of $\widehat{\mathcal{F}}_0$ can be presented as a *finite* linear combination of the infinite series of the form

(2.1.3) $$\sum_{i_1+\cdots+i_m=N} c_{i_1\ldots i_m}\cdot u_{i_1}\ldots u_{i_m},$$

where the u_i's are the Fourier components of $u(t)=\sum_{i\in\mathbb{Z}}u_it^{-i-1}$, $c_{i_1\ldots i_m}$ is a polynomial in $i_1,\ldots,i_m$, and N is an arbitrary integer.

We can consider these elements as lying in a certain completion $\bar{A}$ of the polynomial algebra $A=\mathbb{C}[u_n]_{n\in\mathbb{Z}}$. In order to define this completion, introduce a $\mathbb{Z}$-grading on A by putting $\deg u_n=-n$. We have

$$A=\oplus_{N\in\mathbb{Z},m\geq 0}A_{N,m},$$

where $A_{N,m}$ is the linear span of monomials of degree N and power m. Denote by I^M, $M>0$ the ideal of A generated by $u_n, n\geq M$ and by $I^M_{N,m}$ the intersection $I^M\cap A_{N,m}$. Let $\bar{A}_{N,m}$ be the completion of $A_{N,m}$ with respect to the topology, generated by the open sets $I^M_{N,m}, M>0$. Clearly,

$$\bar{A}=\oplus_{N\in\mathbb{Z},m\geq 0}\bar{A}_{N,m}$$

is a commutative algebra. It consists of finite linear combinations of infinite series of the form (2.1.3), where $c_{i_1,\ldots,i_m}$ is an arbitrary function of $i_1,\ldots,i_m$. In particular, we have an embedding $\widehat{\mathcal{F}}_0\to\bar{A}$.

Now, $u_i, i\in\mathbb{Z}$, are elements of $\widehat{\mathcal{F}}_0$ and $\bar{A}$. We find from formula (2.1.2):

(2.1.4) $$\{u_n,u_m\}=n\delta_{n,-m}.$$

By the Leibnitz rule

(2.1.5) $$\{xy,z\}=\{x,z\}+\{y,z\},$$

we can extend this bracket to a bracket, defined for any pair of monomials in u_n. We can then formally apply this formula by linearity to any pair of elements of $\bar{A}$, using their presentation in the form (2.1.3). This will give us a well-defined bracket $[\cdot,\cdot]$ on $\bar{A}$. One can check directly that the restriction (2.1.4) of this formula to the subspace $\oplus_{n\in\mathbb{Z}}\mathbb{C}u_n$ is antisymmetric and satisfies the Jacobi identity. Therefore, by construction, $[\cdot,\cdot]$ is antisymmetric and satisfies the Jacobi identity on the whole $\bar{A}$. Thus $[\cdot,\cdot]$ defines a Lie algebra structure on $\bar{A}$.

It is shown in [95], §7.23 that the restriction of the Lie bracket $[\cdot,\cdot]$ to $\widehat{\mathcal{F}}_0$ coincides with the bracket (2.1.2) (for instance, this is clear for the subspace $\oplus_{n\in\mathbb{Z}}\mathbb{C}u_n$ of $\widehat{\mathcal{F}}_0$). Therefore $\widehat{\mathcal{F}}_0$ is a Lie algebra, and $\mathcal{F}_0$ is its Lie subalgebra.

2.1.5. *Remark.* A Poisson algebra is usually defined as an object, which carries two structures: associative commutative product and a Lie bracket, which are compatible in the sense that the Leibnitz rule (2.1.5) holds. It is clear that the product of two local functionals is not a local functional, so that $\mathcal{F}_0$ and $\widehat{\mathcal{F}}_0$ are not Poisson algebras in the usual sense, but merely Lie algebras (however, $\bar{A}$ is a Poisson algebra in the usual sense). We could, of course, take the algebra of all polynomials in local functionals and extend our Poisson bracket to it by the Leibnitz rule. This would give us a Poisson algebra in the usual sense. But this would not make any difference for us, because we will never use the product structure, only the Lie bracket. For this reason we will work with $\mathcal{F}_0$ and $\widehat{\mathcal{F}}_0$, but we will still refer to them as Poisson algebras and will call the bracket (2.1.2) the Poisson bracket, because of another meaning of the term "Poisson structure" – as the classical limit of some quantum structure. In § 4 we will define this quantum structure.

2.1.6. *Kirillov-Kostant structure.* The Poisson structure on $\mathcal{F}_0$, defined above, has a nice interpretation as a Kirillov-Kostant structure.

Let us introduce an anti-symmetric scalar product $\langle, \rangle$ on $L\mathfrak{h}$:

$$\langle u(t), v(t) \rangle = \int u(t) dv(t).$$

Note that this scalar product does not depend on the choice of coordinate t on the circle and that its kernel consists of constants. Using this scalar product, we can define a Heisenberg Lie algebra $\widehat{\mathfrak{h}}$ as the central extension of $L\mathfrak{h}$ by the one-dimensional center with generator I. The commutation relations in $\widehat{\mathfrak{h}}$ are

$$[u(t), v(t)] = \langle u(t), v(t) \rangle I, \qquad [u(t), I] = 0.$$

In the natural basis $b_j = t^{-j}, j \in \mathbb{Z}$, of $L\mathfrak{h}$ they can be rewritten as

$$(2.1.6) \qquad [b_n, b_m] = n\delta_{n,-m} I, \qquad [b_n, I] = 0.$$

The (restricted) dual space $\widehat{\mathfrak{h}}^*$ of $\widehat{\mathfrak{h}}$ consists of pairs $(y(t)dt, \mu)$, which define linear functionals on $\widehat{\mathfrak{h}}$ by formula

$$(y(t)dt, \mu)[u(t) + \nu I] = \mu\nu + \int u(t)y(t) dt.$$

One has the Kirillov-Kostant Poisson structure on $\widehat{\mathfrak{h}}^*$. Since I generates the center of $\widehat{\mathfrak{h}}$, we can restrict this structure to a hyperplane $\widehat{\mathfrak{h}}^*_\mu$, which consists of the linear functionals, taking value μ on I.

If we choose a coordinate t on the circle, then we can identify $\widehat{\mathfrak{h}}^*_1$ with the space $L\mathfrak{h}$. This gives us a Poisson bracket on various spaces of functionals on $L\mathfrak{h}$, e.g., on the space A of polynomial functionals or its completion $\bar{A}$. Formulas (2.1.4) and (2.1.6) show that this bracket coincides with the bracket $[\cdot, \cdot]$, defined in § 2.1.4. The restrictions of this Poisson bracket to $\widehat{\mathcal{F}}_0$ and $\mathcal{F}_0$ coincide with the ones, defined by formula (2.1.2).

2.1.7. *The action of $\mathcal{F}_0$ on π_0.* The Poisson bracket $\mathcal{F}_0 \times \mathcal{F}_0 \to \mathcal{F}_0$, defined by formula (2.1.2), can be thought of as the adjoint action of the Lie algebra $\mathcal{F}_0$ on itself. We can lift this action to an action of $\mathcal{F}_0$ on the space π_0 of differential polynomials [70]. Namely, we can rewrite formula (2.1.2) as follows:

$$\{F, G\} = -\int \frac{\delta P}{\delta u} \partial_t \frac{\delta R}{\delta u} dt = -\int \sum_{n \geq 0} (-\partial_t)^n \frac{\partial P}{\partial(\partial^n u)} \partial_t \frac{\delta R}{\delta u} dt =$$

$$-\int \sum_{n \geq 0} \frac{\partial P}{\partial(\partial^n u)} \cdot \partial_t^{n+1} \frac{\delta R}{\delta u} dt,$$

using the fact that the integral of a total derivative is 0. This suggests to define an action of the functional $G = \int R dt$ on the space $\pi_0 \simeq \mathbb{C}[\partial^n u]_{n \geq 0}$ by the vector field

$$(2.1.7) \qquad -\sum_{n \geq 0} \left(\partial_t^{n+1} \frac{\delta R}{\delta u} \right) \frac{\partial}{\partial(\partial^n u)}.$$

The map $\pi_0 \to \pi_0$, given by this formula, manifestly commutes with the action of the derivative ∂, and the projection of this map to a map from $\mathcal{F}_0 = \pi_0/(\partial\pi_0 \oplus \mathbb{C})$ to itself coincides with the adjoint action.

We will use the same notation $\{\cdot, G\}$ for the action of $G \in \mathcal{F}_0$ on π_0. Note that the action of ∂ on π_0 coincides with the action of $\frac{1}{2} \int u^2(t)dt$ and that it commutes with the action of any other $G \in \mathcal{F}_0$.

2.1.8. Now we define another space, $\mathcal{F}_1$, of functionals on $L\mathfrak{h}$ and extend our Poisson bracket. We follow [86].

Let π_1 be the tensor product of the space of differential polynomials π_0 with a one dimensional space $\mathbb{C}v_1$. Let us define an action of ∂ on π_1 as $(\partial + u) \otimes 1$, where u stands for the operator of multiplication by u on π_0. Define the space $\mathcal{F}_1$ as the cokernel of the homomorphism $\partial : \pi_1 \to \pi_1$. We have the exact sequence:

$$0 \longrightarrow \pi_1 \overset{\partial}{\longrightarrow} \pi_1 \longrightarrow \mathcal{F}_1 \longrightarrow 0.$$

To motivate this definition, introduce formally $\phi(t) = \int^t u(s)ds$, so that $\partial_t\phi(t) = u(t)$. Consider the space of functionals on $L\mathfrak{h}$, which have the form

$$(2.1.8) \qquad \int P(u(t), \partial u(t), \dots)e^{\phi(t)}dt.$$

There is a map, which sends $P \otimes v_1 \in \pi_1$ to the functional (2.1.8), so that the action of derivative ∂_t on $Pe^{\phi(t)}$ coincides with the action of ∂ on $P \in \pi_1$. The kernel of this map consists of the elements of π_1, which are total derivatives; therefore $\mathcal{F}_1$ can be interpreted as the space of functionals of the form (2.1.8).

Note that if P is a differential polynomial in u, then

$$-\partial\frac{\delta P}{\delta u} = \frac{\delta P}{\delta\phi}.$$

This formula allows us to extend the Poisson bracket (2.1.2), which was defined for two differential polynomials in u, to the case, when P is a differential polynomial in u, and R depends explicitly on ϕ. In particular, we obtain a well-defined map $\mathcal{F}_0 \times \mathcal{F}_1 \to \mathcal{F}_1$:

$$(2.1.9) \qquad \left\{\int Pdt, \int Re^\phi dt\right\} = \int \frac{\delta P}{\delta u}\frac{\delta[Re^\phi]}{\delta\phi}dt = \int \frac{\delta P}{\delta u}\left[Re^\phi - \partial_t\left(\frac{\delta R}{\delta u}e^\phi\right)\right]dt.$$

This bracket satisfies the Jacobi identity for any triple $F, G \in \mathcal{F}_0, H \in \mathcal{F}_1$. In other words, $\mathcal{F}_1$ is a module over the Lie algebra $\mathcal{F}_0$.

This statement can be proved in the same way as in § 2.1.4. Namely, we can extend the space $\mathcal{F}_1$ to the space $\widehat{\mathcal{F}}_1$ by adjoining elements of the form $\int Re^\phi dt$, where R is a polynomial in $\partial^m u, m \geq 0$, and t. Consider the elements $w_n = \int t^n e^\phi dt \in \widehat{\mathcal{F}}_1$. By formula (2.1.9), the action of $u_m = \int t^m u(t)dt \in \widehat{\mathcal{F}}_0$ (cf. § 2.1.4) on w_n is given by

$$(2.1.10) \qquad \{u_m, w_n\} = w_{n+m}.$$

This formula defines a map $U \times W \to W$, where $U = \oplus_{m\in\mathbb{Z}}\mathbb{C}u_m$ and $W = \oplus_{n\in\mathbb{Z}}w_n$. Recall that U is a Lie algebra, with the commutation relations given by formula (2.1.4). One can check directly that this map defines a structure of a U-module on W.

In the same way as in § 2.1.4 we can define a completion $\bar{B}$ of the space $B = \mathbb{C}[u_m]_{m\in\mathbb{Z}} \otimes W$, which contains $\widehat{\mathcal{F}}_1$ and $\mathcal{F}_1$. Using the Leibnitz rule, we can extend the map $U \times W \to W$ given by (2.1.10) to a map $\bar{A} \times \bar{B} \to \bar{B}$. By construction, this map defines a structure of an $\bar{A}$-module on $\bar{B}$. The restriction $\widehat{\mathcal{F}}_0 \times \widehat{\mathcal{F}}_1 \to \widehat{\mathcal{F}}_1$ of this map coincides with the map

$\{\cdot,\cdot\}$ given by (2.1.9). Therefore it defines on $\hat{\mathcal{F}}_1$ a structure of a module over the Lie algebra $\hat{\mathcal{F}}_0$. Hence it makes $\mathcal{F}_1$ into a module over $\mathcal{F}_0$.

2.1.9. *The Liouville hamiltonian.* We now introduce the hamiltonian H of the Liouville model by the formula $H = \int e^{\phi(t)}dt \in \mathcal{F}_1$. We can rewrite the Liouville equation (1.1.2) in the hamiltonian form as

$$\partial_\tau U(t) = \{U(t), H\}.$$

Here $U(t)$ stands for the delta-like functional on $L\mathfrak{h}$, whose value on a function from $L\mathfrak{h}$ is equal to the value of this function at the point t. We can rewrite it as $U(t) = \int \delta(t-s)u(s)ds$. The formula (2.1.9) can be extended to such functionals as well. Applying this formula, we obtain

$$\{\int \delta(t-s)u(s)ds, \int e^{\phi(s)}ds\} = \int \delta(t-s)e^{\phi(s)}ds = e^{\phi(t)}.$$

2.1.10. *Definition.* The kernel of the linear operator

$$(2.1.11) \qquad \bar{Q} = \{\cdot, \int e^\phi dt\} : \mathcal{F}_0 \to \mathcal{F}_1$$

will be called the space of local integrals of motion of the classical Liouville theory and will be denoted by $I_0(\mathfrak{sl}_2)$.

2.1.11. Our goal is to compute the space $I_0(\mathfrak{sl}_2)$. Note that by the Jacobi identity, it is closed with respect to the Poisson bracket.

It is more convenient to work with the spaces π_0 and π_1, than with $\mathcal{F}_0$ and $\mathcal{F}_1$. We want to define a linear operator $\tilde{Q} : \pi_0 \to \pi_1$, which commutes with the action of ∂ on these spaces, and descends down to the operator $\bar{Q}$.

To define such an operator, we will use the same approach as in § 2.1.7. According to formula (2.1.9), we have:

$$\{\int P dt, \int e^\phi dt\} = \int \sum_{n\geq 0}(-\partial_t)^n \frac{\partial P}{\partial(\partial^n u)} \cdot e^\phi dt = \int \sum_{n\geq 0} \frac{\partial P}{\partial(\partial^n u)} \cdot \partial_t^n e^\phi dt,$$

where we used the fact that the integral of a total derivative is 0. We find: $\partial_t e^\phi = ue^\phi$, $\partial_t^2 e^\phi = (u^2 + \partial_t u)e^\phi$, etc. In general, $\partial_t^n e^\phi = B_n e^\phi$, where the B_n's are certain differential polynomials in $u(t)$, which are connected by the recurrence relation

$$(2.1.12) \qquad B_{n+1} = uB_n + \partial_t B_n.$$

Then we obtain

$$(2.1.13) \qquad \{\int P dt, \int e^\phi dt\} = \sum_{n\geq 0} \int \frac{\partial P}{\partial(\partial^n u)} \cdot B_n e^\phi dt.$$

Therefore we can define a map $\tilde{Q} : \pi_0 \to \pi_1$ as follows:

$$\tilde{Q} \cdot P = \sum_{n\geq 0} B_n \frac{\partial P}{\partial(\partial^n u)} \otimes v_1.$$

It commutes with the action of ∂ and descends down to the operator $\bar{Q} : \mathcal{F}_0 \to \mathcal{F}_1$.

In fact, in the same way we can define for any $G \in \mathcal{F}_1$ a map $\{\cdot, G\} : \pi_0 \to \pi_1$, which commutes with ∂ and descends down to the map $\mathcal{F}_0 \to \mathcal{F}_1$, given by formula (2.1.9):

$$\{P, \int Re^{\phi(t)}dt\} = \sum_{n\geq 0} \frac{\partial P}{\partial(\partial^n u)} \cdot \partial^n(R \otimes v_1) - \partial^{n+1}\left(\frac{\delta}{\delta u}R \otimes v_1\right).$$

We can also extend the action of $G \in \mathcal{F}_0$ on π_0, given by (2.1.7), to an action on π_1 by adding to the vector field (2.1.7) the term $\frac{\delta R}{\delta u} \frac{\partial}{\partial \phi}$, where $\frac{\partial}{\partial \phi}$ acts on π_n by multiplication by n.

It is convenient to pass to the new variables $x_n = \partial^{-n-1} u/(-n-1)!, n < 0$. Then $\pi_0 = \mathbb{C}[x_n]_{n<0}, \pi_1 = \mathbb{C}[x_n]_{n<0} \otimes \mathbb{C}v_1$. In these variables the action of the derivative ∂ on these spaces is given by

$$\partial = -\sum_{n<0} n x_{n-1} \frac{\partial}{\partial x_n} + x_{-1} \frac{\partial}{\partial \phi}.$$

Let $T : \pi_0 \to \pi_1$ be the translation operator, which sends $P \in \pi_0$ to $P \otimes v_1 \in \pi_1$. In the new variables the operator $\tilde{Q} : \pi_0 \to \pi_1$ is given by the formula

(2.1.14)
$$\tilde{Q} = T \sum_{n<0} S_{n+1} \frac{\partial}{\partial x_n},$$

where the polynomials S_n are the Schur polynomials, defined via the generating function:

(2.1.15)
$$\sum_{n\leq0} S_n z^n = \exp(\sum_{m<0} -\frac{x_m}{m} z^m).$$

One can check that $\partial S_n = -x_1 S_n - (n-1)S_{n-1}$, and therefore, by formula (2.1.12), the differential polynomial B_{-n} coincides with S_n in the new variables x_m. We summarize these results in the following Lemma.

2.1.12. **Lemma.** *The operator $\tilde{Q}$, given by formula (2.1.14), commutes with the action of the derivative ∂ and the corresponding operator $\mathcal{F}_0 \to \mathcal{F}_1$ coincides with the operator $\bar{Q}$, given by formula (2.1.11).*

2.1.13. Let us put $\deg 1 = 0, \deg v_1 = 1, \deg x_n = -n$. Since the x_n's generate the spaces π_0 and π_1 from 1 and v_1, respectively, this defines a $\mathbb{Z}$-grading on these spaces such that the homogeneous components are finite-dimensional. The operator ∂ is homogeneous of degree 1, and we can define grading on the spaces $\mathcal{F}_0$ and $\mathcal{F}_1$ by subtracting 1 from the grading on the spaces π_0 and π_1, respectively. The operators $\tilde{Q}$ and $\bar{Q}$ are homogeneous of degree 0 and therefore their kernels and cokernels are $\mathbb{Z}$-graded with finite-dimensional homogeneous components.

2.1.14. To find the space of local integrals of motion of the classical Liouville equation we have to find the 0th cohomology of the complex

$$\mathcal{F}_0 \xrightarrow{\bar{Q}} \mathcal{F}_1.$$

Consider the following double complex.

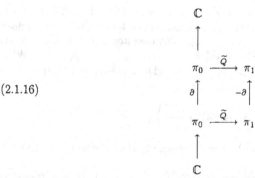

(2.1.16)

We can calculate its cohomology by means of two spectral sequences (as general references on spectral sequences, cf., e.g., [93, 15]). In one of them the 0th differential is vertical. Therefore, in this spectral sequence the 1st term coincides with our complex with the degrees shifted by 1. Hence the 1st cohomology of the double complex coincides with the space of integrals of motion.

In the other spectral sequence the first differential is horizontal. We have two identical complexes

(2.1.17) $$\pi_0 \xrightarrow{\tilde{Q}} \pi_1.$$

Let us calculate the cohomology of this complex.

2.1.15. Proposition. *The operator* $\tilde{Q} : \pi_0 \to \pi_1$ *is surjective, so that its cokernel is equal to 0. The kernel* $\mathcal{W}_0(\mathfrak{sl}_2)$ *of the operator* $\tilde{Q}$ *contains an element* W_{-2} *of degree 2, such that* $\mathcal{W}_0(\mathfrak{g})$ *coincides with the polynomial algebra* $\mathbb{C}[W_n]_{n \leq -2}$, *where* $W_n = \partial^{-n-2} W_{-2}/(-n - 2)!$.

Proof. The operator $Q = T^{-1}\tilde{Q}$ is a linear combination of the vector fields $S_{n+1}\partial/\partial x_n, n < 0$, where $S_0 = 1$ and S_{n+1} is a polynomial in $x_{-1}, \ldots, x_{n+1}, n < -1$. The operator

$$\sum_{-m \leq j \leq -1} S_{j+1} \frac{\partial}{\partial x_j}$$

is therefore a well-defined linear operator $Q(m)$ from $\mathbb{C}[x_j]_{j=-1,\ldots,-m}$ to itself.

The operator $Q(m)$ is surjective. To see that, consider the dual operator $Q(m)^*$, acting on the space dual to $\mathbb{C}[x_j]_{j=-1,\ldots,-m}$. Since our operator is homogeneous, it is sufficient to consider the restricted dual space, which we can identify with itself, choosing the monomials $x_{-j_1}^{k_1} \ldots x_{-j_n}^{k_n}/(k_1! \ldots k_n!)^{1/2}$ as the orthonormal basis. The operator $Q(m)^*$ then has the form

$$Q(m)^* = \sum_{-m \leq j \leq -1} x_j S_{j+1}^*,$$

where S_j^* is obtained from the polynomial S_j by replacing x_i with $\partial/\partial x_i$. We see that the operator $Q(m)^*$ is the sum of multiplication by x_{-1}, which increases the power of any polynomial by 1, and other operators, which do not change or decrease the power. Since the operator of multiplication by x_{-1} has no kernel, the operator $Q(m)^*$ is injective. Therefore, the operator $Q(m)$ is surjective. Hence, the operators Q and $\tilde{Q}$ are also surjective.

For any $n < -1$, there exist polynomials $W_n = (n+1)x_n + W'_n$, where W'_n is a linear combination of terms of power greater then 1 and of degree n, such that $\tilde{Q}\cdot W_n = 0$. Indeed, $Q(n) \cdot x_n = S_{n+1}$ is an element of $\mathbb{C}[x_j]_{j=-1,\ldots,n+1}$. The operator $Q(n+1)$ is surjective on this space. Therefore, there exists such $W'_n \in \mathbb{C}[x_j]_{j=-1,\ldots,n+1}$ that $Q(n+1) \cdot W'_n = -Q(n)\cdot(n+1)x_n$. But then $Q(n)\cdot((n+1)x_n+W'_n) = 0$, and hence $\tilde{Q}\cdot W_n = TQ(n)\cdot W_n = 0$. In the coordinates $x_{-1}, W_n, n < -1$, the operator $\tilde{Q}$ is equal to

$$\tilde{Q} = T\left(\frac{\partial}{\partial x_{-1}} + \sum_{n<-1}(\tilde{Q}\cdot W_n)\frac{\partial}{\partial W_n}\right) = T\frac{\partial}{\partial x_{-1}}.$$

The W_n's are algebraically independent by construction, therefore the kernel of the operator $\tilde{Q}$ coincides with $\mathbb{C}[W_n]_{n<-1}$.

Finally, we can choose as W_n the polynomial $\partial^{-n-2}W_{-2}/(-n-2)!$. Indeed, the latter lies in the kernel of $\tilde{Q}$, because $[\partial, \tilde{Q}] = 0$. Its linear term is equal to $(n+1)x_n$, because the linear term of W_{-2} is $-x_{-2}$, and the derivative preserves the power of a polynomial. The Proposition is proved.

2.1.16. Remark. For any $\mathbb{Z}$-graded vector space

$$V = \oplus_{m\in\mathbb{Z}}V(m)$$

with finite-dimensional homogeneous components, introduce its character as

$$\mathrm{ch}V = \sum_{m\in\mathbb{Z}}\dim V(m)q^m.$$

The Euler character of the complex (2.1.17) is equal to

$$\mathrm{ch}\pi_0 - \mathrm{ch}\pi_1 = \mathrm{ch}\mathrm{Ker}\tilde{Q} - \mathrm{ch}\mathrm{Coker}\tilde{Q} =$$
$$\prod_{n\geq1}(1-q^n)^{-1} - q\prod_{n\geq1}(1-q^n)^{-1} = \prod_{n\geq2}(1-q^n)^{-1} = 1 + q^2 + \ldots.$$

This formula shows that there exists an element of degree 2 in the kernel of $\tilde{Q}$. Denote it by W_{-2}. The operator of multiplication by W_{-2}, acting on π_0 and π_1, commutes with the action of $\tilde{Q}$. Further, since $[\partial, \tilde{Q}] = 0$, the operators of multiplication by $W_n = \partial^{-n-2}W_{-2}/(-n-2)!, n \leq -2$, also commute with $\tilde{Q}$. Therefore any polynomial in the W_n's, constructed this way, lies in the kernel of the operator $\tilde{Q}$. The algebraic independence of these W_n's and the surjectivity of the operator $\tilde{Q}$ allowed us to identify the polynomial algebra in the W_n's with the kernel of $\tilde{Q}$.

2.1.17. Proposition. *The space $I_0(\mathfrak{sl}_2)$ of local integrals of motion of the classical Liouville theory coincides with the quotient of $\mathcal{W}_0(\mathfrak{sl}_2)$ by the total derivatives and constants.*

Proof. We have to prove that the kernel of the operator $\tilde{Q}$ coincides with the quotient of $\mathcal{W}_0(\mathfrak{sl}_2)$ by the total derivatives and constants. As we explained before, this kernel is the same as the 1st cohomology of the double complex (2.1.16). By Proposition 2.1.15, the first term of the spectral sequence, associated to this double complex, looks as follows:

$$\mathbb{C} \longrightarrow \mathcal{W}_0(\mathfrak{sl}_2) \longrightarrow \mathcal{W}_0(\mathfrak{sl}_2) \longrightarrow \mathbb{C}.$$

The 1st cohomology of this complex is equal to the quotient of $\mathcal{W}_0(\mathfrak{sl}_2)$ by the total derivatives and constants.

2.1.18. We can write down explicit formulas for the W_n's as follows:

$$W_{-2} = \frac{1}{2}x_{-1}^2 - x_{-2}, \quad W_n = \frac{1}{(-n-2)!}\partial^{-n-2}W_{-2}, \quad n < -2.$$

Thus, the space $I_0(\mathfrak{sl}_2)$ consists of local functionals, which are defined by differential polynomials, depending on

$$W = \frac{1}{2}u^2 - \partial u.$$

These local functionals constitute a Poisson subalgebra in $\mathcal{F}_0$. It is known that this Poisson subalgebra is isomorphic to the classical Virasoro algebra.

The Virasoro algebra is the central extension of the Lie algebra of vector fields on the circle. Its dual space is equipped with the Kirillov-Kostant Poisson structure. This structure can be restricted to the hyperlane, which consists of the linear functionals, whose value on the central element is 1. If we choose a coordinate on the circle, then this hyperplane can be identified with the space of Laurent polynomials $W(t)$. The local functionals on this space form a Poisson algebra, which is isomorphic to $I_0(\mathfrak{sl}_2)$.

There is a map from a hyperplane in the dual space to the Heisenberg algebra to a hyperplane in the dual space to the Virasoro algebra, which sends $u(t)$ to $W(t) = \frac{1}{2}u^2(t) - \partial_t u(t)$ and preserves the Poisson structure. This map is called the Miura transformation.

In the case of the Liouville theory, which is the simplest Toda field theory, we were able to find explicit formulas for the local integrals of motion and identify their Poisson algebra with the classical Virasoro algebra. However, in general explicit formulas are much more complicated, and we will have to rely on homological algebra to obtain information about the integrals of motion.

2.2. General case.

2.2.1. *Hamiltonian space.* Let $\mathfrak{h}$ be the Cartan subalgebra of $\mathfrak{g}$. It is equipped with the scalar product $(,)$, which is the restriction of the invariant scalar product on $\mathfrak{g}$, normalized as in [81]. In what follows we will identify $\mathfrak{h}$ with its dual by means of this scalar product.

We choose as the Hamiltonian space the space $L\mathfrak{h}$ of Laurent polynomials on the circle with values in $\mathfrak{h}$, that is the space $\mathfrak{h} \otimes \mathbb{C}[t, t^{-1}]$.

Each element $\mathbf{u}(t)$ of $L\mathfrak{h}$ can be represented by its coordinates $(u^1(t), \ldots, u^l(t))$ with respect to the basis of simple roots $\alpha_1, \ldots, \alpha_l$, where l is the rank of $\mathfrak{g}$. Let π_0 be the space of differential polynomials of $\mathbf{u}(t)$, i.e. the space of polynomials in $\partial^n u^i, i = 1, \ldots, l, n \geq 0$.

We define the space $\mathcal{F}_0$ of local functionals as the space of functionals on $L\mathfrak{h}$, which can be represented as formal residues

$$F[\mathbf{u}(t)] = \int P(\partial^n u^i(t))dt,$$

where $P \in \pi_0$. Again, we have the exact sequence

$$0 \longrightarrow \pi_0/\mathbb{C} \overset{\partial}{\longrightarrow} \pi_0/\mathbb{C} \overset{\int}{\longrightarrow} \mathcal{F}_0 \longrightarrow 0.$$

Introduce the Poisson structure on $\mathcal{F}_0$ by the formula

$$(2.2.1) \qquad \{F, G\}[\mathbf{u}(t)] = -\int \left(\frac{\delta P}{\delta \mathbf{u}}, \partial_t \frac{\delta R}{\delta \mathbf{u}} \right) dt.$$

Here $\frac{\delta P}{\delta \mathbf{u}}$ and $\partial_t \frac{\delta R}{\delta \mathbf{u}}$ are vectors in the dual space to $\mathfrak{h}$, and we take their scalar product.

In coordinates, we can rewrite it as

$$\{F,G\}[\mathbf{u}(t)] = -\int \sum_{i,j=1}^{l} (\alpha_i, \alpha_j) \frac{\delta P}{\delta u^i} \partial_t \frac{\delta R}{\delta u^j} dt.$$

This Poisson structure has an interpretation as a Kirillov-Kostant structure on a hyperplane in the dual space to the Heisenberg algebra $\hat{\mathfrak{h}}$, which is the central extension of $L\mathfrak{h}$ (cf. § 2.1.6).

Note that our conventions in the case of $\mathfrak{g} = \mathfrak{sl}_2$ (cf. § 2.1) differ from our conventions in general by a factor of 2, since for $\mathfrak{sl}_2$ we have $(\alpha, \alpha) = 2$.

2.2.2. Let us define other spaces of functionals on $L\mathfrak{h}$. For each element γ of the weight lattice $P \subset \mathfrak{h}^* \simeq \mathfrak{h}$ we define the space $\pi_\gamma = \pi_0 \otimes \mathbb{C}v_\gamma$, equipped with the action of the derivative by the formula $(\partial + \gamma) \otimes 1$, where γ denotes the operator of multiplication by γ (as an element of π_0) on π_0.

Let $\mathcal{F}_\gamma$ be the quotient of π_γ by the image of the operator ∂, i.e. by the total derivatives. We have the exact sequence

(2.2.2)
$$0 \longrightarrow \pi_\gamma \xrightarrow{\partial} \pi_\gamma \longrightarrow \mathcal{F}_\gamma \longrightarrow 0.$$

As in § 2.1.8, we can interpret $\mathcal{F}_\gamma$ as the space of functionals on $L\mathfrak{h}$ of the form

$$\int P(\partial^n u^i(t)) e^{\bar{\gamma}(t)} dt,$$

where $\bar{\gamma}(t)$ is such that $\partial_t \bar{\gamma}(t) = \gamma(t)$.

In the same way as in § 2.1.8, we can extend our Poisson bracket (2.2.1) to a map

$$\mathcal{F}_0 \times \oplus_{\gamma \in P} \mathcal{F}_\gamma \to \oplus_{\gamma \in P} \mathcal{F}_\gamma$$

by the formula

(2.2.3)
$$\left\{ \int P dt, \int R e^{\bar{\gamma}} dt \right\} = \int \left[\left(\frac{\delta P}{\delta \mathbf{u}}, \gamma \right) R e^{\bar{\gamma}} - \left(\frac{\delta P}{\delta \mathbf{u}}, \partial_t \cdot \frac{\delta R}{\delta \mathbf{u}} e^{\bar{\gamma}} \right) \right] dt.$$

One can check that this bracket satisfies the Jacobi identity for any triple $F, G \in \mathcal{F}_0, H \in \mathcal{F}_\gamma$.

2.2.3. *The Toda hamiltonian.* We can now introduce the hamiltonian of the Toda field theory, associated to $\mathfrak{g}$, by the formula

$$H = \frac{1}{2} \sum_{i=1}^{l} \int e^{\phi_i(t)} dt \in \oplus_{i=1}^{l} \mathcal{F}_{\alpha_i},$$

where $\phi_i(t) = \bar{\alpha}_i(t)$. Note that $\int e^{\phi_i(t)} dt$ stands for the image of $v_{\alpha_i} \in \pi_{\alpha_i}$ under the projection $\pi_{\alpha_i} \to \mathcal{F}_{\alpha_i}$. The equation

$$\mathbf{U}(t) = \{\mathbf{U}(t), H\},$$

where $\mathbf{U}(t)$ denotes the delta-like functional as in § 2.1.9, coincides with the Toda equation (1.1.1).

The operator $\bar{Q}_i = \{\cdot, \int e^{\phi_i(t)} dt\}$ is a well-defined linear operator, acting from $\mathcal{F}_0$ to $\mathcal{F}_{\alpha_i}$. We can therefore give the following definition.

2.2.4. Definition. *The kernel of the linear operator*

$$\frac{1}{2}\sum_{i=1}^{l}\bar{Q}_i : \mathcal{F}_0 \to \oplus_{i=1}^{l}\mathcal{F}_{\alpha_i}$$

will be called the space of local integrals of motion of the classical Toda field theory associated to $\mathfrak{g}$ and will be denoted by $I_0(\mathfrak{g})$.

2.2.5. Clearly, $I_0(\mathfrak{g})$ is equal to the intersection of the kernels of the operators $\bar{Q}_i : \mathcal{F}_0 \to \mathcal{F}_{\alpha_i}$. By Jacobi identity, $I_0(\mathfrak{g})$ is a Poisson subalgebra of $\mathcal{F}_0$.

Let us write down an explicit formula for the operator $\bar{Q}_i$.

It is convenient to pass to the new variables $x_n^i = \partial^{-n-1}u^i/(-n-1)!, i = 1,\ldots,l, n < 0$. Then $\pi_\gamma = \mathbb{C}[x_n^i] \otimes \mathbb{C}v_\gamma$ (here $v_0 = 1$).

In these new variables the action of the derivative ∂ can be written as

$$\sum_{i=1}^{l}\left(-\sum_{n<0} nx_{n-1}^i\frac{\partial}{\partial x_n^i} + x_{-1}^i\frac{\partial}{\partial\phi_i}\right),$$

where the action of $\partial/\partial\phi_i$ on π_γ with $\gamma = \sum_{i=1}^{l}l_i\alpha_i$ is given by multiplication by l_i.

Let $T_i : \pi_\gamma \to \pi_{\gamma+\alpha_i}$ be the translation operator, which maps $P\otimes v_\gamma \in \pi_\gamma$ to $P\otimes v_{\gamma+\alpha_i} \in \pi_{\gamma+\alpha_i}$.

Introduce the operators $\tilde{Q}_i : \pi_\gamma \to \pi_{\gamma+\alpha_i}$ by the formula

$$(2.2.4) \qquad \tilde{Q}_i = T_i \sum_{n<0} S_{n+1}^i\partial_n^{(i)},$$

where

$$\partial_n^{(i)} = \sum_{j=1}^{l}(\alpha_i,\alpha_j)\frac{\partial}{\partial x_n^j},$$

and the Schur polynomials S_n^i are given by the generating function

$$\sum_{n\leq 0} S_n^i z^n = \exp(\sum_{m<0} -\frac{x_m^i}{m}z^m).$$

2.2.6. Lemma. *The operator $\tilde{Q}_i : \pi_0 \to \pi_{\alpha_i}$ commutes with the action of derivative ∂ and the corresponding operator $\mathcal{F}_0 \to \mathcal{F}_{\alpha_i}$ coincides with the operator $\bar{Q}_i$.*

Proof. The same as in Lemma 2.1.12.

2.2.7. Our task is to compute the kernel of the operator $\sum_{i=1}^{l}\bar{Q}_i$, or, in other words, the 0th cohomology of the complex

$$(2.2.5) \qquad\qquad \mathcal{F}_0 \longrightarrow \oplus_{i=1}^{l}\mathcal{F}_{\alpha_i}.$$

The cohomology of this complex is very difficult to compute. Indeed, it is clear that the 1st cohomology is very large, because the first group of the complex is "l times larger" than the 0th group. So, we can not use the argument we used in the proof of Proposition 2.1.15. What is even worse is that as was explained in the Introduction, for such a complex it is virtually impossible to prove that the cohomology classes can be quantized.

To fix this situation we will extend this complex further to the right. Clearly, by doing so we will not change the 0th cohomology, but we will be able to kill all higher cohomologies. We will then use the resulting complex to compute the 0th cohomology, and to prove that it can be quantized.

First of all, it is convenient to realize our complex as the double complex

$$\mathbb{C}$$

$$\uparrow$$

$$\pi_0 \xrightarrow{\sum \tilde{Q}_i} \oplus_{i=1}^{l} \pi_{\alpha_i}$$

(2.2.6)
$$\partial \uparrow \qquad -\partial \uparrow$$

$$\pi_0 \xrightarrow{\sum \tilde{Q}_i} \oplus_{i=1}^{l} \pi_{\alpha_i}$$

$$\uparrow$$

$$\mathbb{C}$$

in the same way as in § 2.1.14.

We can compute the cohomology of the double complex (2.2.6) by means of the spectral sequence, whose first term consists of two identical complexes

$$(2.2.7) \qquad\qquad \pi_0 \longrightarrow \oplus_{i=1}^{l} \pi_{\alpha_i}.$$

We will extend both complexes (2.2.7) in such a way that the higher differentials will commute with the derivative ∂. We will then be able to form a double complex, which will give us an extension of the complex (2.2.5) that we are looking for.

The key observation, which will enable us to do that, is as follows. Introduce the operators $Q_i : \pi_0 \to \pi_0$ as $T^{-1}\tilde{Q}_i$. We will use the notation $\mathrm{ad}A \cdot B = [A, B]$.

2.2.8. Proposition. *The operators Q_i satisfy the Serre relations of the Lie algebra $\mathfrak{g}$*

$$(\mathrm{ad}\, Q_i)^{-a_{ij}+1} \cdot Q_j = 0,$$

where $\|a_{ij}\|$ is the Cartan matrix of $\mathfrak{g}$.

Proof. The Proposition follows from the following formula:

$$(\mathrm{ad}Q_i)^m \cdot Q_j = C_m \cdot (-a_{ij} - m + 1) \sum_{n_1,\dots,n_{m+1}<0} S_{n_1}^i \dots S_{n_m}^i S_{n_{m+1}}^j \frac{1}{n_1 \dots n_m} \cdot$$

$$\cdot \left(\sum_{l=1}^{m} \frac{n_l}{n_1 + \dots \widehat{n_l} \dots + n_{m+1}} \partial_{n_1+\dots+n_{m+1}}^{(i)} - \partial_{n_1+\dots+n_{m+1}}^{(j)} \right),$$

where C_m is a constant. This formula can be proved by induction, using the commutation relations

$$[\partial_n^{(i)}, S_m^j] = -(\alpha_i, \alpha_j)\frac{1}{n}S_{m-n}^j$$

(here we put $S_m^j = 0$, if $m > 0$) and the simple identity

$$\frac{1}{a(a+b)} + \frac{1}{b(a+b)} = \frac{1}{ab}.$$

2.2.9. Remark. In the proof of Proposition 2.2.8 we never used the fact that $\|a_{ij}\|$ is the Cartan matrix of a simple Lie algebra. In fact, we could associate the main objects, defined in this section, such as π_γ, $\mathcal{F}_\gamma$, the Poisson structure, and the operators Q_i to any symmetrizable Cartan matrix, so that the results of this section, such as Proposition 2.2.8, remain valid.

2.2.10. Proposition 2.2.8 shows that the operators Q_i generate an action of the nilpotent subalgebra $\mathfrak{n}_+$ of $\mathfrak{g}$ on π_0. In order to extend the complex (2.2.7), we will use the Bernstein-Gelfand-Gelfand (BGG) resolution of the trivial representation of $\mathfrak{g}$ by Verma modules. We will recall the relevant facts about this resolution in the next subsection.

2.3. BGG resolution.

2.3.1. *Verma modules.* Recall that the Lie algebra $\mathfrak{g}$ has the Cartan decomposition $\mathfrak{g} = \mathfrak{n}_- \oplus \mathfrak{h} \oplus \mathfrak{n}_+$. For $\lambda \in \mathfrak{h}^*$ denote by $\mathbb{C}_\lambda$ the corresponding one-dimensional representation of $\mathfrak{h}$. We can extend it trivially to a representation of $\mathfrak{b}_- = \mathfrak{h} \oplus \mathfrak{n}_-$. The induced module over $\mathfrak{g}$,

$$M_\lambda = U(\mathfrak{g}) \otimes_{U(\mathfrak{b}_-)} \mathbb{C}_\lambda,$$

is called the Verma module with lowest weight λ. It is freely generated from the lowest weight vector $1_\lambda = 1 \otimes 1$ by the action of the nilpotent subalgebra $\mathfrak{n}_+$ of $\mathfrak{g}$.

A vector w in M_λ is called a singular vector of weight μ, if it satisfies the properties:

$$\mathfrak{n}_- \cdot w = 0, \quad y \cdot w = \mu(y)w, y \in \mathfrak{h}.$$

In particular, 1_λ is a singular vector of weight λ. A singular vector of weight μ generates a submodule of M_λ, which is isomorphic to the Verma module M_μ.

Consider the Verma module M_0. It is known that the singular vectors of M_0 are labeled by the elements of the Weyl group of $\mathfrak{g}$ [10]. Such a vector w_s, corresponding to an element s of the Weyl group, has the weight $\rho - s(\rho)$, where $\rho \in h^*$ is the half-sum of the positive roots of $\mathfrak{g}$. Let us fix these vectors once and for all.

2.3.2. *The definition of the resolution.* The BGG resolution is a complex, i.e. a $\mathbb{Z}$--graded vector space

$$B_*(\mathfrak{g}) = \oplus_{j \geq 0} B_j(\mathfrak{g}),$$

together with differentials $d_j : B_j(\mathfrak{g}) \to B_{j-1}(\mathfrak{g})$, which are nilpotent: the composition of two consecutive differentials $d_{j-1}d_j$ is equal to 0.

The vector space $B_j(\mathfrak{g})$ is the direct sum of the Verma modules $M_{\rho-s(\rho)}$, where s runs over the set of elements of the Weyl group of length j [11].

Once we fixed the vectors w_s, we have canonical embeddings $M_{\rho-s(\rho)} \to M_0$. Therefore the module $M_{\rho-s(\rho)}$ can be thought of as a submodule of the module M_0, generated by the vector w_s. It is known that the vector $w_{s'}$ belongs to the module $M_{\rho-s(\rho)}$ if and only if $s \preceq s'$ with respect to the Bruhat order on the Weyl group [10]. It is clear that these vectors are singular vectors of the module $M_{\rho-s(\rho)}$ and that there are no other singular vectors in this module. In that case we have an embedding $i_{s',s} : M_{\rho-s'(\rho)} \to M_{\rho-s(\rho)}$, which sends the lowest weight vector of $M_{\rho-s'(\rho)}$ to the singular vector of $M_{\rho-s(\rho)}$ of weight $\rho - s'(\rho)$.

2.3.3. Lemma. [11]

(a) *Let s and s'' be two elements of the Weyl group, such that $s \prec s''$ and $l(s'') = l(s) + 2$. Then there are either two or no elements s', such that $s \prec s' \prec s''$.*

(b) *Let us call a square a set of four elements of the Weyl group, satisfying the conditions of the part (a).*

It is possible to attach a sign $\epsilon_{s',s}$, + or −, to each pair of elements of the Weyl group s, s', such that $s \prec s', l(s') = l(s) + 1$, so that the product of signs over any square is −.

2.3.4. *The differential.* We are now ready to define the differential of the BGG resolution $d_j : B_j(\mathfrak{g}) \to B_{j-1}(\mathfrak{g})$ as

$$(2.3.1) \qquad d_j = \sum_{l(s)=j-1, l(s')=j, s \prec s'} \epsilon_{s',s} \cdot i_{s',s}.$$

In other words, we take the sum of all possible embeddings of the Verma modules, which are direct summands of $B_j(\mathfrak{g})$, with the special choice of signs from Lemma 2.3.3. By definition, these differentials commute with the action of $\mathfrak{g}$.

2.3.5. *Theorem.* [11]

(a) *The differentials* $d_j, j > 0$, *are nilpotent:* $d_{j-1}d_j = 0$, *and so* $B_*(\mathfrak{g})$ *is a complex.*
(b) *The 0th homology of the complex* $B_*(\mathfrak{g})$ *is the trivial one-dimensional representation of* $\mathfrak{g}$, *and all higher homologies vanish.*

Proof. In the notation of § 2.3.2, we have $w_{s'} = P_{s',s} \cdot w_s$, for some element $P_{s',s}$ of $U(\mathfrak{n}_+)$. If we have a square s, s_1', s_2', s'' of elements of the Weyl group, such that $s \prec s_1', s_2' \prec s''$, then we can write: $w_{s''} = P_{s'',s_1'} P_{s_1',s} w_s$ and $w_{s''} = P_{s'',s_2'} P_{s_2',s} w_s$. Therefore we obtain the following identity

$$(2.3.2) \qquad P_{s'',s_1'} P_{s_1',s} = P_{s'',s_2'} P_{s_2',s}$$

in $U(\mathfrak{n}_+)$. By definition, $i_{s',s}(u \cdot 1_{\rho-s'(\rho)}) = (u P_{s',s}) \cdot 1_{\rho-s(\rho)}$. So, we obtain from formula (2.3.2): $i_{s_1',s} \circ i_{s'',s_1'} = i_{s_2',s} \circ i_{s'',s_2'}$. Thus, because of our sign convention (cf. Lemma 2.3.3, (b)), such terms in the composition of two consecutive differentials $d_{j-1}d_j$ will cancel out. This proves part (a) of the Theorem.

The proof of part (b) is rather complicated; it uses the so-called *weak* BGG resolution, which is obtained from the de Rham complex on the big cell of the flag manifold of $\mathfrak{g}$ (cf. [11]).

2.3.6. *Remarks.* (1) One can define analogous resolutions for arbitrary finite-dimensional representations of $\mathfrak{g}$.
(2) There are generalizations of the BGG resolutions to arbitrary symmetrizable Kac-Moody algebras [107]. We will use such resolutions for affine algebras in § 3.

2.4. Extended complex and its cohomology. One of the main applications of the BGG resolution is to computation of the cohomologies of the nilpotent Lie algebra $\mathfrak{n}_+$ of $\mathfrak{g}$. In this subsection we will use this resolution to extend our complex (2.2.7), and to compute the cohomology of the resulting complex.

2.4.1. The jth component $F^j(\mathfrak{g})$ of our extended complex

$$F^*(\mathfrak{g}) = \oplus_{j \geq 0} F^j(\mathfrak{g})$$

will be the direct sum of the spaces $\pi_{\rho-s(\rho)}$, where s runs over the set of elements of the Weyl group of length j.

Now let us define the differentials. The algebra $U(\mathfrak{n}_+)$ is generated by $e_i, i = 1, \dots, l$, which satisfy the Serre relations. So any element of $U(\mathfrak{n}_+)$ can be expressed in terms of e_i. Let $P_{s',s}(Q) : \pi_{\rho-s(\rho)} \to \pi_{\rho-s'(\rho)}$ be the map, obtained by inserting into $P_{s',s} \in U(\mathfrak{n}_+)$ the operators $\tilde{Q}_i$ instead of e_i. We can then introduce the differential $\delta^j : F^{j-1}(\mathfrak{g}) \to F^j(\mathfrak{g})$ of our complex by the formula

$$(2.4.1) \qquad \delta^j = \sum_{l(s)=j-1, l(s')=j, s \prec s'} \epsilon_{s',s} \cdot P_{s',s}(Q).$$

2.4.2. Lemma. *The differentials* $\delta^j, j > 0$, *are nilpotent:* $\delta^{j+1}\delta^j = 0$, *and so* $F^*(\mathfrak{g})$ *is a complex.*

Proof. Just as in the proof of part (a) of Theorem 2.3.5, we have to check that $P_{s'',s_1'}(Q)$ $P_{s_1',s}(Q) = P_{s'',s_2'}(Q)P_{s_2',s}(Q)$. But this follows at once from (2.3.2), since, according to Proposition 2.2.8, the operators Q_i, and therefore the operators $\tilde{Q}_i$, satisfy the defining relations of the algebra $U(\mathfrak{n}_+)$.

2.4.3. We introduce a $\mathbb{Z}$–grading on the complex $F^*(\mathfrak{g})$ by putting $\deg x_n^i = -n$, and $\deg v_{\rho-s(\rho)} = (\rho^\vee, \rho - s(\rho))$, where $\rho^\vee \in h^*$ is defined by the property $(\rho^\vee, \alpha_i) = 1, i = 1, \ldots, l$. Clearly, all homogeneous subspaces of π_γ have finite dimensions. With respect to this grading, the operator ∂ is homogeneous of degree 1, and we can define a grading on the spaces $\mathcal{F}_\gamma$ by subtracting 1 from the grading on the space π_γ. The differentials δ^j are homogeneous of degree 0 with respect to this grading. Therefore our complex $F^*(\mathfrak{g})$ decomposes into a direct sum of finite-dimensional subcomplexes, corresponding to its different graded components.

2.4.4. Example of $\mathfrak{sl}_3$. In this case the Weyl group consists of six elements. It is generated by two reflections: s_1 and s_2 with the relation $s_1s_2s_1 = s_2s_1s_2$. The complex $F^*(\mathfrak{sl}_3)$ is shown on Fig. 1.

The vertices represent the spaces $\pi_{\rho-s(\rho)}$, and arrows represent the maps of the differential. There are four squares. The anti-commutativity of maps, associated to one of them, reads:

$$\tilde{Q}_1^2 \tilde{Q}_2 = -A\tilde{Q}_1.$$

To find a solution A to this equation, let us consider the Serre relation, which the operators $\tilde{Q}_1$ and $\tilde{Q}_2$ satisfy (cf. Proposition 2.2.8):

$$(\text{ad}\tilde{Q}_1)^2 \cdot \tilde{Q}_2 = \tilde{Q}_1^2\tilde{Q}_2 - 2\tilde{Q}_1\tilde{Q}_2\tilde{Q}_1 + \tilde{Q}_2\tilde{Q}_1^2 = 0.$$

Therefore, $A = -2\tilde{Q}_1\tilde{Q}_2 + \tilde{Q}_2\tilde{Q}_1$ is such a solution. Similarly, if we put $B = -2\tilde{Q}_2\tilde{Q}_1 + \tilde{Q}_1\tilde{Q}_2$; then all squares will be anti-commutative.

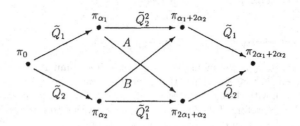

Fig. 1

Our complex $F^*(\mathfrak{sl}_3)$ has four non-trivial groups: $F^0(\mathfrak{sl}_3) = \pi_0, F^1(\mathfrak{sl}_3) = \pi_{\alpha_1} \oplus \pi_{\alpha_2}, F^2(\mathfrak{sl}_3) = \pi_{\alpha_1+2\alpha_2} \oplus \pi_{2\alpha_1+\alpha_2}$, and $F^3(\mathfrak{sl}_3) = \pi_{2\alpha_1+2\alpha_2}$. The differentials have the form: $\delta_1 = \tilde{Q}_1 + \tilde{Q}_2, \delta_2 = \tilde{Q}_1^2 + \tilde{Q}_2^2 + A + B$, and $\delta_3 = \tilde{Q}_1 + \tilde{Q}_2$.

2.4.5. Proposition. *The cohomologies of the complex $F^*(\mathfrak{g})$ are isomorphic to the cohomologies of the Lie algebra $\mathfrak{n}_+$ with coefficients in the module π_0, $H^*(\mathfrak{n}_+, \pi_0)$.*

Proof. The complex $F^*(\mathfrak{g})$ is isomorphic to $\text{Hom}_{\mathfrak{n}_+}(B_*(\mathfrak{g}), \pi_0)$. Indeed, for any λ the module M_λ is isomorphic to a free $\mathfrak{n}_+$–module M with one generator. Therefore the space of $\mathfrak{n}_+$–homomorphisms $\text{Hom}_{\mathfrak{n}_+}(M, \pi_0)$ is canonically isomorphic to π_0. Indeed, any non-zero homomorphism $x \in \text{Hom}_{\mathfrak{n}_+}(M, \pi_0)$ defines a non-zero element in π_0: the image of the lowest weight vector of M. The embedding $i_{s',s}$ of M into itself then induces the homomorphism from π_0 to π_0, which sends $y \in \pi_0$ to $P_{s',s} \cdot y$. This is precisely the homomorphism $P_{s',s}(Q)$. Hence the differentials d_j of the BGG resolution $B_*(\mathfrak{g})$ map to the differentials δ^j of the complex $F^*(\mathfrak{g})$.

According to part (b) of Theorem 2.3.5, the BGG resolution $B_*(\mathfrak{g})$ is the resolution of the trivial $\mathfrak{n}_+$–module by free $\mathfrak{n}_+$–modules. The cohomologies of the $\text{Hom}_{\mathfrak{n}_+}$ of such a resolution to an $\mathfrak{n}_+$–module, are, by definition, the cohomologies of $\mathfrak{n}_+$ with coefficients in this module, cf., e.g., [76]. Therefore the cohomologies of the complex $F^*(\mathfrak{g})$ coincide with $H^*(\mathfrak{n}_+, \pi_0)$.

2.4.6. Proposition. *All higher cohomologies of the complex $F^*(\mathfrak{g})$ vanish.*

Proof. Each of the root generators e_α of $\mathfrak{n}_+$ acts on π_0 by a certain vector field. This vector field has a shift term, which is a linear combination of $\partial/\partial x_n^i$ (cf. the proof of Proposition 3.1.10). It follows from the proof of Proposition 3.2.5 that the shift terms of the root generators of $\mathfrak{n}_+$ are linearly independent. Therefore the dual operators to e_α are equal to the sum of some linear combination of x_n^i and some differential operators which do not change or decrease the power of a polynomial (cf. the proof of Proposition 2.1.15). Hence the dual module to the module π_0 is a free $\mathfrak{n}_+$–module. But then the module π_0 is injective, and so all higher cohomologies of $\mathfrak{n}_+$ with coefficients in π_0 must vanish. Proposition 2.4.5 then implies that all higher cohomologies of the complex $F^*(\mathfrak{g})$ vanish.

2.4.7. Proposition. *There exist elements $W^{(1)}_{-d_1-1}, \dots, W^{(l)}_{-d_l-1}$ of π_0 of degrees $d_1+1, \dots, d_l +1$, where the d_i's are the exponents of $\mathfrak{g}$, such that the 0th cohomology $\mathcal{W}_0(\mathfrak{g})$ of the complex $F^*(\mathfrak{g})$ is isomorphic to the polynomial algebra*

$$\mathbb{C}[W^{(i)}_{n_i}]_{1 \le i \le l, n_i < -d_i},$$

where $W^{(i)}_{n_i} = \partial^{-n_i-d_i-1}W^{(i)}_{-d_i-1}/(-n_i - d_i - 1)!$.

Proof. The algebra π_0 is the inductive limit of the free commutative algebras with the generators x_n^i, $-M \le n \le -1$. Hence the spectrum of π_0 is the inverse limit of the affine spaces R_M, $M > 0$, with the coordinates x_n^i, $-M \le n \le -1$. From the explicit formula for the action of the generators of the Lie algebra $\mathfrak{n}_+$ on π_0 we see that the algebras $\mathbb{C}[R_M]$ are preserved under the action of $\mathfrak{n}_+$ (cf. the proof of Proposition 2.1.15). The infinitesimal action of $\mathfrak{n}_+$ on R_M by vector fields can be integrated to an action of the Lie group N_+ by means of the exponential map $\mathfrak{n}_+ \to N_+$, which is an isomorphism. The action of N_+ commutes with the projections $R_{M+K} \to R_M$.

At each point of the spectrum of π_0 the vector fields of the infinitesimal action of the root generators e_α of $\mathfrak{n}_+$ are linearly independent (cf. the proof of Proposition 2.4.6). Hence the action of N_+ on R_M is free for M large enough. The orbits of this action are isomorphic to the affine space $\mathbb{C}^{\dim N_+}$. Therefore the quotient space R_M/N_+ is also an affine space and the algebra of functions on this space is a free commutative algebra.

Since the projections $R_{M+K} \to R_M$ are compatible with the action of N_+, we can take the inverse limit of the quotient spaces R_M/N_+. The algebra of functions on this inverse limit is the inductive limit of the free polynomial algebras of functions on R_M/N_+ and therefore it is a free polynomial algebra with infinitely many generators. It consists of all N_+-invariant elements of π_0, which are the same as the n_+-invariant elements. Hence this algebra coincides with the 0th cohomology of our complex.

The algebra π_0 is $\mathbb{Z}$-graded and the action of n_+ preserves this grading, if we introduce the principal grading on n_+ by putting $\deg e_i = 1$. Therefore the 0th cohomology of our complex is also $\mathbb{Z}$-graded, and it is easy to compute the degrees of the generators of this algebra by computing its character.

By Proposition 2.4.6, all higher cohomologies of the complex $F^*(\mathfrak{g})$ vanish. Therefore, the character of the 0th cohomology is equal to the Euler character of the complex. The latter is equal to

$$\sum_{j \geq 0} (-1)^j \sum_{l(s)=j} \mathrm{ch}\pi_{\rho - s(\rho)} = \prod_{n > 0} (1 - q^n)^{-l} \sum_s (-1)^{l(s)} q^{(\rho^\vee, \rho - s(\rho))}.$$

From the specialized Weyl character formula we deduce

$$\sum_s (-1)^{l(s)} q^{(\rho^\vee, \rho - s(\rho))} = \prod_{1 \leq i \leq l, 1 \leq n_i \leq d_i} (1 - q^{n_i}).$$

This gives for the character of the 0th cohomology, $\mathcal{W}_0(\mathfrak{g})$,

$$\mathrm{ch}\mathcal{W}_0(\mathfrak{g}) = \prod_{1 \leq i \leq l, n_i > d_i} (1 - q^{n_i})^{-1}.$$

This formula shows that the 0th cohomology is the free commutative algebra with generators $W_{n_i}^{(i)}$ of degree $-n_i$, where $1 \leq i \leq l, n_i < -d_i$. In the same way as in the proof of Proposition 2.1.15 we can see that as the generators $W_{n_i}^{(i)}$ we can take $\partial^{-n_i - d_i - 1} W_{-d_i - 1}^{(i)}/(-n_i - d_i - 1)!$. The Proposition follows.

2.4.8. Lemma. *Let P be a homogeneous element of the algebra $U(n_+)$ of weight γ, such that $P \cdot \mathbf{1}_\lambda$ is a singular vector of the Verma module M_λ of weight $\lambda + \gamma$. Then the operator $P(Q) : \pi_\lambda \to \pi_{\lambda + \gamma}$ commutes with the action of the derivative ∂.*

Proof. The action of the derivative ∂ on π_λ differs from its action on π_0 by the operator of multiplication by λ_{-1}. We have: $[\lambda_{-1}, \tilde{Q}_i] = (\alpha_i, \lambda)T_i$. Therefore, by Lemma 2.2.6, the commutator of the operator $\tilde{Q}_i : \pi_\lambda \to \pi_{\lambda + \alpha_i}$ with ∂ is equal to $(\alpha_i, \lambda)T_i$. Hence, the commutator of the monomial $\tilde{Q}_{i_m} \ldots \tilde{Q}_{i_1} : \pi_\lambda \to \pi_{\lambda + \gamma}$, where $\gamma = \sum_{j=1}^m \alpha_{i_j}$, with ∂ is equal to

$$\sum_{j=1}^m (\alpha_{i_j}, \lambda + \alpha_{i_1} + \ldots + \alpha_{i_{j-1}}) \tilde{Q}_{i_m} \ldots T_{i_j} \ldots \tilde{Q}_{i_1}.$$

This precisely coincides with the action of

$$\sum_{i=1}^l \frac{(\alpha_i, \alpha_i)}{2} f_i,$$

where the $f_i, i = 1, \ldots, l$, are the generators of the Lie algebra n_-, on the vector $e_{i_m} \ldots e_{i_1} \mathbf{1}_\lambda$ of the Verma module M_λ. If $P\mathbf{1}_\lambda$ is a singular vector in M_λ, then

$$\sum_{i=1}^l \frac{(\alpha_i, \alpha_i)}{2} f_i \cdot P\mathbf{1}_\lambda = 0,$$

and so $[\partial, P(Q)] = 0$.

2.4.9. Corollary. *The higher differentials $\delta^j, j > 1$, of the complex $F^*(\mathfrak{g})$ commute with the action of the derivative ∂.*

Proof. Each δ^j is a linear combination of maps $P_{s',s}(Q)$. Since by definition $P_{s',s}$ defines a singular vector, such a map commutes with ∂ by Lemma 2.4.8.

2.4.10. Theorem. *The space $I_0(\mathfrak{g})$ of local integrals of motion of the classical Toda field theory, associated with $\mathfrak{g}$, coincides with the quotient of $\mathcal{W}_0(\mathfrak{g})$ by the total derivatives and constants.*

Proof. Using Corollary 2.4.9, we can construct the double complex $\mathbb{C} \longrightarrow F^*(\mathfrak{g}) \longrightarrow F^*(\mathfrak{g}) \longrightarrow \mathbb{C}$, which is shown on Fig. 2.

By Corollary 2.4.9, the total differential of this complex is nilpotent. If we compute the cohomology of this complex by means of the spectral sequence, in which the 0th differential is vertical, then in the first term we obtain the complex $\tilde{F}^*(\mathfrak{g}) = \oplus_{j \geq 0} \tilde{F}^j(\mathfrak{g})$, where

$$\tilde{F}^j(\mathfrak{g}) = \oplus_{l(s)=j} \mathcal{F}_{\rho - s(\rho)}.$$

By definition, the 0th cohomology of the complex $\tilde{F}^*(\mathfrak{g})$ is the space $I_0(\mathfrak{g})$. Therefore it coincides with the 1st cohomology of our double complex.

But we can compute this cohomology by means of the other spectral sequence, in which the 0th differential is the horizontal one. Then the Theorem follows from Proposition 2.4.7 in the same way as in the proof of Proposition 2.1.17.

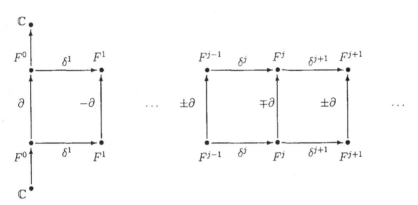

Fig. 2

2.4.11. *The Adler-Gelfand-Dickey algebra.* According to Theorem 2.4.10, the Poisson algebra $I_0(\mathfrak{g})$ of local integrals of motion of the Toda field theory associated to $\mathfrak{g}$ is the algebra of local functionals on a certain hamiltonian space, $H(\mathfrak{g})$.

This Poisson algebra coincides with the Adler-Gelfand-Dickey (AGD) algebra, or the classical $\mathcal{W}$−algebra [71, 2].

The Drinfeld-Sokolov reduction [32, 33] produces the hamiltonian space of the AGD algebra as the result of a hamiltonian reduction of a hyperplane in the dual space to the affinization $\hat{\mathfrak{g}}$ of $\mathfrak{g}$. Following the standard technique of hamiltonian reduction [83], one can obtain this algebra as the 0th cohomology of the corresponding (classical) BRST complex.

It was explained in [48, 56, 57] that the complex $\bar{F}^*(\mathfrak{g})$ appears as the first term of a spectral sequence, associated to the BRST complex of the Drinfeld-Sokolov reduction. Therefore $I_0(\mathfrak{g})$ is precisely the AGD algebra. We also see that higher cohomologies of the BRST complex vanish.

Usually, one constructs a map, which is called the Miura transformation, from the hamiltonian space $L\mathfrak{h}$ to $H(\mathfrak{g})$, which preserves the Poisson structures. The image of the inverse map of the spaces of functionals embeds the AGD algebra into $\mathcal{F}_0$. As we have explained, the image of this map coincides with the algebra of local integrals of motion of the corresponding Toda field theory, and can be characterized in very simple terms as the intersection of the kernels of certain linear operators, acting from $\mathcal{F}_0$ to the spaces $\mathcal{F}_{\alpha_i}$.

For the classical simple Lie algebras explicit formulas for the Miura transformation map are known [33]. They give explicit formulas for the generators $W_n^{(i)}$ of the Poisson algebra $I_0(\mathfrak{g})$.

For example, the AGD hamiltonian space $H(\mathfrak{sl}_n)$ is isomorphic to the space of differential operators on the circle of the form

$$\partial_t^n + \sum_{i=1}^{n-1} W^{(i)}(t)\partial_t^{n-i-1}.$$

The Miura transformation from the space $L\mathfrak{h}$, which consists of functions on the circle with values in the Cartan subalgebra $\mathfrak{h}$ of $\mathfrak{sl}_n$, $\mathbf{u}(t) = (u^1(t), \dots, u^{n-1}(t))$, to $H(\mathfrak{sl}_n)$ can be constructed as follows.

Introduce new variables $v^1(t), \dots, v^n(t)$, such that $\sum_{i=1}^n v^i(t) = 0$, and $u^i(t) = v^i(t) - v^{i+1}(t)$. Then put

$$\partial_t^n + \sum_{i=1}^{n-1} W^{(i)}(t)\partial_t^{n-i-1} = (\partial_t + v^1(t)) \dots (\partial_t + v^n(t)).$$

These formulas allow to express $W^{(i)}(t)$ as a differential polynomial in $u^j(t)$ (cf. § 2.1.15 for the case of $\mathfrak{sl}_2$, when the AGD algebra is isomorphic to the classical Virasoro algebra). One can find other generators $\widetilde{W}^{(i)}(t)$ of $I_0(\mathfrak{sl}_n)$, which transform as tensor fields on the circle under changes of variables [28].

2.4.12. *Integrals of motion in the extended space of local functionals.* As in § 2.1.4, we can extend our Poisson algebra of local functionals $\mathcal{F}_0$ by adjoining all Fourier components of differential polynomials, not only the (-1)st ones. Let $\widehat{\mathcal{F}}_0$ be the space of functionals on $L\mathfrak{h}$, which can be represented as residues of differential polynomials with explicit dependence on t:

$$F[\mathbf{u}(t)] = \int P(\partial^n u^i(t); t)dt.$$

We can define the Poisson structure on $\widehat{\mathcal{F}}_0$ by the same formula (2.2.1). Thus, $\widehat{\mathcal{F}}_0$ is a Poisson algebra, and $\mathcal{F}_0 \subset \widehat{\mathcal{F}}_0$ is its Poisson subalgebra.

Analogously, one can define the spaces $\widehat{\mathcal{F}}_\gamma, \gamma \in P$ by allowing differential polynomials to depend on t.

One has the analogue of the exact sequence (2.2.2):

$$0 \longrightarrow \pi_\gamma \otimes \mathbb{C}[t, t^{-1}] \overset{\partial}{\longrightarrow} \pi_\gamma \otimes \mathbb{C}[t, t^{-1}] \longrightarrow \widehat{\mathcal{F}}_\gamma \longrightarrow 0,$$

where the action of ∂ on $\pi_\gamma \otimes \mathbb{C}[t, t^{-1}]$ is given by $\partial \otimes 1 + 1 \otimes \partial_t$. If $\gamma = 0$, we have to replace the first $\pi_0 \otimes \mathbb{C}[t, t^{-1}]$ by $\pi_0 \otimes \mathbb{C}[t, t^{-1}]/\mathbb{C} \otimes \mathbb{C}$.

The Poisson bracket with $\int e^{\phi_i(t)} dt$ defines a linear operator $\widehat{\mathcal{F}}_0 \to \widehat{\mathcal{F}}_{\alpha_i}$, which we also denote by $\bar{Q}_i$. We can then define the space of integrals of motion of the Toda field theory as the intersection of kernels of the operators $\bar{Q}_i : \widehat{\mathcal{F}}_0 \to \widehat{\mathcal{F}}_{\alpha_i}, i = 1, \ldots, l$.

We can use the methods of this section to compute this space. Indeed, consider the tensor product of the complex $F^*(\mathfrak{g})$ with $\mathbb{C}[t, t^{-1}]$, with the differentials acting on the first factor as δ^j and identically on the second factor. Such differentials commute with the action of the derivative, and therefore we can use the double complex

$$\mathbb{C} \otimes \mathbb{C} \longrightarrow F^*(\mathfrak{g}) \otimes \mathbb{C}[t, t^{-1}] \longrightarrow F^*(\mathfrak{g}) \otimes \mathbb{C}[t, t^{-1}] \longrightarrow \mathbb{C} \otimes \mathbb{C}$$

to compute the space of integrals of motion.

The cohomologies of the complex $F^*(\mathfrak{g}) \otimes \mathbb{C}[t, t^{-1}]$ are equal to the cohomologies of the complex $F^*(\mathfrak{g})$ tensored with $\mathbb{C}[t, t^{-1}]$. We deduce from Proposition 2.4.7 that the space of integrals of motion is isomorphic to the quotient of $\mathcal{W}_0(\mathfrak{g}) \otimes \mathbb{C}[t, t^{-1}]$ by the total derivatives and constants. This Poisson algebra is the algebra of local functionals on the AGD hamiltonian space $H(\mathfrak{g})$, extended in the same way – by adjoining all Fourier components of differential polynomials. It contains $I_0(\mathfrak{g})$ as a Poisson subalgebra. Sometimes it is this algebra, which is called the classical $\mathcal{W}$–algebra.

3. CLASSICAL AFFINE TODA FIELD THEORIES.

3.1. The case of $\widehat{\mathfrak{sl}}_2$ – classical sine-Gordon theory.

3.1.1. *Hamiltonian structure.* The hamiltonian space of the classical sine-Gordon theory is the same as the hamiltonian space of the classical Liouville theory, namely, the space $L\mathfrak{h}$ of polynomial functions on the circle with values in the one-dimensional Cartan subalgebra $\mathfrak{h}$ of $\mathfrak{sl}_2$, cf. § 2.1.1. As the space of functions on this space, we again take the space $\mathcal{F}_0$ of local functionals. The Poisson structure on $\mathcal{F}_0$ is given by formula (2.1.2).

We also define spaces $\pi_n, n \in \mathbb{Z}$, as the tensor products $\pi_0 \otimes \mathbb{C}v_n$, where π_0 is the space of differential polynomials (cf. § 2.1.1), and extend the action of derivative ∂ from π_0 to π_n by the formula $(\partial + n \cdot u) \otimes 1$. We then put: $\mathcal{F}_n = \pi_n/\partial \pi_n$. This space has an interpretation as the space of functionals of the form

$$\int P(u(t), \partial u(t), \ldots) e^{n\phi(t)} dt.$$

As in § 2.1.8, we can extend the Poisson structure on $\mathcal{F}_0$ to well-defined maps $\mathcal{F}_0 \times \mathcal{F}_n \to \mathcal{F}_n$ given by the formula

$$(3.1.1) \quad \left\{ \int P dt, \int R e^{n\phi} dt \right\} = \int \frac{\delta P}{\delta u} \frac{\delta [R e^{n\phi}]}{\delta \phi} dt = \int \frac{\delta P}{\delta u} \left[n R e^{n\phi} - \partial_t \left(\frac{\delta R}{\delta u} e^{n\phi} \right) \right] dt.$$

These brackets satisfy the Jacobi identity for any triple $F, G \in \mathcal{F}_0, H \in \mathcal{F}_n$.

3.1.2. *The sine-Gordon hamiltonian and local integrals of motion.* The hamiltonian H of the sine-Gordon model is given by

$$H = \int e^{\phi(t)} dt + \int e^{-\phi(t)} dt \in \mathcal{F}_1 \oplus \mathcal{F}_{-1}.$$

In other words, it is equal to the sum of the projections of the vectors $v_{\pm 1} \in \pi_{\pm 1}$ to $\mathcal{F}_{\pm 1}$. One can check that the corresponding hamiltonian equation

$$\partial_\tau U(t) = \{U(t), H\}$$

coincides with the sine-Gordon equation (1.1.3).

We then define the space of local integrals of motion $I_0(\widehat{\mathfrak{sl}}_2)$ of the sine-Gordon theory as the kernel of the linear operator

$$\bar{Q}_1 + \bar{Q}_0 : \mathcal{F}_0 \to \mathcal{F}_1 \oplus \mathcal{F}_{-1},$$

where $\bar{Q}_1 = \{\cdot, \int e^{\phi(t)}dt\} : \mathcal{F}_0 \to \mathcal{F}_1$, and $\bar{Q}_0 = \{\cdot, \int e^{-\phi(t)}dt\} : \mathcal{F}_0 \to \mathcal{F}_{-1}$. In other words, $I_0(\widehat{\mathfrak{sl}}_2)$ is the intersection of the kernels of the operators $\bar{Q}_1$ and $\bar{Q}_0$. By the Jacobi identity, $I_0(\widehat{\mathfrak{sl}}_2)$ is closed with respect to the Poisson bracket. Note that the operator $\bar{Q}_1$ coincides with the operator $\bar{Q}$, defined by (2.1.11), and therefore $I_0(\widehat{\mathfrak{sl}}_2)$ is a Poisson subalgebra in the Poisson algebra $I_0(\mathfrak{sl}_2)$ of local integrals of motion of the Liouville theory, that is in the classical Virasoro algebra.

3.1.3. Now introduce the operators $\tilde{Q}_1 : \pi_m \to \pi_{m+1}$ and $\tilde{Q}_0 : \pi_m \to \pi_{m-1}$ by the formulas:

$$\tilde{Q}_1 = T \sum_{n<0} S_{n+1}^+ \frac{\partial}{\partial x_n}, \qquad \tilde{Q}_0 = -T^{-1} \sum_{n<0} S_{n+1}^- \frac{\partial}{\partial x_n},$$

where $T^{\pm 1} : \pi_m \to \pi_{m\pm 1}$ are the translation operators, and $S_n^\pm$ are the Schur polynomials:

$$\sum_{n\leq 0} S_n^\pm z^n = \exp\left(\sum_{m<0} \mp \frac{x_m}{m} z^m\right).$$

We will also need the operators $Q_1 = T^{-1}\tilde{Q}_1$ and $Q_0 = T\tilde{Q}_0$, which are linear endomorphisms of π_0 of degree 1. Note that S_n^+ coincides with S_n, given by formula (2.1.15), and so $\tilde{Q}_1$ coincides with $\tilde{Q}$, given by formula (2.1.14).

In the same way as in Lemma 2.1.12, one shows that the operators $\tilde{Q}_1$ and $\tilde{Q}_0$, acting from π_0 to $\pi_{\pm 1}$ commute with the action of derivative ∂, and that the corresponding operators $\mathcal{F}_0 \to \mathcal{F}_{\pm 1}$ coincide with $\bar{Q}_1$ and $\bar{Q}_0$.

The following Proposition will enable us to compute the space of local integrals of motion $I_0(\widehat{\mathfrak{sl}}_2)$.

3.1.4. *Proposition. The operators Q_1 and Q_0 satisfy the Serre relations of the Lie algebra $\widehat{\mathfrak{sl}}_2$:*

$$(\operatorname{ad} Q_1)^3 \cdot Q_0 = 0, \qquad (\operatorname{ad} Q_0)^3 \cdot Q_1 = 0.$$

Proof follows from Proposition 2.2.8 and Remark 2.2.9.

3.1.5. Thus, operators Q_0 and Q_1 generate an action of the nilpotent subalgebra $\mathfrak{n}_+$ of $\widehat{\mathfrak{sl}}_2$ on the space π_0. We will use this fact and the BGG resolution for $\widehat{\mathfrak{sl}}_2$ to extend the complex

$$\mathcal{F}_0 \longrightarrow \mathcal{F}_1 \oplus \mathcal{F}_{-1},$$

whose 0th cohomology is, by definition, the space $I_0(\widehat{\mathfrak{sl}}_2)$, to a larger complex with nicer cohomologies.

Again, as in the previous section, we will be using the double complex, which consists of the spaces π_n. So, our task is to extend the complex

$$\pi_0 \longrightarrow \pi_1 \oplus \pi_{-1}$$

in such a way that all higher differentials commute with the action of the derivative.

3.1.6. *BGG resolution of* $\widehat{\mathfrak{sl}}_2$. As in the case of finite-dimensional Lie algebras, there exists a BGG resolution $B_*(\widehat{\mathfrak{sl}}_2)$ of $\widehat{\mathfrak{sl}}_2$ [107]. It is a complex, consisting of Verma modules over $\widehat{\mathfrak{sl}}_2$, whose higher homologies vanish, and the 0th homology is one-dimensional.

The jth group $B_j(\widehat{\mathfrak{sl}}_2)$ of this complex is equal to M_0, if $j = 0$, and $M_{2j} \oplus M_{-2j}$, if $j > 0$. Here M_λ stands for the Verma module over $\widehat{\mathfrak{sl}}_2$ of level 0 and lowest weight λ (that is the weight of the Cartan subalgebra of $\mathfrak{sl}_2$, embedded into $\widehat{\mathfrak{sl}}_2$ as the constant subalgebra). Such a module is defined in exactly the same way as a Verma module over a finite-dimensional Lie algebra (cf. § 2.3.1).

The module M_0 contains the singular vectors, labeled by the elements of the Weyl group of $\widehat{\mathfrak{sl}}_2$. Let us denote them by w_0, and $w_j, w_j', j > 0$. The weight of w_j (respectively, w_j') is equal to $2j$ (respectively, $-2j$), and it generates the submodule of M_0, which is isomorphic to M_{2j} (respectively, M_{-2j}). We have $w_1 = Y_1' w_0, w_1' = Y_1 w_0$, and $w_j = X_j w_{j-1}, w_j' = Y_j w_{j-1}, w_j = X_j' w_{j-1}', w_j = Y_j' w_{j-1}'$, for $j > 1$, where X_j, Y_j, X_j' and Y_j' are certain elements from $U(\mathfrak{n}_+)$.

The differential $d_j : B_j^q \to B_{j-1}^q, j > 0$, is given by the alternating sum of the embeddings of M_{2j} and M_{-2j} into $M_{2(j-1)}$ and $M_{-2(j-1)}$, which map the lowest weight vectors to the corresponding singular vectors. The nilpotency of the differential, $d_{j-1}d_j = 0$, is ensured by the commutativity of the embeddings, corresponding to the "squares" in the Weyl group and a special sign convention, analogous to the one from Lemma 2.3.3.

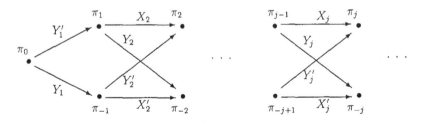

Fig. 3

3.1.7. *Extended complex.* We are ready now to define the extended complex $F^*(\widehat{\mathfrak{sl}}_2)$ (cf. Fig. 3). The jth group $F^*(\widehat{\mathfrak{sl}}_2)$ of this complex is equal to π_0, if $j = 0$, and $\pi_j \oplus \pi_{-j}$, if $j > 0$. The differential $\delta^j : F^{j-1}(\mathfrak{g}) \to F^j(\mathfrak{g})$ is given by the formula $\delta^1 = Y_0(Q) + Y_0'(Q), \delta^j = X_j(Q) - (-1)^j Y_j(Q) - (-1)^j Y_j'(Q) + X_j'(Q), j > 1$, where we insert into these elements of $U(\mathfrak{n}_+)$ the operators $\widetilde{Q}_1$ and $\widetilde{Q}_0$ instead of the generators e_1 and e_0.

The nilpotency of the differential, $\delta^j \delta^{j-1} = 0$, follows from the relations in $U(\mathfrak{n}_+)$: $X_j' Y_{j-1} = Y_j X_{j-1}, X_j Y_{j-1}' = Y_j' X_{j-1}', X_j X_{j-1} = Y_j' Y_{j-1}$, and $X_j' X_{j-1}' = Y_j Y_{j-1}$, and Proposition 3.1.4.

Note that $Y_j'(Q) = \widetilde{Q}_1^{2j-1}, Y_j(Q) = \widetilde{Q}_0^{2j-1}$. Other operators are more complicated, but explicit formulas for them can be obtained in principle from the commutativity relations above, using the Serre relations, in the same way as in § 2.4.4.

Let us introduce a $\mathbb{Z}$-grading on our complex, by putting $\deg v_n = n^2$, and $\deg x_m = -m$. One can check that with respect to this grading the differentials δ^j are homogeneous of degree 0. Therefore our complex is a direct sum of finite-dimensional subcomplexes, corresponding to various graded components.

3.1.8. Proposition. *The operators Q_1 and Q_0 define an action of the nilpotent Lie subalgebra $\mathfrak{n}_+$ of $\widehat{\mathfrak{sl}}_2$ on π_0. The cohomologies of the complex $F^*(\widehat{\mathfrak{sl}}_2)$ are isomorphic to the cohomologies of $\mathfrak{n}_+$ with coefficients in π_0, $H^*(\mathfrak{n}_+, \pi_0)$.*

Proof. The same as in Proposition 2.4.5.

3.1.9. Principal commutative subalgebra. Recall that in the realization of $\widehat{\mathfrak{sl}}_2$ as the central extension of the loop algebra $\mathfrak{sl}_2 \otimes \mathbb{C}[t, t^{-1}]$, the nilpotent subalgebra $\mathfrak{n}_+$ of $\widehat{\mathfrak{sl}}_2$ is identified with the Lie algebra $(\bar{\mathfrak{n}}_+ \otimes 1) \oplus (\mathfrak{sl}_2 \otimes t\mathbb{C}[t])$, where $\bar{\mathfrak{n}}_+$ is the nilpotent subalgebra of $\mathfrak{sl}_2$. Let e, h and f be the standard generators of $\mathfrak{sl}_2$. If y is one of them, we will denote by $y(m)$ the element $y \otimes t^m$ of $\widehat{\mathfrak{sl}}_2$. The basis of $\mathfrak{n}_+$ consists of $e(m), m \geq 0, h(m), m > 0$, and $f(m), m > 0$.

Let $\mathfrak{a}$ be the commutative subalgebra of $\mathfrak{n}_+$, which is linearly generated by $e(m) + f(m+1), m \geq 0$. We call $\mathfrak{a}$ the principal commutative subalgebra.

3.1.10. Proposition. *The cohomologies of the complex $F^*(\widehat{\mathfrak{sl}}_2)$ are isomorphic to the exterior algebra $\bigwedge^*(\mathfrak{a}^*)$ of the dual space to the principal commutative subalgebra of $\mathfrak{n}_+$.*

Proof. If X is an operator on π_0 of the form $\sum_i X_i \partial/\partial x_i$, where X_i are polynomials in x_n, then we can define its shift term as the sum of terms $X_i \partial/\partial x_i$, for which X_i is a constant. According to the definition of the operators Q_1 and Q_0, their shift terms are equal to $\partial/\partial x_{-1}$ and $-\partial/\partial x_{-1}$, respectively. Thus the shift term of the operator $p = e(0) + f(1) = Q_0 + Q_1$ is equal to 0. This operator has the form $\sum_{n<0} A_n \frac{\partial}{\partial x_{n-1}}$, where $A_n = S_n^+ - S_n^-$ is a certain polynomial in x_m; A_n is equal to $-\frac{2}{n}x_n +$ higher power terms.

It is known that the Lie algebra $\mathfrak{n}_+$ splits into the direct sum $\mathrm{Ker}(\mathrm{ad}\,p) \oplus \mathrm{Im}(\mathrm{ad}\,p)$, and that $\mathrm{Ker}(\mathrm{ad}\,p) = \mathfrak{a}$. In the principal grading of $\mathfrak{n}_+$, in which $\deg e(0) = 1, \deg f(1) = 1$, $\mathrm{Im}(\mathrm{ad}\,p)$ is linearly generated by vectors of all positive degrees, and $\mathrm{Ker}(\mathrm{ad}\,p)$ is linearly generated by vectors of all positive odd degrees.

The element $(\mathrm{ad}p)^m \cdot e_1, m \geq 0$, can be chosen as a generator y_{m+1} of $\mathrm{Im}(\mathrm{ad}p)$ of degree $m+1$. By induction one can check that the shift term of the operator $(\mathrm{ad}p)^m \cdot Q_1$ is equal to a non-zero multiple of $\partial/\partial x_{-m-1}$. Indeed, suppose that we have shown this for $m = 0, 1, \ldots, n-1$. Since p does not have a shift term, the shift term of $y_{n+1} = [p, y_n]$ is equal to the commutator of the shift term $\partial/\partial x_{-n}$ (times a constant) of y_n and a linear term of the form $x_{-n}\partial/\partial x_j$ from p. There is only one such summand in p, namely,

$$-\frac{2}{n}x_{-n}\frac{\partial}{\partial x_{-n-1}}.$$

Its commutator with $\partial/\partial x_{-n}$ is equal to $\partial/\partial x_{-n-1}$ up to a non-zero constant. Therefore the shift term of y_{n+1} equals $\partial/\partial x_{-n-1}$ up to a non-zero constant.

In the same way we can show that the shift term of any element of $\mathrm{Ker}(\mathrm{ad}p)$ has to be 0, because otherwise the commutator of this element with p would be non-trivial.

Let us consider the module π_0^* over $\mathfrak{n}_+$, dual to π_0. One can identify π_0 with π_0^* as vector spaces. We can then obtain the formulas for the action of $\mathfrak{n}_+$ on π_0^* from the formulas for its action on π_0 by interchanging x_m's and $\partial/\partial x_m$'s (cf. the proof of Proposition 2.1.15). Since the Lie algebra $\mathfrak{a}$ acts on π_0 by vector fields, which have no shift terms, this Lie algebra acts by 0 on the vector $1^* \in \pi_0^*$, dual to $1 \in \pi_0$. Let L be the $\mathfrak{n}_+$-module, induced from the trivial one-dimensional representation of the Lie subalgebra $\mathfrak{a}$ of $\mathfrak{n}_+$:

$$L = U(\mathfrak{n}_+) \otimes_{U(\mathfrak{a})} \mathbb{C}.$$

Since the vector $1^* \in \pi_0^*$ is $\mathfrak{a}$-invariant, there is a unique $\mathfrak{n}_+$-homomorphism: $L \to \pi_0$, which sends the generating vector of L to 1^*.

Under this homomorphism, a monomial $y_{i_1} \ldots y_{i_m} \otimes 1 \in L$ maps to a vector of π_0^*, which is equal to a non-zero multiple of $x_{-i_1} \ldots x_{-i_m} +$ lower power terms. Therefore this map has no kernel. On the other hand, the character of the module L in the principal gradation is equal to

$$\prod_{n>0} (1 - q^n)^{-1}.$$

This coincides with the character of the module π_0^*. Therefore, π_0^* is isomorphic to L as an $\mathfrak{n}_+$-module.

Going back, we see that the module π_0 is isomorphic to the module L^*, which is the $\mathfrak{n}_+$-module coinduced from the trivial representation of $\mathfrak{a}$.

By "Shapiro's lemma" (cf. [66], §1.5.4, [76], §II.7), $H^*(\mathfrak{n}_+, \pi_0) \simeq H^*(\mathfrak{a}, \mathbb{C})$. But $\mathfrak{a}$ is an abelian Lie algebra, hence $H^*(\mathfrak{a}, \mathbb{C}) = \wedge^*(\mathfrak{a}^*)$. The Proposition now follows from Proposition 3.1.8.

3.1.11. *Theorem. The space $I_0(\widehat{\mathfrak{sl}}_2)$ of local integrals of motion of the sine-Gordon model is linearly generated by mutually commuting local functionals $\mathcal{H}_{2i+1}, i \geq 0$, of all positive odd degrees.*

Proof. According to Proposition 3.1.10, the 1st cohomology of the complex $F^*(\widehat{\mathfrak{sl}}_2)$ is isomorphic to $\mathfrak{a}^*$. As a $\mathbb{Z}$-graded space it is a direct sum of one-dimensional subspaces, generated by some elements h_j of all positive odd degrees. In the same way as in Proposition 2.4.9 one checks that the higher differentials of the complex $F^*(\widehat{\mathfrak{sl}}_2)$ commute with the action of the derivative ∂. Therefore we can form the double complex

$$\mathbb{C} \longrightarrow F^*(\widehat{\mathfrak{sl}}_2) \longrightarrow F^*(\widehat{\mathfrak{sl}}_2) \longrightarrow \mathbb{C}$$

(cf. Fig. 2). The 1st cohomology of this double complex is isomorphic to the space $I_0(\widehat{\mathfrak{sl}}_2)$.

We can calculate the cohomologies of this double complex $F^*(\widehat{\mathfrak{sl}}_2)$ by means of the spectral sequence, in which the 0th differential is horizontal. The first term of this spectral sequence consists of two copies of the cohomologies of the complex $F^*(\widehat{\mathfrak{sl}}_2)$, and the first differential coincides with the action of the derivative on them.

The 0th cohomology of the complex $F^*(\widehat{\mathfrak{sl}}_2)$ is generated by the vector $1 \in \pi_0$. Clearly, the corresponding two cohomology classes in the 1st term of the spectral sequence are canceled by the two spaces $\mathbb{C}$ in the double complex (cf. Fig. 2).

On the other hand, the derivative ∂ acts by 0 on $h_{2j+1} \in H^1(\widehat{\mathfrak{sl}}_2)$, because ∂ has degree 1 and so it should send cohomology classes of odd degrees to cohomology classes of even degrees, which we do not have.

Since we only have two rows in our double complex, the spectral sequence collapses in the first term. Therefore, the 1st cohomology of the double complex is equal to the 1st cohomology H^1 of the complex $F^*(\widehat{\mathfrak{sl}}_2)$, which is equal to $\oplus_{j \geq 0} \mathbb{C} h_{2j+1}$, by Proposition 3.1.10.

Each h_{2j+1} gives rise to a local integral of motion $\mathcal{H}_{2j+1} \in \mathcal{F}_0$ as follows. It is clear that ∂h_{2j+1} is also a cocycle in $F^1(\widehat{\mathfrak{sl}}_2)$, of even degree. Therefore it must be a coboundary:

$\partial h_{2j+1} = \delta^1 H_{2j+1}$, for some element H_{2j+1} from $F^0(\widehat{\mathfrak{sl}}_2) = \pi_0$. This element is not a total derivative, because otherwise h_{2j+1} would also be a coboundary. Hence $\mathcal{H}_{2j+1} = \int H_{2j+1}(t)dt$ is non-zero and it lies in the kernel of $\bar{Q}_1 + \bar{Q}_0$, because $(\tilde{Q}_1 + \tilde{Q}_0)H_{2j+1} = \delta^1 H_{2j+1}$ is a total derivative.

Recall that $I_0(\widehat{\mathfrak{sl}}_2)$ is closed with respect to the Poisson bracket. So, it is a Poisson subalgebra in $\mathcal{F}_0$. One can easily check that the Lie bracket in $\mathcal{F}_0$ is compatible with the $\mathbb{Z}$-grading, introduced in § 2.4.3. Therefore the degree of $\{H_{2i+1}, H_{2j+1}\}$ should be even, and we obtain $\{H_{2i+1}, H_{2j+1}\} = 0$.

3.1.12. *Examples of integrals of motion.* In this subsection we will analyze our complex $F^*(\widehat{\mathfrak{sl}}_2)$ in low degrees and give explicit formulas for local integrals of motion of degrees 1 and 3.

Let us consider the homogeneous components of our complex $F^*(\widehat{\mathfrak{sl}}_2)$ of degrees up to 4. Since $\deg v_n > 4$ for $|n| > 2$ (cf. § 3.1.7), only $F^0(\widehat{\mathfrak{sl}}_2) = \pi_0, F^1(\widehat{\mathfrak{sl}}_2) = \pi_1 \oplus \pi_{-1}$ and $F^2(\widehat{\mathfrak{sl}}_2) = \pi_2 \oplus \pi_{-2}$ contain subspaces of degrees less than or equal to 4, cf. Fig. 4.

On Fig. 4 dots represent basis vectors in $F^0(\widehat{\mathfrak{sl}}_2), F^1(\widehat{\mathfrak{sl}}_2)$, and $F^2(\widehat{\mathfrak{sl}}_2)$, and the dots, corresponding to vectors of the same degree, are situated at the same horizontal level.

We can see that the vector $1 \in \pi_0$ is the only element of degree 0, therefore it is necessarily in the kernel of the differential $\delta^1 : F^0(\widehat{\mathfrak{sl}}_2) \to F^1(\widehat{\mathfrak{sl}}_2)$. It is, as we know, the only cohomology class in the 0th cohomology of our complex.

The second group of the complex $F^2(\widehat{\mathfrak{sl}}_2)$ has degrees greater than or equal to 4. Therefore, all vectors of $\pi_1 \oplus \pi_{-1}$ of degrees $1, 2$ and 3 necessarily lie in the kernel of the differential $\delta^2 : F^1(\widehat{\mathfrak{sl}}_2) \to F^2(\widehat{\mathfrak{sl}}_2)$. But some of them lie in the image of the differential δ^1.

The component of degree 1 of the space π_0 is one-dimensional, and of $\pi_1 \oplus \pi_{-1}$ is two-dimensional, therefore, the 1st cohomology in degree 1 is one-dimensional. As a representative of this cohomology class we can take, for instance, vector $2v_{-1}$. In degree 2 both spaces have two-dimensional components, so there are no cohomologies of degree 2. In degree 3, the component of the space π_0 is three-dimensional, and the component of the space $\pi_1 \oplus \pi_{-1}$ is four-dimensional, so again we have a cohomology class. As a representative we can choose vector $x_{-1}^2 v_{-1}$.

In degree 4 the vectors v_2 and v_{-2} from $F^2(\widehat{\mathfrak{sl}}_2)$ are both in the kernel of the differential $\delta^3 : F^2(\widehat{\mathfrak{sl}}_2) \to F^3(\widehat{\mathfrak{sl}}_2)$. But some linear combination of them is in the image of the differential δ^2. Indeed, we know that the differential δ^1 has no kernel in degree 4. Therefore its image in the subspace of $F^2(\widehat{\mathfrak{sl}}_2)$ of degree 4 is 5-dimensional. But the 1st cohomology of degree 4 is trivial. Hence the kernel of the differential δ^2 is also 5-dimensional. But the space $F^1(\widehat{\mathfrak{sl}}_2)$ has dimension 6 in degree 4. Therefore the image of δ^3 is one-dimensional. Thus the second cohomology is one-dimensional in degree 4. It can be represented by the vector v_{-2}.

In view of Proposition 3.1.10, the cohomology classes that we constructed in $F^1(\widehat{\mathfrak{sl}}_2)$, correspond to the generators of degrees 1 and 3 of the space $\mathfrak{a}^*$, and the class that we constructed in F^2 corresponds to their exterior product. The encircled dots represent these classes on the picture.

Now let us assign to the 1st cohomology classes of degrees 1 and 3 local integrals of motion $\mathcal{H}_1$ and $\mathcal{H}_3$ (cf. Theorem 3.1.11).

The cohomology class h_1 of degree 1 was represented by the vector $2v_1$. According to the general procedure, described in Theorem 3.1.11, we have to take $\partial v_1 = -2x_{-1}v_1$ and find a vector H_1 from π_0, such that $\delta^1 H_1 = -2x_{-1}v_1$. One checks that $\frac{1}{2}x^2_{-1} - x_{-2} \in \pi_0$ is such a vector. Therefore, we can take as the integral of motion

(3.1.2)
$$\mathcal{H}_1 = \int (\frac{1}{2}u(t)^2 - \partial_t u(t))dt.$$

According to § 2.1.7, we can define an action of $\mathcal{H}_1$ on $\oplus_{n\in\mathbb{Z}}\pi_n$, and this action will coincide with the action of ∂.

The cohomology class h_3 of degree 3 was represented by vector $x^2_{-1}v_{-1}$. We have: $\partial h_3 = (-x^3_{-1} + 2x_{-1}x_{-2})v_{-1}$, and one can check that $\delta^1 \cdot \frac{1}{2}(\frac{1}{2}x^2_{-1} - x_{-2})^2 = \partial h_3$. Therefore,

(3.1.3)
$$\mathcal{H}_3 = \frac{1}{2} \int (\frac{1}{2}u(t)^2 - \partial_t u(t))^2 dt$$

is an integral of motion.

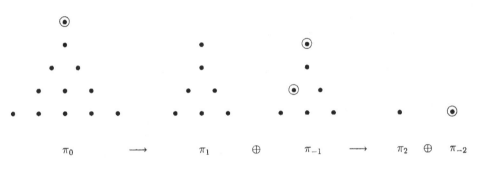

$$\pi_0 \qquad \longrightarrow \qquad \pi_1 \qquad \oplus \qquad \pi_{-1} \qquad \longrightarrow \qquad \pi_2 \qquad \oplus \qquad \pi_{-2}$$

Fig. 4

3.1.13. *Connection with the KdV and mKdV systems.* According to Theorem 3.1.11, the integrals of motion $\mathcal{H}_{2j+1} \in \mathcal{F}_0, j \geq 0$, mutually commute with respect to the Poisson bracket. In particular, we have

$$\{\mathcal{H}_1, \mathcal{H}_{2j+1}\} = 0, \qquad \{\mathcal{H}_3, \mathcal{H}_{2j+1}\} = 0,$$

where $\mathcal{H}_1$ and $\mathcal{H}_3$ are given by formulas (3.1.2) and (3.1.3), respectively. Since $\mathcal{H}_1 = \partial$, the first equation is satisfied automatically for any element of $\mathcal{F}_0$ (cf. § 2.1.7). However, the second equation is non-trivial, and it shows that the higher integrals of motion $\mathcal{H}_{2j+1}, j > 1$, coincide with the higher hamiltonians of the mKdV hierarchy. In fact, this equation can be taken as a definition of these integrals of motion.

This is because $\mathcal{H}_3$ is the hamiltonian of the modified Korteweg-de Vries (mKdV) equation. In other words, the mKdV equation

(3.1.4)
$$\partial_\tau U(t) = \partial_t^3 U(t) - \frac{3}{2}U(t)^2\partial_t U(t)$$

can be written in the hamiltonian form as

$$\partial_\tau U(t) = \{U(t), \mathcal{H}_3\}.$$

The elements of $\mathcal{F}_0$, which commute with $\mathcal{H}_3$ are called the higher mKdV hamiltonians. It is known that they exist precisely for all odd degrees, therefore they coincide with our

$\mathcal{H}_{2j+1}$. Thus, we see that the local integrals of motion of the sine-Gordon model coincide with the hamiltonians of the mKdV hierarchy, and that the property of commutativity with $\mathcal{H}_3$ defines these local integrals of motion uniquely. The mKdV hamiltonians define hamiltonian flows on $L\mathfrak{h}$, which commute with the flow, defined by $\mathcal{H}_3$. Altogether they define the mKdV hierarchy of partial differential equations.

The mKdV hierarchy is closely connected with the KdV hierarchy. Namely, as was explained in § 2.1.18, the kernel of the operator $\bar{Q}_1$ is isomorphic to the classical Virasoro algebra, which is the Poisson algebra of local functionals, depending on $W(t) = \frac{1}{2}u^2(t) - \partial_t u(t)$. Since $I_0(\widehat{\mathfrak{sl}}_2)$ was defined as the intersection of the kernels of the operators $\bar{Q}_1$ and $\bar{Q}_0$, it is a subspace in the classical Virasoro algebra. For example, we can rewrite $\mathcal{H}_1$ and $\mathcal{H}_3$ via $W(t)$ by the formulas

$$\mathcal{H}_1 = \int W(t)dt, \qquad \mathcal{H}_3 = \frac{1}{2}\int W^2(t)dt.$$

The functional $\mathcal{H}_3$ is the KdV hamiltonian, therefore the commutativity condition

$$\{\mathcal{H}_3, \mathcal{H}_{2j+1}\} = 0$$

implies that the $\mathcal{H}_{2j+1}$'s, rewritten in terms of $W(t)$, coincide with the higher KdV hamiltonians.

These hamiltonians define the KdV hierarchy of mutually commuting flows on a hyperplane in the dual space to the Virasoro algebra, cf., e.g., [111]. The Miura transformation (cf. § 2.1.18) maps the hierarchy of these flows to the mKdV hierarchy. In particular, the Miura transformation maps the mKdV equation (3.1.4) to the KdV equation

$$\partial_\tau W(t) = \partial_t^3 W(t) + 3W(t)\partial_t W(t).$$

3.2. General case. Let $\mathfrak{g}$ be an arbitrary affine algebra, twisted or untwisted. We will denote by $\bar{\mathfrak{g}}$ the finite-dimensional simple Lie algebra, whose Dynkin diagram is obtained by deleting the 0th node of the Dynkin diagram of $\mathfrak{g}$.

3.2.1. *Hamiltonian structure.* As the hamiltonian space, the spaces of functionals and the Poisson brackets of the affine Toda field theory, associated to $\mathfrak{g}$, we will take the objects, corresponding to the Toda field theory, associated to $\bar{\mathfrak{g}}$. They were defined in § 2.2.1 and § 2.2.2. Throughout this section we will use notation, introduced in § 2.2.

3.2.2. *The hamiltonian and integrals of motion.* The imaginary root δ of the Lie algebra $\mathfrak{g}$ has the decomposition

$$\delta = \sum_{i=0}^{l} a_i \alpha_i,$$

where $\alpha_i, i = 0, \ldots, l$, are the simple roots of $\mathfrak{g}$, and the a_i's are the labels of the Dynkin diagram of $\mathfrak{g}$ [81]. In particular, $a_0 = 1$ for all affine algebras, except $A_{2n}^{(2)}$, in which case $a_0 = 2$. Somewhat abusing notation, we will introduce a vector α_0 in the Cartan subalgebra $\mathfrak{h}$ of $\mathfrak{g}$ by the formula

$$\alpha_0 = -\frac{1}{a_0}\sum_{i=1}^{l} a_i \alpha_i.$$

We can now define the hamiltonian of the affine Toda field theory, associated to $\mathfrak{g}$, by the formula

$$H = \frac{1}{2}\sum_{i=0}^{l} \int e^{\phi_i(t)}dt \in \oplus_{i=0}^{l}\mathcal{F}_{\alpha_i},$$

where $\phi_i(t) = \bar{\alpha}_i(t)$ (cf. § 2.2.2). The corresponding hamiltonian equation coincides with the Toda equation (1.1.1), associated to $\mathfrak{g}$.

We define the space $I_0(\mathfrak{g})$ of local integrals of motion as the kernel of the operator $\{\cdot, H\} : \mathcal{F}_0 \to \oplus_{i=0}^{l} \mathcal{F}_{\alpha_i}$, or, in other words, as the intersection of the kernels of the operators $\tilde{Q}_i = \{\cdot, \int e^{\phi_i(t)} dt\} : \mathcal{F}_0 \to \mathcal{F}_{\alpha_i}, i = 0, \ldots, l$.

As before, we will use the operators $\tilde{Q}_i : \pi_0 \to \pi_{\alpha_i}, i = 0, \ldots, l$, given by formula (2.2.4), where we put

$$x_n^0 = -\frac{1}{a_0} \sum_{i=1}^{l} a_i x_n^i.$$

These operators commute with the action of the derivative and descend down to the operators $\tilde{Q}_i, i = 0, \ldots, l$. We will also need the operators $Q_i = T_i^{-1} \tilde{Q}_i$, acting on π_0.

According to Proposition 2.2.8 and Remark 2.2.9, the operators $\tilde{Q}_i, i = 0, \ldots, l$, as well as $Q_i, i = 0, \ldots, l$, satisfy the Serre relations of the affine algebra $\mathfrak{g}$. Thus, the operators Q_i generate an action of the nilpotent subalgebra $\mathfrak{n}_+$ of $\mathfrak{g}$ on the space π_0.

In the rest of this subsection we will go through the main steps of § 2 and § 3.1 to describe the space $I_0(\mathfrak{g})$. Most of the proofs are the same as in those sections.

3.2.3. *The complex $F^*(\mathfrak{g})$.* For an affine algebra $\mathfrak{g}$ there also exists a BGG resolution, which is defined in the same way and has the same properties as in the case of finite-dimensional simple Lie algebras [107].

Using this resolution, we can define the complex

$$F^*(\mathfrak{g}) = \oplus_{j \geq 0} F^j(\mathfrak{g})$$

in the same way as in § 2.4, by putting

$$F^j(\mathfrak{g}) = \oplus_{l(s)=j} \pi_{\rho - s(\rho)},$$

where s runs over the affine Weyl group. The differential $\delta^j : F^{j-1}(\mathfrak{g}) \to F^j(\mathfrak{g})$ is given by formula (2.4.1). In the same way as in the case of finite-dimensional simple Lie algebras (cf. Lemma 2.4.2), it follows that this differential is nilpotent.

Introduce a $\mathbb{Z}$-grading on this complex similar to the one introduced in § 2.4.3. Namely, we put $\deg v_{\rho - s(\rho)} = (\rho^\vee, \rho - s(\rho))$, where $\rho^\vee$ is an element in the dual space to the Cartan subalgebra of $\mathfrak{g}$, such that $(\rho^\vee, \alpha_i) = 1, i = 0, \ldots, l$, and $\deg x_n^i = -n$. With respect to this grading, the differentials of our complex have degree 0, and our complex decomposes into a direct sum of finite-dimensional subcomplexes.

One has the analogue of Proposition 2.4.5: the cohomologies of the complex $F^*(\mathfrak{g})$ are isomorphic to the cohomologies of $\mathfrak{n}_+$ with coefficients in π_0.

3.2.4. *Principal commutative subalgebra.* Consider the element $p = \sum_{i=0}^{l} a_i e_i$ in the Lie algebra $\mathfrak{n}_+$. It is known that $\mathfrak{n}_+ = \mathrm{Ker}(\mathrm{ad}\, p) \oplus \mathrm{Im}(\mathrm{ad}\, p)$, where $\mathrm{Ker}(\mathrm{ad}\, p)$ is an abelian Lie subalgebra $\mathfrak{a}$ of $\mathfrak{n}_+$, which we will call the principal commutative subalgebra.

In the principal gradation of $\mathfrak{n}_+$, which is obtained by assigning degree 1 to the generators e_i, the Lie algebra $\mathfrak{a}$ has a basis of homogeneous elements of degrees equal to the exponents $d_i, i = 1, \ldots, l$, of $\mathfrak{g}$ modulo the Coxeter number h.

The space $\mathrm{Ker}(\mathrm{ad}\, p)$ splits into the direct sum $\mathrm{Ker}(\mathrm{ad}\, p) = \oplus_{j>0} \mathfrak{n}_+^j$ of homogeneous components of degree j with respect to the principal grading. Each of these components has dimension l and the operator $\mathrm{ad}\, p : \mathfrak{n}_+^j \to \mathfrak{n}_+^{j+1}$ is an isomorphism.

For the proof of these facts, cf. [80], Proposition 3.8 (b). We will use them in the proof of the following statement, which is a generalization of Proposition 3.1.10.

3.2.5. Proposition. The cohomologies of the complex $F^*(\mathfrak{g})$ are isomorphic to the exterior algebra $\bigwedge^*(\mathfrak{a}^*)$ of the dual space to the pricipal commutative subalgebra $\mathfrak{a}$.

Proof. From the formulas, defining the operators Q_i, one can see that the operator $p = \sum_{i=0}^l a_i Q_i$ does not have a shift term (cf. the proof of Proposition 3.1.10). Therefore, other elements of the Lie subalgebra $\mathfrak{a}$ of $\mathfrak{n}_+$ also do not have shift terms.

Denote by V_j the vector space of operators $\partial/\partial x_{-j}^i, i = 1, \ldots, l, j > 0$. One can check that the operator $\text{ad} p$ isomorphically maps V_j to V_{j+1}. The space of shift terms of the operators from $\mathfrak{n}_+^1 \subset \mathfrak{n}_+$ coincides with V_1. By induction, in as in the proof of Proposition 3.1.10, one can show that the space of shift terms of operators from $\mathfrak{n}_+^j \subset \mathfrak{n}_+$ coincides with V_j.

In the same way as in the proof of Proposition 3.1.10, we deduce from these facts that the $\mathfrak{n}_+$−module π_0 is isomorphic to the module, coinduced from the trivial representation of $\mathfrak{a}$. Therefore the cohomology $H^*(\mathfrak{n}_+, \pi_0)$, are equal, by "Shapiro's lemma", to $H^*(\mathfrak{a}, \mathbb{C}) \simeq \bigwedge^*(\mathfrak{a}^*)$, because $\mathfrak{a}$ is an abelian Lie algebra. But $H^*(\mathfrak{n}_+, \pi_0)$ coincides with the cohomology of the complex $F^*(\mathfrak{g})$. The Proposition is proved.

3.2.6. Theorem. The space $I_0(\mathfrak{g})$ of local integrals of motion of the Toda field theory, associated to an affine algebra $\mathfrak{g}$, is linearly generated by mutually commuting local functionals of degrees equal to the exponents of $\mathfrak{g}$ modulo the Coxeter number.

Proof. If the exponents of $\mathfrak{g}$ are odd and the Coxeter number is even, then the proof simply repeats the proof of Theorem 3.1.11. We can again identify the space $I_0(\mathfrak{g})$ with the 1st cohomology of the double complex $\mathbb{C} \longrightarrow F^*(\mathfrak{g}) \longrightarrow F^*(\mathfrak{g}) \longrightarrow \mathbb{C}$. According to Proposition 3.2.5, the 1st cohomology of the complex $F^*(\mathfrak{g})$ is linearly spanned by elements of certain odd degrees. We can then compute this cohomology using the spectral sequence in the same way as in the proof of Theorem 3.1.11.

The proof in other cases $(A_n^{(1)}, n > 1, D_{2n}^{(1)}, E_6^{(1)}$ and $E_7^{(1)})$ will be published in [51].

3.2.7. Integrals of motion in the extended space of local functionals. We can define the space of integrals of motion of an affine Toda field theory as the intersection of the kernels of operators $\bar{Q}_i : \hat{\mathcal{F}}_0 \to \hat{\mathcal{F}}_{\alpha_0}$, cf. § 2.4.12. We can use the tensor product of the complex $F^*(\mathfrak{g})$ with $\mathbb{C}[t, t^{-1}]$ to compute this space. By repeating the proof of Theorem 3.2.6, we conclude that the space of integrals of motion in the larger space $\hat{\mathcal{F}}_0$ of local functionals coincides with the space $I_0(\mathfrak{g})$ of integrals of motion in the small space $\mathcal{F}_0$.

3.2.8. Remark. One usually defines a conservation law of the Toda field theory (in the light cone coordinates x_+ and x_-) as a pair $(P^-, P^+) \in \pi_0 \oplus (\oplus_i \pi_{\alpha_i})$, such that

$$\frac{\partial P^-}{\partial x_+} = \frac{\partial P^+}{\partial x_-}.$$

In our language, $\partial/\partial x_+ = \sum_i \bar{Q}_i = \delta^1$ is the Poisson bracket with the Toda hamiltonian, and $\partial/\partial x_- = \partial$. Thus, the pair (P^-, P^+), where P^+ is a 1-cocycle of the complex $F^*(\mathfrak{g})$ and P^- is the density of our integral of motion, i.e. the 0-cocycle of the quotient complex $F^*(\mathfrak{g})/\partial F^*(\mathfrak{g})$ such that $\delta^1 P^- = \partial P^+$, is a conservation law. The important observation, which enabled us to find all such conservation laws was that in order to be a component of a conservation law, P^+ should satisfy a certain equation, namely, $\delta^2 \cdot P^+ = 0$.

The space $I_0(\mathfrak{g})$ constitutes a maximal abelian Poisson subalgebra in the classical $\mathcal{W}$−algebra $I_0(\bar{\mathfrak{g}})$. One can show that the first generators of $I_0(\mathfrak{g})$ of degrees $d_i, i =$

$1, \ldots, l$, are equal to $\int W^{(i)}(t)dt$, where the $W^{(i)}$'s can be chosen as the generators of $I_0(\bar{\mathfrak{g}})$, defined in § 2.4.7.

Using this fact, one can identify the local integrals of motion of the classical affine Toda field theory, associated with $\mathfrak{g}$, with the hamiltonians of the corresponding generalized KdV and mKdV hierarchies.

3.2.9. Remark. In the course of proving Proposition 3.2.5, we showed that the $\mathfrak{n}_+$−module π_0 is isomorphic to the module, coinduced from the trivial representation of $\mathfrak{a}$. It means that $\pi_0 \simeq \mathbb{C}[\partial^n u^i]$ is isomorphic to the space of functions on the homogeneous space N_+/A, where N_+ and A are Lie groups of the Lie algebras $\mathfrak{n}_+$ and $\mathfrak{a}$, respectively. In other words, the space N_+/A can be identified with an infinite-dimensional vector space with coordinates $\partial^n u^i$. The local integrals of motion of the corresponding Toda field theory therefore define certain vector fields η_j's on N_+/A (cf. formula (2.1.7)), which commute among themselves. The first of them, η_1, is the vector field ∂.

In our paper [51] we identify the vector field η_j with the infinitesimal right action of a generator of degree $-j$ of the opposite abelian subalgebra $\mathfrak{a}_-$ of $\mathfrak{g}$. Indeed, since $\mathfrak{g}$ infinitesimally acts on N_+ from the right as on the big cell of the flag manifold, and $\mathfrak{a}_-$ commutes with $\mathfrak{a}$, the Lie algebra $\mathfrak{a}_-$ acts on the coset space N_+/A by mutually commuting vector fields. In particular, the action of the vector field ∂ corresponds to the action of the element

$$\sum_{i=0}^{l} \frac{(\alpha_i, \alpha_i)}{2} f_i \in \mathfrak{a}_-,$$

in accordance with Lemma 2.4.8.

In the next section we will show how to quantize the classical integrals of motion.

4. QUANTUM TODA FIELD THEORIES.

4.1. Vertex operator algebras. The theory of vertex operator algebras was started by Borcherds [14] and then further developed by I.Frenkel, Lepowsky and Meurman in [62] (cf. also [63, 75, 91]).

In this section we will give the definition of vertex operator algebras and study some of their properties following closely § 3 of [58]. Ater that we will introduce the vertex operator algebra (VOA) of the Heisenberg algebra (or the VOA of free fields), which we will need in the study of quantum integrals of motion of Toda field theories.

4.1.1. Fields. Let $V = \oplus_{n=0}^{\infty} V_n$ be a $\mathbb{Z}_+$-graded vector space, where $\dim V_n < \infty$ for all n, called the *space of states*. A *field* on V of conformal dimension $\Delta \in \mathbb{Z}$ is a power series $\phi(z) = \sum_{j \in \mathbb{Z}} \phi_j z^{-j-\Delta}$, where $\phi_j \in \text{End } V$ and $\phi_j V_n \subset V_{n-j}$. Note that if $\phi(z)$ is a field of conformal dimension Δ, then the power series $\partial_z \phi(z) = \sum_{j \in \mathbb{Z}} (-j - \Delta)\phi_j z^{-j-\Delta-1}$ is a field of conformal dimension $\Delta + 1$.

If $z \in \mathbb{C}^\times$ is a non-zero complex number, then $\phi(z)$ can be considered as a linear operator $V \to \bar{V}$, where $\bar{V} = \prod_{n=0}^{\infty} V_n$.

We have a natural pairing $\langle , \rangle : V_n^* \times V_n \to \mathbb{C}$. A linear operator $P : V \to \bar{V}$ can be represented by a set of finite-dimensional linear operators $P_i^j : V_i \to V_j, i, j \in \mathbb{Z}$, such that $\langle A, P \cdot B \rangle = \langle A, P_i^j \cdot B \rangle$ for $B \in V_i, A \in V_j^*$. Let P, Q be two linear operators $V \to \bar{V}$. We say that the *composition PQ exists*, if for any $i, k \in \mathbb{Z}, B \in V_i, A \in V_k^*$ the series $\sum_{j \in \mathbb{Z}} \langle A, P_j^k Q_i^j \cdot B \rangle$ converges absolutely.

4.1.2. Definition. *Two fields $\phi(z)$ and $\psi(z)$ are called* local *with respect to each other, if*

- *for any $z, w \in \mathbb{C}^\times$, such that $|z| > |w|$, the composition $\phi(z)\psi(w)$ exists and can be analytically continued to a rational operator-valued function on $(\mathbb{C}^\times)^2 \backslash diagonal$, $R(\phi(z)\psi(w))$;*
- *for any $z, w \in \mathbb{C}^\times$, such that $|w| > |z|$, the composition $\psi(w)\phi(z)$ exists and can be analytically continued to a rational operator-valued function on $(\mathbb{C}^\times)^2 \backslash diagonal$, $R(\psi(w)\phi(z))$;*
- $R(\phi(z)\psi(w)) = R(\psi(w)\phi(z))$.

In other words, fields $\phi(z)$ and $\psi(z)$ are local with respect to each other, if for any $x \in V_n$ and $y \in V_m^*$ both matrix coefficients $\langle y|\phi(z)\psi(w)|x\rangle$ for $|z| > |w|$ and $\langle y|\psi(w)\phi(z)|x\rangle$ for $|z| < |w|$ converge to the same rational function in z and w which has no poles outside the lines $z = 0$, $w = 0$ and $z = w$.

4.1.3. Definition. *A VOA structure on V is a linear map $Y(\cdot, z) : V \longrightarrow \mathrm{End}\, V\,[[z, z^{-1}]]$ which associates to each $A \in V_n$ a field of conformal dimension n (also called a vertex operator) $Y(A, z) = \sum_{j \in \mathbb{Z}} A_j z^{-j-n}$, such that the following axioms hold:*

(A1) (vacuum axiom) *There exists an element $|0\rangle \in V_0$ (vacuum vector) such that $Y(|0\rangle, z) = \mathrm{Id}$ and $\lim_{z \to 0} Y(A, z)|0\rangle = A$.*

(A2) (translation invariance) *There exists an operator $T \in \mathrm{End}\, V$ such that*

$$\partial_z Y(A, z) = [T, Y(A, z)] \quad and \quad T|0\rangle = 0.$$

(A3) (locality) *All fields $Y(A, z)$ are local with respect to each other.*

A VOA V is called conformal *of* central charge *$c \in \mathbb{C}$ if there exists an element $\omega \in V_2$ (called the* Virasoro element*), such that the corresponding vertex operator $Y(\omega, z) = \sum_{n \in \mathbb{Z}} L_n z^{-n-2}$ satisfies the following properties:*

(C) $L_{-1} = T$, $L_0|_{V_n} = n \cdot \mathrm{Id}$, and $L_2\omega = \frac{1}{2}c|0\rangle$.

4.1.4. Proposition. *A VOA V automatically satisfies the associativity property: for any $A, B \in V$,*

$$(4.1.1) \qquad R(Y(A, z)Y(B, w)) = R(Y(Y(A, z - w)B, w)),$$

where the left-hand (resp. right-hand) side is the analytic continuation from the domain $|z| > |w|$ (resp. $|w| > |z - w|$).

Proof which was communicated to us by V. Kac follows from the following two Lemmas.

4.1.5. Lemma. *For any $z \in \mathbb{C}^\times$ and $w \in \mathbb{C}$, such that $|w| < |z|$, the composition $e^{wT}Y(A, z)e^{-Tw}$, where*

$$e^{wT} = \sum_{n=0}^{\infty} \frac{(wT)^n}{n!},$$

exists and can be analytically continued to a rational operator-valued function, which is equal to $Y(A, z + w)$, i.e.

$$R(e^{wT}Y(A, z)e^{-Tw}) = Y(A, z + w).$$

Proof. By axiom **(A2)**, $[T, Y(A, z)] = \partial_z Y(A, z)$. Hence as a formal powers series,

$$e^{wT}Y(A, z)e^{-Tw} = \sum_{n=0}^{\infty} \frac{w^n}{n!} \partial_z^n Y(A, z).$$

The right hand side is the Taylor expansion of $Y(A, z + w)$. Since $Y(A, z)$ is holomorphic everywhere except the origin, $e^{wT}Y(A, z)e^{-Tw}$ converges to $Y(A, z + w)$ if $|w| < |z|$.

4.1.6. *Lemma.* For any $z \in \mathbb{C}^\times$, $Y(A, z)B = e^{zT}Y(B, -z)A$.

Proof. By Lemma 4.1.5,

$$R(e^{(z+w)T}Y(B, -z)A) = R(Y(B, w)e^{(z+w)T}A).$$

We can derive from axioms **(A1)** and **(A2)** that $e^{(z+w)T}A = Y(A, z + w)|0\rangle$. This and the previous formula give:

$$R(e^{(z+w)T}Y(B, -z)A) = R(Y(B, w)Y(A, z + w)|0\rangle).$$

By locality,

$$R(e^{(z+w)T}Y(B, -z)A) = R(Y(A, z + w)Y(B, w)|0\rangle).$$

Both sides of the last formula have well-defined limits when $w \to 0$, hence these limits coincide and we obtain using axiom **(A1)**: $e^{Tz}Y(B, -z)A = Y(A, z)B$.

4.1.7. *Proof of Proposition* 4.1.4. For any vector $C \in V$ we have:

$$Y(A, z)Y(B, w)C = Y(A, z)e^{wT}Y(C, -w)B$$

for $|z| > |w|$, where we applied Lemma 4.1.6 to $Y(B, w)C$. By applying $1 = e^{Tw}e^{-Tw}$ to both sides of this formula (we can do this, because $e^{\pm Tw}$ is a series infinite in only one direction) and using Lemma 4.1.5 in the right hand side of the previous formula we obtain

(4.1.2) $$Y(A, z)Y(B, w)C = e^{wT}Y(A, z - w)Y(C, -w)B$$

for $|z| > |w|$.

On the other hand, consider

$$Y(Y(A, z - w)B, w)C = \sum_{n \in \mathbb{Z}}(z - w)^{-n-\Delta_A}Y(A_n \cdot B, w)C$$

as a formal power series in $(z - w)$. By Lemma 4.1.6,

$$Y(A_n \cdot B, w)C = e^{wT}Y(C, -w)A_nB.$$

Hence

(4.1.3) $$Y(Y(A, z - w)B, w)C = e^{wT}Y(C, -w)Y(A, z - w)B,$$

as formal power series in $(z - w)$. But the right hand side of (4.1.3) converges when $|w| > |z - w|$ and can be analytically continued to a rational function in z and w, by axiom **(A3)**. Therefore the left hand side of (4.1.3) has the same properties.

By applying axiom **(A3)** to the analytic continuations of the right hand sides of (4.1.2) and (4.1.3), we obtain the equality of the analytic continuations of the left hand sides:

$$R(Y(A, z)Y(B, w))C = R(Y(Y(A, z - w)B, w))C$$

for any $C \in V$, and Proposition 4.1.4 follows.

4.1.8. *Operator product expansion.* We may rewrite formula (4.1.1) as

$$Y(A,z)Y(B,w) = \sum_{n \in \mathbb{Z}} (z-w)^{-n-\deg A} Y(A_n \cdot B, w),$$

using the formula $Y(A,z) = \sum_{n \in \mathbb{Z}} A_n z^{-n-\deg A}$. Here and further on to simplify notation we omit $R(\cdot)$ in formulas for analytic continuation of functions. Such an identity is called an operator product expansion (OPE). There exists such $M \in \mathbb{Z}$ that $A_n \cdot B = 0$ for any $n > M$. Therefore, the right hand side of this formula has only finitely many terms with negative powers of $(z-w)$. Combining axioms (3), (4), and the Cauchy theorem we obtain the following identity [62]:

$$\int_{C_w^\rho} \int_{C_z^R} Y(A,z)Y(B,w)f(z,w)\,dzdw - \int_{C_w^\rho} \int_{C_z^r} Y(B,w)Y(A,z)f(z,w)\,dzdw =$$

$$\int_{C_w^\rho} \int_{C_z^\delta(w)} \sum_{n \in \mathbb{Z}} (z-w)^{-n-\deg A} Y(A_n \cdot B, w)f(z,w)\,dzdw,$$

where C^x denotes a circle of radius x around the origin, $R > \rho > r$, $C^\delta(w)$ denotes a small circle of radius δ around w, and $f(z,w)$ is an arbitrary rational function on $(\mathbb{C}^\times)^2 \backslash diagonal$.

This formula can be used in order to compute commutation relations between Fourier components of vertex operators. Indeed, if we choose $f(z,w) = z^{m+\deg A-1}w^{k+\deg B-1}$, then we obtain:

$$[A_m, B_k] = \sum_{n > -\deg A} \int_{C_w^\rho} \frac{dw}{w} w^{k+\deg B} \int_{C_z^\delta(w)} dz \frac{z^{m+\deg A-1}}{(z-w)^{n+\deg A}} Y(A_n \cdot B, w) =$$

$$(4.1.4) \qquad \sum_{-\deg A < n \leq m} \binom{m+\deg A-1}{n+\deg A-1}(A_n \cdot B)_{m+k}.$$

In particular, we see that only the terms in the OPE, which are singular at the diagonal $z = w$, contribute to the commutator. This formula also shows that the commutator of Fourier components of two fields is a linear combination of Fourier components of other fields (namely, the ones, corresponding to the vectors $A_n \cdot B$). Therefore we obtain the following result.

4.1.9. *Theorem. The space of all Fourier components of vertex operators defined by a VOA is a Lie algebra.*

In particular we derive from formula (4.1.1) and axiom (C) that

$$Y(\omega, z)Y(\omega, w) = \frac{c/2}{(z-w)^4} + \frac{2Y(\omega,w)}{(z-w)^2} + \frac{\partial_w Y(\omega,w)}{z-w} + O(1),$$

which implies using (4.1.4) the following commutation relations between the Fourier coefficients $L_n, n \in \mathbb{Z}$, of $Y(\omega, z)$:

$$[L_n, L_m] = (n-m)L_{n+m} + \frac{1}{12}(n^3-n)\delta_{n,-m}c.$$

These are the defining relations of the Virasoro algebra with central charge c.

4.1.10. *Remark.* The axioms of VOA may look rather complicated, but in fact they are quite natural generalizations of the axioms of a $\mathbb{Z}$–graded associative commutative algebra with a unit. Indeed, such an algebra is defined as a $\mathbb{Z}$–graded vector space V along with a linear operator $Y : V \to \mathrm{End}(V)$ of degree 0 and an element $1 \in V$, such that $Y(1) = \mathrm{Id}$. The linear operator Y defines a product structure by the formula $A \cdot B = Y(A)B$. The axioms of commutativity and associativity of this product then read as follows: $Y(A)Y(B) = Y(B)Y(A)$, and $Y(Y(A)B) = Y(A)Y(B)$.

On a VOA V the operator $Y(\cdot, z)$ defines a family of "products" – linear operators $V \to V$, depending (in the formal sense) on a complex parameter z. The axiom **(A3)** and formula (4.1.1) can be viewed as analogues of the axioms of commutativity and associativity (with a proper regularization of the compositions of operators), and the first half of the axiom **(A1)** is the analogue of the axiom of unit. A rather surprising aspect of the theory of VOA is that the "associativity" property (4.1.1) follows from the "commutativity" (locality) **(A3)** together with **(A1)** and **(A2)**.

Another novelty is vector ω. The meaning of this vector is the following. The Fourier components of $Y(\omega, z)$ define on V an action of the Virasoro algebra, which is the central extension of the Lie algebra of vector fields on a punctured disc. The existence of such an element inside V means that all infinitesimal changes of the coordinate z can be regarded as "interior automorphisms" of V.

4.1.11. *Ultralocal fields.* Let us call two fields $\phi(z)$ and $\psi(z)$ *ultralocal* with respect to each other if there exists an integer N, such that for any $v \in V_n$ and $v^* \in V_m^*$, both series $\langle v^* | \phi(z)\psi(w)|v \rangle (z-w)^N$ and $\langle v^* | \psi(w)\phi(z)|v \rangle (z-w)^N$ are equal to the same finite polynomial in $z^{\pm 1}$ and $w^{\pm 1}$. Clearly, ultralocality implies locality. Moreover, in a vertex operator algebra any two vertex operators are automatically ultralocal with respect to each other according to formula (4.1.1) and the fact that the $\mathbb{Z}$–gradation on V is bounded from below.

Given two fields $\phi(z)$ and $\psi(z)$ of conformal dimensions Δ_ϕ and Δ_ψ one defines their *normally ordered product* as the field

$$(4.1.5) \qquad : \phi(z)\psi(z) := \sum_{n \in \mathbb{Z}} \left(\sum_{m < -\Delta_\phi} \phi_m \psi_{n-m} + \sum_{m \geq -\Delta_\phi} \psi_{n-m}\phi_m \right) z^{-n-\Delta_\phi-\Delta_\psi}$$

of conformal dimension $\Delta_\phi + \Delta_\psi$. The Leibniz rule holds for the normally ordered product:

$$(4.1.6) \qquad \partial_z : \phi(z)\psi(z) :=: \partial_z\phi(z)\psi(z) : + : \phi(z)\partial_z\psi(z) : .$$

The following proposition proved in [58], Proposition 3.1, allows one to check easily the axioms of a VOA.

4.1.12. *Proposition.* Let V be a $\mathbb{Z}_+$-graded vector space. Suppose that to some vectors $a^{(0)} = |0\rangle \in V_0, a^{(1)} \in V_{\Delta_1}, \ldots$, one associates fields $Y(|0\rangle, z) = \mathrm{Id}, Y(a^{(1)}, z) = \sum_j a_j^{(1)} z^{-j-\Delta_1}$, $\ldots$, of conformal dimensions $0, \Delta_1, \ldots$, such that the following properties hold:

(1) all fields $Y(a^{(i)}, z)$ are ultralocal with respect to each other;

(2) $\lim_{z\to 0} Y(a^{(i)}, z)|0\rangle = a^{(i)}$;

(3) the space V has a linear basis of vectors

$$(4.1.7) \qquad a_{-j_s-\Delta_{k_s}}^{(k_s)} \cdots a_{-j_1-\Delta_{k_1}}^{(k_1)} |0\rangle, \quad j_1, \ldots, j_s \in \mathbb{Z}_+;$$

(4) *there exists an endomorphism T of V such that*

(4.1.8) $$\left[T, a^{(k)}_{-j-\Delta_k}\right] = (j+1)a^{(k)}_{-j-\Delta_k-1}, \quad T(|0\rangle) = 0.$$

Then letting

(4.1.9) $Y(a^{(k_s)}_{-j_s-\Delta_{k_s}} \cdots a^{(k_1)}_{-j_1-\Delta_{k_1}} |0\rangle, z)$

$$= (j_1! \cdot \ldots \cdot j_s!)^{-1} \cdot\, : \partial_z^{j_s} Y(a^{(k_s)}, z) \ldots \partial_z^{j_2} Y(a^{(k_2)}, z) \partial_z^{j_1} Y(a^{(k_1)}, z) :$$

(where the normal ordering of more than two fields is nested from right to left), gives a well-defined VOA structure on V.

Proof. We define the map $Y(\cdot, z)$ by formula (4.1.9). It is clear that axiom **(A1)** holds. Given two fields $\phi(z)$ and $\psi(z)$, if $[T, \phi(z)] = \partial_z \phi(z)$ and $[T, \psi(z)] = \partial_z \psi(z)$, then from (4.1.5) and (4.1.6) it follows that

(4.1.10) $$[T, : \phi(z)\psi(z) :] = \partial_z : \phi(z)\psi(z) : .$$

Hence the axiom **(A2)** follows inductively from (4.1.9) and (4.1.10).

Using an argument of Dong (cf. [91], Proposition 3.2.7), one can show that if three fields $\chi(z)$, $\phi(z)$ and $\psi(z)$ are ultralocal with respect to each other, then $: \phi(z)\psi(z):$ and $\chi(z)$ are ultralocal.

Indeed, by assumption, there exists such $r \in \mathbb{Z}_+$ that

(4.1.11) $$(w-z)^s \phi(z)\psi(w) = (w-z)^s \psi(w)\phi(z),$$

(4.1.12) $$(u-z)^s \phi(z)\chi(u) = (u-z)^s \chi(u)\phi(z),$$

(4.1.13) $$(u-w)^s \psi(w)\chi(u) = (u-w)^s \chi(u)\psi(w),$$

for any $s \geq r$.

Consider the formal power series in z, w and u:

(4.1.14) $$(w-u)^{3r}\left[(z-w)^{-1}\phi(z)\psi(w) - (z-w)^{-1}\psi(w)\phi(z)\right]\chi(u),$$

where $(z-w)^{-1}$ is considered as a power series in w/z in the first summand and as a power series in z/w in the second summand.

This series is equal to

$$(w-u)^r \sum_{s=0}^{2r} \binom{2r}{s} (w-z)^s (z-u)^{2r-s} \left[(z-w)^{-1}\phi(z)\psi(w) - (z-w)^{-1}\psi(w)\phi(z)\right]\chi(u).$$

The terms with $r < s \leq 2r$ in the last formula vanish by (4.1.11). Hence we can rewrite it as

$$(w-u)^r \sum_{s=0}^{r} \binom{2r}{s} (w-z)^s (z-u)^{2r-s} \left[(z-w)^{-1}\phi(z)\psi(w) - (z-w)^{-1}\psi(w)\phi(z)\right]\chi(u),$$

and further as

$$(w-u)^r \sum_{s=0}^{r} \binom{2r}{s} (w-z)^s (z-u)^{2r-s}\chi(u)\left[(z-w)^{-1}\phi(z)\psi(w) - (z-w)^{-1}\psi(w)\phi(z)\right],$$

using (4.1.12) and (4.1.13). By the same trick as above, we see that (4.1.14) is equal to

(4.1.15) $$(w-u)^{3r}\chi(u)\left[(z-w)^{-1}\phi(z)\psi(w) - (z-w)^{-1}\psi(w)\phi(z)\right].$$

Now it follows from the definition of normal ordering (4.1.5) that the series : $\phi(w)\psi(w)$: is the coefficient of z^{-1} in the series

$$(z-w)^{-1}\phi(z)\psi(w) - (z-w)^{-1}\psi(w)\phi(z),$$

where again $(z-w)^{-1}$ is considered as a power series in w/z in the first summand and as a power series in z/w in the second summand.

Hence if we take the coefficients in front of z^{-1} in formulas (4.1.14) and (4.1.15), we obtain the following equality of formal power series in $w^{\pm 1}$ and $u^{\pm 1}$:

$$(w-u)^{3r} : \phi(w)\psi(w) : \chi(u) = (w-u)^{3r}\chi(u) : \phi(w)\psi(w) : .$$

But since the gradation on V is bounded from below, the matrix coefficients of the left hand side are finite in u^{-1} and w, whereas the matrix coefficients of the right hand side are finite in w^{-1} and in u. Therefore both are polynomials in $w^{\pm 1}$ and $u^{\pm 1}$, and so : $\phi(z)\psi(z)$: and $\chi(z)$ are ultralocal.

It is also clear that if $\phi(z)$ and $\psi(z)$ are ultralocal, then $\partial_z\phi(z)$ and $\psi(z)$ are ultralocal. This implies axiom (A3) and completes the proof.

4.2. The VOA of the Heisenberg algebra.

4.2.1. Let $\mathfrak{h}$ be the Cartan subalgebra of a simple finite-dimensional Lie algebra $\mathfrak{g}$. In § 2.1.6 we defined the Heisenberg algebra $\widehat{\mathfrak{h}}$. It has generators $b_n^i, i = 1, \ldots, l, n \in \mathbb{Z}$, and relations

$$[b_n^i, b_m^j] = n(\alpha_i, \alpha_j)\delta_{n,-m}.$$

In the case $\mathfrak{g} = \mathfrak{sl}_2$ we will have generators $b_n, n \in \mathbb{Z}$, and relations

$$[b_n, b_m] = n\delta_{n,-m}.$$

For $\lambda \in \mathfrak{h}^*$, let π_λ be the Fock representation of $\widehat{\mathfrak{h}}$, which is freely generated by the operators $b_n^i, i = 1, \ldots, l, n \in \mathbb{Z}$, from a vector v_λ, such that

$$b_n^i v_\lambda = 0, n > 0, \qquad b_0^i v_\lambda = (\lambda, \alpha_i)v_\lambda.$$

The fact that we used the same notation π_λ for these representations as for the spaces of differential polynomials in § 2 will be justified later on.

We want to introduce a structure of VOA on π_0.

First we introduce a $\mathbb{Z}$-grading on π_0 by putting $\deg v_0 = 0, \deg b(n) = -n$, so that $\deg b(n_1) \ldots b(n_m)v_0 = -\sum_{i=1}^m n_i$.

Next we introduce a linear map $Y(\cdot, z)$ from π_0 to $\mathrm{End}\pi_0[[z, z^{-1}]]$.

Defining the operator $Y(\cdot, z)$ amounts to assigning to each homogeneous vector $A \in \pi_0$ a formal power series

$$Y(A, z) = \sum_{n\in\mathbb{Z}} A_n z^{-n-\deg A}.$$

In this formula, the Fourier component A_n stands for some linear operator of degree $-n$, acting on π_0.

Let us define these operators by explicit formulas. The degree 0 subspace of the module π_0 is spanned by one vector: v_0. The corresponding operator $Y(v_0, z)$ is equal to the identity operator (times 1, and all other Fourier components are equal to 0).

In degree one we have l linearly independent vectors: $b_{-1}^i v_0, i = 1, \ldots, l$. We assign to them the fields

$$Y(b_{-1}^i v_0, z) = b^i(z) = \sum_{n\in\mathbb{Z}} b_n^i z^{-n-1}.$$

In degree 2 the module π_0 has two types of vectors: $b^i_{-2}v_0$, to which we assign

$$Y(b^i_{-2}v_0, z) = \partial_z b^i(z),$$

where $\partial_z = \partial/\partial z$; and $b^i_{-1}b^j_{-1}v_0$, to which we assign

$$Y(b^i_{-1}b^j_{-1}v_0, z) =: b^i(z)b^j(z) := \sum_{n\in\mathbb{Z}} \left(\sum_{m\in\mathbb{Z}} : b^i(m)b^j(n-m): \right) z^{-n-2}.$$

Here the columns denote the *normal ordering*. If we have a monomial in b^i_n's, its normal ordering is the ordering, in which all "creation" operators $b^i_n, n < 0$, are to the left of all "annihilation" operators $b^i_n, n \geq 0$.

The general formula for a monomial basis element $b^{i_1}_{n_1}\ldots b^{i_m}_{n_m} v_0$ of π_0 is the following:

(4.2.1) $Y(b^{i_1}_{n_1}\ldots b^{i_m}_{n_m} v_0, z) =$

$$\frac{1}{(-n_1-1)!}\cdots\frac{1}{(-n_m-1)!} : \partial_z^{-n_1-1}b^{i_1}(z)\ldots\partial_z^{-n_m-1}b^{i_m}(z): .$$

By linearity, we can extend the map $Y(\cdot, z)$ to any vector of π_0.

Finally, we put $|0\rangle = v_0$ and

(4.2.2) $$T = \frac{1}{2}\sum_{i=1}^{l}\sum_{n\in\mathbb{Z}} : b^i_n b^{i*}_{-n-1} :,$$

where b^{i*}_n are the dual generators of the Heisenberg algebra, i.e. the following commutation relations hold:

$$[b^i_n, b^{j*}_m] = n\delta_{i,j}\delta_{n,-m}.$$

We also have:

$$[\partial, b^i_n] = -nb^i_{n-1}.$$

4.2.2. Theorem. *The map $Y(\cdot, z)$ defined by formula (4.2.1) together with the vector $|0\rangle$ and the operator T define the structure of VOA on π_0.*

Proof. Consider the space π_0 and put $a^{(0)} = v_0, a^{(1)} = b(-1)v_0$. Define also T by formula (4.5.2). Then all conditions of Proposition 4.1.12 will be satisfied. In particular, the fact that $Y(a^{(1)}, z) = b(z)$ is ultralocal with itself follows from the computation in § 4.2.3. Therefore, by Proposition 4.1.12, the map $Y(\cdot, z)$ defined in § 4.2 satisfies the axioms of vertex operator algebra.

The VOA π_0 can be given a structure of conformal VOA with an arbitrary central charge. Define for any $\gamma \in \mathfrak{h}$ the elements ω_γ by the formula

$$\omega_\gamma = \left(\frac{1}{2}\sum_{i=1}^{l} b^i_{-1}b^{i*}_{-1} + \gamma_{-2}\right)v_0.$$

These elements satisfy the axiom (C) with $c = 1 - 12\|\gamma\|^2$.

4.2.3. *Commutation relations.* Now we are going to compute the commutation relations in the Heisenberg algebra $\hat{\mathfrak{h}}$ using formula (4.1.4). First let us compute $b^i(z)b(w)$ for $|z| > |w|$. We have to rewrite this composition as a linear combination of well-defined, i.e. normally ordered, linear operators on π_0. We have

$$b^i(z)b^j(w) = \sum_{n,m\in\mathbb{Z}} b^i(n)b^j(m)z^{-n-1}w^{-m-1} = \sum_{n<0,m\in\mathbb{Z}} b^i(n)b^j(m)z^{-n-1}w^{-m-1}+$$

$$+ \sum_{n\geq 0,m\in\mathbb{Z}} b^i(m)b^j(n)z^{-n-1}w^{-m-1} + \sum_{n\geq 0,m\in\mathbb{Z}} [b^i(n),b^j(m)]z^{-n-1}w^{-m-1} =$$

$$=: b^i(z)b^j(w) : + \sum_{n>0} n(\alpha_i,\alpha_j)\frac{1}{z^2}\left(\frac{w}{z}\right)^{n-1} =: b^i(z)b^j(w) : + \frac{(\alpha_i,\alpha_j)}{(z-w)^2}.$$

We can further rewrite this as

$$(4.2.3) \qquad b^i(z)b^j(w) = \frac{(\alpha_i,\alpha_j)}{(z-w)^2} + \sum_{n\geq 0}\frac{1}{n!}(z-w)^n : \partial_w^n b^i(w)b^j(w) :,$$

by Taylor's formula.

On the other hand, we have: $b^i(n)\cdot b^j(-1)v_0 = 0$, for $n > 1$ or $n = 0$, $b^i(1)\cdot b^j(-1)v_0 = (\alpha_i,\alpha_j)v_0$, and $b^i(n)\cdot b^j(-1)v_0 = b^i(n)b^j(-1)v_0, n < 0$. Hence we obtain

$$(4.2.4)\; Y(Y(b^i(-1)v_0, z - w)\cdot b^j(-1)v_0, w) = \sum_{n\in\mathbb{Z}}(z-w)^{-n-1}Y(b^i(n)\cdot b^j(-1)v_0, w)$$

$$= \frac{(\alpha_i,\alpha_j)}{(z-w)^2} + \sum_{n\geq 0}\frac{1}{n!}(z-w)^n : \partial_w^n b^i(w)b^j(w) :,$$

which coincides with formula (4.2.3), in accordance with formula (4.1.1). Note that the second computation is simpler.

Now we obtain using formula (4.1.4):

$$[b^i(n),b^j(m)] = \int_{C_w^p} dww^m \int_{C_z^q(w)} dzz^n \frac{(\alpha_i,\alpha_j)}{(z-w)^2} = n(\alpha_i,\alpha_j)\int dww^{n+m-1} = n(\alpha_i,\alpha_j)\delta_{n,-m},$$

in accordance with the commutation relations of $\hat{\mathfrak{h}}$.

4.2.4. *Bosonic vertex operators.* Other Fock representations π_λ carry the structure of a module over the VOA π_0 (for general definition, cf. [63]). The Fourier components of the vertex operators $Y(A,z)$, given by formula (4.2.1), define linear operators on any of the modules π_λ, and so we obtain maps $\pi_0 \to \text{End}(\pi_\lambda)[[z,z^{-1}]]$, which satisfy axioms similar to the axioms of VOA.

There is also another structure: a map $\pi_\lambda \to \text{Hom}(\pi_0,\pi_\lambda)[[z,z^{-1}]]$, in other words, to each vector of π_λ we can assign a field, whose Fourier components are linear operators, acting from π_0 to π_λ.

Let us define this map by explicit formulas. To the highest weight vector $v_\lambda \in \pi_\lambda$ we associate the following field, which is called the bosonic vertex operator:

$$(4.2.5) \qquad \tilde{V}_\lambda(z) = \sum_{n\in\mathbb{Z}} V_\lambda(n)z^{-n} = T_\lambda \exp\left(-\sum_{n<0}\frac{\lambda_n z^{-n}}{n}\right)\exp\left(-\sum_{n>0}\frac{\lambda_n z^{-n}}{n}\right),$$

where $T_\lambda : \pi_0 \to \pi_\lambda$ is the shift operator, which maps v_0 to v_λ and commutes with the operators $b_n^i, n < 0$, and $\lambda_n = \lambda \otimes t^n$ is the element of the Heisenberg algebra $\mathfrak{h} = h \otimes \mathbb{C}[t,t^{-1}]$, corresponding to the element $\lambda \in h$ ($\mathfrak{h}$ is identified with $\mathfrak{h}^*$ by means of the scalar product).

The field $\tilde{V}_\lambda(z)$ can be viewed as the normally ordered exponential of

$$\bar{\lambda}(z) = \int^z \lambda(w)dw = -\sum_{n\neq 0} \frac{\lambda_n z^{-n}}{n} + \lambda_0 \log z.$$

We can then define the fields for other elements of π_λ as follows:

$$Y(Pv_\lambda, z) =: Y(Pv_0, z)\tilde{V}_\lambda(z):,$$

where P is a polynomial in $b_n^i, n < 0$.

These fields satisfy the axioms (A1) and (A2). The axiom (A3) and formula (4.1.1) are also satisfied for any pair $A \in \pi_0, B \in \pi_\lambda$, or $B \in \pi_0, A \in \pi_\lambda$.

Using formula (4.1.1), it is easy to compute the commutation relations between b_n^i and $V_\lambda(m)$. Indeed, from the fact that $b_0^i v_\lambda = (\lambda, \alpha_i)v_\lambda$ we derive the following OPE:

$$b^i(z)\tilde{V}_\lambda(w) = \frac{(\lambda, \alpha_i)\tilde{V}_\lambda(w)}{z - w} + \text{regular terms},$$

which gives (using the technique of § 4.2.3) the commutation relations

(4.2.6) $$[b_n^i, V_\lambda(m)] = (\lambda, \alpha_i)V_\lambda(n + m).$$

In the next subsection we will show that the operators $\int \tilde{V}_{\beta\alpha_i}(z)dz$, where β is a deformation parameter, can be viewed as quantizations of the classical operators $\tilde{Q}_i$ from § 2, corresponding to the value $\beta = 0$. We will use these operators to define the space of quantum integrals of motion of Toda field theories and related structures.

4.3. Quantum integrals of motion.

4.3.1. *Quantization of the Poisson algebra.* Let us introduce a deformation parameter β and modify the commutation relations in our Heisenberg algebra by putting

(4.3.1) $$[b_n^i, b_m^j] = n(\alpha_i, \alpha_j)\delta_{n,-m}\beta^2.$$

If $\mathfrak{g} = \mathfrak{sl}_2$, we put

$$[b_n, b_m] = n\delta_{n,-m}\beta^2.$$

We also modify the definition of the Fock module π_λ by putting $b_0^i v_\lambda = \beta(\alpha_i, \lambda)v_\lambda$. Of course, if $\beta \neq 0$, this algebra and these modules are equivalent to the original Heisenberg algebra and Fock modules if we multiply the old generators by β.

According to Theorem 4.1.9, the vector space of all Fourier components of fields from a VOA is a Lie algebra. It contains a Lie subalgebra, which consists of residues of fields, i.e. their (-1)st Fourier components. Indeed, according to formula (4.1.4),

(4.3.2) $$\left[\int Y(A,z)dz, \int Y(B,z)dz\right] = \int Y\left(\int Y(A,w)dw \cdot B, z\right)dz.$$

Let us denote the Lie algebra of all Fourier components of fields from the VOA π_0, by $\hat{\mathcal{F}}_0^\beta$, and its Lie subalgebra of the residues of the fields – by $\mathcal{F}_0^\beta$. They depend on β, because β enters the defining commutation relations (4.3.1). Since the commutation relations in these Lie algebras are polynomial in β^2, we can consider $\hat{\mathcal{F}}_0^\beta$ and $\mathcal{F}_0^\beta$ as Lie algebras over the ring $\mathbb{C}[\beta^2]$, which are free as $\mathbb{C}[\beta^2]$-modules.

The constant term in the commutation relations in $\hat{\mathcal{F}}_0^\beta$ is always equal to 0. We can therefore define the structure of Lie algebra on the space $\hat{\mathcal{F}}_0^\beta/(\beta^2 \cdot \hat{\mathcal{F}}_0^\beta)$, by taking the β^2-linear term in the commutator. That is for any pair $A, B \in \hat{\mathcal{F}}_0^\beta/(\beta^2 \cdot \hat{\mathcal{F}}_0^\beta)$ consider their arbitrary liftings $\tilde{A}, \tilde{B} \in \hat{\mathcal{F}}_0$ and define $\{A, B\}$ as the β^2–linear term in the commutator of $\tilde{A}$ and $\tilde{B}$:

$$[\tilde{A}, \tilde{B}] = \beta^2 \{A, B\} + \beta^4 (\dots).$$

Clearly this bracket does not depend on the liftings of A and B and satisfies the axioms of Lie bracket.

As vector spaces $\hat{\mathcal{F}}_0^\beta / (\beta^2 \cdot \hat{\mathcal{F}}_0^\beta)$ and $\mathcal{F}_0^\beta / (\beta^2 \cdot \mathcal{F}_0^\beta)$ are isomorphic to $\hat{\mathcal{F}}_0$ and $\mathcal{F}_0$, respectively: $\int : P(\partial_z^n b^i(z)) : z^n dz$ is identified with $\int P(\partial_t^n u^i(t)) t^n dt$.

4.3.2. **Lemma.** *The Poisson structure on $\hat{\mathcal{F}}_0^\beta / (\beta^2 \cdot \hat{\mathcal{F}}_0^\beta)$, induced by the Lie algebra structure on $\hat{\mathcal{F}}_0^\beta$, coincides with the Poisson structure on $\hat{\mathcal{F}}_0$, given by formula (2.2.1).*

Proof. Recall that the Poisson structure on $\hat{\mathcal{F}}_0$ is uniquely determined by the Poisson bracket of the functionals $u_n^i = \int u^i(t) t^n dt$, which is equal to

$$\{u_n^i, u_m^j\} = n(\alpha_i, \alpha_j) \delta_{n, -m}.$$

Indeed, any local functional can be represented by an infinite sum

$$\sum_{n_1 + \cdots + n_m = M} c_{n_1 \dots n_m} \cdot u_{n_1}^{i_1} \cdots u_{n_m}^{i_m},$$

and we can compute the Poisson bracket of two such sums term by term, using the Leibnitz rule (cf. § 2.1.4).

Likewise, the commutation relations in the Lie algebra $\hat{\mathcal{F}}_0^\beta$ are uniquely determined by formula (4.3.1), and we can again compute the Lie bracket of two Fourier components term by term. But then we immediately see that the β^2–linear term in the Lie bracket in $\hat{\mathcal{F}}_0^\beta$ coincides with the Poisson bracket in $\hat{\mathcal{F}}_0$.

4.3.3. Thus, the Lie algebras $\hat{\mathcal{F}}_0^\beta$ and $\mathcal{F}_0^\beta$ can be viewed as quantizations of the Poisson algebras $\hat{\mathcal{F}}_0$ and $\mathcal{F}_0$, respectively.

Let $\mathcal{F}_\lambda^\beta$ be the space of linear operators $\pi_0 \to \pi_\lambda$ of the form $\int Y(A, z) dz, A \in \pi_\lambda$. Formula (4.3.2) shows that the Lie algebra $\mathcal{F}_0^\beta$ acts on $\mathcal{F}_\lambda^\beta$ by commutation. In the same way as in Lemma 4.3.2 we can check that this action is a quantization of the action of $\mathcal{F}_0$ on $\mathcal{F}_\lambda$, defined in § 2.2.2. In other words, we can check that the β^2–linear term in the commutator between operators from $\mathcal{F}_0^\beta$ and $\mathcal{F}_\lambda^\beta$ coincides with the Poisson bracket between the corresponding elements of $\mathcal{F}_0$ and $\mathcal{F}_\lambda$.

Moreover, the actions of $\mathcal{F}_\lambda^\beta$ from π_0 to π_λ and of $\mathcal{F}_0^\beta$ from π_λ to π_λ are quantizations of the actions of $\mathcal{F}_\lambda$ from π_0 to π_λ and of $\mathcal{F}_0$ from π_0 to π_0, defined in § 2.1.11. More precisely, let us consider the Fock representation π_λ as a free module π_λ^β over $\mathbb{C}[[\beta^2]]$. The operator $\int Y(A, z) dz$ can be considered over the ring $\mathbb{C}[[\beta^2]]$. It has the form $\int Y(A, z) dz = \beta^2 \cdot Y(A)^{(0)} + \beta^4 \cdot (\dots)$. Hence it induces the operator $Y(A)^{(0)}$ on the quotients $\pi_\lambda^\beta / \beta^2 \cdot \pi_\lambda$. Such a quotient can be identified with the space π_λ of differential polynomials by identifying $b_n^i, n < 0$, with x_n^i. Note that the actions of derivative on both spaces coincide. We have

$$Y(A)^{(0)} dz = \{ \int \bar{A} dt, \cdot \},$$

where $\bar{A} = A \bmod \beta^2$. A particular example of this formula, when $A = v_{\alpha_i}$, will be considered in Lemma 4.3.4.

Consider the map $\pi_\lambda \to \mathcal{F}_\lambda^\beta$, which sends $A \in \pi_\lambda$ to $\int Y(A,z)dz$. From the classical result of § 2.2.2 we derive that the kernel of this map consists of total derivatives, and therefore we have the exact sequences

$$(4.3.3) \qquad\qquad 0 \longrightarrow \pi_\lambda \xrightarrow{\partial} \pi_\lambda \longrightarrow \mathcal{F}_\lambda^\beta \longrightarrow 0$$

(if $\lambda = 0$, then π_0 should be replaced by $\pi_0/\mathbb{C}v_0$).

Analogously, one obtains the exact sequence

$$(4.3.4) \qquad 0 \longrightarrow \pi_\lambda \otimes \mathbb{C}[z,z^{-1}] \xrightarrow{\partial} \pi_\lambda \otimes \mathbb{C}[z,z^{-1}] \longrightarrow \widehat{\mathcal{F}}_\lambda^\beta \longrightarrow 0$$

(if $\lambda = 0$, then the first $\pi_0 \otimes \mathbb{C}[z,z^{-1}]$ should be replaced by $\pi_0 \otimes \mathbb{C}[z,z^{-1}]/\mathbb{C}v_0 \otimes \mathbb{C}$).

Now we can quantize the Toda hamiltonian. Let us define the map $\tilde{Q}_i^\beta : \pi_0 \to \pi_\lambda^\beta$ as $\int \tilde{V}_{\alpha_i}(z)dz$. For any $A \in \pi_\lambda$ the operator $\int Y(A,z)dz$ commutes with the derivative ∂. Hence it defines a map $\mathcal{F}_0 \to \mathcal{F}_\lambda$. Denote the map $\mathcal{F}_0^\beta \to \mathcal{F}_{\alpha_i}^\beta$, corresponding to $\tilde{Q}_i^\beta$, by $\bar{Q}_i^\beta$.

4.3.4. Lemma. *The β^2-linear terms of the operators $\tilde{Q}_i^\beta$ and $\bar{Q}_i^\beta$ coincide with the operators $-\tilde{Q}_i$ and $-\bar{Q}_i$, respectively.*

Proof. According to formula (4.3.1), the operator $b_n^i, n > 0$, acts on π_0 as

$$\beta^2 \sum_{j=1}^l (\alpha_i, \alpha_j) \frac{\partial}{\partial b_{-n}^j}.$$

Thus, we obtain from formula (4.2.5):

$$\tilde{Q}_i^\beta = \int \exp\left(-\sum_{n<0} \frac{b_n^i z^{-n}}{n}\right) \exp\left(-\sum_{n>0} \frac{b_n^i z^{-n}}{n}\right) dz =$$

$$\int \exp\left(-\sum_{n<0} \frac{b_n^i z^{-n}}{n}\right) dz + \int \exp\left(-\sum_{n<0} \frac{b_n^i z^{-n}}{n}\right)\left(-\sum_{n>0} \frac{b_n^i z^{-n}}{n}\right) dz + \ldots =$$

$$-\beta^2 \sum_{n<0} \tilde{V}_i(n+1) \sum_{j=1}^l (\alpha_i, \alpha_j) \frac{\partial}{\partial b_n^j} + \beta^4(\ldots),$$

where $\tilde{V}_i(n+1)$ are the Schur polynomials in $b_n^i, n < 0$. If we replace $b_n^i, n < 0$ by x_n^i, then the linear term in this formula will coincide with formula (2.2.4), which defines the operator $\tilde{Q}_i$, with the sign minus. Therefore, the β^2-linear term in $\bar{Q}_i^\beta$ coincides with the operator $-\bar{Q}_i$.

4.3.5. Definition. *The intersection of kernels of the linear operators $\bar{Q}_i^\beta : \mathcal{F}_0^\beta \to \mathcal{F}_{\alpha_i}^\beta, i \in S$, will be called the space of local integrals of motion of the quantum Toda field theory, associated to a finite-dimensional simple Lie algebra $\mathfrak{g}$ (in this case $S = \{1,\ldots,l\}$), or an affine Lie algebra $\mathfrak{g}$ (in this case $S = \{0,\ldots,l\}$), and will be denoted by $I_\beta(\mathfrak{g})$.*

4.3.6. Definition. *The intersection of kernels of the operators $\tilde{Q}_i^\beta : \pi_0 \to \pi_{\alpha_i}, i = 1,\ldots,l$, will be called the $\mathcal{W}$-algebra of a finite-dimensional simple Lie algebra $\mathfrak{g}$ and will be denoted by $\mathcal{W}_\beta(\mathfrak{g})$.*

4.3.7.   We will say that U is a vertex operator subalgebra of a vertex operator algebra V, if there is an embedding of vector spaces $i : U \to V$, which preserves the $\mathbb{Z}$–gradings, such that $i(|0\rangle_U) = |0\rangle_V$, and that for any $A, B \in U$

(4.3.5) $i \cdot [Y(A, z) \cdot B] = Y(i \cdot A, z) \cdot (i \cdot B)$.

We will say that U is a conformal vertex operator subalgebra of V, if in addition $i(\omega_U) = \omega_V$.

4.3.8. Lemma. *Let $A^{(1)}, \dots, A^{(N)}$ be homogeneous vectors of $\pi_{\lambda_1}, \dots, \pi_{\lambda_N}$, respectively. Then*

(a) *The intersection of kernels of the operators $\int Y(A^{(j)}, z)dz : \pi_0 \to \pi_{\lambda_j}, j = 1, \dots, N$, is a vertex operator subalgebra of π_0;*

(b) *The intersection of kernels of the operators $\int Y(A^{(j)}, z)dz : \mathcal{F}_0 \to \mathcal{F}_{\lambda_j}, j = 1, \dots, N$, is a Lie subalgebra of $\mathcal{F}_0$.*

Proof. (a) Denote by X the intersection of kernels of the operators $\int Y(A^{(j)}, z)dz, j = 1, \dots, N$, and by i the embedding of X into π_0. Since the elements $A^{(j)}$ are homogeneous, these operators are homogeneous, and hence X is $\mathbb{Z}$-graded. Clearly, $v_0 \in \pi_0$ lies in X, so we can put $|0\rangle_X = v_0$. Thus, we have $i(|0\rangle_X) = |0\rangle_{\pi_0}$. We have to show that for any $B, C \in X$ the vector $B_k \cdot C$, where B_k is a Fourier component of the field $Y(B, z)$, lies in X for any $k \in \mathbb{Z}$. In other words, we have to show that

$$\int Y(A^{(j)}, z)dz \cdot (B_k \cdot C) = 0, \qquad j = 1, \dots, N.$$

But this follows from vanishing of the commutators

$$[\int Y(A^{(j)}, z)dz, B_k] = 0, \qquad j = 1, \dots, N,$$

and the fact that $\int Y(A^{(j)}, z)dz \cdot C = 0$. The commutator vanishes according to formula (4.1.4): $\int Y(A^{(j)}, z)dz = A^{(j)}_{-\deg A^{(j)}+1}$, so there is only one term in the commutator, which vanishes, because $A^{(j)}_{-\deg A^{(j)}+1} \cdot B = \int Y(A^{(j)}, z)dz \cdot B = 0$.

(b) follows at once from formula (4.1.4).

Thus we see that $I_\beta(\mathfrak{g})$ is a Lie algebra and $\mathcal{W}_\beta(\mathfrak{g})$ is a VOA.

4.3.9.   A quantum integral of motion can be represented as $x = x^{(0)} + \beta^2 x^{(1)} + \cdots \in \mathcal{F}_0^\beta$. By Lemma 4.3.4, the constant term $x^{(0)}$ should be a classical integral of motion. We will call x a deformation or a quantization of $x^{(0)}$. Clearly, if exists, it is uniquely defined up to adding β^2 times other quantum integrals of motions. In other words, the dimension of the space of quantum integrals of motion of a given degree is less than or equal to the dimension of the space of classical integrals of motion of the same degree. In the rest of this section we will show that the dimension of the space of integrals of motion of a given degree of a Toda field theory for generic values of β is the same as for $\beta = 0$. Since the quantum integrals of motion depend algebraically on β, this will imply that all classical integrals of motion of the Toda field theories can be deformed. The same is true for the VOA $\mathcal{W}_\beta(\mathfrak{g})$.

4.4. Liouville theory.

4.4.1. In the quantum Liouville theory the space of integrals of motion is defined as the kernel of the operator $\bar{Q}^{\beta} : \mathcal{F}_0 \to \mathcal{F}_1$. We will proceed in the same way as in § 2.1, by computing the kernel of the operator $\widetilde{Q}^{\beta} : \pi_0 \to \pi_1$ and then using the spectral sequence. Note that since the operator $\widetilde{Q}^{\beta}$ is a residue of a field, its kernel is a VOA, by Lemma 4.3.8.

Thus, we consider the complex

$$(4.4.1) \qquad\qquad \pi_0 \xrightarrow{\beta^{-2}\widetilde{Q}^{\beta}} \pi_1,$$

where we normalized $\widetilde{Q}^{\beta} = \int \widetilde{V}(z)dz$ by β^{-2}. According to Lemma 4.3.4, this complex makes sense even when $\beta = 0$, when it coincides with the classical complex (2.1.17). We have therefore a family of complexes, depending on a complex parameter β. Moreover, as a vector space our complex does not change, and the differential depends on β polynomially in each homogeneous component. The following simple observation will enable us to prove that all classical integrals of motion can be quantized.

4.4.2. *Lemma. Suppose, one is given a family of finite-dimensional complexes, depending on a complex parameter β, which are the same as vector spaces, and the differentials depend analytically on β. Then for any n the dimension of the nth cohomology group of the complex is the same for generic values of β, and it may only increase for special values of β. In particular, if a certain cohomology group of the complex vanishes, when $\beta = 0$, then it also vanishes for generic values of β.*

Proof. The nth cohomology is the quotient of the kernel of the $(n+1)$st differential by the image of the nth differential. The dimension of the kernel (respectively, the image) of a finite-dimensional linear operator, depending analytically on a parameter β, is the same for generic values of β, and it may only increase (respectively, decrease) for special values of β.

4.4.3. *Corollary. Under the conditions of Lemma 4.4.2, suppose also that all higher cohomologies of the complex vanish, when $\beta = 0$. Then the dimension of the 0th cohomology for generic β is the same as for $\beta = 0$.*

Proof. Vanishing of the higher cohomology groups for generic values of β follows from vanishing for $\beta = 0$ and Lemma 4.4.2. Since the complex does not depend on β as a vector space, its Euler characteristics, which is equal to the alternating sum of the dimensions of the groups of the complex, also does not depend on β. But the Euler characteristics is also equal to the alternating sum of the dimensions of the cohomology groups. Since the dimensions of higher cohomology groups are 0 for generic β and for $\beta = 0$, the Euler characteristics is equal to the dimension of the 0th cohomology group for generic β and for $\beta = 0$. Hence the dimension of the 0th cohomology group for generic β is the same as for $\beta = 0$.

4.4.4. *Proposition. The operator $\widetilde{Q}^{\beta} : \pi_0 \to \pi_1$ has no cokernel for generic β. Its kernel, $\mathcal{W}_{\beta}(\mathfrak{sl}_2)$, is a conformal vertex operator algebra. It contains a Virasoro element $W^{\beta}_{-2}v_0$, and $\mathcal{W}_{\beta}(\mathfrak{sl}_2)$ is freely generated from v_0 under the action of the Fourier components $W^{\beta}_n, n < -1$, of the field*

$$Y(W^{\beta}_{-2}v_0, z) = \sum_{n \in \mathbb{Z}} W^{\beta}_n z^{-n-2}.$$

This VOA is isomorphic to the VOA of the Virasoro algebra with central charge $c = c(\beta) = 13 - 3\beta^2 - 12\beta^{-2}$.

Proof. By Proposition 2.1.15, the operator $-\tilde{Q}$, which, according to Lemma 4.3.4, is the limit of the operator $\beta^{-2}\tilde{Q}^\beta$, when $\beta \to 0$, has no cokernel. In other words, the 1st cohomology of the complex (4.4.1) vanishes, when $\beta = 0$. This complex decomposes into a direct sum of finite-dimensional subcomplexes with respect to the $\mathbb{Z}$–grading. By Corollary 4.4.3, the character of the 0th cohomology for generic β ($=$ the kernel of the operator $\tilde{Q}^\beta$) is the same as for $\beta = 0$:

$$\prod_{n \geq 2} (1 - q^n)^{-1}.$$

This formula means that there is a vector $W^\beta_{-2}v_0$ of degree 2 in the kernel of $\tilde{Q}^\beta$. One can deduce from the axioms of VOA and the fact that the kernel of $\tilde{Q}^\beta$ is a VOA (cf. Lemma 4.3.8) that this vector must be a Virasoro element in π_0. But we can also find an explicit formula for this vector:

$$(4.4.2) \qquad W^\beta_{-2}v_0 = \left(\frac{1}{2\beta^2}b^2_{-1} + (\frac{1}{2} - \frac{1}{\beta^2})b_{-2} \right) v_0.$$

The Fourier components W^β_n of the corresponding field generate an action of the Virasoro algebra with central charge $c(\beta)$ on π_0 and π_1. This action commutes with the action of the differential $\tilde{Q}^\beta$. The operators $W^\beta_n, n < -1$, act freely on π_0 for any value of β, because they act freely for $\beta = 0$ (cf. Proposition 2.1.15). Therefore, by applying these operators to v_0 we obtain a subspace of the kernel, which has the same character as the kernel; hence it coincides with the kernel.

Clearly, this is the vertex operator algebra of the Virasoro algebra, which is defined, e.g., in [44].

4.4.5. Corollary. *The space $I_\beta(\mathfrak{sl}_2)$ of local integrals of motion of the quantum Liouville model is isomorphic to the Lie algebra of residues of fields of the Virasoro vertex operator algebra.*

Proof. By definition, $I_\beta(\mathfrak{sl}_2)$ coincides with the 1th cohomology of the complex $\mathbb{C} \longrightarrow \mathcal{F}^\beta_0 \longrightarrow \mathcal{F}^\beta_1 \longrightarrow \mathbb{C}$. We can use the spectral sequence (2.2.6) to compute this cohomology, in the same way as in the proof of Proposition 2.1.17. The Corollary then follows from Proposition 4.4.4 and the fact that for any VOA V the quotient of V by the total derivatives and constants is isomorphic to the space of residues of fields.

4.4.6. General case. We now want to show that all classical integrals of motion of the classical Toda field theory, associated to a finite-dimensional simple Lie algebra $\mathfrak{g}$, can be deformed. In order to establish that, we will construct a quantum deformation $F^*_\beta(\mathfrak{g})$ of the complex $F^*(\mathfrak{g})$ and then use vanishing of higher cohomologies in the classical limit $\beta = 0$.

Roughly speaking, this can be achieved as follows. We will first construct a resolution $B^q_*(\mathfrak{g})$ over the quantized universal enveloping algebra (quantum group) of $\mathfrak{g}$, $U_q(\mathfrak{g})$, which is a deformation of the standard BGG resolution $B_*(\mathfrak{g})$. The resolution $B^q_*(\mathfrak{g})$ consists of Verma modules $M^q_{\rho-s(\rho)}$ over $U_q(\mathfrak{g})$, and the differentials are given by linear combinations of embeddings of Verma modules, defined by singular vectors $P^q_{s',s} \cdot 1_{\rho-s'(\rho)} \in M^q_{\rho-s(\rho)}$, where $P^q_{s',s} \in U_q(\mathfrak{n}_+)$. In the limit $q \to 1$ this resolution coincides with the BGG resolution of $\mathfrak{g}$.

We will define the quantum deformations of the operators $P_{s',s}(Q) : \pi_{\rho-s(\rho)} \to \pi_{\rho-s'(\rho)}$ as linear combinations of certain multiple integrals of products of bosonic vertex operators

$\widetilde{V}_{\alpha_i}(z)$, corresponding to $P^q_{s',s}$. It turns out that the operators $\widetilde{Q}^\beta_i$ satisfy the q–deformed Serre relations, which are the relations in the quantized universal enveloping algebra of the nilpotent subalgebra of $\mathfrak{g}$, $U_q(\mathfrak{n}_+)$, with $q = \exp(\pi i \beta^2)$. This will allow us to define differentials on the quantum complex.

Using the quantum complex, we will show that all classical integrals of motion can be quantized.

4.5. Quantum groups and quantum BGG resolutions.

4.5.1. *Quantum group $U_q(\mathfrak{g})$.* Let $\mathfrak{g}$ be a Kac-Moody Lie algebra associated to a symmetrizable Cartan matrix $\|a_{ij}\|, i, j \in S$. The quantum group $U_q(\mathfrak{g})$ [31, 78], where $q \in \mathbb{C}^\times, q \neq \pm 1$, is the associative algebra over $\mathbb{C}$ with generators $e_i, f_i, K_i, K_i^{-1}, i \in S$, and the relations:

$$K_i K_j = K_j K_i, \qquad K_i K_i^{-1} = K_i^{-1} K_i = 1,$$
$$K_i e_j K_i^{-1} = q^{(\alpha_i, \alpha_j)} e_j, \qquad K_i f_j K_i^{-1} = q^{-(\alpha_i, \alpha_j)} f_j,$$
$$[e_i, f_j] = \delta_{i,j} \frac{K_i - K_i^{-1}}{q^{(\alpha_i, \alpha_i)/2} - q^{-(\alpha_i, \alpha_i)/2}},$$

and the so-called q-Serre relations, which can be defined as follows.

Introduce a grading on the free algebra with generators $e_i, i = 1, \ldots, l$, with respect to the weight lattice P, by putting $\deg e_i = \alpha_i$. If x is a homogeneous element of this algebra of weight γ, put

$$\mathrm{ad}_q e_i \cdot x = e_i x - q^{(\alpha_i, \gamma)} x e_i.$$

Likewise, we can introduce operators $\mathrm{ad}_q f_i$ on the free algebra with generators $f_i, i = 1, \ldots, l$. Then the q-Serre relations read:

$$(4.5.1) \qquad (\mathrm{ad}_q e_i)^{-a_{ij}+1} \cdot e_j = 0, \qquad (\mathrm{ad}_q f_i)^{-a_{ij}+1} \cdot f_j = 0.$$

If we put

$$h_i = \frac{K_i - K_i^{-1}}{q^{(\alpha_i, \alpha_i)/2} - q^{-(\alpha_i, \alpha_i)/2}},$$

then in the limit $q \to 1$ these relations coincide with the standard relations of $\mathfrak{g}$ in terms of $e_i, f_i, h_i, i \in S$.

Denote by $U_q(\mathfrak{n}_+)$ the subalgebra of $U_q(\mathfrak{g})$, generated by $e_i, i = 1, \ldots, l$, with the relations (4.5.1). This algebra is a quantum deformation of the universal enveloping algebra of the nilpotent subalgebra $\mathfrak{n}_+$ of $\mathfrak{g}$. Let $U_q(\mathfrak{b}_-)$ be the subalgebra of $U_q(\mathfrak{g})$, generated by $f_i, K_i, K_i^{-1}, i = 1, \ldots, l$.

4.5.2. *Verma modules.* Verma modules over $U_q(\mathfrak{g})$ are defined in the same way as Verma modules over $\mathfrak{g}$ (cf. § 2.3.1).

Let $\mathbb{C}_\lambda$ be the one-dimensional representation of $U_q(\mathfrak{b}_-)$, which is spanned by vector 1_λ, such that

$$f_i \cdot 1_\lambda = 0, \qquad K_i \cdot 1_\lambda = q^{(\lambda, \alpha_i)} 1_\lambda, \qquad i = 1, \ldots, l.$$

The Verma module M^q_λ over $U_q(\mathfrak{g})$ of lowest weight λ is the module induced from the $U_q(\mathfrak{b}_-)$-module $\mathbb{C}_\lambda$:

$$M^q_\lambda = U_q(\mathfrak{g}) \otimes_{U_q(\mathfrak{b}_-)} \mathbb{C}_\lambda.$$

4.5.3. *Singular vectors.* We want to construct a resolution $B_*^q(\mathfrak{g})$ over $U_q(\mathfrak{g})$, which would coincide with the standard BGG resolution in the limit $q \to 1$. Therefore as a vector space, the jth group $B_j^q(\mathfrak{g})$ of this resolution should be the direct sum of Verma modules

$$B_j^q(\mathfrak{g}) = \oplus_{l(s)=j} M_{\rho-s(\rho)}^q.$$

Now we have to construct the differentials.

We want to show first that the structure of singular vectors in the modules $M_{\rho-s(\rho)}^q$ for generic q is the same as the structure of the singular vectors for $q = 1$, which is described in § 2.3.2. We will then proceed in the same way as in the case $q = 1$.

The existence of these singular vectors can be derived from the determinant formula for the Shapovalov form on M_λ^q. This formula has been established in [26] (cf. formula (1.9.3)) for finite-dimensional simple Lie algebras. In [79] (cf. Lemma 6.4) the irreducible factors of the formula were found for arbitrary symmetrizable Kac-Moody algebras. This is sufficient for our purposes.

The above cited results show that the Shapovalov form on M_λ^q at weight η vanishes, if one of the following equations is satisfied:

$$(4.5.2) \qquad (\lambda - \rho, \gamma) + \frac{m}{2}(\gamma, \gamma) = 0,$$

where γ runs over the set of positive roots of $\mathfrak{g}$, and m runs over the set of positive integers, such that $m\gamma < \eta$. Note that in this formula we have signs different from those in [26, 79], because we work with modules of *lowest weight.*

Consider now the module $M_{\rho-s(\rho)}^q$. It is known that for any s', which satisfies $s \prec s'$ and $l(s') = l(s) + 1$, there exist γ and m, such that $\rho - s'(\rho) = \rho - s(\rho) + m\gamma$ and $(s(\rho), \gamma) - \frac{m}{2}(\gamma, \gamma) = 0$. Therefore the equation (4.5.2) is satisfied and hence the determinant is equal to 0 at weight $\rho - s'(\rho)$.

4.5.4. *Remark.* In fact, in order to prove this statement, we do not need the exact formula for the determinant. We can proceed along the lines of [82], using the Casimir operators, constructed in [34] or [79] and the limit $q \to 1$ of the determinant formula from [82].

4.5.5. On the other hand, the determinant is not equal to 0 at any level $\eta < \rho - s'(\rho)$ for $q = 1$ [82], and hence for generic q. Therefore in $M_{\rho-s(\rho)}^q$ there should exist a singular vector of weight $\rho - s'(\rho)$.

This singular vector defines a map

$$i_{s',s}^q : M_{\rho-s'(\rho)}^q \to M_{\rho-s(\rho)}^q$$

by sending the lowest weight vector $1_{\rho-s'(\rho)}^q$ of $M_{\rho-s'(\rho)}^q$ to the singular vector $P_{s',s}^q \cdot 1_{\rho-s(\rho)}^q$ of $M_{\rho-s(\rho)}^q$ of weight $\rho - s'(\rho)$. This map is an embedding of Verma modules for generic q, because it is an embedding for $q = 1$.

We can map singular vectors, constructed this way, inductively to M_0^q. Thus, we obtain in M_0^q singular vectors of weights $\rho - s(\rho)$ for arbitrary elements s of the Weyl group.

We know that for $q = 1$ there is only one singular vector of weight $\rho - s(\rho)$ in M_0. But the dimension of the space of singular vectors of a Verma module M_λ^q of a certain weight for generic q is greater than or equal to that for $q = 1$. Therefore there is only one singular vector of such weight in M_0^q for generic q.

Uniqueness implies the relation

$$(4.5.3) \qquad P_{s'',s_1'}^q P_{s_1',s}^q = P_{s'',s_2'}^q P_{s_2',s}^q$$

in $U_q(\mathfrak{n}_+)$. Indeed, both $P^q_{s'',s'_1}P^q_{s'_1,s}1^q_{\rho-s(\rho)}$ and $P^q_{s'',s'_2}P^q_{s'_2,s}1^q_{\rho-s(\rho)}$ are singular vectors in $M_{\rho-s(\rho)}$ of weight $\rho - s''(\rho)$. Therefore they coincide for generic q.

4.5.6. *The differential.* Now we can construct a differential on the complex $B^q_*(\mathfrak{g})$. For any pair s, s' of elements of the Weyl group of $\mathfrak{g}$, such that $s \prec s''$, we have the embeddings of $\mathfrak{g}$-modules $i^q_{s',s} : M^q_{\rho-s'(\rho)} \to M^q_{\rho-s(\rho)}$. They satisfy: $i^q_{s'_1,s} \circ i^q_{s'',s'_1} = i^q_{s'_2,s} \circ i^q_{s'',s'_2}$, according to (4.5.3). We define the differential $d^q_j : B^q_j(\mathfrak{g}) \to B^q_{j-1}(\mathfrak{g})$ by the formula

$$(4.5.4) \qquad d^q_j = \sum_{l(s)=j-1, l(s')=j, s \prec s'} \epsilon_{s',s} \cdot i^q_{s',s}$$

(cf. formula (2.3.1)). By construction, this differential is nilpotent (cf. Theorem 2.3.5).

Note that since higher cohomologies of $B^q_*(\mathfrak{g})$ vanish for $q = 1$, they also vanish for generic q, by Lemma 4.4.2. Therefore the 0th cohomology is one-dimensional for generic q, by Corollary 4.4.3. Thus, for generic q, $B_*(\mathfrak{g})$ is a free resolution of the trivial representation of $U_q(\mathfrak{n}_+)$.

4.5.7. *Remark.* In the same way we can q-deform the BGG resolution of an arbitrary integrable representation $V_\lambda, \lambda \in P$, of a Kac-Moody algebra. Such a resolution has also been constructed in [94] by other methods.

Vanishing of higher cohomologies of this resolution for $q = 1$ implies that they vanish for generic q. But then the 0th cohomology for generic q is a module over $U_q(\mathfrak{g})$, which is irreducible, since V_λ is irreducible, and whose character is the same as the character of V_λ for $q = 1$, by Lemma 4.4.3. This gives an alternative proof of the fact that any integrable representation of $\mathfrak{g}$ can be q-deformed, previously proved by Lusztig [92] by other methods.

4.6. Toda field theories associated to finite-dimensional simple Lie algebras.

4.6.1. Now we can use the resolution $B_*(\mathfrak{g})$ to define the quantum complex $F^*_\beta(\mathfrak{g}) = \oplus_{j \geq 0} F^j_\beta(\mathfrak{g})$. The jth group $F^j_\beta(\mathfrak{g})$ of the complex is the same as the jth group of the classical complex $F^*(\mathfrak{g})$:

$$F^j_\beta(\mathfrak{g}) = \oplus_{l(s)=j} \pi_{\rho-s(\rho)}.$$

To define the differentials of the complex, we have to quantize the operators $P_{s',s}(Q) : \pi_{\rho-s(\rho)} \to \pi_{\rho-s'(\rho)}$. In order to do that we should first learn how to compose the operators $\tilde{Q}^\beta_i$.

4.6.2. Let $p = (p_1, \ldots, p_m)$ be a permutation of the set $(1, 2, \ldots, m)$. We define a contour of integration C_p in the space $(\mathbb{C}^\times)^m$ with the coordinates $z_1, \ldots, z_m$ as the product of one-dimensional contours along each of the coordinates, going counterclockwise around the origin starting and ending at the point $z_i = 1$, and such that the contour of z_{p_i} is contained inside the contour of z_{p_j} for $i > j$.

Denote by $\mathbf{i} = (i_1, \ldots, i_m)$ a sequence of numbers from 1 to l, such that $i_1 \leq i_2 \leq \ldots \leq i_m$. We can apply a permutation p to this sequence to obtain another sequence $\mathbf{j} = (j_1, \ldots, j_m) = (i_{p_1}, \ldots, i_{p_m})$. Let us define an operator $V^\beta_\mathbf{j}$ as the integral

$$\int_{C_p} dz_1 \ldots dz_m \prod_{1 \leq k < l \leq m} (z_k - z_l)^{\beta^2(\alpha_{i_k}, \alpha_{i_l})} \prod_{1 \leq k \leq m} z_k^{\beta^2(\lambda, \alpha_{i_k})} : \tilde{V}_{\alpha_{i_1}}(z_1) \ldots \tilde{V}_{\alpha_{i_m}}(z_m) :=$$

$$(4.6.1) \qquad \sum_{n_1, \ldots, n_m \in \mathbb{Z}} \Gamma^{n_1, \ldots, n_m}_\mathbf{j} : V_{\alpha_{i_1}}(n_1) \ldots V_{\alpha_{i_m}}(n_m) :;$$

where the coefficient $\Gamma_{\mathbf{j}}^{n_1,\dots,n_m}$ is given by

$$(4.6.2) \qquad \Gamma_{\mathbf{j}s}^{n_1,\dots,n_m} = \int_{C_p} dz_1 \dots dz_m \prod_{1 \le k < l \le m} (z_k - z_l)^{\beta^2(\alpha_{i_k},\alpha_{i_l})} \prod_{1 \le k \le m} z_k^{\beta^2(\lambda,\alpha_{i_k})-n_k}.$$

In the integrals above, we choose the branch of the power function, which takes real values for real values of the z_i's, such that $z_{j_1} > z_{j_2} > \dots > z_{j_m}$. Thus, C_p should be viewed as an element of the group of relative m-chains in $(\mathbb{C}^\times)^m$ modulo the diagonals, with values in the one-dimensional local system ξ_i, which is defined by the multivalued function

$$\prod_{1 \le k < l \le m} (z_k - z_l)^{-\beta^2(\alpha_{i_k},\alpha_{i_l})} \prod_{1 \le k \le m} z_k^{-\beta^2(\lambda,\alpha_{i_k})}.$$

The integral of the type (4.6.2) over any such relative chain is well-defined for generic values of β. Indeed, the integral

$$\int_C dz_1 \dots dz_m \prod_{1 \le k < l \le m} (z_k - z_l)^{\mu_{kl}} \prod_{1 \le k \le m} z_k^{\nu_k}$$

over such a chain C converges in the region $\mathrm{Re}\mu_{kl} \ge 0$, and can be uniquely analytically continued to other values of μ_{kl}, which do not lie on certain hyperplanes, cf. [114], especially, Theorem (10.7.7), for details. These hyperplanes are defined by setting some linear combinations of μ_{kl}'s, with integral coefficients, to negative integers. For generic β the exponents in our integral do not lie on those hyperplanes, therefore the integral is well-defined.

Note that the integral in (4.6.2) depends on $\mathbf{j}$ and not on p. For any $N \in \mathbb{Z}$ the operator

$$\sum_{n_1+\dots+n_m=N} \Gamma_{\mathbf{j}}^{n_1,\dots,n_m} : V_{\alpha_{i_1}}(n_1) \dots V_{\alpha_{i_m}}(n_m) :,$$

is a well-defined homogeneous operator acting from π_λ to $\pi_{\lambda+\gamma}$, where $\gamma = \sum_{j=1}^m \alpha_{i_j}$. Therefore, the operator $V_{\mathbf{j}}^\beta$ is a linear operator from π_λ to the completion $\bar\pi_{\lambda+\gamma}$ of $\pi_{\lambda+\gamma}$.

4.6.3. One can interpret the operator $V_{\mathbf{j}}^\beta$ as a suitably defined composition operator $\tilde{Q}_{j_1}^\beta \dots \tilde{Q}_{j_m}^\beta$. Indeed, let us define the bosonic vertex operator $V_{\alpha_i}(z)$, acting from π_λ to $\pi_{\lambda+\alpha_i}$ as $z^{\beta^2(\lambda,\alpha_i)} \tilde{V}_{\alpha_i}(z)$. One has (cf., e.g., [62], formulas (8.4.25) and (A.2.9))

$$(4.6.3) \qquad V_{\alpha_{j_1}}(z_1) \dots V_{\alpha_{j_m}}(z_m) = \prod_{1 \le k < l \le m} (z_k - z_l)^{\beta^2(\alpha_{j_k},\alpha_{j_l})} : V_{\alpha_{j_1}}(z_1) \dots V_{\alpha_{j_m}}(z_m) :,$$

for $|z_1| > \dots > |z_m|$. Note that unlike the Fourier components of $V_{\alpha_{j_1}}(z_1) \dots V_{\alpha_{j_m}}(z_m)$, the Fourier components of $: V_{\alpha_{j_1}}(z_1) \dots V_{\alpha_{j_m}}(z_m) :$ are well-defined linear operators, analytically depending on $z_1, \dots, z_m$ in the region $\mathbb{C}^m \backslash diag$, which do not change under the permutations of coordinates.

We have for $\mathbf{j}_1 = (j_1^1,\dots,j_m^1)$ and $\mathbf{j}_2 = (j_1^2,\dots,j_n^2)$:

$$(4.6.4) \qquad V_{\mathbf{j}_1}^\beta V_{\mathbf{j}_2}^\beta = V_{(\mathbf{j}_1,\mathbf{j}_2)}^\beta.$$

where $V_{\mathbf{j}_2}^\beta : \pi_\lambda \to \bar\pi_{\lambda+\gamma_2}$ and $V_{\mathbf{j}_1}^\beta : \pi_{\lambda+\gamma_2} \to \bar\pi_{\lambda+\gamma_1+\gamma_2}$. Indeed, we can choose the contour C_p in such a way that $|z_{p_i}| \ge 1$ for $i = 1,\dots,m$ and $|z_{p_i}| \le 1$ for $i = m+1,\dots,m+n$. Then the composition in the left hand side of (4.6.4) is well-defined, and concides with the right hand side.

According to Lemma 4.3.4, the β^2-linear term of the operator $\tilde{Q}_j^\beta$ coincides with the operator $-\tilde{Q}_j$, therefore the leading (β^{2m}th) term of $V_{\mathbf{j}}^\beta$ coincides with $(-1)^m \tilde{Q}_{j_1} \dots \tilde{Q}_{j_m}$.

The operators $\tilde{Q}_i$ satisfy the Serre relations. We want to show that the operators V_i^β satisfy the q-Serre relations, where $q = \exp(\pi i \beta^2)$.

4.6.4. Consider a free algebra A with generators $g_i, i = 1, \ldots, l$. We can assign to each monomial $g_{j_1} \ldots g_{j_m}$ the contour C_j and hence the operator V_j^β. This gives us a map Δ from A to the space of linear combinations of such contours. Given such a linear combination C, we define V_C^β as the linear combination of the corresponding operators $V_{C_j}^\beta$.

Consider the two-sided ideal S_q in A, which is generated by the q-Serre relations $(\text{ad} g_i)_q^{-a_{ij}+1} \cdot g_j, i \neq j$, where $q = \exp(\pi i \beta^2)$.

4.6.5. *Lemma. If C belongs to $\Delta(S_q)$, then $V_C^\beta = 0$.*

Proof is given in [19]. It is based on rewriting the integrals over the contours C_j as integrals over the contours, where all variables are on the unit circle with some ordering of their arguments.

We have to prove that $(\text{ad} \tilde{Q}_i)^{-a_{ij}+1} \cdot \tilde{Q}_j = 0$. Let $\mathcal{V}_{p(i)}^\beta$ be the operator, defined by formula (4.6.1), where the contour C_p is replaced by contour

$$C_p = \{(z_1, \ldots, z_m) | \, |z_i| = 1, 0 < \arg z_{p_1} < \ldots < \arg z_{p_m} < 2\pi\}.$$

By induction one can prove [19] that

$$(\text{ad} \tilde{Q}_i)^m \cdot \tilde{Q}_j = I_m \mathcal{V}_{(i,\ldots,i,j)},$$

where

$$I_m = \prod_{j=0}^{m-1} \frac{(1 - q^{j(\alpha_i, \alpha_i) + 2(\alpha_i, \alpha_j)})(1 - q^{(j+1)(\alpha_i, \alpha_i)})}{1 - q^{(\alpha_i, \alpha_i)}},$$

and Lemma follows.

Thus, we obtain a well-defined map, which assigns to each element P of the algebra $U_q(\mathfrak{n}_+) \simeq A/S_q$ the operator V_P^β.

4.6.6. *Lemma. Let $P \in U_q(\mathfrak{n}_+)$ be such that $P \cdot 1_\lambda$ is a singular vector of M_λ^q of weight $\lambda + \gamma$. Then the operator V_P^β is a homogeneous linear operator $\pi_\lambda \to \pi_{\lambda+\gamma}$.*

Proof. It is known that under the conditions of the Lemma, the contour $\Delta(P)$ is a cycle in the group $H^m((\mathbb{C}^\times)^m, diag; \xi_i)$, cf. [17, 18, 19, 109, 114]. Therefore, only the degree 0 integrands (with respect to the grading $\deg z_i = \deg dz_i = 1$) of the integral over $\Delta(P)$ give non-zero results. In other words, all coefficients

$$\int_{\Delta(P)} dz_1 \ldots dz_n \prod_{1 \leq k < l \leq m} (z_k - z_l)^{\beta^2(\alpha_{i_k}, \alpha_{i_l})} \prod_{1 \leq k \leq m} z_k^{\beta^2(\lambda, \alpha_{i_k}) - n_k}$$

vanish, unless $\sum_{i=1}^m n_i = -m$. It means that in the operator V_P^β all homogeneous components, except one, vanish, and Lemma follows.

404

4.6.7. *Remark.* The statements of Lemma 4.6.5 and Lemma 4.6.6 follow from more general results on the remarkable and not yet fully understood correspondence between quantum groups and local systems of the type ξ_i on configuration spaces, established in the works of Schechtman and Varchenko [109, 114, 110] (cf. also [54]).

Let us also remark that in [114] Varchenko constructed explicitly certain *absolute* chains in $(\mathbb{C}^\times)\backslash diag$, which have all the nice properties of the *relative* chains C_p, including the factorization property (4.6.4) (cf. §14 of [114]). One can use these chains instead of C_p in the definition of our operators.

4.6.8. *Quantum differentials.* We are ready now to define the differentials $\delta_\beta^j : F_\beta^{j-1}(\mathfrak{g}) \to F_\beta^j(\mathfrak{g})$ of the quantum complex $F_\beta^*(\mathfrak{g})$. Recall that for any pair s, s' of elements of the Weyl group there exists a singular vector $P_{s',s}^q \cdot 1_{\rho-s(\rho)}^q \in M_{\rho-s(\rho)}$ of weight $\rho - s'(\rho)$, where $P_{s',s}^q \in U_q(\mathfrak{n}_+)$. We put:

$$(4.6.5) \qquad \delta_\beta^j = \sum_{l(s)=j-1,l(s')=j,s\prec s'} \epsilon_{s',s} \cdot V_{P_{s',s}^q}^\beta,$$

where $q = \exp(\pi i \beta^2)$. By Lemma 4.6.6, the differentials δ_β^j are well-defined homogeneous linear operators. From the nilpotency of the differential of the quantum BGG resolution and Lemma 4.6.5 we deduce that these differentials are nilpotent.

Thus, we obtain a family of complexes $F_\beta^*(\mathfrak{g})$, depending on the parameter β. As was explained above, we can rescale the differentials by some powers of β, so that when $\beta = 0$ the complex $F_\beta^*(\mathfrak{g})$ becomes the classical complex $F^*(\mathfrak{g})$.

4.6.9. *Theorem.* For generic β higher cohomologies of the complex $F_\beta^*(\mathfrak{g})$ vanish. The 0th cohomology, $\mathcal{W}_\beta(\mathfrak{g})$, is a conformal vertex operator algebra. There exist elements $W_{-d_1-1}^{(1),\beta}v_0, \ldots, W_{-d_l-1}^{(l),\beta}v_0$ of $\mathcal{W}_\beta(\mathfrak{g})$ of degrees $d_1 + 1, \ldots, d_l + 1$, where the d_i's are the exponents of $\mathfrak{g}$, such that $\mathcal{W}_\beta(\mathfrak{g})$ is freely generated (in the sense explained below) from v_0 under the action of the Fourier components $W_{n_i}^{(i),\beta}, 1 \leq i \leq l, n_i < -d_i$, of the corresponding fields

$$\mathbf{W}_i^\beta(z) = Y(W_{-d_i-1}^{(i),\beta}v_0, z) = \sum_{n\in\mathbb{Z}} W_n^{(i),\beta}z^{-n-d_i-1}.$$

Moreover, $W_{-2}^{(1),\beta}v_0$ is the Virasoro element, and the other elements $W_{-d_i-1}^{(i),\beta}v_0$ can be chosen so as to be annihilated by the corresponding Virasoro generators $W_n^{(1),\beta}, n > 0$.

Proof. Lemma 4.4.2 together with Proposition 2.4.7 lead us to conclude that for generic β all higher cohomologies of the complex $F_\beta^*(\mathfrak{g})$ vanish.

Therefore, by Corollary 4.4.3, the 0th cohomology, $\mathcal{W}_\beta(\mathfrak{g})$, of the complex $F_\beta^*(\mathfrak{g})$ for generic β has the same character as the 0th cohomology of the complex $F^*(\mathfrak{g})$ (cf. the proof of Proposition 2.4.7):

$$(4.6.6) \qquad \prod_{1\leq i\leq l,n_i>d_i} (1 - q^{n_i})^{-1}.$$

This formula shows that there is a vector of degree 2, $W_{-2}^{(1),\beta}v_0$, in $\mathcal{W}_\beta(\mathfrak{g})$, which is a Virasoro element. It is given by the following formula:

$$W_{-2}^{(1),\beta}v_0 = \left(\frac{1}{2\beta^2}\sum_{i=1}^l b_{-1}^i b_{-1}^{i*} + \rho_{-2} - \frac{1}{\beta^2}\rho_{-2}^\vee\right)v_0.$$

The Fourier components $L_n = W_n^{(1),\beta}, n \in \mathbb{Z}$, of the field corresponding to $W_{-2}^{(1),\beta} v_0$, generate an action of the Virasoro algebra on π_λ. They commute with the action of the differential δ_β^1 and hence they preserve the space $\mathcal{W}_\beta(\mathfrak{g})$, which is the kernel of δ_β^1.

Recall that a singular vector is a vector, which is annihilated by the positive Virasoro generators $L_n, n > 0$. Verma module is a module over the Virasoro algebra, which is freely generated from a singular vector by $L_n, n < 0$. The degree of the singular vector, from which it is generated is called highest weight. The vacuum Verma module is the module, freely generated by $L_n, n < -1$, from a vector, which is annihilated by $L_n, n \geq -1$. From the structural theory of Verma modules over the Virasoro algebra [52] we know that Verma modules of integral non-zero highest weight and the vacuum Verma module are irreducible for generic central charge. Moreover, there can be no non-trivial extensions between such modules for generic central charge.

Therefore for generic β each singular vector of π_0, except for v_0, generates an irreducible Verma module of positive integral highest weight under the free action of the generators $L_n, n < 0$. The vector v_0 generates the irreducible vacuum Verma module under the free action of the generators $L_n, n < -1$. As a module over the Virasoro algebra, π_0 is a direct sum of the vacuum Verma module generated from v_0 and some Verma modules of positive integral highest weights. Since $\mathcal{W}_\beta(\mathfrak{g})$ is a submodule of π_0, it is also a direct sum of such modules. In particular, we see that the character of the space of singular vectors in $\mathcal{W}_\beta(\mathfrak{g})$ is equal to

$$(4.6.7) \qquad (1-q) \prod_{2 \leq i \leq l, n_i > d_i} (1 - q^{n_i})^{-1} + q.$$

From this fact and the character formula (4.6.6) we can derive that there exist singular vectors $W_{-d_i-1}^{(i),\beta} v_0$ of degrees $d_i + 1$ in the 0th cohomology, which in the limit $\beta = 0$ can be chosen as polynomial generators $W_{-d_i-1}^{(i)}$ of $\mathcal{W}(\mathfrak{g})$ from Proposition 2.4.7.

This can be proved by induction. Suppose, we have proved this fact for $i < j$. Thus, we have constructed singular vectors $W_{-d_i-1}^{(i),\beta} v_0, i = 1, \ldots, j-1$, satisfying the conditions above. Let $\mathcal{W}(\mathfrak{g})'$ be the subspace of $\mathcal{W}(\mathfrak{g})$, which consists of all polynomials in $W_{n_i}^{(i)}, n_i < -d_i, i = 1, \ldots, j-1$. Consider the component $\mathcal{W}(\mathfrak{g})_{d_j+1}$ of $\mathcal{W}(\mathfrak{g})$ of degree d_j+1. In this component the subspace $\mathcal{W}(\mathfrak{g})'_{d_j+1} = \mathcal{W}(\mathfrak{g})' \cap \mathcal{W}(\mathfrak{g})_{d_j+1}$ has codimension 1. Now consider β^2 as a formal variable and the space $W_\beta(\mathfrak{g})$ as a free module over the ring $\mathbb{C}[[\beta^2]]$ as in § 4.3.1. Denote by $S_{d_j+1}^\beta$ the space of singular vectors of $\mathcal{W}_\beta(\mathfrak{g})$ of degree $d_j + 1$. We have a natural projection $\mathcal{W}_\beta(\mathfrak{g}) \to \mathcal{W}_\beta(\mathfrak{g})/\beta^2 \cdot \mathcal{W}_\beta(\mathfrak{g}) \simeq \mathcal{W}(\mathfrak{g})$, the classical limit. We will show that the image of S_{j+1}^β in $\mathcal{W}(\mathfrak{g})_{d_j+1}$ is not contained in $\mathcal{W}(\mathfrak{g})'_{d_j+1}$.

Indeed, note that the action of the operators $L_n, n \geq -1$, on π_0 is well-defined in the limit $\beta = 0$, and denote the corresponding operators by $L_n^{(0)}$. These are derivations of π_0, which generate a Lie subalgebra of the Virasoro algebra. The action of the operator L_{-1} does not depend on β and coincides with the action of the derivative ∂. From our inductive assumption we already know the commutation relations between L_n and $W_{n_i}^{(i),\beta}, i = 1, \ldots, j-1$. They are given by formula (4.6.9) below. These relations give us in the limit $\beta = 0$:

$$L_n^{(0)} \cdot W_{n_i}^{(i)} = (nd_i - n_i)W_{n+n_i}^{(i)},$$

where we put $W_m^{(i)} = 0$, if $m \geq -d_i$. From these formulas it is clear that the polynomial algebra $A^{(j)}$, generated by $W_{n_i}^{(i)}, i = 2, \ldots, j-1$, is preserved by the action of $L_n^{(0)}, n \geq -1$. But the operator $L_{-1}^{(0)} = \partial$ acts freely on $\pi_0/\mathbb{C}$ and hence on $A^{(j)}/\mathbb{C}$. If X is a singular

406

vector in $\mathcal{W}(\mathfrak{g})'$, i.e. if it is annihilated by the operators $L_n^{(0)}, n > 0$, it should lie in $A^{(j)}$. But if X is a singular vector, then $L_{-1}^{(0)} X$ can not be a singular vector. Therefore the character of the space of singular vectors, contained in $\mathcal{W}(\mathfrak{g})'$, is less then or equal to the character of the quotient of $A^{(j)}$ by the total derivatives, which is equal to

$$(4.6.8) \qquad (1-q) \prod_{2 \le i \le j-1, n_i > d_i} (1-q^{n_i})^{-1} + q$$

(in fact, it can be shown that it is equal to (4.6.8)).

The image of $S_{j+1}^\beta \subset \mathcal{W}_\beta(\mathfrak{g})_{d_j+1}$ in $\mathcal{W}(\mathfrak{g})_{d_j+1}$ lies in the space of singular vectors of $\mathcal{W}(\mathfrak{g})_{d_j+1}$. Formulas (4.6.7) and (4.6.8) show that the dimension of the space S_{j+1}^β is greater than the dimension of the space of singular vectors of $\mathcal{W}(\mathfrak{g})'_{d_j+1}$. Hence there should exist a singular vector in $\mathcal{W}_\beta(\mathfrak{g})_{d_j+1}$, whose image in $\mathcal{W}(\mathfrak{g})_{d_j+1}$ is linearly independent from the subspace $\mathcal{W}(\mathfrak{g})'_{d_j+1}$. Denote such a vector by $W_{-d_j-1}^{(j),\beta} v_0$. By Proposition 2.4.7, its image in $\mathcal{W}(\mathfrak{g})$ is algebraically independent from the previously constructed $W_{-d_i-1}^{(i)}, i < j$, and hence it can be chosen as a generator $W_{-d_j-1}^{(j)}$ of $\mathcal{W}(\mathfrak{g})$. This completes our inductive argument.

The Fourier components $W_{n_i}^{(i),\beta}, n_i < -d_i$, of the fields corresponding to the singular vectors $W_{-d_i-1}^{(i),\beta} v_0$ act on π_λ and commute with the differential δ_β^1. Therefore they act on $\mathcal{W}_\beta(\mathfrak{g})$. In the limit $\beta = 0$ their action coincides with the action by multiplication by the polynomials $W_{n_i}^{(i)}$.

The polynomials $W_{n_i}^{(i)}, n_i < -d_i$, were shown in Proposition 2.4.7 to be algebraically independent. Therefore monomials

$$W_{n_{i_1}}^{(i_1)} \dots W_{n_{i_m}}^{(i_m)} \in \mathcal{W}(\mathfrak{g}),$$

where $i_1 \le \dots \le i_m$ and $n_{i_j} < n_{i_{j+1}}$ for $i_j = i_{j+1}$ are linearly independent. But these monomials are images (classical limits) of monomial elements

$$W_{n_{i_1}}^{(i_1,\beta)} \dots W_{n_{i_m}}^{(i_m,\beta)} v_0 \in \mathcal{W}_\beta(\mathfrak{g})$$

ordered so that $i_1 \le \dots \le i_m$ and $n_{i_j} < n_{i_{j+1}}$ for $i_j = i_{j+1}$. Therefore the latters are linearly independent in $\mathcal{W}_\beta(\mathfrak{g})$.

Hence such monomials linearly span a subspace in the $\mathcal{W}_\beta(\mathfrak{g})$, whose character is given by (4.6.6). But we know that this is the character of $\mathcal{W}_\beta(\mathfrak{g})$. Hence these monomials form a basis of $\mathcal{W}_\beta(\mathfrak{g})$. By analogy with the case of universal enveloping algebras, where one can choose the Poincare-Birkhoff-Witt basis, one can say that the operators $W_{n_i}^{(i),\beta}, 1 \le i \le l, n_i < -d_i$, *freely generate* $\mathcal{W}_\beta(\mathfrak{g})$ from v_0.

4.6.10. *Remark.* A vector in a VOA, which is a singular vector with respect to the action of the Virasoro algebra, gives rise to a field, which is called a *primary field* of conformal dimension equal to the degree of this vector.

Suppose, A is such a vector, of degree Δ. Then we have the following OPE:

$$Y(T,z)Y(A,w) = \frac{\Delta Y(A,z)}{(z-w)^2} + \frac{\partial_z Y(A,z)}{z-w} + \text{regular terms}.$$

This gives us the following formula for the commutation relations between the generators $L_n, n \in \mathbb{Z}$, of the Virasoro algebra, and the Fourier components, A_m, of the field $Y(A,w) = \sum_{m \in \mathbb{Z}} A_m z^{-m-\Delta}$:

$$[L_n, A_m] = (n(\Delta - 1) - m)A_{n+m}.$$

These commutation relations show that the Fourier components of a primary field of conformal dimension Δ behave with respect to the generators $L_n = t^{-n+1}\partial_t$ as the $(1 - \Delta)$-differentials on the circle $t^{-m+\Delta-1}dt^{1-\Delta}$.

Theorem 4.6.9 shows that the VOA $\mathcal{W}_\beta(\mathfrak{g})$ is "generated" by l fields: $\mathbf{W}_i^\beta(z), i = 1, \ldots, l$. The first of them, $\mathbf{W}_1^\beta(z)$, is the Virasoro field, and the others are primary fields of conformal dimensions equal to the exponents of $\mathfrak{g}$ plus 1 with respect to this Virasoro field for generic β. This means that we have the following commutation relations for generic β:

$$(4.6.9) \qquad [L_n, W_m^{(i),\beta}] = (nd_i - m)W_{n+m}^{(i),\beta}.$$

The fields $\mathbf{W}_i^\beta(z), i = 1, \ldots, l$, generate the VOA $\mathcal{W}_\beta(\mathfrak{g})$ in the sense that the OPE of any two of these fields can be expressed through normally ordered expressions of the same fields and their derivatives. This is equivalent to the property that $\mathcal{W}_\beta(\mathfrak{g})$ is freely generated from v_0 under the action of the non-negative Fourier components of the fields $\mathbf{W}_i^\beta(z)$.

The vertex operator algebras $\mathcal{W}_\beta(\mathfrak{g})$ were constructed by Fateev and Zamolodchikov in the case of $\mathfrak{g} = \mathfrak{sl}_3$ [43], and by Fateev and Lukyanov in the cases $\mathfrak{g} = \mathfrak{sl}_n$ [40] and $\mathfrak{so}_{2n}$ [41]. They found explicit expressions of $\mathbf{W}_i^\beta(z)$ through the free fields and then verified that the OPE closes. It was conjectured that such vertex operator algebras exist for arbitrary finite-dimensional simple Lie algebras. In Theorem 4.6.9 we prove this conjecture (this was first announced in [57]). A similar construction for $\mathfrak{g} = \mathfrak{sl}_n$ was proposed in [99].

One can also define $\mathcal{W}$-algebra $\mathcal{W}_\beta(\mathfrak{g})$ through the *quantum Drinfeld-Sokolov reduction* of the affine algebra $\widehat{\mathfrak{g}}$ of level k [45, 48, 57]. In this setting, $\mathcal{W}_\beta(\mathfrak{g})$ with $\beta = -(k+h^\vee)^{-1/2}$ is the 0th cohomology of the corresponding BRST complex. We have shown in [48], §4, and [57], §3 that the complex $F_\beta^\bullet(\mathfrak{g})$ appears as the first term of a spectral sequence of this BRST complex for generic k.

One can also use the opposite spectral sequence of the BRST complex to prove the existence of $\mathcal{W}$-algebras, cf. [25].

4.6.11. Theorem. *The Lie algebra $I_\beta(\mathfrak{g})$ of local integrals of motion of the quantum Toda field theory, associated to $\mathfrak{g}$, is isomorphic to the Lie algebra of residues of fields from the $\mathcal{W}$-algebra $\mathcal{W}_\beta(\mathfrak{g})$.*

Proof. As explained in Lemma 4.6.6, the differentials of the complex $F_\beta^\bullet(\mathfrak{g})$ are integrals over cycles. Therefore, the differentials of the complex $F_\beta^\bullet(\mathfrak{g})$ commute with the derivative, and we can form the double complex

$$\mathbb{C} \longrightarrow F_\beta^\bullet(\mathfrak{g}) \longrightarrow F_\beta^\bullet(\mathfrak{g}) \longrightarrow \mathbb{C}$$

with $\pm\partial$ as the vertical differentials. By definition, the space $I_\beta(\mathfrak{g})$ coincides with the 1st cohomology of this double complex. The Theorem now follows from Theorem 4.6.9 and the analogue of the exact sequence (4.3.3) for the VOA $\mathcal{W}_\beta(\mathfrak{g})$. Thus, the Lie algebra $I_\beta(\mathfrak{g})$ is a quantum deformation of the Poisson algebra $I_0(\mathfrak{g})$ in the same sense as before: the Lie brackets in $I_\beta(\mathfrak{g})$ in the β^2-expansion have no constant term, and the linear term coincides with the Poisson bracket in $I_0(\mathfrak{g})$.

In the same way we can show that the space of integrals of motion in the larger space $\widehat{\mathcal{F}}_0^\beta$ coincides with the Lie algebra of all Fourier components of fields from $\mathcal{W}_\beta(\mathfrak{g})$.

4.7. Affine Toda field theories. In this subsection we will extend the methods of the previous subsection to the quantum affine Toda field theories.

4.7.1. Our task is again to construct the deformed complex $F_\beta^*(\mathfrak{g})$, which becomes $F^*(\mathfrak{g})$ in the limit $\beta \to 0$, for an affine Lie algebra $\mathfrak{g}$.

According to § 4.5.6 the quantum BGG resolution $B_*^q(\mathfrak{g})$ exists for the quantized affine algebra $U_q(\mathfrak{g})$. Lemma 4.6.5 and Lemma 4.6.6 also hold in the affine case. Thus, we can define the complex $F_\beta^*(\mathfrak{g})$ in the same way as in the case of finite-dimensional simple Lie algebras.

As a vector space, the jth group of the complex $F_\beta^j(\mathfrak{g})$ is the direct sum of the modules $\pi_{\rho-s(\rho)}$, where s runs over the set of elements of the Weyl group of $\mathfrak{g}$ of length j. The differential $\delta_\beta^j : F_\beta^{j-1}(\mathfrak{g}) \to F_\beta^j(\mathfrak{g})$ is given by formula (4.6.5). Note that this complex is $\mathbb{Z}$–graded with finite-dimensional homogeneous components and the differentials are homogeneous of degree 0.

After proper rescaling of the differentials, we obtain a family of complexes, defined for generic $\beta \in \mathbb{C}$, such that for $\beta = 0$ we obtain our classical complex $F^*(\mathfrak{g})$. Let us restrict ourselves with the affine algebras, whose exponents are odd and the Coxeter number is even. It turns out that using the result of Proposition 3.2.5, in which the cohomologies of the complex $F^*(\mathfrak{g})$ were computed, and the fact that the Euler characteristics of the cohomologies does not depend on β, we can prove that the cohomologies of the complex $F_\beta^*(\mathfrak{g})$ for generic β are the same as for $\beta = 0$.

4.7.2. *Proposition. For generic β the cohomologies of the complex $F_\beta^*(\mathfrak{g})$ are isomorphic to the exterior algebra $\bigwedge^*(\mathfrak{a}^*)$ of the dual space to the principal commutative subalgebra $\mathfrak{a}$ of $\mathfrak{n}_+$.*

Proof. Let $F_\beta^j(\mathfrak{g})_m$ and $H_\beta^j(\mathfrak{g})_m$ be the mth homogeneous components of the jth group of the complex $F_\beta^*(\mathfrak{g})$ and of its jth cohomology group, respectively. The Euler characteristics of the mth homogeneous component of the complex $F_\beta^*(\mathfrak{g})$,

$$\sum_{j \geq 0}(-1)^j \dim F_\beta^j(\mathfrak{g})_m = \sum_{j \geq 0}(-1)^j \dim H_\beta^j(\mathfrak{g})_m,$$

does not depend on β.

From Proposition 3.2.5 we know that the Euler character of the complex $F^*(\mathfrak{g})$ is equal to that of $\bigwedge^*(\mathfrak{a}^*)$:

$$\prod_{n \equiv d_i \bmod h} (1 - q^n),$$

where $d_1, \ldots, d_l$ are the exponents of $\mathfrak{g}$ and h is the Coxeter number. In fact, one can compute the Euler character of our complex by a different method. By definition, it is equal to

$$\sum_{j \geq 0}(-1)^j \mathrm{ch} F_\beta^j(\mathfrak{g}) = \prod_{n > 0}(1 - q^n)^{-l} \sum_s (-1)^{l(s)} q^{\langle \rho - s(\rho) | \rho^\vee \rangle}.$$

Using the Weyl-Kac character formula for the trivial representation of $\mathfrak{g}$ in the principal gradation [80], we can reduce it to the product formula above.

We restrict ourselves with the case when all the exponents of $\mathfrak{g}$ are odd and the Coxeter number is even (the general proof is technically more complicated and it will be published separately). Then the subcomplex $F^*(\mathfrak{g})_m$ has cohomologies of only even degrees, if m is even, and of only odd degrees, if m is odd. It means, according to Lemma 4.4.2, that the same property holds for the subcomplex $F_\beta^*(\mathfrak{g})_m$ for generic values of β.

But then we have:

$$\sum_{j \geq 0} \dim H_\beta^{2j}(\mathfrak{g})_m = \sum_{j \geq 0} \dim H_0^{2j}(\mathfrak{g})_m$$

for even m and

$$\sum_{j\geq 0} \dim H_\beta^{2j-1}(\mathfrak{g})_m = \sum_{j\geq 0} \dim H_\beta^{2j-1}(\mathfrak{g})_m$$

for odd m. If the value $\beta = 0$ of the parameter were not generic, then for generic β we would have $\dim H_\beta^l(\mathfrak{g})_m \geq \dim H_0^l(\mathfrak{g})_m$ for all l, and there would exist such i that $\dim H_\beta^i(\mathfrak{g})_m > \dim H_0^i(\mathfrak{g})_m$. But this would contradict the equalities above. Therefore for any m and j

$$\dim H_\beta^j(\mathfrak{g})_m = \dim H_0^j(\mathfrak{g})_m,$$

and Proposition follows.

4.7.3. Theorem. *All local integrals of motion of the classical affine Toda field theory can be deformed, and so the space $I_\beta(\mathfrak{g})$ of quantum integrals of motion is linearly generated by mutually commuting elements of degrees equal to the exponents of $\mathfrak{g}$ modulo the Coxeter number.*

Proof. The same as in Theorem 3.2.5, in particular, since the Lie bracket preserves the grading, from the fact that they all have odd degrees follows that they commute with each other.

4.8. Concluding remarks.

4.8.1. The duality $\beta \to -(r^\vee)^{\frac{1}{2}}/\beta$ and the limit $\beta \to \infty$. There is a remarkable duality in $\mathcal{W}$-algebras [48, 56, 49]. Let $\mathfrak{g}$ be a simple Lie algebra and $\mathfrak{g}^L$ be the Langlands dual Lie algebra, whose Cartan matrix is the transpose of the Cartan matrix of $\mathfrak{g}$. Let $r^\vee$ be the maximal number of edges, connecting two vertices of the Dynkin diagram of $\mathfrak{g}$.

For generic values of β the vertex operator algebra $\mathcal{W}_\beta(\mathfrak{g})$ is isomorphic to the vertex operator algebra $\mathcal{W}_{\beta^L}(\mathfrak{g}^L)$, where $\beta^L = -(r^\vee)^{\frac{1}{2}}/\beta$.

Accordingly, under these conditions $I_\beta(\mathfrak{g}) \simeq I_{\beta^L}(\mathfrak{g}^L)$.

Clearly, $\mathfrak{g} \simeq \mathfrak{g}^L$, unless $\mathfrak{g}$ is of types B_n or C_n, in which case they are dual to each other. The duality means then that there is only one family of $\mathcal{W}$-algebras associated to the Lie algebras B_n and C_n.

The proof of this duality [48, 56] is based on the explicit computation in the rank one case, which follows from the proof of Proposition 4.4.4. Indeed, it is clear that the Virasoro element in $\mathcal{W}_\beta(\mathfrak{sl}_2)$ given by formula (4.4.2) is invariant under the transformation $\beta \to -2/\beta$. Therefore for generic β we have the isomorphism $\mathcal{W}_\beta(\mathfrak{sl}_2) \simeq \mathcal{W}_{-2/\beta}(\mathfrak{sl}_2)$.

General case can be reduced to the case of $\mathfrak{sl}_2$. Let $\pi_0^{(i)}$ be the subspace in π_0, which is generated from v_0 by the operators $b_n^{j*}, n < 0, j \neq i$. These operators commute with $b_m^i, m \in \mathbb{Z}$, and hence with $\tilde{Q}_i^\beta = \int V_{\beta\alpha_i}(z)dz$. Therefore the kernel of the operator $\tilde{Q}_i^\beta$ on π_0 coincides with the tensor product of $\pi_0^{(i)}$ and the kernel of the operator $\tilde{Q}_i^\beta$ on the subspace of π_0, generated from v_0 by the operators $b_n^i, n < 0$. But the latter is isomorphic to $\mathcal{W}_{\beta\|\alpha_i\|}(\mathfrak{sl}_2)$, and hence does not change, if we replace β by $-2/(\beta\|\alpha_i\|^2)$.

Thus we see that the kernel of the operator $\tilde{Q}_i^\beta = \int V_{\beta\alpha_i}(z)dz$ coincides with the kernel of the operator $\int V_{-\alpha_i^\vee/\beta}(z)dz$, where $\alpha_i^\vee = 2\alpha_i/(\alpha_i, \alpha_i)$. It is known [81] that the scalar product $(\alpha_i^\vee, \alpha_j^\vee)$ in $\mathfrak{h}^* \subset \mathfrak{g}$ equals $r^\vee$ times the scalar product (α_i^L, α_j^L) of the simple roots $\alpha_i^L \in \mathfrak{h}^L \subset \mathfrak{g}^L$ of the Langlands dual Lie algebra $\mathfrak{g}^L$. We can therefore identify the Heisenberg algebras $\hat{\mathfrak{h}}$ and $\hat{\mathfrak{h}}^L$ by identifying $\alpha_i^\vee$ with $(r^\vee)^{\frac{1}{2}}\alpha_i^L$. But then the operator $\int V_{-\alpha_i^\vee/\beta}(z)dz$ becomes the operator $\tilde{Q}_i^{\beta^L}$, where $\beta^L = -(r^\vee)^{\frac{1}{2}}/\beta$. Therefore for generic values of β the kernel of the operator $\tilde{Q}_i^\beta$ coincides with the kernel of the operator $\tilde{Q}_i^{\beta^L}$ of the Langlands dual Lie algebra. Thus, $\mathcal{W}_\beta(\mathfrak{g}) \simeq \mathcal{W}_{\beta^L}(\mathfrak{g}^L)$.

This duality has a remarkable limit when $\beta \to \infty$. Of course, in this case the operator $\tilde{Q}_i^\beta$ is not well-defined and has to be regularized.

It suffices describe this regularization in the rank one case. To this end, consider the Heisenberg algebra with generators $b_n, n \in \mathbb{Z}$ with the commutation relations

$$[b_n, b_m] = n\delta_{n,-m}.$$

We will introduce bases in the Fock spaces π_0 and π_β in such a way that the matrix elements of the operator $\tilde{Q}^\beta = \int V_\beta(z)dz : \pi_0 \to \pi_\beta$ are well-defined when $\beta \to \infty$.

As the basis elements in π_0 we will take monomials in $b'_n = b_n/\beta, i < 0$. Denote $W_N = \int V_\beta z^{N-1}dz, N \in \mathbb{Z}$. The basis in π_β consists of elements

(4.8.1) $\qquad b'_{n_1} \ldots b'_{n_m} w_N, \qquad n_1 \leq n_2 \leq \ldots \leq n_m < -1, \quad N \leq 0,$

where for finite β we put $w_N = W_N \cdot v_0$. Here v_0 is the vacuum vector of π_0. In particular, w_0 is the vacuum vector v_1 of π_β.

In order to find out how $\tilde{Q}^\beta$ acts on a monomial basis element of π_0, we can use the commutation relations

$$[W_N, b'_n] = -W_{N+n},$$

which follow from formula (4.2.6). So, when we apply $\tilde{Q}^\beta = W_1$ to a monomial in b'_n's, we obtain a linear combination of terms of the form (4.8.1), but with $n_m \leq -1$. However, we can re-express elements $P \cdot (b'_{-1})^k w_N$, where P is a polynomial in $b'_n, n < -1$, in terms of elements of the form (4.8.1), using the identity

(4.8.2) $\qquad \displaystyle\sum_{n+M=N, n<0, M\leq 0} b'_n w_M = -\frac{1}{\beta^2} N w_N, \quad N < 0.$

This identity can be obtained by applying to v_0 the negative Fourier components of the formula

$$: b(z)V_\beta(z) := \frac{1}{\beta}\frac{\partial}{\partial z}V_\beta(z).$$

But then see that the matrix elements of the operator $\tilde{Q}^\beta$ in this new basis are polynomials in β^{-2}, and therefore they define a certain linear operator, when $\beta^{-2} = 0$. We will denote this operator by $\tilde{Q}^\infty$.

We can show that the kernel of the operator $\tilde{Q}^\infty$ coincides with the $\beta \to \infty$ limit of the kernel of the operator $\tilde{Q}^\beta$ for generic β [56]. We can then check that this kernel is isomorphic to the $\beta \to 0$ limit of the kernel of the operator $\tilde{Q}^\beta$ for generic β, which is described in Proposition 2.1.15.

By extending this result to higher rank case in the same way as for generic β, we can prove that $\mathcal{W}_\infty(\mathfrak{g}) \simeq \mathcal{W}_0(\mathfrak{g}^L)$, where $\mathcal{W}_\infty(\mathfrak{g})$ denotes the intersection of kernels of the operators $\tilde{Q}_i^\infty$ associated to $\mathfrak{g}$ [48, 56, 49].

The quotient $I_\infty(\mathfrak{g})$ of $\mathcal{W}_\infty(\mathfrak{g})$ by total derivatives and constants has a Poisson bracket, which is equal to the β^{-2}-linear term in the commutator in $I_\beta(\mathfrak{g})$. This isomorphism implies that $I_\infty(\mathfrak{g}) \simeq I_0(\mathfrak{g}^L)$.

In [48, 56] we showed that $I_\infty(\mathfrak{g})$ is isomorphic to the center $Z(\hat{\mathfrak{g}})$ of a certain completion of the universal enveloping algebra of the affine algebra $\hat{\mathfrak{g}}$ at the critical level. Thus, we see that the center $Z(\hat{\mathfrak{g}})$ is isomorphic to the classical $\mathcal{W}$-algebra of the Langlands dual Lie algebra $\mathfrak{g}^L$.

In the same way we can prove the isomorphism $I_\infty(\mathfrak{g}) \simeq I_0(\mathfrak{g}^L)$, where $\mathfrak{g}$ is an affine Kac-Moody algebra and $\mathfrak{g}^L$ is the affine algebra, whose Cartan matrix is obtained by transposing the Cartan matrix of $\mathfrak{g}$.

4.8.2. *Explicit formulas.* It is an interesting problem to find explicit formulas for the quantum integrals of motion of the affine Toda field theories.

Explicit formulas for the classical ones are known in many cases. For instance, there are many effective methods to compute explicitly the KdV hamiltonians, which are the local integrals of motion of the sine-Gordon model. However, it seems none of those methods can be used to produce the quantum integrals, cf. e.g. [85]. So far, only partial results have been obtained in this direction.

First of all, a few quantum integrals of motion of the sine-Gordon theory of low degrees are known for any value of the central charge of the Virasoro algebra, cf. e.g. [108]. When the central charge is equal to $1 - 3(2n - 1)^2/(2n + 1)$ (the $(2, 2n + 1)$ minimal model of the Virasoro algebra), it is known that the quantum integral of motion of degree $2n - 1$ can be obtained as the residue of the field, corresponding to the singular vector of degree $2n$ in the vacuum Verma module of the Virasoro algebra [65, 36, 29]. An explicit formula is known for this singular vector, and this allows one to write down the corresponding integral of motion for this value of the deformation parameter. A similar phenomenon has been observed in other theories [29].

Finally, the quantum integrals of motion are known for the central charge $c = -2$, which corresponds to $\beta = 2$. The reason for that is that in this case the operators $\tilde{Q}_1$ and $\tilde{Q}_0$ have a simple realization in terms of the Clifford algebra with the generators $\psi_i, \psi_i^*, i \in \mathbb{Z}$ and the anti-commutation relations

$$[\psi_i, \psi_j^*]_+ = \delta_{i,-j}.$$

Indeed, let $\Lambda^* = \oplus_{n \in \mathbb{Z}} \Lambda^n$ be the Fock representation of this algebra with the vacuum vector v, satisfying

$$\psi_i v = 0, i \geq 0, \quad \psi_i^* v = 0, i > 0.$$

This representation is $\mathbb{Z}$-graded in accordance with the convention $\deg \psi_i = 1, \deg \psi_i^* = -1$. One can introduce an action of this Clifford algebra on the space $\oplus_{n \in \mathbb{Z}} \pi_n$ using the vertex operators by the formulas

$$\psi(z) = \sum_{m \in \mathbb{Z}} \psi_m z^{-m-1} = V_1(z), \qquad \psi^*(z) = \sum_{m \in \mathbb{Z}} \psi_m^* z^{-m} = V_{-1}(z).$$

Since β is an integer, all Fourier components of these vertex operators are well-defined on any of the modules π_n. This boson-fermion correspondence allows us to identify our complex $F_2^*(\hat{\mathfrak{sl}}_2) \simeq \oplus_{n \in \mathbb{Z}} \pi_{2n}$ with the even part of Λ^*. The operators $\tilde{Q}_1$ and $\tilde{Q}_0$ then become

(4.8.3)

$$\tilde{Q}_1 = \int V_2(z) dz = \int \psi(z) \partial_z \psi(z) dz, \qquad \tilde{Q}_0 = \int V_{-2}(z) dz = \int \psi^*(z) \partial_z \psi^*(z) dz.$$

It is not difficult to prove directly that they satisfy the Serre relations with $q = 1$. Recall that for generic values of β only the compositions of the operators $\tilde{Q}_1^\beta$ and $\tilde{Q}_0^\beta$, corresponding to the singular vectors in the Verma modules over the quantum group, are well-defined as linear operators acting between the spaces π_n. The operators (4.8.3) are always well-defined and they generate an action of the nilpotent Lie subalgebra $\mathfrak{n}_+$ of $\hat{\mathfrak{sl}}_2$ on Λ^*. In fact, this action can be extended to an action of the whole Lie algebra $\hat{\mathfrak{sl}}_2$ [89].

It is possible to write down explicit formulas for the integrals of motion for $\beta = 2$ in terms of the fermions $\psi(z), \psi^*(z)$ [30]:

$$\mathcal{H}_{2n+1} = \int \psi(z)\partial_z^{2n-1}\psi^*(z)dz.$$

These formulas can be converted into nice formulas in terms of the generators of the Virasoro algebra, which first appeared in [108]. Other features of the case $\beta = 2$ have been studied in [89].

Similar, but more complicated is the case when $\beta = N$, a positive integer. Then the operators $\tilde{Q}_1$ and $\tilde{Q}_0$ can be written in terms of $\psi(z)$ and $\psi^*(z)$ as follows:

$$\tilde{Q}_1 = \int \psi(z)\partial_z\psi(z)\ldots\partial_z^{N-1}\psi(z)dz,$$

$$\tilde{Q}_0 = \int \psi^*(z)\partial_z\psi^*(z)\ldots\partial_z^{N-1}\psi^*(z)dz.$$

In these cases explicit formulas for the integrals of motion are still lacking. Finding such formulas for infinitely many integers N would lead to an independent proof of the existence of quantum integrals of motion.

4.8.3. *Special values of β for finite-dimensional $\mathfrak{g}$.* So far, we have only been interested in the generic values of the deformation parameter β. In this subsection we will discuss briefly what happens for special values of β, that is the values, for which the kernel of the operator $\sum_i \tilde{Q}_i^\beta$ on $\mathcal{F}_0^\beta$ becomes larger.

Let us first look at the case of finite-dimensional $\mathfrak{g}$. In the simplest case of $\mathfrak{g} = \mathfrak{sl}_2$ our complex is $\pi_0 \longrightarrow \pi_1$, and we have proved that for generic values of β the first cohomology of this complex is trivial. In fact, this statement can be proved directly, using the description of the structure of the modules π_0 and π_1 over the Virasoro algebra from [52]. According to this description, the module π_1 is irreducible for generic β, while the module π_0 contains an irreducible submodule, such that the quotient by this submodule is isomorphic to π_1. If β^2 is a positive rational number, however, the modules π_0 and π_1 may become highly reducible, and that leads to the appearance of the first cohomology and the enlargement of the 0th cohomology.

The most interesting situation occurs when $\beta^2 = 2p/q$, where $p, q > 1$ are two relatively prime integers. The corresponding central charge $c = 1 - 6(p-q)^2/pq$ is the central charge of the (p,q) minimal model [8]. In that case the composition structure of the modules π_0 and π_1 becomes very complicated [52], and our complex has very large cohomology groups. It turns out, however, that one can extend this complex to an infinite two-sided complex, whose cohomologies are concentrated in one dimension and are isomorphic to the irreducible representation of the Virasoro algebra of highest weight 0. This representation is the quotient of the vacuum Verma module of the Virasoro algebra by its submodule, generated by a unique singular vector, which it contains. Note that this irreducible representation is at the same time the VOA of the corresponding minimal model.

Such a complex was constructed by Felder [53]. It has one Fock space π_λ with an appropriate λ in each group, and the only cohomology occurs in the 0th group of the complex.

The situation with the $\mathcal{W}$−algebras, associated to general finite-dimensional Lie algebras, is apparently very similar. The vacuum Verma module $\mathcal{W}_\beta(\mathfrak{g})$, freely generated from the vacuum vector by the operators $W_{n_i}^{(i),\beta}, n_i < -d_i$ (cf. Theorem 4.6.9), which is irreducible for generic values of β, may contain singular vectors, if β^2 is a positive rational number. The quotient $L_\beta(\mathfrak{g})$ of $\mathcal{W}_\beta(\mathfrak{g})$ by the submodule, generated by these singular vectors, is irreducible. One should be able to construct two-sided complexes, which consist of the Fock spaces π_λ, labeled by elements of the *affine* Weyl group of $\mathfrak{g}$, in which the 0th cohomology would be isomorphic to $L_\beta(\mathfrak{g})$ and all other cohomologies would be trivial. Such complexes have been conjectured in [59], Conjecture 3.5.2. These complexes should appear [59] as the result of the quantum Drinfeld-Sokolov reduction of similar complexes (two-sided BGG resolutions) over the corresponding affine Kac-Moody algebra $\hat{\mathfrak{g}}$ [47, 9, 18]. This has been proved for $\mathfrak{g} = \mathfrak{sl}_2$ in [46, 12].

These complexes are closely connected with certain complexes [18] of modules over the quotient of the quantum group $U_q(\mathfrak{g})$ with q a root of unity, $q = \exp(2\pi i p/q)$, by a big central subalgebra, cf. [26].

There is also another interesting value of β, namely, $\beta = 1$ for a *simply-laced* Lie algebra $\mathfrak{g}$ [60, 6, 18, 58]. In this case $q = -1$, and slightly redefined operators $\bar{Q}_i^\beta$ generate the nilpotent Lie algebra $\mathfrak{n}_+$. This nilpotent subalgebra lies in the constant subalgebra of the whole affine algebra $\hat{\mathfrak{g}}$ acting on the direct sum of the Fock modules π_λ where the summation is over the root lattice Q of $\mathfrak{g}$, by vertex operators [61]. The $\mathcal{W}$-algebra $\mathcal{W}_1(\mathfrak{g})$ can then be interpreted as the space of invariants of the constant subalgebra of $\hat{\mathfrak{g}}$ in $\pi = \oplus_{\lambda \in Q} \pi_\lambda$. This implies that $I_1(\mathfrak{g})$ (for which the central charge is the rank of $\mathfrak{g}$) is the commutant of $\mathfrak{g}$ in $\hat{\mathcal{F}}_0^1$. A version of $I_1(\mathfrak{g})$ was defined for the first time by I. Frenkel in [60].

It was proved in [58], Theorem 4.2 (cf. also [16, 20]), that $\mathcal{W}_1(\mathfrak{g})$ has the same character as $\mathcal{W}_\beta(\mathfrak{g})$ for generic β, i.e. that the intersection of kernels of the operators $\bar{Q}_i^\beta$ does not increase at the point $\beta = 1$. Let us show that the higher cohomologies of the complex $F_1^*(\mathfrak{g})$ vanish. This has been conjectured (and proved for $\mathfrak{g} = \mathfrak{sl}_2$) in [60].

Consider the complex $\tilde{F}^*(\mathfrak{g})$, in which the jth group consists of $\#\{w|l(w) = j\}$ copies of π, and the differentials are given by the same formulas as the differentials of the complex $F_1^*(\mathfrak{g})$. The cohomology of the complex $\tilde{F}^*(\mathfrak{g})$ it is the cohomology of the Lie algebra $\mathfrak{n}_+$ with coefficients in π. The complex $F_1^*(\mathfrak{g})$ is a subcomplex of $\tilde{F}^*(\mathfrak{g})$. It is easy to show that its cohomology is the subspace of the cohomology of $\tilde{F}^*(\mathfrak{g})$ of weight 0 with respect to the Cartan subalgebra of $\mathfrak{g}$ acting on $\tilde{F}^*(\mathfrak{g})$ and commuting with the differentials. But π is a direct sum of finite-dimensional $\mathfrak{g}$-modules. By Borel-Weil-Bott-Kostant theorem, weight 0 cohomology classes in $\tilde{F}^*(\mathfrak{g})$ can occur only in dimension 0 and those are the invariants of $\mathfrak{g}$ in π. Therefore the higher cohomologies of $F_1^*(\mathfrak{g})$ vanish. This implies that the character of $\mathcal{W}_1(\mathfrak{g})$ coincides with the character of $\mathcal{W}_\beta(\mathfrak{g})$ for generic β, and thus we obtain an alternative proof of Theorem 4.2 of [58].

It was shown in [58] that $I_1(\mathfrak{sl}_N)$ is the quotient of the local completion of the universal enveloping algebra of the Lie algebra $\mathcal{W}_\infty$ with central charge $N - 1$.

4.8.4. *Special values of β for affine $\mathfrak{g}$.* Now let us turn to the space $I_\beta(\mathfrak{g})$ of the integrals of motion of the affine Toda field theory associated to an affine algebra $\mathfrak{g}$. The space of these integrals was defined as the intersection of kernels of the operators $\bar{Q}_i^\beta, i = 0, \ldots, l$, on $\mathcal{F}_0^\beta$. For generic values of β the intersection of kernels of the operators $\bar{Q}_i^\beta$ with $i = 1, \ldots, l$ coincides with the $\mathcal{W}$-algebra $I_\beta(\bar{\mathfrak{g}})$, and so the space $I_\beta(\mathfrak{g})$ can be defined as the kernel of the operator $\bar{Q}_0^\beta$ on $I_\beta(\bar{\mathfrak{g}})$.

Recall that the $\mathcal{W}$-algebra $I_\beta(\bar{\mathfrak{g}})$ is the quotient of the vacuum Verma module $\mathcal{W}_\beta(\bar{\mathfrak{g}})$ by the total derivatives and constants. If $\mathfrak{g}$ is untwisted, the operator $\bar{Q}_0^\beta$ can be interpreted as the residue $\int \Phi_{1,1,Adj}(z)dz$ of a certain primary field $\Phi_{1,1,Adj}(z)$, acting from $\mathcal{W}_\beta(\bar{\mathfrak{g}})$ to another module $M_\beta(\bar{\mathfrak{g}})$ over the $\mathcal{W}$-algebra [120, 42, 35, 77], so that the operator $\bar{Q}_0^\beta$ is the corresponding operator on the quotients by the total derivatives and constants. Therefore, for generic β the space $I_\beta(\mathfrak{g})$ consists of the elements $P^- \in \mathcal{W}_\beta(\bar{\mathfrak{g}})$, for which $\tilde{Q}_0^\beta \cdot P^-$ is a total derivative in $M_\beta(\bar{\mathfrak{g}})$:

$$(4.8.4) \qquad \int \Phi_{1,1,Adj}(z)dz \cdot P^- = \partial P^+.$$

This equation shows that the pair (P^-, P^+) can be interpreted as a conservation law (compare with Remark 3.2.8) in the deformation of the corresponding conformal field theory obtained by adding $\lambda \int \Phi_{1,1,Adj}(z)dz$, where λ is a parameter of deformation, to the action [120].

When β^2 is a positive rational number, the module $\mathcal{W}_\beta(\bar{\mathfrak{g}})$ may become reducible. Because of that, the cohomologies of the complex $F^*(\mathfrak{g})$ increase. In such a case it is appropriate to redefine integrals of motion as elements P^- of the *irreducible* module $L_\beta(\bar{\mathfrak{g}})$, which satisfy the equation (4.8.4) [120]. This may result in dropping out of some of the "generic" integrals of motion. At the same time some new ones may appear.

For instance, for $\mathfrak{g} = \widehat{\mathfrak{sl}}_2, \beta^2 = 4/(2n+1)$ (the $(2, 2n+1)$ model) the density of the integral of motion of the quantum sine-Gordon theory of degree $2n-1$ coincides with the field, corresponding to the singular vector of degree $2n$. Since we take the quotient by the submodule, generated by this vector, this integral of motion drops out (cf. the previous section). It has been argued that the integrals of motion of degrees, which are divisible by $2n-1$, also drop out in this case [65, 84, 36].

Another example of dropping out of integrals of motion is (in our terminology) the Toda field theory associated to the twisted algebra $A_2^{(2)}$ for the value $\beta^2 = 3/2$. The integrals of motion of this Toda theory for generic values of β have all positive integral degrees, which are not divisible by 2 and 3. They are elements of the Virasoro algebra $I_\beta(\mathfrak{sl}_2)$, because for $\mathfrak{g} = A_2^{(2)}$, $\bar{\mathfrak{g}} = \mathfrak{sl}_2$. They have the property (4.8.4) with the field $\Phi_{(1,3)}(z) = \Phi_{1,1,Adj}(z)$ replaced by $\Phi_{(1,2)}(z)$ [120]. The value $\beta^2 = 3/2$ corresponds to the Ising model $(3, 4)$ with central charge $c = 1/2$. It was found in [120, 44] that the integral of motion of degree 5 drops out for this special value of parameter. It was conjectured that the degrees of the integrals of motion which should occur are relatively prime with 30, so that they are the exponents of $E_8^{(1)}$ modulo the Coxeter number.

On the other hand, in the same theory with $\beta^2 = 8/5$ (the Ising tri-critical point, the $(4, 5)$ minimal model) the appearance of an integral of motion of degree 9 was observed [44, 24] and it was conjectured that there should also be integrals of motion of degrees $9n$, where n is an arbitrary positive odd integer.

In these examples, the dropping out or appearance of new integrals of motion is caused by the existence of a larger vertex operator algebra of symmetries of the model. For instance, it is known that the $(2, 2n+1)$ minimal model of the Virasoro algebra coincides with $(2n-1, 2n+1)$ minimal model of the $\mathcal{W}$-algebra $\mathcal{W}(\mathfrak{sl}_{2n-1})$ [84], and that the Ising model has a hidden symmetry of $\mathcal{W}(E_8)$ [21]. Therefore one should expect that the degrees of integrals of motion in such a model should satisfy "exclusion rules" of the larger symmetry algebra as well. It is interesting whether there are other reasons for dropping out or appearance of new integrals of motion.

It seems plausible that for the special values of β one can construct a two-sided complex, consisting of the modules π_λ, whose first cohomology would give the space of integrals of motion, corresponding to the irreducible representation $L_\beta(\bar{\mathfrak{g}})$. We have constructed a candidate for such a complex for the $(2, 2n + 1)$ model. The computation of the Euler character of this complex suggests that its first cohomology is indeed generated by elements of all odd degrees, which are not divisible by $2n - 1$. We will discuss this complex elsewhere.

4.8.5. Spectrum of the integrals of motion. Our integrals of motion, both classical and quantum, act on the spaces π_λ. They are not diagonalizable, since they are all of negative degrees (in particular, the first of them is the operator of derivative ∂). However, one can define a transformation on the space $\hat{\mathcal{F}}_0^\beta$ of Fourier components of fields [98], which maps the set of integrals of motion to a set of mutually commuting elements of degree 0 (for example, the first integral of motion, $\mathcal{H}_1 = \partial = L_{-1}$ maps to $L_0 - c/24$), cf. [101].
It would be very interesting to find the spectrum of these operators on the modules π_λ.

Acknowledgments. E.F. would like to thank the organizers of the C.I.M.E. Summer School, M. Francaviglia and S. Greco, for giving him the opportunity to present these lectures, and for creating a very stimulating atmosphere at the School.
The main part of this work was done during our visits to the Research Institute for Mathematical Sciences and Yukawa Institute of Kyoto University in 1991-1993. We express our deep gratitude to these institutions, and especially to T. Inami, M. Kashiwara, and T. Miwa, for financial support and generous hospitality.
We thank J. Bernstein, V. Kac, D. Kazhdan, F. Smirnov, and A. Varchenko for their interest in this work and valuable comments.
E.F. was supported by a Junior Fellowship from the Society of Fellows of Harvard University and in part by NSF grant DMS-9205303.

REFERENCES

1. M. Ablowitz, D. Kaup, A. Newell, H. Segur, Stud. Appl. Math. 53 (1974) 249-315
2. M. Adler, Inv. Math. 50 (1979) 219-248
3. O. Babelon, D. Bernard, Comm. Math. Phys. 149 (1992) 279-306
4. O. Babelon, D. Bernard, Int. J. Math. Phys. A8 (1993) 507-543
5. O. Babelon, L. Bonora, Phys. Lett. B244 (1990) 220
6. F.A. Bais, P. Bouwknegt, K. Schoutens and M. Surridge, Nucl. Phys. B304 (1988) 348-370
7. J. Balog, L. Feher, L. O'Raifeartaigh, P. Forgacs, A. Wipf, Ann. Phys. (NY) 203 (1990) 76
8. A. Belavin, A. Polyakov, A. Zamolodchikov, Nucl. Phys. B241 (1984) 333-380
9. D. Bernard, G. Felder, Comm. Math. Phys. 127 (1990) 145
10. J.N. Bernstein, I.M. Gelfand, S.I. Gelfand, Funct. Anal. Appl. 5 (1971) 1-8
11. J.N. Bernstein, I.M. Gelfand, S.I. Gelfand, in *Representations of Lie groups*, ed. I.Gelfand, 21-64, Wiley, New York 1975
12. M. Bershadsky, H. Ooguri, Comm. Math. Phys. 126 (1989) 49
13. A. Bilal, J.-L. Gervais, Phys. Lett. 206B (1988) 412; Nucl. Phys. B314 (1989) 646; B318 (1989) 579
14. R. Borcherds, Proc. Natl. Acad. Sci. USA, 83 (1986) 3068-3071
15. R. Bott, L.W. Tu, *Differential forms in algebraic topology*, Springer, New York 1982
16. P. Bouwknegt, in *Infinite-dimensional Lie algebras and groups*, ed. V. Kac, Adv. Ser. in Math. Phys. 7, 527-555, World Scientific, Singapore 1989
17. P. Bouwknegt, J. McCarthy, K. Pilch, Comm. Math. Phys. 131 (1990) 125-155
18. P. Bouwknegt, J. McCarthy, K. Pilch, Progr. Theor. Phys. Suppl. 102 (1990) 67

19. P. Bouwknegt, J. McCarthy, K. Pilch, in *Strings and Symmetries*, eds. N. Berkovits, e.a., 407-422. World Scientific, 1992
20. P. Bouwknegt, K. Schoutens, Phys. Reports 223 (1993) 183-276
21. P. Bowcock, P. Goddard, Nucl. Phys. B305 (1988) 685
22. P. Bowcock, G.M.T. Watts, Nucl. Phys. B379 (1992) 63-95
23. H.W. Braden, E. Corrigan, P.E. Dorey, R. Sasaki, Nucl. Phys. B338 (1990) 698
24. P. Christe, G. Mussardo, Nucl. Phys. B330 (1990) 465-487
25. J. de Boer, T. Tjin, Preprint THU-93/01, ITFA-02-93
26. C. De Concini, V. Kac, in *Operator algebras, unitary representations, enveloping algebras, and invariant theory*, eds. A.Connes e.a., Progress in Math. 92, 471-506, Birkhäuser 1990
27. L.A. Dickey, *Soliton Equations and Hamiltonian Systems*, Adv. Ser. in Math. Phys. 12, World Scientific, Singapore, 1990
28. P. Di Francesco, C. Itzykson, J.-B. Zuber, Comm. Math. Phys. 140 (1991) 543-567
29. P. Di Francesco, P. Mathieu, Phys. Lett. B278 (1992) 79-84
30. P. Di Francesco, P. Mathieu, D. Sénéchal, Mod. Phys. Lett. A7 (1992) 701
31. V.G. Drinfeld, Sov. Math. Dokl. 32 (1985) 254
32. V.G. Drinfeld, V.V. Sokolov, Sov. Math. Dokl. 23 (1981) 457-462
33. V.G. Drinfeld, V.V. Sokolov, J. Sov. Math. 30 (1985) 1975-2035
34. V.G. Drinfeld, Algebra i Analiz (Leningrad Math. J.) 1, N2 (1989) 30-46
35. T. Eguchi, S.-K. Yang, Phys. Lett. 224B (1989) 373
36. T. Eguchi, S.-K. Yang, Phys. Lett. 235B (1989) 282
37. L.D. Faddeev, L.A. Takhtajan, *Hamiltonian Methods in the Theory of Solitons*, Springer, 1987
38. L.D. Faddeev, L.A. Takhtajan, in Lect. Notes in Phys. 246, 66. Springer, 1986
39. L.D. Faddeev, A. Volkov, Phys. Lett. 315B (1993) 311-318
40. V. Fateev, S. Lukyanov, Int. J. of Mod. Phys. A3 (1988) 507
41. V. Fateev, S. Lukyanov, Sov. J. Nucl. Phys. 49 (1989) 925
42. V. Fateev, S. Lukyanov, Kiev Preprints ITF-88-74R, ITF-88-75R, ITF-88-76R, 1988 (English translation: Sov. Sci. Rev. A Phys. 15 (1990) 1)
43. V. Fateev, A. Zamolodchikov, Nucl. Phys. B280 (1987) 644
44. V. Fateev, A. Zamolodchikov, in *Physics and Mathematics of Strings*, Memorial Volume for Vadim Knizhnik, eds. L. Brink, D. Friedan, A.M. Polyakov, 245-270, World Scientific, 1990
45. B. Feigin, E. Frenkel, Phys. Lett. B246 (1990) 75-81
46. B. Feigin, E. Frenkel, Lett. Math. Phys. 19 (1990) 307-317
47. B. Feigin, E. Frenkel, Comm. Math. Phys. 128 (1990) 161-189
48. B. Feigin, E. Frenkel, Int. J. Mod. Phys. A7, Supplement 1A (1992) 197-215
49. B. Feigin, E. Frenkel, Duke Math. Journal 64, IMRN 6 (1991) 75-82
50. B. Feigin, E. Frenkel, Phys. Lett. B276 (1992) 79-86
51. B. Feigin, E. Frenkel, *Kac-Moody groups and integrability of soliton equations*, Preprint RIMS-970, hep-th/9311171, to appear in Invent. Math.
52. B. Feigin, D. Fuchs, in *Representations of Lie Groups and Related Topics*, eds. A.M.Vershik and D.P.Zhelobenko, 465-554, Gordon and Breach, 1990
53. G. Felder, Nucl. Phys. B324 (1989) 548
54. G. Felder, C. Wieczerkowski, Comm. Math. Phys. 138 (1991) 583-605
55. J.M. Figueroa-O'Farrill, Nucl. Phys. B343 (1990) 450
56. E. Frenkel, *Affine Kac-Moody algebras at the critical level and quantum Drinfeld-Sokolov reduction*, Ph.D. Thesis, Harvard University, 1991
57. E. Frenkel, in *"New Symmetry Principles in QFT"*, eds. J. Fröhlich, e.a., 433-447, Plenum Press, 1992
58. E. Frenkel, V. Kac, A. Radul, W. Wang, $\mathcal{W}_{1+\infty}$ and $\mathcal{W}(gl_N)$ *with central charge* N, Preprint hep-th/9405121, to appear in Comm. Math. Phys.
59. E. Frenkel, V. Kac, M. Wakimoto, Comm. Math. Phys. 147 (1992) 295-328
60. I. Frenkel, in *Applications of Group Theory in Physics and Mathematical Physics*, eds. M. Flato, P. Sally, G. Zuckerman, Lect. in Appl. Math. 21, 325-353. AMS, Providence, 1985
61. I. Frenkel and V. Kac, Invent. Math. 62 (1980), 23-66
62. I. Frenkel, J. Lepowsky, A. Meurman, *Vertex Operator Algebras and the Monster*, Academic Press

417

1988
63. I. Frenkel, Y.-Z. Huang, J. Lepowsky, Memoirs of the AMS, 594 (1993)
64. I. Frenkel, Y. Zhu, Duke Math. Journal 66 (1992) 123-168
65. P.G.O. Freund, T.R. Klassen, E. Melzer, Phys. Lett. B229 (1989) 243
66. D.B. Fuchs, *Cohomology of Infinite-dimensional Lie Algebras*, Plenum 1988
67. C.S. Gardner, J. Math. Phys. 12 (1971) 1548-1551
68. C.S. Gardner, J.M. Green, M.D. Kruskal, R.M. Miura, Phys. Rev. Lett. 19 (1967) 1095-1097
69. I.M. Gelfand, L.A. Dikii, Russ. Math. Surv. 30, N5 (1975) 77-113
70. I.M. Gelfand, L.A. Dikii, Funct. Anal. Appl. 10 (1976) 16-22
71. I.M. Gelfand, L.A. Dikii, *A family of Hamiltonian structures connected with integrable non-linear differential equations*, Preprint IPM AN SSSR, Moscow, 1978
72. J.-L. Gervais, Phys. Lett. B160 (1985) 277
73. J.-L. Gervais, Phys. Lett. B160 (1985) 279
74. J.-L. Gervais, A. Neveu, Nucl. Phys. B 224 (1983) 329
75. P. Goddard, in *Infinite-dimensional Lie algebras and groups*, ed. V. Kac, Adv. Ser. in Math. Phys. 7, 556-587, World Scientific, Singapore 1989.
76. A. Guichardet, *Cohomologie des Groupes Topologiques et des Algebres de Lie*, Cedic/Fernand Nathan, Paris 1980
77. T. Hollowood, P. Mansfield, Phys. Lett. 226B (1989) 73
78. M. Jimbo, Lett. Math. Phys. 10 (1985) 63
79. A. Joseph, G. Letzter, *Rosso's form and quantized Kac-Moody algebras*, Preprint, 1993
80. V. Kac, Adv. Math. 30 (1978) 85
81. V. Kac, *Infinite-dimensional Lie Algebras*, 3rd Edition, Cambridge University Press 1990
82. V. Kac, D. Kazhdan, Adv. Math. 34, 97-108 (1979)
83. B. Kostant, S. Sternberg, Ann. Phys. 176 (1987) 49
84. A. Kuniba, T. Nakanishi, J. Suzuki, Nucl. Phys. B356 (1991) 750-774
85. B. Kupershmidt, P. Mathieu, Phys. Lett. 227B (1989) 245
86. B. Kupershmidt, G. Wilson, Comm. Math. Phys. 81 (1981) 189-202
87. B. Kupershmidt, G. Wilson, Inv. Math. 62 (1981) 403-436
88. P. Lax, Lect. Appl. Math. 15 (1974) 85-96
89. A. LeClair, Preprint CLNS 93/1220, hep-th/9305110, May 1993
90. A.N. Leznov, M.V. Saveliev, Lett. Math. Phys. 3 (1979) 498-494
91. H. Li, *Local systems of vertex operators, vertex superalgebras and modules*, Preprint
92. G. Lusztig, Adv. Math. 70 (1988) 237
93. S. MacLane, *Homology*, Springer 1963
94. F. Malikov, Int. J. Mod. Phys. A7, Supplement 1B (1992) 623-643
95. Yu.I. Manin, J. Sov. Math. 11 (1978) 1-122
96. D. McLaughlin, J. Math. Phys. 16 (1975) 96
97. A. Mikhailov, M. Olshanetsky, V. Perelomov, Comm. Math. Phys. 79 (1981) 473-488
98. W. Nahm, Int. J. Mod. Phys. A6 (1991) 2837-2845
99. M. Niedermaier, Comm. Math. Phys. 148 (1992) 249-281
100. M. Niedermaier, Preprint DESY 92-105 (1992)
101. M. Niedermaier, in *"New Symmetry Principles in QFT"*, eds. J. Fröhlich, e.a., 493-503, Plenum Press, 1991
102. D. Olive, N. Turok, Nucl. Phys. B220 (1983) 491
103. D. Olive, N. Turok, Nucl. Phys. B265 (1985) 469
104. D. Olive, N. Turok, J. Underwood, Preprint, April 1993
105. A. Pressley, G. Segal, *Loop Groups*, Clarendon Press, Oxford 1986
106. N. Reshetikhin, F. Smirnov, Comm. Math. Phys. 131 (1990) 157-177
107. A. Rocha-Caridi, Trans. AMS 262 (1980) 335
108. R. Sasaki, I. Yamanaka, Adv. Stud. in Pure Math. 16, 271-296 (1988)
109. V. Schechtman, A. Varchenko, in ICM-90 Satellite Conference Proceedings *Algebraic Geometry and Analytic Geometry*, 182-191, Springer 1991
110. V. Schechtman, Duke Math. J., Int. Math. Res. Not. 2 (1992) 39-49; 10 (1992) 307-315
111. G. Segal, Int. J. Mod. Phys. A6 (1991) 2859-2869

418

112. E. Sklyanin, L. Takhtajan, L. Faddeev, Theor. Math. Phys. 40 (1980) 688
113. L. Takhtadjian, L. Faddeev, Teor. Mat. Fiz. 21 (1974) 680
114. A. Varchenko, *Multidimensional hypergeometric functions and representation theory of Lie algebras and quantum groups*, Adv. Ser. in Math. Phys. 21, World Scientific 1995
115. A. Volkov, Phys. Lett. 167A (1992) 345
116. G. Wilson, Ergod. Th. and Dynam. Syst. 1 (1981) 361
117. V.E. Zakharov, L.D. Faddeev, Funct. Anal. Appl. 5 (1971) 280-287
118. A. Zamolodchikov, Al. Zamolodchikov, Ann. Phys. 120 (1979) 253-291
119. A. Zamolodchikov, Theor. Math. Phys. 65 (1985) 1205
120. A. Zamolodchikov, Adv. Stud. in Pure Math. 19, 641-674 (1989)

LANDAU INSTITUTE FOR THEORETICAL PHYSICS, 2 KOSYGINA ST, MOSCOW 117940, RUSSIA AND R.I.M.S., KYOTO UNIVERSITY, KYOTO 606, JAPAN

DEPARTMENT OF MATHEMATICS, HARVARD UNIVERSITY, CAMBRIDGE, MA 02138, USA

Seventy Years of Spectral Curves: 1923–1993

Emma Previato*

Boston University, Boston, MA 02215

The unifying theme of this series of lectures is the use of explicit (and elementary) methods of algebraic geometry, chiefly the theory of curves and vector bundles, for studying integrable systems. Each lecture tapers into an open direction where the dictionary between algebra and differential analysis needs to be further refined. This paper is expository: the results and proofs are taken from the literature, with the only addition of remarks and examples.

Lecture 1: The Burchnall-Chaundy problem [BC1, 1923].

Before stating (in 1.3) this most simple and attractive problem, we must go through a considerable amount of notation.

1.1 Definitions (i) The ring of formal pseudodifferential operators Ψ is the set

$$\{\sum_{j=-\infty}^{N} u_j(x)\partial^j, \ u_j \text{ analytic in some connected neighborhood of } x = 0\}.$$

If we think of these symbols as acting on functions of x by multiplication and differentiation: $(u(x)\partial)f(x) = u\frac{d}{dx}f$, and formally integrate by parts: $\int(uf) = u\int f - \int(u'f)$, we can motivate the composition rules:

$$\partial u = u\partial + u'$$
$$\partial^{-1}u = u\partial^{-1} - u'\partial^{-2} + u''\partial^{-3} - \dots$$

and easily check an extended Leibnitz rule [M3, p.183ff] for $A, B \in \Psi$:

$$A \circ B = \sum_{i=0}^{\infty} \frac{1}{i!}\tilde{\partial}^i A * \partial^i B$$

where $\tilde{\partial}$ is a partial differentiation w.r.t. the symbol ∂ and $*$ has the effect of bringing all functions to the left and powers of ∂ to the right.

More rigorously, this multiplication law can be motivated by introducing a free, rank 1 Ψ-module M = space of formal expressions $f = e^{xz}\tilde{f}$ where $\tilde{f} = \sum_{-\infty}^{N} f_j(x)z^j$ (cf. [SW§4]), with generator e^{xz}. Indeed, scalar multiplication $\partial e^{xz}\tilde{f} = e^{xz}(\partial + z)\tilde{f}$ can be inverted:

$$\partial^{-1}e^{xz}\tilde{f} = e^{xz}z^{-1}(z^{-1}\partial + 1)^{-1}\tilde{f} = e^{xz}z^{-1}(1 - z^{-1} + z^{-2}\partial^2 - \dots)\tilde{f}.$$

* Supported in part by NSF Grant DMS-9105221

(ii) Ψ contains the subring $\mathcal{D}$ of differential operators $A = \sum_0^N u_j \partial^j$ and we denote by $(\ )_+$ the projection $B_+ = \sum_0^N u_j \partial^j$ where $B = \sum_{-\infty}^N u_j \partial^j$.

1.2 Remarks. (i) The only reason for requiring that the functions $u_j(x)$ in the definition of Ψ be analytic near $x = 0$ is that the algebro geometric constructions preserve this restriction; many of the statements would be true for larger classes of functions. (ii) The much studied (cf. [Di] for example) Weyl algebra in two generators $\mathbf{C}[p,q]$ with multiplication rule defined by the commutator $[p,q] = 1$ can be viewed as a subring of $\mathcal{D}$, by letting $p = \partial$ and $q = x$.

1.3 Definition. The Burchnall-Chaundy (hereafter BC for short) problem asks to find and classify all commutative subrings of $\mathcal{D}$. If we denote by $\mathcal{C}_{\mathcal{D}}(L)$ the centralizer in $\mathcal{D}$ of an element $L \in \mathcal{D}$, we see that the polynomial ring $\mathbf{C}[L]$ is always contained in $\mathcal{C}_{\mathcal{D}}(L)$. We also see that if L has order $n > 0$ and its leading coefficient is regular, i.e. $u_n(0) \neq 0$, then L can be brought to standard form:

$$L = \partial^n + u_{n-2}(x)\partial^{n-2} + u_{n-3}(x)\partial^{n-3} + \ldots + u_0(x)$$

by using change of variable and conjugation by a function, which are the only two automorphisms of $\mathcal{D}$; we shall always assume L to be in standard form, and define a BC solution to be such an L for which $\mathcal{C}_{\mathcal{D}}(L)$ is not a polynomial ring $\mathbf{C}[M]$, $M \in \mathcal{D}$.

1.4 Remark. Notice that any translation in $x: x \mapsto x - a$, transforms a BC solution L into another solution L_a. We refer to this operation as the "x-flow" and typically never specify how small $|a|$ must be so that the analyticity of the coefficients is preserved; on the contrary, we shall often treat classical examples whose coefficients have a pole at $x = 0$ (cf. 1.5): strictly speaking, a translation in x would be necessary for them to fit the definitions 1.1, 1.3.

To illustrate the intricate beauty of the BC problem, we immediately give some examples, which will be reprised and explained in the course of the lectures.

1.5 Examples.
(i) As [BC1] points out, for general $A, B \in \mathcal{D}$ of orders n, m the commutator in general has order $n + m - 1$, so the commutativity condition imposes $n + m$ equations on $n + m + 2$ arbitrary functions. For example, for $n = 2, m = 3$, $A = \partial^2 + u$, $B = \partial^3 + a\partial + b$, the condition becomes: $a' = \frac{3}{2}u'$, $b' = \frac{3}{4}u''$, $b'' = u''' + au'$, so that $-\frac{u'^2}{2} = (u+c)^3 + du + e$ where c, d, e are constants of integration. Now it is not surprising that we find the solution $L = \partial^2 - 2\wp(x)$, where $\wp(x) = \frac{1}{x^2} + \sum_{\omega \in \Omega \setminus \{0\}} \left(\frac{1}{(x-\omega)^2} - \frac{1}{\omega^2} \right)$ is the Weierstrass $\wp$-function with lattice Ω and equation $\wp'^2 = 4\wp^3 - g_2\wp - g_3$. Moreover, $\mathcal{C}_{\mathcal{D}}(L) = \mathbf{C}[L, B]$ is the affine ring of the curve $\mu^2 = 4\lambda^3 - g_2\lambda - g_3$. The limiting cases in which one (simply periodic functions, nodal curve $\mu^2 = \lambda^2(\lambda - 1)$) or both (cuspidal curve $\mu^2 = \lambda^3$) generators of Ω go to infinity also yield solutions (cf. [AMcKM]); the latter was the first example found by BC, as part of the family:

$$A = x^{-n}\delta(\delta - m)(\delta - 2m)\ldots(\delta - (n-1)m)$$
$$B = x^{-m}\delta(\delta - n)(\delta - 2n)\ldots(\delta - (m-1)n)$$

where $\delta = x\partial$, the greatest common divisor $\gcd(n,m) = 1$, and for $n = 2, m = 3$

$$x^{-1}Ax = x^{-1}\delta(\delta - 3)x = \partial^2 - \frac{2}{x^2}, \quad x^{-1}Bx = x^{-3}(\delta + 1)(\delta - 1)(\delta - 3) = \partial^3 - \frac{3}{x^2}\partial + \frac{3}{x^3}.$$

(ii) If $c \in \mathbf{C}$ is a constant, $L = \partial^2 - c\wp(x)$ is a BC solution iff $c = n(n+1)$ with n an integer greater than zero (cf. [AMcKM]); if this is the case, the centralizer $C_{\mathcal{D}}(L)$ is the affine ring of a hyperelliptic curve of genus n, given by an equation: $\mu^2 = \lambda^{2n+1}+$ lower order. [TV3] gives the following generalization: if $\omega_i(0 \leq i \leq 3)$ are the half periods of $\wp, c_i \in \mathbf{C}$ are constants, $u = -2\sum_{i=0}^{3} c_i\wp(x - \omega_i)$, and $L = \partial^2 + u$, then L is a BC solution iff each c_i is a triangular number $a_i(a_i + 1)/2$ for a_i some positive integers.

(iii) In the Weyl algebra of 1.2(ii), define $u = p^3 + q^2 + \alpha$, $v = \frac{1}{2}p$, $L = u^2 + 4v$, $B = u^3 + 3(uv - vu)$; then $C(L) = \mathbf{C}[L,B]$ and $B^2 - L^3 = -\alpha$, as shown in [Di]. By the assignment $p = \partial$, $q = x$ we obtain $L, B \in \mathcal{D}$ of order 6,9, but the automorphism $\partial \mapsto -x$, $x \mapsto \partial$ will turn the orders into 4,6. Moreover, it will still be true that $C_{\mathcal{D}}(L) = \mathbf{C}[L,B]$, the affine ring of the curve $\mu^2 = \lambda^3 - \alpha$; in particular, L is a BC solution.

We would now like to: illustrate the significance of these examples; give a summary of the BC theory; and indicate what problems remain open, by postponing the details until Lemma 1.7. We need just one more word to address the problem fully:

1.6 Definition. The rank of a subset of $\mathcal{D}$ is the greatest common divisor of the orders of all the elements of $\mathcal{D}$.

1.6.1 Remark. The rank is a somewhat tricky concept; for example, the rank of $C_{\mathcal{D}}(L)$ in the example $L = \partial^2 + u$ is 1 or 2, according as L is a BC solution or not. However, for the BC solution of order 4 in 1.5(iii) the rank of $C_{\mathcal{D}}(L)$ is 2 (and this was morally the first known explicit example of a solution whose centralizer has rank greater than 1). On the other hand, if L, B are of order 2,3 and satisfy $B^2 = 4L^3 - g_2 L - g_3$ (cf. 1.5(i)), then $\mathbf{C}[L, L^2 + B]$ has rank 1 even if the generators have order 2,4. Lastly, $\mathbf{C}[L]$ has rank ord L, which shows that an algebra of rank 1 cannot be of type $\mathbf{C}[L]$ except for the trivial (normalized) $L = \partial$.

This is then a description of the theory: rather than BC solutions, we seek to describe commutative subalgebras of $\mathcal{D}$, of nontrivial type, namely containing some normalized element L of order > 0, and excluding $\mathbf{C}[M]$. It will follow from the next 2 lemmas that centralizers $C_{\mathcal{D}}(L)$ are maximal-commutative subalgebras of $\mathcal{D}$. How large can they be? Not very: since their quotient fields are function fields of 1 variable, they are affine rings of curves, and in a formal sense these are indeed spectral curves. Different rings may correspond to the same curve, and this is where the notion of rank enters. BC's theory gives a complete and explicit classification for rank 1: the algebras that correspond to a fixed curve make up the (generalized) Jacobian of that curve, and the x flow is a holomorphic vector field on it. We may (formally) view this as a "direct" spectral problem; the "inverse" spectral problem allows us to reconstruct the coefficients of the operators (in terms of theta functions) from the data of a point on the Jacobian. To insert a brief commentary on the theory of integrable PDE's, the x-flow is tangent to the Abel image of the curve in its Jacobian, at a specific point. The higher osculating flows form a sequence (essentially finite): $x = t_1, t_2, \ldots, t_s, \ldots$ and the corresponding operators depend on these parameters in such a way as to satisfy the KP hierarchy. Finally, the higher-rank algebras are still much of a mystery. In this lecture we will sketch a very concrete account of the BC theory. In Lecture II we will turn to an interpretation of the flows in the theory of completely integrable Hamiltonian systems, and in Lecture III we will address the higher-rank case.

1.7 Lemma (cf. [Ve]) *If* $A = a_n \partial^n + a_{n-1} \partial^{n-1} + \ldots + a_0$, $B = b_m \partial^m + b_{m-1} \partial^{m-1} + \ldots + b_0 \in \mathcal{D}$ *are such that* $n > 0$ *and* ord $[A, B] < n+m-1$, *then* $\exists \alpha \in \mathbf{C}$ *s.t.* $b_m^n = \alpha a_n^m$. *Moreover, if* a_n *and* b_m *are constant and* ord $[A, B] < n + m - 2$, *then* $\exists \alpha, \beta \in \mathbf{C}$ *s.t.* $b_{m-1} = \alpha a_{n-1} + \beta$.
(The proof is straightforward.)

1.8 Lemma (cf. [Ve]) *If* $\mathcal{A}$ *is a commutative subalgebra of* $\mathcal{D}$ *and* $M \in \mathcal{D}$, $\exists p \in \mathbf{Z} \cup \{-\infty\}$ *s.t.* $\forall L \in \mathcal{A}$, ord $L > 0$, ord $[M, L] = p + $ord L.

Proof. Let $p(L) = $ ord $[M, L] - $ ord L. Since "ad_M"$=[M, \ ]$ is a derivation, for $s \in \mathbf{N}$ $[M, L^s] = \sum_{i=0}^{s-1} L^i [M, L] L^{s-i-1} = sL^{s-1}[M, L]+$ lower order, so that $p(L^s) = p(L)$. Now let $p = \sup p(L)$ for $L \in \mathcal{A}$, ord $L > 0$ and let $K \in \mathcal{A}$, ord $K = k > 0$, $p(K) = p$ (the sup is attained because $p(L) \leq $ ord $M - 1$). Now $\forall L \in \mathcal{A}$ with ord $L = l > 0$ by Lemma 1.7 $\exists 0 \neq \alpha \in \mathbf{C}$, $L^k = \alpha K^l + R$, ord $R < kl$. Then ord $[M, R] = p(R)+$ ord $R < p + kl = $ ord $[M, K^l]$; from $[M, L^k] = \alpha[M, K^l] + [M, R]$ we finally get ord $[M, L^k] = $ ord $[M, K^l]$ so that $p(L) = p(L^k) = p(K^l) = p(K) = p$. ∎

1.9 Corollary *If* ord $L > 0$ *and* $A, B \in \mathcal{D}$ *both commute with* L, *then* $[A, B] = 0$; *in particular,* $\mathcal{C}_\mathcal{D}(L)$ *is commutative, hence every maximal-commutative subalgebra of* $\mathcal{D}$ *is a centralizer.*

1.9.1 Remark. The analog of 1.9 is not true for operators on finite-dimensional spaces. It is easy to find two noncommuting matrices that commute with a third one. A stronger concept was introduced by C. Chevalley (cf. [V], 3.1.3-13): in a finite-dimensional vector space V, the "replicas" of an endomorphism L are the endomorphisms M such that for any $r, s \geq 0$ and element u of the tensor product of r copies of V and s copies of V^*, $Lu = 0 \Rightarrow Mu = 0$, with the natural action of L and M on u. Then the replicas of a matrix L are the matrices $p(L)$, where p is a polynomial with zero constant term, so if N is a replica of L and M is a replica of N then M is a replica of L.

In Ψ any (normalized) L has a unique nth root, $n = $ord L, of the form $\mathcal{L} = \partial + u_{-1}(x)\partial^{-1} + u_{-2}(x)\partial^{-2} + \ldots$.

1.10 Theorem (I. Schur) $\mathcal{C}_\mathcal{D}(L) = \{\sum_{-\infty}^{N} c_j \mathcal{L}^j, c_j \in \mathbf{C}\} \cap \mathcal{D}$.
(The proof is a dimension count together with 1.7.)

We could use 1.10 to show that the quotient field of $\mathcal{C}_\mathcal{D}(L)$ is a function field of one variable; indeed, a B which commutes with L must satisfy an algebraic equation $f(L, B) = 0$ (identically in x), by a dimension count as sketched in [M2], moreover the degree of f in B is bounded; but we shall follow the more algorithmic idea of BC to see this and find the spectral curve.

1.11 **Lemma** [BC1]. *If* $[L, B] = 0$ *then there exists a polynomial in two variables* $f(\lambda, \mu) \in \mathbf{C}[\lambda, \mu]$ *such that* $f(L, B) \equiv 0$, *if we assign "weight"* $na + mb$ *to a monomial* $\lambda^a \mu^b$ *where* $n = \operatorname{ord} L$, $m = \operatorname{ord} B$, *then the terms of highest weight in* f *are* $\alpha \lambda^m + \beta \mu^n$ *for some constants* α, β.

Proof and Construction: The idea is that by commutativity B acts on V_λ, the n-dimensional vector space of solutions $y(x)$ of $Ly = \lambda y$ (L is regular); $f(\lambda, \mu)$ is the characteristic polynomial of this operator; to see that $f(L, B) \equiv 0$ it is enough to remark that $f(\lambda, \mu) = 0$ iff L, B have a "common eigenfunction": $\begin{cases} Ly = \lambda y \\ By = \mu y \end{cases}$ hence $f(L, B)$ would have an infinite-dimensional kernel (eigenfunctions belonging to distinct eigenvalues $\lambda_1, \ldots, \lambda_k$ are independent by a Vandermonde argument). But what brings out the algebraic structure of the problem, and of the polynomial f, is the construction of the "BC matrix" M: if

$$L - \lambda = u_{0,0} + u_{0,1}\partial + \ldots + \partial^n \quad (0 = u_{0,n+1} = u_{0,n+2} = \ldots)$$
$$\partial \circ (L - \lambda) = u_{1,0} + u_{1,1}\partial + \ldots$$

$$\cdots$$

$$\partial^{m-1} \circ (L - \lambda) = u_{m-1,0} + \ldots$$
$$B - \mu = u_{m,0} + u_{m,1}\partial + \ldots + \partial^m$$

$$\cdots$$

$$\partial^{n-1} \circ (B - \mu) = u_{m+n-1,0} + \ldots$$

then $M = [m_{ij}]$ with $m_{ij} = u_{i-1,j-1}$ ($i = 1, \ldots, m + n$; $j = 1, \ldots, m + n$) is such that $\det M = f(\lambda, \mu)$. Let us justify this and calculate a simple example. To see why M works, let's recall that two polynomials in one variable (over an integral domain) $F = a_0 + \ldots + a_n t^n$, $G = b_0 + \ldots + b_m t^m$ have a common zero if and only if their resultant $R(F, G) = 0$ where:

$$R = \det \begin{bmatrix} a_0 & a_1 & a_n & 0 & \cdots & & 0 \\ 0 & a_0 & & a_n & \cdots & & \\ \vdots & & & & & & \\ 0 & & & & & & a_n \\ b_0 & b_1 & \cdots & & b_m & \cdots & 0 \\ 0 & b_0 & & & & & \ddots \\ & & & & & & b_m \end{bmatrix} = \det M_{alg}.$$

The matrix M_{alg} is constructed from $F, tF, \ldots, t^{m-1}F$; $G, tG, \ldots, t^{n-1}G$ much as the above M, and the main tool in the proofs (which are standard) is the Euclidean algorithm, which holds in $\mathcal{D}$: there are unique $Q, R \in \mathcal{D}$ with $\operatorname{ord} R < \operatorname{ord} L$ and

$B - \mu = Q(L - \lambda) + R$. We obtain the characteristic polynomials from M as follows. Subdivide either matrix M_{alg} or M into blocks:

$$\begin{array}{cc} G_1 & G_2 \\ G_3 & G_4 \end{array} \begin{array}{l} \}n \\ \}m \end{array}$$
$$\underbrace{}_{n} \ \underbrace{}_{m}$$

and consider, in the algebraic case, the basis $1, t, \ldots, t^{n-1}$ of the vector space $V_F = C[t]/(F(t))$ and the vectors $X_n = [1, t, \ldots, t^{n-1}]^T$ and $X_m = [t^n, t^{n+1}, \ldots, t^{n+m-1}]^T$; in the differential case, the basis $y_0(x), \ldots, y_{n-1}(x)$ of V_λ normalized at $x = 0$ and the vectors $Y_n = [y, y', \ldots, y^{(n-1)}]^T$, $Y_m = [y^{(n)}, \ldots, y^{(n+m-1)}]^T$, $B_n = [(B - \mu)y, ((B - \mu)y)', \ldots, ((B - \mu)y)^{(n-1)}]^T$. Now by construction,

$$G_1 X_n + G_2 X_m = F(t) X_m \qquad G_1 Y_n + G_2 Y_m = 0$$
$$G_3 X_n + G_4 X_n = G(t) X_n \qquad G_3 Y_n + G_4 Y_m = B_n$$

so that in the algebraic case the matrix $(G_3 - G_4 G_2^{-1} G_1)^T$ represents multiplication by $G(t)$ in the chosen basis of V_F; and in the differential case, since:

$$\begin{bmatrix} (B - \mu)y_1 & \cdots & (B - \mu)y_n \\ ((B - \mu)y_1)' & \cdots & ((B - \mu)y_n)' \\ \vdots & & \vdots \\ ((B - \mu)y_1)^{(n-1)} & \cdots & ((B - \mu)y_n)^{(n-1)} \end{bmatrix} = \begin{bmatrix} y_1 & \cdots & y_n \\ y_1' & & y_n' \\ \vdots & & \vdots \\ y_1^{(n-1)} & & y_n^{(n-1)} \end{bmatrix}$$

$$(M_B^T - \mu I) = (G_3 - G_4 G_2^{-1} G_1) \begin{bmatrix} y_1 & \cdots & y_n \\ y_1' & \cdots & y_n' \\ \vdots & & \vdots \\ y_1^{(n-1)} & \cdots & y_n^{(n-1)} \end{bmatrix}$$

where M_B represents the action of B on V_λ in the given basis, by evaluation at $x = 0$,

$$M_B - \mu I = (G_3 - G_4 G_2^{-1} G_1)^T \Big|_{x=0}.$$

Finally, the identity

$$\begin{bmatrix} G_1 & G_2 \\ G_3 & G_4 \end{bmatrix} \begin{bmatrix} 0 & I \\ G_2^{-1} & -G_2^{-1} G_1 \end{bmatrix} = \begin{bmatrix} I & 0 \\ G_4 G_2^{-1} & G_3 - G_4 G_2^{-1} G_1 \end{bmatrix}$$

gives the conversion between $\det M$, $\det M_{alg}$ and the characteristic polynomials. ∎

1.11.1 Remark. Since the algebra $C[L, B]$ has no zero-divisors, it can be viewed as the affine ring $C[X, Y]/(h)$ of a plane curve, with $h(X, Y)$ an irreducible polynomial. The BC curve $= \{(\lambda, \mu) \mid L, B, \text{ have a joint eigenfunction } Ly = \lambda y, By = \mu y\}$ is included in the curve Spec $C[L, B]$ and since the latter is irreducible, they must coincide; this shows in particular that the BC polynomial is some power of an irreducible polynomial h: $f(\lambda, \mu) = h^{r_1}$.

1.12 **Proposition** (cf. [W], Appendix, for a rigorous proof). *Let* $r_2 = \text{rank } \mathbf{C}[L, B]$, *and* $r_3 = \dim V_{(\lambda,\mu)}$ *where* $V_{(\lambda,\mu)}$ *is the vector space of common eigenfunctions at any smooth point* (λ, μ) *of the BC curve. Then* r_1 *(defined in 1.11.1)* $= r_2 = r_3$. *Moreover, this integer is the order of* $G = gcd(L - \lambda, B - \mu)$, *the operator (found by the Euclidean algorithm) of highest order for which* $B - \mu = T_1 G$, $L - \lambda = T_2 G$.

Unless differently specified, the rank of the commutative pair L, B (cf. 1.12) is assumed to be 1 for the rest of this section.

1.12.1 **Remark.** It is clear from the form of the BC matrix M that its term of highest weight is of the form $(-\lambda)^m + (-1)^{mn}\mu^n$. Let us define the semigroup of weights $W = \{an + bm \mid a, b \text{ nonnegative integers}\}$. By analyzing the general solution $(a + cm)n + (b - cn)m$, it is easy to prove the following useful statements [BC2]: (i) every number in the closed interval $[(m - 1)(n - 1), mn - 1]$ belongs to W and exactly half the numbers in the closed interval $[1, (m - 1)(n - 1)]$ do not; (ii) in this range, a solution (a, b) to $an + bm = k$ is unique. To explain the significance of the weight, we compactify the BC curve following $[M2]$ to $X = \text{Proj } R$, where R is the graded ring $\oplus_{s=0}^{\infty} A_s$, $A_s = \{A \in \mathcal{A} \mid ord\ A \le s\}$ and the operator 1 is represented by an element $e \in A_1$ (in our case the commutative algebra $\mathcal{A}$ is $\mathbf{C}[L, B]$, but the construction holds in general). That the point P_∞ which we added is smooth can be seen as follows: the affine open $e \neq 0$ is $\text{Spec }(R[\frac{1}{e}]_0) = \text{Spec } \mathcal{A}$ (the subscript 0 signifies the degree zero component); the affine open where $L \neq 0$ is $\text{Spec }(R[\frac{1}{L}]_0)$ and the completion of this ring in the e-adic topology is $\mathbf{C}[[z]]$ if z corresponds to $L^i B^j / L^k$ with $in + jm = kn - 1$ (basically we are using $\mathcal{L}^{-1}$ as a local parameter, with $\mathcal{L} = L^{1/n}$). Thus, the weight is the valuation at P_∞ of a function in $\mathcal{A}$, W is the Weierstrass semigroup and the number of gaps $g = \frac{(m-1)(n-1)}{2}$ is the genus of X if there are no finite singular points.

We have learned that a commutative algebra $\mathcal{A} \subset \mathcal{D}$ is the affine ring of a spectral curve X, in such a way that a point of X gives a homomorphism $\phi: \mathcal{A} \to \mathbf{C}$, a (generically) unique common eigenfunction: $Ay = \phi(A)y$, and the order of pole at P_∞ of $A \in \mathcal{A}$ equals ord A. But this correspondence is not one-to-one, and we introduce now the other spectral data, which will allow us to solve the inverse problem, in a completely elementary way at first, through an example: let

$$L = \partial^2 - \frac{2}{x^2}$$

$$B = \partial^3 - \frac{3}{x^2}\partial + \frac{3}{x^3}$$

(nothing changes if you prefer to replace all x by $x + c$). It is fascinating to watch all x-dependence disappear in the determinant of the BC matrix:

$$\det \begin{bmatrix} -\lambda - \frac{2}{x^2} & 0 & 1 & 0 & 0 \\ \frac{4}{x^3} & -\lambda - \frac{2}{x^2} & 0 & 1 & 0 \\ -\frac{12}{x^4} & \frac{8}{x^3} & -\lambda - \frac{2}{x^2} & 0 & 1 \\ -\mu + \frac{3}{x^3} & -\frac{3}{x^2} & 0 & 1 & 0 \\ -\frac{9}{x^4} & -\mu + \frac{9}{x^3} & -\frac{3}{x^2} & 0 & 1 \end{bmatrix} = \mu^2 - \lambda^3.$$

This curve has a singularity at $(0,0)$, but the number $g = (n-1)(m-1)/2 = 1$; we'll see that there are $g = 1$ points naturally associated to L, B, in two equivalent ways: (i) the matrix $G_3 - G_4 G_2^{-1} G_1$ is

$$
\begin{bmatrix} -\frac{1}{x^3} - \mu & -\frac{1}{x^2} + \lambda \\ \frac{1}{x^4} + \frac{\lambda}{x^2} + \lambda^2 & \frac{1}{x^3} - \mu \end{bmatrix} = \begin{bmatrix} -V(\lambda) - \mu & U(\lambda) \\ W(\lambda) & V(\lambda) - \mu \end{bmatrix}
$$

with U, W monic polynomials in λ of degrees 1,2 and V of degree 0 (these would become $g, g+1, g-1$ resp. if you had used example 1.5(ii) for $n = g$, or 1.5(i) for $n = 2$, $m = 2g + 1$); the fact that M_B is traceless should not be surprising: since $B + \Sigma c_j L^j$ also commutes with L, such a renormalization is always possible. Now the equations $U(\lambda) = V(\lambda) - \mu = 0$ define g points on the curve (for fixed $x = x_0$), which we call the divisor of the algebra $\mathbf{C}[L, B]$. The fact that the algebra can be recovered from the divisor, and that each divisor corresponds to an algebra, will be demonstrated below by introducing the BC construction of adjoint eigenfunctions. For now, let us just explain the significance of the divisor. Since $M_B - \mu$ is given by the transpose of the matrix $G_3 - G_4 G_2^{-1} G_1$, the eigenvectors of this matrix will give the common eigenfunctions: $-U y_0 + (-V - \mu) y_1 \propto (-V + \mu) y_0 + W y_1$. Now there is a generically 1-dimensional space of common eigenfunctions, and to normalize them we require that the value at the reference point $x = 0$ be 1. Then $\psi(x) = y_0 + \frac{V+\mu}{U} y_1$, and the only point of the curve where this normalization fails is $U\Big|_{x=0} = 0 \neq (V + \mu)\Big|_{x=0}$, but since to any given λ there correspond $\pm \mu$ on the curve, this means $V - \mu\Big|_{x=0} = 0$. To read this off in our case we had better replace x by $x + c$, with a non-zero constant c, and our divisor becomes $(\lambda, \mu) = (\frac{1}{c^2}, \frac{1}{c^3})$.

(ii) The other way to find this divisor is by calculating the gcd of $L - \lambda$, $B - \mu$:

$$
B - \mu = \partial(L - \lambda) + (\lambda - \frac{1}{x^2})\partial - \frac{1}{x^3} - \mu
$$

$$
L - \lambda = ((\lambda - \frac{1}{x^2})^{-1}\partial - (\frac{1}{x^3} - \mu)(\lambda - \frac{1}{x^2})^{-2})((\lambda - \frac{1}{x^2})\partial - \frac{1}{x^3} - \mu)
$$

$$
+ (\mu^2 - \lambda^3)(\lambda - \frac{1}{x^2})^{-2}
$$

so on the curve gcd $(L - \lambda, \ B - \mu) = G = (\lambda - \frac{1}{x^2})\partial - \frac{1}{x^3} - \mu = U(\lambda)\partial - (V(\lambda) + \mu)$, and again this has singular points exactly when $U(\lambda) = 0$, $V(\lambda) - \mu = 0$ (for the $x = 0$ normalization as above). Before we get to the crux of the BC construction, we give yet another way to look at the divisor. The most efficient way to do it is to introduce some formal tools, in the vein of 1.1.

1.13 Definitions. (i) In Ψ, it is possible to conjugate any $\mathcal{L} = \partial + u_{-1}(x)\partial^{-1} + \cdots$ into ∂ by a $K \in \Psi$, $K = 1 + v_{-1}(x)\partial^{-1} + \cdots$, determined up to elements of $\mathbf{C}[\partial] = \mathcal{C}_{\mathcal{D}}(\partial)$. From now on we assume that $K^{-1}\mathcal{L}K = \partial$. (ii) We define a formal Baker function for $\mathcal{L}$ as the element of the module M (cf. 1.1) such that $\mathcal{L}\psi = z\psi$; notice that $\psi = K e^{xz}$. (iii) We say that the formal adjoint A^+ of a (formal pseudo) differential operator $A = \sum_{j=-\infty}^{N} u_j(x)\partial^j$ is $A^+ = \sum_{j=-\infty}^{N}(-\partial)^j u_j(x)$, and that the dual Baker function ψ^+ to ψ is the Baker function of $\mathcal{L}^+$; the operator which corresponds to K in (i) is $(K^+)^{-1}$, i.e. $(K^+)^{-1}\mathcal{L}^+K^+ = -\partial$.

Next we take a closer look at the special case in which L has order $n = 2$ (hence B has odd order $m = 2g + 1$). We will come back to this important class of examples in Lecture II, but a few remarks now will afford us a much clearer view of the BC construction.

1.14 The hyperelliptic case. The reason for the nomenclature is that the (normalized) BC curve is $\mu^2 = \prod_{j=1}^{2g+1}(\lambda - e_j)$, which can be viewed as a 2:1 covering of $\mathbf{P}^1$ by the function $\lambda: X \to \mathbf{P}^1$, with branchpoints $e_1, \ldots, e_{2g+1}, P_\infty$. Since L is (formally) self-adjoint, and B can be chosen skew-adjoint, in this case the adjoint of the Baker function $\psi(x, P)$ is simply $\phi(x, P) = \psi(x, \iota P)$, where ι is the hyperelliptic involution: $\iota P = \iota(\lambda, \mu) = (\lambda, -\mu)$. Then ψ, ϕ is a basis of V_λ except at the branchpoints $(e_i, 0)$ where $\psi = \phi$. But then the normalized basis of V_λ is linked to ψ, ϕ by a constant matrix: $\begin{bmatrix} y_0 \\ y_1 \end{bmatrix} = C \begin{bmatrix} \psi \\ \phi \end{bmatrix}$, while $B \begin{bmatrix} \psi \\ \phi \end{bmatrix} = \begin{bmatrix} \mu & 0 \\ 0 & -\mu \end{bmatrix} \begin{bmatrix} \psi \\ \phi \end{bmatrix}$ so that $\begin{bmatrix} -V & U \\ W & V \end{bmatrix}^T = M_B = C \begin{bmatrix} \mu & 0 \\ 0 & -\mu \end{bmatrix} C^{-1}$. By evaluating at $x = 0$, you find: $C = \frac{1}{Wr} \begin{bmatrix} \phi' & -\psi' \\ -\phi & \psi \end{bmatrix} \Big|_{x=0}$ with $Wr = \psi\phi' - \psi'\phi$. Finally, we calculate: $C \begin{bmatrix} \mu & 0 \\ 0 & -\mu \end{bmatrix} C^{-1} = \frac{\mu}{Wr} \begin{bmatrix} \psi\phi' + \psi'\phi & 2\psi'\phi' \\ -2\psi\phi & -(\psi\phi' + \psi'\phi) \end{bmatrix}$ so that $U(\lambda) = \psi\phi' + \psi'\phi$, $V(\lambda) = -2\psi\phi$, $W(\lambda) = 2\psi'\phi'$, and the fact that $UW + V^2$ does not depend on x expresses the fact that $Wr = $ constant.

We are finally ready to absorb the BC construction: the product of the Baker function and its adjoint becomes algebraic and by an inductive argument the divisor is recovered; the x flow is made explicit on the Jacobian, and the inverse spectral problem is solved thereby.

1.15 Proposition. *Let (λ, μ) be a point of the spectral curve and write, by Euclidean division,*

$$f(L, \mu) = T(L - \lambda)(B - \mu) + (a(x)\partial^{n-1} + a_1(x)\partial^{n-2} + \ldots + a_{n-1}(x))(B - \mu)$$

$$f(L^+, \mu) = S(L^+ - \lambda)(B^+ - \mu) + (b(x)\partial^{n-1} + b_1(x)\partial^{n-2} + \ldots + b_{n-1}(x))(B^+ - \mu).$$

Then $\psi\phi = a = b$, the intersections of $a(x)$ (which is polynomial in λ, μ) with the finite part of the curve X consists of $2g$ points, half of which (generically in x) belong to the curve $a_1(x), (\alpha_s, \beta_s)$ say, $s = 1, \ldots, g$; and the other half to the curve $b_1(x), (\gamma_s, \delta_s)$ say, $s = 1, \ldots, g$.

1.15.1 Example. If the equation is $\mu^2 = \lambda^3 - \frac{g_2}{4}\lambda - \frac{g_3}{4}$ as in 1.4(ii), $f(L, \mu) = L^3 - \frac{g_2}{4}L - \frac{g_3}{4} - \mu^2 = (B + \mu)(B - \mu) = T(L - \lambda + G)(B - \mu)$ where G is gcd $(L, B + \mu)$ so that $a(x) = \lambda + \frac{u}{2} = U(\lambda)$, $b_1(x) = V(\lambda)$. More generally, the left-hand side is a product $\prod(B - \mu_j)$ where the μ_j depend algebraically on λ, and we divide by $B - \mu_1$, say.

1.16 Theorem *Let the g functions of x $\omega(\alpha_s, \beta_s)$ be defined by the g linear equations*

$$\begin{cases} \sum_{s=1}^{g} \alpha_s^a \beta_s^b \omega(\alpha_s, \beta_s) = 0 & \text{if } an + bm < 2g - 2 \\ \sum_{s=1}^{g} \alpha_s^a \beta_s^b \omega(\alpha_s, \beta_s) = -1 & \text{if } an + bm = 2g - 2. \end{cases}$$

Then:

(i) $\omega(\alpha_s, \beta_s) = \dfrac{-\alpha_s'}{f_\mu(\alpha_s, \beta_s)} = \dfrac{\beta_s'}{f_\lambda(\alpha_s, \beta_s)} = \dfrac{-a'(\alpha_s, \beta_s)}{\partial(f, a)/\partial(\alpha_s, \beta_s)};$

(ii) $\prod\limits_{r=1}^{n} \psi_r = \tau \prod\limits_{s=1}^{g} (\lambda - \alpha_j)$ *where ψ_i give a normalized basis of eigenfunctions for L;*
here τ may be a function of x but not λ;
(iii) conversely, if the $\omega(\alpha_s, \beta_s)$ are defined as above and the g points (α_s, β_s) obey the differential equations in (i), which give g free parameters as constants of integration, then there can be constructed a pair L, B of commuting operators.

Rather than providing the proofs, which are elementary in nature and occupy the entire [BC2], we provide a little more interpretation than [BC2], especially in view of the open problem in which we would like to mention at the end of the Lecture.

1.16.1 Comments. The formulas in 1.16 are given by the residue theorem on X. Indeed, assume for simplicity that X is smooth and hence of genus g. Then $\{\lambda^a \mu^b \frac{d\lambda}{f_\mu} = -\lambda^a \mu^b \frac{d\mu}{f_\lambda}, \ na + mb \leq 2g - 2\}$ give a basis of holomorphic differentials $\omega_1, \ldots, \omega_g$ say, and the function $\frac{b_1(x)}{a(x)}$ has poles exactly at the (α_s, β_s) $s = 1, \ldots, g$, so that the residue theorem is being applied to $\frac{b_1}{a} \omega_j$ $j = 1, \ldots, g$ if $\omega(\alpha_s, \beta_s) = \mathrm{Res}_{(\alpha_s, \beta_s)} \frac{b_1}{a} \frac{d\lambda}{f_\mu}$. Next, equations (i) give the abelian flow explicitly, since they say that the Abel map of the divisor $\sum_{s=1}^{g} (\alpha_s, \beta_s)$, differentiated with respect to x, is $[0, \ldots, 0, 1]$. We can actually check our observation that $\omega(\alpha_s, \beta_s) = \mathrm{Res}_{(\alpha_s, \beta_s)} \frac{b_1}{a} \frac{d\lambda}{f_\mu}$ against equations (i), as follows; we must use $a'(\alpha_s \beta_s) = -b_1(\alpha_s \beta_s)$, which BC derive form the equations. Now we want to show $\mathrm{Res}_{(\alpha_s, \beta_s)} \frac{a'}{a} \frac{d\lambda}{f_\mu} = \frac{\alpha_s'}{f_\mu(\alpha_s, \beta_s)}$, which is straightforward, by the chain rule, and the fact that a has a zero at α_s, β_s.

1.17 The Mumford viewpoint [M2]. This is a more modern version of the BC theory; we highlight some points of the construction which give information of more analytic nature on the Baker function and lead immediately into the KP hierarchy. Mumford presents a dictionary between the following two sets of data:
{curve X, smooth point P_∞, local parameter z^{-1} near P_∞ and rank 1, torsion-free sheaf $\mathcal{F}$ on X with $h^0(\mathcal{F}) = h^1(\mathcal{F}) = 0$} and
{$\mathcal{A} \subset \mathcal{D}$ commutative subalgebra of rank 1 together with a normalized element L}
The ingredients:
(i) for the inverse spectral problem one constructs a formal deformation of $\mathcal{F}$ to a sheaf $\mathcal{F}^*$ over $X^* = X \times_{\mathbb{C}} \mathbb{C}[[x]]$, by twisting it with the transition function e^{xz} on a punctured neighborhood of P_∞. Then differentiation w.r.t. x gives an operator $\nabla \colon \mathcal{F}^*(n, P_\infty) \to \mathcal{F}^*((n+1)P_\infty)$, $n > 0$, and the choice of a generator σ_0 of $H^0(X^*, \mathcal{F}^*(P)) \cong H^0(X^*, \mathcal{F}^*(P)/\mathcal{F}^*) \cong \mathbb{C}[[x]]$ permits to associate to any function

$\lambda \in H^0(X\backslash P_\infty, \mathcal{O}_X)$ a differential operator $\Lambda = l_n \nabla^n + \sum_{i=0}^{n-1} l_i(x)\nabla^i$ such that $\lambda \sigma_0 = \Lambda \sigma_0$. The choice of σ_0 is only unique up to multiplication by $a(x) \in \mathbf{C}[[x]]$, $a(0) \neq 0$, but the choices we made allow us to fix that.

(ii) The direct spectral problem: we already got curve, point and local parameter such that $z = L^{1/n}$. The sheaf is given by the dual space of the space of joint eigenfunctions at each finite point; near enough P_∞, the assignment $y \mapsto y(0)$ is a regular section and to extend the sheaf to X one trivializes it consistently on this neighborhood of P_∞. But more algebraically, the sheaf is defined by $M = \mathbf{C}[\partial]$ viewed as a module over $\mathbf{C}[[x]] \otimes_{\mathbf{C}} \mathcal{A}$ (with natural multiplication from the left and right); and for each point $\phi: \mathcal{A} \to \mathbf{C}$ the pairing $M/m_\phi M \times V_\phi \to \mathbf{C}$ (where m_ϕ is the maximal ideal corresponding to ϕ and V the space of common eigenfunctions) is an isomorphism.

1.18 Krichever's Construction. It turns out that the formal Baker function of 1.12 for these algebraic solutions has a beautiful expression in terms of theta functions, which makes the inverse spectral problem very clean; we only quote the result:

$$\psi(x, P) = \exp(x \int \eta - xa) \frac{\vartheta(Ux + A(P) - A(D) - \Delta)\vartheta(A(D) + \Delta)}{\vartheta(A(P) - A(D) - \Delta)\vartheta(Ux - A(D) - \Delta)}$$

where $A(-)$ is the Abel map, Δ is the Riemann constant, $U \in \mathbf{C}^g$ a suitable vector, D the BC divisor, η a differential of the second kind and a is an intriguing, cf. [Du], constant depending on the curve.

1.19 Applications. (i) If in 1.17 we had defined the deformation of $\mathcal{F}$ by transition functions $e^{\sum_{j=1}^\infty t_j z^j}$, we would have got differential operators (depending on the parameters t_j) satisfying the KP hierarchy (cf. [SW]); $\mathcal{L}$ is then referred to as a KP solution;

$$\partial_{t_j} \mathcal{L} = [(\mathcal{L}^j)_+, \mathcal{L}].$$

Notice that if $\mathcal{L} = L^{1/n}$ then $\partial_{t_n} \mathcal{L} = [\mathcal{L}^n, \mathcal{L}] = 0$ and all integral multiples of n also give stationary flows; these solutions are referred to as "generalized KdV" solutions, because for $n = 2$ they solve the KdV equation. [G] gives the explicit calculation of these higher flows for the example that precedes 1.13, using the high-brow recipe of 1.17.

(ii) To obtain the coefficients of $\mathcal{L}$ explicitly, 1.18 is used or else (but with more difficulty) trace formulas for the squared eigenfunctions which we encountered in 1.14, 1.15. For example, for $L = \partial^2 + u$, $(\partial^3 + 4u\partial + 2u')\psi\phi = 4\lambda(\psi\phi)'$, so by replacing $\psi\phi$ with $\prod_{s=1}^g (\lambda - \alpha_s)$ and comparing powers of λ you get $u = \sum_{i=1}^{2g+1} e_i - 2 \sum_{i=1}^g \alpha_i$.

(iii) Contrary to common (?) belief, Burchnall and Chaundy had not missed out on the higher KP flows. In [BC1] before solving the inverse problem by using the divisor of $\psi\phi$ they manage to span the hyperelliptic Jacobian by producing inductively all the operators which "semi-commute" with L and which are nothing but $(L^{j+1/2})_+$, $j = 1, \ldots, g$.

(iv) The use of the adjoint operator is quite classical and Darboux realized its remarkable algebraic properties [D, L.IV C.V]. His viewpoint is that of differential algebra in that he treats the dependent variable u (a function of x, say) and its derivatives $u', u'' \ldots$ as indeterminates. The *adjoint equation* $g(v)$ of an equation $f(u) =$

$su + s_1u' + \ldots + s_nu^{(n)}$ is then characterized by the fact that $vf(u) - ug(v)$ is a perfect derivative $= \frac{d}{dx}\Psi(u,v)$ where Ψ is bilinear in u, v and their derivatives up to the $(n-1)$st. A fundamental system of solutions of $f, u_1, \ldots, u_n$, defines uniquely a dual system for g:

$$v_i = \frac{1}{\lambda_n}\frac{\partial \Delta}{\partial u_i^{(n-1)}} \text{ (it turns out that } u_i = \frac{(-1)^{n-1}}{\lambda_n}\frac{\partial \log \Delta_1}{\partial v_i^{(n-1)}}), i = 1, \ldots, n$$

where $\Delta = \det \begin{bmatrix} u_1 & u_1' & \cdots & u_1^{(n-1)} \\ u_2 & u_2' & & u_2^{(n-1)} \\ \vdots & \vdots & & \vdots \\ u_n & u_n' & & u_n^{(n-1)} \end{bmatrix}$ and Δ_1 is obtained by replacing u with v. As a

consequence, the eigenfunctions and the dual eigenfunctions satisfy quadratic equations:

$$\begin{cases} v_1^{(i)}u_1^{(k)} + v_2^{(i)}u_2^{(k)} + \ldots + v_n^{(i)}u_n^{(k)} = 0, \ i + k < n - 1 \\ v_1^{(i)}u_1^{(k)} + v_2^{(i)}u_2^{(k)} + \ldots + v_n^{(i)}u_n^{(k)} = \frac{(-1)^i}{\lambda_n}, \ i + k = n - 1. \end{cases}$$

But now if f is skew adjoint of order $2n - 1$, say, some further calculation shows that $v_i = (-1)^{i-1}u_{2n-i}$ so that the solutions of f in fact lie on a quadric!

$$\begin{cases} 2u_1u_{2n-1} - 2u_2u_{2n-2} + \ldots + (-1)^{n-1}u_n^2 = 0 \\ 2u_1^{(i)}u_{2n-1}^{(i)} - 2u_2^{(i)}u_{2n-2}^{(i)} + \ldots + (-1)^{n-1}(u_n^{(i)})^2 \ i < n - 1 \\ 2v_1^{(n-1)}u_{2n-1}^{(n-1)} - 2u_2^{(n-1)}u_{2n-2}^{(n-1)} + \ldots + (-1)^{n-1}(u_n^{(n-1)})^2 = \frac{(-1)^{n-1}}{\lambda_{2n-1}} \end{cases}$$

[BC1] remarks that a B of order $2n - 1 (= 2g + 1)$ which commutes with a second-order L can be taken to be skew-adjoint, but a basis of Baker-function solutions $y_1, \ldots, y_{2g-1}$ of $By = 0$ (notice that they correspond to the branchpoints $(e_i, 0)$ of the hyperelliptic curve) lie on $g + 1$ quadrics! Indeed, recall that $\psi(x, e_i) = \phi(x, e_i) = y_i$ from 1.13 but also that $y_i^2 = e_i^g + s_1 e_i^{g-1} + \ldots + s_g$, with s_j functions of x but independent of i (cf. 1.15(ii)). Then if Δ_r is the minor of e_r^{2g} (and Δ the determinant) in the matrix:

$$\begin{bmatrix} 1 & e_1 & e_1^2 & \cdots & e_1^{2g} \\ 1 & e_2 & e_2^2 & \cdots & e_2^{2g} \\ \vdots & & & & \\ 1 & e_{2g+1} & e_{2g+1}^2 & \cdots & e_{2g+1}^{2g} \end{bmatrix}$$

we have $g + 1$ relations

$$\begin{cases} F_{j-1}(y_1, \ldots, y_{2g+1}) = \sum_{r=1}^{2g+1} e_r^{j-1}\Delta_r y_r^2 = 0 \ j = 1, 2, \ldots, g \\ F_g(y_1, \ldots, y_{2g+1}) = \sum_{r=1}^{2g+1} e_r^g \Delta_r y_r^2 = \Delta \end{cases}$$

and by differentiating and using $Ly_r = e_r y_r$ we obtain $g - j + 1$ quadratic relations among the jth derivatives, until $F_0(y_1^{(g)}, \ldots, y_{2g+1}^{(g)}) = (-1)^g\Delta$, which must be Darboux' quadric.

(v) Open problems. For a (smooth) hyperelliptic curve X with affine equation $\mu^2 = \prod_{i=1}^{2g+1}(\lambda - e_i)$ there is a marvelous geometric model in $\mathbf{P}^{2g+1}$ which provides information on X, Jac X, and moduli spaces of higher-rank bundles on X. This construction, which generalizes classical results for genus 1 and 2 (the quadratic line complex for the 4-dimensional Grassmannian Gr $(2,4)$, cf. [GH]) is due to A. Weil and M. Reid, cf. [DR]; it is being used at present to investigate the Verlinde formulas, namely the equations for projective images of moduli spaces of vector bundles, cf. [BS]. It all starts with the intersection of the two quadrics: $\sum_{i=0}^{2g+1} x_i^2 = 0$ and $\sum_{i=1}^{2g+1} e_i x_i^2 = x_0^2$, which in $\mathbf{P}^3$ is the elliptic curve. More generally, let V be the intersection of the 2 quadrics: the largest linear subspaces contained in V are $\mathbf{P}^{g-1}$'s. The curve X is recovered as follows: for fixed $l \subset V, F_l(V) = $ the closure of the set $\{l' \in F(v) | \dim(l \cap l') = g - 2\}$ is isomorphic to X; the 2:1 map to the pencil of quadrics spanned by the two given ones assigns to an $l' \neq l$ the quadric which contains the $\mathbf{P}^n$ spanned by l and l'; this map is branched exactly over the $2g + 2$ singular quadrics of the pencil. For $g = 2$ again, by intersecting V with a third quadric $\sum_{i=1}^{2g+1} e_i^2 x_i^2 = x_0^2$ a surface birational to the Kummer surface is obtained. There are 2 problems worth investigating: one is an interpretation of the BC quadrics of (iv) in these projective spaces, especially in view of the fact that the sets $(y_1, \ldots, y_{2g+1})$ correspond to points of Jac X; further geometric properties of the KdV flows acting on $(y_1, \ldots, y_{2g+1})$ are brought out by Knörrer [K] via a hamiltonian system which we recall in Lecture II. The other problem is the geometric analog of Reid's Jacobian for nonhyperelliptic curves.

Lecture 2: Integrable Systems

Given the wealth of this subject, and the depth of the links (between dynamics and representation theory, to mention one only) which were found to explain several mysteries over the past 20 years, I will set three very simple goals, and use mainly the Neumann system as a prototype, because it has proved attractive to many people and the salient features have been worked out beautifully in the literature. Goal 1 is to show the link between finite-dimensional completely integrable systems and PDE's of KP type discussed in Lecture 1. Goal 2 is to generalize the construction to the case of nonhyperelliptic spectral curves. Goal 3 is to reprise a certain duality (which will be explained in the course of reaching Goal 2) and view it as a stroboscope, thanks to an important tool, the "transference," which had also been introduced by BC and which will be put to a different use in Lecture 3.

2.1 **The Neumann system.** It seems best to just give the facts of this well-known theory; for the history and all the proofs we refer to [M3]. Recall the hyperelliptic curve X of 1.14 and the 3 polynomials (U, V, W) which define a divisor on it. In fact these polynomials can be used to parametrize Jac X, an idea that goes back to Jacobi:

2.1.1 **Theorem** *There is an isomorphism between the (affine) subvariety of $\mathbf{C}^{3g+1}$ given by the coefficients of (U, V, W) satisfying the equation $UW + V^2 = f(\lambda) = \prod_{i=1}^{2g+1}(\lambda - e_i)$, and Jac $X\backslash\Theta$, where Θ is a theta divisor (given by $\sum_{i=1}^{g} P_i$ for which $h^0(\sum_{i=1}^{g} P_i - P_\infty) > 0$).*

2.1.2 Theorem *The prescriptions*

$$
\left.
\begin{aligned}
D_k U_l &= \Sigma(V_i U_j - V_j U_i) \\
D_k V_l &= \frac{1}{2}(\Sigma(U_i W_j - U_j W_i) - U_k U_l) \\
D_k W_l &= \sum(W_i V_j - W_j V_i) + U_k V_l
\end{aligned}
\right\}
\quad \text{with}
\quad
\begin{aligned}
i + j &= k + l - 1 \\
i &\geq \max(k, l) \\
j &\leq \min(k, l) - 1
\end{aligned}
$$

for $1 \leq k \leq g$ give g commuting vector fields on $\mathbf{C}^{3g+1}$ with $2g+1$ polynomial invariants a_i (defined by $\lambda^{2g+1} + a_{2g}\lambda^{2g} + \ldots + a_0 = UW + V^2$) and integral manifolds Jac $X\backslash\Theta$, where X varies over all hyperelliptic curves [here we named $U = \lambda^g + U_1\lambda^{g-1} + \ldots + U_g$ and V, W similarly].

2.1.3 Lemma *The following map from the (complexified) tangent bundle TS^g of the unit sphere to $\mathbf{C}^{3g+1}$: $\pi(x, y) = (U, V, W)$ with*

$$
U(\lambda) = f_1(\lambda)\sum_{k=1}^{g+1} \frac{x_k^2}{\lambda - e_k}, \quad V(\lambda) = \sqrt{-1}f_1(\lambda)\sum \frac{x_k y_k}{\lambda - e_k}
$$

$$
W(\lambda) = f_1(\lambda)(\sum_{k=1}^{g+1} \frac{y_k^2}{\lambda - e_k} + 1) \text{ with } f_1(\lambda) = \prod_{k=1}^{g+1}(\lambda - e_k)
$$

is surjective and a $2^{g+1}:1$ covering, unramified outside the subvariety where $U(e_k) = V(e_k) = W(e_k) = 0$ for some k.

We define the Neumann system as the system of harmonic oscillators $\ddot{x}_k = e_k x_k$ constrained to TS^g; it is Hamiltonian with $H = \frac{1}{2}(\Sigma y_k^2 + \Sigma e_k x_k^2)$ and has integrals $F_k = x_k^2 + \sum_{l\neq k} \frac{(x_k y_l - x_l y_k)^2}{e_k - e_l}$; moreover if $f_2(\lambda) = \sum_k \prod_{i\neq k}(\lambda - e_i)F_k$, then $f_1 f_2 = UW + V^2$, so that the coefficients of $f(\lambda)$ are constants of motion. To put the final touch to the algebro-geometric integration, we show that the image of the hamiltonian vector fields under the map π of 2.1.3 spans the invariant vector fields of Jac X; to do that, we define combinations of the vector fields of 2.1.2 (which are nothing but the invariant vector fields determined by the tangent directions to the image of X in Jac X, with Abel map normalized at P_∞, at some point P: $D_P = \sum_{k=1}^g \lambda(P)^{g-k}D_k$.)

2.1.4 Theorem $\pi_*(X_{F_k}) = c_k D_k$ *where* $c_k = 4\sqrt{-1}\prod_{l\neq k}(e_k - e_l)^{-1}$ *and* $\pi_*(X_H) = -2\sqrt{-1}D_\infty$.

If we put this together with the BC matrix of 1.14, we see that the KdV potential $u = 2U_1 = -2\sum_{k=1}^{g+1}(\sum_{l\neq k} e_l)x_k^2$ does indeed evolve according to the flow $\pi_*(X_H)$ (or a constant multiple of it). The higher KdV flows are linear combinations of the D_k's.

We have reached Goal 1, but we want to go a little further and report the theta formulas given by Mumford for the Neumann coordinates, and his amazing use of the Neumann system to characterize hyperelliptic Jacobians.

2.1.5 Theorem *After choosing a suitable symplectic homology basis for X, define the Riemann theta function ϑ and the Riemann vector $\vec{\Delta} = \Omega[\frac{1}{2}\cdots\frac{1}{2}]^T + [\frac{g}{2}\frac{g-1}{2}\cdots 1\frac{1}{2}]^T$*

and theta characteristics η_k corresponding to the points e_k; $\vec{e}$ is the vector in $\mathbf{C}^g$ corresponding to the vector field D_∞. Then the functions

$$x_k(t) = \frac{\vartheta[\eta_{2k-1}](0)\vartheta[\eta_{2k-1}](\vec{z}_0 - \vec{\Delta} + 2\sqrt{-1}t\vec{e})}{\vartheta[0](0)\vartheta[0](\vec{z}_0 - \vec{\Delta} + 2\sqrt{-1}t\vec{e})}$$

(where $\vec{z}_0$ is a fixed point corresponding to (U, V, W)) are defined on the torus $\mathbf{C}^g/\Omega'$ where $\Omega' = \{\Omega n + m | n, m \in \mathbf{Z}^g$ and $m_1, m_2 + n_1, \ldots, m_g + n_1 + \ldots + n_{g-1}, n_1 + \ldots + n_g$ are even$\}$ and evolve according to the t-flow X_H of the Neumann system.

Now fix a set B with $2g + 2$ elements and a subset U of B with $g + 1$ elements, and any bijection: $T \leftrightarrow [\eta'_T, \eta''_T]$ between the even subsets of B (modulo $T \sim (B \backslash T)$) and $\frac{1}{2}\mathbf{Z}^{2g}/\mathbf{Z}^{2g}$ such that $\#(T_1 \cap T_2) \equiv 2\eta'_{T_1} \cdot 2\eta''_{T_2} - 2\eta'_{T_2} \cdot 2\eta''_{T_1} \bmod 2$ and $\frac{\#(T \circ U) - (g+1)}{2} \equiv 2\eta'_T \cdot 2\eta''_T \bmod 2$ (where the circle denotes symmetric difference).

2.1.6 Theorem If a matrix Ω in the Siegel upper half g space satisfies $\vartheta[\eta_T] = 0 \Leftrightarrow \#(T \circ U) \neq g + 1$, then Ω is the period matrix of a smooth hyperelliptic curve of genus g (conversely, if B is the set of branch points of the hyperelliptic curve, the condition is satisfied).

This theorem is actually proved by constructing a Neumann system on a 2^{g+1} cover of $\mathbf{C}^g/\Omega$!

2.2 Moser's systems. Now we turn to Goal 2. The solution of Neumann's system appeared out of the blue, and we justify it now by reporting the two constructions of an equivalent Lax pair, which defines the curve very naturally. Moser showed that the Neumann system together with other important classical examples were special cases of "rank 2 perturbations" which preserve the spectrum of a matrix

$$2.2.1 \qquad L = A + ax \otimes x + bx \otimes y + cy \otimes x + dy \otimes y$$

(A is a fixed constant matrix which we can take diagonal diag $(e_1, \ldots, e_{g+1})$, det $\begin{bmatrix} a & b \\ c & d \end{bmatrix}$ $\neq 0$ and $u \otimes v$ denotes the matrix $[u_i v_j]$). Indeed, the hamiltonian flow of $H = a < Bx, x > +(b+c) < Bx, y > +d < By, y > - \frac{ad-bc}{2} \sum_{i \neq j} \frac{b_i - b_j}{a_i - a_j}(x_i y_j - x_j y_i)^2$, where $B = $ diag $(b_1, \ldots, b_n)$ is any fixed matrix, is equivalent to the Lax-pair equation $\dot{L} = [M, L]$ where M is a suitable matrix

$$M = \frac{1}{2}(b - c)[b_i \delta_{ij}] + (ad - bc)\left[\frac{b_i - b_j}{a_i - a_j + \delta_{ij}}(x_i y_j - x_j y_i)\right]$$

(Unless otherwise specified, the symplectic structure is the standard $\omega = \Sigma dx_i \wedge dy_i$ so that the hamiltonian H defines a flow $\dot{x}_i = \frac{\partial H}{\partial y_i}, \dot{y}_i = -\frac{\partial H}{\partial x_i}$, and $\frac{\partial G}{\partial t} = \{H, G\} = -\Sigma \frac{\partial H}{\partial x_i} \frac{\partial G}{\partial y_i} + \frac{\partial G}{\partial x_i} \frac{\partial H}{\partial y_i}$). This result still does not provide a spectral curve, but if we use the following key formula called "Weinstein-Aronszajn" in [Mos]

$$\det(I_n - \sum_{i=1}^{r} \xi_i \otimes \eta_i) = \det(I_r - [< \xi_i, \eta_j >])$$

(where $\xi_1, ..., \xi_r, \eta_1, ... \eta_r$ are n-vectors) then we find for the spectral invariants

$$\frac{l(\lambda)}{a(\lambda)} = \frac{\det(\lambda - L)}{\det(\lambda - A)} = \det(I - ((\lambda - A)^{-1}x) \otimes (ax + by) - ((\lambda - A)^{-1}y) \otimes (cx + dy))$$

$$= \det(I_2 - W_\lambda(x, y))$$

with

$$W_\lambda(x, y) = \begin{bmatrix} < (\lambda - A)^{-1}x, x > & < (\lambda - A)^{-1}x, y > \\ < (\lambda - A)^{-1}x, y > & < (\lambda - A)^{-1}y, y > \end{bmatrix} \begin{bmatrix} a & b \\ c & d \end{bmatrix}$$

and $\det(I - W_\lambda(x, y)) = 1 - tr\ W_\lambda + \det W_\lambda = 1 - \phi_\lambda(x, y)$.
Moser also showed that the system is completely integrable and linearizes on the (generalized) Jacobian of the curve $\mu^2 = \phi_\lambda(x, y)$ and if you let $a = 1$, $b = -c = 1$, $d = 0$ which gives the Neumann system you recognize $a^2(\lambda)\phi_\lambda(x, y) = UW + V^2$, our Neumann hyperelliptic curve.

That's not all, however: if you play with the dilation $x \mapsto \lambda x$ then $A \mapsto A + \lambda^2 x \otimes x + \lambda(x \otimes y - y \otimes x)$ and you get a Lax pair with a parameter, which perhaps makes the spectral curve look a bit more natural:

2.2.2 [AvM, Theorem 4.4] The Neumann flow is equivalent to the Lax-pair: $\dot{L}_1 = [M_1, L_1]$ where $L_1 = A\mu^2 + \mu(x \otimes y - y \otimes x) + x \otimes x$ and $M_1 = A\mu + x \otimes y - y \otimes x$. Moreover, the hamiltonian(s) are of "Adler-Kostant-Symes" (AKS) type, namely projections (with respect to an ad-invariant inner product) of gradients of orbit-invariant functions to half of the splitting of a Lie algebra. Specifically, $\{\sum_{-\infty}^{N} A_j \mu^j | A_j \in gl(n, \mathbf{C})\} = K \oplus N$ with $K = \{\sum_0^N A_j \mu^j\}$ and $N = \{\sum_{-\infty}^{-1} A_j \mu^j\}$; if the inner product is $< A, B > = \sum_{i+j=-1} tr A_i B_j$; then the dual of N can be identified with $K = K^\perp$ and the hamiltonian for the Neumann flow can be taken to be $H = < \frac{1}{2}(L_1\mu^{-2})^2, \mu^3 I_n >$ under the Lie-Poisson brackets and suitable reduction. The flows linearize on the Jacobian of the (hyperelliptic) curve $\det (L_1 - \eta) = 0$. (This is the same curve we got in the 2×2 model, but if you are wondering whether different representations for the same system can give rise to nonisomorphic curves, you are right on the mark; Donagi's lectures in this volume are likely to address this question.)

In view of 2.2.1, 2.2.2, we pose two puzzles: one, how do we generalize Moser's rank 2 perturbations to nonhyperelliptic curves (and higher-order KdV equations); two, how do we explain the 2×2 versus $(g + 1) \times (g + 1)$ different Lax-pair models for the spectral curve, and generalize such pattern? The answer to the first is to be found in [AHP] and to the second in [AHH]. We sketch the picture roughly first, then give the formulas together with some proofs which turn out to have appealing simplicity. First we define the symplectic manifold of (generically) rank r perturbations:

2.2.3 **Definition.** We let $M_{n,r}$ denote the space of $n \times r$ complex matrices, with $n \geq r$ and give $M = M_{n,r} \times M_{n,r}$ the symplectic structure $\omega(F, G) = tr(dF \wedge dG^T)$ for $F, G \in M$. A rank r perturbation of a fixed $n \times n$ matrix A is $L = A + FG^T$.

2.2.4 **Definition.** We split the formal loop algebra $\widetilde{gl(r)} = \widetilde{gl(r)}^+ \oplus \widetilde{gl(r)}^-$ where $\widetilde{gl(r)}^+$ consists of $r \times r$ matricial polynomials in λ and $\widetilde{gl(r)}^-$ of strictly negative formal power series. Under the pairing $< X(\lambda), Y(\lambda) >= tr(X(\lambda)Y(\lambda))_-$ (where the subscript $-$ means the coefficient of λ^{-1}) the dual of $\widetilde{gl(r)}^+$ is identified with $\widetilde{gl(r)}^-$, which therefore admits a Lie-Poisson structure.

Roughly then, we consider an action on M whose moment map (see below) lands in $\widetilde{gl(r)}^-$; we check that the AKS flows on $\widetilde{gl(r)}^-$ correspond to isospectral deformations of $L = A + FG^T$ for flows on M_A (cf. 2.2.9); finally we perform a Marsden-Weinstein reduction for an (equivariant) $GL(r)$ action to obtain a completely integrable system on a symplectic leaf, whose flows are linear on the Jacobian of the spectral curve. We recall very briefly the general definitions.

2.2.5 **Moment map.** (i) A smooth group-action of G on a symplectic manifold (M, ω) is said to be hamiltonian if there exists a "moment map" $J: M \to g^*$ such that the Hamiltonian vector field associated to J and a fixed element $\xi \in g$ is the same as the infinitesimal action associated to ξ. We will, however, be content with an infinitesimal definition because in our formal setup the group of a Lie algebra will often remain in the background. We just recall that (ii) the Lie-Poisson structure of g^* is defined by:

$$\{\phi, \psi\}_{g^*}(\mu) = < \mu, [d\phi(\mu), d\psi(\mu)] > \text{ for } \phi, \psi \in C^\infty(g^*), \ \mu \in g^*$$

where $d\phi: g^* \to g^{**}$ (which in our situations will always be identified with g) is defined by $< d\phi(\mu), \nu >= \frac{d}{dt}\phi(\mu + t\nu)\Big|_{t=0}$, $\mu, \nu \in g^*$. Now we say that $J: M \to g^*$ is a moment map if (iii) its linear dual $j: g \to C^\infty(M)$ is a Lie-algebra homomorphism; or if (iv) it is a Poisson map w.r.t. the Lie-Poisson structure: $\phi, \psi \in C^\infty(g^*) \Rightarrow \{J^*\phi, J^*\psi\} = J^*\{\phi, \psi\}_{g^*}$. In case we do have a hamiltonian G-action, then the subspace $C_G^\infty(M)$ of G-invariant functions is a Lie subalgebra of $C^\infty(M)$. If G acts freely and properly on M, then M/G is a manifold with a Poisson structure inherited from the one on M through the identification $C^\infty(M/G) \cong C_G^\infty(M)$. The symplectic leaves of M/G have the form $M_\mu = J^{-1}(\mu)/G_\mu = J^{-1}(\mathcal{O}_\mu)/G$ where $\mu \in g^*$, G_μ is the isotropy group of μ in G and $\mathcal{O}_\mu$ is the G-orbit through μ. The reduced manifold M_μ has a natural symplectic structure ω_μ such that $i^*\omega = \pi^*\omega_\mu$, where $i: J^{-1}(\mu) \to M$ is inclusion and $\pi: J^{-1}(\mu) \to M_\mu$ is the natural projection taking points to their G_μ-orbits.

2.2.6 **The R matrix.** Given a linear map $R: g \to g$, the alternating bilinear form $[X, Y]_R = \frac{1}{2}([RX, Y] + [X, RY])$ satisfies the Jacobi identity $\Leftrightarrow$ some quadratic conditions on R are satisfied. Assuming they are, for all pairs of invariant functions I, J on g^* we have $\{I, J\}_R = 0$ (where $\{ \}_R$ is the attendant (Lie-Poisson) structure).

Proof. $\{I, J\}_R(\mu) = < [dI(\mu), dJ(\mu)]_R, \mu >$
$= \frac{1}{2} < [RdI(\mu), dJ(\mu)], \mu > + \frac{1}{2} < [dI(\mu), RdJ(\mu)], \mu >$,
but e.g. $< [RdI(\mu), dJ(\mu)], \mu >=< RdI(\mu), ad^*dJ(\mu)(\mu) >= 0$. ∎

2.2.7 **Remark.** As is clear from the proof, our definition of invariant need only be infinitesimal, i.e. $f \in I(g^*)$ iff $< \mu, [df(\mu), X] >= 0 \ \forall \mu \in g^*$, $X \in g$. Of course when we have a corresponding Lie group the invariants are the functions which are invariant under the natural action, such as the symmetric functions of the eigenvalues of a matrix.

2.2.8 The AKS flows. For a splitting $g = K \oplus N$, as in the situation of 2.2.2 or 2.2.4, with $g^* = N^* \oplus K^*$, an example of R-matrix (2.26) is given by $R(X) = X_+ - X_-$ (where $+, -$ denote projection to K, N): the Jacobi identity is straightforward to check. As a consequence, invariants on g^* are in involution w.r.t. $\{ \ \}_R$ and these are called AKS flows: $\dot{X} = [df(\tilde{X})_+, X] = -[df(\tilde{X})_-, X]$, which we give for the case in which we can identify K with K^* and $\tilde{X}$ is the element in K^* that corresponds to $X \in K$. Moreover, we can further restrict the flows to any $k \times n$, if k and n are orbits of a symplectic action, a special case of which is $K \times \{\text{point}\}$: this shifts the flows by a character χ of N, namely a $\chi \in N^*: < \chi, [N, N] > = 0$. To apply this to our situation, if $\phi \in I(g^*)$ we define $\phi_\lambda = \phi(\mu + \chi)$, $\mu \in g^*$ and let $\overline{\phi}_\chi$ be the restriction of ϕ_χ to K^*; then $\{\overline{\phi}_\chi, \overline{\psi}_\chi\}_{K^*} = 0$ for all $\phi, \psi \in I(g^*)$ and $\dot{X} = [d\phi((\widetilde{X + \chi}))_+, X + \chi]$ for $X \in K^*$, with $(\widetilde{X + \chi})$ the element of g^* corresponding to $X + \chi \in g$. In the situation of 2.2.4 χ could be any constant matrix.

We now proceed to the appropriate moment maps. We generalize the constant matrix A by allowing multiple eigenvalues α_i of multiplicities $n_i \leq r, n_1 + \ldots + n_k = n$, so that $\det (A - \lambda I) = \prod_{i=1}^{k}(\alpha_i - \lambda)^{n_i}$. Let $a(\lambda) = \prod_{i=1}^{k}(\alpha_i - \lambda)$. We split an $n \times r$ matrix F into k blocks F_i accordingly.

2.2.9 Definition/Statement (i) $J_r^n(F, G)(x_1, \ldots, x_n) = -\sum_{j=1}^{n} tr(F_j X_j G_j^T)$ is the moment map of the action $[(g_1, \ldots, g_n)(F, G)]_i = (F_i g_i^{-1}, G_i g_i^T)$ where $g_i \in GL(r)$ so that under standard identifications $J_r^n(F, G) = -(G_1^T F_1, \ldots, G_n^T F_n)$ and restricting the action to the diagonal subgroup $\{(g, \ldots, g)\}$, $J_r(F, G) = -G^T F$.

(ii) For $X(\lambda) \in \widetilde{gl(r)}^+$ we define $\alpha(X(\lambda)) = (X(\alpha_1), \ldots, X(\alpha_r))$ and obtain the exact sequence

$$0 \to a(\lambda)\widetilde{gl(r)}^+ \xrightarrow{\iota} \widetilde{gl(r)}^+ \xrightarrow{\alpha} g_r^n \to 0$$

By dualizing, and identifying g_r^n to its dual by using the trace componentwise, we get $\alpha^*(Y_1, \ldots, Y_n) = \sum_{i=1}^{k} \frac{Y_i}{\lambda - \alpha_i}$ and finally check that $\tilde{J}_r = \alpha^* \circ J_r^n$ is a moment map. By combining (i) and (ii), we get a moment map $\tilde{J}_r(F, G) = \sum_{i=1}^{k} \frac{G_i^T F_i}{\alpha_i - \lambda} = G^T(A - \lambda)^{-1}F$, which becomes injective on $\mathcal{M}/H$, where $\mathcal{M}$ is a suitable open submanifold of M and $H = GL(n_1) \times \ldots \times GL(n_k)$ acts blockwise by $(h_i F_i, h_i^{-1T} G_i)$.

(iii) We also notice that the "Moser space" $M_A = \{A + FG^T | F, G \in \mathcal{M}\}$ of rank r perturbations can be identified with the orbit space $\mathcal{M}/G_r, G_r = GL(r)$ acting as in (i).

We are now going to turn on the obvious AKS flows on $\widetilde{gl(r)}^-$: the key observation is that they are isospectral for the rank r perturbation $A + FG^T$; let's consider more generally the matrix $L_A(\tau) = A + F\tau G^T$, for some fixed $\tau \in GL(r)$ (this occurs in the classical examples); if $Y \in gl(r)$ is such that $I_r + Y$ is invertible and $\tau = (I + Y)^{-1}$, we use Y for the character-shift of 2.2.8. Since

$$
\begin{aligned}
\det(A + F\tau G^T - \lambda I) &= \det(A - \lambda I)\det(I + G^T(A - \lambda I)^{-1}F\tau) \\
&= \det(A - \lambda I)\det(\tau^{-1} + G^T(A - \lambda I)^{-1}F)\det\tau \\
&= \det(A - \lambda I)\det(I + G^T(A - \lambda I)^{-1}F + Y)\det\tau
\end{aligned}
$$

we see that the Poisson-commutative ring $\mathcal{F}_+^Y$ of projected, Y-shifted invariants (cf. 2.2.8) defines by composition with $\tilde{J}_r$ a Poisson-commutative ring $\mathcal{F}^Y$ of isospectral flows on $M_{n,r} \times M_{n,r}$. The final reduction in the presence of Y should be taken w.r.t. the stabilizer of τ in $Gl(r)$, which leaves the functions of $\mathcal{F}^Y$ invariant. To show that the symplectic leaves are completely integrable systems, we calculate their dimensions, as well as the genus of the spectral curve, and invoke [RS] which gives the link between AKS and Jacobian flows.

2.2.10 **Examples:** (i) The Neumann system ($r = 2$) requires an extra reduction from $\widetilde{gl(2)}$ to $\widetilde{sl(2)}$ and a choice of $Y = \begin{bmatrix} 0 & -1 \\ 0 & 0 \end{bmatrix}$; the reduction requires $J_H = 0$ where H is multiplication by 2×2 diagonal matrices, so that $G_i \cdot F_i = 0$ and we can let $(x, y) = F, (y, -x) = G$.

We rediscover the spectral curve of 2.2, $\det\left[\sum_{i=1}^n \frac{G_i^T F_i}{\alpha_i - \lambda} + \lambda Y - \mu \begin{bmatrix} 1 & 0 \\ 0 & 1 \end{bmatrix}\right] = 0$; the group $SL(2)_Y$ is one-dimensional, so that the conserved quantities (i.e. moment map = constant) are given by the restriction of $J_{SL(2)}(x, y)$ to $sl(2)_Y$,

i.e. $tr \begin{bmatrix} x \cdot y & y \cdot y \\ -x \cdot x & -x \cdot y \end{bmatrix} \begin{bmatrix} 0 & -1 \\ 0 & 0 \end{bmatrix} = x \cdot x$ which we can take to equal 1.

(ii) The Boussinesq equation $3u_{yy} + u_{xxxx} + 12(uu_x)_x = 0$ is a consequence of the Lax pairs $\frac{\partial}{\partial x}\mathcal{L} = [A, \mathcal{L}]$ and $\frac{\partial}{\partial y}\mathcal{L} = [B, \mathcal{L}]$, where

$$
A = \lambda \begin{bmatrix} 0 & 0 & 0 \\ 0 & 0 & 0 \\ 1 & 0 & 0 \end{bmatrix} + \begin{bmatrix} 0 & 1 & 0 \\ 0 & 0 & 1 \\ -3u_x - v & -3u & 0 \end{bmatrix},
$$

$$
B = \lambda \begin{bmatrix} 0 & 0 & 0 \\ 1 & 0 & 0 \\ 0 & 1 & 0 \end{bmatrix} + \begin{bmatrix} 2u & 0 & 1 \\ -u_x - 3v & -u & 0 \\ -u_{xx} - 3v_x & -2u_x - 3v & -u \end{bmatrix}
$$

and $\mathcal{L}$ is a matricial polynomial in $\lambda^{-1}, \mathcal{L} = \mathcal{L}_0 + \mathcal{L}_1\lambda^{-1} + \ldots + \mathcal{L}_n\lambda^{-n}$. We produce this system by rank 3 perturbations and constrained $\widetilde{sl(3)}$ flows. We use the hamiltonians

$$
\phi_k = (\mathcal{N}(\lambda)) = \frac{1}{2}tr\left(\frac{a(\lambda)}{\lambda^n}\lambda^k\mathcal{N}^2(\lambda)\right)_0 \quad \text{where } \mathcal{N}(\lambda) \in \widetilde{sl(3)}_-
$$

$$
\psi_k(\mathcal{N}(\lambda)) = \frac{1}{3}tr\left(\left(\frac{a(\lambda)}{\lambda^n}\right)^2 \lambda^k\mathcal{N}^3(\lambda)\right)_0
$$

and ϕ_1, ψ_2 will give us our x and y flows. For $(d\phi_1)_+$ and $(d\psi_2)_+$ to be matrices of the form displayed above, with $\mathcal{L}(\lambda) = \frac{a(\lambda)}{\lambda^n}\mathcal{N}(\lambda)$, we must impose some constraints, besides prescribing a number of constants of motion. To produce the rank 3 perturbation, we choose each F_i, G_i to be 3-vectors p_i, q_i, with $\|p_i\| = 1$ and $p_i \cdot q_i = 0$, and obtain $\tilde{J}_3 = \mathcal{N}(\lambda) = \sum_{i=1}^n \frac{p_i^T q_i}{\alpha_i - \lambda}$.

Exercise: Find a Boussinesq model with a character Y and identify the constraints with the appropriate reduction w.r.t. $Gl(3)_Y$.

This completes the discussion of one puzzle. The answer to the second is provided by the beautiful diagram (found in [AHH]):

$$\widetilde{gl(n)}^{-} \quad\overset{J_n^Y}{\longleftarrow}\quad M \quad\overset{J_r^A}{\longrightarrow}\quad \widetilde{gl(r)}^{-}$$

$$\overset{J_{n,0}^Y}{\nwarrow} \qquad \nearrow \qquad \searrow \qquad \overset{J_{r,0}^A}{\nearrow}$$

$$\downarrow \quad M/GL(r)_Y \qquad\qquad\qquad M/GL(n)_A \quad \downarrow$$

$$\searrow \qquad \swarrow$$

$$\widetilde{gl(n)}_-/GL(n)_A \longleftarrow M/(GL(r)_Y \times GL(n)_A) \longrightarrow \widetilde{gl(r)}^{-}/GL(r)_Y$$

where much of the notation can be guessed from the previous, but specifically:
$J_r^A(F,G) = -G^T(A - \lambda)^{-1}F$, $J_n^Y(F,G) = F(Y - z)^{-1}G^T$;
$GL(n): M \to M$ by $g: (F,G) \mapsto (gF, (g^T)^{-1}G)$,
$GL(r): M \to M$ by $g: (F,G) \to (Fg^{-1}, Gg^T)$; $GL(n)_A$ and $GL(r)_Y$ denote the stabilizers under conjugation of $A \in gl(n)$ and $Y \in gl(r)$, resp., and the meaning of the diagram is that we restrict when necessary to open dense submanifolds (to have, for example, $Gl(n)_A, G(r)_Y$ act freely). The relevant function rings on M which give hamiltonians are:

$$\mathcal{F}^Y = \{\psi \in C^\infty(M)|\psi(F,G) = \phi(Y + G^T(A - \lambda)^{-1}F), \phi \in I(\widetilde{gl(r)}^*)\}$$

$$\mathcal{F}^A = \{\psi \in C^\infty(M)|\psi(F,G) = \phi(A + F(Y - z)^{-1}G), \phi \in I(\widetilde{gl(n)}^*)\}$$

and the final statement on the AKS flows (which can be viewed by reduction as flows on the symplectic leaves occurring at the bottom of the diagram, where the arrows are 1:1 Poisson maps) is the following:

2.2.11 Theorem. *The two rings $\mathcal{F}^Y$ and $\mathcal{F}^A$ are equal, their elements Poisson commute and their Hamiltonian flows preserve the spectrum of $N(\lambda) = Y + G^T(A - \lambda)^{-1}F$ and $M(z) = A + F(Y - z)^{-1}G$. If $\psi \in \mathcal{F}^Y = \mathcal{F}^A$ is of the form $\psi(F,G) = \phi_1(N) = \phi_2(M)$, the integral curves of the corresponding hamiltonian flow satisfy*

$$\frac{dN}{dt} = [(d\phi_1)_+, N], \quad \frac{dM}{dt} = [(d\phi_2)_+, M].$$

Note that the equality of the rings follows from the usual Weinstein-Aronszajn identity:

$$\det(A - \lambda)\det(Y + G^T(A - \lambda)^{-1}F - z) = \det(Y - z)\det(A + F(Y - z)^{-1}G^T - \lambda),$$

since generators of $\mathcal{F}^Y(\mathcal{F}^A)$ will appear as coefficients in the expansion of the left (right)-hand side in λ and z; the formula also shows that the two spectral curves are birational.

2.3 Flaschka's matrix and stroboscopes.

Now that we accomplished Goal 2 we turn to Goal 3. The work I am presenting is due to H. Flaschka and collaborators [Fl,EF,S]; they viewed the significance of this viewpoint in two directions: extrinsic geometric information on the spectral curve and surface (=total space of the line bundle which corresponds to the BC divisor); and generalization of the Neumann system to the nonhyperelliptic case; I would add as a third item the stroboscope algorithms and their potential generalizations, and think that this viewpoint deserves much further work and systematization. As usual, before providing precise formulas, I would like to discuss, if loosely, the general idea. We saw in Lecture 1 that two commuting ODO's L, B of orders n, m give rise to a curve X and a divisor; the data can be encoded in an $n \times n$ matrix, say the action of B on $V_\lambda = \lambda$-eigenspace of L. We saw in Lecture 2 that this matrix was the image of a moment map and could be made to flow according to the AKS system; all the motion is linear on Jac X. But some duality (2.2.11) provided us with a related matrix; specifically, if $n = 2$ and $m = 2g + 1$ (Neumann) this other matrix has size $g + 1$: what is its significance? Let me point out something important. To write the BC matrix we only needed the curve, the point P_∞, and the divisor (plus some choices of normalization). Whereas to write the $(g+1) \times (g+1)$ matrix we also need to choose $g+1$ branchpoints, $e_1, \ldots, e_{g+1}$ say; it is this choice which will afford us an integrable system with $2g$-dimensional phase space–the 'angle variables' live on Jacobians whereas the 'action variables', as it were, are moduli of curves obtained by varying the remaining g finite branchpoints. It is this choice of points that gives us our clue: the "other" matrix is part of another way of representing meromorphic functions on the curve. By Krichever's construction, if we fix an algebraic function h with $g + 1$ simple zeroes at $e_{g+2}, \ldots, e_{2g+1}$, P_∞ and poles at $e_1, \ldots, e_{g+1}$, $h = \mu / \prod_{i=1}^{g+1}(\lambda - e_i)$, say, then we can find a unique first-order matrix differential operator L_1 such that $L_1 \psi = h\psi$, where ψ is a suitable Baker *vector* with essential singularities at $e_1, \ldots, e_{g+1}$; through this way of representing the ring $H^0(X \backslash \{e_1, \ldots, e_{g+1}\}, \mathcal{O}_X)$ we obtain a second-order matrix differential operator B_1 corresponding to the function λh^2 and commuting with L_1: the "other model" for the Neumann Lax pair is the action of this B_1 on the $(g + 1)$-dimensional kernel of $L_1 - h$. We now sketch a few proofs because they are highly representative of methods in the subject.

First of all, the link we gave between Neumann and KdV (2.1.4) was based on the BC construction; there is an alternative, more direct proof which uses eigenfunctions and residues:

2.3.1 **Definitions.** (i) We give a more geometric description, due to Krichever, of the Baker function introduced in 1.13: $\psi(x, P)$ is the unique function with the following properties: for $|x|$ sufficiently small it is meromorphic on $X \backslash \{P_\infty\}$, with pole divisor bounded by $\delta = P_1 + \ldots + P_g$, independent of x, such that $h^0(\delta - P_\infty) = 0$, and near P_∞ $\psi(x, P)e^{-xz} = 1 + O(z^{-1})$ is holomorphic, with z chosen to be $\lambda^{1/2}$ in our case.

(ii) We let Ω be the unique meromorphic differential with zeroes on δ and a double pole of the form $(-\lambda + \text{holomorphic})dz^{-1}$ at P_∞. Note (1): Riemann-Roch shows that Ω is unique, $\dim\{\varphi|(\varphi) + \delta - 2P_\infty \geq 0\} - \dim\{\omega|(\omega) \geq \delta - 2\infty\} = (g-2) - g + 1$. (2) We get another characterization of the dual Baker function as meromorphic on $X\backslash\{P_\infty\}$ with poles bounded by δ' and behavior $e^{-xz}(1 + O(z^{-1}))$ near P_∞, where $\delta + \delta'$ are the $2g$ zeroes of Ω. (3) $\Omega = d\lambda/Wr(\psi, \phi)$.

2.3.2 Theorem. *Let* $\rho_j = \text{Res}_{e_j}h\Omega$, $x_j = \sqrt{\rho_j}\psi(x, e_j)$, $y_j = \sqrt{\rho_j}\phi(x, e_j)$; *then* $\sum_{j=1}^{g+1} x_j^2 = 1$, $\sum_{j=1}^{g+1} x_j y_j = 0$, $\sum_{j=1}^{g+1}(e_j x_j^2 + y_j^2) = u(x)$ *and* $\{x_j, y_j\}$ *satisfy the Neumann system.*

Proof. The constraints follow from the Residue Theorem applied to the differential $h\Omega\psi\phi$ (it has a residue of -1 at P_∞); the differential equations $\ddot{x}_j = e_j x_j - u x_j$ follow from the assumption $L\psi = \lambda\psi$. ∎

Next we report a result of Flaschka's which explains why $g+1$ is the 'dual' number of 2; before stating it we observe that ψ is a quotient of (normalized) sections of $\mathcal{L}_x$ and $\mathcal{L}(\delta)$, where $\mathcal{L}_x = \mathcal{L}(\delta) \otimes \mathcal{L}'_x = \mathcal{L}(\zeta(x))$ is the bundle $\mathcal{L}(\delta)$ twisted by e^{-xz} near P_∞ (cf. 1.16) and $\zeta(x)$ are the zeroes of the Baker function; we denote a similar twist at a point P_i by $\mathcal{L}_x^{(i)}$ and define the vector Baker function $\underline{\psi}$ in analogy to 2.3.1, with pole divisor bounded by Δ of degree $g + l - 1$ in the case of l points $P_1, \ldots, P_l$, and exponential behavior $\psi_\alpha(x, P) = e^{C_\beta hx}(\delta_{\alpha\beta} + O(h^{-1}))$ near P_β, with $C_1, \ldots, C_l$ distinct constants and h a function with simple poles at $P_1, \ldots, P_l$.

2.3.3 Proposition. *With notation as above,* $\mathcal{L}_x \otimes \mathcal{L}(Z)$ *is isomorphic to the bundle of a vector Baker function, i.e.* $\mathcal{L}(\delta) \otimes \mathcal{L}'_x \otimes \mathcal{L}(Z) \cong \mathcal{L}(\delta + Z) \otimes \mathcal{L}_x^{(1)} \otimes \ldots \otimes \mathcal{L}_x^{(l)}$ *for any generic divisor* Z *of degree* g, *but for no* Z *of smaller degree.*

Proof. Assuming the isomorphism exists, we have a function h with simple poles at $P_1 + \ldots + P_l$, and a nowhere vanishing holomorphic section σ of $\mathcal{L}_x^{(1)} \otimes \ldots \otimes \mathcal{L}_x^{(l)} \otimes \mathcal{L}_x^{-1}$, which defines a nowhere vanishing function F meromorphic on $X\backslash\{P_1, \ldots, P_l, P_\infty\}$ and with local behavior $e^{C_\alpha hx}\sigma_{P_\alpha}(P)$, $e^{-zt}\sigma_\infty(P)$. Then $G = \log F$ is well defined, meromorphic on X, with simple poles at $P_1, \ldots, P_l, P_\infty$ (and conversely if such G exists the statement of the proposition holds). Assume that $l \leq g$; then $P_1 + \ldots + P_l$ is a special divisor (by Riemann-Roch), so that it is made out of pairs $Q_j + \iota Q_j$; but the product $(\lambda - \lambda(Q_j))$ G then has only a pole of order $\leq l + 1$ at P_∞ and μ is the lowest-odd-order-pole function at P_∞, so that $2g + 1 \leq l + 1 \leq g + 1$, a contradiction. For $l = g+1$, and h a function with simple poles at a nonspecial divisor $P_1 + \ldots + P_{g+1}(P_j \neq P_\infty)$, we normalize $h = \frac{1}{z} + \ldots$ at P_∞ and $G(P) = -\lambda h$ satisfies the requirement. ∎

2.3.4 Vaguely Formulated Conjecture: For this hyperelliptic case, the choice of "augmented divisor" (in Flaschka's terminology) $\Delta = \delta + Z$ was given a lower-bound in 2.3.3, which is intimately related with the dynamical system; in the currently studied generalizations of integrable systems to situations where $X \to \mathbf{P}^1$ is replaced by $\Gamma \to X$ and the Hamiltonians (coefficients of the spectral curve) by invariants of a twisted endomorphism $E \to E \otimes L$, E a vector bundle and L a line bundle over X, there have appeared [Bo,M] lower bounds on L for the construction of a symplectic leaf that hosts the flows: 2.3.3 must be a special example of these bounds.

2.3.5 The Baker vector. Let $\chi_i(x,P) = \frac{\sqrt{\rho_i}}{h} \frac{Wr(\psi(x,P),\phi(x,e_i))}{\lambda - e_i}$ $i = 1,\ldots,g+1$ and $\psi_i(x,P) = \chi_i(x,P)e^{-h\lambda x}$. Then

$$\dot{\psi} = (-Xh + y \otimes x - x \otimes y)\psi$$

where $X = \text{diag}\,(e_1,\ldots,e_{g+1})$; we call the matrix $-Xh + y \otimes x - x \otimes y = T$; and if we define L_2 by

$$L_2\underline{\psi} = (-Xh^2(y \otimes x - x \otimes y)h - x \otimes x)\underline{\psi},$$

then $L_2\underline{\psi} = -\lambda h^2\underline{\psi}$ and $\dot{L}_2 = [T, L_2]$ is equivalent to the Neumann system.

Proof. We sketch another important technique due to Krichever: it is straightforward to compute the poles and exponential behavior of all the functions in sight; to show for example that $\dot{\psi} = T\underline{\psi}$, one checks that $\dot{\psi} - T\underline{\psi}$ has behavior $O(h^{-1})e^{-h\lambda t}$ near each P_i, so when expressed in terms of the normalized Baker basis of the $(g+1)$-dimensional eigenspace it must have all coefficients $= 0$. ∎

2.3.6 Stroboscopes. The operation of "transference" will be introduced in detail in the next lecture. Here we only indicate a possible link with the vector-Baker map of 2.3.5. (i) For a commuting pair of differential operators, L, B as in Lecture 1, the Baker function gives a common divisor: $L - \lambda = T_1(\partial - \frac{\psi'}{\psi})$, $B - \mu = T_2(\partial - \frac{\psi'}{\psi})$ and the transference (at λ, μ) gives a new pair $\tilde{L} - \lambda = (\partial - \frac{\psi'}{\psi})T_1$, $\tilde{B} - \mu = (\partial - \frac{\psi'}{\psi})T_2$. At the divisor level, [BC2] show that this corresponds to the Abelian sum: $\delta(\tilde{L}, \tilde{B}) \sim \delta(L, B) - (\lambda, \mu) + P_\infty$ (linearly equivalent). What of the Baker function? The formula in general is only slightly more complicated, but for $n = 2$ if we recall that the dual Baker function ϕ also solves $(L - \lambda)\phi = 0$, we see that $(\partial - \frac{\psi'}{\psi})\phi$ is in Ker $(\tilde{L} - \lambda)$ and so is $Wr(\phi, \psi)$, which needs however to be normalized. (ii) motivated by (i), [EF] create the Baker vector by a sequence of $g + 1$ modified transferences at the points $e_1,\ldots,e_{g+1}$ (recall that the augmented divisor is $\delta + e_{g+2} + \ldots + e_{2g+1} \sim \delta - e_1 - \ldots - e_{g+1} + (2g+1)P_\infty$). More precisely, if $\chi(x,P) = \frac{Wr(\psi(x,P),\phi(x,P_0))}{\lambda(P) - \lambda(P_0)}$, then the Baker function after transference at P_0 becomes $\tilde{\psi}(x,P) = \chi(x,P)\psi(0,\tau P_0)/\chi(0,P)\psi(x,\tau P_0)$ whereas the Baker vector is given by $\psi_j(x,P) = \sqrt{\rho_j}\frac{Wr(\psi(x,P),\psi(x,e_j))}{h(\lambda(P) - e_j)}$.

(iii) The QR algorithm is used in numerical analysis to approximate a matrix by a diagonal one: it consists in applying a sequence of steps: $A_0 = Q_0R_0$ (orthogonal times upper-triangular), $A_1 = R_0Q_0 = Q_1R_1$, etc. It is obviously redolent of transferences, since it is obtained by conjugation, which does not destroy the spectrum, and indeed it can be produced by applying a stroboscope to the Toda flow [Sy]: if $L_0 = L(0)$ is the tridiagonal matrix corresponding to the initial condition of the Toda lattice, and if $e^{L(k)} = Q_kR_k$, then $e^{L(k+1)} = R_kQ_k$. Moreover, a suitable flow that commutes with Toda exhibits the QR algorithm at integral times [DLT].

2.3.7 Open Problems

(i) The intent of the observations 2.3.6 is to suggest finding a closer link between PDE's of KP type and finite-dimensional integrable systems by stroboscope algorithms.

(ii) The significance of the Neumann system in moduli (2.1.6) calls for exploration in the case of generalized Neumann systems. One place to start would be the moduli of "$\mathbf{Z}_N$-curves", of kind $\mu^N = \prod_{i=1}^{h}(\lambda - \alpha_i)^{r_i}$, $1 \leq r_i \leq N - 1$, investigated by [BR], who generalized certain classical hyperelliptic theta equations (Thomae's formulas, of fundamental importance in the proof of 2.1.6) to the curves, by means of conformal field theory!

Lecture 3: Higher rank.

At the time of writing, this is the one area covered in these notes where most work still needs to be done. Let's distinguish at the outset between two applications of higher (than 1) rank bundles over curves to integrable systems: one, which we address in this lecture, is to deformations of KP type; the other, which we skirt in lectures 4-6 while doing other things, has to do with moduli spaces/equivalence classes of semistable bundles and finite-dimensional hamiltonian systems. The latter area is immensely active at present, the impetus having been provided by Hitchin's work, cf. [Hi1], and we believe Donagi's lectures in this volume will be devoted to it. As for the former area, we refer to [PW1] for a 'historical' survey; there are two approaches, one fairly general which we'll discuss first: it has the advantage of ensuring the construction in principle and the disadvantage of lacking explicitness, as will be pointed out; the other, very concrete and ad hoc, has the disadvantage of being only adapted to extremely particular situations, and the advantage of yielding some pleasant surprises–we will give three examples, 1. the decomposable bundles over an elliptic curve; 2. the elliptic rank 2 transference; 3. the maximality of certain abelian subalgebras of $\mathcal{D}$. Finally, while we saw in Lectures 1,2 that the two applications of line bundles to integrability (KP and finite hamiltonians) come together, there seems to be no such structure to the general higher-rank situation.

3.1 General Construction. This part is not entirely self-contained and we refer to the book [PS] and the paper [SW] for the definition of the (analytic) grassmannian Gr and the construction of the KP flows; we will recall the facts we need. While the statements are more complete and clean in this setting, we emphasize that the underlying idea is again the Krichever construction [KN].

3.1.1 KP flows on Gr. We recall that an element of $GrH^{(r)}$ is a subspace $W \subseteq L^2(S^1, \mathbf{C}^r)$ such that the orthogonal projections $\pi_+ \colon W \to H_+^{(r)}, \pi_- \colon W \to H_-^{(r)}$ are respectively Fredholm and compact (the entries of H_+ have nonnegative exponents in the orthonormal basis z^j of $L^2(S^1, \mathbf{C})$, those of H_- negative exponents); we shall always tacitly assume that W belongs to the connected component where dim Ker $\pi_+ -$ dim Coker$\pi_+ = 0$. We also recall the operation of "interleaving Fourier coefficients" which gives an isometry between $H^{(r)}$ and $H^{(1)}$:

$$[f_0(z), \ldots, f_{r-1}(z)] \mapsto f(z) = \sum_{i=0}^{r} z^i f_i(z^r),$$

with inverse:

$$f(z) \mapsto f_i(z) = \frac{1}{r} \sum_{\zeta^r = z} \zeta^{-i} f(\zeta)$$

Example: $r = 2$, $[a_0 + a_1 z + a_2 z^2 + \ldots, b_0 + b_1 z + b_2 z^2 + \ldots] \mapsto a_0 + b_0 z + a_1 z^2 + b_1 z^3 + \ldots$
$c_0 + c_1 z + c_2 z^2 + \ldots \mapsto$
$[c_0 + c_1 z^{1/2} - c_1 z^{1/2} + c_2 z + c_3 z^{3/2} - c_3 z^{3/2} + \ldots, c_0 z^{1/2} - c_0 z^{1/2} + c_1 + \ldots]$

$W \in GrH^{(1)}$ is called "transverse" if $\pi_+ : W \to H_+$ is an isomorphism and such a W contains a unique function $\psi(0, z) = \pi_+^{-1}(1)$ which has the form $1 + O(z^{-1})$.

Notice that the isometry we just described preserves the gradation by making the standard basis z^j correspond to the basis $\{\underline{e}_i\}_{i \in \mathbb{Z}}$ where if $i = ar + b$ with $a \in \mathbb{Z}$, $0 \le b \le r - 1$, $\underline{e}_i$ is the vector with b-th entry z^a and all other entries zero; thus the vector corresponding to $\psi(0, z)$ is $\underline{e}_0 + \sum_{i=1}^{\infty} a_i(0)\underline{e}_{-i}$. The matrix

$$\zeta = \begin{bmatrix} 0 & 1 & 0 & \ldots & \\ 0 & 0 & 1 & \ldots & 0 \\ \vdots & & & & \\ 0 & 0 & \ldots & & 1 \\ z & 0 & \ldots & & 0 \end{bmatrix}$$

gives the shift $\underline{e}_i = \underline{e}_0 \zeta^i$.

It is easy to check that the vector space M of 1.1 is a free module of rank 1 over Ψ and this fact enters crucially in the proof that the flows that we are going to construct satisfy the KP hierarchy. Define two subgroups of $H^{(1)}$:

$$\Gamma_+ = \{\sum_{j \ge 0} c_j z^j, \ c_0 = 1\} \text{ (extend analytically to } D_0 = \{|z| \le 1\})$$

$$\Gamma_- = \{\sum_{j \le 0} c_j z^j, \ c_0 = 1\} \text{ (extend analytically to } D_\infty = \{|z| \ge 1\})$$

Recall that these abelian groups act on Gr [PS] (and commute with each other), and if W is transverse $g = e^{\Sigma t_j z^j} \in \Gamma_+$, $|t|$ small, then $g^{-1}W$ is still transverse [SW]. Define $g^{-1}\psi(\underline{t}, z) = \pi_+^{-1}(1)$ via $\pi_+ : g^{-1}W \to H_+$; then $\psi(\underline{t}, z) = e^{\Sigma t_j z^j}(1 + O(z^{-1}))$ is called the Baker function, belongs to W for all time, and if $\psi = Ke^{xz}$ (cf. 1.13) then $\mathcal{L} = K^{-1}\partial K$ is a solution of the KP hierarchy. This correspondence W- $\to$ solutions is not 1:1, and it is not hard to show that the ambiguity is provided exactly by the action of Γ_-. Let's notice that the construction of 3.1.1 didn't involve any geometric data; the link with Lecture 1 is the following:

(A) Given a quintuple $(X, P_x, z, \mathcal{L}, \phi)$ as in Lecture 1, with ϕ a local trivialization on D_∞ of the line bundle $\mathcal{L}$, which here is obliged to have $h^0(\mathcal{L}) = 1, h^0(\mathcal{L} \otimes \mathcal{O}(-P_\infty)) = 0$, the space W is defined as the Hilbert closure of the space of holomorphic sections of $\mathcal{L}$ over $X \backslash D_\infty$ (viewed as elements of $L^2(S^1, \mathbf{C})$ via the trivialization over $|z| = 1$). It is shown in [PS] that $W \in Gr$; the KP flow corresponds precisely to the one we studied geometrically.

(B) Given a $W \in GrH^{(1)}$, we associate to it the ring A_W of analytic functions $a(z)$ which have a finite number of positive Fourier coefficients and are such that $a(z)W^{alg} \subset$

W^{alg}; then W comes from the construction in (A) precisely if A_W contains functions of any sufficiently high order; the BC ring is given by $\{a(\mathcal{L})\}$, where $\mathcal{L}$ is the KP solution constructed in 3.1.1 and evaluated at $t_j = 0$ for $j > 1$.

Again, (A) and (B) do not provide a 1:1 correspondence, but if $\mathcal{L}$ is maximal in a geometric sense (cf. [SW]) then they do.

For the higher-rank construction, we shall first illustrate the corresponding geometry: for motivation, we remark that to a rank r BC algebra $\mathcal{A}$ there are associated a curve X with a smooth point P_∞, a local parameter z^{-1}, a rank r vector bundle with a local trivialization near P_∞ (cf. 1.12.1) given by completing over P_∞ the basis $s_j: y \mapsto y^{(j)}(0)$ $j = 0, \ldots, r - 1$ valid for finite P near P_∞. This will give us a $W \in Gr H^{(r)}$ by the usual recipe (completion of sections on $X \backslash D_\infty$); more generally,

3.1.2 Lemma. *The subspace W corresponding to the quintuple $(X, P_\infty, z, \mathcal{L}, \phi)$ belongs to the big cell if and only if (i) $h^0(\mathcal{L}) = r, h^1(\mathcal{L}) = 0$, (ii) the values of the global (holomorphic) sections span the fibre of $\mathcal{L}$ over P_∞.*

However, the point $W \in Gr H^{(r)}$ alone does not give sufficiently many interesting KP solutions; the generalization of the rank 1 construction is the following: choose an arbitrary r-th order operator of the form $L_0 = \partial^r + u_{r-2}(x)\partial^{r-2} + \ldots + u_0(x)$ and define its normalized (at $x = 0$) Wronskian matrix $\Psi_0(x, z)$, so that the first row $\underline{\psi}_0 = (\psi_0, \ldots, \psi_{r-1})$ is a basis of solutions of $L_0\psi = z\psi$, and the jth row is given by $\underline{\psi}_0^{(j-1)}$. For

this, $\frac{\partial \Psi_0}{\partial x} = (\zeta - U)\Psi_0$ where $U = \begin{bmatrix} 0 & \cdots & & 0 \\ \cdots & & & \\ 0 & \cdots & & 0 \\ u_0 & \cdots & u_{r-2} & 0 \end{bmatrix}$. Notice that if the coefficients

of L_0 also depend on a sequence of time parameters $\underline{t}$ in such a way that $L_0^{1/r}$ is a solution of the KP hierarchy, then we obtain a matrix $\Psi_0(\underline{t}, z)$ for the eigenvalue problem

$$
\begin{cases}
L_0\psi = z\psi \\
\partial_{t_j}\psi = (L_0^{j/r})_+\psi.
\end{cases}
$$

Now we introduce the analog of the module M of 1.1: let $M(\Psi_0)$ be the space of formal expressions $(\sum_{i=-\infty}^{N} b_i(x)\underline{e}_i)\Psi_0(x, z)$; it is easy to see by using the gradation that the (obvious, left-) action of ∂ on $M(\Psi_0)$ is invertible and that $M(\Psi_0)$ is a free Ψ-module of rank 1 (as a generator, we can take any element with invertible leading coefficient b_N, for instance, $\underline{\psi}_0 = \underline{e}_0\Psi_0$). At last, the construction is quite simple:

3.1.3 Theorem. *(i) If we define (for $|\underline{t}|$ small enough) the Baker vector $\underline{\xi}(\underline{t}, z)$ of (W, L_0) by $\pi_+^{-1}|_W \Psi_0^{-1}(\underline{e}_0)$ i.e. by the conditions that $\underline{\psi} = \underline{\xi}\Psi_0$ has the form $(\underline{e}_0 + \sum_{i=1}^{\infty} a_i(\underline{t})\underline{e}_i)\Psi_0(\underline{t}, z)$ and belongs to W for each fixed time, and if K is the unique element of Ψ such that $\underline{\psi} = K\underline{\psi}_0$, then $L = KL_0K^{-1} \in \Psi$ is such that $L\underline{\psi} = z\underline{\psi}$ and $L^{1/r}$ is a solution of the KP hierarchy.*

(ii) $W \in Gr H^{(r)}$ comes from geometric data as in 3.1.2 exactly when its ring A_W contains a function of any sufficiently large order (notice that the corresponding $A_{\tilde{W}}$ of

the interleaved image $\tilde{W} \in Gr H^{(1)}$ has then rank r since $z \mapsto z^r$); in this case, by setting all higher times $= 0$, we obtain a rank r commutative subalgebra of $\mathcal{D} = \{a(L), a(z) \in A_W\}$ (where L is defined in (i)), which is isomorphic to $H^0(X \setminus \{P_\infty\}, \mathcal{O}_X)$.

3.1.4 Remarks. (i) The choice $L_0 = \partial^r$ reproduces the KP flows of 3.1.1 under the interleaving isometry $W \mapsto \tilde{W}$. However, the general L_0 will not give rise to a "flow" on Gr, in the sense that $\psi(x_1 + x_2, z)$ can be reached in time x_2 from $\psi(x_1, z)$ by an operation that involves W_{x_1} only; the reason for this is that, unlike for $r = 1$, $\Psi_0(x_1 + x_2, z) \neq \Psi_0(x_1, z)\Psi_0(x_2, z)$. What is a flow is the operation on pairs $(W, L_0) \to (W_{x_1}, L_{0,x_1})$; but here the geometry is (irreparably?) lost. (ii) part of the problem with (i) is that in higher rank the condition that tells whether (W, L_0) and (W', L_0') give the same solution is very inexplicit (although conceptually clear). (iii) On the plus side, let's remark that, when we deform the algebra in time, the determinant of the associated BC bundle does not move in Jac X, because it is given by a meromorphic function (the exponential singularities of the r eigenfunctions add up to zero).

3.2 Examples. We try to show by 3 examples why the situation is so far from explicit.

(1) First of all, there are no formulas for the coefficients of the solutions; there are formulas for the x-dependence only, and for the case of genus $X = 1$ only; they were found by direct computations by three schools: the Krichever-Novikov school expresses the solution in terms of the $\wp$ function; [De], who worked on this side project upon Verdier's suggestion, describes the motion in terms of plane geometry, and Grünbaum in terms of a set of differential equations: the formulas are complete for rank 2 and only partial for rank 3, [Mo] being the most advanced paper to date. It is possible that the reason why explicit formulas can be given in $g = 1$ is that the moduli space of bundles with fixed determinant is of dimension $r - 1$, and since there are $r - 1$ free parameters in the construction of the algebra, these can be taken to be coordinates for the moduli space (albeit by a very nontrivial recipe; Dehornoy's formulas give the simplest form). The findings I would like to describe are the following: the general construction of 3.1 guarantees that there will be a rank 2 BC algebra corresponding to any (sufficiently generic, see 3.1.2) bundle over X; however [De] does find the general rank 2 BC algebra over X, but claims that bundles of type $\mathcal{O}(P) \oplus \mathcal{O}(P)$ cannot occur (we are going to use the words sheaf and line bundle interchangeably for the corresponding objects). This would contradict 3.1.3, and prompted us to look for an explicit way to identify the bundle associated to (any) of the explicit solutions; we did find $\mathcal{O}(P) \oplus \mathcal{O}(P)$, and also a plausible reason why [De] overlooked it; and a pleasant relationship between the numerology of the operator L and the bundle.

(2) There is no known explicit KP solution whatsoever, coming from higher-rank data, except when the curve X is singular. The problem for genus 1 (only!) has been reduced to an equivalent equation in x and $t = t_3$ only, the Krichever-Novikov equation (KN), to which however no solutions are known except in the singular case. The calculation of a transference, in the case of the known solution, gives an alternative way to derive the equation, and more effectively, an explanation of the link [SSY] between KdV and the singular KN equation.

(3) A rank r algebra could be contained in a commutative algebra of smaller rank, even rank 1. In this case, it should have been possible to produce it much more easily by using a line bundle over a covering curve. There seems to be however no way to tell! Let's say that the rank is "fake" if the algebra is not maximal-commutative. Dixmier [Di] gave an elegant argument based on gradations to show that example 1.5(iii) is maximal-commutative. We check by transference that a "new" example (suggested by BC!) is maximal-commutative (unfortunately the curve is singular).

We give only a sketch of the main features entering the work of example (1). The pair (L, B) will have orders 4,6 and we may assume $B = (L^{3/2})_+$. We find it most convenient to write L in Grünbaum's form:

3.2.1 The solution.
$L = (\partial^2 + \frac{1}{2}c_2(x))^2 + 2c_1(x)\partial + c_1'(x) + c_0(x)$, so that L is (formally) self-adjoint if and only if $c_1 \equiv 0$. We assume rank $\mathbf{C}[L, B] = 2$; then:

Case I. If $c_1 \equiv 0$ then $c_0' \not\equiv 0$ and c_0 can play the role of the arbitrary function.

$$c_2 = \frac{K_2}{c_0'^2} + \frac{2K_3c_0}{c_0'^2} + \frac{c_0^3}{c_0'^2} - \frac{c_0'''}{c_0'} + \frac{1}{2}(\frac{c_0''}{c_0'})^2$$

and $B^2 = L^3 + \frac{K_3}{2}L - \frac{K_2}{8}$.

Case II. If $c_1 \not\equiv 0$, we let g be an arbitrary function, $c_1 = g'$,

$$c_0 = -g^2 + K_{11} + K_{12}$$

$$c_2 = \frac{K_{14} + 6g^2 K_{12} + 2g^3 K_{11} - 2gK_{10} - g^4 + g''^2 - 2g'g'''}{2g'^2}$$

and $B^2 = L^3 + \frac{K_{14} - 3K_{12}^2 - K_{10}K_{11}}{4}L + \frac{K_{10}^2 - 2K_{10}K_{11}K_{12} - 4K_{12}^3 - K_{14}(K_{11}^2 + 4K_{12})}{16}$

3.2.2 The gcd.
We call E the dual of the bundle of common eigenfunctions of the algebra $\mathbf{C}[L, B]$. First we notice that the sections s_j give a preferred holomorphic section of the line bundle det E, namely $s_0 \wedge s_1$, and the divisor D of this section is effective of degree rg (= 2 in our case). Then we normalize the greatest common divisor of $L - \lambda$ and $B - \mu$ (for (λ, μ) a point of the curve, otherwise the gcd is 1) to be $R = \partial^r +$ lower order, with coefficients that are meromorphic functions of P (and analytic in x with possible poles), and whose kernel at each point $P \in X \backslash \{P_\infty\}$ is the fibre of the dual bundle to E. This says that if P is not a point of the divisor D, the coefficients of R are regular at $x = 0$, whereas if P appears in D then R has a regular singular point at $x = 0$; the singular point is regular because Ker R is a subspace of the kernel of the regular operator $L - \lambda$, so that furthermore the exponents of R are distinct integers ρ_i satisfying the inequalities $0 \leq \rho_i < \text{ord } L$ (the exponents are the r numbers ρ_i such that the equation $R\psi = 0$ has a solution of the form $\psi(x) = x^{\rho_i}\psi_0(x)$ with ψ_0 regular and nonvanishing at $x_0 = 0$). Lastly, near P_∞ R is of the form $R(x, z) = \partial^r + v_{r-2}(x)\partial^{r-2} + \ldots + v_0(x) - z + O(z^{-1})$.

We remark in passing that it follows easily from 3.1.2 that our bundles are semistable; also, by the well-known classification of vector bundles over an elliptic curve (due to M.F. Atiyah) they are not stable, and they are either decomposable, $\mathcal{O}(P) \oplus \mathcal{O}(Q)$, or the unique indecomposable extension E_P of $\mathcal{O}(P)$ by itself.

3.2.3 Lemma. *(i) All the global sections of the bundle $\mathcal{O}(P) \oplus \mathcal{O}(P)$ vanish at P. (ii) The bundle E_P has a section that does not vanish at P.*

Proof. (i) is obvious; to prove (ii), assume that both sections s_0, s_1 of a basis vanish at P. Then they are sections of the degree-zero bundle $E(-P)$ which span the fibres at every point, except perhaps P but then at P too, or the degree would be less than zero. Thus $E(-P)$ is the trivial bundle and E is $\mathcal{O}(P) \oplus \mathcal{O}(P)$. ∎

The following result is remarkable because it gives a link between the nature of the eigenfunctions as functions of x and of the sections as functions of (λ, μ); it says that the knowledge of the exponents at one of the singular points, P say, determines the type of the bundle E (and also the exponents at Q if $P \neq Q$); there are 5 possibilities for the exponents because $(0,1)$ gives a regular point.

3.2.4 Theorem *(i) If the exponents at P are $(0,2)$ then E is $\mathcal{O}(P) \oplus \mathcal{O}(Q)$ if $P \neq Q$, or E_P if $P = Q$; in the former case the exponents at Q are also $(0,2)$. (ii) If the exponents at P are $(1,2)$, then $P \neq Q$ and the exponents at Q are $(0,3)$; (iii) If the exponents at P are $(0,3)$, then $P \neq Q$ and the exponents at Q are $(1,2)$. (iv) If the exponents at P are $(1,3)$, then $P = Q$ and the bundle E is E_P. (v) If the exponents at P are $(2,3)$, then $P = Q$ and the bundle E is $\mathcal{O}(P) \oplus \mathcal{O}(P)$.*

Proof. We will not go through all the cases but do some, to give an idea of the method. First let us observe that all the eigenfunctions together with their first derivative vanish at P if and only if both sections s_0 and s_1 vanish at P, so in view of 3.2.3 (v) is proved, and is also the only case when $E = \mathcal{O}(P) \oplus \mathcal{O}(P)$ can occur. For the other cases, we use the asymptotic form of R (3.2.2) and derive the following asymptotic behavior for the section $s_2(y) = y''(0)$: $s_2(z) = (z - v_0(0) + O(z^{-1}))s_0(z) + O(z^{-1})s_1(z)$. This implies that $s_0 \wedge s_2$ is a holomorphic section of $\det E$ that vanishes at P_∞; and $s_1 \wedge s_2$ is a meromorphic section of $\det E$ whose only singularity is a simple pole at P_∞.

3.2.5 Lemma. *The section $s_0 \wedge s_2$ is identically zero unless the exponents at P are $(0,2)$ in which case it is not.*

Proof of Lemma. Let's tabulate the values of the sections on a basis in the $(0,2)$ case:

	y_0	y_1
s_0	1	0
s_1	?	0
s_2	?	1

which shows that $s_0 \wedge s_2$ is not zero. Now if $s_0 \wedge s_2$ does vanish at P then its zero divisor (unless it's identically zero) must be $P + P_\infty$, and this must be linearly equivalent to $P + Q$, impossible. ∎

This lemma allows us to prove (i), because if $P \neq Q$ $s_0 \wedge s_1$ will imply $(0,2)$ at Q and if $P = Q$ the bundle is E_P as remarked at the beginning. The other cases are similar. ∎

3.2.6 Corollary. *If we let* $\mu(P) = \rho_1 + \rho_2 - 2$, *then* $\mu(P) = \mu(Q) = \mu$ *and* μ *can be 0,1,2, or 3. If* $\mu = 0$ *then the bundle* E *may be either* $\mathcal{O}(P) \oplus \mathcal{O}(Q)$ *for* $P \neq Q$, *or* E_P. *But if* $\mu > 0$, *the type of the bundle is determined as follows:*
(i) If $\mu = 1$, *the bundle is* $\mathcal{O}(P) \oplus \mathcal{O}(Q)$ *with* $P \neq Q$;
(ii) If $\mu = 2$, *the bundle is* E_P;
(iii) If $\mu = 3$, *the bundle is* $\mathcal{O}(P) \oplus \mathcal{O}(P)$.

Before we state the main, and surprising, result which allows us to read off the bundle from the solution 3.2.1, we hazard a guess as to why [De] overlooked case (iii) of 3.2.6: it will be shown that if (iii) occurs, the denominator of the solution vanishes at $x = 0$; we (and Dehornoy) do exclude pairs L, B with poles at 0; but there is a possibility, and we give an example below, that the numerator vanishes at $x = 0$ as well and the solution is still regular.

3.2.7 Theorem. *The number* μ *in 3.2.6 is equal to the order of vanishing at* $x = 0$ *of the coefficient* c_1 *in 3.2.1, or of* c'_0 *if* L *is self-adjoint.*

Proof. Two pages of calculations, using 3.2.1 and the fact that $\rho_1 + \rho_2 - 1 = \operatorname{Res}_{x=0} a_1$, if $R = \partial^2 - a_1\partial + a_2$ (since ρ_i are the roots of the indicial equation of R). ∎

3.2.8 Example. By taking $c_0 = x^4, K_2 = 0, K_3 = 12$ we obtain the solution $L = (\partial^2 + \frac{1}{32}x^6)^2 + x^4$ whose curve is $\mu^2 = \lambda^3 + 6\lambda$ and bundle $E = \mathcal{O}(P) \oplus \mathcal{O}(P)$ where P is the origin, in view of what we proved.

3.2.9 Exercise. Prove or disprove that the algebra $\mathbf{C}[L, B]$ is maximal abelian; my idea is that since L belongs to the Weyl algebra you can use the same method as Dixmier [Di] to show that it is.

When we deal with example (2) of 3.2, it turns out that the Krichever-Novikov formulas are more geometrically transparent, so we give the dictionary between these and the above 3.2.1:

3.2.10 The solution. $L = (\partial^2 + u)^2 + 2c'(\wp(\gamma_2) - \wp(\gamma_1))\partial + c'(\wp(\gamma_2) - \wp(\gamma_1))' - \wp(\gamma_2) - \wp(\gamma_1)$, with:

$$u = -\frac{1}{4c'^2} + \frac{1}{4}\frac{(c'')^2}{(c')^2} + 2\Phi(\gamma_1, \gamma_2)c' - \frac{c'''}{2c'} + c'^2(\Phi_c(\gamma_0 + c, \gamma_0 - c) - \Phi^2(\gamma_1, \gamma_2)),$$

where: $\Phi(\gamma_1, \gamma_2) = \zeta(\gamma_2 - \gamma_1) + \zeta(\gamma_1) - \zeta(\gamma_2)$ ($\zeta' = -\wp$ are the classical functions of Weierstrass), $\gamma_1(x) = \gamma_0 + c(x), \gamma_2(x) = \gamma_0 - c(x)$.
Notice that γ_1, γ_2 give two points which move in a pencil since $\gamma_1 + \gamma_2 = \text{const.}$ and $c(x)$ is the arbitrary function.
The dictionary betwen 3.2.1 and 3.2.10 is given in [G1] for the case $c_1 \not\equiv 0$:

$$K_{12} = -2\wp(\gamma_0)$$
$$K_{11} = \wp''(\gamma_0)/\wp'(\gamma_0)$$
$$K_{14} = 0$$
$$K_{10} = -2\wp'(\gamma_0)$$
$$g(x) = \wp'(\gamma_0)/(\wp(\gamma_0) - \wp(c(x))),$$

and can be formulated as

$$c_0 = -\wp(\gamma_1) - \wp(\gamma_2)$$

$$c_1 = c'(\wp(\gamma_2) - \wp(\gamma_1))$$

$$c_2 = -(\frac{c''}{c'} - 2c'\Phi)' - \frac{1}{2}(\frac{c''}{c'} - 2c'\phi)^2 - \frac{1}{2c'^2}$$

so as to get the self-adjoint formula in the limit:

$$\lim_{\gamma_0 \to 0} \Phi = 2\zeta(c) - \zeta(2c) = -\frac{1}{2}\frac{\wp''(c)}{\wp'(c)}$$

$$c_2 = -\left(\frac{c''}{c'} + c'\frac{\wp''(c)}{\wp'(c)}\right)' - \frac{1}{2}\left(\frac{c''}{c'} + c'\frac{\wp''(c)}{\wp'(c)}\right)^2 - \frac{1}{2c'^2}.$$

3.2.11 Definition. Let L, B be a commutative pair; a Darboux transformation of L is a conjugation $ALA^{-1} = \tilde{L}$ where $A = \partial - \psi'/\psi$ and $L\psi = \lambda\psi$; this guarantees that L is divisible by A on the right and $\tilde{L} \in \mathcal{D}$ (as opposed to just Ψ). A Darboux transformation of the pair (L, B) is conjugation by means of a joint eigenfunction: $L\psi = \lambda\psi$, $B\psi = \mu\psi$, so that $\tilde{L}, \tilde{B}$ both are in $\mathcal{D}$. A transference for the pair (L, B) is conjugation by $R = gcd(L - \lambda, B - \mu)$; it is the transference that has geometric meaning. In the rank 1 case, BC showed that transference for chosen $P = (\lambda, \mu)$ corresponds to the abelian sum for the divisor of the algebra, as previewed in 2.3.6, which they checked by the use of their plane-curve description of the divisors; this gave them their idea that the algebras corresponding to a fixed curve should constitute the Jacobian of that curve, and in fact they showed how to reach a pair from any other pair attached to the same curve by a sequence of g transferences. In higher rank, where the geometry is much more mysterious, we would like to know: (1) does the transference have the same geometric effect on the divisor $\det E$ where E is the rank r bundle? (yes in principle); (2) do the Darboux transformations allow us to produce all bundles from a given one (it is not clear that there are enough parameters, for the gcd $(L - \lambda, B - \mu)$ has an r-dimensional kernel while the moduli space of bundles with fixed determinant in $g > 1$ has dimension $r^2(g - 1) - g + 1$, but iterated transferences may increase the parameter count, as happens in rank 1 where the choice of a point N times, $N \geq g$, does indeed give g parameters); (3) do the Darboux transformations detect, by having different effects, whether the algebra has fake rank? Again, we have only looked at the $g = 1$ case and have sparse results, of which we give a sample below. First we reproduce the statement for rank 1 in a specific case, which is interesting in its own right and will make our path easier to follow.

3.2.12 Rank 1 case. (i) If $L = \partial^2 - 2\wp(x - e)$, then the divisor of the algebra $\mathcal{C}_{\mathcal{D}}(L)$ is $P_1 = (\wp(-e), \frac{1}{2}\wp'(-e))$ and transference at the point $P = (\wp(f), \frac{1}{2}\wp'(f))$ has the effect $e \mapsto e - f$ at the $\wp$-function level, and $\mathcal{O}(P_1) \mapsto \mathcal{O}(P_1) \otimes \mathcal{O}(-P) \otimes \mathcal{O}(P_\infty)$ at the sheaf level. (ii) If (L, B) is a hyperelliptic pair such that $B^2 = f(L)$ with order $B = 2g + 1$ the smallest odd-order in $\mathcal{C}_{\mathcal{D}}(L)$, and if y is a solution of $Ly - ay$, then: *Case I:* if y is not an eigenfunction for B, the new potential $\tilde{u}$ in $\tilde{L} = \partial^2 + \tilde{u}$ corresponds to the singular curve $\mu^2 = (\lambda - a)^2 f(\lambda)$; *Case II:* If y is a joint eigenfunction, the bundle transforms as in (i); *Case III:* if $(\lambda - a)^2 | f(\lambda)$ so the curve was singular, then transference at $(c, 0)$ by a common eigenfunction undoes the singularity: $\mu^2 = (\lambda - a)^{-2} f(\lambda)$.

The first statement is a calculation and (ii) is proved in [EK], the idea in Case I being that if B is not divisible by A on the right, then the ring of the new curve is $C[ALA^{-1}, ABLA^{-1}] = C[\tilde{L}, \tilde{B}]$, where now $\tilde{B}^2 = (\tilde{L} - a)^2 f(\tilde{L})$; Case III performs the inverse operation. It is very interesting that by a sequence of Case I operations beginning with the curve $\mathbf{P}^1$ and the trivial solution $u = 0$, all the rational KdV solutions (vanishing for $x \to \infty$) can be obtained, as shown originally by [AM].

3.2.13 Rank 2 case (i) If transference is performed on 3.2.10 with respect to $P = (\wp(\rho), \frac{1}{2}\wp'(\rho))$ then the new solution is given by $\tilde{\gamma}_1 = \gamma_0 + \rho + c(x)$, $\tilde{\gamma}_2 = \gamma_0 + \rho - c(x)$.
(ii) It is known that the (rj) KP deformations of rank r algebra have the effect of twisting by $\exp(z^j t_{rj})$ the transition matrix of the bundle. Over an elliptic curve, for $j = 1$ and $r = 2$, this has the effect of translating by t_2 the abelian coordinate of the corresponding divisor.

The proofs of 3.2.13 are direct calculations. About (i) we remark that it confirms the rank 1 behavior of tensoring the bundle by $\mathcal{O}(-P) \otimes \mathcal{O}(P_\infty)$; indeed, to find det E, namely the zeroes of $s_0 \wedge s_1$, we need the poles of $\partial Wr(\psi_0, \psi_1) = a_1$ (notations as in 3.2.7), which can be calculated as $-\partial \log(\frac{\sigma(\gamma_1)\sigma(\gamma_2)}{\sigma(\gamma_1 + \rho)\sigma(\gamma_2 + \rho)})$ so that one gets $\rho_0 = -\gamma_1$ and $\rho_0 = -\gamma_2$ (evaluated at $x = 0$). As for (ii) we invoke Grünbaum's formula for the t_2 flow of KP, which he calculates directly: $g(x, t_2, t_3) = \wp'(t_2)/(\wp(t_2) - \wp(c(x, t_3)))$.

3.2.14 The BC example. Now we would like to explain how BC found their first solutions. Recall that they wanted to exclude the trivial case $\mathcal{C}_D(L) = \mathbf{C}[M]$; this is automatically achieved by a rank 1 algebra unless it contains ∂ (this is the reason for the requirement gcd $(n, m) = 1$ in Example 1.5(i)). But, how did they find these examples? Consider $\mathbf{C}[\partial]$; if we make a Darboux transformation of the pair (∂^2, ∂^3) by an eigenfunction of ∂^2 which is not an eigenfunction of ∂, such as x (so that $A = \partial - \frac{1}{x}$) we obtain indeed $(\partial^2 - \frac{2}{x^2}, \partial^3 - \frac{3}{x^2}\partial + \frac{3}{x^3})$! We want to use this trick to produce interesting higher-rank examples, but found it even more useful to actually undo the singularity. Before stating our result, we introduce the KN equation:

3.2.15 The Krichever-Novikov equation. (i) The solution in 3.2.10 satisfies the first three evolution equations of the KP hierarchy if and only if $\gamma_0 = t_2$ and $c(x, t_3)$ obeys the KN equation:

$$\frac{\partial c}{\partial t_3} = \frac{1}{4}\frac{\partial^3 c}{\partial x^3} + \frac{3}{8}\frac{1 - (\partial^2 c / \partial x^2)^2}{\partial c / \partial x} - \frac{3}{2}\wp(2c)(\frac{\partial c}{\partial x})^3$$

(ii) [SSY] shows that the $\wp$-function can be removed from the KN equation by the transformation $v = \wp(c)$, which gives:

$$\frac{v_{t_3}}{v_x} = \frac{1}{4}\{v, x\} + \frac{3}{2}\frac{v^3 - \frac{1}{4}g_2 v - \frac{1}{4}g_3}{v_x^2},$$

where $\{v, x\}$ denotes the Schwarzian derivative: $v_{xxx}v_x^{-1} - \frac{3}{2}v_{xx}^2 v_x^{-2}$ and $\wp$ satisfies $\wp'^2 = 4\wp^3 - g_2\wp - g_3$. Moreover, they show that the KN equation is equivalent to KdV if and only if the cubic is singular, in which case a linear fractional transformation of v lowers the degree of the cubic; we will only need the example $g_2 = g_3 = 0$, $w = \frac{1}{v}$, which gives

$$\frac{w_{t_3}}{w_x} = \frac{1}{4}\{w, x\} + \frac{3}{2}\frac{w}{w_x^2}.$$

Finally, letting $\phi = -\frac{w_{xxx}}{w_x} + \frac{1}{2}\frac{w_{xx}^2}{w_x^2} - \frac{2w}{w_x^2}$ gives a solution of KdV: $\phi_{t_3} = \frac{1}{4}\phi_{xxx} + \frac{3}{4}\phi\phi_x$.

We explain the KN $\leftrightarrow$ KdV correspondence of 3.2.15(ii) as follows: if the curve of the pair L, B (of orders 4,6) is a singular cubic, we can undo the singularity by a transference; the new algebra $\mathbb{C}[\tilde{L}, \tilde{B}]$ then will not be maximal, because $\mathcal{C}_D(\tilde{L})$ will actually contain an operator T of order 2. Not only does this explain the occurrence of the KdV equation (KP hierarchy for the initial condtion $T^{1/2}$), it also gives an explicit expression for T, which allows us to check whether the rank ($= 2$) of the original solution L was true or fake. Indeed, it is simple to show that a Darboux transformation does not affect rank-fakery, and if rank $\mathcal{C}_D(T)$ is 1 then the potential of T is of hyperelliptic known form. We give the precise statements in 3.2.16 below (proofs in [LP]); notice that we can get by just with self-adjoint operators which are simpler to calculate with; the reason for this is that L is self-adjoint $\Leftrightarrow$ the divisor of $s_0 \wedge s_1$ belongs to the hyperelliptic series; this can always be achieved by transference at a suitable point (you have to solve the equation $P + Q = 2P_0$, but this you can do at the level of line bundles, and in $g = 1$ the curve equals its Jacobian).

3.2.16 Theorem. *Any solution L such that the associated rank 2 algebra has singular spectrum can be transferenced to the (translate of the) square of a second-order operator, as follows: let $\mu^2 = \lambda^3 - \frac{1}{4}g_2\lambda - \frac{1}{4}g_3 = (\lambda - e_1)(\lambda - e_2)(\lambda - e_3) = f(\lambda)$, and L self-adjoint. Transference by $(e_i, 0)$ gives*

$$\tilde{c}_2 = c_2 + 4\partial^2\log(e_i + \frac{1}{2}c_0)$$

$$\tilde{c}_0 = -2e_i + \frac{2f'(e_i)}{e_i + \frac{1}{2}c_0}$$

so that when $f'(e_i) = 0$, say $\tilde{L} = (\partial^2 + \frac{1}{2}\tilde{c}_2)^2 - 2e_i$; to corroborate the transformation 3.2.15(ii), we can check that by letting $c_0 = -2v$,

$$\tilde{c}_2 = -\frac{v_{xxx}}{v_x} + \frac{1}{2}\frac{v_{xx}^2}{v_x^2} - 2\frac{f(v)}{v_x^2} + 4\partial_x^2\log(v - e_i)$$

(from 3.2.1), and if $\phi = \tilde{c}_2$ (which explains the strange formula for ϕ in 3.2.15(ii)),

$$-\frac{1}{4}\phi_{xxx} - \frac{3}{4}\phi\phi_x + \phi_{t_3} = 6\frac{f'(e_i)v_x}{(v - e_i)^2}$$

whenever v solves the transformed KN: $v_{t_3} = \frac{1}{4}(v_{xxx} - \frac{3}{2}\frac{v_{xx}^2}{v_x}) + \frac{3}{2}\frac{f(v)}{v_x}$, which gives KdV for ϕ when $f'(e) = 0$.

Example 3.2 (3) is now a curiosity: I took a non-coprime case of the B (example 1.4(i)): $n = 4, m = 6$, $L = x^{-4}\delta(\delta - 6)(\delta - 12)(\delta - 18)$, $L = x^{-6}\delta(\delta - 4)(\delta - 8)(\delta - 12)(\delta - 16)(\delta - 20)$ and I found a sequence of transferences which brings L into $\tilde{L} = T^2$ with $T = \partial^2 - \frac{1}{x}\partial - \frac{35}{x^2}$. I renormalized this into $\partial^2 - \frac{11 \cdot 13}{4x^2}$, and since $\frac{11 \cdot 13}{4} \neq n(n+1)$ I proved that the rank is not fake, by [AMcKM] and by the argument given above. The general (n, m) case is treated in [P3].

For open problems, see the first of "Three Questions on Vector Bundles," [P2].

Lecture 4: Elliptic parameter.

Almost ten years ago as a fresh Ph.D. I was telling B. Dwork "all about" the Burchnall-Chaundy-Krichever-Mumford dictionary and writing L's and λ's all over his blackboard when he stopped me with his gentle diffidence to ask "have you thought of letting λ live on, instead of $\mathbf{P}^1$, an elliptic curve say?" In this lecture I am hoping to demonstrate the beauty of that idea, even though I for one don't know how to encase it within principles as general as those that work for the $\mathbf{P}^1$ case. Let's review what those principles do. We concentrate on 3 features: 1. a dictionary between commutative algebras of ODO's and function fields of one variable, with KP "isospectral deformations" which are Jacobian flows; 2. completely integrable hamiltonian systems, also given by Jacobian flows, and by a choice of Poisson structure which we haven't completely clarified (cf. 2.3.4), but which–after undergoing many revisions–have come to be described most clearly as follows: they are given by Lax-pairs for a matrix that describes (locally) a twisted endomorphism $E \to E \otimes F$, where $E = \mu_*\Delta$, Δ is a line bundle on the spectral curve Γ, and Γ comes with a function $\pi: \Gamma \to \mathbf{P}^1$ (our λ) and the endomorphism is given as multiplication by another function (our μ–recall how the Neumann matrix was the action of B on the λ-eigenspace of L); the twist F is given by a suitable line bundle over $\mathbf{P}^1$ (to offset the poles of μ at ∞, say); 3. flows on a homogenous space, the manifold Gr: this model now becomes especially important because special classes of solutions are geometrically classified. The submanifolds $Gr_0 = \{W \in Gr: \exists q \in \mathbf{N} \text{ for which } z^q H_+ \subset W \subset z^{-q} H_+\} \subseteq Gr_1 = \{W \in Gr: \exists p(z), q(z) \in \mathbf{C}[z] \text{ for which } p(z)H_+ \subset W \subset q^{-1}(z)H_+\}$ were shown [SW§7] to correspond to curves with one cusp, rational curves resp. (under suitable normalization of the choices). By another of the (for me) mysterious interactions between x and z (cf. 3.2.4), elements of Gr_0 give rise to KP solutions which are rational in the time variables [SW§8], while the "soliton" solutions (there isn't a unique definition of these but any one will do) which come from geometry will correspond to elements of Gr_1. What sense to make then, of an "elliptic parameter" λ? Here are a few suggestions, to be justified in the course of the lecture–as I said, no complete theory exists, at least to my understanding. Aspect 1 should require that the coefficients of all the operators in the algebra $\mathcal{A} \subset \mathcal{D}$ be elliptic functions of x, this ties in with aspect 3, since the limit of elliptic functions as one period goes to infinity will live on a cylinder, namely be made up of sines, cosines, etc. (solitons) and as both periods go to infinity they will become rational; and we suggest that the corresponding points of the Grassmannian be called Gr_2, but do not know how to characterize them explicitly. As for aspect 2, as we mentioned in passing (2.3.4), there is indeed a growing theory of spectral curves $\Gamma \to X$ where genus

453

$X > 0$, but that does not guarantee, for instance, that the corresponding coefficients of the differential operators enjoy a given periodicity as functions of x; the best we can do is to present the beautiful theory of the Krichever system, which seems as suited to the elliptic parameter as the Neumann system is suited to the rational parameter. Historically the theory began with the Calogero-Moser system, which corresponds to the rational solutions of Gr_0 and part of Gr_1; this case inspired Krichever and can now be viewed as a limit; the explanation of why it works is the following beautiful geometric accident: the first KP flow is the tangent to the spectral curve Γ in its Jacobian at the point P_∞; for some curves ("tangential covers," see 4.2) the closure of this 1-parameter group in Jac Γ is an elliptic curve. We give a brief account of the hamiltonian systems first, then the geometric theory due to Treibich-Verdier, and some related results.

4.1 **The Krichever system.** We follow the survey [OP] and encode several cases in a single formulation: let (x_i, y_i), $1 \le i \le n$ be the customary phase space, v be an even function and consider the hamiltonian

$$H = \frac{1}{2} \sum_{i=1}^n y_i^2 + \sum_{i<j} v(x_i - x_j)$$

4.1.1 **Theorem** *If L, M are $n \times n$ matrices with entries:*

$$L_{jk} = y_j \delta_{jk} + \sqrt{-1}(1 - \delta_{jk})u(x_j - x_k)$$
$$M_{jk} = \delta_{jk} \sum_{l \ne j} w(x_j - x_l) - (1 - \delta_{jk})z(x_j - x_k)$$

where u is odd, w and z are even, then the equation

$$\sqrt{-1}\dot{L} = [M, L]$$

entails $z = -\dot{u}$, $v = u^2 +$ const., and a functional equation for u and w, which has solutions $u = \frac{1}{x}, \frac{1}{\sin x}, \frac{1}{\operatorname{sn} x}$; in each case, the invariants

$$F_k = \frac{1}{k} tr(L^k) \quad k = 1, \ldots, n$$

are (generically) functionally independent and in involution, so that the system is completely integrable.

Recall that the KdV, resp. Boussinesq equation, are both reductions of the KP equation:

$$u_{yy} = (u_t + u_{xxx} - 6uu_x)_x$$

the former when the solution is y-independent, the latter when it is t-independent.

These particle systems were linked to the motion of poles of KP for the first time in [AMcKM] for the rational case $u = \frac{1}{x}$.

4.1.2 **Theorem.** *The function $f(x,t) = -2\sum_{j=1}^n (x - x_j(t))^{-2}$ is a solution of the KdV equation if and only if the set $(x_i(t_0), y_i(t_0))$ is an equilibrium point for the hamiltonian F_2; moreover, the t dependence of $x_j(t)$ is the flow of the hamiltonian F_3.*

The analogous statements hold for the function $f(x,y) = -2\sum_{j=1}^{n}(x - x_j(y))^{-2}$ and the Boussinesq equation if we interchange F_2 and F_3 (also a linear change of variables in t and y is needed for Boussinesq to be normalized as above).

This naturally leads to two questions: 1. Is there a link with the higher KP reductions and does the link persist for the other types of potentials; 2. Are the equilibrium loci nonempty? Krichever answered the first question by combining the (x-independent) entries of an eigenvector of the matrix L against functions that have a pole for $x = x_i$, and obtaining a Baker function whose corresponding KP solution $\mathcal{L}$ gives potential $u_{-1} = -\sum_{i=1}^{n} v(x - x_i(y,t))$; for example, in the rational case

$$\psi = (1 + \sum_{j=1}^{n} a_j(y,tz)(x - x_j(y,t))^{-1})e^{zx+z^2y+z^3t}$$

and Krichever [Kr2] also gives explicit formulas for the potential in terms of the divisor and the (rational, with arithmetic genus n) curve. In the elliptic case, Krichever introduced a new Lax pair, also equivalent to the system, which contains an elliptic parameter and yields the spectral curve as an n-sheeted cover of the elliptic curve (corresponding to $\wp$), generically of genus n:

$$L_{ij}(\alpha) = y_i\delta_{ij} + 2(1 - \delta_{ij})\Phi(x_i - x_j, \alpha)$$

$$T_{ij}(\alpha) = \delta_{ij}(-\wp(\alpha) + 2\sum_{k\neq i} \wp(x_i - x_k)) + 2(1 - \delta_{ij})\Phi'(x_i - x_j, \alpha)$$

where $\Phi(x,\alpha) = -\frac{\sigma(x-\alpha)}{\sigma(\alpha)\sigma(x)}e^{\zeta(\alpha)x}$, $\sigma(z) = z\prod_{m,n\neq 0}(1 - \frac{z}{\omega_{mn}})\exp[\frac{z}{\omega_{mn}} + \frac{1}{2}(\frac{z}{\omega_{mn}})^2]$ and $\psi = \sum_{i=1}^{n} a_i(y,k,\alpha)\Phi(x-x_i,\alpha)e^{kx+k^2y+k^3t}$. The spectral equations are: $(L(\alpha)+2k)\underline{a} = 0$, $(\frac{\partial}{\partial y} + T)\underline{a} = 0$. The upshot is:

4.1.3 Theorem [Kr3]. *The solutions of the system 4.1.1 give solutions $u_{-1} = \sum_{j=1}^{n} v(x - x_j(y,t))$ of the KP equation, which are elliptic in x in the case $v = \wp$. Conversely [TV2] if $u(x,y,t)$ is a solution to the KP equation defined and meromorphic on $\mathbb{C} \times U, U \subset \mathbb{C} \times \mathbb{C}$ an open domain and doubly periodic in x with period lattice Ω, then up to a linear change of variables $u = 2\sum_{i=1}^{n} \wp(x - x_i(y,t))$.*

As for question 2, in the rational case [AMcKM] proved the following:

4.1.4 Definition. Let the set of equilibrium points for the hamiltonian F_k be $\mathcal{V}_k(\frac{1}{x^2})$ or $\mathcal{V}_k(\wp(x))$, depending on the potential. The equations defining $\mathcal{V}_2$ are: x_i pairwise distinct, $y_i = 0, \sum_{k\neq j}(x_j - x_k)^{-3} = 0$ ($\sum_{k\neq j} \wp'(x_j - x_k) = 0$, resp.), $j = 1,\ldots,n$, so that $\mathcal{V}_2$ can be viewed as a sublocus of the symmetric product $\mathbb{C}^{(n)}, X^{(n)}$, resp.; let $\overline{\mathcal{V}}_2$ be the closure of that locus.

4.1.5 Theorem [AMckM] *Let $v(x) = \frac{1}{x^2}$. The locus $\mathcal{V}_2$ is nonempty iff $n = \frac{g(g+1)}{2}$, g any positive integer; in this case $\overline{\mathcal{V}}_2$ is diffeomorphic to $\mathbb{C}^g$. The locus $\mathcal{V}_3$ is nonempty for $n = 2$ and is a manifold of dimension 2.*

The numbers $g(g+1)/2$ had already appeared in 1.5(ii); however, after introducing a considerable amount of auxiliary machinery, [TV1] showed that $\overline{V}_2(\wp(x))$ is nonempty for all n, its connected components are closures of generalized Jacobians and unions of closures of KdV orbits and may have different dimensions but all $\leq \gamma(n) = \max\{p \in \mathbb{N} | \frac{p(p+1)}{2} \leq n\}$.

4.2 Tangential covers.

I must make a disclaimer. What I'll try to do is to convey the idea which is quite simple and attractive; and I will try to survey the results and the methods. A thorough discussion of the technical details would unbalance the spirit of these notes; in omitting it, I will be unable to do justice to two major themes that were developed: (1) the very concrete function-theoretic investigation of the spectral curve Γ and the surface S; (2) the delicate analysis of singularities; this is one case where it isn't good to say "let's assume that everything is smooth because the results would be the same apart from requiring more effort"; it's precisely the singular cases that contain the numbers we want. I refer to the series of papers by [TV] for the more sophisticated aspects (primitive tangential covers, for one).

4.2.1 Definition. (i) Let Γ be a (projective, integral) curve of (arithmetic) genus $> 0, p \in \Gamma$ a smooth point and $\pi \colon (\Gamma, p) \to (X, q)$ a finite pointed morphism to an elliptic curve. π is said to be a tangential cover if $\pi^* (\text{Jac } X \cong X) \subset \text{Jac} \Gamma$ is tangent to $A_\Gamma(\Gamma)$ at the origin of Jac Γ, where A_Γ is the Abel map (based at the point p and defined on the smooth part of Γ). (ii) A tangential cover is said to be minimal if it cannot be factored through another tangential cover $(\Gamma', p') \to (X, q)$ in such a way that $(\Gamma, p) \to (\Gamma', p')$ is a birational morphism other than an isomorphism; to each tangential cover there corresponds a unique minimal one.

[TV2] proves rigorously that the data of an elliptic KP solution (4.1.3) is equivalent to the data of a tangential cover (4.2.1); for delicate questions of one-to-oneness, cf. [TV2]. To link up the tangential cover with Krichever's systems, [T] proves the following.

4.2.2 Definition. Let $\mathbf{C}(X)$ be the function field of the elliptic curve X, T an indeterminate over $\mathbf{C}(X)$ and w a local parameter near $q \in X$. A monic polynomial of degree $n \geq 1$, $P(t) \in \mathbf{C}(X)[T]$ is said to be a tangential polynomial (with respect to w) if all the coefficients of $wP(T - w^{-1})$ are holomorphic at q: this condition depends only on the tangent vector $\frac{d}{dw}$; we denote by $\mathcal{T}(n, x\frac{d}{dw})$ the set of tangential polynomials w.r.t. w.

4.2.3 Theorem $\mathcal{T}(n, X, \frac{d}{dw})$ *is an affine space of dimension n. To every tangential cover $(\Gamma, p) \to (X, q)$ together with a choice of $\frac{d}{dw}$ we can associate a unique meromorphic function on Γ whose minimal polynomial $P(T)$ over $\mathbf{C}(T)$ belongs to $\mathcal{T}(n, X, \frac{d}{dw})$; conversely, any $P(t) \in \mathcal{T}(n, X, \frac{d}{dw})$ gives rise to a minimal tangential cover.*

This explicit criterion is very practicable, for instance it gives a Boussinesq tangential cover:

4.2.4 Example. [PV] A smooth hyperelliptic curve with affine equation $\mu^2 = \lambda^6 + a\lambda^2 + b$ covers 2:1 an elliptic curve $\nu^2 = \eta^3 + a\eta + b$ (a, b are suitable parameters), is tangential at the point p where $\frac{\mu}{\lambda^3} - 1$ has a zero with tangential polynomial $T^2 - \eta$ and parameter $w = \frac{n}{\omega}$, thus gives a 2-dimensional manifold of elliptic Boussinesq solutions. In order to obtain moduli for tangential covers, [TV1] introduces a ruled (over X) surface S; roughly speaking, S represents the tangent vector $\frac{d}{dw}$ and the automorphisms of S/X correspond exactly to changing $\frac{d}{dw}$.

4.2.5 Remarks and Definitions. 1. The image α of the first coboundary in the exact sequence $O \to \mathcal{O} \to \mathcal{O}_X(q) \to \mathbf{C}_q \to 0$ is the image of the differential of the Abel map at q. 2. If W is the rank 2 bundle over X associated to the element $\alpha \in H^1(\mathcal{O}_X) \cong \mathrm{Ext}^1(\mathcal{O}_X, \mathcal{O}_X)$, we let S be the ruled surface $= \mathrm{Proj}\ W$ and $\pi : S \to X$ the natural projection. 3. The curves on S can be described quite explicitly; the following are standard facts [H]: π has a unique section C_0 with self intersection $C_0 \cdot C_0 = 0$; if S_q is the fibre over q, $C_0 \cdot S_q = 1$ and $S_q \cdot S_q = 0$. The divisors up to linear equivalence form the group $\mathrm{Pic}\ S \cong \mathbf{Z} \oplus \pi^* \mathrm{Pic}\ X$ and the divisors up to algebraic (here $=$ numerical) equivalence form the group $NS(S)$ free abelian over C_0 and S_q. 4. $S \backslash C_0$ is an additive group Δ which can be viewed as the generalized Jacobian of X with a cusp $2q$: $0 \to \mathbf{G}_a \to \Delta \to X \to 0$, $\mathbf{G}_a$ ($=$additive group of $\mathbf{C}$) acts as affine transformations on the fibres and gives $\mathrm{Aut}\ (S/X)$. 5. S is obtained from $\mathbf{P}^2 \times X$ by a sequence of two elementary transformations, the first with center (∞, q) (this is what you always did to put the spectral curve $f(\lambda, \mu) = 0$ on the total space of the line bundle corresponding to the divisor of Lecture 2), the second with center the point of slope -1 in the fibre over q, which corresponds to the choice of a local parameter w (this singles out the tangential point at $k = \infty$ in Krichever's polynomial, from among the n points that cover q).

4.2.6 The moduli space [TV1]

(i) Given a tangential cover $\pi_n : (\Gamma, p) \to (X, q)$ there exists a morphism $\iota : \Gamma \to S$ of degree 1 that sends p to $p = C_0 \cap S_q$ and Γ to a member of the linear system $|nC_0 + S_q|$. Any two such morphisms ι_1, ι_2 are related by $\iota_1 = \sigma \circ \iota_2, \sigma \in \mathrm{Aut}\ (S/X)$.

(ii) (Γ, p, π_n) is a minimal tangential cover $\Leftrightarrow \iota$ is an immersion.

(iii) (Γ, p, π_n) is a minimal tangential cover $\Leftrightarrow$ the arithmetic genus $g_a(\Gamma) = n$ (in general, $g_a \leq n$).

(iv) Each irreducible divisor in $|nC_0 + S_q|$ is a minimal tangential cover of degree n (with π_n induced by π).

(v) The linear system $|nC_0 + S_q|$ has dimension n and only one fixed point p; a divisor in this system is irreducible if and only if it cuts C_0 transversally at p. The set of irreducible divisors $V(n, X)$ in the linear system is an n-dimensional affine space and its smooth members form a Zariski-open dense subset.

(vi) In summary, there is a 1:1 correspondence between isomorphism classes of minimal tangential covers of degree n and points of the affine space of dimension $n - 1$: $V(n, X)/\mathrm{Aut}(S/X)$.

To treat the second question formulated after 4.1.2, we introduce "hyperelliptic" and "Boussinesq" tangential covers, but will see that the theory developed so far is not powerful enough, except for the case when S has a symmetry that acts as a $\mathbf{Z}/2\mathbf{Z}, \mathbf{Z}/3\mathbf{Z}$ resp. on $H^1(\mathcal{O}_X)$; this is always the case for hyperelliptic Γ's, but is special for Boussinesq.

4.2.7 Definitions: (i) Let σ be the natural symmetry of S induced by the inversion on Δ. We say that a tangential cover (Γ, p, π) has a symmetry if Γ has an involution that fixes p and covers the natural one on X. We say that Γ is a hyperelliptic tangential cover (HTC) if there is an involution σ of Γ that fixes p and such that $\Gamma/\sigma \cong \mathbf{P}^1$, in particular, a HTC has a symmetry. (ii) The tangential cover (Γ, p, π) is said to be a Galois Boussinesq tangential cover (GBTC) if the curve Γ has an automorphism χ of order 3 such that $\Gamma/\chi \cong \mathbf{P}^1$ and p is a fixed point for χ. Equivalently, there exists a rational function $f \in \mathbf{C}(\Gamma)$ that has a pole of order 3 at p, is regular elsewhere, and the field extension associated to $f \colon \Gamma \to \mathbf{P}^1$, $\mathbf{C}(\Gamma)/\mathbf{C}(\mathbf{P}^1)$, is Galois (cyclic). For a primitive tangential cover, meaning that it does not factor through a nontrivial isogeny $(X', q') \to (X, q)$, the curve X and the surface S inherit an automorphism of order 3, thus X is unique up to isomorphism (e.g. $X \colon \nu^2 = \eta^3 + 1$) and we implicitly fix it from now on when we mention GBTC.

We sketch the complicated and rich classification of HTC obtained in [TV1]; the arguments are mainly cohomological and arithmetical.

4.2.8 Theorem *Any symmetric tangential cover admits a unique $\iota \colon \Gamma \to S$ such that $\iota(\Gamma) \in V(n, X)^\sigma$ which is a space of dimension $[\frac{n}{2}]$. For the generic $C \in V(n, X)^\sigma$, C/σ is the smooth of genus $[\frac{n}{2}]$.*

4.2.9 Definitions. (i) The surface S has 8 points fixed by σ, 4 of them on C_0; s_0, s_1, s_2, s_3 and 4 on $\Delta \colon r_0, r_1, r_2, r_3$; s_i and r_i belong to the same fibre and s_0, r_0 to S_q. We identify a symmetric tangential cover Γ with its image $\iota(\Gamma) \in V(n, X)^\sigma$ and let $\mu_i = $ multiplicities of Γ at r_i. (ii) Let $e \colon S^\perp \to S$ be the blow up of S at the 8 points s_i, r_i and $s_i^\perp, r_i^\perp$ the corresponding exceptional divisors and let $\phi \colon S^\perp \to \tilde{S}$ be the identification mod σ. Then $\iota \colon \Gamma \to S$ can be lifted uniquely to $S^\perp$ because the image of Γ in S is symmetric; in fact $\iota^\perp(\Gamma)$ is the strict transform of $\iota(\Gamma)$. Now if we call $\rho(\Gamma)$ the reduced image of $\iota^\perp(\Gamma)$ in $\tilde{S}$, we can recover Γ from $\rho(\Gamma)$ if Γ was a minimal HTC (which in general is not a minimal TC). We note that $\tilde{S}$ is a smooth rational surface and Pic $\tilde{S}$ has rank 10. (iii) We let $\lambda(n, \mu)$ be the unique element of Pic $\tilde{S}$ such that $\phi^*(\lambda(n, \mu)) = e^*(nC_0 + S_q) - s_0^\perp - \sum_{i=0}^{3} \mu_i r_i^\perp$. The μ_i necessarily satisfy the congruences: $\mu_i + \epsilon_i \equiv n \bmod 2$, $\epsilon_0 = 1, \epsilon_1 = \epsilon_2 = \epsilon_3 = 0$.

4.2.10 Theorem *There is a 1:1 correspondence between minimal-hyperelliptic tangential covers and rational (i.e., of geometric genus 0) curves in $|\lambda(n, \mu)|$ given by $\Gamma \mapsto \rho(\Gamma)$, and*

$$g_a(\Gamma) = (\sum_{i=0}^{3} \mu_i - 1)/2, \quad g_a(\rho(\Gamma)) = 2n + 1 - \sum_{i=0}^{3} \mu_i^2/4.$$

4.2.11 Theorem *The following are equivalent:*
(i) $g_a(\rho(\Gamma)) = 0$
(ii) $\rho(\Gamma)$ is an exceptional divisor.
(iii) $\iota^\perp$ is an immersion.

4.2.12 Theorem *(i) For fixed X and n, there is only a finite number of isomorphism classes of HTC of X of degree n and their arithmetic genus is bounded by $\gamma(n) = \sup\{l \in \mathbf{N} | \frac{l(l+1)}{2} \leq n\}$.*
(ii) If we let $\chi(n) =$ number of exceptional HTC of degree n and $\psi(n) =$ number of those that in addition have $g_a = \gamma(n)$, theen $\chi(n)$ is an increasing function of the order $O(n \log \log n)$, whereas the behavior of $\psi(n)$ is 'chaotic': the function $n \mapsto \sup\{\psi(l)|0 < l \leq n\}$ grows like $o(n^{1/4} \log^2 n)$ but $\psi(n) = 1$ iff n is a triangular number $(= \gamma(n)(\gamma(n)+1)/2)$.

The GBTC case is treated with similar techniques; but S is blown up once at 3 points and twice at 3 further points, so that the geometry and the diophantine equations become more intricate. However, the reward is the following:

4.2.13 Theorem [PV]. *There exists a finite nonzero number of exceptional GBTC of degree n for every positive n (exceptional refers to the divisor $\rho(\Gamma)$ defined in analogy to 4.2.9, 4.2.11).*

4.3 Related results. If we go back to the introduction to this lecture, we notice that Gr_0 is dense in Gr simply because when thought of as graphs $H_+ \to H_-$, the corresponding W are given by the matrices that have a finite number of nonzero entries. In [CPP] we show that Gr_2 is dense in Gr, although not in this "holomorphic" sense but rather as $\mathbf{Q}$ is dense in $\mathbf{R}$, and indeed by methods of Hodge structure.

4.3.1 Theorem *Tangential covers of genus g are dense in $\mathcal{M}_g$ for the natural topology over the complex numbers. Thus, Gr_2 is dense in Gr.*

What of tangential covers $\pi: \Gamma \to X$ with genus $X > 1$? (The definition is the natural one, $A_\Gamma(\Gamma)$ and $(\pi^* \circ A_X)(X)$ tangent at a point.) [T] shows that they can only occur in a very special situation, and we refer to [DP] not because we add any new information but because we display a fibered diagram which will motivate what follows:

4.3.2 Theorem *$(\Gamma, p) \to (X, q)$ is a tangential cover with genus $X > 1$ iff Γ is a fibered product $X \times_{\mathbf{P}^1} X'$, with X' of genus zero branched at say $\{0, \infty\}$, X hyperelliptic branched at $\{e_1, \ldots, e_{2g+2}\}$ and X, X' have one or two branchpoints in common.*

Because of this negative result, in [DP] we propose a different generalization of the "elliptic solitons" (this is the name [TV1] gives the elliptic KP solutions of 4.1.3): intuitively speaking, we ask that an abelian variety of dimension $n < g$ contain the first osculating vectors to the curve Γ in Jac Γ at some point p; in particular, these could be the first n KP flows of another curve for which $\pi: \Gamma \to X$ defines the natural map $X \to$ JacΓ.

4.3.3 Definition. A pointed curve (Γ, p) (p smooth) is said to be an abelian soliton of dimension n if there exists an abelian subvariety A of Jac Γ such that the first n (nontrivial) flows of the KP hierarchy for (Γ, p) are tangent to A.

4.3.4 Example. Coelliptic abelian solitons. Let Γ be an arbitrary cover of degree n of an elliptic curve E with one ramification point of index $g-1$ and $g-1$ points of index 1 (generic case; they might coincide). This gives a g-parameter family of abelian solitons since Jac $\Gamma \sim E \oplus A$, A an abelian variety of dimension $g-1$, and the Prym-Abel map $\Gamma \to A$ has differential that vanishes to order $g-1$ at least.

We were surprised that none of the fibered-product situations we analyzed yielded abelian solitons, apart from 4.3.2: cf. [DP] for a group-theoretic explanation of this negative result.

4.4 Questions
4.4.1 Exercise. Interpret Krichever's matrices for the elliptic system by the use of transferences (cf. 2.3.6 and [TV2] where Krichever's matrix is given geometric interpretation).

4.4.2 Research exercise. The triangular numbers $n(n+1)/2$ that enter the KdV equation (4.1.5) should be generalized to higher "generalized KdV flows"; their significance is representation-theoretic and geometric, for they are dimensions of cells in Gr. Notice that the simplest case of a curve with Burchnall-Chaundy ring (cf. 1.11.1) generated by 2 functions of weight 3 and $m \equiv 1$ ($\equiv 2$, resp.) mod 3 has Weierstrass gap number $2\frac{m-1}{3} + \frac{m+2}{3} - 1$ ($2\frac{m-2}{3} + \frac{m+1}{3}$, resp.), cf. also [S]; for the two Boussinesq examples in [Kr2] the number of particles is $\frac{N(N+2)}{4}$ (N even), $\frac{(N+1)^2}{4}$ (N odd). For a start, in [LP2] we produce rational Boussinesq solutions by transference from the trivial, as [AM] does for KdV. Another strategy is to study the multiplicity of intersection of X and Θ_Γ as in [F2] and [TV3]. Pertaining to the former direction, in [K] the part of Gr_1 that corresponds to rational KP solutions is given a stratification in finite-dimensional Grassmannians and some characterizing properties of the corresponding tau function are listed.

Lecture 5: Verlinde formulas: Geometry of vector bundles.

These last two lectures are (even) closer to the flavor of classical algebraic geometry. We feel that we need something of an excuse for presenting this material here, and that is provided by the current large amount of research devoted to generalizing the "Hitchin systems," cf. [Bo][M]. In that process, a more explicit and precise knowledge of the projective images of the moduli spaces of bundles (cf. 5.1 for a more precise definition) is needed and is being uncovered; the Seshadri 'school' who had a ground-breaking role in the theory in the 1960s and 1970s has produced a large number of important papers of which we cite only [BNR,DN]. At a somewhat opposite end, research in conformal field theory, which is also related to the themes of this conference, produced a stunning set of numbers ($g > 1$):

$$N_{0,k} = \sum_{j=1}^{k+1} \left(\frac{k+2}{2\sin^2 \frac{j\pi}{k+2}} \right)^{g-1}$$

(Verlinde Formulas)

$$(5.0) \qquad N_{1,k} = \sum_{j=1}^{2k+1} (-1)^{j+1} \left(\frac{k+1}{\sin^2 \frac{j\pi}{2k+2}} \right)^{g-1}$$

(which are integers), and linked them to a certain geometric object (nonabelian theta divisor) which will be recalled below. In analyzing some of these numbers for small k, Hitchin's spectral curve [BNR] played an important role; moreover, the $|2\Theta|$ map, which enters deeply the projective models mentioned above, provided auxiliary numbers and threw light on some special cases of the formula. It is on this aspect that we are going to report, and more specifically on the (somewhat unexpected) link of $N_{0,4}$ with the Schottky problem, although in so doing we will also sketch known facts on the $|2\Theta|$ map to provide a more complete backdrop.

5.1 **Meaning of the numbers.** In this paragraph we collect some generalities on vector bundles. Of the several approaches, the one we most draw on is algebraic; references to the original proofs can be found within the bibliographies of [B1,2], [DR], [Hi1], [NR1,2], [N], [Tu]. We fix a smooth curve X of genus $g > 1$. The degree of a vector bundle E over X is defined as the degree of the line bundle $\det E$, the slope of E is the rational number $m(E) = \deg E / \operatorname{rank} E$, and a bundle is said to be semistable (stable) if $m(L) \leq m(E)$ $(<)$ for any subbundle L of E. This definition is intended to avoid situations like $\mathcal{O}(-(n+1)P) \oplus \mathcal{O}(nP)$ which has many sections but negative degree. Line bundles are all stable and tensoring with a line bundle does not affect stability. Riemann-Roch for vector bundles of rank r becomes: $h^0(E) - h^1(E) = \deg E - r(g-1)$. From now on we will always assume $g > 1$. For elliptic curves $(g = 1)$, M.F. Atiyah gave a complete classification of vector bundles by using extensions of line bundles; there are no stable bundles of rank r and fixed determinant of degree d unless $(r, d) = 1$ in which case there is a unique one. However, [Tu] shows that when suitably interpreted the Verlinde numbers reflect the geometry in this case as well. D. Mumford proved in the 1960s that for fixed rank and degree there is a moduli space of stable bundles, a smooth quasiprojective variety of dimension $r^2(g-1)+1$ (the dimension count is given by Kodaira-Spencer as $\dim H^1(X, E^* \otimes E)$). C.S. Seshadri showed that the semistable bundles can be given an equivalence relation (two are equivalent when $\oplus_{j=1}^n F_j$ are isomorphic, F_j being stable Jordan-Hölder blocks which make up the bundle; of course there may be nonisomorphic bundles which are equivalent, we saw two in Lecture 3, E_P and $\mathcal{O}(P) \oplus \mathcal{O}(P)$) in such a way that the quotient space is a projective variety, in fact its singular locus corresponds exactly to the semistable bundles unless $g = 2 = r$, deg $= 0$. When gcd $(r, d) = 1$ the semistable bundles are all stable. The standard notation for these projective spaces is $U_X(r, d)$, and $SU_X(r, L)$ when $\det E = L$ is a fixed line bundle; the notation originated with Seshadri's result that stable bundles correspond to irreducible unitary representations of $\pi_1(X)$. Since we'll be concerned with $r = 2$ only, and by tensoring with a line bundle we have an isomorphism between the spaces corresponding to degrees which are congruent mod 2, we shall adopt the simpler notation $\mathcal{M}_0 = SU_X(2, \mathcal{O}_X)$ ("even case") and $\mathcal{M}_1$ (or sometimes $\mathcal{M}_p$)$= SU_X(2, \mathcal{O}_X(p))$ ("odd case"). Drezet and Narasimhan showed [DN] that the projective variety $SU(r, L)$ has Picard group isomorphic to $\mathbf{Z}$, so that we can choose an ample generator which we call

$\mathcal{L}_0, \mathcal{L}_1 = \mathcal{L}_p$ resp. in our case. We define the numbers $D_{\epsilon,k} = \dim H^0(M_\epsilon, \mathcal{L}_\epsilon^{\otimes k})$ with $\epsilon = 0, 1$. By a cohomological argument (Euler characteristic) the numbers $D_{\epsilon,k}$ do not depend on the curve, whereas the spaces $\mathcal{M}_0$, $\mathcal{M}_1$ do. In fact, a Torelli-type theorem holds: just as in the line-bundle case Jac X determines the curve X, so does $\mathcal{M}_1$, as proved in [MN]; as for the "even case" $\mathcal{M}_0$, the same is true with the exception of genus 2 when $\mathcal{M}_0 \cong \mathbf{P}^3$. Indeed, the singular locus of $\mathcal{M}_0$ is the Kummer variety of X and there is a unique (indecomposable) ppav with given Kummer (for example, an argument can be given using the moduli space with (2,4) level structure, mapped to $\mathbf{P}^{2^g-1}$, and [SM]). Finally, let's draw an analogy with the classical case: the moduli space of rank 1, degree zero bundles is Jac X, whose Picard group is also $\mathbf{Z}$ generated by $[\Theta]$ if Θ is a theta divisor. Then $\dim H^0$ (Jac $X, [\Theta]^{\otimes k}) = k^g$, and the sections of the bundle are called theta functions of order k; the corresponding rank r spaces are referred to as nonabelian theta functions. Now we can explain the significance of the Verlinde numbers, which is given by the equality: $D_{\epsilon,k} = N_{\epsilon,k}$. This has now been proved by several methods: [BS], with collaboration by D. Zagier, used hyperelliptic curves and Atiyah-Bott fixed-point theorems; [F] used a degeneration argument (to singular curves) which provides a completely general result for any rank. For small k, [B1,2] gives a very classical interpretation of the numbers, and this method is closer in spirit to what we want to do, so that we'll survey his results as we set up our method.

5.2 Some numbers.

We recall briefly some theta-function notation, referring to [F1,I]. A theta characteristic $\kappa \in \mathrm{Pic}^{g-1}(X)$ is a line bundle for which $\kappa^{\otimes 2} \cong \Omega_X^1$, the canonical bundle. We say that κ is even resp. odd if $h^0(X, \kappa)$ is an even resp. odd number. There are 2^{2g} theta characteristics in all $=\# J[2]$ (the group of points of order divisor of 2 in J=Jac X), $2^{g-1}(2^g+1)$ of which are even and $2^{g-1}(2^g-1)$ odd. Now we display a few Verlinde numbers:

$$N_{0,0} = 1$$
$$N_{0,1} = 2^g \qquad\qquad N_{1,0} = 1$$
$$N_{0,2} = 2^{g-1}(2^g+1) \qquad\qquad N_{1,1} = 2^{g-1}(2^g-1)$$
$$N_{0,3} = 2(5+\sqrt{5})^{g-1} + 2(5-\sqrt{5})^{g-1} \qquad N_{1,2} = 3^{g-1}2^{2g-1} - 2^{2g-1} + 3^{g-1}$$
$$N_{0,4} = (3^g+1)/2 + (2^{2g}-1)(3^{g-1}+1)/2$$

The first three nontrivial numbers are related to the dimension of the space $V = H^0$ $(J, \mathcal{O}(2\Theta))$, which is 2^g and $V \otimes V = S^2 V + \Lambda V = \dim H^0(J, \mathcal{O}(4\Theta))_+ + \dim(J, \mathcal{O}(4\Theta))_-$ (even and odd theta functions). Our theme will be to explore this link between 'nonabelian' and 'abelian' theta functions, and provide a similar explanation for the number $N_{0,4}$; our proof will show $D_{0,4} \geq N_{0,4}$ and also draw some geometric conclusion assuming the fact, which is proved by other methods, $D_{0,4} \leq N_{0,4}$.

We introduce more notation. Each theta characteristic defines a symmetric divisor $\Theta_\kappa = \Theta_X - \kappa$ in J (where Θ_X is the canonical theta divisor in $\mathrm{Pic}^{g-1} X$), and all effective symmetric divisors giving the principal polarization are obtained in this way. For any integer s, the divisors $2s\Theta_\kappa$ are linearly equivalent on J; we simply write $2s\Theta$ for this equivalence class. The Mumford group $\mathcal{G}$ is the group of automorphisms of $\mathcal{O}(2\Theta)$ which lift the action of an element of $J[2]$ on $V = \{(x, \phi) | \phi : T_x^* \mathcal{O}(2\Theta) \xrightarrow{\simeq} \mathcal{O}(2\Theta)\}$. We fix a theta structure, namely an isomorphism between $\mathcal{G}$ and the Heisenberg group:

$$H = H_g = \mathbf{C}^* \times (\mathbf{Z}/2)^g \times \mathrm{Hom}((Z/2)^g, \mathbf{C}^*)$$

with multiplication $(s, \alpha, \alpha^*)(t, \beta, \beta^*) = (st\beta^*(\alpha), \alpha + \beta, \alpha^*, \beta^*)$. Using the theta structure, the action on V can be described as follows: there is a basis $\{X_\sigma\}$ ($\sigma \in (\mathbf{Z}/2\mathbf{Z})^g$) for which $(t, \alpha, \alpha^*)X_\sigma = t\alpha^*(\alpha + \sigma)X_{\sigma+\alpha}$.

Remark: the choice of a symplectic homology basis for X with corresponding period matrix τ determines a theta structure and the basis $\{X_\sigma\}$ of V is given by $X_\sigma = \vartheta \begin{bmatrix} \sigma \\ 0 \end{bmatrix} (2z, 2\tau)$. The action of the Heisenberg group on V induces an action on $V \otimes V$, which will be a direct sum of nonisomorphic one-dimensional representations, each inducing a duality $V \to V^*$. The representations in $\mathrm{Sym}^2 V$ (resp $\Lambda^2 V$) correspond to even (resp. odd) theta characteristics. We denote the corresponding duality by ξ_κ.

Let us also recall the definition of a Prym variety: for any finite covering of (smooth, irreducible) curves $\pi : Y \to X$, the Prym variety $\mathrm{Prym}(Y/X)$ is the connected component of the identity of the kernel of the "norm map" $Nm : \mathrm{Jac}\, Y \to \mathrm{Jac}\, X$ (which at the level of divisors can be obtained simply as the image under π).

5.3 Hyperplanes and quadrics. We start describing more concretely the bundles $\mathcal{L}_\epsilon$ and their sections; we are more interested in conveying the geometric significance of the constructions than in reproducing all the technical arguments necessary to check e.g. that a divisor is reduced, or a given locus on a singular variety corresponds to a Cartier divisor, etc., which are contained in the proofs of [B1,2,BNR]. Set theoretically, then, $\Theta_n = \{E \in \mathcal{S}\,\mathcal{U}(n, g-1) | h^0(X, \epsilon) \geq 1\}$ is a divisor associated to the line bundle $\mathcal{L}$ (for the moment the rank is arbitrary and $\mathcal{L}$ is the ample generator of $\mathrm{Pic}\,\mathcal{S}\,\mathcal{U}(n, g-1)$); but we can similarly define subvarieties of J, and obtain the following:

5.3.1 Theorem [BNR] *The (rational) map which associates to any point e of $\mathcal{S}\,\mathcal{U}(n, \mathcal{O})$ the divisor $D_E = \{\xi \in \mathrm{Pic}^{g-1} X : \Gamma(E \otimes \xi) \neq 0\}$, where E is a semistable bundle in the class of e, induces an isomorphism of $\Gamma(\mathrm{Pic}^{g-1} X, \mathcal{O}(n, \Theta_X))$ with $\Gamma(\mathcal{S}\,\mathcal{U}(n, \mathcal{O}), \mathcal{L})^*$. In particular:*

$$\dim \Gamma(\mathcal{S}\,\mathcal{U}(n, \mathcal{O}), \mathcal{L}) = n^g.$$

This theorem is proven by the theory of spectral curves: there exist n-sheeted covers $\pi : \tilde{X} \to X$ such that the generic semistable rank n bundle over X is the direct image of a line bundle over $\tilde{X}$; by comparing sections of theta divisors, restricting to a Prym variety to obtain bundles in $\mathcal{S}\,\mathcal{U}(n, \mathcal{O})$, and using equivariance under the Mumford group, the theorem is proven. But the result also gives a map $\phi_\mathcal{L}$ of $\mathcal{S}\,\mathcal{U}(n, \mathcal{O})$ to $\mathrm{P}\Gamma(\mathrm{Pic}^{g-1} X, \mathcal{O}(n\Theta_X))$, which we from now on study in the case $n = 2$; we can also take the image of the Jacobian in this projective space, and it is intuitively clear how it is contained in the image of $\mathcal{M}_0$: consider

$$j : J \to \mathcal{M}_0 \quad j : L \mapsto L \oplus L^{-1};$$

then j factors over the Kummer variety $K = J/\pm$; all these maps fit in the following diagram:

5.3.2 Theorem [B1]

(i) The isomorphism $\rho: H^0(\mathcal{M}_0, \mathcal{L}_0) \to H^0(J, 2\Theta_X)^*$ of 5.3.1 is induced by the morphism

$$\delta: \mathcal{M}_0 \to |2\Theta_X| \quad E \mapsto \Delta_E = \{\xi \in \mathrm{Pic}^{g-1}X : h^0(E \otimes \xi) > 0)\}.$$

(ii) We have $j^*\mathcal{L}_0 \cong \mathcal{O}(2\Theta)$ and the map $j^*: H^0(\mathcal{M}_0, \mathcal{L}_0) \to H^0(J, \mathcal{O}(2\Theta))$ is an isomorphism.

(iii) For any theta characteristic $\kappa \in \mathrm{Pic}^{g-1}X$ we have an isomorphism $|2\Theta_X| \cong_\kappa |2\Theta|$, $D \mapsto D - \kappa$ and the diagram

$$
\begin{array}{ccc}
 & |2\Theta| \xrightarrow{\cong_\kappa} |2\Theta_X| & \\
 {}^{\delta_J}\nearrow & {}^{\delta}\nearrow & \\
J \xrightarrow{\;j\;} \mathcal{M}_0 & & \downarrow \xi_\kappa \\
{}^{\phi_{2\Theta}}\searrow & \searrow & \\
 & PV \cong |2\Theta|^* &
\end{array}
$$

is commutative, where $\delta_J: L \mapsto T_L^*\Theta \oplus T_{L^{-1}}^*\Theta$.

So far we talked about sections of $\mathcal{L}_0$, which can be viewed as hyperplanes in PV. To study $H^0(\mathcal{M}_0, \mathcal{L}_0^{\otimes 2})$ we restrict to $\delta(\mathcal{M}_0)$ the quadrics of PV, which amounts to the 'multiplication map' m_2 of the diagram below (the kernel of m_2 are the quadrics that vanish on $\delta(\mathcal{M}_0)$, which we again compare with an 'abelian' multiplication map n_2:

$$
\begin{array}{ccc}
S^2V & \xrightarrow{m_2} & H^0(\mathcal{M}_0, \mathcal{L}_0^{\otimes 2}) \\
 & & \\
{}^{n_2}\searrow & & \downarrow j^* \\
 & & \\
 & H^0(J, \mathcal{O}(4\Theta))_+ &
\end{array}
$$

Since $\dim S^2V = 2^{g-1}(2^g + 1) = \dim H^0(H, \mathcal{O}(4\Theta))_+ = N_{0,2}$, it would be nice if all the arrows in the diagram were isomorphisms; Beauville showed that this is so for the general curve, and constructed an explicit basis for $H^0(\mathcal{M}_0, \mathcal{L}_0^{\otimes 2})$ and $H^0(\mathcal{M}_0, \mathcal{L}_p)$; let $\mathcal{E}nd_0E$ be the sheaf of trace-zero endomorphisms:

5.3.3 Theorem [B2] Let κ be an even (resp. odd) theta characteristic and let D_κ be the reduced subvariety of $\mathcal{M}_0$ (resp $\mathcal{M}_1$) given by

$$D_\kappa = \{E \in \mathcal{M}_\epsilon : h^0(\mathcal{E}nd_0E \otimes \kappa) > 0\}.$$

(i) There is a section $d_\kappa \in H^0(\mathcal{M}_0, \mathcal{L}_0^{\otimes 2})$ (resp. $H^0(\mathcal{M}_1, \mathcal{L}_1)$) whose divisor is D_κ and these sections give bases of these spaces.

(ii) Let $\vartheta_\kappa \in H^0(J, \mathcal{O}(\Theta_\kappa))$ be a section defining Θ_κ. Then a basis for $H^0(J, \mathcal{O}(4\Theta))_+$ is given by the $[2]^*\vartheta_\kappa$ with κ even, where $[2]: J \to J$, $x \mapsto 2x$ (similarly, $H^0(J, \mathcal{O}(4\Theta))_-$ is spanned by the $[2]^*\vartheta_\kappa$ with κ odd).

464

(iii) For any even κ there is a $c_\kappa \in \mathbf{C}$ such that:

$$j^* d_\kappa = c_\kappa([2]^* \vartheta_\kappa) \text{ and } c_\kappa = 0 \Leftrightarrow h^0(X, \kappa) > 0.$$

In particular, j^* is an isomorphism iff $h^0(X, \kappa) \neq 0$ for all κ, i.e. J has no vanishing thetanulls.
(iv) The map m_2 is an isomorphism iff J has no vanishing thetanulls (in fact $m_2^* d_\kappa$ lies in the representative space corresponding to κ, and zero iff $h^0(X, \kappa) = 0$).

It is of interest here to mention the methods of [B2]: for the even case, Pryms are used but not the large ones used in 5.3.1; in fact unramified 2:1 covers of X and their $(g-1)$ dimensional Pryms suffice; we define these in the next section because of their fundamental role in our proof. For the odd case, the Hecke correspondence is used; this plays a fundamental role in [BS] as well, in that it allows one to lift geometric statements for $\mathcal{M}_0$ to a $\mathbf{P}^1$ bundle $\mathcal{P}$ over it and deduce corresponding statements for $\mathcal{M}_p$; the following diagram defines $\mathcal{P}$:

$$(E, u) \in \mathcal{P}$$

$$pr \swarrow \qquad \searrow q$$

$$\mathcal{M}_p \qquad \mathcal{M}_0$$

$0 \to \text{Ker } u \to E \to \mathbf{C}_p \to 0$; the fibre of q at a point F of $\mathcal{M}_0$ is the line $\mathbf{P}(\text{Ext}^1_{\mathcal{O}_x}(\mathbf{C}, F))$.

5.3.4 Theorem [B2]. *If $l \in J$ does not have order 2 (i.e. $l \not\cong l^{-1}$), then we can define a bundle $F_l \in \mathcal{M}_p$ which fits in an exact sequence $0 \mapsto l \oplus l^{-1} \to F_l \to \mathbf{C}_p \to 0$; the map $l \mapsto F_l$ can be extended to a morphism: $j_p \colon \hat{J} \to \mathcal{M}_p$ which factors through the involution to define a morphism of $\hat{K}$ to $\mathcal{M}_p$. The circumflexes indicate blow-up at the points of order 2. If the fibre of pr is mapped to $\mathcal{M}_0$ by sending $(E, u) \mapsto \text{Ker } u$, by following this with $\mathcal{M}_0 \xrightarrow{\phi_\zeta} \mathbf{P}(V)$ we get a line, namely a point in $\Lambda^2 V$; let this define $\phi_p \colon \mathcal{M}_p \to \mathbf{P}(\Lambda^2 V)$; the composite $\phi_p \circ j_p$ is the Gauss map associated to the tangent vector to X at p. The diagram*

$$\Lambda^2 V \xrightarrow{\phi_p^*} H^0(\mathcal{M}_p, \mathcal{L}_p)$$

$$w_{D_p} \searrow \qquad \qquad \downarrow j_p^*$$

$$H^0(J, L^2)_- = H^0(\hat{J}, e^* L^2(-\mathcal{E}))_-$$

(where $e \colon \hat{J} \to J$ is the blow-up of J at J_2 and $\mathcal{E}$ the exceptional divisor) is commutative; to define w_{D_p}, one applies the vector field D_p on J, gotten from the tangent vector to $X \subset J$ at p, to the Wahl homomorphism $(s, t) \mapsto t^{\otimes 2} d(\frac{s}{t})$

$$w \colon \Lambda^2 H^0(J, L) \to H^0(J, L^2 \otimes \Omega^1_J).$$

Finally, if κ is an odd theta characteristic, $j_p^* d_\kappa = [2]^* \vartheta_\kappa$ if $h^0(\kappa(-p)) = 0$, while $j_p^* d_\kappa = 0$ if $h^0(\kappa(-p)) > 0$. In particular, ϕ_p^* and w_{D_p} have the same kernel and are isomorphisms iff, for all odd theta characteristics κ of C, $h^0(\kappa(-p)) = 0$.

5.4 Schottky-Prym relations via bundles. We review in some detail the geometric configuration in $\mathbf{P}V$ corresponding to all the Prym varieties P_x of the unramified double covers $\pi_x : X_x \to X$, which correspond to the $2^{2g} - 1$ points of order 2 of J. The crucial part of the theory is found in [M1]. $P_x = (\ker Nm_x)^0$ comes with a principal polarization Θ_{P_x}. The norm map Nm_x is a surjective homomorphism; superscript 0 means the connected component containing $0 \in \operatorname{Jac} X_x$. The picture to have in mind is that not only do the $|2\Theta_{P_x}|$ maps of the P_x fit in the projective space $|2\Theta|$ (equivariantly under the action of the corresponding Heisenbergs); they also factor through the operation of producing rank 2 bundles over X via π_{x*} and applying δ to $\mathcal{M}_0$ (Proposition 5.4.1). The details are as follows:

The (projective) Heisenberg action on V can also be viewed as the action of $J[2]$ on $\mathcal{M}_0$ by tensorization, since the pull-back of $\mathcal{L}_0$ under such action must be isomorphic to itself, and so defines

$$U : J[2] \to \operatorname{Aut} \mathbf{P}H^0(\mathcal{M}_0, \mathcal{L}_0); \quad U(x)\delta(E) = \delta(E \otimes x).$$

Notice that the Kummer K is stable under this action and $U(x)j(z) = j(z \otimes x)$. For $x \neq 0$, $U(x)$ has two eigenspaces in $\mathbf{P}V$, each of dimension 2^{g-1}, and since the Weilparing $E : J[2] \times J[2] \to \mathbf{Z}/2$ corresponds to the commutator in H, we have:

$$U(x)U(y) = (-1)^{E(x,y)} U(y)U(x).$$

Thus if $E(x,y) = 1$, the eigenspaces of $U(x)$ and $U(y)$ do not intersect. On an eigenspace $\mathbf{P}V_x$ of $U(x)$ the group

$$x^\perp := \{y \in J[2] : E(x,y) = 0\}$$

acts, and since $U(x)$ acts trivially on an eigenspace, the action factors over $x^\perp / < x > \cong (\mathbf{Z}/2)^{2(g-1)}$. This action is the projectivization of the Schrödinger representation of H_{g-1} on V_x.

When $E(x,y) = 0$, the space $\mathbf{P}V_{x,y} := \mathbf{P}V_x \cap \mathbf{P}V_y$, is an eigenspace of $U(\bar{y})$,

$$\bar{y} := y + < x > \in x^\perp / < x >$$

so $\dim \mathbf{P}V_{x,y} = 2^{g-2} - 1$, the group $(x^\perp \cap y^\perp) / < x, y > \cong (\mathbf{Z}/2)^{2(g-2)}$ acts on it and the action is the projectivization of the Schrödinger representation of H_{g-2} on $V_{x,y}$.

For any ppav (A, Θ_A), with symmetric Θ_A, the map $\phi_A : A \to \mathbf{P}H^0(A, L_A)$, with $L_A := \mathcal{O}_A(2\Theta_A)$ is equivariant under the action of $A[2]$ and factors over $A/\pm$. If the decomposition of this ppav into indecomposable ppav's is given by:

$$(A, \Theta_A) \cong \prod_i (A_i, \Theta_{A_i}), \text{ then } A \xrightarrow{\phi} K(A) := \prod(A_i/\pm) \subset \mathbf{P}H^0(A, L_A)$$

and we call $K(A)$ the Kummer variety of A. In case A is a Jacobian, then $K(A) \cong A/\pm$.

For non-zero $x \in A[2]$, let:

$$S_x := \{z \in A: z^{\otimes 2} \cong x\}.$$

The set S_x is a principal homogeneous space for $A[2]$, and since for $z \in S_x$, $\phi(z) = \phi(z^{-1}) = \phi(z \otimes x)$, the image of S_x lies in the union of the two eigenspaces of $U(x)$ in $\mathbf{P}H^0(A, L_A)$, and each eigenspace gets half the points. If A is indecomposable, then $K(A) \cap \mathbf{P}V_x$ consists exactly of the image of one half of S_x (since then $\phi(z) = \phi(z \otimes x)$ is equivalent to $z \in S_x$) and since ϕ is 2:1, that is a set of 2^{2g-2} points.

At the level of abelian varieties, one has $\ker(\pi_x^*: \operatorname{Jac} X \to \operatorname{Jac}(X_x)) \cong < x >$, π_x^* induces an isomorphism:

$$\pi_x^*: x^\perp / < x > \xrightarrow{\cong} P_x[2] \text{ and } \ker(Nm_x) = P_x + (P_x + \pi_x^*y),$$

for (any) $y \in J[2]$ with $E(x, y) = 1$. Since $Nm_x(p \otimes \pi_x^*z) = Nm_x(p) \otimes z^{\otimes 2}$, the group $J[2]$ acts on $\ker(Nm_x)$ via $p \mapsto p \otimes \pi_x^*y$, $y \in J[2]$ and this action factors over $J[2]/ < x > \cong (\mathbf{Z}/2)^{2g-1}$.

Since $\pi_{x*}\mathcal{O}_{X_x} \cong \mathcal{O}_X \oplus x$ we have $\det \pi_*\mathcal{O}_{X_x} \cong x$. For any $p \in \ker(Nm_x)$, the rank two bundle $\pi_{x*}p$ on X then also has determinant x. Fixing a $z_x \in J$ with $z_x^{\otimes 2} \cong x$, we get a map:

$$\psi_x: \ker(Nm_x) \cong P_x \cup P_x \to \mathcal{M}_0, \ p \mapsto (\pi_{x*}p) \otimes z_x.$$

Note that the actual map depends on the choice of z_x, but that if also $z_x'^{\otimes 2} \cong x$, then $z_x' \cong z_x \otimes y$ for some $y \in J[2]$. Thus if ψ_x' is the map defined by z_x', then we have $\psi_x'(p) = U(y)\psi_x(p)$ for all $p \in \ker(Nm_x)$.

By the projection formula we have, for any $p \in \ker(Nm_x)$ and $z \in X$:

$$\pi_{x*}(p) \otimes z \cong \pi_{x*}(p \otimes \pi_x^*s) \text{ thus } \psi_x(p) \otimes y = \psi_x(p \otimes \psi_x^*y) \ (\forall y \in J[2])$$

and so the map ψ_x is equivariant for the actions of $J[2]$ on $\ker(Nm_x)$ and $\mathcal{M}_0$. Moreover, the choice of z_x in the definition of ψ_x does not affect the image of ψ_x (i.e. $\operatorname{Im}(\psi_x) = \operatorname{Im}(\psi_x')$).

Since both ψ_x and ϕ are equivariant for the $J[2]$-action, and since $< x >$ acts trivially on $\ker(Nm_x)$, we have, for all $p \in \ker(Nm_x)$:

$$U(x)\phi(\psi_x(p)) = \phi(\psi_x(p)), \text{ thus } \psi_x(\ker(Nm_x)) \subset \mathbf{P}V_x^+ \cup \mathbf{P}V_x^-,$$

the union of the two eigenspaces of $U(x)$ in $\mathbf{P}V$. Since $U(y)$, with $E(x, y) = 1$, interchanges the two components of $\ker(Nm_x)$ and the two eigenspaces, there is one component of $\ker(Nm_x)$ in each eigenspace. We can restrict $\phi \circ \psi_x$ to get a map:

$$\phi_x: P_x \to \mathbf{P}V_x$$

where $\mathbf{P}V_x$ is the appropriate eigenspace of $U(x)$. The description of ϕ combined with results of Mumford's [M1] allows us to identify ϕ_x.

5.4.1 **Proposition.** *The map $\phi_x: P_x \to \mathbf{P}V_x$ is the natural map*

$$P_x \to K(P_x) \subset \mathbf{P}H^0(P_x, L_{P_x}) \cong \mathbf{P}V_x \text{ with } L_{P_x} := \mathcal{O}(2\Theta_{P_x}).$$

Proof. Let $\Theta_{g-1} := \{z \in \operatorname{Pic}^{g-1}(X): H^0(X,z) \neq 0\}$ be the theta divisor of X. Then there is a unique $w \in \operatorname{Pic}^{g-1}(X)$, in fact a theta characteristic, such that $\Theta = \{z \otimes w^{-1}: z \in \Theta_{g-1}\}$. The map $\phi: \mathcal{M}_0 \to PV \cong |2\Theta| \cong |2\Theta_{g-1}|$ is given by:

$$\phi: E \mapsto D_E \in |2\Theta_{g-1}|, \quad D_E := \{z \in \operatorname{Pic}^{g-1}(X): H^0(X, E \otimes z) \neq 0\}.$$

For $p \in \ker(Nm_x)$ we have:

$$H^0(X, (\pi_{x*}p) \otimes z_x \otimes z) = H^0(X, \pi_{x*}(p \otimes \pi_x^*(z_x \otimes z))) = H^0(X_x, p \otimes \pi_x^*(z_x \otimes z)).$$

The degree of $\pi_x^* z$ is $2(g-1) = g_x - 1$ with g_x the genus of X_x, therefore

$$D_{\pi_{x*}(p) \otimes z_x} = D_p := \{z \in \operatorname{Pic}^{g-1}(X): \pi_x^*(z) \in p^{-1} \otimes (\pi_x^* z_x)^{-1} \otimes \tilde{\Theta}_{gx-1}\},$$

with $\tilde{\Theta}_{gx-1} \subset \operatorname{Pic}^{g_x-1}(X_x)$ the theta divisor of X_x, and Mumford [M1, p. 334] proved that $p \mapsto D_p$ gives the natural map $P_x \to PH^0(P_x, L_{P_x})$. $\blacksquare$

From now on, we simply write K and $K(P_x)$ for the images of the maps $\phi_K := \phi \circ \psi_K$ and ϕ_x respectively.

5.4.2 Proposition. *For any curve X and any $x, y \in J[2] - \{0\}$ with $E(x,y) = 0$ we have:*

(i) The SJ (Schottky-Jung) relations:

(SJ) $$K \cap PV_x = K(P_x[2]).$$

(ii) The Donagi relations:

(D) $$\phi_y(S_{\bar{x}}) \cap PV_{x,y} = \phi_x(S_{\bar{y}}) \cap PV_{x,y}$$

with $\bar{x} := x+ < y > \in y^\perp / < y > \cong P_y[2]$ and $\bar{y} := y+ < x > \in x^\perp / < x > \cong P_x[2]$.

Proof. Because of the equivariance under the action of $x^\perp$ on PV_x and the identification $\pi_x^*: x^\perp / < x > \to P_x[2]$, to prove (SJ) it is enough to find one point in the intersection. In view of 5.4.1 there is a very natural choice, namely the zero divisor of X_x; indeed, $\pi_*(\mathcal{O}_{X_x}) \cong \mathcal{O}_X \oplus x$ which, after a translation to get trivial determinant, implies that $\phi_K(y) = \phi_x(0)$ for some $y \in J[2]$. More precisely, let $\phi_K(z_x) \in K \cap PV_x$, then $z_x^{\otimes 2} \cong x$ and $K \cap PV_x = \{\phi_K(z_x \otimes y): y \in x^\perp\}$. Let z_x define $\phi_x: P_x \to PV_x$ then:

$$\phi_K(z_x) := \phi(z_x \oplus z_x^{-1}) = \phi((\mathcal{O}_X \oplus x) \otimes z_x) = \phi_x(\mathcal{O}_{X_x}).$$

As y runs over $x^\perp$:

$$
\begin{aligned}
K \cap PV_x &= \{\phi_K(x_x \otimes y)\} = \{U(y)\phi_K(z_x)\} \\
&= \{U(y)\phi_x(\mathcal{O}_{X_x})\} = \{\phi_x(\pi_x^* y)\} = K(P_x[2]).
\end{aligned}
$$

For the Donagi relations, we explicitly construct a unitary rank 2 vector bundle E on X defined by the representation:

$$\rho_E \colon \pi_1(X) \to SU(2) \begin{cases} \alpha_k \mapsto I,\ 1 \le k \le g,\ \beta_l \mapsto I,\ 3 \le l \le g \\[2mm] \beta_1 \mapsto \begin{pmatrix} i & 0 \\ 0 & -i \end{pmatrix},\ \beta_2 \mapsto \begin{pmatrix} 0 & i \\ i & 0 \end{pmatrix}, \end{cases}$$

where the α_i, β_j are standard generators of $\pi_1(X)$. To prove the Donagi relations, we will first show that $E \cong \pi_{x*}q_x \cong \pi_{y*}q_y$, for $x, y \in J[2], q_x \in \operatorname{Jac}(X_x), q_y \in \operatorname{Jac}(X_y)$, which is equivalent to proving that

$$\rho_E \sim \operatorname{Ind}_{\pi_1(X_x)}^{\pi_1(X)}(\rho_x) \sim \operatorname{Ind}_{\pi_1(X_y)}^{\pi_1(X)}(\rho_y)$$

with $\rho_x \colon \pi_1(X_x) \to U(1)$ the representation corresponding to q_x etc., then we check that these q_x, q_y describe the sets $S_{\overline{y}}, S_{\overline{x}}$, resp.

Let X_x be the unramified 2:1 cover of X defined by the subgroup $\ker(\epsilon_x)$ of $\pi_1(X)$ with

$$\epsilon_x \colon \pi_1(X) \to U(1),\ \epsilon_x(\beta_1) = -1,\ \epsilon_x(\alpha_i) = \epsilon_x(\beta_j) = 1,$$

with $1 \le i \le g, 2 \le j \le g$. Similarly, we define X_y by a character ϵ_y with $\epsilon_y(\beta_2) = -1$, and ϵ_y trivial on the other generators.

Note that $\pi_{x*} \colon \pi_1(X_x) \overset{\cong}{\to} \ker(\epsilon_x) \subset \pi_1(X)$, and that $\pi_1(X_x)$ is the fundamental group of a Riemann surface of genus $2g - 1$. For suitable generators γ_i, δ_i or $\pi_1(X_x)$ the homomorphism π_{x*} is given by:

$\gamma_1,$	$\gamma_2,$	$\ldots,$	$\gamma_g,$	$\gamma_{g+1},$	$\cdots$	$\gamma_{2g-1},$	$\delta_1,$	δ_2	$\ldots,$	$\delta_g,$	$\delta_{g+1},$	$\ldots,$	δ_{2g-1}
$\downarrow$	$\downarrow$		$\downarrow$	$\downarrow$		$\downarrow$	$\downarrow$	$\downarrow$		$\downarrow$	$\downarrow$		$\downarrow$
$\alpha_1,$	$\alpha_2',$	$\ldots,$	$\alpha_g',$	$\alpha_2,$	$\ldots,$	$\alpha_g,$	$\beta_1^2,$	β_2'	$\cdots$	$\beta_g',$	$\beta_2,$	$\ldots,$	β_g

with $\lambda' := \beta_1 \lambda \beta_1^{-1}$ and π_{y*} is given by a similar prescription, with the roles of 1 and 2 interchanged.

Then it is easy to check that the restriction of ρ_E to $\pi_1(X_x) \subset \pi_1(X)$ is reducible; invariant subspaces are $W_x := < \binom{1}{1} >$ and $W_x^\perp$. Since

$$\pi_1(X) = \pi_1(X_x) \cup \beta_1 \pi_1(X_x) \text{ and } \mathbf{C}^2 = W_x \oplus W_x^\perp = W_x \oplus \rho_E(\beta_1) W_x,$$

it follows that $\rho_E = \operatorname{Ind}_{\pi_1(X_x)}^{\pi_1(X)}(\rho_x)$, with $\rho_x \colon \pi_1(X_x) \to U(W_x) \cong U(1)$ the restriction of ρ_E. Similarly, the restriction to $\pi_1(X_y)$ has invariant subspaces $W_y := (< \binom{1}{0} >)$ and $W_y^\perp = \rho_E(\beta_2) W_y$, and $\rho_E = \operatorname{Ind}_{\pi_1(X_y)}^{\pi_1(X)}(\rho_y)$.

Let $z_x \in J$ define the map ψ_x and let $p_x := q_x \otimes \pi_x^* z_x$, then $p_x \in \ker(Nm_x)$ and, after tensoring by some $y \in J[2]$ if necessary, we get $p_x \in P_x$. Then:

$$\psi_x(p_x) := \pi_{x*}(p_x) \otimes z_x \cong \pi_{x*}(q_x \otimes \pi_x^* z_x \otimes \pi_x^* z_x) \cong \pi_{x*}(q_x) \cong E$$

since $z_x^{\otimes 2} = x$ implies $(\pi_x^* z_x)^{\otimes 2} \cong \pi_x^* x \cong O_{X_x}$. Since $p_x^{\otimes 2} \cong q_x^{\otimes 2}$ is the line bundle defined by the character ρ_x^2 of $\pi_1(X_x)$ and since $\rho_x^2 = \epsilon_y$, the character defining X_y when restricted to $\pi_1(X_x)$ (easy verification), we have that $p_x \in S_{\overline{y}}$. Using a similar argument for y and using the equivariance of ψ_x, ψ_y for the action of $x^{\perp} \cap y^{\perp}$, the stabilizer of $\mathbf{P}V_{x,y}$ in $J[2]$, we get the desired result. ∎

5.5 Overview and Problems.

Before we continue, in the next section, our investigation of quartic hypersurfaces in $|2\Theta|$, we collect here some fascinating facts on the $|2\Theta|$ map, for two reasons. The first is the role of this map in the "Schottky problem" which, despite much progress, remains profound and challenging. The second is the amount of classical algebraic geometry that might be reinterpreted in terms of vector bundles (via spectral curves!) in the spirit of 5.4.2. For instance, Klein's quadratic complex enters the description of the Hitchin flows [vGP2], cf. 6.11; given the modern applications of the quadratic complex (Hodge theory, intermediate Jacobians, cf. [GH] Ch. 6), this direction is surely worth pursuing.

5.5.1 The Schottky problem. We don't presume to survey this long-studied question, but only highlight those features that are somewhat linked to quartics in $|2\Theta|$. The Schottky problem loosely stated asks for a geometric characterization of the Jacobian locus in $\mathcal{A}_g$ (=moduli space of ppav=$\mathcal{H}_g/Sp(2g, \mathbf{Z})$); the original question was asked in terms of modular forms. Notice that, while the Jacobian locus has dimension $3g - 3$ within the $\frac{g(g+1)}{2}$ deformation space of abelian varieties, the number of moduli of $\mathcal{SU}(r,d)$ is the same as that of the curve X when $(r,d) = 1$.

5.5.2 (i) It was proved by Wirtinger [Wi §§17-23] that $K(J) = \phi_{2\Theta}(J)$ can be defined by quartics and the coefficients of these quartics are algebraic expressions in the thetanulls.

The importance of (order k) theta functions is due precisely to this double-nature: as functions of $z \in \mathbf{C}^g$ they give sections of the bundle $k\Theta$; as functions of the period matrix τ (e.g., the thetanulls at $z = 0$) they map the moduli space $\mathcal{A}_g$ (with some level structure) to $|k\Theta|$. The link between z and τ is provided by the heat equation:

$$\nabla_{\frac{\partial}{\partial \tau_{kl}}} \vartheta(z, \tau) = 0, \quad \text{with } \nabla_{\frac{\partial}{\partial \tau_{kl}}} := \frac{\partial}{\partial \tau_{kl}} - 2\pi i(1 + \delta_{kl}) \frac{\partial^2}{\partial z_k \partial z_l}.$$

Merely as a pointer: this flat connection on the global sections of the theta bundles, over the moduli space of curves, has been extended by N. Hitchin [Hi2] to higher rank and to the bundles $\mathcal{L}$, but the explicit calculation of the holonomy is still an open problem (in rank 1 it reduces to the monodromy on the set of theta structures).

The heat equation was given a geometric interpretation in $|2\Theta|$ space in [vGvdG]:

(ii) By virtue of the Heisenberg group action, there is a natural map

$$Th : \mathcal{H}_g \longrightarrow \mathbf{P}V, \quad \tau \mapsto (\ldots, \vartheta[\begin{smallmatrix} r \\ 0 \end{smallmatrix}](\tau, 0), \ldots),$$

which factors over $\mathcal{A}_g(2,4) := \mathcal{H}_g/\Gamma_g(2,4)$.

As a consequence of the heat equation, the intersection of the tangent space $T_{K(0_J)}$ to $Th(\mathcal{H}_g)$ at the point $K(0_J) \in Th(\mathcal{H}_g)$ with $K(J)$ is given as follows: identifying $\mathbf{P}V \cong |2\Theta|^*$, the hyperplanes in $\mathbf{P}V$ passing through $K(0_J)$ are those $D \in |2\Theta|$ with $0_J \in D$, and the hyperplanes which contain $T_{K(0_J)}$ correspond to the divisors wich have multiplicity at least 4 in 0_J; we denote that space by:

$$|2\Theta|_{00} := \{D \in |2\Theta| : \; mult_0(D) \geq 4\}.$$

(iii) The above-illustrated link between points of $|2\Theta|$, points of $\mathcal{A}_g$ and points of $K(J)$ is precisely the object of the Schottky-Jung relations:

$$\vartheta \begin{bmatrix} \epsilon & 0 \\ \epsilon' & 0 \end{bmatrix} \vartheta \begin{bmatrix} \epsilon & 0 \\ \epsilon' & 1 \end{bmatrix} (\tau, 0) = \vartheta \begin{bmatrix} \epsilon \\ \epsilon' \end{bmatrix}^2 (\pi_x, 0)$$

In the next lecture these equations are used to connect the vanishing of a quartic in $|2\Theta|$ on $K(J)$ and on the Pryms, in the light of the Verlinde fomula for $k = 4$, cf. 6.2,6.3,6.9.

(iv) In the case of a nonhyperelliptic curve of genus 3, $\delta \mathcal{M}_0$ is a quartic hypersurface, given in [C §33 (5)]. More is true: if we denote by D_i the closure in $\mathcal{SU}(2, K_X)$ (here K_X is the canonical bundle) of the set of stable bundles E such that $h^0(E) \geq i$, then: D_1 is a hyperplane section of this quartic; D_2, given in [C §49 (4)], is a cone over the Veronese embedding of $\mathbf{P}^2$ in $\mathbf{P}^5$; its vertex is D_3 and it corresponds to the unique bundle E with 3 independent sections; this bundle can be obtained in the following canonical way: its dual E^* fits in the exact sequence:

$$0 \to E^* \to H^0(X, K_X) \otimes_{\mathbf{C}} \mathcal{O}_X \to K_X \to 0$$

(in other words E is the normal bundle to the curve in canonical space). The same bundle was studied in [PR] in connection with a conjecture of M. Green.

(v) The Schottky problem is linked in another surprising fashion with (iv) above. Recall the linear series $|2\Theta|_{00}$ from (ii); this turns out to have significance to the Schottky problem, because if a ppav A is the Jacobian of a curve, one knows that the intersection of $Th(\mathcal{H}_g)$ with $K(A)$, the image of A in $\mathbf{P}V$, contains a surface but for general A it seems likely the the intersection consists just of points, in fact the set $K(A[2])$ (note that, with 0_A the identity element of A one has $K(0_A) = Th(\tau)$, with τ a suitable period matrix of A). Writing $X - X := \{a - b \in J : a, b \in X\}$, which is a surface in J, a well known result of Welters gives (except for $g = 4$, when there can be two additional points):

$$\cap_{D \in |2\Theta|_{00}} D = X - X, \quad \text{so} \quad T_{K(0_J)} \cap K(J) = K(X - X).$$

As of yet, little is known about the divisors in $|2\Theta|_{00}$, but the theory of rank 2 bundles provides a way of producing elements in this space, as demonstrated in [L], where for the bundle E described in (vi) it is shown:

$$\Delta_E = \{x \in J \mid h^0(E \otimes x) > 0\} = X - X,$$

the unique element in $|2\Theta|_{00}$ for $g = 3$, X nonhyperelliptic.

Lecture 6. Theta functions: The multiplication maps.

Finally, we study the degree 4 multiplication maps, which we again compare in the abelian and nonabelian case, according to the diagram below: we need some notation first.

The spaces $S^k V$ can be explicitly decomposed into irreducible pieces for the H action; when k is even the action factors through the abelian quotient $H/H' \cong \mathbf{C}^* \times$

$(\mathbf{Z}/2)^g \times \mathrm{Hom}((\mathbf{Z}/2)^g, \mathbf{C}^*)$; denote by $Ch(H)$ the group of characters of $(\mathbf{Z}/2)^g \times \mathrm{Hom}((\mathbf{Z}/2)^g, \mathbf{C}^*)$ and by 0 the trivial character; then

$$S^4 V = \oplus_{\chi \in Ch(H)} S^4_\chi V, \quad \dim S^4_0 V = d(g), \quad \dim S^4_\chi V = d(g-1)$$

with $d(g) = (2^g + 1)(2^{g-1} + 1)/3$, cf. [vG].

Notice that by restricting forms of degree k to $\delta(\mathcal{M}_0)$ we will have a kernel (degree k hypersurfaces which contain $\delta(\mathcal{M}_0)$) and if we can bound from above the dimension of that kernel then we have a lower bound for $D_{0,k}$:

$$\ker m_k \to S^k V \overset{m_k}{\to} H^0(\mathcal{M}_0, \mathcal{L}_0^{\otimes k})$$

More precisely, using the ϕ_x and j, we restrict sections of $\mathcal{L}_0^{\otimes 4}$ to the union of all Pryms and J (that is, to their images in $\mathcal{M}_0$). It will be convenient to write $P_0 := J$ and to agree that $0 \in J[2]$ corresponds to the trivial character $0 \in Ch(H)$.

The following diagram exhibits the maps we will be interested in; note that we write $m = \oplus m_\chi$, $n = \oplus_\chi n_\chi$ for the multiplication maps originating from $S^4 V = \oplus_\chi S^4_\chi V$.

$$
\begin{array}{ccc}
S^4_\chi V & \overset{m_\chi}{\to} & H^0(\mathcal{M}_0, \mathcal{L}_0^{\otimes 4}) \\
& \searrow n_\chi & \downarrow res \\
& & \oplus_{x \in J[2]} H^0(P_x, \mathcal{O}(8\Xi_x))_+
\end{array}
$$

The diagram again relates multiplication maps for nonabelian theta functions (the m_χ) with those of abelian theta functions (the n_χ). Since the Heisenberg group acts on both $S^4 V$ and on $H^0(\mathcal{M}_0, \mathcal{L}_0^{\otimes 4})$ and m is equivariant, we have that $\ker m = \oplus_\chi \ker m_\chi$. The diagram implies that $\dim \ker m_\chi \leq \dim \ker n_\chi$, and thus we obtain the following lower bound for the fourth Verlinde number

$$
\begin{aligned}
N_4 &= \dim H^0(\mathcal{M}_0, \mathcal{L}_0^{\otimes 4}) \\
&\geq \dim S^4 V - \dim \ker m \\
&\geq \sum_\chi (\dim S^4_\chi V - \dim \ker n_\chi)
\end{aligned}
$$

The main result of [vGP1], explained in the remainder of this section, is that for a generic curve one has $\dim S^4_\chi V - \dim \ker n_\chi = e(g)$ if $\chi = 0$ and $= e(g-1)$ if $\chi \neq 0$ with $e(g) := (3^g + 1)/2$. The lower bound for N_4 thus obtained actually coincides with the value of N_4 predicted by Verlinde, and again the non-abelian theta functions are 'identified' with certain abelian theta functions:

6.1 **Theorem.** *Let X be a curve with no vanishing thetanulls. Then the following hold:*

(i) The multiplication map $m : S^4 V \to H^0(\mathcal{M}_0, \mathcal{L}^{\otimes 4})$ is surjective.

(ii) $\ker m_\chi = \ker n_\chi$ for all χ, thus also $\ker m = \ker n$.

(iii) The map $res : H^0(\mathcal{M}_0, \mathcal{L}^{\otimes 4}) \to \oplus_{x \in J[2]} H^0(P_x, \mathcal{O}(8\Xi_x))$ is injective.

The map n_0 was already studied in [vG], for the other n_χ we need the results of [vGP1]. It is a curious phenomenon that quartics in the kernel of the n_χ's must satisfy a certain condition on their singular locus:

6.2 Theorem (a weaker version of Theorem 1 in [vG], less technical to state). *Let* $F \in S_0^4 V$. *Then F vanishes on all Pryms (i.e. $K(P_x) \subset Z(F)$ for all $x \in J[2]$) if and only if $K(J) := \phi_{\mathcal{L}}(j(P_0)) \subset SingZ(F)$.*

6.3 Theorem (Theorem 1 in [vGP1]). *Let $F \in S_\chi^4 V$, $\chi \neq 0$, and let $x \in J[2]$ correspond to χ. Then F vanishes on all Pryms if and only if $K(P_x) \subset SingZ(F)$.*

The proofs of these theorems are discussed in (6.8) below; they imply, with χ corresponding to x:

$$\ker(n_\chi) = \{F \in S_\chi^4 V : K(P_x) \subset SingZ(F)\}.$$

6.4 Therefore for $F \in S_\chi^4$ to be in $\ker(n_\chi)$ it is necessary and sufficient that $\frac{\partial F}{\partial X_\sigma}$ be a (cubic) equation for $K(P_x)$ for all $\sigma \in (\mathbf{Z}/2\mathbf{Z})^g$ (in view of the Heisenberg action, it suffices in fact that $\frac{\partial F}{\partial X_\sigma}$ be such an equation because: (case 0), $(1, \alpha, 0) \in H$ leaves K invariant, while $X_0 \mapsto X_\alpha$, $F \mapsto \chi(\alpha)F$; (case χ), same argument for the $x^\perp$ part of $H = x^\perp \oplus < y >$, but if $E(x, y) = 1$ then $F \equiv 0$ on $\mathbf{P}V_y$). There are interesting relations between cubics and quartics under the action of H. We have an isomorphism

$$\frac{1}{4}\frac{\partial}{\partial X_0} : S_0^4 V \longrightarrow S_0^3 V, \quad \text{with } S_0^3 V := \{G \in S^3 V : (1, 0, \alpha^*)G = G, \ \forall \alpha^*\}.$$

and homomorphisms

$$M(\chi) : S_0^3 V \to S_\chi^4 V, \quad G \mapsto \sum_\sigma \alpha^*(\sigma)X_{\sigma+\alpha}((1, \sigma, 0)G),$$

where $\chi = (\alpha, \alpha^*) \in Ch(H)$, $\sigma \in (\mathbf{Z}/2\mathbf{Z})^g$; $M(0)$ is the inverse of $\frac{1}{4}\frac{\partial}{\partial X_0}$. Moreover, if χ corresponds to x then the restrictions of $M(0)F$ and $M(\chi)F$ to V_x, the eigenspace of x in V, differ by a non-zero multiplicative constant.

Using the isomorphism $\frac{1}{4}\frac{\partial}{\partial X_0} : S_0^4 V \to S_0^3 V$, we see that $\ker n_0 = \ker(S_0^3 V \to H^0(J, \mathcal{O}(6\Theta))_+)$, and the well-known fact that the multiplication map $S^3 V \to H^0(J, \mathcal{O}(6\Theta))_+$ is surjective for any X with no vanishing thetanull then implies that dim $\ker n_0 = d(g) - e(g)$, as desired.

In case $\chi \neq 0$ one proceeds in a rather similar way to derive dim $\ker n_\chi = d(g - 1) - e(g - 1)$, replacing $J = P_0$ by P_x with x corresponding to χ; for the details, see [vGP1].

6.5 The proofs of the theorems (6.2) and (6.3) involve a comparison between the multiplication maps (restricted to subspaces of) $S^4 V_x \to H^0(P_x, \mathcal{O}(8\Xi_x))$ for the various x. The main point is that, w.r.t. suitable bases, the entries of the matrices defining these maps are theta constants which become equal in virtue of the Schotty-Jung (SJ) (when comparing $\chi = 0$ with a nonzero x) and the Donagi (D) relations (when comparing two orthogonal x).

The advertised comparison between multiplication maps is given by explicit matrices: a choice of theta structure gives a period matrix, hence as in [vG] explicit bases for the spaces involved. For a suitable choice of bases, Riemann's formula gives (cf. [I], IV.1):

6.6 Proposition.

The multiplication $S^2V \xrightarrow{n_2} H^0(J, L^{\otimes 2})$ where $L = \mathcal{O}(2\Theta)$ is given by

$$n_2(Q\begin{bmatrix} \epsilon \\ \epsilon' \end{bmatrix}(\dots\vartheta_{\frac{\epsilon}{2},0}(2\tau,2z)\dots)) = \vartheta_{\frac{\epsilon}{2},\frac{\epsilon'}{2}}(\tau,0)\vartheta_{\frac{\epsilon}{2},\frac{\epsilon'}{2}}(\tau,2z) \text{ where } \vartheta_{\frac{\epsilon}{2},0}(2\tau,2z)$$

$(\sigma \in (\mathbf{Z}/2)^g)$ is a basis of $H^0(J, L)$ and $\vartheta_{\frac{\epsilon}{3},\frac{\epsilon'}{2}}(\tau,2z)(^t\epsilon\epsilon' = 0)$ of $H^0(J, L^{\otimes 2})_+$,

and $Q\begin{bmatrix} \epsilon \\ \epsilon' \end{bmatrix} = \sum_\sigma (-1)^{t_{\sigma\epsilon'}} X_\sigma X_{\sigma+\epsilon}$ $(^t\epsilon\epsilon' = 0)$ is a basis of eigenvectors in S^2V. As a consequence n_2 is an isomorphism $\Leftrightarrow X$ has no vanishing thetanulls.

Next, given a non-trivial $\chi \in Ch(H)$, we can choose an automorphism ϕ of H, which is the identity on $\mathbf{C}^*$, such that

$$\phi_\chi = \chi_1 := ((\alpha_1, \dots, \alpha_g, \alpha_1^*, \dots, \alpha_g^*) \mapsto (-1)^{\alpha_g}).$$

The element $\bar{h} \in Ch(H)/\mathbf{C}^*$ corresponding to χ_1, that is the class of a (t, h) for which $[(t, h), (s, l)] = (\chi_1(l), 0)$, is then

$$\bar{h} = ((0, \dots, 0), (0, \dots, 0, 1)) \in (\mathbf{Z}/2)^g \times \mathrm{Hom}((\mathbf{Z}/2)^g, \mathbf{C}^*)$$

since the commutator in H is: $[(t, \alpha, \alpha^*), (s, \beta, \beta^*)] = (\alpha^*(\beta)\beta^*(\alpha), 0, 1)$.

In S_0^4V, we choose the following basis:

$$P_I = \sum_\sigma X_\sigma X_{\sigma+\rho} X_{\sigma+\nu} X_{\sigma+\rho+\nu}$$

for all $I = \{0, \rho, \nu, \rho+\nu\} \subset (\mathbf{Z}/2)^g$

and in $S_{\chi_1}^4 V$:

$$R_J = \sum_\sigma X_{(\delta 0)} X_{(\delta+\beta 0)} X_{(\delta+\gamma 0)} X_{(\delta+\beta+\gamma 0)} - X_{(\delta 1)} X_{(\delta+\beta 1)} X_{(\delta+\gamma 1)} X_{(\delta+\beta+\gamma 1)}$$

for all $J = \{0, \beta, \gamma, \beta+\gamma\} \subset (\mathbf{Z}/2)^{g-1}$

and in the following subspace of $H^0(X, L^{\otimes 4})$, which contains the image $n_4(S_\chi^4V)$: $[2]^*H^0(X, L_\chi)_+$, where $L_\chi = T_y^* L$, $2y = x$, and $x \in J[2]$ corresponds to χ (the subscript $+$ means even thetas, note all elements of V are even!):

$$\theta_{\frac{\sigma}{2}0}(2\tau, 4z), \quad \sigma \in (\mathbf{Z}/2)^g, \quad \text{for } \chi = 0$$

$$\theta_{(\frac{\alpha}{2}0)(0\frac{1}{2})}(2\tau, 4z), \quad \alpha \in (\mathbf{Z}/2)^{g-1}, \quad \text{for } \chi = \chi_1.$$

Using Riemann's formula [I, IV.1], the matrices of the multiplication map n_4 are computed in [vG, Proposition 4]:

(i) $\qquad n_4: S_0^4(X, L) \to [2]^* H^0(J, L_\chi)_+, \quad P_I \xrightarrow{n_4} \sum_\lambda C_{I,\lambda} \theta_{\frac{\lambda}{2}0}(2\tau, 4z)$

where $\lambda \in (\mathbf{Z}/2)^g$ and

$$C_{I,\lambda}(\tau) = (\theta_{\frac{\lambda+\rho}{2}0} \theta_{\frac{\lambda+\nu}{2}0} \theta_{\frac{\lambda+\rho+\nu}{2}0}(2\tau, 0);$$

(ii)
$$n_4 \colon S_{\chi_1}^4 V \to [2]^* H^0(J, L_\chi)_+, \ R_J \overset{p_4}{\longrightarrow} \sum_\sigma D_{J,\delta} \theta_{(\frac{\delta}{2}0)(0\frac{1}{2})}(2\tau, 4z)$$

where $\delta \in (\mathbf{Z}/2)^{g-1}$ and

$$D_{J,\delta}(\tau) = \left(\theta_{(\frac{\delta+\beta}{2}0)(0\frac{1}{2})} \theta_{(\frac{\delta+\gamma}{2}0)(0\frac{1}{2})} \theta_{(\frac{\delta+\beta+\gamma}{2}0)(0\frac{1}{2})} \right)(2\tau, 0).$$

In [vG], the (SJ) relations are shown to be equivalent to $\theta_{(\frac{\delta}{2}0)(0\frac{1}{2})}(2\tau, 0) = \theta_{\frac{\delta}{2}0}(2\pi_x, 0)$, for suitable period matrices τ of X and π_x of P_x thus, (ibid., Proposition 6), the matrices of the multiplication maps n_4 in the above bases differ by a non-zero multiplicative constant, where the maps are:

$$n_4 \colon S_{\chi_1}^4 V \to [2]^* H^0(J, L_\chi)_+ \text{ and } n_4 \colon S_0^4 V_{P_x} \to [2]^* H^0(P_x, L_{P_x})_+.$$

Note that the identification $S_{\chi_1}^4 V \cong S_0^4 V_{P_x}$ maps $R_J \in S_{\chi_1}^4 V$, $J = \{0, \rho, \nu, \rho + \nu\} \subset (\mathbf{Z}/2)^{g-1}$ to $P_J \in S_0^4 V_{P_x}$, which is now a polynomial in the 2^{g-1} variables, $X_\delta, \delta \in (\mathbf{Z}/2)^{g-1}$. The identification is thus obtained by restricting R_J to the eigenspace V_x defined by $X_{(\delta1)} = 0$ and then putting $X_{(\delta0)} = X_\delta$. More intrinsically, we use the action of H_{g-1} to identify V_x and $H^0(P_x, L_{P_x})$ and take Sym^4 of this identification.

Next we proceed to compare multiplication maps for two Pryms:

$$
\begin{array}{ccc}
\ker(n_{x,\bar\mu}) & \longrightarrow & S_{\bar\mu}^4 V_x \overset{n_{x,\bar\mu}}{\longrightarrow} [2]^* H^0(P_x, L_{P_x,\bar\mu})_+ \\
\cong \downarrow & & \cong \downarrow \\
W & \longrightarrow & S_0^4 V_{x,y} \\
\cong \uparrow & & \cong \uparrow \\
\ker(n_{y,\bar\chi}) & \longrightarrow & S_{\bar\chi}^4 V_y \overset{n_{y,\bar\chi}}{\longrightarrow} [2]^* H^0(P_y, L_{P_y,\bar\chi})_+.
\end{array}
$$

By a suitable identification $J[2] \cong H/\mathbf{C}^*$ we can assume that χ and μ correspond to the points $x = e_{2g}, y = e_{2g-1}$ in the "standard basis" $\{e_j\}$ of $J[2]$. The linearized eigenspaces of $U(x), U(y)$ are coordinatized by:

$$[X_0, \ldots, X_{2^{g-1}-1}, 0, \ldots, 0] \text{ and } [Y_0, ..., Y_{2^{g-2}-1}, 0, ..., 0; Y_{2^{g-1}}, \ldots, Y_{2^{g-1}+2^{g-2}-1}, 0, \ldots, 0]$$

and their intersection, which is a $\mathbf{P}^{2^{g-2}-1}$, has coordinates $[Z_0, \ldots, Z_{2^{g-2}-1}]$ which correspond to $X_{(\sigma''00)}, \sigma'' \in (\mathbf{Z}/2)^{g-2}$. The $X_{(\sigma'0)}, \sigma' \in (\mathbf{Z}/2)^{g-1}$, and $Y_{(\sigma''0\epsilon)}, \epsilon = 0, 1$, are acted on by the obvious Heisenbergs H_{g-1} in the obvious way.

Thus we can use ([vG Proposition 4], see also above) to express the matrices of the multiplication maps n_4 on $S_{\bar\mu}^4 V_\chi$ and $S_{\bar\chi}^4 V_\mu$, resp. $(\bar\mu, \bar\chi \in H_{g-1})$, where we use the (analog of the) basis R_J on the left $\theta_{(\sigma''0)(0\frac{1}{2})}(2\pi_x, 2z)$, and $\theta_{(\sigma''0)(0\frac{1}{2})}(2\pi_y, 2z)$ resp. on the right. The theta constants in the matrices are the coordinates of 4-torsion points on the Pryms, and the Donagi relations say precisely that the coefficients of the multiplication matrices are the same up to a (nonzero) multiplicative constant.

Lastly, note that the R_J's from $S_{\bar\mu}^4 V_\chi$ and $S_{\bar\chi}^4 V_\mu$ restrict to the same elements in $S_0^4 V_{x,y}$, which are polynomials in the $Z_i, 0 \le i \le 2^{g-2} - 1$. This proves the:

6.7 Lemma.

(i)(SJ) For any non-zero $x \in J[2]$ corresponding to χ, the restriction map gives an isomorphism $S_\chi^4 V \cong S_0^4 V_x$, and the multiplication maps $S_\chi^4 V \to H^0(J, \mathcal{O}(8\Theta))$ and $S_0^4 V_x \to H^0(P_x, \mathcal{O}(8\Xi_x))$ are defined (when restricted to their image) by the same matrices up to a (nonzero) constant.

(ii)(D) Let x, y be orthogonal elements of $J[2] \setminus \{0\}$, corresponding to characters χ, μ, and let $\overline{\chi}, \overline{\mu}$ be the characters of H_{g-1} induced on $(P_x)[2]$, $(P_y)[2]$. Then under the isomorphism given by the restriction:

$$S_{\overline{\chi}}^4 V_x \cong S_0^4 V_{x,y} \cong S_{\overline{\chi}}^4 V_y,$$

the multiplication maps $n_{4,x,\overline{\mu}}$ and $n_{4,y,\overline{\chi}}$ differ by a nonzero multiplicative constant.

6.8 Sketch of proof of (6.2), (6.3). We indicate the argument used to obtain (6.2). Let $F \in S_0^4 V$ and define $G := M(0)^{-1} F$; $F_\chi := M(\chi)G \in S_\chi^4 V$ (note $F_0 = F$).

If $K(J) \subset Sing(Z(F))$, then $G = \frac{1}{4} \frac{\partial}{\partial X} F$ is a cubic equation for $K(J)$, and, for all $\alpha \in (\mathbf{Z}/2\mathbf{Z})^g$, $G_\alpha := (1, \alpha, 0)G$ is also an equation for $K(J)$ since $K(J)$ is invariant under the action of H. Therefore each $F_\chi \in S_\chi^4 V$ is a quartic equation for $K(J)$. Using (6.7)(i), we see that F_χ also vanishes on $K(P_x) \subset PV_x$ (with x corresponding to χ). Since F and F_χ have the same restriction to PV_x, F also vanishes on $K(P_x)$. This shows that for all non-zero x F vanishes on $K(P_x)$ and since by assumption $K(J) \subset Z(F)$, F vanishes on all Pryms, i.e. $F \in \ker n_0$.

Conversely, assume that $F = F_0 \in \ker n_0$, i.e. F_0 vanishes on all Pryms, in particular F_0 vanishes on $K(J)$. We will first show that all other F_χ's, $\chi \neq 0$, also vanish on $K(J)$. For this we restrict F_χ to PV_x, with x corresponding to χ. Since F and F_χ coincide, up to scalar multiple, and since F vanishes by assumption on all Pryms, in particular on $P_x \in PV_x$, we see that F_χ vanishes on P_x. Using (6.7)(i) again, we see that for any non-trivial χ, F_χ is also an equation for $K(J)$. To finish the argument, we take suitable linear combinations of the various F_χ, and we find that $K(J) \subset Z(X_\alpha \frac{\partial}{\partial X_\alpha} F)$. Since X_α is not zero on $K(J)$, each $\frac{\partial}{\partial X_\alpha} F$ is an equation for $K(J)$, so we get $K(J) \subset Sing(Z(F))$.

The same principles, with the role of (SJ) being played by (D) give the proof of (6.3); for details see [vGP1].

6.9 Corollary. *For a curve with no vanishing thetanulls, a quartic vanishes on $\phi(\mathcal{M}_0)$ iff it vanishes on $\phi_x(P_x)$ for all Pryms of C.*

The "only if" part had been proved, and the "if" conjectured, in[vG].

6.10 Comments and problems:

6.10.1 It is perhaps not immediate to perceive that the Verlinde numbers in (5.0) are integers. D. Zagier, in a letter to R. Bott (4/26/1990) interpreted them as traces over $\mathbf{Q}$ of certains roots of unity and gave estimates on their generating functions in k and g.

6.10.2 It goes without saying that a geometric interpretation of the other Verlinde numbers would advance the theory of theta functions (abelian and non). We have not attempted the $k = 2$, odd-determinant case (the odd case for $k = 1$, was treated together with the even case, $k = 2$, in 5.3.3) but we should mention that in that case the natural map $P_x \to \mathbf{P}(\wedge^2 V)$ is an embedding [NR1]: the images of the two components of $\ker Nm_x$ overlap. However, the $k = 4$, even case was spectacularly put together with the $k = 2$, odd case by W.M. Oxbury [O]: *for every curve there exist canonical isomorphisms*

$$H^0(\mathcal{M}_0, \mathcal{L}_0^4)^* \cong \oplus_{x \in J[2]} H_+^0(P_x, 3\Theta_x), \quad H^0(\mathcal{M}_1, \mathcal{L}_1^2)^* \cong \oplus_{x \in J[2]} H_-^0(P_x, 3\Theta_x).$$

6.10.3 "Easy" exercise. It follows from 5.3.3(iv) that $\delta(\mathcal{M})$ is projectively normal in degree 2 iff X has no vanishing thetanulls. We don't know if the analogous statement holds for degree 4, in fact it's unlikely; what 6.1 implies is that for X with no vanishing thetanulls $\delta(\mathcal{M})$ is projectively normal in degree 4. One might amuse oneself by checking what happens for $g = 3$; in that case, the Coble quartic L^4 [C §33 (5)] defines $\delta(\mathcal{M})$ in the nonhyperelliptic case, while when X is hyperelliptic a double quartic (set-theoretically smooth) gives $\mathcal{M}/\iota$ [NR2,B2].

6.11 **Back to integrable systems.** Hitchin used his theory of spectral curves to show that the cotangent bundle to the moduli space of stable bundles is a completely integrable system [Hi1], and a concrete description of his system has been sought-after ever since. In [vGP2] the genus 2, even determinant case is analyzed using classical constructions for the curves Δ_E of 5.3.2 and the Hamiltonians are written as polynomials in the coordinates of the cotangent bundle of $\mathcal{M}_0 = \mathbf{P}^3$. **Note I.** The $g = 2$ case of the Hitchin system is rich in special phenomena; we mention one of them to settle the question, "can this system be one of those encountered in Lecture 2"? Since genus $X = 2$, pull-back via the hyperelliptic involution gives an isomorphism $H^0(X, K_X^{\otimes 2}) \cong H^0(\mathbf{P}^1, \mathcal{O}_{\mathbf{P}^1}(2))$, thus the spectral curve $\tilde{X}$ is a fibred product of two hyperelliptic curves, X branched at $b_1, ..., b_6$ say, and C branched at $b_1, ...b_6, b_7, b_8$ say. According to Hitchin, the flows (for the fixed-determinant case) linearize on the Prym P_π of the ramified cover $\pi : \tilde{X} \to X$; this is isomorphic to Jac $C/ < \alpha >$ where α is the point of order 2 of Jac C that corresponds to the unramified cover $\tau : \tilde{X} \to C$. This shows that the Hitchin system is not a Neumann-type system, since those linearize on a g-dimensional family of g-dimensional hyperelliptic Jacobians obtained by varying g of the branchpoints of the curve, whereas the curve C has 6 fixed, 2 moving branchpoints. **Note II.** In the introduction to Lecture 3 we expressed pessimism as regards a dictionary between higher-rank KP and the Hitchin systems. Here we want to acknowledge recent work [LM] in that direction, which Donagi also explained in his lectures. However, we would like to highlight one difference between this work and the rank 1 dictionary. In higher rank, [LM] embeds Hitchin's spectral Prym in the cotangent bundle to a vector Grassmannian (speaking loosely) and thus identify the Hitchin flows with (B)KP flows. Yet, the general higher-rank KP flows of 3.1 (corresponding to maximal-commutative Burchnall-Chaundy subalgebras of $\mathcal{D}$ of nonfake rank > 1) are not likely to be among these, because of the infinite dimensionality of the geometric data under the flow $(W, L_0) \to (W_x, L_{0,x})$ of 3.1.4(i).

Acknowledgements. I would like to express special gratitude to Prof. Mauro Francaviglia for proposing the topic of this CIME course and my own participation. Thanks also to the other organizers Proff. S. Greco and F. Magri and to the CIME secretary Prof. P. Zecca for providing a most gracious environment and much opportunity for unencumbered interaction. Lastly, remark 5.5(iv) and 6.10.2 are largely due to Prof. S. Ramanan's explanations; I thank him for his kindness and NSA for funding his visit at Boston University (10/15-11/2, 1992) under grant MDA904-92-H-3032.

References.

[AHH] M.R. Adams, J. Harnad and J. Hurtubise, Dual moment maps into loop algebras, *Lett. Math. Phys.* **20** (1990), 299-308.

[AHP] M.R. Adams, J. Harnad and E. Previato, Isospectral Hamiltonian flows in finite and infinite dimension, *Comm. Math. Phys.* **117** (1988), 451-500.

[AvM] M. Adler and P. van Moerbeke, Completely integrable systems, euclidean Lie algebras and curves. Linearization of Hamiltonian systems, Jacobian varieties and representation theory, *Adv. in Math.* **38** (1980), 267-379.

[AM] M. Adler and J. Moser, On a class of polynomials connected with the Korteweg de Vries equations, *Comm. Math. Phys.* **61** (1978), 1-30.

[AMcKM] H. Airault, H.P. McKean and J. Moser, Rational and elliptic solutions of the Korteweg-de Vries equation and a related many-body problem, *Comm. Pure Appl. Math.* **30** (1977), 95-148.

[B1] A. Beauville, Fibrés de rang 2 sur une courbe, fibré déterminant et fonctions thêta, *Bull. Soc. Math. France* **116** (1988), 431-448.

[B2] A. Beauville, Fibrés de rang 2 sur une courbe, fibré déterminant et fonctions thêta, II, *Bull. Soc. Math. France* **119** (1991), 259-291.

[B3] A. Beauville, Jacobiennes des courbes spectrales et systèmes hamiltoniens complètement intégrables *Acta Math.* **164** (1990), 211-235.

[BNR] A. Beauville, M.S. Narasimhan and S. Ramanan, Spectral Curves and the Generalized Theta Divisor, *J. Reine Angew. Math.* **398** (1989), 169-179.

[BR] M. Bershadsky and A. Radul, Fermionic fields on Z_N curves, *Comm. Math. Phys.* **116** (1988), 689-700.

[BS] A. Bertram and A. Szenes, Hilbert polynomials of moduli spaces of rank 2 vector bundles II, *Topology* **32** (1993), 599-609.

[Bo] F. Bottacin, Symplectic geometry on moduli spaces of stable pairs, preprint Univ. Paris Sud, 1992.

[BC1] J.L. Burchnall and T.W. Chaundy, Commutative ordinary differential operators, *Proc. London Math. Soc.* **211** (1923), 420-440.

[BC2] J.L. Burchnall and T.W. Chaundy, Commutative ordinary differential operators, *Proc. Roy. Soc. London Ser. A* **118** (1928), 557-583.

[C] A.B. Coble, *Algebraic Geometry and Theta Functions*, AMS Colloquium Publications X, 1961.

[CPP] E. Colombo, G.P. Pirola and E. Previato, Density of elliptic solitons, *J. Reine Angew. Math.* **451** (1994), 161-169.

[D] G. Darboux, Théorie Générale des Surfaces II, Chelsea 1972.

[De] P. Dehornoy, Opérateurs différentiels et courbes elliptiques, *Compositio Math.* **43** (1981), 71-99.

[DLT] P. Deift, L.C. Li and C. Tomei, Matrix factorizations and integrable systems, *Comm. Pure Appl. Math.* **42** (1989), 443-521.

[DR] J.V. Desale and S. Ramanan, Classification of vector bundles of rank 2 on hyperelliptic curves, *Invent. Math.* **38** (1976), 161-185.

[Di] J. Dixmier, Sur les algèbres de Weyl, *Bull. Soc. Math. France* **96** (1968), 209-242.

[Do] R. Donagi, Non-Jacobians in the Schottky loci, *Ann. of Math.* **126** (1987), 193-217.

[DP] R. Donagi and E. Previato, Abelian solitons, in preparation.

[DN] J.-M. Drezet and M.S. Narasimhan, Groupe de Picard des variétés de modules de fibrés semi-stables sur les courbes algébriques, *Invent. Math.* **97** (1989), 53-94.

[Du] B.A. Dubrovin, The KP equation and the relations between the periods of holomorphic differentials on Riemann surfaces, *Math. USSR-Izv.* **19** (1982), 285-296.

[EK] F. Ehlers and H. Knörrer, An algebro-geometric interpretation of the Bäcklund transformation for the Korteweg-de Vries equation, *Comment. Math. Helv.* **57** (1982), 1-10.

[EF] N. Ercolani and H. Flaschka, The geometry of the Hill equation and of the Neumann system, *Philos. Trans. Roy. Soc. London Ser. A* **315** (1985), 405-422.

[F] G. Faltings, A proof of the Verlinde formula, preprint 1992.

[F1] J. Fay, Theta Functions, *Lecture Notes in Math* **352**, Springer, Berlin, 1973.

[F2] J. Fay, On the even-order vanishing of Jacobian theta functions, *Duke Math. J.* **51** (1984), 109-132.

[Fl] H. Flaschka, Towards an algebro-geometric interpretation of the Neumann system, *Tôhoku Math. J.* **36** (1984), 407-426.

[vG] B. van Geemen, Schottky-Jung relations and vectorbundles on hyperelliptic curves, *Math. Ann.* **281** (1988), 431-449.

[vGvdG] B. van Geemen and G. van der Geer, Kummer varieties and the moduli spaces of abelian varieties, *Amer. J. Math.* **108** (1986), 615-641.

[vGP1] B. van Geemen and E. Previato, Prym varieties and the Verlinde formula, *Math. Annalen* **294** (1992), 741-754.

[vGP2] B. van Geemen and E. Previato, On the Hitchin system MSRI Preprint No. 057-94.

[G] T. Gneiting, Kommutierende Differentialoperatoren und Solitonen, Diplomarbeit, Universität Bayreuth 1993.

[GrP] S. Greco and E. Previato, Spectral curves and ruled surfaces: projective models, Vol. 8 of "The curves seminar at Queen's," *Queen's Papers in Pure and Appl. Math.*, No. 88 (1991), pp. F1-F33.

[GH] P. Griffiths and J. Harris, Principles of Algebraic Geometry, Wiley, New York, 1978.

[G1] F.A. Grünbaum, Commuting pairs of linear ordinary differential operators of orders four and six, *Phys. D* **31** (1988), 424-433.

[G2] F.A. Grünbaum, The KP equation: an elementary approach to the "rank 2" solutions of Krichever and Novikov, *Phys. Lett. A* **139** (1989), 146-150.

[G3] F.A. Grünbaum, Darboux's method and some "rank two" explicit solutions of the KP equation, in *Nonlinear evolution equations: integrability and spectral methods*, eds. A. Degasperis et al., Manchester Univ. Press, Manchester, UK, 1990, pp. 271-277.

[H] R. Hartshorne, Algebraic Geometry, Springer-Verlag, New York, 1977.

[Hi1] N. Hitchin, Stable bundles and integrable systems, *Duke Math. J.* **54** (1987), 91-114.

[Hi2] N. Hitchin, Flat connections and geometric quantization, *Comm. Math. Phys.* **131** (1990), 347-380.

[I] J.-I. Igusa, Theta functions, Springer-Verlag, Berlin, 1972.

[K] A. Kasman, Rank-r KP solutions with singular rational spectral curves, Thesis, Boston Univ. 1995.

[Kn] H. Knörrer, Geodesics on the ellipsoid, *Invent. Math.* **59** (1980), 119-143.

[Kr1] I.M. Krichever, The integration of non-linear equations by methods of algebraic geometry, *Functional Anal. Appl.* **12** (1977), 12-26.

[Kr2] I.M. Krichever, Rational solutions of the Zakharov-Shabat equations and completely integrable systems of N particles on a line, *J. Soviet Math.* **21** (1983), 335-345.

[Kr3] I.M. Krichever, Elliptic solutions of the KP equation and integrable systems of particles, *Functional Anal. Appl.* **14** (1980), 15-31.

[KN] I.M. Krichever and S.P. Novikov, Holomorphic bundles over algebraic curves and nonlinear equations, *Russian Math. Surveys* **35** (1980), 53-79.

[L] Y. Laszlo, Un théorème de Riemann pour les diviseurs thêta sur les espaces des modules de fibrés stables sur une courbe, *Duke Math. J.* **64** (1991), 333-347.

[LP] G. Latham and E. Previato, Higher rank Darboux transformations, in *Singular Limits of Dispersive Waves*, eds. N.M Ercolani et al., Plenum Press, New York, 1994, pp. 117-134.

480

[LP2] G.A. Latham and E. Previato, Darboux transformations for higher rank Kadomtsev-Petviashvili and Krichever-Novikov equations, in *KdV '95*, eds. M. Hazewinkel *et al.*, Kluwer Acad., Dordrecht, 1995, pp. 405-433.

[LM] Y. Li and M. Mulase, The Hitchin system and the KP equations, preprint 1994.

[M] E. Markman, Spectral curves and integrable systems, *Compositio Math.* **93** (1994), 255-290.

[Mo] I.O. Mokhov, Commuting differential operators of rank 3 and nonlinear differential equations, *Math. USSR-Izv.* **35** (1990), 629-655.

[Mos] J. Moser: I. Three integrable hamiltonian systems connected with isospectral deformations, *Adv. Math.* **16** (1975), 197-220; II. Geometry of quadrics and spectral theory, *The Chern Symposium*, Springer-Verlag, 1980, pp. 147-188.

[M1] D. Mumford, Prym varieties I, in *Contributions to Analysis*, ed. L.V. Ahlfors, AcadOA. Press New York, 1974, pp. 325-350.

[M2] D. Mumford, An algebro-geometric construction of commuting operators and of solutions to the Toda lattice equation, Korteweg de Vries equation and related non-linear equations. Intl. Sympos. on Algebraic Geometry, Kyoto (1977), 115-153.

[M3] D. Mumford, Tata lectures on theta II, *Progr. Math.* **43**, Birkhäuser, Boston, 1984.

[MN] D. Mumford and P. Newstead, Periods of a moduli space of bundles on curves, *Amer. J. Math.* **90** (1968), 1200-1208.

[NR1] M.S. Narasimhan and S. Ramanan, Generalised Prym varieties and fixed points, *J. Indian Math. Soc.* **39** (1975), 1-19.

[NR2] M.S. Narasimhan and S. Ramanan, 2Θ linear systems on Abelian varieties, *Vector bundles on algebraic varieties*, Oxford Univ. Press 1987, pp. 415-427.

[N] P. Newstead, Introduction to moduli problems and orbit spaces, TIFR, Bombay, 1978.

[OP] M.A. Olshanetsky and A.M. Perelomov, Classical Integrable finite-dimensional systems related to Lie algebras, *Phys. Rep.* **71** (1981), 313-400.

[O] W.M. Oxbury, Anticanonical Verlinde spaces as theta spaces on Pryms, preprint 1994.

[PR] K. Paranjape and S. Ramanan, On the canonical ring of a curve, Nagata volume, 1987, pp. 503-516.

[PS] A. Pressley and G.B. Segal, Loop Groups, Oxford 1986.

[P1] E. Previato, Generalized Weierstrass ℘-functions and KP flows in affine space, *Comment. Math. Helvetici* **62** (1987), 292-310.

[P2] E. Previato, Three questions on vector bundles, in *Complex Analysis and Geometry*, eds. V. Ancona and A. Silva, Plenum Press 1993, pp. 395-398.

[P3] E. Previato, Burchnall-Chaundy bundles, to appear, Proc. Europroj '94, Barcelona.

[PV] E. Previato and J.-L. Verdier, Boussinesq elliptic solitons: the cyclic case, in *Proc. Indo-French Conference on Geometry* (Bombay, 1989), Hindustan Book Agency, Delhi 1993, pp. 173-185.

[PW1] E. Previato and G. Wilson, Vector bundles over curves and solutions of the KP equations, *Proc. Sympos. Pure Math.* **49** (1989), 553-569.

[PW2] E. Previato and G. Wilson, Differential operators and rank two bundles over elliptic curves, *Compositio Math.* **81** (1992), 107-119.

[RS] A. Reymann, M. Semenov-Tian-Shansky: Reduction of Hamiltonian systems, affine Lie algebras, and Lax equations I and II, *Invent. Math.* **54** (1979), 81-100 and **63** (1981), 423-432.

[SM] R. Salvati Manni, Modular varieties with level 2 theta structure, *Amer. J. Math.* **116** (1994), 1489-1511.

[S] R. Schilling: Generalizations of the Neumann system: a curve theoretical approach I, II and III, *Comm. Pure Appl. Math.* **40** (1987), 455-522; **42** (1989), 409-422; and **45** (1992), 775-820.

[SW] G.B. Segal and G. Wilson, Loop groups and equations of KdV type, *Publ. Math. Inst. Hautes Études Sci.* **61** (1985), 5-65.

[SSY] S.I. Svinolupov, V. Sokolov and R. Yamilov, On Bäcklund transformations for integrable evolution equations, *Soviet Math. Dokl.* **28** (1983), 165-168.

[Sy] W.W. Symes, Systems of Toda type, inverse spectral problems, and representation theory, *Invent. Math.* **59** (1980), 13-51.

[Tr] A. Treibich, Tangential polynomials and elliptic solitons, *Duke Math. J.* **59** (1989), 611-627.

[TV1] A. Treibich and J.-L. Verdier, Solitons Elliptiques, The Grothendieck Festschrift III, Birkhäuser 1990, pp. 437-480.

[TV2] A. Treibich and J.-L. Verdier, Variétés de Kritchever des solitons elliptiques de KP, *Proc. Indo-French Conf. on Geometry* (Bombay, 1989), Hindustan Book Agency, Delhi, 1993, pp. 187-232.

[TV3] A. Treibich and J.-L. Verdier, Revêtements tangentiels et sommes de 4 nombres triangulaires, *C.R. Acad. Sci. Paris Sér. I Math.*, **311** (1990), 51-54.

[Tu] L. Tu, Semistable bundles over an elliptic curve, *Adv. Math.* **98** (1993), 1-26.

[V] V.S. Varadarajan, Lie Groups, Lie Algebras, and Their Representations, Springer 1984

[Ve] J.-L. Verdier, Équations différentielles algébriques, *Lecture Notes in Math.* **710**, Springer 1979, pp. 101-122.

[W] G. Wilson, Algebraic curves and soliton equations, in *Geometry Today*, eds. E. Arbarello *et al.*, Birkhäuser, Boston, 1985, 303-329.

[Wi] W. Wirtinger, Untersuchungen über Thetafunctionen, Teubner, Leipzig, 1895.

C.I.M.E. Session on "Integrable Systems and Quantum Groups"

List of Participants

M. ADAMS, Dept. of Math., Univ. of Georgia, Athens, GA 30601, USA

M. BERGVELT, Dept. of Math., Univ. of Illinois, 273 Altgeld Hall, 1409 West Green Street, Urbana, ILL 61801, USA

A. BOROWIEC, Inst. for Theor. Phys., Univ. of Wroclaw, 50204 Wroclaw, Poland

P. CASATI, Dip. di Mat., Univ., Via Carlo Alberto 10, 10123 Torino, Italy

B. CASCIARO, Dip. di Mat., Univ., Via Orabona 4, 70121 Bari, Italy

A. CASSA, Dip. di Mat., Univ. di Trento, 38050 Povo, Trento, Italy

E. COLOMBO, Dip. di Mat., Univ., Via C. Saldini 50, 20133 Milano, Italy

V. DRAGOVIC, Math. Inst., Knez-Mihailova 35, 11000 Belgrado, Yugoslavia

G. FALQUI, Dip. di Mat., Univ., Via C. Saldini 50, 20133 Milano, Italy

M. FERRARIS, Dip. di Mat., Via Ospedale 72, 09124 Cagliari

L. GATTO, Dip. di Mat. del Polit., Corso Duca degli Abruzzi 24, 10129 Torino, Italy

T. KRIECHERBAUER, Courant Inst. of Math. Sci., New York Univ., 251 Mercer St. New York, NY 10012, USA

G. MAGNANO, Istituto di Fisica Matematica, Via C. Alberto 10, 10123 Torino, Italy

K. B. MARATHE, Brooklyn College, Dept. of Math. Cuny, NY 11210, USA

M. NIEDERMAIER, Max-Planck-Institute f. Physik, Fohringer Ring 6, Postfach 401212, 8000 Munchen 40, Germany

G. PARESCHI, Dip. di Mat., Via Machiavelli 35, 44100 Ferrara, Italy

M. PEDRONI, Dip. di Mat., Univ., Via C. Saldini 50, 20133 Milano, Italy

G.P. PIROLA, Dip. di Mat., Strada Nuova 65, 27100 Pavia, Italy

N. RAMACHANDRAN, Dept. of Math., Brown Univ., Providence, RI 02917, USA

C. REINA, SISSA, Via Beirut 2-4, 34100 Trieste, Italy

M. ROTHSTEIN, Dept. of Math., Univ. of Georgia, Athens, GA 30602, USA

R. SCOGNAMILLO, S.N.S., Piazza dei Cavalieri 7, 56126 Pisa, Italy

A. STOLIN, Dept. of Math., Royal Inst. of Techn., 100 44 Stockholm, Sweden

B. VAN GEEMEN, Dip. di Mat., Strada Nuova 65, 27100 Pavia, Italy

LIST OF C.I.M.E. SEMINARS   Publisher

1954 - 1. Analisi funzionale   C.I.M.E.
        2. Quadratura delle superficie e questioni connesse   "
        3. Equazioni differenziali non lineari   "

1955 - 4. Teorema di Riemann-Roch e questioni connesse   "
        5. Teoria dei numeri   "
        6. Topologia   "
        7. Teorie non linearizzate in elasticità, idrodinamica,aerodinamica   "
        8. Geometria proiettivo-differenziale   "

1956 - 9. Equazioni alle derivate parziali a caratteristiche reali   "
       10. Propagazione delle onde elettromagnetiche   "
       11. Teoria della funzioni di più variabili complesse e delle
           funzioni automorfe   "

1957 - 12. Geometria aritmetica e algebrica (2 vol.)   "
       13. Integrali singolari e questioni connesse   "
       14. Teoria della turbolenza (2 vol.)   "

1958 - 15. Vedute e problemi attuali in relatività generale   "
       16. Problemi di geometria differenziale in grande   "
       17. Il principio di minimo e le sue applicazioni alle equazioni
           funzionali   "

1959 - 18. Induzione e statistica   "
       19. Teoria algebrica dei meccanismi automatici (2 vol.)   "
       20. Gruppi, anelli di Lie e teoria della coomologia   "

1960 - 21. Sistemi dinamici e teoremi ergodici   "
       22. Forme differenziali e loro integrali   "

1961 - 23. Geometria del calcolo delle variazioni (2 vol.)   "
       24. Teoria delle distribuzioni   "
       25. Onde superficiali   "

1962 - 26. Topologia differenziale   "
       27. Autovalori e autosoluzioni   "
       28. Magnetofluidodinamica   "

1963 - 29. Equazioni differenziali astratte   "
       30. Funzioni e varietà complesse   "
       31. Proprietà di media e teoremi di confronto in Fisica Matematica   "

1964 - 32. Relatività generale   "
       33. Dinamica dei gas rarefatti   "
       34. Alcune questioni di analisi numerica   "
       35. Equazioni differenziali non lineari   "

1965 - 36. Non-linear continuum theories   "
       37. Some aspects of ring theory   "
       38. Mathematical optimization in economics   "

1966 - 39. Calculus of variations   Ed. Cremonese, Firenze
       40. Economia matematica   "
       41. Classi caratteristiche e questioni connesse   "
       42. Some aspects of diffusion theory   "

1967 - 43. Modern questions of celestial mechanics   "
       44. Numerical analysis of partial differential equations   "
       45. Geometry of homogeneous bounded domains   "

1968 - 46. Controllability and observability   "
       47. Pseudo-differential operators   "
       48. Aspects of mathematical logic   "

1969 - 49. Potential theory   "
       50. Non-linear continuum theories in mechanics and physics
           and their applications   "
       51. Questions of algebraic varieties   "

1970 - 52. Relativistic fluid dynamics   "
       53. Theory of group representations and Fourier analysis   "
       54. Functional equations and inequalities   "
       55. Problems in non-linear analysis   "

1971 - 56. Stereodynamics   "
       57. Constructive aspects of functional analysis (2 vol.)   "
       58. Categories and commutative algebra   "

1972 - 59. Non-linear mechanics   "
         60. Finite geometric structures and their applications   "
         61. Geometric measure theory and minimal surfaces   "

1973 - 62. Complex analysis   "
         63. New variational techniques in mathematical physics   "
         64. Spectral analysis   "

1974 - 65. Stability problems   "
         66. Singularities of analytic spaces   "
         67. Eigenvalues of non linear problems   "

1975 - 68. Theoretical computer sciences   "
         69. Model theory and applications   "
         70. Differential operators and manifolds   "

1976 - 71. Statistical Mechanics   Ed Liguori, Napoli
         72. Hyperbolicity   "
         73. Differential topology   "

1977 - 74. Materials with memory   "
         75. Pseudodifferential operators with applications   "
         76. Algebraic surfaces   "

1978 - 77. Stochastic differential equations   "
         78. Dynamical systems   Ed Liguori, Napoli and Birhäuser Verlag

1979 - 79. Recursion theory and computational complexity   "
         80. Mathematics of biology   "

1980 - 81. Wave propagation   "
         82. Harmonic analysis and group representations   "
         83. Matroid theory and its applications   "

1981 - 84. Kinetic Theories and the Boltzmann Equation (LNM 1048) Springer-Verlag
         85. Algebraic Threefolds   (LNM 947)   "
         86. Nonlinear Filtering and Stochastic Control (LNM 972)   "

1982 - 87. Invariant Theory   (LNM 996)   "
         88. Thermodynamics and Constitutive Equations   (LN Physics 228)   "
         89. Fluid Dynamics   (LNM 1047)   "

1983 -	90. Complete Intersections	(LNM 1092)	Springer-Verlag
	91. Bifurcation Theory and Applications	(LNM 1057)	"
	92. Numerical Methods in Fluid Dynamics	(LNM 1127)	"
1984 -	93. Harmonic Mappings and Minimal Immersions	(LNM 1161)	"
	94. Schrödinger Operators	(LNM 1159)	"
	95. Buildings and the Geometry of Diagrams	(LNM 1181)	"
1985 -	96. Probability and Analysis	(LNM 1206)	"
	97. Some Problems in Nonlinear Diffusion	(LNM 1224)	"
	98. Theory of Moduli	(LNM 1337)	"
1986 -	99. Inverse Problems	(LNM 1225)	"
	100. Mathematical Economics	(LNM 1330)	"
	101. Combinatorial Optimization	(LNM 1403)	"
1987 -	102. Relativistic Fluid Dynamics	(LNM 1385)	"
	103. Topics in Calculus of Variations	(LNM 1365)	"
1988 -	104. Logic and Computer Science	(LNM 1429)	"
	105. Global Geometry and Mathematical Physics	(LNM 1451)	"
1989 -	106. Methods of nonconvex analysis	(LNM 1446)	"
	107. Microlocal Analysis and Applications	(LNM 1495)	"
1990 -	108. Geoemtric Topology: Recent Developments	(LNM 1504)	"
	109. H$_\infty$ Control Theory	(LNM 1496)	"
	110. Mathematical Modelling of Industrial Processes	(LNM 1521)	"
1991 -	111. Topological Methods for Ordinary Differential Equations	(LNM 1537)	"
	112. Arithmetic Algebraic Geometry	(LNM 1553)	"
	113. Transition to Chaos in Classical and Quantum Mechanics	(LNM 1589)	"
1992 -	114. Dirichlet Forms	(LNM 1563)	"
	115. D-Modules, Representation Theory, and Quantum Groups	(LNM 1565)	"
	116. Nonequilibrium Problems in Many-Particle Systems	(LNM 1551)	"

1993 - 117.	Integrable Systems and Quantum Groups	(LNM 1620)	Springer-Verlag
118.	Algebraic Cycles and Hodge Theory	(LNM 1594)	
119.	Phase Transitions and Hysteresis	(LNM 1584)	"
1994 - 120.	Recent Mathematical Methods in	to appear	"
	Nonlinear Wave Propagation		
121.	Dynamical Systems	(LNM 1609)	"
122.	Transcendental Methods in Algebraic	to appear	"
	Geometry		
1995 - 123.	Probabilistic Models for Nonlinear PDE's	to appear	"
	and Numerical Applications		
124.	Viscosity Solutions and Applications	to appear	"
125.	Vector Bundles on Curves. New Directions	to appear	"